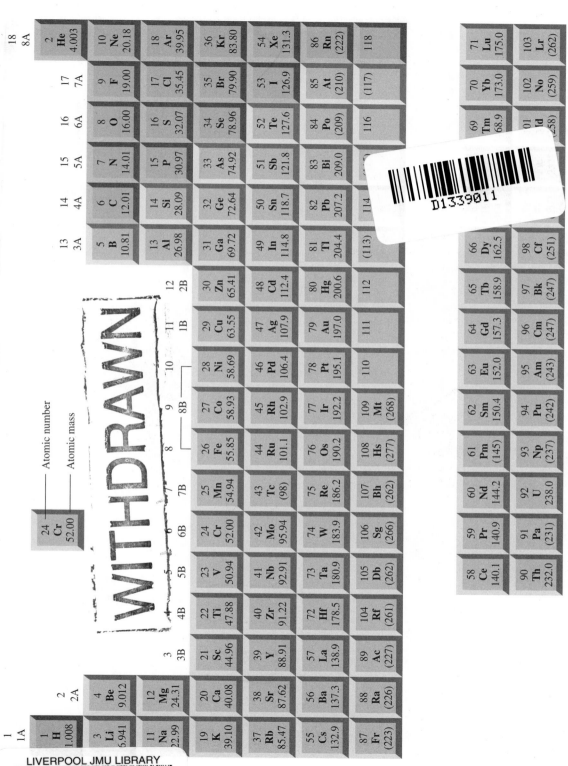

The 1–18 group designation has been recommended by the International Union of Pure and Applied Chemistry (IUPAC) but is not yet in wide use. No names have been assigned for elements 110–112, 114, 116, and 118. Elements 113, 115, and 117 have not yet been synthesized.

Source: Chang, R. *Chemistry*, 7th ed. Copyright © 2002 The McGraw-Hill Companies, Inc. New York. Reproduced with permission.

INTRODUCTION TO ENVIRONMENTAL ENGINEERING

The McGraw-Hill Series in Civil and Environmental Engineering

INTRODUCTION TO ENVIRONMENTAL ENGINEERING

Fifth Edition

Mackenzie L. Davis, Ph.D., P.E., BCEE
Michigan State University

David A. Cornwell, Ph.D., P.E., BCEE
Environmental Engineering & Technology, Inc.

The McGraw·Hill Companies

INTRODUCTION TO ENVIRONMENTAL ENGINEERING, FIFTH EDITION
International Edition 2013

Exclusive rights by McGraw-Hill Education (Asia), for manufacture and export. This book cannot be re-exported from the country to which it is sold by McGraw-Hill. This International Edition is not to be sold or purchased in North America and contains content that is different from its North American version.

Published by McGraw-Hill, a business unit of The McGraw-Hill Companies, Inc., 1221 Avenue of the Americas, New York, NY, 10020. Copyright © 2013 by McGraw-Hill Companies, Inc. All rights reserved. Previous editions © 2008. No part of this publication may be reproduced or distributed in any form or by any means, or stored in a database or retrieval system, without the prior written consent of The McGraw-Hill Companies, Inc., including, but not limited to, in any network or other electronic storage or transmission, or broadcast for distance learning.
Some ancillaries, including electronic and print components, may not be available to customers outside the United States.

10 09 08 07 06 05 04 03 02
20 15 14 13 12
CTP   COS

All credits appearing on page or at the end of the book are considered to be an extension of the copyright page.

When ordering this title, use ISBN 978-007-132624-7 or MHID 007-132624-3

Printed in Singapore

www.mhhe.com

To Elaine, my critic, my cheerleader, my wife . . . lo these 50 years, and my love . . . forever

—*Mackenzie L. Davis*

To my wife Nancy, who not only puts up with me in life, but has to put up with me in business too . . . without you neither I nor EE&T would be the same. Thank you for being my wife, companion, and partner.

—*David A. Cornwell*

ABOUT THE AUTHORS

Mackenzie L. Davis, Ph.D., P.E., BCEE, is an Emeritus Professor of Environmental Engineering at Michigan State University. He received all his degrees from the University of Illinois. From 1968 to 1971 he served as a Captain in the U.S. Army Medical Service Corps. During his military service he conducted air pollution surveys at Army ammunition plants. From 1971 to 1973 he was Branch Chief of the Environmental Engineering Branch at the U.S. Army Construction Engineering Research Laboratory. His responsibilities included supervision of research on air, noise, and water pollution control and solid waste management for Army facilities. In 1973 he joined the faculty at Michigan State University. He has taught and conducted research in the areas of air pollution control and hazardous waste management.

In 1987 and 1989–1992, under an intergovernmental personnel assignment with the Office of Solid Waste of the U.S. Environmental Protection Agency, Dr. Davis performed technology assessments of treatment methods used to demonstrate the regulatory requirements for the land disposal restrictions ("land ban") promulgated under the Hazardous and Solid Waste Amendments.

Dr. Davis is a member of the following professional organizations: American Chemical Society, American Institute of Chemical Engineers, American Society for Engineering Education, American Meteorological Society, American Society of Civil Engineers, American Water Works Association, Air & Waste Management Association, Association of Environmental Engineering and Science Professors, and the Water Environment Federation.

His honors and awards include the State-of-the-Art Award from the ASCE, Chapter Honor Member of Chi Epsilon, Sigma Xi, election as a Fellow in the Air & Waste Management Association, and election as a Diplomate in the American Academy of Environmental Engineers with certification in hazardous waste management. He has received teaching awards from the American Society of Civil Engineers Student Chapter, Michigan State University College of Engineering, North Central Section of the American Society for Engineering Education, Great Lakes Region of Chi Epsilon, and the Amoco Corporation. In 1998, he received the Lyman A. Ripperton Award for distinguished achievement as an educator from the Air & Waste Management Association. In 2007, he was recognized as the Educational Professional of the Year by the Michigan Water Environment Association. He is a registered professional engineer in Michigan.

Dr. Davis is the author of a student and professional edition of *Water and Wastewater Engineering* and Co-author of *Principles of Environmental Engineering* with Dr. Susan Masten.

In 2003, Dr. Davis retired from Michigan State University.

David A. Cornwell, Ph.D., P.E., BCEE, is a registered professional engineer in 19 states and is the founder and president of the consulting firm Environmental Engineering & Technology, Inc. (EE&T) headquartered in Newport News, VA. He attended the University of Florida in Gainesville, FL, where he received his Ph.D. in civil/environmental engineering and has remained a loyal Gator fan ever since, serving as a Bull Gator and on the President's Council. He was an associate professor in the Civil Environmental Engineering Department at Michigan State University prior to entering the consulting field. Many of Dr. Cornwell's students now are active members of the water profession.

During his career as a consultant, Dr. Cornwell has provided process, design, and operational troubleshooting services to water utilities around the world. He has lectured and written on many aspects of water treatment, including over 50 peer-reviewed technical articles and reports. Much of his work has included the development of new and optimized water treatment processes. He has won three *JAWWA* Division best paper awards and the overall *JAWWA* publication award. Dr. Cornwell has an extensive record of service to the water profession. He has been an active member of American Water Works Association (AWWA) since the early 1970s and has served on numerous committees in that organization. He has chaired the Research Division and the Technical and Education Council, and served on the board of directors and executive committee of AWWA.

In 2005, Dr. Cornwell was the recipient of the A.P. Black Research Award given by AWWA to recognize excellence in water treatment research, recognizing his contributions to bridging the gap between research and application. Dr. Cornwell has been a principal investigator on over 20 Water Research Foundation research projects.

PREFACE

Following the format of previous editions, the fifth edition of *Introduction to Environmental Engineering* is designed for use in an introductory sophomore-level environmental engineering course with sufficient depth to allow its use in more advanced courses. We assume that the book will be used in one of the first environmental engineering courses encountered by the student. As such, it provides the fundamental science and engineering principles that instructors in more advanced courses may assume are common knowledge for an advanced undergraduate.

The fifth edition has been reorganized. New chapters on risk assessment, water chemistry, and sustainability and green engineering are included. The risk assessment chapter includes material that was formerly in the chapter on hazardous waste management. It now includes introductory material on probability and risk that is used both in this chapter and in the chapters on water resources engineering and sustainability.

The water chemistry material that was formerly included with water treatment is now presented as a separate chapter. More environmental chemistry concepts are introduced at the beginning of the chapters in which they are relevant. This format integrates the chemistry fundamentals with their application to the subject matter of the chapter. It provides the student with the tools to analyze and understand the environmental engineering issues described in the chapter as well as providing an immediate feedback of the relevance of the basic chemistry. There are over 100 end-of-chapter chemistry-related problems spread throughout the text.

A new chapter on sustainability and green engineering focuses on water resources and energy. Population growth and climate change implications are discussed. This chapter uses water resource engineering concepts presented in Chapter 4 to discuss floods and droughts. Energy conservation measures in building design, water, and wastewater treatment are highlighted.

The Fundamentals of Engineering (FE) examination for civil and environmental engineering has been highlighted as a focal point in this edition. Seventy percent of the topics included in the environmental engineering specific Fundamentals of Engineering (FE) examination are covered in *Introduction to Environmental Engineering*. These include the following subject areas: ethics in Chapter 1; mass balance in Chapter 2; hydrology and watershed processes in Chapter 4; water and wastewater engineering in Chapters 6, 7, and 8; air quality engineering in Chapter 9; the noise pollution aspects of occupational and health safety in Chapter 10; solid and hazardous waste engineering in Chapters 11 and 12; radiological health, safety, and waste management in Chapter 13. We have identified equations in *Introduction to Environmental Engineering* that also appear in the *Fundamentals of Engineering Supplied-Reference Handbook*; these are identified with the FE flag icon. A website has been developed to assist the students in locating similar equations in the *Handbook*. In addition, at the end of each chapter we have supplied typical FE exam formatted problems for the students to work.

Because the FE exam uses both SI units and U.S. Customary System (USCS) units, USCS units are introduced in Chapter 1 and then utilized in numerous example problems as well as the FE exam formatted problems. A conversion factor table is presented in Appendix C.

The concept of materials and energy balance as a tool for understanding environmental processes and solving environmental engineering problems is carried through the text. This concept is introduced in a stand-alone chapter and then applied for conservative systems in hydrology (hydrologic cycle, development of the rational formula, and reservoir design). This theme is expanded to include sludge mass balance in water treatment (Chapter 6), and the DO sag curve in Chapter 7. The design equations for a completely mixed activated sludge system and a more elaborate sludge mass balance are developed in Chapter 8. Mass balance is used to account for the production of sulfur dioxide from the combustion of coal and in the development of absorber design equations in Chapter 9. In Chapter 12, a mass balance approach is used for waste audit. There are over 100 materials and energy balance end-of-chapter problems spread throughout the text.

Each chapter concludes with a list of review items, the traditional end-of-chapter problems, and, perhaps less traditional, discussion questions and FE formatted problems. The review items have been written in the "objective" format of the Accreditation Board for Engineering and Technology (ABET). Instructors will find this particularly helpful for directing student review for exams, for assessing continuous quality improvement for ABET, and for preparing documentation for ABET curriculum review. We have found the discussion questions useful as a "minute check" or spot quiz item to see if the students understand concepts as well as number crunching.

The fifth edition has been thoroughly revised and updated. With the addition of 58 new end-of-chapter problems, there are now a total of over 720 problems. Sixty-six of the problems have been set up for spreadsheet solutions. Two *Solver*® example problems are demonstrated in Chapter 8.

When the senior author was teaching, the course bearing the title of this book provided the foundation for four follow-on senior-level environmental engineering courses. The initial portions of selected chapters (materials and energy balances, water resource engineering, water treatment, water pollution, wastewater treatment, air pollution, noise pollution, and solid waste) were included in the introductory course. Advanced material, including most of the design concepts, were covered in the upper-level courses (water resources engineering, water and wastewater treatment plant design, solid and hazardous waste management). Some of the material is left for the students to pursue on their own (environmental legislation, ionizing radiation). This book provides an ideal foundation for the senior author's book on *Water and Wastewater Engineering: Design Principles and Practice*.

An instructor's manual and set of PowerPoint® slides are available online for qualified instructors. Please inquire with your McGraw-Hill representative for the necessary access password. The instructor's manual includes sample course outlines, solved example exams, and detailed solutions to the end-of-chapter problems. In addition, there are suggestions for using the pedagogic aids in the text.

Numerous Michigan State University alumni have indicated that *Introduction to Environmental Engineering* is an excellent text for review and preparation for the

Professional Engineers examination. It is both readable for self-study as well as a good source of sufficient example problems and data for practical application in the exam. Many have taken it to the exam as one of their reference resources. And they have used it!

As always, we appreciate any comments, suggestions, corrections, and contributions for future revisions.

Mackenzie L. Davis
David A. Cornwell

Acknowledgments

As with any other text, the number of individuals who have made it possible far exceeds those whose names grace the cover. At the hazard of leaving someone out, we would like to explicitly thank the following individuals for their contribution.

Over the many years of the five editions, the following students helped to solve problems, proofread text, prepare illustrations, raise embarrassing questions, and generally make sure that other students could understand the material: Shelley Agarwal, Stephanie Albert, Deb Allen, Mark Bishop, Aimee Bolen, Kristen Brandt, Jeff Brown, Amber Buhl, Nicole Chernoby, Rebecca Cline, Linda Clowater, Shauna Cohen, John Cooley, Ted Coyer, Marcia Curran, Talia Dodak, Kimberly Doherty, Bobbie Dougherty, Lisa Egleston, Karen Ellis, Craig Fricke, Elizabeth Fry, Beverly Hinds, Edith Hooten, Brad Hoos, Kathy Hulley, Geneva Hulslander, Lisa Huntington, Angela Ilieff, Alison Leach, Gary Lefko, Lynelle Marolf, Lisa McClanahan, Tim McNamara, Becky Mursch, Cheryl Oliver, Kyle Paulson, Marisa Patterson, Lynnette Payne, Jim Peters, Kristie Piner, Christine Pomeroy, Susan Quiring, Erica Rayner, Bob Reynolds, Laurene Rhyne, Sandra Risley, Carlos Sanlley, Lee Sawatzki, Stephanie Smith, Mary Stewart, Rick Wirsing, Glenna Wood, and Ya-yun Wu. To them a hearty thank you!

We would also like to thank the following individuals for their many helpful comments and suggestions in bringing out the first four editions of the book: Wayne Chudyk, Tufts University; John Cleasby, Iowa State University; Michael J. Humenick, University of Wyoming; Tim C. Keener, University of Cincinnati; Paul King, Northeastern University; Susan Masten, Michigan State University; R. J. Murphy, University of South Florida; Thomas G. Sanders, Colorado State University; and Ron Wukasch, Purdue University. Myron Erickson, P. E., Clean Water Plant, City of Wyoming, MI; Thomas Overcamp, Clemson University; James E. Alleman, Iowa State University; Janet Baldwin, Roger Williams University; Ernest R. Blatchley, III, Purdue University; Amy B. Chan Hilton, Florida A&M University-Florida State University; Tim Ellis, Iowa State University; Selma E. Guigard, University of Alberta; Nancy J. Hayden, University of Vermont; Jin Li, University of Wisconsin-Milwaukee; Mingming Lu, University of Cincinnati; Taha F. Marhaba, New Jersey Institute of Technology; Alexander P. Mathews, Kansas State University; William F. McTernan, Oklahoma State University; Eberhard Morgenroth, University of Illinois at Urbana-Champaign; Richard J. Schuhmann, The Pennsylvania State University; Michael S. Switzenbaum, Marquette University; Derek G. Williamson, University of Alabama.

The following reviewers provided many helpful comments and useful suggestions for the fifth edition: Gregory Boardman, Virginia Polytechnic Institute & State

University; Shankar Chellam, University of Houston; Cynthia Coles, Memorial University of New Foundland; Timothy Ellis, Iowa State University; Enos Inniss, University of Missouri-Columbia; Edward Kolodziej, University of Nevada, Reno; Taha Marhaba, New Jersey Institute of Technology; Alexander Mathews, Kansas State University.

To John Eastman, our esteemed friend and former colleague, we offer our sincere appreciation. His contribution to the initial work of Chapter 5 in the first edition, as well as constructive criticism and "independent" testing of the material was exceptionally helpful. Kristin Erickson, Radiation Safety Officer, Office of Radiation, Chemical and Biological Safety, Michigan State University, contributed to the Chapter 11 revisions for the third edition. To her we offer our hearty thanks. We especially want to thank Dave's wife, Nancy McTigue, for all her work on making revisions to the Solid Waste Management chapter and her help in reviewing the Water Treatment chapter.

And last, but certainly not least, we wish to thank our families, who have put up with the nonsense of book writing.

CONTENTS

CHAPTER
1

INTRODUCTION

1-1   WHAT IS ENVIRONMENTAL ENGINEERING?

Environmental engineering is a profession that applies mathematics and science to utilize the properties of matter and sources of energy in the solution of problems of environmental sanitation. These include the provision of safe, palatable, and ample public water supplies; the proper disposal of or recycle of wastewater and solid wastes; the adequate drainage of urban and rural areas for proper sanitation; and the control of water, soil, and atmospheric pollution, and the social and environmental impact of these solutions. Furthermore it is concerned with engineering problems in the field of public health, such as control of arthropod-borne diseases, the elimination of industrial health hazards, and the provision of adequate sanitation in urban, rural, and recreational areas, and the effect of technological advances on the environment (ASCE, 1973, 1977).

Environmental engineering is not concerned primarily with heating, ventilating, or air conditioning (HVAC), nor is it concerned primarily with landscape architecture. Neither should it be confused with the architectural and structural engineering functions associated with built environments, such as homes, offices, and other workplaces.

Historically, environmental engineering has been a specialty area of civil engineering. Today it is still primarily associated with civil engineering in academic curricula. However, especially at the graduate level, students may come from a multitude of other disciplines, such as chemical, bio-systems, electrical, and mechanical engineering as well as biochemistry, microbiology, and soil science.

Professional Development

The beginning of professional development for environmental engineers is the successful attainment of the baccalaureate degree. For continued development, a degree in engineering from a program accredited by the Accreditation Board for Engineering and Technology (ABET) provides a firm foundation for professional growth. Other steps in the progression of professional development are:

- Achievement of the title "Engineer in Training" by successful completion of the Fundamentals of Engineering (FE) examination

- Achievement of the title "Professional Engineer" by successful completion of four years of applicable engineering experience and successful completion of the Principles and Practice of Engineering (PE) exam

- Achievement of the title "Board Certified Environmental Engineer" (BCEE) by successful completion of 8 years of experience and successful completion of a written certification examination or 16 years of experience and successful completion of an oral examination

The FE exam and the PE exam are developed and administered by the National Council of Examiners for Engineering and Surveying (NCEES). The BCEE exams are administered by the American Academy of Environmental Engineering (AAEE). Typically, the FE examination is taken in the last semester of undergraduate academic work.

It is noteworthy that this edition of *Introduction to Environmental Engineering* has been written by Board Certified Environmental Engineers. In addition we note that we have made a special effort to flag equations that appear in the NCEES *FE Fundamentals of Engineering Supplied-Reference Handbook*.

Professions

Environmental engineers are professionals. Being a professional is more than being in or of a profession. True professionals are those who pursue their learned art in a spirit of public service (ASCE, 1973). True professionalism is defined by the following characteristics:

1. Professional decisions are made by means of general principles, theories, or propositions that are independent of the particular case under consideration.

2. Professional decisions imply knowledge in a specific area in which the person is expert. The professional is an expert only in his or her profession and not an expert at everything.

3. The professional's relations with his or her clients are objective and independent of particular sentiments about them.

4. A professional achieves status and financial reward by accomplishment, not by inherent qualities such as birth order, race, religion, sex, or age or by membership in a union.

5. A professional's decisions are assumed to be on behalf of the client and to be independent of self-interest.

6. The professional relates to a voluntary association of professionals and accepts only the authority of those colleagues as a sanction on his or her own behavior (Schein, 1968).

A professional's superior knowledge is recognized. This puts the client into a very vulnerable position. The client retains significant authority and responsibility for decision making. The professional supplies ideas and information and proposes courses of action. The client's judgment and consent are required. The client's vulnerability has necessitated the development of a strong professional code of ethics. The code of ethics serves to protect not only the client but the public. Codes of ethics are enforced through the professional's peer group.

1-2 PROFESSIONAL CODES OF ETHICS

Civil engineering, from which environmental engineering is primarily, but not exclusively, derived, has an established code of ethics that embodies these principles. The code is summarized in Figure 1-1. The *FE Fundamentals of Engineering Supplied-Reference Handbook*, published by the National Council of Examiners for Engineering and Surveying (NCEES) includes *Model Rules of Professional Conduct*. The NCEES amplifies the principles of the code of ethics in the *Handbook*. It is available on line at www.ncees.org/Exams/Study_materials/Download_FE_supplied-Reference_Handbook.php

**AMERICAN SOCIETY OF CIVIL ENGINEERS
CODE OF ETHICS**

Fundamental Principles

Engineers uphold and advance the integrity, honor and dignity of the engineering profession by:

1. using their knowledge and skill for the enhancement of human welfare and the environment;
2. being honest and impartial and serving with fidelity the public, their employers and clients;
3. striving to increase the competence and prestige of the engineering profession; and
4. supporting the professional and technical societies of their disciplines.

Fundamental Canons

1. Engineers shall hold paramount the safety, health and welfare of the public and shall strive to comply with the principles of sustainable development in the performance of their professional duties.
2. Engineers shall perform services only in areas of their competence.
3. Engineers shall issue public statements only in an objective and truthful manner.
4. Engineers shall act in professional matters for each employer or client as faithful agents or trustees, and shall avoid conflicts of interest.
5. Engineers shall build their professional reputation on the merit of their services and shall not compete unfairly with others.
6. Engineers shall act in such a manner as to uphold and enhance the honor, integrity, and dignity of the engineering profession.
7. Engineers shall continue their professional development throughout their careers, and shall provide opportunities for the professional development of those engineers under their supervision.

FIGURE 1-1

American Society of Civil Engineers code of ethics. (ASCE, 2005. Reprinted with permission.)

1-3 ENVIRONMENTAL ETHICS

The birth of environmental ethics as a force is partly a result of concern for our own long-term survival, as well as our realization that humans are but one form of life, and that we share our earth with other forms of life (Vesilind, 1975).

Although it seems a bit unrealistic for us to set a framework for a discussion of environmental ethics in this short introduction, we have summarized a few salient points in Table 1-1.

TABLE 1-1
An Environmental Code of Ethics

1. Use knowledge and skill for the enhancement and protection of the environment.
2. Hold paramount the health, safety, and welfare of the environment.
3. Perform services only in areas of personal expertise.
4. Be honest and impartial in serving the public, your employers, your clients, and the environment.
5. Issue public statements only in an objective and truthful manner.

Although these few principles seem straightforward, real-world problems offer distinct challenges. Here is an example for each of the principles listed:

- The first principle may be threatened when it comes into conflict with the need for food for a starving population and the country is overrun with locusts. Will the use of pesticides enhance and protect the environment?

- The EPA has stipulated that wastewater must be disinfected where people come into contact with the water. However, the disinfectant may also kill naturally occurring beneficial microorganisms. Is this consistent with the second principle?

- Suppose your expertise is water and wastewater chemistry. Your company has accepted a job to perform air pollution analysis and asks you to perform the work in the absence of a colleague who is the company's expert. Do you decline and risk being fired?

- The public, your employers, and your client believe that dredging a lake to remove weeds and sediment will enhance the lake. However, the dredging will destroy the habitat for muskrats. How can you be impartial to *all* these constituencies?

- You believe that a new regulation proposed by EPA is too expensive to implement but you have no data to confirm that opinion. How do you respond to a local newspaper reporter asking for your opinion? Do you violate the fifth principle even though it is "your opinion" that is being sought?

We think it is important to point out that many environmentally related decisions such as those described above are much more difficult than the problems presented in the remaining chapters of this book. Frequently these problems are related more to ethics than to engineering. The problems arise when there are several courses of action with no *a priori* certainty as to which is best. Decisions related to safety, health, and welfare are easily resolved. Decisions as to which course of action is in the best interest of the public are much more difficult to resolve. Furthermore, decisions as to which course of action is in the best interest of the environment are at times in conflict with those that are in the best interest of the public. Whereas decisions made in the public interest are based on professional ethics, decisions made in the best interest of the environment are based on environmental ethics.

Ethos, the Greek word from which "ethic" is derived, means the character of a person as described by his or her actions. This character was developed during the evolutionary process and was influenced by the need for adapting to the natural environment. Our ethic is our way of doing things. Our ethic is a direct result of our natural environment. During the latter stages of the evolutionary process, *Homo sapiens* began to modify the environment rather than submit to what, millennia later, became known as Darwinian natural selection. As an example, consider the cave dweller who, in the chilly dawn of prehistory, realized the value of the saber-toothed tiger's coat and appropriated it for personal use. Inevitably a pattern of appropriation developed, and our ethic became more self-modified than environmentally adapted. Thus, we are no longer adapted to our natural environment but rather to our self-made environment. In the ecological context, such maladaptation results in one of two consequences: (1) the organism (*Homo sapiens*) dies out; or (2) the organism evolves to a form and character that is once again compatible with the natural environment (Vesilind, 1975). Assuming that we choose the latter course, how can this change in character (ethic) be brought about? Each individual must change his or her character or ethic, and the social system must change to become compatible with the global ecology.

The acceptable system is one in which we learn to share our exhaustible resources—to regain a balance. This requires that we reduce our needs and that the materials we use must be replenishable. We must treat all of the earth as a sacred trust to be used so that its content is neither diminished nor permanently changed; we must release no substances that cannot be reincorporated without damage to the natural system. The recognition of the need for such adaptation (as a means of survival) has developed into what we now call the *environmental ethic* (Vesilind, 1975).

1-4 ENGINEERING DIMENSIONS AND UNITS

The *FE Fundamentals of Engineering Supplied-Reference Handbook* uses the metric system of units. Ultimately, the FE examination will be entirely metric. However, currently some of the FE examination problems use both metric and U.S. Customary System (USCS) units. This text uses the metric system of units. Because the FE examination has some problems in U.S. Customary units, we have included some example problems and some FE formatted end-of-chapter problems in U.S. Customary units.

Our experience is that U.S. students are very familiar with the metric system of units and have an adequate knowledge of fundamental U.S. Customary System (USCS) units such as feet per second (ft/s or fps), miles per hour (mph), pounds mass (lb_m), and gallons (gal) that we need not elaborate more than this brief reminder. However, there are a small number of units and abbreviations that are particular to environmental engineering that we feel should be addressed here. They will be used without further elaboration in the following chapters. At appropriate places, we will provide examples of the use of handy equivalences.

The following are USCS definitions:

acre-ft (or ac-ft): a volume of water that has a surface area of one acre and a depth of one foot or an equivalent volume by other measurements, for example, an area of ½ acre and a depth of 2 feet or an area of 2 acres and a depth of ½ foot.

Btu: British thermal unit

cfs: cubic feet per second

gal: U.S. gallon(s)

gpm: U.S. gallon(s) per minute

gpcd: U.S. gallons per capita per day

hp: horsepower

MGD (or sometimes mgd): million U.S. gallons per day

ppb: parts (mass) per billion parts of fluid; the fluid is understood to be water. Alternatively it may be parts (mass) per billion parts of soil. ppb is equivalent to μg/kg.

TABLE 1-2
U.S. Customary System conversions factors

Multiply	by	To obtain
acre (ac)	43,560	square feet (ft^2)
acre-ft	325,851	U.S gallons
Btu	2.930×10^{-4}	kW-hour
Btu/min	0.02358	hp
Btu/min	0.01758	kW
ft^3 of water	62.4	lb_m of water
ft^3 of water	7.48	U.S. gallons of water
gal of water	0.1337	ft^3 of water
gal of water	8.34	lb_m of water
gpd/ft^2	0.04074	$m^3/d \cdot m^2$
gpm/ft^2	2.445	$m^3/h \cdot m^2$
hp	0.7457	kW
psi	2.307	ft of water
$lb_m/ft^2 \cdot d$	0.2048	$kg/m^2 \cdot d$
lb_m/U.S. ton	0.4999	g/kg
U.S. short tons	2,000	lb_m
U.S. tons/acre	0.2242	kg/ha

ppm: parts (mass) of substance per million parts of the fluid. Alternatively it may be parts (mass) per billion parts of soil. ppm is equivalent to mg/kg.

ppm(v/v): volume of substance per million volumes of fluid; the fluid is understood to be air

psi: a pressure; pounds force per square inch of surface area

sf: square feet

U.S. ton: 2000 lb_m

Conversion factors for the USCS are given in Table 1-2 and in Appendix C. Conversions from SI units to USCS units are given inside the back cover of this book.

1-5 ENVIRONMENTAL SYSTEMS OVERVIEW

Systems

Before we begin in earnest, we thought it worth taking a look at the problems to be discussed in this text in a larger perspective. Engineers like to call this the "systems approach," that is, looking at all the interrelated parts and their effects on one another. In environmental systems it is doubtful that mere mortals can ever hope to identify all the interrelated parts, to say nothing of trying to establish their effects on one another. The first thing the systems engineer does, then, is to simplify the system to a tractable size that behaves in a fashion similar to the real system. The simplified model does not behave in detail as the system does, but it gives a fair approximation of what is going on.

We have followed this pattern of simplification in our description of three environmental systems: the water resource management system, the air resource management system, and the solid waste management system. Pollution problems that are confined to one of these systems are called single-medium problems if the medium is either air, water, or soil. Many important environmental problems are not confined to one of these simple systems but cross the boundaries from one to the other. These problems are referred to as *multimedia* pollution problems.

Water Resource Management System

Water Supply Subsystem. The nature of the water source commonly determines the planning, design, and operation of the collection, purification, transmission, and distribution works.* The two major sources used to supply community and industrial needs are referred to as *surface water* and *groundwater*. Streams, lakes, and rivers are the surface water sources. Groundwater sources are those pumped from wells.

Figure 1-2 depicts an extension of the water resource system to serve a small community. The source in each case determines the type of collection works and the type of treatment works. The pipe network in the city is called the distribution system. The

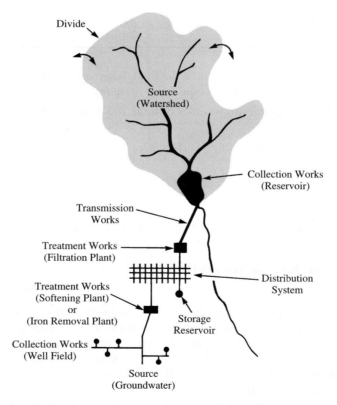

FIGURE 1-2
An extension of the water supply resource system.

Works is a noun used in the plural to mean "engineering structures." It is used in the same sense as *art works.*

pipes themselves are often referred to as *water mains*. Water in the mains generally is kept at a pressure between 200 and 860 kilopascals (kPa). Excess water produced by the treatment plant during periods of low *demand** (usually the nighttime hours) is held in a storage reservoir. The storage reservoir may be elevated (the ubiquitous water tower), or it may be at ground level. The stored water is used to meet high demand during the day. Storage compensates for changes in demand and allows a smaller treatment plant to be built. It also provides emergency backup in case of a fire.

Population and water-consumption patterns are the prime factors that govern the quantity of water required and hence the source and the whole composition of the water resource system. One of the first steps in the selection of a suitable water-supply source is determining the demand that will be placed on it. The essential elements of water demand include average daily water consumption and peak rate of demand. Average daily water consumption must be estimated for two reasons: (1) to determine the ability of the water source to meet continuing demands over critical periods when surface flows are low or groundwater tables are at minimum elevations, and (2) for purposes of estimating quantities of stored water that would satisfy demands during these critical periods. The peak demand rates must be estimated in order to determine plumbing and pipe sizing, pressure losses, and storage requirements necessary to supply sufficient water during periods of peak water demand.

Many factors influence water use for a given system. For example, the mere fact that water under pressure is available stimulates its use, often excessively, for watering lawns and gardens, for washing automobiles, for operating air-conditioning equipment, and for performing many other activities at home and in industry. The following factors have been found to influence water consumption in a major way:

1. Climate

2. Industrial activity

3. Meterage

4. System management

5. Standard of living

The following factors also influence water consumption but to a lesser degree: extent of sewerage, system pressure, water price, and availability of private wells.

If the demand for water is measured on a *per capita*† basis, climate is the most important factor influencing demand. This is shown dramatically in Table 1-3. The average annual precipitation for the "wet" states is about 100 cm per year while the average annual precipitation for the "dry" states is only about 25 cm per year. Of course, the dry states are also considerably warmer than the wet states.

Demand is the use of water by consumers. This use of the word derives from the economic term meaning "the desire for a commodity." The consumers express their desire by opening the faucet or flushing the water closet (W.C.).
†Per capita is a Latin term that means "by heads." Here it means "per person." This assumes that each person has one head (on the average).

TABLE 1-3
Total fresh water withdrawals for public supply[a]

State	Withdrawal (Lpcd)[b]
"Wet"	
Connecticut	680
Michigan	598
New Jersey	465
Ohio	571
Pennsylvania	543
Average	571
"Dry"	
Nevada	1,450
New Mexico	698
Utah	926
Average	1,025

[a]Compiled from Kenny et al., 2009.
[b]Lpcd = liters per capita per day.

The influence of industry is to increase per capita water demand. Small rural and suburban communities will use less water per person than industrialized communities.

The third most important factor in water use is whether individual consumers have water meters. Meterage imposes a sense of responsibility not found in unmetered residences and businesses. This sense of responsibility reduces per capita water consumption because customers repair leaks and make more conservative water-use decisions almost regardless of price. For residential consumers, water is so inexpensive, price is not much of a factor in water use. Water price is extremely important for industrial and farming operations that use large volumes of water.

Following meterage closely is the aspect called system management. If the water distribution system is well managed, per capita water consumption is less than if it is not well managed. Well-managed systems are those in which the managers know when and where leaks in the water main occur and have them repaired promptly.

Climate, industrial activity, meterage, and system management are more significant factors controlling water consumption than the standard of living. The rationale for the last factor is straightforward. Per capita water use increases with an increased standard of living. Highly developed countries use much more water than the less developed nations. Likewise, higher socioeconomic status implies greater per capita water use than lower socioeconomic status.

The total U.S. water withdrawal for all uses (agricultural, commercial, domestic, mining, and thermoelectric power), including both fresh and saline water, was estimated to be approximately 5,100 liters per capita per day (Lpcd) in 2005 (Kenny, et al., 2009). The amount for U.S. public supply (domestic, commercial, and industrial use) was estimated to be 550 Lpcd in 2005 (Kenny, et al., 2009). The American Water

TABLE 1-4
Examples of variation in per capita water consumption

Location	Lpcd	Percent of per capita consumption		
		Industry	Commercial	Residential
Lansing, MI	512	14	32	54
East Lansing, MI	310	0	10	90
Michigan State University	271	0	1	99

Data from local treatment plants, 2004.

Works Association estimated that the average daily household water use in the United States was 1,320 liters per day in 1999 (AWWA, 1999). For a family of three, this would amount to about 440 Lpcd. The variation in demand is normally reported as a factor of the average day. For metered dwellings the factors are as follows: maximum day = 2.2 × average day; peak hour = 5.3 × average day (Linaweaver et al., 1967). Some mid-Michigan average daily use figures and the contribution of various sectors to demand are shown in Table 1-4.

International per capita domestic water use has been estimated by the Pacific Institute for Studies in Development, Environment, and Security (Pacific Institute, 2000). For example, they report the following (all in Lpcd): Australia, 1,400; Canada, 430; China, 60; Ecuador, 85; Egypt, 130; Germany, 270; India, 30; Mexico, 130; Nigeria, 25.

Wastewater Disposal Subsystem.   Safe disposal of all human wastes is necessary to protect the health of the individual, the family, and the community and also to prevent the occurrence of certain nuisances. To accomplish satisfactory results, human wastes must be disposed of so that:

1. They will not contaminate any drinking water supply.

2. They will not give rise to a public health hazard by being accessible to *vectors* (insects, rodents, or other possible carriers) that may come into contact with food or drinking water.

3. They will not give rise to a public health hazard by being accessible to children.

4. They will not cause violation of laws or regulations governing water pollution or sewage disposal.

5. They will not pollute or contaminate the waters of any bathing beach, shellfish-breeding ground, or stream used for public or domestic water-supply purposes, or for recreational purposes.

6. They will not give rise to a nuisance due to odor or unsightly appearance.

These criteria can best be met by the discharge of domestic sewage to an adequate public or community sewerage system (U.S. PHS, 1970). Where no community sewer system exists, on-site disposal by an approved method is mandatory.

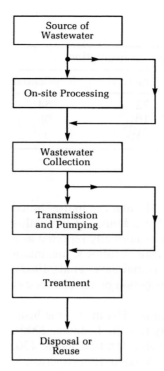

FIGURE 1-3
Wastewater management subsystem. (Linsley and Fanzini, 1979)

In its simplest form the wastewater management subsystem is composed of six parts (Figure 1-3). The source of wastewater may be either industrial wastewater or domestic sewage or both.* Industrial wastewater may be subject to some pretreatment on site if it has the potential to upset the municipal wastewater treatment plant (WWTP). Federal regulations refer to municipal wastewater treatment systems as publicly owned treatment works, or POTWs.

The quantity of sewage flowing to the WWTP varies widely throughout the day in response to water usage. A typical daily variation is shown in Figure 1-4. Most of the water used in a community will end up in the sewer. Between 5 and 15 percent of the water is lost in lawn watering, car washing, and other consumptive uses. In warm, dry climates, consumptive use out of doors may be as high as 60 percent. Consumptive use may be thought of as the difference between the average rate that water flows into the distribution system and the average rate that wastewater flows into the WWTP (excepting the effects of leaks in the pipes).

The quantity of wastewater, with one exception, depends on the same factors that determine the quantity of water required for supply. The major exception is that underground water (groundwater) conditions may strongly affect the quantity of water in the system because of leaks. Whereas the drinking water distribution system is under pressure

---

*Domestic sewage is sometimes called sanitary sewage, although it is far from being sanitary!

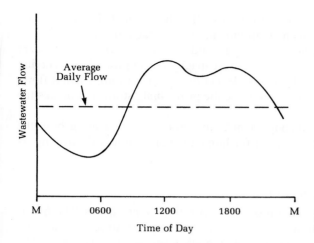

FIGURE 1-4
Typical variation in daily wastewater flow.

and is relatively tight, the sewer system is gravity operated and is relatively open. Thus, groundwater may *infiltrate,* or leak into, the system. When manholes lie in low spots, there is the additional possibility of *inflow* through leaks in the manhole cover. Other sources of inflow include direct connections from roof gutters and downspouts, as well as sump pumps used to remove water from basement footing tiles. *Infiltration* and *inflow* (I & I) are particularly important during rainstorms. The additional water from I & I may hydraulically overload the sewer causing sewage to back up into houses as well as to reduce the efficiency of the WWTP. New construction techniques and materials have made it possible to reduce I & I to insignificant amounts.

Sewers are classified into three categories: sanitary, storm, and combined. *Sanitary sewers* are designed to carry municipal wastewater from homes and commercial establishments. With proper pretreatment, industrial wastes may also be discharged into these sewers. *Storm sewers* are designed to handle excess rainwater to prevent flooding of low areas. While sanitary sewers convey wastewater to treatment facilities, storm sewers generally discharge into rivers and streams. *Combined sewers* are expected to accommodate both municipal wastewater and stormwater. These systems were designed so that during dry periods the wastewater is carried to a treatment facility. During rain storms, the excess water is discharged directly into a river, stream, or lake without treatment. Unfortunately, the storm water is mixed with untreated sewage. The U.S. Environmental Protection Agency (EPA) has estimated that 40,000 overflows occur each year. Combined sewers are no longer being built in the United States. Many communities are in the process of replacing the combined sewers with separate systems for sanitary and storm flow.

When gravity flow is not possible or when sewer trenches become uneconomically deep, the wastewater may be pumped. When the sewage is pumped vertically to discharge into a higher-elevation gravity sewer, the location of the sewage pump is called a *lift station.*

Sewage treatment is performed at the WWTP to stabilize the waste material, that is, to make it less *putrescible.* The *effluent* from the WWTP may be discharged into an ocean, lake, or river (called the receiving body). Alternatively, it may be discharged onto (or into) the ground, or be processed for reuse. The by-product sludge from the WWTP also must be disposed of in an environmentally acceptable manner.

Whether the waste is discharged onto the ground or into a receiving body, care must be exercised not to overtax the assimilative capacity of the ground or receiving body. The fact that the wastewater effluent is cleaner than the river into which it flows does not justify the discharge if it turns out to be the proverbial "straw that breaks the camel's back."

In summary, water resource management is the process of managing both the quantity and the quality of the water used for human benefit without destroying its availability and purity.

Air Resource Management System

Our air resource differs from our water resource in two important aspects. The first is in regard to quantity. Whereas engineering structures are required to provide an adequate water supply, air is delivered free of charge in whatever quantity we desire. The second aspect is in regard to quality. Unlike water, which can be treated before we use it, it is impractical to go about with a gas mask on to treat impure air and with ear plugs in to keep out the noise.

The balance of cost and benefit to obtain a desired quality of air is termed *air resource management*. Cost-benefit analyses can be problematic for at least two reasons. First is the question of what is desired air quality. The basic objective is, of course, to protect the health and welfare of people. But how much air pollution can we stand? We know the tolerable limit is something greater than zero, but tolerance varies from person to person. Second is the question of cost versus benefit. We know that we don't want to spend the entire Gross Domestic Product to ensure that no individual's health or welfare is impaired, but we do know that we want to spend some amount. Although the cost of control can be reasonably determined by standard engineering and economic means, the cost of pollution is still far from being quantitatively assessed.

Air resource management programs are instituted for a variety of reasons. The most defensible reasons are that (1) air quality has deteriorated and there is a need for correction, and (2) the potential for a future problem is strong.

In order to carry out an air resource management program effectively, all of the elements shown in Figure 1-5 must be employed. (Note that with the appropriate substitution of the word *water* for *air,* these elements apply to management of water resources as well.)

Solid Waste Management System

In the past, solid waste was considered a resource, and we will examine its current potential as a resource. Generally, however, solid waste is considered a problem to be solved as cheaply as possible rather than a resource to be recovered. A simplified block diagram of a solid waste management system is shown in Figure 1-6.

While typhoid and cholera epidemics of the mid-1800s spurred water resource management efforts, and while air pollution episodes have prompted better air resource management, we have yet to feel the impact of material or energy shortages

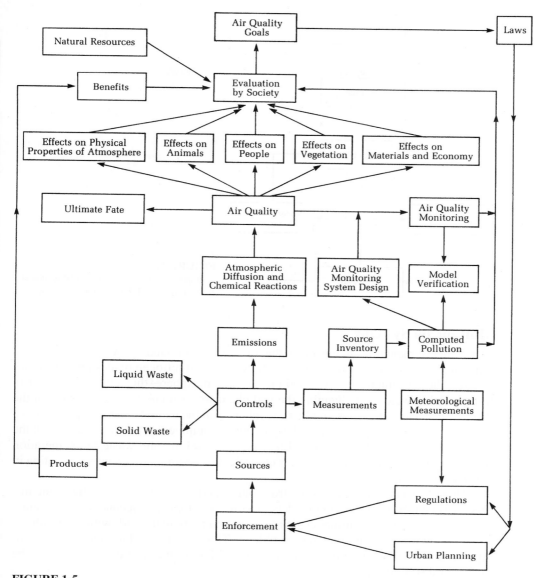

FIGURE 1-5

A simplified block diagram of an air resource management system.

severe enough to encourage modern solid waste management. The landfill "crisis" of the 1980s appears to have abated in the early 1990s due to new or expanded landfill capacity and to many initiatives to reduce the amount of solid waste generated. By 1999, more than 9,000 curbside recycling programs served roughly half of the U.S. population (U.S. EPA, 2005a).

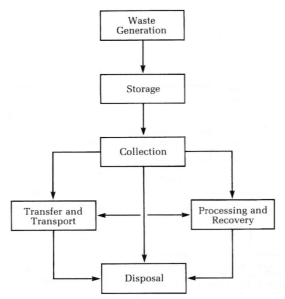

FIGURE 1-6

A simplified block diagram of a solid waste management system. (Tchobanoglous et al., 1977)

Multimedia Systems

Many environmental problems cross the air-water-soil boundary. An example is acid rain that results from the emission of sulfur oxides and nitrogen oxides into the atmosphere. These pollutants are washed out of the atmosphere, thus cleansing it, but in turn polluting water and changing the soil chemistry, which ultimately results in the death of fish and trees. Thus, our historic reliance on the natural cleansing processes of the atmosphere in designing air-pollution-control equipment has failed to deal with the multimedia nature of the problem. Likewise, disposal of solid waste by incineration results in air pollution, which in turn is controlled by scrubbing with water, resulting in a water pollution problem.

Three lessons have come to us from our experience with multimedia problems. First, it is dangerous to develop models that are too simplistic. Second, environmental engineers must use a multimedia approach and, in particular, work with a multidisciplinary team to solve environmental problems. Third, the best solution to environmental pollution is waste minimization—if waste is not produced, it does not need to be treated or disposed of.

1-6 ENVIRONMENTAL LEGISLATION AND REGULATION

The requirements of environmental legislation and regulation are major considerations in the selection of technology and the design of treatment processes to remove contaminants. The following paragraphs provide a brief introduction to the process leading to the establishment of regulations and the terms used to identify the location of information about bills, laws, and regulations. This discussion is restricted to the federal process and nomenclature.

Acts and Laws

A proposal for a new law, called a *bill,* is introduced in either the Senate or the House of Representatives (House). The bill is given a designation, for example S. 2649 in the Senate or H.R. 5959 in the House. Bills often have "companions" in that similar bills may be started in both the Senate and House at the same time. The bill is given a title, for example, the "Safe Drinking Water Act" which implies an "act" of Congress. The act may be listed under one "Title" or it may be divided into several Titles. References to the Titles of the act are given by roman numeral. For example, Title III of the Clean Air Act Amendments establishes a list of hazardous air pollutants. Frequently a bill directs some executive branch of the government such as the EPA to carry out an action such as setting limits for contaminants. On occasion, such a bill includes specific numbers for limits on contaminants. If the bills successfully pass the committee to which they are assigned, they are "reported out" to the full Senate (for example, Senate Report 99-56) or to the full House (House Report 99-168). The first digits preceding the dash refer to the session of Congress during which the bill is reported out. In this example, it is the 99th Congress. If bills pass the full Senate/House they are taken by a joint committee of senators and congressional representatives (conference committee) to form a single bill for action by both the Senate and House. If the bill is adopted by a majority of both houses, it goes to the president for approval or veto. When the president signs the bill it becomes a *law* or *statute*. It is then designated, for example, as Public Law 99-339 or PL 99-339. This means it is the 339th law passed by the 99th Congress. The law or statute approved by the president's signature may alternatively be called an act that is referred to by the title assigned the bill in Congress.

The Office of the Federal Register prepares the *United States Statutes at Large* annually. This is a compilation of the laws, concurrent resolutions, reorganization plans, and proclamations issued during each congressional session. The statutes are numbered chronologically. They are not placed in order by subject matter. The shorthand reference is, for example, 104 Stat. 3000.

The *United States Code* is the compiled written set of laws in force on the day before the beginning of the current session of Congress (U.S. Code, 2005). Reference is made to the U.S. Code by Title and Section number (for example, 42 USC 6901 or 42 U.S.C. §6901). Table 1-5 gives a sample of titles and sections of environmental interest. Note that Titles of the U.S. Code do not match the Titles of the Acts of Congress.

Regulations

In carrying out the directives of the Congress to develop a *regulation* or *rule,* the EPA or other executive branch of the government follows a specific set of formal procedures in a process referred to as *rule making*. The government agency (EPA, Department of Energy, Federal Aviation Agency, etc.) first publishes a *proposed rule* in the *Federal Register*. The *Federal Register* is, in essence, the government's newspaper. It is published every day that the federal government is open for business. The agency provides the logic for the rule making (called a *preamble*) as well as the proposed rule and requests comments. The preamble may be several hundred pages in length for a rule that is only

TABLE 1-5
U.S. Code title and section numbers of environmental interest

Title	Sections	Statute
7	136 to 136y	Federal Insecticide, Fungicide, and Rodenticide Act
16	1531 to 1544	Endangered Species Act
33	1251 to 1387	Clean Water Act
33	2701 to 2761	Oil Pollution Act
42	300f to 300j-26	Safe Drinking Water Act
42	4321 to 4347	National Environmental Policy Act
42	4901 to 4918	Noise Control Act
42	6901 to 6922k	Solid Waste Disposal Act
42	7401 to 7671q	Clean Air Act (includes noise at §7641)
42	9601 to 9675	Comprehensive Environmental Response, Compensation, and Liability Act
42	11001 to 11050	Emergency Planning and Community Right-to-Know Act
42	13101 to 13109	Pollution Prevention Act
46	3703a	Oil Pollution Act
49	2101	Aviation Safety and Noise Abatement Act[a]
49	2202	Airport and Airway Improvement Act[a]
49	47501 to 47510	Airport Noise Abatement Act

[a] At U.S. Code Annotated (U.S.C.S.A.)

a few lines long or a single page table of allowable concentrations of contaminants. Prior to the issuance of a final rule, the agency allows and considers public comment. The time period for submitting public comments varies. For rules that are not complex or controversial it may be a few weeks. For more complicated rules, the comment period may extend for as long as a year. The reference citation to *Federal Register* publications is in the following form: 59 FR 11863. The first number is the volume number. Volumes are numbered by year. The last number is the page number. Pages are numbered sequentially beginning with page 1 on the first day of business in January of each year. From the number shown, this rule making starts on page 11,863! Although one might assume this is late in the year, it may not be if a large number of rules have been published. This makes the date of publication very useful in searching for the rule.

Once a year, on July 1, the rules that have been finalized in the past year are *codified*. This means they are organized and published in the *Code of Federal Regulations* (CFR, 2005). Unlike the *Federal Register,* the *Code of Federal Regulations* is a compilation of the rules/regulations of the various agencies without explanation of how the government arrived at its decision. The explanation of how the rule was developed may be found only in the *Federal Register.* The notation used for *Code of Federal Regulations* is as follows: 40 CFR 280. The first number is the Title number. The second number in the citation refers to the part number. Unfortunately, this title number has no relation to either the title number in the Act or the United States Code title number. The CFR title numbers and subjects of environmental interest are shown in Table 1-6. A detailed discussion of the development of environmental legislation is available at the text website: www.mhhe.com/davis.

TABLE 1-6
Code of Federal Regulations **title numbers of environmental interest**

Title number	Subject
7	Agriculture (soil conservation)
10	Energy (Nuclear Regulatory Commission)
14	Aeronautics and Space (noise)
16	Conservation
23	Highways (noise)
24	Housing and Urban Development (noise)
29	Labor (noise)
30	Mineral Resources (surface mining reclamation)
33	Navigation and Navigable Waters (wet lands and dredging)
40	Protection of the Environment (Environmental Protection Agency)
42	Public Health and Welfare
43	Public Lands: Interior
49	Transportation (transporting hazardous waste)
50	Wildlife and Fisheries

1-7   CHAPTER REVIEW

When you have completed studying this chapter, you should be able to do the following without the aid of your textbook or notes:

1. Sketch and label a water resource system including (*a*) source; (*b*) collection works; (*c*) transmission works; (*d*) treatment works; and (*e*) distribution works.

2. State the proper general approach to treatment of a surface water and a groundwater (see Figure 1-2).

3. Define the word "demand" as it applies to water.

4. List the five most important factors contributing to water consumption and explain why each has an effect.

5. State the rule-of-thumb water requirement for an average city on a per-person basis and calculate the average daily water requirement for a city of a stated population.

6. Define the acronyms WWTP and POTW.

7. Explain why separate storm sewers and sanitary sewers are preferred over combined sewers.

8. Explain the purpose of a lift station.

1-8   PROBLEMS

1-1.   Estimate the total daily water withdrawal (in m^3/d) including both fresh and saline water for all uses for the United States in 2000. The population was 281,421,906.

Answer: 1.52×10^9 m^3/d

1-2.   Estimate the per capita daily water withdrawal for public supply in the United States in 2005 (in Lpcd). Use the following population data (McGeveran, 2002) and water supply data (Kenny et al., 2009):

Year	Population	Public supply withdrawal, m^3/d
1950	151,325,798	5.30×10^7
1960	179,323,175	7.95×10^7
1970	203,302,031	1.02×10^8
1980	226,542,203	1.29×10^8
1990	248,709,873	1.46×10^8
2000	281,421,906	1.64×10^8

(Note: This problem may be worked by hand calculation and then plotted on graph paper to extrapolate to 2005, or it may be worked by using a spreadsheet to perform the calculations, plot the graph, and extrapolate to 2005.)

1-3.   A residential development of 280 houses is being planned. Assume that the American Water Works Association average daily household consumption applies, and that each house has three residents. Estimate the additional average daily water production in L/d that will have to be supplied by the city.

Answer: 3.70×10^5 L/d

1-4.   Repeat Problem 1-3 for 320 houses, but assume that low-flush valves reduce water consumption by 14 percent.

1-5.   Using the data in Problem 1-3 and assuming that the houses are metered, determine what additional demand will be made at the peak hour.

Answer: 1.96×10^6 L/d

1-6.   If a faucet is dripping at a rate of one drop per second and each drop contains 0.150 milliliters, calculate how much water (in liters) will be lost in 1 year.

1-7.   Savabuck University has installed standard pressure-operated flush valves on its water closets. When flushing, these valves deliver 130.0 L/min. If the delivered water costs $0.45 per cubic meter, what is the monthly cost of not repairing a broken valve that flushes continuously?

Answer: $2,527.20, or $2,530/mo

1-8.   The American Water Works Association estimates that 15 percent of the water that utilities process is lost each day. Assuming that the loss was from public supply withdrawal in 2000 (Problem 1-2), estimate the total value of the lost water if delivered water costs $0.45 per cubic meter.

1-9. Water delivered from a public supply in western Michigan costs $0.45 per cubic meter. A 0.5-L bottle of water purchased from a dispensing machine costs $1.00. What is the cost of the bottled water on a per cubic meter basis?

Answer: $2,000/m^3

1-10. Using U.S. Geological Survey Circular 1268 (http://usgs.gov), estimate the daily per capita domestic withdrawal of fresh water in South Carolina in Lpcd. (Note: conversion factors inside the back cover of this book may be helpful.)

1-11. Using the Pacific Institute for Studies in Development, Environment, and Security website (http://www.worldwater.org/table2.html), determine the lowest per capita domestic water withdrawal in the world in Lpcd and identify the country in which it occurs.

1-9 DISCUSSION QUESTIONS

1-1. Would you expect the demand for water to drop in half if the price ($/L) doubled? Explain your reasoning.

1-2. The water supply for the city of Peoria is from wells. Other than disinfection, no water treatment is provided. A filtration plant would be appropriate to improve the quality of the water. True or False? If the answer is false, revise the statement so that it is true.

1-3. The water treatment plant for the town of Gettysburg was built 20 years ago. Over the last few years, there has been difficulty in maintaining water pressure in the system over the 4th of July weekend. In some parts of town only a trickle of water flows from the tap during early morning and late evening hours. There are no problems during the remainder of the year. Explain why the town may be having water pressure problems.

1-4. The town of West Lafayette is considering two proposals for a new water-treatment plant. West Lafayette's average daily demand is 11,400 m^3/d. Proposal A is to build a plant that will produce 475 m^3/h and a storage reservoir to hold 2,520 m^3 of water. Proposal B is to build a plant that will produce 1,425 m^3/h but no water storage reservoir will be provided. Which proposal do you recommend? Explain why.

1-5. Homeowners in the town of Rolla have connected their downspouts and the sump pumps from their footing drains to the sanitary sewer system. The rainwater and sump water entering the sewer is called (choose one):

(a) Infiltration

(b) Inflow

These connections to the sanitary sewer, in effect, make it a (choose one):

(c) Storm sewer

(d) Combined sewer

Explain why you have made your choices.

1-6. The Shiny Plating Company is using about 2,000.0 kg/wk of organic solvent for vapor degreasing of metal parts before they are plated. The Air Pollution Engineering and Testing Company (APET) has measured the air in the workroom and in the stack that vents the degreaser. APET has determined that 1,985.0 kg/wk is being vented up the stack and that the workroom environment is within occupational standards. The 1,985.0 kg/wk is well above the allowable emission rate of 11.28 kg/wk.

Elizabeth Fry, the plant superintendent, has asked J.R. Injuneer, the plant engineer, to review two alternative control approaches offered by APET and to recommend one of them.

The first method is to purchase a pollution control device to put on the stack. This control system will reduce the solvent emission to 1.0 kg/wk. Approximately 1,950.0 kg of the solvent which is captured each week can be recycled back to the degreaser. Approximately 34.0 kg of the solvent must be discharged to the wastewater treatment plant (WWTP). J. R. has determined that this small amount of solvent will not adversely affect the performance of the WWTP. In addition, the capital cost of the pollution control equipment will be recovered in about two years as a result of savings from recovering lost solvent.

The second method is to substitute a solvent that is not on the list of regulated emissions. The price of the substitute is about 10 percent higher than the solvent currently in use. J. R. has estimated that the substitute solvent loss will be about 100.0 kg/wk. The substitute collects moisture and loses its effectiveness in about a month's time. The substitute solvent cannot be discharged to the WWTP because it will adversely affect the WWTP performance. Consequently, about 2,000 kg must be hauled to a hazardous waste disposal site for storage each month. Because of the lack of capital funds and the high interest rate for borrowing, J. R. recommends that the substitute solvent be used. Do you agree with this recommendation? Explain your reasoning.

1-7. Ted Terrific is the manager of a leather tanning company. In part of the tanning operation a solution of chromic acid is used. It is company policy that the spent chrome solution is put in 0.20-m^3 drums and shipped to a hazardous waste disposal facility.

On Thursday the 12th, the day shift miscalculates the amount of chrome to add to a new batch and makes it too strong. Since there is not enough room in the tank to adjust the concentration, Abe Lincoln, the shift supervisor, has the tank emptied and a new one prepared and makes a note to the manager that the bad batch needs to be reworked.

On Monday the 16th, Abe Lincoln looks for the bad batch and cannot find it. He notifies Ted Terrific that it is missing. Upon investigation, Ted finds that Rip Van-Winkle, the night-shift supervisor, dumped the batch into the sanitary sewer at 3:00 a.m. on Friday the 13th. Ted makes discreet inquiries at the wastewater plant and finds that they have had no process upsets. After Ted severely disciplines Rip, he should: (Choose the correct answer and explain your reasoning.)

A. Inform the city and state authorities of the illegal discharge as required by law even though no apparent harm resulted.

B. Keep the incident quiet because it will cause trouble for the company without doing the public any good. No harm was done and the shift supervisor has been punished.

C. Advise the president and board of directors and let them decide whether to follow A or B.

1-10 FE EXAM FORMATTED PROBLEMS

1-1. A residential development of 320 houses is being planned. For planning purposes, an average daily consumption of 120 gpcd is used. Estimate the average daily volume of water that must be supplied to this development if each house is occupied by 4 people.

    a. 153,600 gallons per day   b. 38,400 gallons per day
    c. 145,300 gallons per day   d. 581,400 gallons per day

1-2. The maximum day demand for a current population of people is 90 MGD. The maximum day demand 20 years from the current estimate is 120 MGD. The following assumptions were made for the estimate: peaking factor $= 1.5$; average day demand remains constant at 150 gpcd over the 20 year period; population growth is exponential. What annual rate of population growth was used for the estimate?

    a. 15.0 percent   b. 6.67 percent
    c. 1.44 percent   d. 0.0144 percent

1-3. State University has a resident population of 43,000. If the average day demand is 300 Lpcd, what average flow rate per day (in m^3/d) must be supplied to the campus?

    a. $1.29 \times 10^7 \, m^3/d$ b. $1.29 \times 10^4 \, m^3/d$
    c. $3.41 \times 10^6 \, m^3/d$ d. $3.41 \times 10^3 \, m^3/d$

1-4. If 60 percent of the average household water use in a dry climate is used outside of the house, what is the estimated wastewater flow rate (in m^3/d) for a community of 20,100 that has an average day demand of 960 Lpcd.

    a. $7.72 \times 10^6 \, m^3/d$ b. $1.16 \times 10^7 \, m^3/d$
    c. $7.72 \times 10^3 \, m^3/d$ d. $1.16 \times 10^4 \, m^3/d$

1-11 REFERENCES

ASCE (1973) *Official Record,* American Society of Civil Engineers, New York.

ASCE (1977) *Official Record,* Environmental Engineering Division, Statement of Purpose, American Society of Civil Engineers, New York.

ASCE (2005) http://www.asce.org/inside/codeofethics.cfm.

AWWA (1999) American Water Works Association, "Stats on Tap," Denver, http://www.awwa.org/Advocacy/pressroom/STATS.cfm.

CFR (2005) U.S. Government Printing Office, Washington, DC, http://www.gpoaccess.gov/ecfr/ (in January 2005 this was a beta test site for searching the CFR).

Kenny, J. F., N. L. Barker, S. S. Hutson, et al. (2009) *Estimated Use of Water in the United States in 2005,* U.S. Geological Survey Circular 1344, Washington, DC. http://www.usgs.gov

Linaweaver, F. P., J. C. Geyer, and J. B. Wolff (1967) "Summary Report on the Residential Water Use Research Project," *Journal of the American Water Works Association,* vol. 59, p. 267.

Linsley, R. K., and J. B. Fanzini (1979) *Water Resources Engineering,* McGraw-Hill, New York, p. 546.

McGeveran, W. A. (editorial director) (2002) *The World Almanac and Book of Facts: 2002,* World Almanac Books; New York, p. 377.

Pacific Institute (2000) Pacific Institute for Studies in Development, Environment, and Security, Oakland, CA. http://www.worldwater.org/table2.html.

Schein, E. H. (1968) "Organizational Socialization and the Profession of Management," 3rd Douglas Murray McGregor Memorial Lecture to the Alfred P. Sloan School of Management, Massachusetts Institute of Technology.

Tchobanoglous, G., H. Theisen, and R. Eliassen (1977) *Solid Wastes,* McGraw-Hill, New York, p. 21.

U.S. Code (2005) House of Representatives, Washington, DC, http://uscode.house.gov/.

U.S. EPA (2005a) "Municipal Solid Waste: Reduce, Reuse, Recycle," U.S. Environmental Protection Agency, Washington, DC, http://www.epa.gov/epaoswer/non-hw/muncpl/reduce.htm.

U.S. PHS (1970) *Manual of Septic Tank Practice,* Public Health Service Publication No. 526, Department of Health, Education and Welfare, Washington, DC.

Vesilind, P. A. (1975) *Environmental Pollution and Control,* Ann Arbor Science, Ann Arbor, MI, p. 214.

CHAPTER
2

MATERIALS AND ENERGY BALANCES

2-1   INTRODUCTION

Materials and energy balances are key tools in achieving a quantitative understanding of the behavior of environmental systems. They serve as a method of accounting for the flow of energy and materials into and out of environmental systems. Mass balances provide us with a tool for modeling the production, transport, and fate of pollutants in the environment. Energy balances likewise provide us with a tool for modeling the production, transport, and fate of energy in the environment.

Applications

Examples of the application of mass balances include prediction of rainwater runoff (Chapter 4), oxygen balance in streams (Chapter 7), and audits of hazardous waste production (Chapter 12). Energy balances predict the temperature rise in a stream from the discharge of cooling water from a power plant (Chapter 7), and the temperature rise due to global warming (Chapter 9).

2-2   UNIFYING THEORIES

Conservation of Matter

The *law of conservation of matter* states that (without nuclear reaction) matter can neither be created nor destroyed. This is a powerful theory. It means that if we observe an environmental process carefully, we should be able to account for the "matter" at any point in time. It does not mean that the form of the matter does not change nor, for that matter, the properties of the matter. Thus, if we measure the volume of a fresh glass of water on the counter on Monday and measure it again a week later and find the volume to be less, we do not presume magic has occurred but rather that matter has changed in form. The law of conservation of matter says we ought to be able to account for all the mass of the water that was originally present, that is, the mass of water remaining in the glass plus the mass of water vapor that has evaporated equals the mass of water originally present. The mathematical representation of this accounting system is called a *materials balance* or *mass balance.*

Conservation of Energy

The *law of conservation of energy* states that energy cannot be created or destroyed. Like the law of conservation of matter, this theory means that we should be able to account for the "energy" at any point in time. It also does not mean that the form of the energy does not change. Thus, we should be able to trace the energy of food through a series of organisms from green plants through animals. The mathematical representation of the accounting system we use to trace energy is called an *energy balance.*

Conservation of Matter and Energy

At the turn of the 20th century, Albert Einstein hypothesized that matter could be transformed to energy and vice versa. The birth of the nuclear age proved his hypothesis correct, so today we have a combined *law of conservation of matter and energy* that states that the total amount of energy and matter is constant. A nuclear change produces new materials by changing the identity of the atoms themselves. Significant amounts of matter are converted to energy in nuclear explosions. Exchange between mass and energy is not an issue in environmental applications. Thus, there are generally two separate balances for mass and energy.

2-3 MATERIALS BALANCES

Fundamentals

In its simplest form a materials balance or mass balance may be viewed as an accounting procedure. You perform a form of material balance each time you balance your checkbook.

$$\text{Balance} = \text{deposit} - \text{withdrawal} \tag{2-1}$$

For an environmental process, the equation would be written

$$\text{Accumulation} = \text{input} - \text{output} \tag{2-2}$$

where accumulation, input, and output refer to the mass quantities accumulating in the system or flowing into or out of the system. The "system" may be, for example, a pond, river, or a pollution control device.

The Control Volume.   Using the mass balance approach, we begin solving the problem by drawing a flowchart of the process or a conceptual diagram of the environmental subsystem. All of the known inputs, outputs, and accumulation are converted to the same mass units and placed on the diagram. Unknown inputs, outputs, and accumulation are also marked on the diagram. This helps us define the problem. System boundaries (imaginary blocks around the process or part of the process) are drawn in such a way that calculations are made as simple as possible. The system within the boundaries is called the *control volume*.

We then write a materials balance equation to solve for unknown inputs, outputs, or accumulations or to demonstrate that we have accounted for all of the components by demonstrating that the materials balance "closes," that is, the accounting balances. Alternatively, when we do not have data for all inputs or outputs, we can assume that the mass balance closes and solve for the unknown quantity. The following example illustrates the technique.

Example 2-1.   Mr. and Mrs. Green have no children. In an average week they purchase and bring into their house approximately 50 kg of consumer goods (food, magazines, newspapers, appliances, furniture, and associated packaging). Of this amount, 50 percent is consumed as food. Half of the food is used for biological maintenance and ultimately released as CO_2; the remainder is discharged to the sewer system. The Greens recycle approximately 25 percent of the solid waste that is generated. Approximately 1 kg accumulates in the house. Estimate the amount of solid waste they place at the curb each week.

Solution. Begin by drawing a mass balance diagram and labeling the known and unknown inputs and outputs. There are, in fact, two diagrams: one for the house and one for the people. However, the mass balance for the people is superfluous for the solution of this problem.

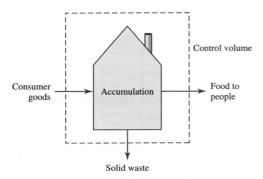

Write the mass balance equation for the house.

Input = accumulation in house + output as food to people + output as solid waste

Now we need to calculate the known inputs and outputs.

One half of input is food = $(0.5)(50 \text{ kg}) = 25 \text{ kg}$

This is output as food to the people. The mass balance equation is then rewritten as

$$50 \text{ kg} = 1 \text{ kg} + 25 \text{ kg} + \text{output as solid waste}$$

Solving for the mass of solid waste gives

Output as solid waste = $50 - 1 - 25 = 24 \text{ kg}$

The mass balance diagram with the appropriate masses may be redrawn as shown below:

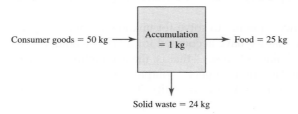

We can estimate the amount of solid waste placed at the curb by performing another mass balance around the solid waste as shown in the following diagram:

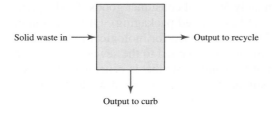

The mass balance equation is

Solid waste in = output to recycle + output to curb

Because the recycled amount is 25 percent of the solid waste

Output to recycle = (0.25)(24 kg) = 6 kg

Substituting into the mass balance equation for solid waste and solving for output to curb;

24 kg = 6 kg + output to curb

Output to curb = 24 − 6 = 18 kg

Comment: This type of analysis is used in *waste audits* to identify opportunities for waste reduction or to measure the effectiveness of a recycling program.

Time as a Factor

For many environmental problems time is an important factor in establishing the degree of severity of the problem or in designing a solution. In these instances, Equation 2-2 is modified to the following form:

Mass rate of accumulation = mass rate of input − mass rate of output   (2-3)

where *rate* is used to mean "per unit of time." In the calculus this may be written as

$$\frac{dM}{dt} = \frac{d(\text{in})}{dt} - \frac{d(\text{out})}{dt} \qquad (2\text{-}4)$$

where M refers to the mass accumulated and (in) and (out) refer to the mass flowing in or out of the control volume. As part of the description of the problem, a convenient time interval that is meaningful for the system must be chosen.

Example 2-2.   Truly Clearwater is filling her bathtub but she forgot to put the plug in. If the volume of water for a bath is 0.350 m^3 and the tap is flowing at 1.32 L/min and the drain is running at 0.32 L/min, how long will it take to fill the tub to bath level? Assuming Truly shuts off the water when the tub is full and does not flood the house, how much water will be wasted? Assume the density of water is $1{,}000 \text{ kg/m}^3$

Solution.   The mass balance diagram is shown here.

Control volume

$Q_{in} = 1.32 \text{ L/min}$ $V_{\text{accumulation}}$ $Q_{out} = 0.32 \text{ L/min}$

Because we are working in mass units, we must convert the volumes to masses. To do this, we use the density of water.

$$\text{Mass} = (\text{volume})(\text{density}) = (\Psi)(\rho)$$

where the

$$\text{Volume} = (\text{flow rate})(\text{time}) = (Q)(t)$$

So for the mass balance equation, noting that $1.0 \text{ m}^3 = 1{,}000 \text{ L}$, $0.350 \text{ m}^3 = 350 \text{ L}$,

$$\text{Accumulation} = \text{mass in} - \text{mass out}$$
$$(\Psi_{\text{ACC}})(\rho) = (Q_{\text{in}})(\rho)(t) - (Q_{\text{out}})(\rho)(t)$$
$$\Psi_{\text{ACC}} = (Q_{\text{in}})(t) - (Q_{\text{out}})(t)$$
$$\Psi_{\text{ACC}} = 1.32t - 0.32t$$
$$350 \text{ L} = (1.00 \text{ L/min})(t)$$
$$t = 350 \text{ min}$$

The amount of water wasted is

$$\text{Wasted water} = (0.32 \text{ L/min})(350 \text{ min}) = 112 \text{ L}$$

Comment: With an assumption of the duration of rainfall (t), this type of analysis forms the basis for the design of the volume of a rainwater retention basin.

More Complex Systems

A key step in the solution of mass balance problems for systems that are more complex than the previous examples is the selection of an appropriate control volume. In some instances, it may be necessary to select multiple control volumes and then solve the problem sequentially using the solution from one control volume as the input to another control volume. For some complex processes, the appropriate control volume may treat all of the steps in the process as a "black box" in which the internal process steps are not required and therefore are hidden in a black box. The following example illustrates a case of a more complex system and a method of solving the problem.

Example 2-3.   A storm sewer network in a small residential subdivision is shown in the following sketch. The storm water flows by gravity in the direction shown by the pipes. Storm water only enters the storm sewer on the east–west legs of pipe. No storm water enters on the north–south legs. The flow rate for each section of pipe is also shown by the arrows at each section of pipe. The capacity of each pipe is 0.120 m³/s. During large rainstorms, River Street floods below junction number 1 because the flow of water exceeds the capacity of the storm sewer pipe. To alleviate this problem and to provide extra capacity for expansion, it is proposed to build a retention pond to hold the storm water until the storm is over and then gradually release it. Where in the pipe network should the retention pond be built to provide approximately 50 percent extra capacity (0.06 m³/s) in the remaining system?

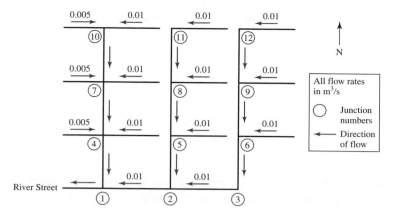

Solution.   This is an example of a balanced flow problem. That is Q_{out} must equal Q_{in}. Although this problem can almost be solved by observation, we will use a sequential mass balance approach to illustrate the technique. Starting at the upper end of the system at junction number 12, we draw the following mass balance diagram:

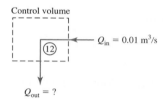

The mass balance equation is

$$\frac{dM}{dt} = \frac{d(in)}{dt} - \frac{d(out)}{dt}$$

Because no water accumulates at the junction

$$\frac{dM}{dt} = 0$$

and

$$\frac{d(in)}{dt} = \frac{d(out)}{dt}$$

$$(\rho)(Q_{in}) = (\rho)(Q_{out})$$

Because the density of water remains constant, we may treat the mass flow rate in and out as directly proportional to the flow rate in and out.

$$Q_{in} = Q_{out}$$

So the flow rate from junction 12 to junction 9 is 0.01 m³/s.

At junction 9, we can draw the following mass balance diagram.

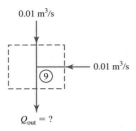

Again using our assumption that no water accumulates at the junction and recognizing that the mass balance equation may again be written in terms of flow rates,

$$Q_{\text{from junction 9}} = Q_{\text{from junction 12}} + Q_{\text{in the pipe connected to junction 9}}$$
$$= 0.01 + 0.01 = 0.02 \text{ m}^3/\text{s}$$

Similarly

$$Q_{\text{from junction 6}} = Q_{\text{from junction 9}} + Q_{\text{in the pipe connected to junction 6}}$$
$$= 0.02 + 0.01 = 0.03 \text{ m}^3/\text{s}$$

and, noting that storm water enters on the east–west legs of pipe

$$Q_{\text{from junction 3}} = Q_{\text{from junction 6}} + Q_{\text{in the pipe connecting junction 3 with junction 2}}$$
$$= 0.03 + 0.01 = 0.04 \text{ m}^3/\text{s}$$

By a similar process for all the junctions, we may label the network flows as shown in the following diagram.

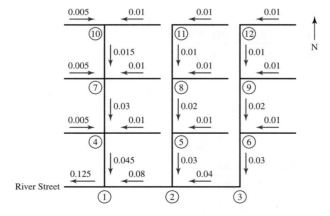

It is obvious that the pipe capacity of 0.12 m³/s is exceeded just below junction 1. By observation we can also see that the total flow into junction 2 is 0.07 m³/s and that a retention pond at this point would require that the pipe below junction 1 carry only 0.055 m³/s. This meets the requirement of providing approximately 50 percent of the capacity for expansion.

Comments:

1. While the construction of storm water retention ponds is now common for new developments, the addition of a retention pond in a built-up community is rare because of the lack of available land.

2. An alternative solution to the retention pond is to replace the River Street storm sewer below junction 1 with a larger pipe. This may be more economical as well as more practical.

Efficiency

The effectiveness of an environmental process in removing a contaminant can be determined using the mass balance technique. Starting with Equation 2-4,

$$\frac{dM}{dt} = \frac{d(\text{in})}{dt} - \frac{d(\text{out})}{dt}$$

The mass of contaminant per unit of time [$d(\text{in})/dt$ and $d(\text{out})/dt$] may be calculated as

$$\frac{\text{Mass}}{\text{Time}} = (\text{concentration})(\text{flow rate})$$

For example,

$$\frac{\text{Mass}}{\text{Time}} = (\text{mg/m}^3)(\text{m}^3\text{/s}) = \text{mg/s}$$

This is called a *mass flow rate*. In concentration and flow rate terms, the mass balance equation is

$$\frac{dM}{dt} = C_{\text{in}}Q_{\text{in}} - C_{\text{out}}Q_{\text{out}} \tag{2-5}$$

where dM/dt = rate of accumulation of contaminant in the process
$C_{\text{in}}, C_{\text{out}}$ = concentrations of contaminant into and out of the process
$Q_{\text{in}}, Q_{\text{out}}$ = flow rates into and out of the process

The ratio of the mass that is accumulated in the process to the incoming mass is a measure of how effective the process is in removing the contaminant

$$\frac{dM/dt}{C_{\text{in}}Q_{\text{in}}} = \frac{C_{\text{in}}Q_{\text{in}} - C_{\text{out}}Q_{\text{out}}}{C_{\text{in}}Q_{\text{in}}} \tag{2-6}$$

For convenience, the fraction is multiplied by 100%. The left-hand side of the equation is given the notation η. Efficiency (η) is then defined as

$$\eta = \frac{\text{mass in} - \text{mass out}}{\text{mass in}}(100\%) \tag{2-7}$$

If the flow rate in and the flow rate out are the same, this ratio may be simplified to

$$\eta = \frac{\text{concentration in} - \text{concentration out}}{\text{concentration in}}(100\%) \tag{2-8}$$

The following example illustrates a multistep solution using efficiency as part of the solution technique.

Example 2-4.   The air pollution control equipment on a municipal waste incinerator includes a fabric filter particle collector (known as a *baghouse*). The baghouse contains 424 cloth bags arranged in parallel, that is 1/424 of the flow goes through each bag. The gas flow rate into and out of the baghouse is 47 m³/s, and the concentration of particles entering the baghouse is 15 g/m³. In normal operation the baghouse particulate discharge meets the regulatory limit of 24 mg/m³. During preventive maintenance replacement of the bags, one bag is inadvertently not replaced, so only 423 bags are in place.

Calculate the fraction of particulate matter removed and the efficiency of particulate removal when all 424 bags are in place and the emissions comply with the regulatory requirements. Estimate the mass emission rate when one of the bags is missing and recalculate the efficiency of the baghouse. Assume the efficiency for each individual bag is the same as the overall efficiency for the baghouse.

Solution.   The mass balance diagram for the baghouse in normal operation is shown here.

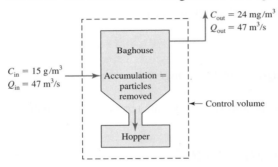

In concentration and flow rate terms, the mass balance equation is

$$\frac{dM}{dt} = C_{in}Q_{in} - C_{out}Q_{out}$$

The mass rate of accumulation in the baghouse is

$$\frac{dM}{dt} = (15{,}000 \text{ mg/m}^3)(47 \text{ m}^3/\text{s}) - (24 \text{ mg/m}^3)(47 \text{ m}^3/\text{s}) = 703{,}872 \text{ mg/s}$$

The fraction of particulates removed is

$$\frac{703{,}872 \text{ mg/s}}{(15{,}000 \text{ mg/m}^3)(47 \text{ mg/s})} = \frac{703{,}872 \text{ mg/s}}{705{,}000 \text{ mg/s}} = 0.9984$$

The efficiency of the baghouse is

$$\eta = \frac{15{,}000 \text{ mg/m}^3 - 24 \text{ mg/m}^3}{15{,}000 \text{ mg/m}^3}(100\%)$$

$$= 99.84\%$$

Note that the fraction of particulate matter removed is the decimal equivalent of the efficiency.

To determine the mass emission rate with one bag missing, we begin by drawing a mass balance diagram. Because one bag is missing, a portion of the flow (1/424 of Q_{out}) effectively bypasses the baghouse. The "Bypass" line around the baghouse is drawn to show this.

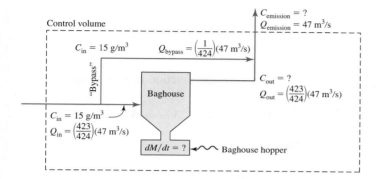

A judicious selection of the control volume aids in the solution of this problem. As shown in the diagram, a control volume around the overall baghouse and bypass flow yields three unknowns: the mass flow rate out of the baghouse, the rate of mass accumulation in the baghouse hopper, and the mass flow rate of the mixture. A control volume around the baghouse alone reduces the number of unknowns to two:

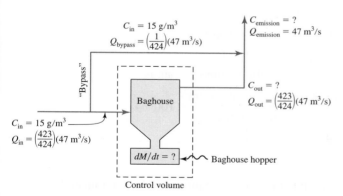

Because we know the efficiency and the influent mass flow rate, we can solve the mass balance equation for the mass flow rate out of the filter.

$$\eta = \frac{C_{in}Q_{in} - C_{out}Q_{out}}{C_{in}Q_{in}}$$

Solving for $C_{out}Q_{out}$

$$
\begin{aligned}
C_{out}Q_{out} &= (1 - \eta)C_{in}Q_{in} \\
&= (1 - 0.9984)(15{,}000 \text{ mg/m}^3)(47 \text{ m}^3/\text{s})(423/424) = 1{,}125 \text{ mg/s}
\end{aligned}
$$

This value can be used as an input for a control volume around the junction of the bypass, the effluent from the baghouse and the final effluent.

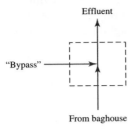

Effluent

"Bypass"

From baghouse

A mass balance for the control volume around the junction may be written as

$$\frac{dM}{dt} = C_{in}Q_{in \; from \; bypass} + C_{in}Q_{in \; from \; baghouse} - C_{emission}Q_{emission}$$

Because there is no accumulation in the junction

$$\frac{dM}{dt} = 0$$

and the mass balance equation is

$$C_{emission}Q_{emission} = C_{in}Q_{in \; from \; bypass} + C_{in}Q_{in \; from \; baghouse}$$
$$= (15{,}000 \text{ mg/m}^3)(47 \text{ m}^3/\text{s})(1/424) + 1{,}125 = 2788 \text{ mg/s}$$

The concentration in the effluent is

$$\frac{C_{emission}Q_{emission}}{Q_{out}} = \frac{2{,}788 \text{ mg/s}}{47 \text{ m}^3/\text{s}} = 59 \text{ mg/m}^3$$

The overall efficiency of the baghouse with the missing bag is

$$\eta = \frac{15{,}000 \text{ mg/m}^3 - 59 \text{ mg/m}^3}{15{,}000 \text{ mg/m}^3}(100\%)$$
$$= 99.61\%$$

Comments:

1. The efficiency is still very high but the control equipment does not meet the allowable emission rate of 24 mg/m^3.

2. It is not likely that a baghouse would ever operate with a missing bag because the unbalanced gas flows would be immediately apparent. However, many small holes in a number of bags could yield an effluent that did not meet the discharge standards but would otherwise appear to be functioning correctly. To prevent this situation, the bags undergo periodic inspection and maintenance and the effluent stream is monitored continuously.

The State of Mixing

The state of mixing in the system is an important consideration in the application of Equation 2-4. Consider a coffee cup containing approximately 200 mL of black coffee (or another beverage of your choice). If we add a dollop (about 20 mL) of cream and immediately take a sample (or a sip), we would not be surprised to find that the cream was not evenly distributed throughout the coffee. If, on the other hand, we mixed the coffee and cream vigorously and then took a sample, it would not matter if we sipped from the left or right of the cup, or, for that matter, put a valve in the bottom and sampled from there, we would expect the cream to be distributed evenly. In terms of a mass balance on the coffee cup system, the cup itself would define the system boundary for the control volume. If the coffee and cream were not mixed well, then the place we take the sample from would strongly affect the value of $d(\text{out})/dt$ in Equation 2-4. On the other hand, if the coffee and cream were instantaneously well mixed, then any place we take the sample from would yield the same result. That is, any output would look exactly like the contents of the cup. This system is called a completely mixed system. A more formal definition is that *completely mixed systems* are those in which every drop of fluid is homogeneous with every other drop, that is, every drop of fluid contains the same concentration of material or physical property (e.g., temperature). If a system is completely mixed, then we may assume that the output from the system (concentration, temperature, etc.) is the same as the contents within the system boundary. Although we frequently make use of this assumption to solve mass balance problems, it is often very difficult to achieve in real systems. This means that solutions to mass balance problems that make this assumption must be taken as approximations to reality.

If completely mixed systems exist, or at least systems that we can approximate as completely mixed, then it stands to reason that some systems are completely unmixed or approximately so. These systems are called *plug-flow systems*. The behavior of a plug-flow system is analogous to that of a train moving along a railroad track (Figure 2-1). Each car in the train must follow the one preceding it. If, as in Figure 2-1b, a tank car is inserted in a train of box cars, it maintains its position in the train until it arrives at its destination. The tank car may be identified at any point in time as the train travels down the track. In terms of fluid flow, each drop of fluid along the direction of flow remains unique and, if no reactions take place, contains the same concentration of material or physical property that it had when it entered the plug-flow system. Mixing may or may not occur in the radial direction. As with the completely mixed systems, ideal plug-flow systems don't happen very often in the real world.

When a system has operated in such a way that the rate of input and the rate of output are constant and equal, then, of course, the mass rate of accumulation is zero (i.e., in Equation 2-4 $dM/dt = 0$). This condition is called *steady state*. In solving mass balance problems, it is often convenient to make an assumption that steady-state conditions have been achieved. We should note that steady state does not imply *equilibrium*. For example, water running into and out of a pond at the same rate is not at equilibrium, otherwise it would not be flowing. However, if there is no accumulation in the pond, then the system is at steady state.

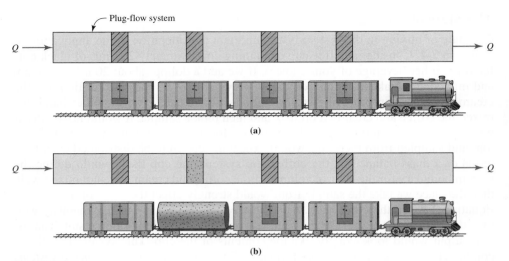

FIGURE 2-1

(*a*) Analogy of a plug-flow system and a train. (*b*) Analogy when a pulse change in influent concentration occurs.

Example 2-5 demonstrates the use of two assumptions: complete mixing and steady state.

Example 2-5.   A storm sewer is carrying snow melt containing 1.200 g/L of sodium chloride into a small stream. The stream has a naturally occurring sodium chloride concentration of 20 mg/L. If the storm sewer flow rate is 2,000 L/min and the stream flow rate is 2.0 m³/s, what is the concentration of salt in the stream after the discharge point? Assume that the sewer flow and the stream flow are completely mixed, that the salt is a conservative substance (it does not react), and that the system is at steady state.

Solution.   The first step is to draw a mass balance diagram as shown here.

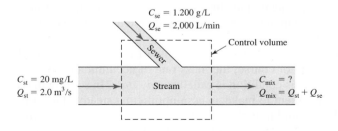

Note that the mass flow of salt may be calculated as

$$\frac{\text{Mass}}{\text{Time}} = (\text{concentration})(\text{flow rate})$$

or

$$\frac{\text{Mass}}{\text{Time}} = (\text{mg/L})(\text{L/min}) = \text{mg/min}$$

Using the notation in the diagram, where the subscript "st" refers to the stream and the subscript "se" refers to the sewer, the mass balance may be written as

$$\text{Rate of accumulation of salt} = [C_{st}Q_{st} + C_{se}Q_{se}] - C_{mix}Q_{mix}$$

where $Q_{mix} = Q_{st} + Q_{se}$.

Because we assume steady state, the rate of accumulation equals zero and

$$C_{mix}Q_{mix} = [C_{st}Q_{st} + C_{se}Q_{se}]$$

Solving for C_{mix}

$$C_{mix} = \frac{[C_{st}Q_{st} + C_{se}Q_{se}]}{Q_{st} + Q_{se}}$$

Before substituting in the values, the units are converted as follows:

$$C_{se} = (1.200 \text{ g/L})(1000 \text{ mg/g}) = 1,200 \text{ mg/L}$$
$$Q_{st} = (2.0 \text{ m}^3/\text{s})(1000 \text{ L/m}^3)(60 \text{ s/min}) = 120,000 \text{ L/min}$$
$$C_{mix} = \frac{[(20 \text{ mg/L})(120,000 \text{ L/min})] + [(1,200 \text{ mg/L})(2,000 \text{ L/min})]}{120,000 \text{ L/min} + 2,000 \text{ L/min}}$$
$$= 39.34 \text{ or } 39 \text{ mg/L}$$

Comments:

1. The final salt concentration is not very high but, in effect, it doubles the upstream concentration. This may have severe effects on the biota in the stream.
2. Real-world lakes, lagoons, and rivers are seldom completely mixed without the addition of external mixing systems. For discharge into rivers, dispersion jets from piping along the bottom may be used to enhance mixing. For lagoons, mechanical mixers may be employed. The use of these systems increases mixing but the ideal theoretical assumption of complete mixing cannot be achieved in natural systems. It is rarely, if ever, achieved in tank systems.

Including Reactions and Loss Processes

Equation 2-4 is applicable when no chemical or biological reaction takes place and no radioactive decay occurs of the substances in the mass balance. In these instances the substance is said to be *conserved*. Examples of conservative substances include salt in water and argon in air. An example of a nonconservative substance (i.e., one that reacts or settles out) is decomposing organic matter. Particulate matter that is settling from the air is considered a loss process.

In most systems of environmental interest, transformations occur within the system: by-products are formed (e.g., CO_2) or compounds are destroyed (e.g., ozone). Because many environmental reactions do not occur instantaneously, the time dependence of the reaction must be taken into account. Equation 2-3 may be written to account for time-dependent transformation as follows:

$$\text{Accumulation rate} = \text{input rate} - \text{output rate} \pm \text{transformation rate} \quad (2\text{-}9)$$

Time-dependent reactions are called *kinetic reactions*. The rate of transformation, or reaction rate (r), is used to describe the rate of formation or disappearance of a substance or chemical species. With reactions, Equation 2-4 may become

$$\frac{dM}{dt} = \frac{d(\text{in})}{dt} - \frac{d(\text{out})}{dt} + r \quad (2\text{-}10)$$

The reaction rate is often some complex function of temperature, pressure, the reacting components, and products of reaction.

$$r = -kC^n \quad (2\text{-}11)$$

where k = reaction rate constant, s^{-1} or d^{-1}
C = concentration of substance
n = exponent or reaction order

The minus sign before reaction rate, k, indicates the disappearance of a substance or chemical species.

In many environmental problems, for example the oxidation of organic compounds by microorganisms (Chapter 7) and radioactive decay (Chapter 14), the reaction rate, r, may be assumed to be directly proportional to the amount of material remaining, that is the value of $n = 1$. This is known as a *first-order reaction*. In first-order reactions, the rate of loss of the substance is proportional to the amount of substance present at any given time, t.

$$r = -kC = \frac{dC}{dt} \quad (2\text{-}12)$$

The differential equation may be integrated to yield either

$$\ln\frac{C}{C_o} = -kt \qquad \text{FE} \;\; (2\text{-}13)$$

or

$$C = C_o e^{-kt} = C_o \exp(-kt) \quad (2\text{-}14)$$

where C = concentration at any time t
C_o = initial concentration
ln = logarithm to base e
e = exp = exponential e = 2.7183 raised to the $-kt$ power

For simple completely mixed systems with first-order reactions, the total mass of substance (M) is equal to the product of the concentration and volume ($C\forall$) and, when $\forall$ is a constant, the mass rate of decay of the substance is

$$\frac{dM}{dt} = \frac{d(C\forall)}{dt} = \forall\frac{d(C)}{dt} \quad (2\text{-}15)$$

Because first-order reactions can be described by Equation 2-12, we can rewrite Equation 2-10 as

$$\frac{dM}{dt} = \frac{d(\text{in})}{dt} - \frac{d(\text{out})}{dt} - kC\mathcal{V} \tag{2-16}$$

Example 2-6. A well-mixed sewage lagoon (a shallow pond) is receiving 430 m³/d of untreated sewage. The lagoon has a surface area of 10 ha (hectares) and a depth of 1.0 m. The pollutant concentration in the raw sewage discharging into the lagoon is 180 mg/L. The organic matter in the sewage degrades biologically (decays) in the lagoon according to first-order kinetics. The reaction rate constant (decay coefficient) is 0.70 d⁻¹. Assuming no other water losses or gains (evaporation, seepage, or rainfall) and that the lagoon is completely mixed, find the steady-state concentration of the pollutant in the lagoon effluent.

Solution. We begin by drawing the mass balance diagram.

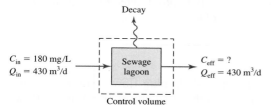

The mass balance equation may be written as

$$\text{Accumulation} = \text{input rate} - \text{output rate} - \text{decay rate}$$

Assuming steady-state conditions, that is, accumulation = 0, then

$$\text{Input rate} = \text{output rate} + \text{decay rate}$$

This may be written in terms of the notation in the figure as

$$C_{\text{in}}Q_{\text{in}} = C_{\text{eff}}Q_{\text{eff}} + kC_{\text{lagoon}}\mathcal{V}$$

Solving for C_{eff}, we have

$$C_{\text{eff}} = \frac{C_{\text{in}}Q_{\text{in}} - kC_{\text{lagoon}}\mathcal{V}}{Q_{\text{eff}}}$$

Now calculate the values for terms in the equation. The input mass rate $(C_{\text{in}}Q_{\text{in}})$ is

$$(180 \text{ mg/L})(430 \text{ m}^3/\text{d})(1{,}000 \text{ L/m}^3) = 77{,}400{,}000 \text{ mg/d}$$

With a lagoon volume of

$$(10 \text{ ha})(10^4 \text{ m}^2/\text{ha})(1 \text{ m}) = 100{,}000 \text{ m}^3$$

and the decay coefficient of 0.70 d⁻¹, the decay rate is

$$k C\mathcal{V} = (0.70 \text{ d}^{-1})(100{,}000 \text{ m}^3)(1{,}000 \text{ L/m}^3)(C_{\text{lagoon}}) = (70{,}000{,}000 \text{ L/d})(C_{\text{lagoon}})$$

Now using the assumption that the lagoon is completely mixed, we assume that $C_{\text{eff}} = C_{\text{lagoon}}$. Thus,

$$k C\mathcal{V} = (70{,}000{,}000 \text{ L/d})(C_{\text{eff}})$$

Substituting into the mass balance equation

$$\text{Output rate} = 77{,}400{,}000 \text{ mg/d} - 70{,}000{,}000 \text{ L/d} \times C_{\text{eff}}$$

or

$$C_{\text{eff}}(430 \text{ m}^3/\text{d})(1{,}000 \text{ L/m}^3) = 77{,}400{,}000 \text{ mg/d} - 70{,}000{,}000 \text{ L/d} \times C_{\text{eff}}$$

Solving for C_{eff}, we have

$$C_{\text{eff}} = \frac{77{,}400{,}000 \text{ mg/d}}{70{,}430{,}000 \text{ L/d}} = 1.10 \text{ mg/L}$$

Comments:

1. The reaction rate constant is quite high. In warm climates, it might be as high as 0.2 d^{-1}. Typically, it will be on the order of 0.05 d^{-1} or less.
2. The flow pattern through a lagoon is more closely approximated by arbitrary flow (intermediate between completely mixed and plug-flow) than completely mixed flow (Thirumurthi, 1969).

Plug-Flow with Reaction. As noted in Figure 2-1, in plug-flow systems, the tank car, or "plug" element of fluid, does not mix with the fluid ahead or behind it. However, a reaction can take place in the tank car element. Thus, even at steady state, the contents within the element can change with time as the plug moves downstream. The control volume for the mass balance is the plug or differential element of fluid. The mass balance for this moving plug may be written as

$$\frac{dM}{dt} = \frac{d(\text{in})}{dt} - \frac{d(\text{out})}{dt} + V\frac{d(C)}{dt} \tag{2-17}$$

Because no mass exchange occurs across the plug boundaries (in our railroad car analogy, there is no mass transfer between the box cars and the tank car), $d(\text{in})$ and $d(\text{out}) = 0$. Equation 2-17 may be rewritten as

$$\frac{dM}{dt} = 0 - 0 + V\frac{d(C)}{dt} \tag{2-18}$$

As noted earlier in Equation 2-12, for a first-order decay reaction, the right-hand term may be expressed as

$$\frac{dC}{dt} = -kC \tag{2-19}$$

The total mass of substance (M) is equal to the product of the concentration and volume (CV) and, when V is a constant, the mass rate of decay of the substance in Equation 2-18 may be expressed as

$$V\frac{dC}{dt} = -kCV \tag{2-20}$$

where the left-hand side of the equation $= dM/dt$. The steady-state solution to the mass balance equation for the plug-flow system with first-order kinetics is

$$\ln\frac{C_{out}}{C_{in}} = -kt_o \tag{2-21}$$

or

$$C_{out} = (C_{in})e^{-kt_o} \tag{2-22}$$

where k = reaction rate constant, s^{-1}, min^{-1}, or d^{-1}
$\quad t_o$ = residence time in plug-flow system, s, min, or d

In a plug-flow system of length L, each plug travels for a period $= L/u$, where u = the speed of flow. Alternatively, for a cross-sectional area A, the residence time is

$$t_o = \frac{(L)(A)}{(u)(A)} = \frac{\text{V}}{Q} \tag{2-23}$$

where V = volume of the plug-flow system, m^3
$\quad Q$ = flow rate, m^3/s

Thus, for example, Equation 2-21, may be rewritten as

$$\ln\frac{C_{out}}{C_{in}} = -k\frac{L}{u} = -k\frac{\text{V}}{Q} \tag{2-24}$$

where L = length of the plug-flow segment, m
$\quad u$ = linear velocity, m/s

Although the concentration within a given plug changes over time as the plug moves downstream, the concentration at a fixed point in the plug-flow system remains constant with respect to time. Thus, Equation 2-24 has no time dependence.
    Example 2-7 illustrates an application of plug-flow with reaction.

Example 2-7. A wastewater treatment plant must disinfect its effluent before discharging the wastewater to a nearby stream. The wastewater contains 4.5×10^5 fecal coliform colony-forming units (CFU) per liter. The maximum permissible fecal coliform concentration that may be discharged is 2,000 fecal coliform CFU/L. It is proposed that a pipe carrying the wastewater be used for disinfection process. Determine the length of pipe required if the linear velocity of the wastewater in the pipe is 0.75 m/s. Assume that the pipe behaves as a steady-state plug-flow system and that the reaction rate constant for destruction of the fecal coliforms is 0.23 min^{-1}.

Solution. The mass balance diagram is sketched here. The control volume is the pipe itself.

$C_{in} = 4.5 \times 10^5$ CFU/L
$u = 0.75$ m/s

$L = ?$

$C_{out} = 2,000$ CFU/L
$u = 0.75$ m/s

Using the steady-state solution to the mass balance equation, we obtain

$$\ln\frac{C_{out}}{C_{in}} = -k\frac{L}{u}$$

$$\ln\frac{2{,}000\,\text{CFU/L}}{4.5 \times 10^5\,\text{CFU/L}} = -0.23\ \text{min}^{-1}\frac{L}{(0.75\ \text{m/s})(60\ \text{s/min})}$$

Solving for the length of pipe, we have

$$\ln(4.44 \times 10^{-3}) = -0.23\ \text{min}^{-1}\frac{L}{45\ \text{m/min}}$$

$$-5.42 = -0.23\ \text{min}^{-1}\frac{L}{45\ \text{m/min}}$$

$$L = 1{,}060\,\text{m}$$

Comment: A little over 1 km of pipe is needed to meet the discharge standard. For most wastewater treatment systems this would be an exceptionally long discharge. Another alternative such as a baffled reactor (discussed in Chapter 6) would be investigated. The baffled reactor would have a much larger cross-sectional area. As a result, the velocity would be much lower.

Reactors

The tanks in which physical, chemical, and biochemical reactions occur, for example, in water softening (Chapter 6) and wastewater treatment (Chapter 8), are called *reactors*. These reactors are classified based on their flow characteristics and their mixing conditions. With appropriate selection of control volumes, ideal chemical reactor models may be used to model natural systems.

Batch reactors are of the fill-and-draw type: materials are added to the tank (Figure 2-2a), mixed for sufficient time to allow the reaction to occur (Figure 2-2b), and then drained (Figure 2-2c). Although the reactor is well mixed and the contents are uniform at any instant in time, the composition within the tank changes with time as the reaction proceeds. A batch reaction is unsteady. Because there is no flow into or out of a batch reactor

$$\frac{d(\text{in})}{dt} = \frac{d(\text{out})}{dt} = 0$$

For a batch reactor Equation 2-16 reduces to

$$\frac{dM}{dt} = -kC\forall \tag{2-25}$$

As we noted in Equation 2-15

$$\frac{dM}{dt} = \forall\frac{dC}{dt}$$

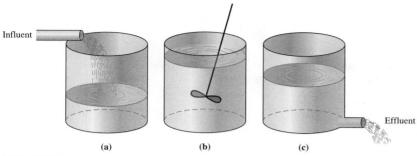

FIGURE 2-2

Batch reactor operation. (*a*) Materials added to the reactor. (*b*) Mixing and reaction. (*c*) Reactor is drained. *Note:* There is no influent or effluent during the reaction.

So that for a first-order reaction in a batch reactor, Equation 2-25 may be simplified to

$$\frac{dC}{dt} = -kC \qquad (2\text{-}26)$$

Flow reactors have a continuous type of operation: material flows into, through, and out of the reactor at all times. Flow reactors may be further classified by mixing conditions. The contents of a *completely mixed flow reactor* (CMFR), also called a *continuous-flow stirred tank reactor* (CSTR), ideally are uniform throughout the tank. A schematic diagram of a CSTR and the common flow diagram notation are shown in Figure 2-3. The composition of the effluent is the same as the composition in the tank. If the mass input rate into the tank remains constant, the composition of the effluent remains constant. The mass balance for a CSTR is described by Equation 2-16.

In *plug-flow reactors* (PFR), fluid particles pass through the tank in sequence. Those that enter first leave first. In the ideal case, it is assumed that no mixing occurs in the lateral direction. Although composition varies along the length of the tank, as long as the flow conditions remain steady, the composition of the effluent remains constant.

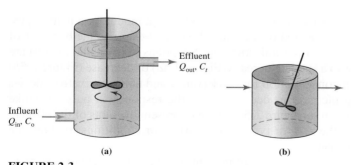

FIGURE 2-3

Schematic diagram of (*a*) continuous-flow stirred tank reactor (CSTR) and (*b*) the common diagram. The propeller indicates that the reactor is completely mixed.

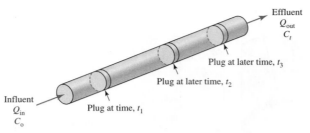

FIGURE 2-4
Schematic diagram of a plug-flow reactor (PFR). *Note:* $t_3 > t_2 > t_1$.

A schematic diagram of a plug-flow reactor is shown in Figure 2-4. The mass balance for a PFR is described by Equation 2-18, where the time element (*dt*) is the time spent in the PFR as described by Equation 2-23. Real continuous-flow reactors are generally something in between a CSTR and PFR.

For time-dependent reactions, the time that a fluid particle remains in the reactor obviously affects the degree to which the reaction goes to completion. In ideal reactors the average time in the reactor (*detention time* or *retention time* or, for liquid systems, *hydraulic detention time* or *hydraulic retention time*) is defined as

$$t_o = \frac{\forall}{Q} \qquad\qquad \text{(2-27)}$$

where t_o = theoretical detention time, s
$\forall$ = volume of fluid in reactor, m³
Q = flow rate into reactor, m³/s

Real reactors do not behave as ideal reactors because of density differences due to temperature or other causes, short circuiting because of uneven inlet or outlet conditions, and local turbulence or dead spots in the tank corners. The detention time in real tanks is generally less than the theoretical detention time calculated from Equation 2-27.

Reactor Analysis

The selection of a reactor either as a treatment method or as a model for a natural process depends on the behavior desired or recognized. We will examine the behavior of batch, CSTR, and PFR reactors in several situations. Situations of particular interest are the response of the reactor to a sudden increase (Figure 2-5a) or decrease (Figure 2-5b) in the steady-state influent concentration for conservative and nonconservative species (commonly called a step increase or decrease) and the response to pulse or spike change in influent concentration (Figure 2-5c). We will present the plots of the effluent concentration for each of the reactor types for a variety of conditions to show the response to these influent changes.

For nonconservative substances, we will present the analysis for first-order reactions. The behavior of zero-order and second-order reactions will be summarized in comparison at the conclusion of this discussion.

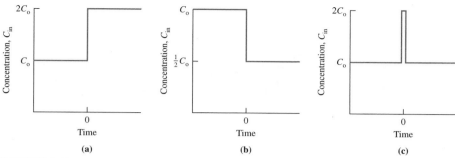

FIGURE 2-5

Example influent graphs of (*a*) step increase in influent concentration, (*b*) step decrease in influent concentration, and (*c*) a pulse or spike increase in influent concentration. *Note:* The size of the change is for illustration purposes only.

Batch Reactor.   Laboratory experiments are often conducted in batch reactors because they are inexpensive and easy to build. Industries that generate small quantities of wastewater (less than 150 m³/d) use batch reactors because they are easy to operate and provide an opportunity to check the wastewater for regulatory compliance before discharging it.

Because there is no influent to or effluent from a batch reactor, the introduction of a conservative substance into the reactor either as a step increase or a pulse results in an instantaneous increase in concentration of the conservative substance in the reactor. The concentration plot is shown in Figure 2-6.

Because there is no influent or effluent, for a nonconservative substance that decays as a first-order reaction, the mass balance is described by Equation 2-26. Integration yields

$$\frac{C_t}{C_o} = e^{-kt} \qquad (2\text{-}28)$$

The final concentration plot is shown in Figure 2-7*a*. For the formation reaction, where the sign in Equation 2-28 is positive, the concentration plot is shown in Figure 2-7*b*.

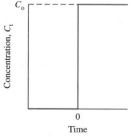

FIGURE 2-6

Batch reactor response to a step or pulse increase in concentration of a conservative substance. C_o = mass of conservative substance/volume of reactor.

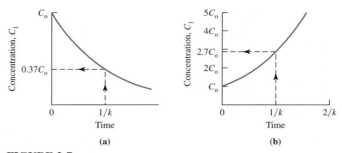

FIGURE 2-7
Batch reactor response for (*a*) decay of a nonconservative substance and
(*b*) for a formation reaction.

Example 2-8.   A contaminated soil is to be excavated and treated in a completely
mixed aerated lagoon at a Superfund site. To determine the time it will take to treat the
contaminated soil, a laboratory completely mixed batch reactor is used to gather
the following data. Assuming a first-order reaction, estimate the rate constant, k, and
determine the time to achieve 99 percent reduction in the original concentration.

Time (d)	Waste Concentration (mg/L)
1	280
16	132

Solution.   The rate constant may be estimated by solving Equation 2-28 for k. Using
the 1st and 16th day, the time interval $t = 16 - 1 = 15$ d,

$$\frac{132 \text{ mg/L}}{280 \text{ mg/L}} = \exp[-k(15 \text{ d})]$$

$$0.4714 = \exp[-k(15)]$$

Taking the logarithm (base e) of both sides of the equation, we obtain

$$-0.7520 = -k(15)$$

Solving for k, we have

$$k = 0.0501 \text{ d}^{-1}$$

To achieve 99 percent reduction the concentration at time t must be $1 - 0.99$ of the
original concentration:

$$\frac{C_t}{C_o} = 0.01$$

The estimated time is then

$$0.01 = \exp[-0.05(t)]$$

Taking the logarithm of both sides and solving for t, we get

$$t = 92 \text{ days}$$

Comment: It is not unusual for biological treatment of hazardous wastes to take much longer than this to achieve acceptable levels of residual chemical concentrations.

CSTR. A batch reactor is used for small volumetric flow rates. When water flow rates are greater than 150 m^3/d, a CSTR may be selected for chemical mixing. Examples of this application include equalization reactors to adjust the pH, precipitation reactors to remove metals, and mixing tanks (called *rapid mix* or *flash mix tanks*) for water treatment. Because municipal wastewater flow rates vary over the course of a day, a CSTR (called an *equalization basin*) may be placed at the treatment plant influent point to level out the flow and concentration changes. Some natural systems such as a lake or the mixing of two streams or the air in a room or over a city may be modeled as a CSTR as an approximation of the real mixing that is taking place.

For a step increase in a conservative substance entering a CSTR, the initial level of the conservative substance in the reactor is C_o prior to $t = 0$. At $t = 0$, the influent concentration (C_{in}) instantaneously increases to C_1 and remains at this concentration (Figure 2-8a). With balanced fluid flow ($Q_{in} = Q_{out}$) into the CSTR and no reaction, the mass balance equation for a step increase is

$$\frac{dM}{dt} = C_t Q_{in} - C_{out} Q_{out} \tag{2-29}$$

where $M = C\forall$. The solution is

$$C_t = C_o\left[\exp\left(-\frac{t}{t_o}\right)\right] + C_1\left[1 - \exp\left(-\frac{t}{t_o}\right)\right] \tag{2-30}$$

where C_t = concentration in the reactor at any time t
C_o = concentration in reactor prior to step change
C_1 = concentration in influent after instantaneous increase
t = time after step change
t_o = theoretical detention time = $\forall/Q$

Figure 2-8b shows the effluent concentration plot.

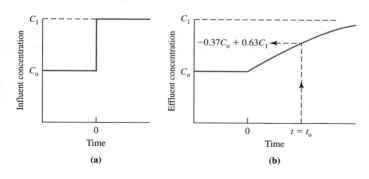

FIGURE 2-8
Response of a CSTR to (*a*) a step increase in the influent concentration of a conservative substance from concentration C_o to a new concentration C_1. (*b*) Effluent concentration.

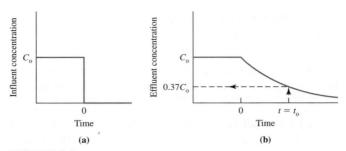

FIGURE 2-9
Flushing of CSTR resulting from (*a*) a step decrease in influent concentration
of a conservative substance from C_o to 0. (*b*) Effluent concentration.

Flushing of a nonreactive contaminant from a CSTR by a contaminant-free fluid is
an example of a step change in the influent concentration (Figure 2-9*a*). Because $C_{in} = 0$
and no reaction takes place, the mass balance equation is

$$\frac{dM}{dt} = -C_{out}Q_{out} \tag{2-31}$$

where $M = C\forall$. The initial concentration is

$$C_o = \frac{M}{\forall} \tag{2-32}$$

Solving Equation 2-31 for any time $t \geq 0$, we obtain

$$C_t = C_o \exp\left(-\frac{t}{t_o}\right) \tag{2-33}$$

where $t_o = \forall/Q$ as noted in Equation 2-27. Figure 2-9*b* shows the effluent concentration plot.

Example 2-9. Before entering an underground utility vault to do repairs, a work
crew analyzed the gas in the vault and found that it contained 29 mg/m³ of hydrogen
sulfide. Because the allowable exposure level is 14 mg/m³, the work crew began ventilating the vault with a blower. If the volume of the vault is 160 m³ and the flow rate of
contaminant-free air is 10 m³/min, how long will it take to lower the hydrogen sulfide
level to a level that will allow the work crew to enter? Assume the manhole behaves as
a CSTR and that hydrogen sulfide is nonreactive in the time period considered.

Solution. This is a case of flushing a nonreactive contaminant from a CSTR. The
theoretical detention time is

$$t_o = \frac{\forall}{Q} = \frac{160 \text{ m}^3}{10 \text{ m}^3/\text{min}} = 16 \text{ min}$$

The required time is found by solving Equation 2-33 for t

$$\frac{14 \text{ mg/m}^3}{29 \text{ mg/m}^3} = \exp\left(-\frac{t}{16 \text{ min}}\right)$$

$$0.4828 = \exp\left(-\frac{t}{16 \text{ min}}\right)$$

Taking the logarithm to the base e of both sides

$$-0.7282 = -\frac{t}{16 \text{ min}}$$

$t = 11.6$ or 12 min to lower the concentration to the allowable level

Comments:

1. Because the odor threshold for H_2S is about 0.18 mg/m^3, the vault will still have quite a strong odor after 12 min.

2. A precautionary note is in order here. H_2S is commonly found in confined spaces such as manholes. It is a very toxic poison and has the unfortunate property of deadening the olfactory senses. Thus, you may not smell it after a few moments even though the concentration has not decreased. Each year a few individuals in the United States die because they have entered a confined space without taking stringent safety precautions. Some precautions include continuous ventilation and the use of an H_2S detector with an alarm.

Because a CSTR is completely mixed, the response of a CSTR to a step change in the influent concentration of a reactive substance results in an immediate corresponding change in the effluent concentration. For this analysis we begin with the mass balance for a balanced flow ($Q_{in} = Q_{out}$) CSTR operating under steady-state conditions with first-order decay of a reactive substance.

$$\frac{dM}{dt} = C_{in}Q_{in} - C_{out}Q_{out} - kC_{out}V \qquad (2\text{-}34)$$

where $M = CV$. Because the flow rate and volume are constant, we may divide through by V and simplify to obtain

$$\frac{dC}{dt} = \frac{1}{t_o}(C_{in} - C_{out}) - kC_{out} \qquad (2\text{-}35)$$

where $t_o = V/Q$ as noted in Equation 2-27. Under steady-state conditions $dC/dt = 0$ and the solution for C_{out} is

$$C_{out} = \frac{C_o}{1 + kt_o} \qquad \text{FE} \ (2\text{-}36)$$

where $C_o = C_{in}$ immediately after the step change. Note that C_{in} can be nonzero before the step change. For a first-order reaction where material is produced, the sign of the reaction term is positive and the solution to the mass balance equations is

$$C_{out} = \frac{C_o}{1 - kt_o} \qquad (2\text{-}37)$$

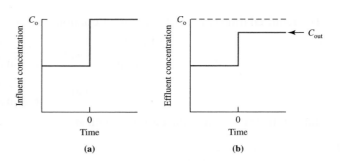

FIGURE 2-10
Steady-state response of CSTR to
(a) a step increase in influent concentration of a reactive substance.
(b) Effluent concentration. *Note:*
Steady-state conditions exist prior to
$t = 0$.

The behavior of the CSTR described by Equation 2-36 is shown diagrammatically in Figure 2-10. The steady-state effluent concentration (C_{out} in Figure 2-10b) is less than the influent concentration because of the decay of a reactive substance. From Equation 2-36, you may note that the effluent concentration is equal to the influent concentration divided by $1 + kt_o$.

A step decrease in the influent concentration to zero ($C_{in} = 0$) in a balanced flow ($Q_{in} = Q_{out}$) CSTR operating under non-steady-state conditions with first-order decay of a reactive substance may be described by rewriting Equation 2-34:

$$\frac{dM}{dt} = 0 - C_{out}Q_{out} - kC_{out}V \tag{2-38}$$

where $M = CV$. Because the volume is constant, we may divide through by V and simplify to obtain

$$\frac{dC}{dt} = \left(\frac{1}{t_o} + k\right)C_{out} \tag{2-39}$$

where $t_o = V/Q$ as noted in Equation 2-27. The solution for C_{out} is

$$C_{out} = C_o \exp\left[-\left(\frac{1}{t_o} + k\right)t\right] \tag{2-40}$$

where C_o is the effluent concentration at $t = 0$.

The concentration plots are shown in Figure 2-11.

CSTRs in Series. Some biological wastewater kinetics are inherently sequential. The treatment process may be designed with a number of CSTRs in series to take advantage

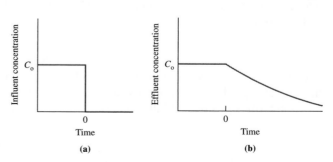

FIGURE 2-11
Non-steady-state response of CSTR
to (a) step decrease from C_o to 0
of influent C_o reactive substance.
(b) Effluent concentration.

of this fact. For the same reactor volume, reactors in series can provide greater treatment efficiency than a single CSTR.

For a conservative substance, the final effluent concentration from a series of CSTRs may be calculated as

$$C_n = C_o \left\{ \frac{\left(\frac{t}{t_o}\right)^{n-1}}{(n-1)!} \exp\left(\frac{t}{t_o}\right) \right\} \qquad \text{🏳 (2-41)}$$

Where C_n = concentration from the nth reactor
$\quad C_o$ = influent concentration or concentration in the first reactor
$\quad t$ = time
$\quad t_o$ = detention time of one reactor = volume of a single reactor/Q
$\quad Q$ = flow rate through the CSTR

For a first-order reaction with material destroyed,

$$C_n = \frac{C_o}{\left[1 + k\left(\dfrac{V_o}{Q}\right)\right]^n} \qquad \text{🏳 (2-42)}$$

Where k = rate constant
$\quad V_o$ = volume of a single reactor
$\quad t_o$ = detention time of one reactor = V_o/Q

PFR. Pipes and long narrow rivers approximate the ideal conditions of a PFR. Biological treatment in municipal wastewater treatment plants is often conducted in long narrow tanks that may be modeled as a PFR.

A step change in the influent concentration of a conservative substance in a plug-flow reactor results in an identical step change in the effluent concentration at a time equal to the theoretical detention time in the reactor as shown in Figure 2-12.

Equation 2-21 is a solution to the mass balance equation for a steady-state first-order reaction in a PFR. The concentration plot for a step change in the influent concentration is shown in Figure 2-13.

A pulse entering a PFR travels as a discrete element as illustrated in Figure 2-14 for a pulse of green dye. The passage of the pulse through the PFR and the plots of concentration with distance along the PFR are also shown.

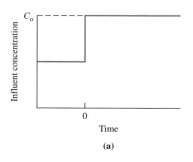

(a)

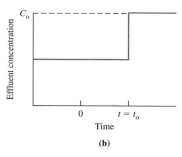

(b)

FIGURE 2-12
Response of a PFR to (a) a step increase in the influent concentration of a conservative substance. (b) Effluent concentration.

For reaction orders greater than or equal to one, an ideal PFR will always require less volume than a single ideal CSTR to achieve the same percent destruction.

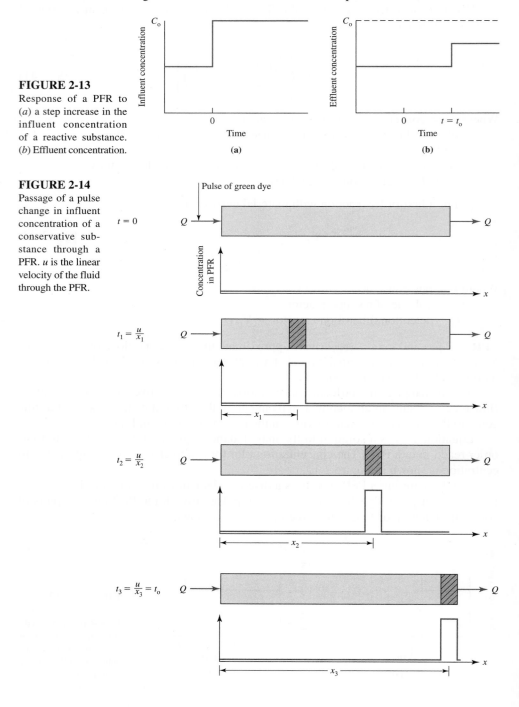

FIGURE 2-13

Response of a PFR to (a) a step increase in the influent concentration of a reactive substance. (b) Effluent concentration.

FIGURE 2-14

Passage of a pulse change in influent concentration of a conservative substance through a PFR. u is the linear velocity of the fluid through the PFR.

TABLE 2-1
Comparison of steady-state mean retention times for decay reactions of different order*

| Reaction order | r | Equations for mean retention times (t_o) | | |
		Ideal batch	Ideal plug flow	Ideal CSTR
Zero[†]	$-k$	$\dfrac{(C_o - C_t)}{k}$	$\dfrac{(C_o - C_t)}{k}$	$\dfrac{(C_o - C_t)}{k}$
First	$-kC$	$\dfrac{\ln(C_o/C_t)}{k}$	$\dfrac{\ln(C_o/C_t)}{k}$	$\dfrac{(C_o/C_t) - 1}{k}$
Second	$-kC^2$	$\dfrac{(C_o/C_t) - 1}{kC_o}$	$\dfrac{(C_o/C_t) - 1}{kC_o}$	$\dfrac{(C_o/C_t) - 1}{kC_t}$

*C_o = initial concentration or influent concentration; C_t = final condition or effluent concentration; units of k: for zero-order reactions — mass/volume · time; for first-order reactions — 1/time; for second-order reactions — volume/mass · time
[†]Expressions are valid for $kt_o \leq C_o$; otherwise $C_t = 0$.

TABLE 2-2
Comparison of steady-state performance for decay reactions of different order*

| Reaction order | r | Equations for C_t | | |
		Ideal batch	Ideal plug flow	Ideal CSTR
Zero[†] $t \leq C_o/k$ $t > C_o/k$	$-k$	$C_o - kt$ 0	$C_o - kt_o$	$C_o - kt_o$
First	$-kC$	$C_o[\exp(-kt)]$	$C_o[\exp(-kt_o)]$	$\dfrac{C_o}{1 + kt_o}$
Second	$-kC^2$	$\dfrac{C_o}{1 + ktC_o}$	$\dfrac{C_o}{1 + kt_oC_o}$	$\dfrac{(4kt_oC_o + 1)^{1/2} - 1}{2kt_o}$

*C_o = initial concentration or influent concentration; C_t = final condition or effluent concentration.
[†]Time conditions are for ideal batch reactor only.

Reactor Comparison. Although first-order reactions are common in environmental systems, other reaction orders may be more appropriate. Tables 2-1, 2-2, and 2-3 compare the reactor types for zero-, first-, and second-order reactions.

Example 2-10. A chemical degrades in a flow-balanced, steady-state CSTR according to first-order reaction kinetics. The upstream concentration of the chemical is 10 mg/L and the downstream concentration is 2 mg/L. Water is being treated at a rate of 29 m³/min. The volume of the tank is 580 m³. What is the rate of decay? What is the rate constant?

Solution. From Equation 2-11, we note that for a first-order reaction, the rate of decay, $r = -kC$. To find the rate of decay, we must solve Equation 2-34 for kC to determine the reaction rate.

TABLE 2-3
Comparison of volumes for steady-state performance for decay reactions of different order*

| | CSTR | | |
Reaction order	Single reactor	n reactors	Plug flow reactor
Zero	$\mathrm{V} = \dfrac{Q}{k}(C_o - C_e)$	$\mathrm{V} = \dfrac{Q}{k}(C_o - C_n)$	$\mathrm{V} = \dfrac{Q}{k}(C_o - C_e)$
First	$\mathrm{V} = \dfrac{Q}{k}\left(\dfrac{C_o}{C_e} - 1\right)$	$\mathrm{V} = \dfrac{nQ}{k}\left[\left(\dfrac{C_o}{C_e}\right)^{1/n} - 1\right]$	$\mathrm{V} = \dfrac{Q}{k}\ln\dfrac{C_o}{C_e}$
Second	$\mathrm{V} = \dfrac{Q}{k}\left(\dfrac{C_o}{C_e} - 1\right)\dfrac{1}{C_e}$	complex	$\mathrm{V} = \dfrac{Q}{k}\left(\dfrac{1}{C_e} - \dfrac{1}{C_o}\right)$

*V = reactor volume; Q = flow rate; k = reaction rate constant; C_o = influent concentration; C_e = effluent concentration; n = number of CSTRs in series.

$$\frac{dM}{dt} = C_{in}Q_{in} - C_{out}Q_{out} - kC_{out}\mathrm{V}$$

For steady-state there is no mass accumulation, so $dM/dt = 0$. Because the reactor is flow-balanced, $Q_{in} = Q_{out} = 29 \text{ m}^3/\text{min}$. The mass balance equation may be rewritten

$$kC_{out}\mathrm{V} = C_{in}Q_{in} - C_{out}Q_{out}$$

Solving for the reaction rate, we obtain

$$r = kC = \frac{C_{in}Q_{in} - C_{out}Q_{out}}{\mathrm{V}}$$

$$= kC = \frac{(10 \text{ mg/L})(29 \text{ m}^3/\text{min}) - (2 \text{ mg/L})(29 \text{ m}^3/\text{min})}{580 \text{ m}^3} = 0.4$$

The rate constant, k, can be determined by using the equations given in Table 2-1. For a first-order reaction in a CSTR

$$t_o = \frac{(C_o/C_t) - 1}{k}$$

The mean hydraulic detention time (t_o) is

$$t_o = \frac{\mathrm{V}}{Q} = \frac{580 \text{ m}^3}{29 \text{ m}^3/\text{min}} = 20 \text{ min}$$

Solving the equation from the table for the rate constant, k, we get

$$k = \frac{(C_o/C_t) - 1}{t_o}$$

and

$$k = \frac{\left(\dfrac{10 \text{ mg/L}}{2 \text{ mg/L}} - 1\right)}{20 \text{ min}} = 0.20 \text{ min}^{-1}$$

Reactor Design. Volume is major design parameter in reactor design. In general, the influent concentration of material, the flow rate into the reactor, and the desired effluent concentration are known. As noted in Equation 2-27, the volume is directly related to the theoretical detention time and the flow rate into the reactor. Thus, the volume can be determined if the theoretical detention time can be determined. The equations in Table 2-1 may be used to determine the theoretical detention time if the decay rate constant, k, is available. The rate constant must be determined from the literature or laboratory experiments.

2-4   ENERGY BALANCES

First Law of Thermodynamics

The *first law of thermodynamics* states that (without nuclear reaction) energy can be neither created nor destroyed. As with the law of conservation of matter, it does not mean that the form of the energy does not change. For example, the chemical energy in coal can be changed to heat and electrical power. *Energy* is defined as the capacity to do useful work. *Work* is done by a force acting on a body through a distance. One *joule* (J) is the work done by a constant force of one newton when the body on which the force is exerted moves a distance of one meter in the direction of the force. *Power* is the rate of doing work or the rate of expanding energy. The first law may be expressed as

$$Q_H = U_2 - U_1 + W \tag{2-43}$$

where Q_H = heat absorbed, kJ
U_1, U_2 = internal energy (or thermal energy) of the system in states 1 and 2, kJ
W = work, kJ

Fundamentals

Thermal Units of Energy.   Energy has many forms, among which are thermal, mechanical, kinetic, potential, electrical, and chemical. Thermal units were invented when heat was considered a substance (caloric), and the units are consistent with conservation of a quantity of substance. Subsequently, we have learned that energy is not a substance but mechanical energy of a particular form. With this in mind we will still use the common metric thermal unit of energy, the calorie.* One *calorie* (cal) is the amount of energy required to raise the temperature of one gram of water from 14.5°C to 15.5°C. In SI units 4.186 J = 1 cal.

The *specific heat* of a substance is the quantity of heat required to increase a unit mass of the substance one degree. Specific heat is expressed in metric units as kcal/kg · K and in SI units as kJ/kg · K where K is kelvins and 1 K = 1°C.

Enthalpy is a thermodynamic property of a material that depends on temperature, pressure, and the composition of the material. It is defined as

$$H = U + P V \tag{2-44}$$

---

*In discussing food metabolism, physiologists also use the term *Calorie*. However, the food Calorie is equivalent to a *kilocalorie* in the metric system. We will use the units cal or kcal throughout this text.

where H = enthalpy, kJ
  U = internal energy (or thermal energy), kJ
  P = pressure, kPa
  Ψ = volume, m^3

Think of enthalpy as a combination of thermal energy (U) and flow energy ($P\Psi$) Flow energy should not be confused with kinetic energy ($\frac{1}{2}Mv^2$). Historically, H has been referred to as a system's "heat content." Because *heat* is correctly defined only in terms of energy transfer across a boundary, this is not considered a precise thermodynamic description and enthalpy is the preferred term.

When a non-phase-change process* occurs without a change in volume, a change in internal energy is defined as

$$\Delta U = Mc_v\Delta T \tag{2-45}$$

where ΔU = change in internal energy
  M = mass
  c_v = specific heat at constant volume
  ΔT = change in temperature

When a non-phase-change process occurs without a change in pressure, a change in enthalpy is defined as

$$\Delta H = Mc_p\Delta T \tag{2-46}$$

where ΔH = enthalpy change
  c_p = specific heat at constant pressure

Equations 2-45 and 2-46 assume that the specific heat is constant over the range of temperature (ΔT). Solids and liquids are nearly incompressible and therefore do virtually no work. The change in $P\Psi$ is zero, making the changes in H and U identical. Thus, for solids and liquids, we can generally assume $c_v = c_p$ and $\Delta U = \Delta H$, so the change in energy stored in a system is

$$\Delta H = Mc_v\Delta T \tag{2-47}$$

Specific heat capacities for some common substances are listed in Table 2-4.

When a substance changes *phase* (i.e., it is transformed from solid to liquid or liquid to gas), energy is absorbed or released without a change in temperature. The energy required to cause a phase change of a unit mass from a solid to a liquid at constant pressure is called the *latent heat of fusion* or *enthalpy of fusion*. The energy required to cause a phase change of a unit mass from a liquid to a gas at constant pressure is called the *latent heat of vaporization* or *enthalpy of vaporization*. The same amounts of energy are released in condensing the vapor and freezing the liquid. For water the enthalpy of fusion at 0°C is 333 kJ/kg and the enthalpy of vaporization is 2,257 kJ/kg at 100°C. The enthalpy of condensation is 2,490 kJ/kg at 0°C. Figure 2-15 illustrates the heat required to convert 1 kg of ice to steam.

---

*Non-phase-change means, for example, water is not converted to steam.

TABLE 2-4
Specific heat capacities for common substances

Substance	c_p (kJ/kg · K)
Air (293.15 K)	1.00
Aluminum	0.95
Beef	3.22
Cement, portland	1.13
Concrete	0.93
Copper	0.39
Corn	3.35
Dry soil	0.84
Human being	3.47
Ice	2.11
Iron, cast	0.50
Steel	0.50
Poultry	3.35
Steam (373.15 K)	2.01
Water (288.15 K)	4.186
Wood	1.76

Adapted from Guyton (1961), Hudson (1959), Masters (1998), Salvato (1972).

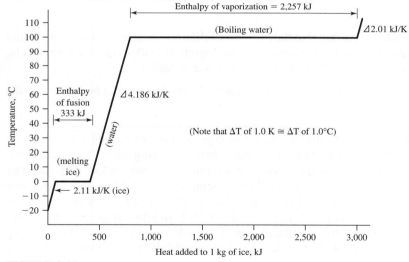

FIGURE 2-15
Heat required to convert 1 kg of ice to steam.

Example 2-11. Standard physiology texts (Guyton, 1961) report that a person weighing 70.0 kg requires approximately 2,000 kcal for simple existence, such as eating and sitting in a chair. Approximately 61 percent of all the energy in the foods we eat becomes heat during the process of formation of the energy-carrying molecule adenosine tri-phosphate (ATP) (Guyton, 1961). Still more energy becomes heat as it is transferred to functional systems of the cells. The functioning of the cells releases still more energy so that ultimately "all the energy released by metabolic processes eventually becomes heat" (Guyton, 1961). Some of this heat is used to maintain the body at a normal temperature of 37°C. What fraction of the 2,000 kcal is used to maintain the body temperature at 37°C if the room temperature is 20°C? Assume the specific heat of a human is 3.47 kJ/kg · K.

Solution. The change in energy stored in the body is

$$\Delta H = (70 \text{ kg})(3.47 \text{ kJ/kg} \cdot \text{K})(37°C - 20°C) = 4{,}129.3 \text{ kJ}$$

Converting the 2,000 kcal to kJ gives

$$(2{,}000 \text{ kcal})(4.186 \text{ kJ/kcal}) = 8{,}372.0 \text{ kJ}$$

So the fraction of energy used to maintain temperature is approximately

$$\frac{4{,}129.3 \text{ kJ}}{8{,}372.0 \text{ kJ}} = 0.49, \text{ or about } 50\%$$

The remaining energy must be removed if the body temperature is not to rise above normal. The mechanisms of removing energy by heat transfer are discussed in the following sections.

Energy Balances. If we say that the first law of thermodynamics is analogous to the law of conservation of matter, then energy is analogous to matter because it too can be "balanced." The simplest form of the energy balance equation is

$$\text{Loss of enthalpy of hot body} = \text{gain of enthalpy by cold body} \tag{2-48}$$

Example 2-12. The Rhett Butler Peach Co. dips peaches in boiling water (100°C) to remove the skin (a process called blanching) before canning them. The wastewater from this process is high in organic matter and it must be treated before disposal. The treatment process is a biological process that operates at 20°C. Thus, the wastewater must be cooled to 20°C before disposal. Forty cubic meters (40 m³) of wastewater is discharged to a concrete tank at a temperature of 20°C to allow it to cool. Assuming no losses to the surroundings, and that the concrete tank has a mass of 42,000 kg and a specific heat capacity of 0.93 kJ/kg · K, what is the equilibrium temperature of the concrete tank and the wastewater?

Solution. Assuming the density of the water is 1,000 kg/m³, the loss in enthalpy of the boiling water is

$$\Delta H = (1{,}000 \text{ kg/m}^3)(40 \text{ m}^3)(4.186 \text{ kJ/kg} \cdot \text{K})(373.15 - T) = 62{,}480{,}236 - 167{,}440T$$

where the absolute temperature is 273.15 + 100 = 373.15 K.

The gain in enthalpy of the concrete tank is

$$\Delta H = (42{,}000 \text{ kg})(0.93 \text{ kJ/kg} \cdot \text{K})(T - 293.15) = 39{,}060T - 11{,}450{,}439$$

The equilibrium temperature is found by setting the two equations equal and solving for the temperature.

$$(\Delta H)_{\text{water}} = (\Delta H)_{\text{concrete}}$$

$$62{,}480{,}236 - 167{,}440T = 39{,}060T - 11{,}450{,}439$$

$$T = 358 \text{ K or } 85°C$$

Comment: This is not very close to the desired temperature, without considering other losses to the surroundings. Convective and radiative heat losses discussed later on also play a role in reducing the temperature, but a cooling tower may be required to achieve 20°C.

For an open system, a more complete energy balance equation is

Net change in energy = energy of mass entering system − energy of
mass leaving system ± energy flow into or out of system   (2-49)

For many environmental systems the time dependence of the change in energy (i.e., the rate of energy change) must be taken into account. Equation 2-49 may be written to account for time dependence as follows:

$$\frac{dH}{dt} = \frac{d(H)_{\text{mass in}}}{dt} + \frac{d(H)_{\text{mass out}}}{dt} \pm \frac{d(H)_{\text{energy flow}}}{dt} \qquad (2\text{-}50)$$

If we consider a region of space where a fluid flows in at a rate of dM/dt and also flows out at a rate of dM/dt, then the change in enthalpy due to this flow is

$$\frac{dH}{dt} = c_{\text{p}}M\frac{dT}{dt} + c_{\text{p}}T\frac{dM}{dt} \qquad (2\text{-}51)$$

where dM/dt is the mass flow rate (e.g., in kg/s) and ΔT is the difference in temperature of the mass in the system and the mass outside of the system.

Note that Equations 2-49 and 2-50 differ from the mass balance equation in that there is an additional term: "energy flow." This is an important difference for everything from photosynthesis (in which radiative energy from the sun is converted into plant material) to heat exchangers (in which chemical energy from fuel passes through the walls of the tubes of the heat exchanger to heat a fluid inside). The energy flow into (or out of) the system may be by conduction, convection, or radiation.

Conduction.   Conduction is the transfer of heat through a material by molecular diffusion due to a temperature gradient. Fourier's law provides an expression for calculating energy flow by conduction.

$$\frac{dH}{dt} = -h_{\text{tc}}A\frac{dT}{dx} \qquad \text{🏴} (2\text{-}52)$$

TABLE 2-5
Thermal conductivity for some common materials[a]

Material	h_{tc} (W/m · K)
Air	0.023
Aluminum	221
Brick, fired clay	0.9
Concrete	2
Copper	393
Glass-wool insulation	0.0377
Steel, mild	45.3
Wood	0.126

[a]Note that the units are equivalent to J/s · m · K.
Adapted from Kuehn et al. (1998), Shortley and Williams (1955).

where dH/dt = rate of change of enthalpy, kJ/s or kW
$\qquad h_{tc}$ = thermal conductivity, kJ/s · m · K or kW/m · K
$\qquad A$ = surface area, m^2
$\qquad dT/dx$ = change in temperature through a distance, K/m

Note that 1 kJ/s = 1 kW. The average values for thermal conductivity for some common materials are given in Table 2-5.

Convection.   Forced convective heat transfer is the transfer of thermal energy by means of large-scale fluid motion such as a flowing river or aquifer or the wind blowing. The convective heat transfer between a fluid at a temperature, T_f, and a solid surface at a temperature, T_s, can be described by Equation 2-53.

$$\frac{dH}{dt} = h_c A (T_f - T_s)$$

(2-53)

where h_c = convective heat transfer coefficient, kJ/s · m^2 · K
$\qquad A$ = surface area, m^2

Radiation.   Although both conduction and convection require a medium to transport energy, radiant energy is transported by electromagnetic radiation. The radiative transfer of heat involves two processes: the absorption of radiant energy by an object and the radiation of energy by that object. The change in enthalpy due to the radiative heat transfer is the energy absorbed minus the energy emitted and can be expressed as

$$\frac{dH}{dt} = E_{abs} - E_{emitted}$$

(2-54)

Thermal radiation is emitted when an electron moves from a higher energy state to a lower one. Radiant energy is transmitted in the form of waves. Waves are cyclical or *sinusoidal* as shown in Figure 2-16. The waves may be characterized by their wavelength

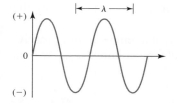

FIGURE 2-16
Sinusoidal wave. The wavelength λ is the distance between two peaks or troughs.

(λ) or their frequency (ν). The wavelength is the distance between successive peaks or troughs. Frequency and wavelength are related by the speed of light (c).

$$c = \lambda v \qquad (2\text{-}55)$$

Planck's law relates the energy emitted to the frequency of the emitted radiation.

$$E = h\nu \qquad (2\text{-}56)$$

where h = Planck's constant = 6.63×10^{-34} J $\cdot$ s

The electromagnetic wave emitted when an electron makes a transition between two energy levels is called a *photon*. When the frequency is high (small wavelengths), the energy emitted is high. Planck's law also applies to the absorption of a photon of energy. A molecule can only absorb radiant energy if the wavelength of radiation corresponds to the difference between two of its energy levels.

Every object emits thermal radiation. The amount of energy radiated depends on the wavelength, surface area, and the absolute temperature of the object. The maximum amount of radiation that an object can emit at a given temperature is called *blackbody radiation*. An object that radiates the maximum possible intensity for every wavelength is called a *blackbody*. The term *blackbody* has no reference to the color of the body. A blackbody can also be characterized by the fact that all radiant energy reaching its surface is absorbed.

Actual objects do not emit or absorb as much radiation as a blackbody. The ratio of the amount of radiation an object emits to that a blackbody would emit is called the emissivity (ε). The energy spectrum of the sun resembles that of a blackbody at 6,000 K. At normal atmospheric temperatures, the emissivity of dry soil and woodland is approximately 0.90. Water and snow have emissivities of about 0.95. A human body, regardless of pigmentation, has an emissivity of approximately 0.97 (Guyton, 1961). The ratio of the amount of energy an object absorbs to that which a blackbody would absorb is called absorptivity (α). For most surfaces, the absorptivity is the same value as the emissivity.

Integration of Planck's equation over all wavelengths yields the radiant energy of a blackbody.

$$E_{\mathrm{B}} = \sigma T^4 \qquad (2\text{-}57)$$

where E_{B} = blackbody emission rate, W/m^2
σ = Stephan–Boltzmann constant = 5.67×10^{-8} W/m$^2 \cdot$ K^4
T = absolute temperature, K

For other than blackbodies, the right-hand side of the equation is multiplied by the emissivity.

For a body with an emissivity ε and an absorptivity α, at a temperature T_b receiving radiation from its environment which is a blackbody of temperature $T_{environ}$, we can express the change in enthalpy as

$$\frac{dH}{dt} = A(\varepsilon \sigma T_b^4 - \alpha \sigma T_{environ}^4) \qquad \text{🏴} \quad (2\text{-}58)$$

where A = surface area of the body, m^2.

The solution to thermal radiation problems is highly complex because of the "re-radiation" of surrounding objects. In addition, the rate of radiative cooling will change with time as the difference in temperatures changes; initially the change per unit of time will be large because of the large difference in temperature. As the temperatures approach each other the rate of change will slow. In the following problem we use an arithmetic average temperature as a first approximation to the actual average temperature.

Example 2-13.   As mentioned in Example 2-12, heat losses due to convection and radiation were not considered. Using the following assumptions, estimate how long it will take for the wastewater and concrete tank to come to the desired temperature (20°C) if radiative cooling and convective cooling are considered. Assume that the average temperature of the water and concrete tank while cooling between 85°C (their combined temperature from Example 2-12) and 20°C is 52.5°C. Also assume that the mean radiant temperature of the surroundings is 20°C, that both the cooling tank and the surrounding environment radiate uniformly in all directions, that their emissivities are the same (0.90), that the surface area of the concrete tank including the open water surface is 56 m^2, and that the convective heat transfer coefficient is 13 $J/s \cdot m^2 \cdot K$.

Solution.   The required change in enthalpy for the wastewater is

$$\Delta H = (1{,}000 \text{ kg/m}^3)(40 \text{ m}^3)(4.186 \text{ kJ/kg} \cdot \text{K})(325.65 - 293.15) = 5{,}441{,}800 \text{ kJ}$$

where the absolute temperature of the wastewater is $273.15 + 52.5 = 325.65$ K.
    The required change in enthalpy of the concrete tank is

$$\Delta H = (42{,}000 \text{ kg})(0.93 \text{ kJ/kg} \cdot \text{K})(325.65 - 293.15) = 1{,}269{,}450 \text{ kJ}$$

For a total of $5{,}441{,}800 + 1{,}269{,}450 = 6{,}711{,}250$ kJ, or 6,711,250,000 J
    In estimating the time to cool down by radiation alone, we note that the emissivities are the same for the tank and the environment and that the net radiation is the result of the difference in absolute temperatures.

$$
\begin{aligned}
E_B &= \varepsilon \sigma (T_c^4 - T_{environ}^4) \\
&= \varepsilon \sigma T^4 = (0.90)(5.67 \times 10^{-8} \text{ W/m}^2 \cdot \text{K}^4)[(273.15 + 52.5)^4 - (273.15 + 20)^4] \\
&= 197 \text{ W/m}^2
\end{aligned}
$$

The rate of heat loss is

$$(197 \text{ W/m}^2)(56 \text{ m}^2) = 11,032 \text{ W, or } 11,032 \text{ J/s}$$

From Equation 2-53, the convective cooling rate may be estimated.

$$\frac{dH}{dt} = h_c A(T_f - T_s)$$
$$= (13 \text{ J/s} \cdot \text{m}^2 \cdot \text{K})(56 \text{ m}^2)[(273.15 + 52.5) - (273.15 + 20)]$$
$$= 23,660 \text{ J/s}$$

The time to cool down is then

$$\frac{6,711,250,000 \text{ J}}{11,032 \text{ J/s} + 23,660 \text{ J/s}} = 193,452 \text{ s, or } 2.24 \text{ days}$$

Comment: This is quite a long time. If land is not at a premium and several tanks can be built, then the time may not be a relevant consideration. Alternatively, other options must be considered to reduce the time. One alternative is to utilize conductive heat transfer and build a heat exchanger. As an energy-saving measure, the heat exchanger could be used to heat the incoming water needed in the blanching process.

Overall Heat Transfer.   Most practical heat transfer problems involve multiple heat transfer modes. For these cases, it is convenient to use an overall heat transfer coefficient that incorporates multiple modes. The form of the heat transfer equation then becomes

$$\frac{dH}{dt} = h_o A(\Delta T) \qquad \qquad \text{FE} \quad (2\text{-}59)$$

where h_o = overall heat transfer coefficient, kJ/s $\cdot$ m^2 $\cdot$ K
  ΔT = temperature difference that drives the heat transfer, K

Among their many responsibilities, environmental scientists (often with a job title of environmental sanitarian) are responsible for checking food safety in restaurants. This includes ensuring proper refrigeration of perishable foods. The following problem is an example of one of the items, namely the electrical rating of the refrigerator, that might be investigated in a case of food poisoning at a family gathering.

Example 2-14.   In evaluating a possible food "poisoning," Sam and Janet Evening evaluated the required electrical energy input to cool food purchased for a family reunion. The family purchased 12 kg of hamburger, 6 kg of chicken, 5 kg of corn, and 20 L of soda pop. They have a refrigerator in the garage in which they stored the food until the reunion. The specific heats of the food products (in kJ/kg $\cdot$ K) are hamburger: 3.22; chicken: 3.35; corn: 3.35; beverages: 4.186. The refrigerator dimensions are 0.70 m $\times$ 0.75 m $\times$ 1.00 m. The overall heat transfer coefficient for the refrigerator is 0.43 J/s $\cdot$ m^2 $\cdot$ K. The temperature in the garage is 30°C. The food must be kept at 4°C to prevent spoilage. Assume that it takes 2 h for the food to reach a temperature of 4°C, that the meat has risen to 20°C in the time it takes to get it home from the store, and that the soda pop and corn have risen to 30°C. What electrical energy input (in kilowatts) is required during the first 2 h the food is in the refrigerator? What is the energy input required to maintain the temperature for the second 2 h. Assume the refrigerator interior is at 4°C when the food is placed in it and that the door is not opened during the 4-h period. Ignore the energy required to heat the air in the refrigerator, and assume that all the electrical energy is used to remove heat. If the refrigerator is rated at 875 W, is poor refrigeration a part of the food poisoning problem?

Solution.   The energy balance equation is of the form

$$\frac{dH}{dt} = \frac{d(H)_{\text{mass in}}}{dt} + \frac{d(H)_{\text{mass out}}}{dt} \pm \frac{d(H)_{\text{energy flow}}}{dt}$$

where dH/dt = the enthalpy change required to balance the input energy
$d(H)_{\text{mass in}}$ = change in enthalpy due to the food
$d(H)_{\text{energy flow}}$ = the change in enthalpy to maintain the temperature at 4°C

There is no $d(H)_{\text{mass out}}$.
    Begin by computing the change in enthalpy for the food products.

Hamburger

$$\Delta H = (12 \text{ kg})(3.22 \text{ kJ/kg} \cdot \text{K})(20°C - 4°C) = 618.24 \text{ kJ}$$

Chicken

$$\Delta H = (6 \text{ kg})(3.35 \text{ kJ/kg} \cdot \text{K})(20°C - 4°C) = 321.6 \text{ kJ}$$

Corn

$$\Delta H = (5 \text{ kg})(3.35 \text{ kJ/kg} \cdot \text{K})(30°C - 4°C) = 435.5 \text{ kJ}$$

Beverages
Assuming that 20 L = 20 kg

$$\Delta H = (20 \text{ kg})(4.186 \text{ kJ/kg} \cdot \text{K})(30°C - 4°C) = 2,176.72 \text{ kJ}$$

The total change in enthalpy = 618.24 kJ + 321.6 kJ + 435.5 kJ + 2,176.72 kJ = 3,552.06 kJ
    Based on a 2-h period to cool the food, the rate of enthalpy change is

$$\frac{3,552.06 \text{ kJ}}{(2 \text{ h})(3,600 \text{ s/h})} = 0.493, \quad \text{or } 0.50 \text{ kJ/s}$$

The surface area of the refrigerator is

$$0.70 \text{ m} \times 1.00 \text{ m} \times 2 = 1.40 \text{ m}^2$$
$$0.75 \text{ m} \times 1.00 \text{ m} \times 2 = 1.50 \text{ m}^2$$
$$0.75 \text{ m} \times 0.70 \text{ m} \times 2 = 1.05 \text{ m}^2$$

for a total of 3.95 m^2

The heat loss through walls of the refrigerator is then

$$\frac{dH}{dt} = (4.3 \times 10^{-4} \text{ kJ/s} \cdot \text{m}^2 \cdot \text{K})(3.95 \text{ m}^2)(30°\text{C} - 4°\text{C}) = 0.044 \text{ kJ/s}$$

Comments:

1. In the first 2 h, the electrical energy required is 0.044 kJ/s + 0.50 kJ/s = 0.54 kJ/s Because 1 W = 1 J/s, the electrical requirement is 0.54 kW, or 540 W. It does not appear that the refrigerator is part of the food poisoning problem.

   In the second 2 h, the electrical requirement drops to 0.044 kW, or 44 W.

2. Note that at the beginning of this example we put "poisoning" in quotation marks because illness from food spoilage may be the result of microbial infection, which is not poisoning in the same sense as, for example, that caused by arsenic.

The result of Example 2-14 is based on an assumption of 100 percent efficiency in converting electrical energy to refrigeration. This is, of course, not possible and leads us to the second law of thermodynamics.

Second Law of Thermodynamics

The second law of thermodynamics states that energy flows from a region of higher concentration to one of lesser concentration, not the reverse, and that the quality degrades as it is transformed. All natural, spontaneous processes may be studied in the light of the second law, and in all such cases a particular one-sidedness is found. Thus, heat always flows spontaneously from a hotter body to a colder one; gases seep through an opening spontaneously from a region of higher pressure to a region of lower pressure. The second law recognizes that order becomes disorder, that randomness increases, and that structure and concentrations tend to disappear. It foretells elimination of gradients, equalization of electrical and chemical potential, and leveling of contrasts in heat and molecular motion unless work is done to prevent it. Thus, gases and liquids left by themselves tend to mix, not to unmix; rocks weather and crumble; iron rusts.

The degradation of energy as it is transformed means that enthalpy is wasted in the transformation. The fractional part of the heat which is wasted is termed unavailable energy. A mathematical expression called the *change in entropy* is used to express this unavailable energy.

$$\Delta s = Mc_p \ln\frac{T_2}{T_1} \tag{2-60}$$

where Δs = change in entropy
M = mass
c_p = specific heat at constant pressure
T_1, T_2 = initial and final absolute temperature
ln = natural logarithm

By the second law, entropy increases in any transformation of energy from a region of higher concentration to a lesser one. The higher the degree of disorder, the higher the entropy. Degraded energy is entropy, dissipated as waste products and heat.

Efficiency (η) or, perhaps, lack of efficiency is another expression of the second law. Sadi Carnot (1824) was the first to approach the problem of the efficiency of a heat engine (e.g., a steam engine) in a truly fundamental manner. He described a theoretical engine, now called a Carnot engine. Figure 2-17 is a simplified representation of a Carnot engine. In his engine, a material expands against a piston that is periodically brought back to its initial condition so that in any one cycle the change in internal energy of this material is zero, that is $U_2 - U_1 = 0$, and the first law of thermodynamics (Equation 2-43) reduces to

$$W = Q_2 - Q_1 \tag{2-61}$$

where Q_1 = heat rejected or exhaust heat
Q_2 = heat input

Thermal efficiency is the ratio of work output to heat input. The output is mechanical work. The exhaust heat is not considered part of the output.

$$\eta = \frac{W}{Q_2} \tag{2-62}$$

where W = work output
Q_2 = heat input

or, from Equation 2-61

$$\eta = \frac{Q_2 - Q_1}{Q_2} \tag{2-63}$$

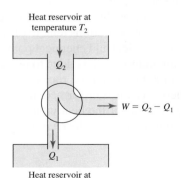

FIGURE 2-17
Schematic flow diagram of a Carnot
heat engine.
(Richards et al., 1960.)

Carnot's analysis revealed that the most efficient engine will have an efficiency of

$$\eta_{max} = 1 - \frac{T_1}{T_2} \tag{2-64}$$

where the temperatures are absolute temperatures (in kelvins). This equation implies that maximum efficiency is achieved when the value of T_2 is as high as possible and the value for T_1 is as low as possible.

Another way to evaluate efficiency is to estimate the power output for a given heat input per unit of time. This technique is used to evaluate the efficiency of electric power generation in fossil fuel power plants. The technique is illustrated in Example 2-15.

Example 2-15.   The Michigan State University power plant is rated at 61 MW. Estimate the efficiency of power generation if anthracite coal with a heating value of 32 MJ/kg is fired at a rate of 5.45 kg/s. Use one hour as the time interval for computation.

Solution.   For 1.0 hour the heat input is

$$(32 \text{ MJ/kg})(5.45 \text{ kg/s})(3,600 \text{ s/h})(1\text{h}) = 6.28 \times 10^5 \text{ MJ}$$

From the table inside of the back cover of this book, find that 1 kilowatt-hour = 3.6 MJ. Therefore, the heat input is equivalent to

$$\frac{6.28 \times 10^5 \text{MJ}}{3.6 \text{ MJ/kW-h}} = 1.74 \times 10^5 \text{ kW-h}$$

or $(1.74 \times 10^5 \text{ kW-h})(10^{-3} \text{ MW/kW}) = 174$ MW-h

With an electricity production of 61 MW-h the efficiency is then

$$\eta = \left(\frac{61 \text{ MW-h}}{174 \text{ MW-h}}\right)(100\%) = 34.98, \text{ or about } 35\%$$

Comments:

1. This efficiency is typical of coal fired power plants built in the last century.
2. The Michigan State University power plant is a *co-generation* plant. That is, it uses the steam that has passed through the electricity generator to heat buildings on campus. The use of this "waste heat" improves the overall efficiency of the power plant's use of the heating value of the coal to about 60 percent.

A refrigerator may be considered to be a heat engine operated in reverse (Figure 2-18). From an environmental point of view, the best refrigeration cycle is one that removes the greatest amount of heat (Q_1) from the refrigerator for the least expenditure of mechanical work. Thus, we use the *coefficient of performance* rather than efficiency.

$$\text{C.O.P} = \frac{Q}{W} = \frac{Q_1}{Q_2 - Q_1} \tag{2-65}$$

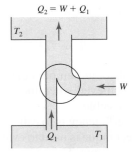

FIGURE 2-18
Schematic flow diagram of a refrigerator.
(Richards et al., 1960.)

By analogy to the Carnot efficiency,

$$\text{C.O.P} = \frac{T_1}{T_2 - T_1} \tag{2-66}$$

Example 2-16.   What is the coefficient of performance of the refrigerator in Example 2-14?

Solution.   The C.O.P. is calculated directly from the temperatures.

$$\text{C.O.P} = \frac{273.15 + 4}{[(273.15 + 30) - (273.15 + 4)]} = 10.7$$

Comments:

1. Note that in contrast to the heat engine, the performance increases if the temperatures are close together.

2. The normal ambient temperature for refrigerators located in household kitchens in the United States is about 20°C. At this temperature, the C.O.P. would be about 17.

2-5   CHAPTER REVIEW

When you have completed studying this chapter, you should be able to do the following without the aid of your textbook or notes:

1. Define the law of conservation of matter (mass).

2. Explain the circumstances under which the law of conservation of matter is violated.

3. Draw a materials-balance diagram given the inputs, outputs, and accumulation or the relationship between the variables.

4. Define the following terms: rate, conservative pollutants, reactive chemicals, steady-state conditions, equilibrium, completely mixed systems, and plug-flow systems.

5. Explain why the effluent from a completely mixed system has the same concentration as the system itself.

6. Define the first law of thermodynamics and provide one example.

7. Define the second law of thermodynamics and provide one example.

8. Define energy, work, power, specific heat, phase change, enthalpy of fusion, enthalpy of evaporation, photon, and blackbody radiation.

9. Explain how the energy balance equation differs from the materials balance equation.

10. List the three mechanisms of heat transfer and explain how they differ.

11. Explain the relationship between energy transformation and entropy.

With the aid of this text, you should be able to do the following:

12. Write and solve mass balance equations for systems with and without transformation.

13. Write the mathematical expression for the decay of a substance by first-order kinetics with respect to the substance.

14. Solve first-order reaction problems.

15. Compute the change in enthalpy for a substance.

16. Solve heat transfer equations for conduction, convection, and radiation individually and in combination.

17. Write and solve energy balance equations.

18. Compute the change in entropy.

19. Compute the Carnot efficiency for a heat engine.

20. Compute the coefficient of performance for a refrigerator.

2-6   PROBLEMS

2-1.   A sanitary landfill has available space of 16.2 ha at an average depth of 10 m. Seven hundred sixty-five (765) cubic meters of solid waste are dumped at the site 5 days per week. This waste is compacted to twice its delivered density. Draw a mass balance diagram and estimate the expected life of the landfill in years.

> *Answer:* 16.29, or 16 years

2-2.   Each month the Speedy Dry Cleaning Company buys 1 barrel (0.160 m^3) of dry cleaning fluid. Ninety percent of the fluid is lost to the atmosphere and 10 percent remains as residue to be disposed of. The density of the dry cleaning fluid is 1.5940 g/mL. Draw a mass balance diagram and estimate the monthly mass emission rate to the atmosphere in kg/mo.

2-3. Congress banned the production of the Speedy Dry Cleaning Company's dry cleaning fluid in 2000. Speedy is using a new cleaning fluid. The new dry cleaning fluid has one-sixth the volatility of the former dry cleaning fluid (Problem 2-2). The density of the new fluid is 1.6220 g/mL. Assume that the amount of residue is the same as that resulting from the use of the old fluid and estimate the mass emission rate to the atmosphere in kg/mo. Because the new dry cleaning fluid is less volatile, the company will have to purchase less per year. Estimate the annual volume of dry cleaning fluid saved (in m^3/y).

2-4. Gasoline vapors are vented to the atmosphere when an underground gasoline storage tank is filled. If the tanker truck discharges into the top of the tank with no vapor control (known as the splash fill method), the emission of gasoline vapors is estimated to be 2.75 kg/m^3 of gasoline delivered to the tank. If the tank is equipped with a pressure relief valve and interlocking hose connection and the tanker truck discharges into the bottom of the tank below the surface of the gasoline in the storage tank, the emission of gasoline vapors is estimated to be 0.095 kg/m^3 of gasoline delivered (Wark et al., 1998). Assume that the service station must refill the tank with 4.00 m^3 of gasoline once a week. Draw a mass balance diagram and estimate the annual loss of gasoline vapor (in kg/y) for the splash fill method. Estimate the value of the fuel that is captured if the vapor control system is used. Assume the density of the condensed vapors is 0.800 g/mL and the cost of the gasoline is $1.06 per liter.

2-5. The Rappahannock River near Warrenton, VA, has a flow rate of 3.00 m^3/s. Tin Pot Run (a pristine stream) discharges into the Rappahannock at a flow rate of 0.05 m^3/s. To study mixing of the stream and river, a conservative tracer is to be added to Tin Pot Run. If the instruments that can measure the tracer can detect a concentration of 1.0 mg/L, what minimum concentration must be achieved in Tin Pot Run so that 1.0 mg/L of tracer can be measured after the river and stream mix? Assume that the 1.0 mg/L of tracer is to be measured after complete mixing of the stream and Rappahannock has been achieved and that no tracer is in Tin Pot Run or the Rappahannock above the point where the two streams mix. What mass rate (kg/d) of tracer must be added to Tin Pot Run?

Answer: 263.52 or 264 kg/d

2-6. The Clearwater water treatment plant uses sodium hypochlorite (NaOCl) to disinfect the treated water before it is pumped to the distribution system. The NaOCl is purchased in a concentrated solution (52,000 mg/L) that must be diluted before it is injected into the treated water. The dilution piping scheme is shown in Figure P-2-6. The NaOCl is pumped from the small tank (called a "day tank") into a small pipe carrying a portion of the clean treated water (called a "slip stream") to the main service line. The main service line carries a flow rate of 0.50 m^3/s. The slip stream flows at 4.0 L/s. At what rate of flow (in L/s) must the NaOCl from the day tank be pumped into the slip stream to achieve a concentration of 2.0 mg/L of NaOCl in the main service line? Although it is reactive, you may assume that NaOCl is not reactive for this problem.

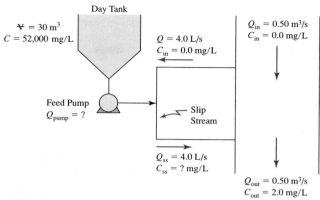

Day Tank

$V = 30\ m^3$
$C = 52{,}000\ mg/L$

$Q = 4.0\ L/s$
$C_{in} = 0.0\ mg/L$

$Q_{in} = 0.50\ m^3/s$
$C_{in} = 0.0\ mg/L$

Feed Pump
$Q_{pump} = ?$

Slip
Stream

$Q_{ss} = 4.0\ L/s$
$C_{ss} = ?\ mg/L$

$Q_{out} = 0.50\ m^3/s$
$C_{out} = 2.0\ mg/L$

FIGURE P-2-6
Dilution piping scheme.

2-7. The Clearwater design engineer cannot find a reliable pump to move the NaOCl from the day tank into the slip stream (Problem 2-6). Therefore, she specifies in the operating instructions that the day tank be used to dilute the concentrated NaOCl solution so that a pump rated at 1.0 L/s may be used. The tank is to be filled once each shift (8 h per shift). It has a volume of 30 m³. Determine the concentration of NaOCl that is required in the day tank if the feed rate of NaOCl must be 1,000 mg/s. Calculate the volume of concentrated solution and the volume of water that is to be added for an 8-hour operating period. Although it is reactive, you may assume that NaOCl is not reactive for this problem.

2-8. In water and wastewater treatment processes a filtration device may be used to remove water from the sludge formed by a precipitation reaction. The initial concentration of sludge from a softening reaction (Chapter 4) is 2 percent (20,000 mg/L) and the volume of sludge is 100 m³. After filtration the sludge solids concentration is 35 percent. Assume that the sludge does not change density during filtration, and that liquid removed from the sludge contains no sludge. Using the mass balance method, determine the volume of sludge after filtration.

2-9. The U.S. EPA requires hazardous waste incinerators to meet a standard of 99.99 percent destruction and removal of organic hazardous constituents injected into the incinerator. This efficiency is referred to as "four nines DRE." For especially toxic waste the DRE must be "six nines." The efficiency is to be calculated by measuring the mass flow rate of organic constituent entering the incinerator and the mass flow rate of constituent exiting the incinerator stack. A schematic of the process is shown in Figure P-2-9. One of the difficulties of assuring these levels of destruction is the ability to measure the contaminant in the exhaust gas. Draw a mass balance diagram for the process and determine the allowable quantity of contaminant in the exit stream if the incinerator is burning 1.0000 g/s of hazardous constituent. (*Note:* the number of significant figures is very important in this calculation.)

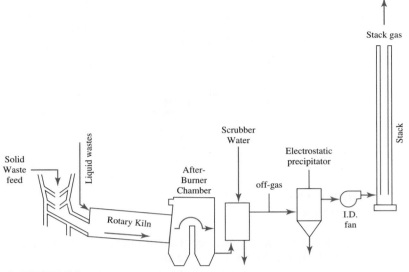

FIGURE P-2-9
Schematic of hazardous waste incinerator.

If the incinerator is 90 percent efficient in destroying the hazardous constituent, what scrubber efficiency is required to meet the standard?

2-10. A new high-efficiency air filter has been designed to be used in a secure containment facility to do research on detection and destruction of anthrax. Before the filter is built and installed it needs to be tested. It is proposed to use ceramic microspheres of the same diameter as the anthrax spores for the test. One obstacle in the test is that the efficiency of the sampling equipment is unknown and cannot be readily tested because the rate of release of microspheres cannot be sufficiently controlled to define the number of microspheres entering the sampling device. The engineers propose the test apparatus shown in Figure P-2-10. The sampling filters capture the microspheres on a membrane filter that allows microscopic

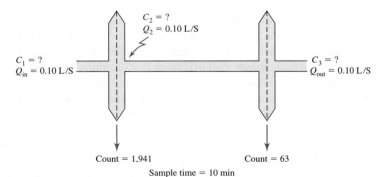

FIGURE P-2-10
Filtration test apparatus.

counting of the captured particles. At the end of the experiment, the number of particles on the first filter is 1,941 and the number on the second filter is 63. Assuming each filter has the same efficiency, estimate the efficiency of the sampling filters. (*Note:* this problem is easily solved by using the particle counts (C_1, C_2, C_3) and efficiency η.)

2-11. To remove the solution containing metal from a part after metal plating the part is commonly rinsed with water. This rinse water is contaminated with metal and must be treated before discharge. The Shiny Metal Plating Co. uses the process flow diagram shown in Figure P-2-11. The plating solution contains 85 g/L of nickel. The parts drag out 0.05 L/min of plating solution into the rinse tank. The flow of rinse water into the rinse tank is 150 L/min. Write the general mass balance equation for the rinse tank and estimate the concentration of nickel in the wastewater stream that must be treated. Assume that the rinse tank is completely mixed and that no reactions take place in the rinse tank.

Answer: $C_n = 28.3$, or 28 mg/L

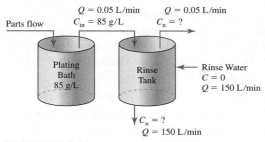

FIGURE P-2-11
Plating rinse water flow scheme.

2-12. Because the rinse water flow rate for a nickel plating bath (Problem 2-11) is quite high, it is proposed that the countercurrent rinse system shown in Figure P-2-12 be used to reduce the flow rate. Assuming that the C_n concentration remains the same at 28 mg/L, estimate the new flow rate. Assume that the rinse tank is completely mixed and that no reactions take place in the rinse tank.

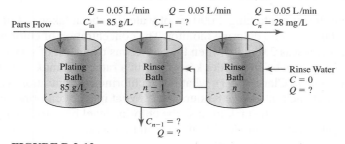

FIGURE P-2-12
Countercurrent rinse water flow scheme.

2-13. The Environmental Protection Agency (U.S. EPA, 1982) offers the following equation to estimate the flow rate for counter current rinsing (Figure P-2-12):

$$Q = \left[\left(\frac{C_{in}}{C_n} \right)^{1/n} + \frac{1}{n} \right] q$$

where Q = rinse water flow rate, L/min
C_{in} = concentration of metal in plating bath, mg/L
C_n = concentration of metal in nth rinse bath, mg/L
n = number of rinse tanks
q = flow rate of liquid dragged out of a tank by the parts, L/min

Using the EPA equation and the data from Problem 2-11, calculate the rinse water flow rate for one, two, three, four, and five rinse tanks in series using a computer spreadsheet you have written. Use the spreadsheet graphing function to plot a graph of rinse water flow rate versus number of rinse tanks.

2-14. If biodegradable organic matter, oxygen, and microorganisms are placed in a closed bottle, the microorganisms will use the oxygen in the process of oxidizing the organic matter. The bottle may be treated as a batch reactor and the decay of oxygen may be treated as a first-order reaction. Write the general mass balance equation for the bottle. Using a computer spreadsheet program you have written, calculate and then plot the concentration of oxygen each day for a period of 5 days starting with a concentration of 8 mg/L. Use a rate constant of 0.35 d^{-1}.

Answer: day 1 = 5.64, or 5.6 mg/L; day 2 = 3.97, or 4.0 mg/L

2-15. In 1908, H. Chick reported an experiment in which he disinfected anthrax spores with a 5 percent solution of phenol (Chick, 1908). The results of his experiment are tabulated below. Assuming the experiment was conducted in a completely mixed batch reactor, determine the decay rate constant for the die-off of anthrax.

Concentration of Survivors (number/mL)	**Time (min)**
398	0
251	30
158	60

2-16. A water tower containing 4,000 m^3 of water has been taken out of service for installation of a chlorine monitor. The concentration of chlorine in the water tower was 2.0 mg/L when the tower was taken out of service. If the chlorine decays by first-order kinetics with a rate constant $k = 1.0$ d^{-1} (Grayman and Clark, 1993), what is the chlorine concentration when the tank is put back in service 8 hours later? What mass of chlorine (in kg) must be added to the tank to raise the chlorine level back to 2.0 mg/L? Although it is not completely mixed, you may assume the tank is a completely mixed batch reactor.

2-17. The concept of "half-life" is used extensively in environmental engineering and science. For example, it is used to describe the decay of radioisotopes, elimination of poisons from people, self-cleaning of lakes, and the disappearance of pesticides from soil. Starting with the mass balance equation, develop an expression that describes the half-life ($t_{1/2}$) of a substance in terms of the reaction rate constant k, assuming the decay reaction takes place in a batch reactor.

2-18. If the initial concentration of a reactive substance in a batch reactor is 100 percent, determine the amount of substance remaining after 1, 2, 3, and 4 half-lives if the reaction rate constant is 6 mo^{-1}.

2-19. Liquid hazardous wastes are blended in a CSTR to maintain a minimum energy content before burning them in a hazardous waste incinerator. The energy content of the waste currently being fed is 8.0 MJ/kg (megajoules/kilogram). A new waste is injected in the flow line into the CSTR. It has an energy content of 10.0 MJ/kg. If the flow rate into and out of the 0.20 m^3 CSTR is 4.0 L/s, how long will it take the effluent from the CSTR to reach an energy content of 9 MJ/kg?

  Answer: t = 34.5, or 35 s

2-20. Repeat Problem 2-19 with a new waste having an energy content of 12 MJ/kg instead of 10 MJ/kg.

2-21. An instrument is installed along a major water distribution pipe line to detect potential contamination from terrorist threats. A 2.54-cm-diameter pipe connects the instrument to the water distribution pipe. The connecting pipe is 20.0 m long. Water from the distribution pipe is pumped through the instrument and then discharged to a holding tank for verification analysis and proper disposal. If the flow rate of the water in the sample line is 1.0 L/min, how many minutes will it take a sample from the distribution pipe to reach the instrument. Use the following relationship to determine the speed of the water in the sample pipe:

$$u = \frac{Q}{A}$$

where u = speed of water in pipe, m/s
  Q = flow rate of water in pipe, m^3/s
  A = area of pipe, m^2

If the instrument uses 10 mL for sample analysis, how many liters of water must pass through the sampler before it detects a contaminant in the pipe?

2-22. A bankrupt chemical firm has been taken over by new management. On the property they found a 20,000 m^3 brine pond containing 25,000 mg/L of salt. The new owners propose to flush the pond into their discharge pipe leading to the Atlantic ocean, which has a salt concentration above 30,000 mg/L. What flow rate of freshwater (in m^3/s) must they use to reduce the salt concentration in the pond to 500 mg/L within one year?

  Answer: Q = 0.0025 m^3/s

2-23. A 1,900-m^3 water tower has been cleaned with a chlorine solution. The vapors of chlorine in the tower exceed allowable concentrations for the work crew to enter and finish repairs. If the chlorine concentration is 15 mg/m^3 and the allowable concentration is 0.0015 mg/L, how long must the workers vent the tank with clean air flowing at 2.35 m^3/s? Although chlorine is a reactive substance, you may consider it nonreactive for this problem.

2-24. A railroad tank car is derailed and ruptured. It discharges 380 m^3 of pesticide into the Mud Lake drain. As shown in Figure P-2-24, the drain flows into Mud Lake which has a liquid volume of 40,000 m^3. The water in the creek has a flow rate of 0.10 m^3/s, a velocity of 0.10 m/s, and the distance from the spill site to the pond is 20 km. Assume that the spill is short enough to treat the injection of the pesticide as a pulse, that the pond behaves as a flow balanced CSTR, and that the pesticide is nonreactive. Estimate the time it will take to flush 99 percent of the pesticide from the pond.

Answer: Time to reach the pond = 2.3 days;
time to flush = 21.3, or 21 days

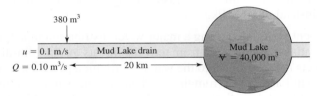

FIGURE P-2-24
Drain flow into Mud lake.

2-25. During a snowstorm the fluoride feeder in North Bend runs out of feed solution. As shown in Figure P-2-25, the rapid-mix tank is connected to a 5-km-long distribution pipe. The flow rate into the rapid-mix tank is 0.44 m^3/s and the volume of the tank is 2.50 m^3. The velocity in the pipe is 0.17 m/s. If the fluoride concentration in the rapid-mix tank is 1.0 mg/L when the feed stops, how long will it be until the concentration of fluoride is reduced to 0.01 mg/L at the end of the distribution pipe? The fluoride may be considered a nonreactive chemical.

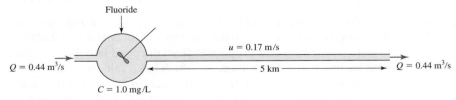

FIGURE P-2-25
Fluoride feeder.

2-26. A sewage lagoon that has a surface area of 10 ha and a depth of 1 m is receiving 8,640 m^3/d of sewage containing 100 mg/L of biodegradable contaminant. At steady state, the effluent from the lagoon must not exceed 20 mg/L of biodegradable contaminant. Assuming the lagoon is well mixed and that there are no losses or gains of water in the lagoon other than the sewage input, what biodegradation reaction rate coefficient (d^{-1}) must be achieved for a first-order reaction?

Answer: $k = 0.3478$ or 0.35 d^{-1}

2-27. Repeat Problem 2-26 with two lagoons in series (see Figure P-2-27). Each lagoon has a surface area of 5 ha and a depth of 1 m.

$C_{in} = 100$ mg/L
$Q_{in} = 8,640$ m^3/d

Lagoon 1
Area = 5 ha

Lagoon 2
Area = 5 ha

$C_{out} = 20$ mg/L
$Q_{out} = 8,640$ m^3/d

FIGURE P-2-27
Two lagoons in series.

2-28. Using a spreadsheet program you have written, determine the effluent concentration if the process producing sewage in Problem 2-26 shuts down ($C_{in} = 0$). Calculate and plot points at 1 day intervals for 10 days. Utilize the graphing function of the spreadsheet to construct your plot.

2-29. A 90-m^3 basement in a residence is found to be contaminated with radon coming from the ground through the floor drains. The concentration of radon in the room is 1.5 Bq/L under steady-state conditions. The room behaves as a CSTR and the decay of radon is a first-order reaction with a decay rate constant of 2.09×10^{-6} s^{-1}. If the source of radon is closed off and the room is vented with radon-free air at a rate of 0.14 m^3/s, how long will it take to lower the radon concentration to an acceptable level of 0.15 Bq/L?

2-30. An ocean outfall diffuser that discharges treated wastewater into the Pacific ocean is 5,000 m from a public beach. The wastewater contains 10^5 coliform bacteria per milliliter. The wastewater discharge flow rate is 0.3 m^3/s. The coliform first-order death rate in seawater is approximately 0.3 h^{-1} (Tchobanoglous and Schroeder, 1985). The current carries the wastewater plume toward the beach at a rate of 0.5 m/s. The ocean current may be approximated as a pipe carrying 600 m^3/s of seawater. Determine the coliform concentration at the beach. Assume that the current behaves as a plug-flow reactor and that the wastewater is completely mixed in the current at the discharge point.

2-31. For the following conditions, determine whether a CSTR or a PFR is more efficient in removing a reactive compound from the waste stream under steady-state conditions with a first-order reaction: reaction volume = 280 m^3, flow rate = 14 m^3/d, and reaction rate coefficient = 0.05 d^{-1}.

Answer: CSTR $\eta = 50\%$; PFR $\eta = 63\%$

2-32. Compare the reactor volume required to achieve 95 percent efficiency for a CSTR and a PFR for the following conditions: steady-state, first-order reaction, flow rate $= 14 \text{ m}^3/\text{d}$, and reaction rate coefficient $= 0.05 \text{ d}^{-1}$.

2-33. The discharge pipe from a sump pump in the dry well of a sewage lift station did not drain properly, and the water at the discharge end of the pipe froze. A hole has been drilled into the ice and a 200-W electric heater has been inserted in the hole. If the discharge pipe contains 2 kg of ice, how long will it take to melt the ice? Assume all the heat goes into melting the ice.

Answer: 55.5, or 56 min

2-34. As noted in Examples 2-12 and 2-13, the time to achieve the desired temperature using the cooling tank is quite long. An evaporative cooler is proposed as an alternative means of reducing the temperature. Estimate the amount of water (in m^3) that must evaporate each day to lower the temperature of the 40 m^3 of wastewater from 100°C to 20°C. (*Note:* While the solution to this problem is straightforward, the design of an evaporative cooling tower is a complex thermodynamic problem made even more complicated in this case by the contents of the wastewater that would potentially foul the cooling system.)

2-35. The water in a biological wastewater treatment system must be heated from 15°C to 40°C for the microorganisms to function. If the flow rate of the wastewater into the process is 30 m^3/d, at what rate must heat be added to the wastewater flowing into the treatment system? Assume the treatment system is completely mixed and that there are no heat losses once the wastewater is heated.

Answer: 3.14 GJ/d

2-36. The lowest flow in the Menominee River in July is about 40 m^3/s. If the river temperature is 18°C and a power plant discharges 2 m^3/s of cooling water at 80°C, what is the final river temperature after the cooling water and the river have mixed? Ignore radiative and convective losses to the atmosphere as well as conductive losses to the river bottom and banks.

2-37. The flow rate of the Seine in France is 28 m^3/s at low flow. A power plant discharges 10 m^3/s of cooling water into the Seine. In the summer the river temperature upstream of the power plant reaches 20°C. The temperature of the river after the power plant discharge mixes with the river is 27°C (Goubet, 1969). Estimate the temperature of the cooling water before it is mixed with the river water. Ignore radiative and convective losses to the atmosphere as well as conductive losses to the river bottom and banks.

2-38. An aerated lagoon (a sewage treatment pond that is mixed with air) is being proposed for a small lake community in northern Wisconsin. The lagoon must be designed for the summer population but will operate year-round. The winter population is about half of the summer population. The volume of the proposed lagoon, based on these design assumptions, is 3,420 m^3. The daily volume of sewage in the winter is estimated to be 300 m^3.

In January, the temperature of the lagoon drops to 0°C but it is not yet frozen. If the temperature of the wastewater flowing into the lagoon is 15°C, estimate the temperature of the lagoon at the end of a day. Assume the lagoon is completely mixed and that there are no losses to the atmosphere or the lagoon walls or floor. Also assume that the sewage has a density of 1,000 kg/m^3 and a specific heat of 4.186 kJ/kg · K.

2-39. Using the data in Problem 2-38 and a spreadsheet program you have written, estimate the temperature of the lagoon at the end of each day for a period of 7 days. Assume that the flow leaving the lagoon equals the flow entering the lagoon and that the lagoon is completely mixed.

2-40. A cooling water pond is to be constructed for a power plant that discharges 17.2 m^3/s of cooling water. Estimate the required surface area of the pond if the water temperature is to be lowered from 45.0°C at its inlet to 35.5°C at its outlet. Assume an overall heat transfer coefficient of 0.0412 kJ/s · m^2 · K (Edinger et al., 1968). (*Note:* the cooling water will be mixed with river water after it is cooled. The mixture of the 35°C water and the river water will meet thermal discharge standards.)

> *Answer:* 174.76 or 175 ha

2-41. A small building that shelters a water supply pump measures 4 m × 6 m × 2.4 m. It is constructed of 1-cm-thick wood having a thermal conductivity of 0.126 W/m · K. The inside walls are to be maintained at 10°C when the outside temperature is −18°C. How much heat must be supplied each hour to maintain the desired temperature? How much heat must be supplied if the walls are lined with 10 cm of glass-wool insulation having a thermal conductivity of 0.0377 W/m · K? Neglect the wood in the second calculation.

2-42. Because the sewage in the lagoon in Problem 2-38 is violently mixed, there is a good likelihood that the lagoon will freeze. Estimate how long it will take to freeze the lagoon if the temperature of the wastewater in the lagoon is 15°C and the air temperature is −8°C. The pond is 3 m deep. Although the aeration equipment will probably freeze before all of the wastewater in the lagoon is frozen, assume that the total volume of wastewater freezes. Use an overall heat transfer coefficient of 0.5 kJ/s · m^2 · K (Metcalf & Eddy, 2003). Ignore the enthalpy of the influent wastewater.

2-43. Bituminous coal has a heat of combustion* of 31.4 MJ/kg. In the United States, the average coal-burning utility produces an average of 2.2 kWh of electrical energy per kilogram of bituminous coal burned. What is the average overall efficiency of this production of electricity?

---

*Heat of combustion is the amount of energy released per unit mass when the compound reacts completely with oxygen. The mass does not include the mass of oxygen.

2-7 DISCUSSION QUESTIONS

2-1. A piece of limestone rock ($CaCO_3$) at the bottom of Lake Superior is slowly dissolving. For the purpose of calculating a mass balance, you can assume:

(a) The system is in equilibrium.

(b) The system is at steady state.

(c) Both of the above.

(d) Neither of the above.

Explain your reasoning.

2-2. A can of a volatile chemical (benzene) has spilled into a small pond. List the data you would need to gather to calculate the concentration of benzene in the stream leaving the pond using the mass balance technique.

2-3. In Table 2-3, specific heat capacities for common substances, the values for c_p for beef, corn, human beings, and poultry are considerably higher than those for aluminum, copper and iron. Explain why.

2-4. If you hold a beverage glass whose contents are at 4°C, "You can feel the cold coming into your hand." Thermodynamically speaking, is this statement true? Explain.

2-5. If you walk barefoot across a brick floor and a wood floor, the brick floor will feel cooler even though the room temperature is the same for both floors. Explain why.

2-8 FE EXAM FORMATTED PROBLEMS

2-1. The decay of chlorine in a distribution system follows first-order decay with a rate constant of $0.360 \ d^{-1}$. If the concentration of chlorine in a well-mixed water storage tank is 1.00 mg/L at time zero, what will the concentration be one day later? Assume no water flows out of the tank.

a. 0.360 mg/L   b. 0.368 mg/L

c. 0.500 mg/L   d. 0.698 mg/L

2-2. A 350 m^3 retention pond that holds rainwater from a shopping mall is empty at the beginning of a rainstorm. The flow rate out of the retention pond must be restricted to 320 L/min to prevent downstream flooding from a 6-hour storm. What is the maximum flow rate (in L/min) into the pond from a 6-hour storm that will not flood it?

a. 5,860 L/min   b. 321 L/min

c. 1,290 L/min   d. 7,750 L/min

2-3. A pipeline carrying 0.50 MGD of a 35,000 mg/L brine solution (NaCl) across a creek ruptures. The flow rate of the creek is 2.80 MGD. If the salt concentration in the creek is 175 mg/L, what is the concentration of salt in the creek after the pipeline discharge mixes completely with the creek water?

    a. 1.80×10^4 mg/L   b. 1.75×10^2 mg/L
    c. 5.45×10^3 mg/L   d. 6.43×10^3 mg/L

2-4. A wastewater treatment plant has experienced a power outage due to a winter storm. A treatment facility (anaerobic digester) contains 2,120 m^3 of wastewater at 37°C. If the wastewater temperature falls below 30°C, the digester will fail. The digester is made of concrete with a thermal conductivity of 2 W/m · K. The surface area of the digester is 989.6 m^2. The concrete walls and ceiling are 30 cm thick. If the outside temperature is 0°C and there is no wind or sunshine, how long will the operator have to get the heating system back into operation before the digester fails? Assume the specific heat capacity of the wastewater is the same as that of water and ignore the lack of mixing.

    a. 16 d   b. 3 d
    c. 22 min   d. 2 h

2-9   REFERENCES

Chick, H. (1908) "An Investigation of the Laws of Disinfection," *Journal of Hygiene*, p. 698.

Edinger, J. E., D. K. Brady, and W. L. Graves (1968) "The Variation of Water Temperatures Due to Steam-Electric Cooling Operations," *Journal of Water Pollution Control Federation* 40, no. 9, pp. 1637–1639.

Goubet, A. (1969) "The Cooling of Riverside Thermal-Power Plants," in F. L. Parker and P. A. Krenkel (eds.), *Engineering Aspects of Thermal Pollution*, Vanderbilt University Press, Nashville, p. 119.

Grayman, W. A., and R. M. Clark (1993) "Using Computer Models to Determine the Effect of Storage on Water Quality," *Journal of the American Water Works Association* 85, no. 7, pp. 67–77.

Guyton, A. C. (1961) *Textbook of Medical Physiology,* 2nd ed., W. B. Saunders, Philadelphia, pp. 920–921, 950–953.

Hudson, R. G. (1959) *The Engineers' Manual*, John Wiley & Sons, New York, p. 314.

Kuehn, T. H., J. W. Ramsey, and J. L. Threkeld (1998) *Thermal Environmental Engineering*, Prentice Hall, Upper Saddle River, NJ, pp. 425–427.

Masters, G. M. (1998) *Introduction to Environmental Engineering and Science*, Prentice Hall, Upper Saddle River, NJ, p. 30.

Metcalf & Eddy, Inc. (2003) revised by G. Tchobanoglous, F. L. Burton, and H. D. Stensel, *Wastewater Engineering, Treatment and Reuse*, McGraw-Hill, Boston, p. 844.

Richards, J. A., F. W. Sears, M. R. Wehr, and M. W. Zemansky, *Modern University Physics*, Addison-Wesley, Reading, MA, 1960, pp. 339, 344.

Salvato, Jr., J. A. (1972) *Environmental Engineering and Sanitation*, 2nd ed., Wiley-Interscience, New York, pp. 598–599.

Shortley, G., and D. Williams (1955) *Elements of Physics*, Prentice-Hall, Engelwood Cliffs, NJ, p. 290.

Tchobanoglous, G., and E. D. Schroeder (1985) *Water Quality*, Addison-Wesley, Reading, MA, p. 372.

Thirumurthi, D. (1969) "Design of Waste Stabilization Ponds," *Journal of Sanitary Engineering Division*, American Society of Civil Engineers, vol. 95, p. 311.

U.S. EPA (1982) *Summary Report: Control and Treatment Technology for the Metal Finishing Industry, In-Plant Changes*, U.S. Environmental Protection Agency, Washington, DC, Report No. EPA 625/8-82-008.

Wark, K., C. F. Warner, and W. T. Davis (1998) *Air Pollution: Its Origin and Control*, 3rd ed., Addison-Wesley, Reading, MA, p. 509.

CHAPTER
3

RISK ASSESSMENT

3-1 INTRODUCTION

The concepts of risk and hazard are inextricably intertwined. *Hazard* implies a probability of adverse effects in a particular situation. *Risk* is a measure of the probability. In some instances the measure is subjective, or perceived risk. Engineers and scientists use models to calculate an estimated risk. In some instances actual data may be used to estimate the risk.

Applications

We make estimates of risk for a wide range of environmental phenomena. Examples include the risk of tornadoes, hurricanes, floods, droughts, landslides, and forest fires. In this chapter we have chosen to limit our discussion to the risk to human health from chemicals released to the environment. In Chapter 4 we will use the probability concepts developed in this chapter as a forecasting tool for the risk of floods and droughts.

In the last three decades an attempt has been made to bring more rigor to the estimation of risk exposure to chemicals released to the environment. Today this process is called *quantitative risk assessment,* or more simply *risk assessment.* The use of the results of a risk assessment to make policy decisions is called *risk management.* Chapters 6–13 discuss alternative measures for managing risk by reducing the amount of contaminants in the environment.

3-2 PROBABILITY AND RISK

Types of Probability

Probability may be discussed as theory based on a subjective determination of fair odds, or an equally likely set of events, or as relative frequencies.

Subjective Probability.   Risk perception is a form of subjective probability. When people feel that they have enough experience, they assess the "odds" that their hypothesis about the risk of an environmental contaminant might be true. The amount of information or experience is the basis for assessing the odds. For example, 1:10 odds indicate a different level of belief or information about a situation than 1:2 odds do. Odds of 1:10 indicate an assessment of probability as 1/11. Odds of 1:2 indicate a probability of 1/3. Changes in information may change subjective probability.

Subjective probability is not based on a quantitative analysis. It is not appropriate for estimating risk to the public. However, risk perception often makes it difficult to achieve social change when quantitative analysis is at odds with subjective probability. It is discussed in this chapter under the heading "Risk Perception."

Relative Frequency.   For a long series of coin tosses, we generally expect heads or tails to be equally likely. For a very long sequence of trials, we may think of probability as the relative frequency of particular events. Thus, we think of the probability of a particular coin falling heads as the ratio of the number of heads occurring to the total number of tosses in a sequence. This form of probability is the one most frequently used in the determination of risk to the public from environmental hazards. Examples are probability of a flood or drought. This type of risk assessment will be addressed in Chapter 4.

Equally Likely. If we select one person from a well-mixed group of 500,000 people, we may consider any of the 500,000 people equally likely to be chosen. This sense of probability only exists if there is a finite set of likely possibilities. An example is when people are chosen for a public opinion poll. This is also the case if we are examining the consequences from exposure to a chemical for a certain class of people. The set of people is carefully limited. An example is children in the age group 1 to 5 at an abandoned hazardous waste site. The axioms and theorems for finite sets are quantitative. They are appropriate for estimating risk to the public for a well-defined set of people exposed to a specific environmental contaminant. They are the basis for the discussion in this chapter.

Risk Perception

There is an old political saying: "Perception is reality." This is no less true for environmental concerns than it is for politics. People respond to the hazards they perceive. If their perceptions are faulty, risk management efforts to improve environmental protection may be misdirected.

Some risks are well quantified. For example, the frequency and severity of automobile accidents are well documented. In contrast, other hazardous activities such as the use of alcohol and tobacco are more difficult to document. Their assessment requires complex epidemiological studies (Slovic et al., 1979).

When lay people (and some experts for that matter) are asked to evaluate risk, they seldom have ready access to the statistics. In most cases, they rely on inferences based on their experience. People are likely to judge an event as likely or frequent if instances of it are easy to imagine or recall. Also, it is evident that acceptable risk is inversely related to the number of people participating in the activity. In addition, recent events such as a disaster can seriously distort risk judgments.

Figure 3-1 illustrates different perceptions of risk. Four different groups were asked to rate thirty activities and technologies according to the present risk of death from each. Three of the groups were from Eugene, Oregon. They included 30 college students, 40 members of the League of Women Voters (LOWV), and 25 business and professional members of the "Active Club." The fourth group was composed of 15 people selected from across the United States because of their professional involvement in risk assessment. The groups were asked to estimate the mean fatality for the same group of activities and technologies given the fact that the annual death toll from motor vehicle accidents in the United States was 50,000. The lines in Figure 3-1 are the lines of best fit to the results. If the lines were at 45 degrees, the estimate would be perfect. The steeper slope of the line for the experts' risk judgments shows that they are more closely associated with the actual annual fatality rates than those of the lay groups.

Putting risk perception into perspective, we can calculate the risk of death from some familiar causes. To begin, we recognize that we all will die at some time. So, as a trivial example, the lifetime risk of death from all causes is 100 percent, or 1.0. In the United States, in 2001, there were about 3.9 million deaths from all causes. Of these, 541,532 were cancer related (National Center for Health statistics, 2004). In 2001, of those dying, the probability that an individual died of cancer was about

$$\frac{541{,}532}{3.9 \times 10^6} = 0.14$$

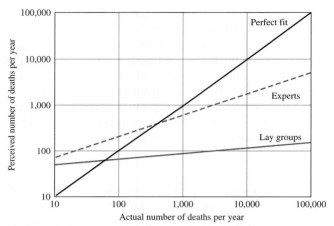

FIGURE 3-1
Judgments of perceived risk for experts and lay people plotted against the best technical estimates of annual fatalities for 25 technologies and activities. The lines are the straight lines that best fit the points. The experts' risk judgments are seen to be more closely associated with annual fatality rates than are the lay judgments. (*Source:* Davis and Masten, 2009.)

In 2006, 559,888 people died of cancer. The population of the United States is estimated to have been about 300 million. Without considering age factors, the probability, or *risk,* of a person dying of cancer in the United States in 2008 is estimated to be

$$\frac{559,888}{3.0 \times 10^8} = 1.9 \times 10^{-3}, \text{ or } 0.0019, \text{ or 19 cancer deaths per 10,000 people}$$

The annual risk (assuming a 75-year life expectancy and again ignoring age factors) is about

$$\frac{0.0019}{75} = 2.5 \times 10^{-5}$$

For comparison, Table 3-1 summarizes the risk of dying from various causes.

You may note that there are no units for the risk calculation. In addition, the description of the character of the risk is an important qualifier. Some example qualifiers are whether or not the risk is an annual risk or a lifetime risk, or whether the risk is an average over the whole population or for a particular set of people.

In addition, we understand that if the risk of dying in one year is increased, the risk of dying from another cause in a later year is decreased. Because accidents often occur early in life, a typical accident may shorten life by 30 years. In contrast, diseases, such as cancer typically cause death later in life, and life is shortened by about 15 years. Therefore, a risk of 10^{-6} shortens life on the average of 30×10^{-6} years, or 15 minutes, for an accident. The same risk for a fatal illness shortens life by about 8 minutes. It has been noted that smoking a cigarette takes 10 minutes and shortens life by 5 minutes (Wilson, 1979).

Incremental probabilities are used in environmental risk assessments. The United States Environmental Protection Agency uses an incremental lifetime cancer risk to

TABLE 3-1
Annual risk of death from selected common human activities

Cause of death	Number of deaths in representative year	Individual risk/year
Black lung disease (coal mining)	1,135	8×10^{-3} or 1/125
Heart attack	631,636	2.1×10^{-3} or 1/450
Cancer	559,888	1.9×10^{-3} or 1/525
Coal mining accident	180	1.3×10^{-3} or 1/770
Fire fighting		3×10^{-4} or 1/3,300
Motor cycle driving	5,154	6.7×10^{-4} or 1/500
Motor vehicle	45,315	1.5×10^{-4} or 1/6,600
Truck driving	802	1.1×10^{-4} or 1/9,500
Falls	20,823	7×10^{-5} or 1/14,000
Football (averaged over participants)		4×10^{-5} or 1/25,000
Home accidents	25,000	1.2×10^{-5} or 1/83,000
Bicycling (one person per bicycle)	700	6.8×10^{-6} or 1/150,000
Air travel: one transcontinental trip/year		2×10^{-6} or 1/500,000

(*Sources:* Heron, Hoyert, Murphy, et al., 2009; NHTSA, 2005, 2009; Hutt, 1978; and Rodricks and Taylor, 1983.

the most exposed members of the public to determine toxic chemical exposure limits. In setting standards, the EPA often selects a lifetime risk in the range of 10^{-6} to 10^{-4} (1 additional cancer per million people to 1 additional cancer per 10,000 people) for a given toxic chemical or combination of chemicals.

3-3 RISK ASSESSMENT

In 1989, the EPA adopted a formal process for conducting a baseline risk assessment (U.S. EPA, 1989). This process includes data collection and evaluation, toxicity assessment, exposure assessment, and risk characterization. Risk assessment is considered to be site-specific. Each step is described briefly below.

Data Collection and Evaluation

Data collection and evaluation includes gathering and analyzing site-specific data relevant to human health concerns for the purpose of identifying substances of major interest. This step includes gathering background and site information as well as the preliminary identification of potential human exposure through sampling, and development of a sample collection strategy.

When collecting background information, it is important to identify the following:

1. Possible contaminants on the site

2. Concentrations of the contaminants in key sources and media of interest, characteristics of sources, and information related to the chemical's release potential

3. Characteristics of the environmental setting that could affect the fate, transport, and persistence of the contaminants

The review of the available site information determines basic site characteristics such as groundwater movement or soil characteristics. With these data, it is possible to initially identify potential exposure pathways and exposure points important for assessing exposure. A conceptual model of pathways and exposure points can be formed from the background data and site information. This conceptual model can then be used to help refine data needs.

Toxicity Assessment

Toxicity assessment is the process of determining the relationship between the exposure to a contaminant and the increased likelihood of the occurrence or severity of adverse effects to people. This procedure includes hazard identification and dose-response evaluation. *Hazard identification* determines whether exposure to a contaminant causes increased adverse effects towards humans and to what level of severity. *Dose-response* evaluation uses quantitative information on the dose of the contaminant and relates it to the incidence of adverse health in an exposed population. Toxicity values can be determined from this quantitative relationship and used in the risk characterization step to estimate different occurrences of adverse health effects based on various exposure levels.

The single factor that determines the degree of harmfulness of a compound is the dose of that compound (Loomis, 1978). *Dose* is defined as the mass of chemical received by the animal or exposed individual. Dose is usually expressed in units of milligrams per kilogram of body mass (mg/kg). Some authors use parts per million (ppm) instead of mg/kg. Where the dose is administered over time, the units may be mg/kg · d. It should be noted that dose differs from the concentration of the compound in the medium (air, water, or soil) to which the animal or individual is exposed.

For toxicologists to establish the "degree of harmfulness" of a compound, they must be able to observe a quantitative effect. The ultimate effect manifested is death of the organism. Much more subtle effects may also be observed. Effects on body weight, blood chemistry, and enzyme inhibition or induction are examples of *graded responses.* Mortality and tumor formation are examples of *quantal* (all-or-nothing) responses. If a dose is sufficient to alter a biological mechanism, a harmful consequence will result. The experimental determination of the range of changes in a biologic mechanism to a range of doses is the basis of the dose-response relationship.

The statistical variability of organism response to dose is commonly expressed as a cumulative-frequency distribution known as a dose-response curve. Figure 3-2 illustrates the method by which a common toxicological measure, namely the LD_{50}, or lethal dose for 50 percent of the animals, is obtained. The assumption inherent in the plot of the dose-response curve is that the test population variability follows a Gaussian distribution and, hence, that the dose-response curve has the statistical properties of a Gaussian cumulative-frequency curve.

Toxicity is a relative term. That is, there is no absolute scale for establishing toxicity; one may only specify that one chemical is more or less toxic than another. Comparison of different chemicals is uninformative unless the organism or biologic mechanism is the same and the quantitative effect used for comparison is the same. Figure 3-2 serves to illustrate how a toxicity scale might be developed. Of the two curves in the figure, the LD_{50} for compound B is greater than that for compound A. Thus, for the test animal represented by the graph, compound A is more toxic than compound B as measured by lethality. There are many difficulties in establishing toxicity relationships. Species

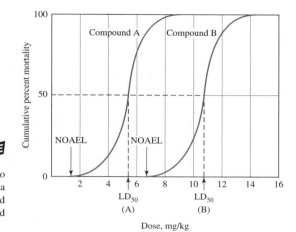

FIGURE 3-2

Hypothetical dose-response curves for two chemical agents (A and B) administered to a uniform population. NOAEL = no observed adverse effect level. (*Source:* Davis and Masten, 2009.)

respond differently to toxicants so that the LD_{50} for a mouse may be very different than that for a human. The shape (slope) of the dose-response curve may differ for different compounds so that a high LD_{50} may be associated with a low "no observed adverse effect level" (NOAEL) and vice versa.

The nature of a statistically obtained value, such as the LD_{50}, tends to obscure a fundamental concept of toxicology: that there is no fixed dose that can be relied on to produce a given biologic effect in every member of a population. In Figure 3-2, the mean value for each test group is plotted. If, in addition, the extremes of the data are plotted as in Figure 3-3, it is apparent that the response of individual members of the population may vary widely from the mean. This implies not only that single point comparisons, such as the LD_{50}, may be misleading, but that even knowing the slope of the average dose-response curve may not be sufficient if one wishes to protect hypersensitive individuals.

Organ toxicity is frequently classified as an acute or subacute effect. Carcinogenesis, teratogenesis, reproductive toxicity, and mutagenesis have been classified as chronic effects. A glossary of these toxicology terms is given in Table 3-2. It is self-evident that an organ may exhibit acute, subacute, and chronic effects and that this system of classification is not well bounded.

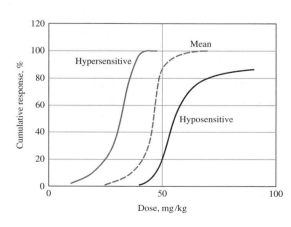

FIGURE 3-3

Hypothetical dose-response relationships for a chemical agent administered to a uniform population. (*Source:* Davis and Masten, 2004.)

TABLE 3-2
Glossary of toxicological terms

Acute toxicity	An adverse effect that has a rapid onset, short course, and pronounced symptoms.
Cancer	An abnormal growth process in which cells begin a phase of uncontrolled growth and spread.
Carcinogen	A cancer-producing substance.
Carcinomas	Cancers of epithelial tissues. Lung cancer and skin cancer are examples of carcinomas.
Chronic toxicity	An adverse effect that frequently takes a long time to run its course and initial onset of symptoms may go undetected.
Genotoxic	Toxic to the genetic material (DNA).
Initiator	A chemical that starts the change in a cell that irreversibly converts the cell into a cancerous or precancerous state. Needs to have a promoter to develop cancer.
Leukemias	Cancers of white blood cells and the tissue from which they are derived.
Lymphomas	Cancers of the lymphatic system. An example is Hodgkin's disease.
Metastasis	Process of spreading/migration of cancer cells throughout the body.
Mutagenesis	Mutagens cause changes in the genetic materal of cells. The mutations may occur either in somatic (body) cells or germ (reproductive) cells.
Neoplasm	A new growth. Usually an abnormally fast-growing tissue.
Oncogenic	Causing cancers to form.
Promoter	A chemical that increases the incidence to a previous carcinogen exposure.
Reproductive toxicity	Decreases in fertility, increases in miscarriages, and fetal or embryonic toxicity as manifested in reduced birth weight or size.
Sarcoma	Cancer of mesodermal tissue such as fat and muscle.
Subacute toxicity	Subacute toxicity is measured using daily dosing during the first 10 percent of the organism's normal life expectancy and checking for effects throughout the normal lifetime.
Teratogenesis	Production of a birth defect in the offspring after maternal or paternal exposure.

Virtually all of the data used in hazard identification and, in particular, hazard quantification, is derived from animal studies. Aside from the difficulty of extrapolating from one species to another, the testing of animals to estimate low-dose response is difficult. Example 3-1 illustrates the problem.

Example 3-1.   An experiment was developed to ascertain whether a compound has a 5 percent probability of causing a tumor. The same dose of the compound was administered to 10 groups of 100 test animals. A control group of 100 animals was, with the exception of the test compound, exposed to the same environmental conditions for the same period of time. The following results were obtained:

Group	No. of Tumors
A	6
B	4
C	10
D	1
E	2
F	9
G	5
H	1
I	4
J	7

No tumors were detected in the controls (not likely in reality).

Solution.   The average number of excess tumors is 4.9 percent. These results tend to confirm that the probability of causing a tumor is 5 percent.

Comments:

1. If, instead of using 1,000 animals (10 groups × 100 animals), only 100 animals were used, it is fairly evident from the data that, statistically speaking, some very anomalous results might be achieved. That is, we might find a risk from 1 percent to 10 percent.
2. Note that a 5 percent risk (probability of 0.05) is very high in comparison to the EPA's objective of achieving an environmental contaminant risk of 10^{-7} to 10^{-4}.

Animal studies are only capable of detecting risks on the order of 1 percent. To extrapolate the data taken from animals exposed to high doses to humans who will be exposed to doses several orders of magnitude lower, toxicologists employ mathematical models.

One of the most controversial aspects of toxicological assessment is the method chosen to extrapolate the carcinogenic dose-response curve from the high doses actually administered to test animals to the low doses that humans actually experience in the environment. The conservative worst-case assessment is that one event capable of altering

DNA will lead to tumor formation. This is called the *one-hit hypothesis.* From this hypothesis, it is assumed that there is no threshold dose below which the risk is zero, so that for carcinogens, there is no NOAEL and the dose-response curve passes through the origin.

Many models have been proposed for extrapolation to low doses. The selection of an appropriate model is more a policy decision than a scientific one since there are no data to confirm or refute any model. The *one-hit model* is frequently used:

$$P(d) = 1 - \exp(-q_0 - q_1 d) \tag{3-1}$$

where
$$P(d) = \text{lifetime risk (probability) of cancer}$$
$$d = \text{dose}$$
$$q_0 \text{ and } q_1 = \text{parameter to fit data}$$

This model corresponds to the simplest mechanistic model of carcinogenesis, namely that a single chemical hit will induce a tumor.

The background rate of cancer incidence, $P(0)$, may be represented by expanding the exponential as

$$\exp(x) = 1 + x + \frac{x^2}{2!} + \cdots + \frac{x^n}{n!} \tag{3-2}$$

For small values of x, this expansion is approximately

$$\exp(x) \simeq 1 + x \tag{3-3}$$

Assuming the background rate for cancer is small, then

$$P(0) = 1 - \exp(-q_0) \simeq 1 - [1 + (-q_0)] = q_0 \tag{3-4}$$

This implies that q_0 corresponds to the background cancer incidence. For small dose rates, the one-hit model can then be expressed as

$$P(d) \simeq 1 - [1 - (q_0 + q_1 d)] = q_0 + q_1 d = P(0) + q_1 d \tag{3-5}$$

For low doses, the additional cancer risk above the background level may be estimated as

$$A(d) = P(d) - P(0) = (P(0) + q_1 d) - P(0) \tag{3-6}$$

or

$$A(d) = q_1 d \tag{3-7}$$

This model, therefore, assumes that the excess lifetime probability of cancer is linearly related to dose.

Some authors prefer a model that is based on an assumption that tumors are formed as a result of a sequence of biological events. This model is called the *multistage model:*

$$P(d) = 1 - \exp[-(q_0 + q_1 d + q_2 d^2 + \cdots + q_n d^n)] \tag{3-8}$$

where q_i values are selected to fit the data. The one-hit model is a special case of the multistage model.

EPA has selected a modification of the multistage model for toxicological assessment. It is called the *linearized multistage model.* This model assumes that we can extrapolate from high doses to low doses with a straight line. At low doses, the slope of the dose-response curve is represented by a *slope factor* (SF). It has units of *risk per unit dose* or *risk* (kg · d/mg).

The EPA maintains a toxicological data base called IRIS (*Integrated Risk Information System*) that provides background information on potential carcinogens. IRIS includes suggested values for the slope factor. A list of slope factors for several compounds is shown in Table 3-3.

In contrast to the carcinogens, it is assumed that for noncarcinogens there is a dose below which there is no adverse effect; that is, there is an NOAEL. The EPA has estimated the acceptable daily intake, or *reference dose* (RfD), that is likely to be without appreciable risk. The RfD is obtained by dividing the NOAEL by safety factors to account for the transfer from animals to humans, sensitivity, and other uncertainties in developing the data. A list of several compounds and their RfD values is given in Table 3-4.

TABLE 3-3
Slope factors for potential carcinogens[a]

Chemical	CPS_0, kg · d/mg	RfC per $\mu g/m^3$	CPS_i, kg · d/mg
Arsenic	1.5	4.3×10^{-3}	15.1
Benzene	0.015	2.2×10^{-6}	0.029
Benzo(a)pyrene	7.3	Not available	
Cadmium	N/A	1.8×10^{-3}	6.3
Carbon tetrachloride	0.13	1.5×10^{-5}	0.0525
Chloroform	0.0061	2.3×10^{-5}	0.08
Chromium VI	N/A	1.2×10^{-2}	42.0
DDT	0.34	9.7×10^{-5}	0.34
1,1-Dichloroethylene	0.6	2×10^{-1}	0.175
Dieldrin	16.0	4.6×10^{-3}	16.1
Heptachlor	4.5	1.3×10^{-3}	4.55
Hexachloroethane	0.014	4.0×10^{-6}	0.014
Methylene chloride	0.0075	4.7×10^{-7}	0.00164
Pentachlorophenol	0.02	Not available	
Polychlorinated biphenyls	0.04	1×10^{-4}	
2,3,7,8-TCDD[b]	1.5×10^5		1.16×10^5
Tetrachloroethylene[b]	0.052		0.002
Trichloroethylene[c]			0.006
Vinyl chloride[b]	0.072	4.4×10^{-6}	

CPS_0 = cancer potency slope, oral; RfC = reference air concentration unit risk; CPS_i = cancer potency slope, inhalation, derived from RfC.
[a]Values are frequently updated. Refer to IRIS for current data.
[b]From Health Effects Assessment Summary Tables (HEAST), 1994.
[c]From U.S. EPA–NCEA Regional Support—provisional value, http://www.epa.gov/ncea.
(*Source:* with exceptions noted above: U.S. Environmental Protection Agency, IRIS database, September 2005.)

TABLE 3-4
RfDs for chronic noncarcinogenic effects for selected chemicals[a]

Chemical	Oral RfD, mg/kg · d
Acetone	0.9
Barium	0.2
Cadmium	0.0005
Chloroform	0.01
Cyanide	0.02
1,1-Dichloroethylene	0.05
Hydrogen cyanide	0.02
Methylene chloride	0.06
Pentachlorophenol	0.03
Phenol	0.3
PCB	
Aroclor 1016	7.0×10^{-5}
Aroclor 1254	2.0×10^{-5}
Silver	0.005
Tetrachloroethylene	0.01
Toluene	0.2
1,2,4-Trichlorobenzene	0.01
Xylenes	0.2

[a]Values are frequently updated. Refer to IRIS for current data.
(*Source:* U.S. Environmental Protection Agency IRIS database, 2005.)

Limitations of Animal Studies. No species provides an exact duplicate of human response. Certain effects that occur in common lab animals generally occur in people. Many effects produced in people can, in retrospect, be produced in some species. Notable exceptions are toxicities dependent on immunogenic mechanisms. Most sensitization reactions are difficult if not impossible to induce in lab animals. The procedure in transferring animal data to people is then to find the "proper" species and study it in context. Observed differences are then often quantitative rather than qualitative.

Carcinogenicity as a result of application or administration to lab animals is often assumed to be transposable to people because of the seriousness of the consequence of ignoring such evidence. However, slowly induced, subtle toxicity—because of the effects of ancillary factors (environment, age, etc.)—is difficult at best to transfer. This becomes even more difficult when the incidence of toxicity is restricted to a small hypersensitive subset of the population.

Limitations of Epidemiological Studies. There are four difficulties in epidemiological studies of toxicity in human populations. The first is that large populations are required to detect a low frequency of occurrence of a toxicological effect. The second difficulty is that there may be a long or highly variable latency period between the exposure to the toxicant and a measurable effect. Competing causes of the observed toxicological response make it difficult to attribute a direct cause and effect. For

TABLE 3-5
Potential contaminated media and corresponding routes of exposure

Media	Routes of potential exposure
Groundwater	Ingestion, dermal contact, inhalation during showering
Surface water	Ingestion, dermal contact, inhalation during showering
Sediment	Ingestion, dermal contact
Air	Inhalation of airborne (vapor phase) chemicals (indoor and outdoor)
	Inhalation of particulates (indoor and outdoor)
Soil/dust	Incidental ingestion, dermal contact
Food	Ingestion

example, cigarette smoking; the use of alcohol or drugs; and personal characteristics such as sex, race, age; and prior disease states tend to mask environmental exposures. The fourth difficulty is that epidemiological studies are often based on data collected in specific political boundaries that do not necessarily coincide with environmental boundaries such as those defined by an aquifer or the prevailing wind patterns.

Exposure Assessment

The objective of this step is to estimate the magnitude of exposure to chemicals of potential concern. The magnitude of exposure is based on chemical intake and pathways of exposure. The most important route (or pathway) of exposure may not always be clearly established. Arbitrarily eliminating one or more routes of exposure is not scientifically sound. The more reasonable approach is to consider an individual's potential contact with all contaminated media through all possible routes of entry. These are summarized in Table 3-5.

Bioconcentration.   For the "Food Ingestion" route listed in Table 3-5, there is an especially important additional consideration for the consumption of contaminated fish. Fish that live in contaminated water tend to accumulate contaminants so that concentration of contaminants in the edible tissue is much higher than in the water they live in. Because it is much more difficult to obtain comprehensive data on concentrations of contaminants in fish than it is to obtain concentrations of contaminants in water, EPA has developed *bioconcentration factors*. The bioconcentration factor is an estimate of the concentration of contaminant in the edible portion of the fish. It is based on measurements of the equilibrium concentration of contaminant in fish that have lived in water containing a known concentration of contaminant. With the bioconcentration factor, the concentration in fish may be estimated as

$$\text{Concentration in fish} = (\text{CW}) \times (\text{BCF}) \qquad (3\text{-}9)$$

where   Concentration in fish is in mg/kg
$\quad\quad$ CW = concentration in water, mg/L
$\quad\quad$ BCF = bioconcentration factor, L/kg

TABLE 3-6
Selected list of chemical bioconcentration factors

Chemical	Bioconcentration factor, L/mg
Aldrin	28
Arsenic and its compounds	44
Benzene	5.2
Cadmium and its compounds	81
Chlordane	14,000
Chromium III and VI and their compounds	200
Copper	51,000
DDT	54,000
Dieldrin	4,760
Heptachlor	15,700
Nickel and its compounds	47
Polychlorinated biphenyls	100,000
Trichloroethylene	10.6

Source: U.S. EPA, 1986.

Note that the parameters CW and BCF are those used by EPA. They are *not* products of terms, that is CW is the parameter. It is not the product of C and W. Some examples of bio concentration factors are given in Table 3-6.

Total Exposure Assessment.   The evaluation of all major sources of exposure is known as total exposure assessment (Butler et al., 1993). After reviewing the available data, it may be possible to decrease or increase the level of concern for a particular route of entry to the body. Elimination of a pathway of entry can be justified if:

1. The exposure from a particular pathway is less than that of exposure through another pathway involving the same media at the same exposure point.

2. The magnitude of exposure from the pathway is low.

3. The probability of exposure is low and incidental risk is not high.

There are two methods of quantifying exposure: point estimate methods and probabilistic methods. The EPA utilizes the point estimate procedure by estimating the *reasonable maximum exposure* (RME). Because this method results in very conservative estimates, some scientists believe probabilistic methods are more realistic (Finley and Paustenbach, 1994).

RME is defined as the highest exposure that is reasonably expected to occur and is intended to be a conservative estimate of exposure within the range of possible exposures. Two steps are involved in estimating RME: first, exposure concentrations are predicted using a transport model such as the Gaussian plume model for atmospheric dispersion (Chapter 9), then pathway-specific intakes are calculated using these exposure concentration estimates. The following equation is a generic intake equation*:

$$CDI = C\left[\frac{(CR)(EDF)}{BW}\right]\left(\frac{1}{AT}\right) \qquad \text{⚑} \qquad (3\text{-}10)$$

where CDI = chronic daily intake, (mg/kg body weight · day)
       C = chemical concentration, contacted over the exposure period
          (e.g., mg/L water)
     CR = contact rate, the amount of contaminated medium contacted per unit
          time or event (e.g., L/day)
   EFD = exposure frequency and duration, describes how long and how often
          exposure occurs. Often calculated using two terms (EF and ED):
          EF = exposure frequency (days/year)
          ED = exposure duration (years)
    BW = body weight, the average body weight over the exposure period (kg)
    AT = averaging time, period over which exposure is averaged (days)

For each different media and corresponding route of exposure, it is important to note that additional variables are used to estimate intake. For example, when calculating intake for the inhalation of airborne chemicals, an inhalation rate and exposure time are required. Specific equations for media and routes of exposure are given in Table 3-7. Standard values for use in the intake equations are shown in Table 3-8.

The EPA has made some additional assumptions in calculating exposure that are not noted in the tables. In the absence of other data, the exposure frequency (EF) for residents is generally assumed to be 350 days/year to account for absences from the residence for vacations. Similarly, for workers, EF is generally assumed to be 250 days/year based on a 5-day workweek over 50 weeks per year.

Because the risk assessment process is considered to be an *incremental risk* for cancer, the exposure assessment calculation is based on an assumption that cancer effects are cumulative over a lifetime and that high doses applied over a short time are equivalent to short doses over a long time. Although the validity of this assumption may be debated, the standard risk calculation incorporates this assumption by using a lifetime exposure duration (ED) of 75 years and an averaging time (AT) of 27,375 d (that is 365 days/year × 75 years). For noncarcinogenic effects, the averaging time (AT) is assumed to be the same as the exposure duration (ED) (Nazaroff and Alverez-Cohen, 2001).

---

*The notation in Equation 3-10 and subsequent equations follows that used in EPA guidance documents. The abbreviation CDI does not imply multiplication of three variables C, D, and I. CDI, CR, EFD, BW, and so on are the notation for the variables. They do not refer to product of terms.

TABLE 3-7

Residential exposure equations for various pathways[a]

Ingestion in drinking water

$$CDI = \frac{(CW)(IR)(EF)(ED)}{(BW)(AT)} \tag{3-11}$$

Ingestion while swimming

$$CDI = \frac{(CW)(CR)(ET)(EF)(ED)}{(BW)(AT)} \tag{3-12}$$

Dermal contact with water

$$AD = \frac{(CW)(SA)(PC)(ET)(EF)(ED)(CF)}{(BW)(AT)} \tag{3-13}$$

Ingestion of chemicals in soil

$$CDI = \frac{(CS)(IR)(CF)(FI)(EF)(ED)}{(BW)(AT)} \tag{3-14}$$

Dermal contact with soil

$$AD = \frac{(CS)(CF)(SA)(AF)(ABS)(EF)(ED)}{(BW)(AT)} \tag{3-15}$$

Inhalation of airborn (vapor phase) chemicals

$$CDI = \frac{(CA)(IR)(ET)(EF)(ED)}{(BW)(AT)} \tag{3-16}$$

Ingestion of contaminated fruits, vegetables, fish, and shellfish

$$CDI = \frac{(CF)(IR)(FI)(EF)(ED)}{(BW)(AT)} \tag{3-17}$$

where ABS = absorption factor for soil contaminant, unitless
AD = absorbed dose, mg/kg · d
AF = soil-to-skin adherence factor, mg/cm^2
AT = averaging time, d
BW = body weight, kg
CA = contaminant concentration in air, mg/m^3
CDI = chronic daily intake, mg/kg · d
CF = volumetric conversion factor for water = 1 L/1,000 cm^3
= conversion factor for soil = 10^{-6} kg/mg
CR = contact rate, L/h
CS = chemical concentration in soil, mg/kg
CW = chemical concentration in water, mg/L
ED = exposure duration, y
EF = exposure frequency, d/y or events/y
ET = exposure time, h/d or h/event
FI = fraction ingested, unitless
IR = ingestion rate, L/d or mg soil/d or kg/meal
= inhalation rate, m^3/h
PC = chemical-specific dermal permeability constant, cm/h
SA = skin surface area available for contact, cm^2

(*Source:* U.S. EPA, 1989.)

TABLE 3-8
EPA recommended values for estimating intake[a, b]

Parameter	Standard value
Body weight, adult female	65.4 kg
Body weight, adult male	78 kg
Body weight, child	
6–11 months	9 kg
1–5 y	16 kg
6–12 y	33 kg
Amount of water ingested daily, adult[c]	2.3 L
Amount of water ingested daily, child[c]	1.5 L
Amount of air breathed daily, adult female	11.3 m^3
Amount of air breathed daily, adult male	15.2 m^3
Amount of air breathed daily, child (3–5 y)	8.3 m^3
Amount of fish consumed daily, adult	6 g/d
Water swallowing rate, swimming	50 mL/h
Skin surface available, adult female	1.69 m^2
Skin surface available, adult male	1.94 m^2
Skin surface available, child	
3–6 y (avg for male and female)	0.720 m^2
6–9 y (avg for male and female)	0.925 m^2
9–12 y (avg for male and female)	1.16 m^2
12–15 y (avg for male and female)	1.49 m^2
15–18 y (female)	1.60 m^2
15–18 y (male)	1.75 m^2
Soil ingestion rate, children 1 to 6 y	100 mg/d
Soil ingestion rate, persons > 6 y	50 mg/d
Skin adherence factor, gardeners	0.07 mg/cm^2
Skin adherence factor, wet soil	0.2 mg/cm^2
Exposure duration	
Lifetime	75 y
At one residence, 90th percentile	30 y
National median	5 y
Averaging time	(ED)(365 d/y)
Exposure frequency (EF)	
Swimming	7 d/y
Eating fish and shell fish	48 d/y
Exposure time (ET)	
Bath or shower, 90th percentile	30 min
Bath or shower, 50th percentile	15 min

([a]*Sources:* U.S. EPA, 1989; U.S. EPA, 1997; U.S. EPA, 2004.)
[b]Average value unless otherwise noted.
[c]90th percentile.

Example 3-2. Estimate the lifetime average chronic daily intake of benzene from exposure to a city water supply that contains a benzene concentration equal to the drinking water standard. The allowable drinking water concentration (maximum contaminant level, MCL) is 0.005 mg/L. Assume the exposed individual is an adult male who consumes water at the adult rate for 63 years* and is an avid swimmer who swims in a local pool (supplied with city water) three days a week for 30 minutes from the age of 30 until he is 75-years-old. As an adult, he takes a long (30-minute) shower every day. Assume that the average air concentration of benzene during the shower is 5 μg/m³ (McKone, 1987). From the literature, it is estimated that the dermal uptake from water is 0.0020 m³/m²/h (This is PC in Table 3-7. PC also has units of m/h or cm/h.) Direct dermal absorption during showering is no more than 1 percent of the available benzene because most of the water does not stay in contact with skin long enough (Byard, 1989).

Solution. From Table 3-5, we note that five routes of exposure are possible from the drinking water medium: (1) ingestion, dermal contact while (2) showering and (3) swimming, (4) inhalation of vapor while showering, and (5) ingestion while swimming.

We begin by calculating the CDI for ingestion (Equation 3-11):

$$CDI = \frac{(0.005 \text{ mg/L})(2.3 \text{ L/d})(365 \text{ d/y})(63 \text{ y})}{(78 \text{ kg})(75 \text{ y})(365 \text{ d})}$$

$$= 1.24 \times 10^{-4} \text{ mg/kg} \cdot \text{d}$$

The chemical concentration (CW) is the MCL for benzene. As noted in the problem statement and footnote, the man ingests water at the adult rate for a duration (ED) equal to his adult years. The ingestion rate (IR) and body weight (BW) were selected from Table 3-8. The exposure averaging time of 365 days/year for 75 years is the EPA's generally accepted value as discussed on page 99.

Equation 3-13 may be used to estimate absorbed dose while showering:

$$AD = \frac{(0.005 \text{ mg/L})(1.94 \text{ m}^2)(0.0020 \text{ m/h})(0.50 \text{ h/event})}{(78 \text{ kg})(75 \text{ years})}$$

$$\times \frac{(1 \text{ event/d})(365 \text{ d/y})(63 \text{ y})(10^3 \text{ L/m}^3)}{(365 \text{ d/y})}$$

$$= 1.04 \times 10^{-4} \text{ mg/kg} \cdot \text{d}$$

As in the previous calculation, CW is the MCL for benzene. SA is the adult male surface area. PC is given in the problem statement. A "long shower" is assumed to be the 90th percentile value in Table 3-8 and the 63 years is derived as noted in the previous calculation.

Only about 1 percent of this amount is available for adsorption in a shower because of the limited contact time, so the actual adsorbed dose by dermal contact is

$$AD = (0.01)(1.04 \times 10^{-4} \text{ mg/kg} \cdot \text{day}) = 1.04 \times 10^{-6} \text{ mg/kg} \cdot \text{day}$$

---

*This is based on Table 3-8 values of a lifetime of 75 years minus a childhood that is assumed to last until age 12.

The adsorbed dose for swimming is calculated in the same fashion:

$$AD = \frac{(0.005 \text{ mg/L})(1.94 \text{ m}^2)(0.0020 \text{ m/h})(0.5 \text{h/event})}{(78 \text{ kg})(75 \text{ years})}$$

$$\times \frac{(3 \text{ events/wk})(52 \text{ weeks/y})(45 \text{ years})(10^3 \text{ L/m}^3)}{(365 \text{ d/y})}$$

$$= 3.19 \times 10^{-5} \text{ mg/kg} \cdot \text{d}$$

In this case, because there is virtually total body immersion for the entire contact period and because there is virtually an unlimited supply of water for contact, there is no reduction for availability. The value of ET is computed from the swimming time (30 minutes = 0.5 h/event. The exposure frequency is computed from the number of swimming events per week and the number of weeks in a year. The exposure duration (ED) is calculated from the lifetime and beginning time of swimming = 75 years − 30 years = 45 y.

The inhalation rate from showering is estimated from Equation 3-16:

$$CDI = \frac{(5 \text{ }\mu\text{g/m}^3)(10^{-3} \text{ mg/g})(0.633 \text{ m}^3\text{/h})(0.50 \text{ h/event})(1 \text{ event/d})(365 \text{ d/y})(365 \text{ d/y})(63 \text{ y})}{(78 \text{ kg})(75 \text{ years})(365 \text{ d/y})}$$

$$= 1.71 \times 10^{-5} \text{ mg/kg} \cdot \text{d}$$

The inhalation rate (IR) is taken from Table 3-8 and converted to an hourly basis, i.e., $15.2 \text{ m}^3/24 \text{ h} = 0.633 \text{ m}^3\text{/h}$. The values for ET and EF were given in the problem statement. As assumed previously, adulthood occurs from age 12 until age 75.

For ingestion while swimming, we apply Equation 3-12:

$$CDI = \frac{(0.005 \text{ mg/L})(50 \text{ mL/h})(10^{-3}\text{L/mL})(0.5 \text{h/event})(3 \text{ events/wk})(52 \text{ wk/y})(45 \text{ y})}{(78 \text{ kg})(75 \text{ y})(365 \text{ d/y})}$$

$$= 4.11 \times 10^{-7} \text{ mg/kg} \cdot \text{d}$$

The contact rate (CR) is the water swallowing rate. It was determined from Table 3-8. Other values were obtained in the same fashion as those for dermal contact while swimming.

The total exposure would be estimated as

$$CDI_T = 1.24 \times 10^{-4} + 1.04 \times 10^{-6} + 3.19 \times 10^{-5} + 1.71 \times 10^{-5} + 4.11 \times 10^{-7}$$
$$= 1.74 \times 10^{-4} \text{ mg/kg} \cdot \text{d}$$

Comment: From these calculations, it becomes readily apparent that, in this case, drinking the water dominates the intake of benzene.

Risk Characterization

In the risk characterization step, all data collected from exposure and toxicity assessments are reviewed to corroborate qualitative and quantitative conclusions about risk. The risk for each media source and route of entry is calculated. This includes the evaluation of compounding effects due to the presence of more than one chemical contaminant and the combination of risk across all routes of entry.

For low-dose cancer risk (risk below 0.01), the quantitative risk assessment for a single compound by a single route is calculated as:

$$\text{Risk} = (\text{Intake})(\text{Slope Factor}) \qquad \text{(3-18)}$$

where intake is calculated from one of the equations in Table 3-7 or a similar relationship. The slope factor is obtained from IRIS (see, for example, Table 3-3). For high carcinogenic risk levels (risk above 0.01), the one-hit equation is used:

$$\text{Risk} = 1 - \exp[-(\text{Intake})(\text{Slope Factor})] \qquad \text{(3-19)}$$

The measure used to describe the potential for noncarcinogenic toxicity to occur in an individual is not expressed as a probability. Instead, EPA uses the noncancer hazard quotient, or hazard index (HI):

$$\text{HI} = \frac{\text{Intake}}{\text{RfD}} \qquad \text{(3-20)}$$

These ratios are not to be interpreted as statistical probabilities. A ratio of 0.001 does *not* mean that there is a one in one thousand chance of an effect occurring. If the HI exceeds unity, there may be concern for potential noncancer effects. As a rule, the greater the value above unity, the greater the level of concern.

To account for multiple substances in one pathway, EPA sums the risks for each constituent:

$$\text{Risk}_T = \sum \text{Risk}_i \qquad \text{(3-21)}$$

For multiple pathways

$$\text{Total Exposure Risk} = \sum \text{Risk}_{ij} \qquad \text{(3-22)}$$

where i = the compounds and j = pathways.

In a like manner, the hazard index for multiple substances and pathways is estimated as

$$\text{Hazard Index}_T = \sum \text{HI}_{ij} \qquad \text{(3-23)}$$

In EPA's guidance documents, they recommend segregation of the hazard index into chronic, subchronic, and short-term exposure.

Although some research indicates that the addition of risks is reasonable (Silva et al., 2002), there is some uncertainty in taking this approach. Namely, should the risk from a carcinogen that causes liver cancer be added to the risk from a compound that causes stomach cancer? The conservative approach is to add the risks.

Example 3-3.   Using the results from Example 3-2, estimate the risk from exposure to drinking water containing the MCL for benzene.

Solution.   Equation 3-22 in the form

$$\text{Total Exposure Risk} = \sum \text{Risk}_j$$

may be used to estimate the risk. Because the problem is to consider only one compound, namely benzene, $i = 1$ and others do not need to be considered. Because the total exposure from Example 3-2 included both oral and inhalation routes and there are different slope factors for each route in Table 3-4, the risk from each route is computed and summed. Because we do not have a slope factor for contact, we have assumed that it is the same as for oral ingestion. The risk is

$$\text{Risk} = (1.57 \times 10^{-4} \text{ mg/kg} \cdot \text{day})(1.5 \times 10^{-2} \text{mg/kg} \cdot \text{d})^{-1}$$
$$+ (1.70 \times 10^{-5} \text{mg/kg} \cdot \text{day}^{-1})(2.9 \times 10^{-2} \text{ mg/kg} \cdot \text{d})^{-1}$$
$$= 2.85 \times 10^{-5} \text{ or } 2.9 \times 10^{-6}$$

This is the total lifetime risk (75 years) for benzene in drinking water at the MCL. Another way of viewing this is to estimate the number of people that might develop cancer. For example, in a population of 2 million,

$$(2 \times 10^6)(2.85 \times 10^{-6}) = 5.7, \text{ or } 6, \text{ people might develop cancer.}$$

Comment: This risk falls within the EPA guidelines of 10^{-4} to 10^{-7} risk. It, of course, does not account for all sources of benzene by all routes. None the less, the risk, compared to some other risks in daily life, appears to be quite small.

3-4   RISK MANAGEMENT

Though some might wish it, it is clear that establishment of zero risk cannot be achieved. There are risks in all societal decisions from driving a car to drinking water with benzene at the MCL concentration. Even banning the production of chemicals, as was done for PCBs, for example, does not remove those that already permeate our environment. Risk management is performed to decide the magnitude of risk that is tolerable in specific circumstances (NRC, 1983). This is a policy decision that weighs the results of the risk assessment against costs and benefits as well as the public acceptance. The risk manager recognizes that if a very high certainty in avoiding risk (that is, a very low risk, for example, 10^{-7}) is required, the costs in achieving low concentrations of the contaminant are likely to be high.

Unfortunately, there is very little guidance that can be provided to the risk manager. We know that people are willing to accept a higher risk for things that they expose themselves to voluntarily than for involuntary exposures, and, hence, insist on lower levels of risk, regardless of cost, for involuntary exposure. We also know that people are willing to accept risk if it approaches that for disease, that is, a fatality rate of 10^{-6} people per person-hour of exposure (Starr, 1969).

3-5   CHAPTER REVIEW

When you have completed studying this chapter, you should be able to do the following without the aid of your textbook or notes:

  1. Define and differentiate between risk and hazard.

  2. List the four steps in risk assessment and explain what occurs in each step.

3. Define the terms dose, LD_{50}, NOAEL, slope factor, RfD, CDI, and IRIS.

4. Explain why it is not possible to establish an absolute scale of toxicity.

5. Explain why an average dose-response curve may not be an appropriate model to develop environmental protection standards.

6. Explain the concept of bioconcentration.

7. Identify routes of potential exposure for the release of contaminants in multiple media.

8. Explain how risk management differs from risk assessment and the role of risk perception in risk management.

With the aid of this text, you should be able to do the following:

9. Calculate lifetime risk using the one-hit or multistage model.

10. Calculate chronic daily intake or other variables given the media and values for remaining variables.

11. Perform a risk characterization calculation for carcinogenic and noncarcinogenic threats by multiple contaminants and multiple pathways.

3-6 PROBLEMS

3-1.   The recommended time-weighted average air concentration for occupational exposure to water-soluble hexavalent chromium (Cr VI) is 0.05 mg/m^3. This concentration is based on an assumption that the individual is generally healthy and is exposed for 8 hours per day over a working lifetime (that is from age 18 to 65 years). Assuming a body weight of 70 kg and an inhalation rate of 20 m^3/h over the working life of the individual, what is the lifetime (70 y) CDI?

   Answer: 2.2×10^{-3} mg/kg · d

3-2.   The National Ambient Air Quality Standard for sulfur dioxide is 80 mg/m^3. Assuming a lifetime exposure (24 h/d, 365 d/y) for an adult male of average body weight, what is the estimated CDI for this concentration?

3-3.   Children are one of the major concerns of environmental exposure. Compare the CDIs for a 1-year-old child and an adult female drinking water contaminated with 10 mg/L of nitrate (as N). Assume a 1-year averaging time.

3-4.   Agricultural chemicals such as 2,4-D (2,4-dichlorophenoxyacetic acid) may be ingested by routes other than food. Compare the CDIs for ingestion of a soil contaminated with 10 mg/kg of 2,4-D by a 3-year-old child and an adult. Assume a 1-year averaging time. Also assume that the fraction of 2,4-D ingested is 0.10.

3-5.   Estimate the chronic daily intake of toluene from exposure to a city water supply that contains a toluene concentration equal to the drinking water standard of 1 mg/L. Assume the exposed individual is an adult female who consumes water at the adult rate for 70 years, that she abhors swimming, and that she takes a long (20-min) bath every day. Assume that the average air concentration of toluene during the bath is 1 μg/m^3. Assume the dermal uptake from water (PC) is 9.0×10^{-6} m/h and that direct dermal absorption during bathing is no more than 80 percent of the available toluene because she is not completely submerged.

> *Answer:* 3.5×10^{-2} mg/kg $\cdot$ d

3-6. Estimate the chronic daily intake of 1,1,1-trichloroethane from exposure to a city water supply that contains a 1,1,1-trichloroethane concentration equal to the drinking water standard of 0.2 mg/L. Assume the exposed individual is a child who consumes water at the child rate for 5 years, that she swims once a week for 30 min, and that she takes a short (10-min) bath every day. Assume her average age over the exposure period is 8. Assume that the average air concentration of 1,1,1-trichloroethane during the bath is 1 μg/m^3. Assume the dermal uptake from water (PC) is 0.0060 m/h and that direct dermal absorption during bathing is no more than 50 percent of the available l,1,1-trichloroethane because she is not completely submerged.

3-7. Estimate the risk from occupational inhalation exposure to hexavalent chromium. (See Problem 3-1 for assumptions.)

> *Answer:* Risk $= 8.83 \times 10^{-2}$ or 0.09

3-7   DISCUSSION QUESTIONS

3-1. What was the outcome of the hazardous waste episode at Times Beach? (*Hint:* You will need to do an Internet search.)

3-2. It has been stated that, on the basis of LD$_{50}$, 2,3,7,8-TCDD is the most toxic chemical known. Why might this statement be misleading? How would you rephrase the statement to make it more scientifically correct?

3-3. Which of the following individuals is at greater risk from inhalation of an airborne contaminant: a 1-year-old child; an adult female; an adult male? Explain your reasoning.

3-4. Which of the following individuals is at greater risk from ingestion of a soil contaminant: a 1-year-old child; an adult female; an adult male? Explain your reasoning.

3-5. A hazard index of 0.001 implies:

a. Risk $= 10^{-3}$

b. The probability of hazard is 0.001

c. The RfD is small compared to the CDI

d. There is little concern for potential health effects

3-8   FE EXAM FORMATTED PROBLEMS

3-1.   A 4-year-old child has been playing in soil contaminated with 500 mg/kg of tetrachloroethylene. It is estimated that he ingested 5 g/d of soil over a single 3-month period. The exposure frequency was 7 days per week. What is the estimated chronic daily intake of tetrachloroethylene for the child over the 3-month period?

a. 0.00016 mg/kg · d   b. 0.0756 mg/kg · d
c. 0.0312 mg/kg · d   d. 0.156 mg/kg · d

3-2.   Characterize the risk of an oral chronic daily intake of 50 ppb of arsenic.

a. 0.072   b. 0.075
c. 0.033   d. 0.047

3-3.   Characterize the hazard of an oral chronic daily intake of 0.005 mg/L of cadmium.

a. 0.032   b. 0.031
c. 10.0   d. 1.00

3-4.   Estimate the concentration of heptachlor in fish if the concentration in water is 5 ppb. The bioconcentration factor is 15,700.

a. 78.5 mg/kg   b. 78,500 mg/kg
c. 3,140 mg/kg   d. 3.14 mg/kg

3-9   REFERENCES

Butler, J. P., A. Greenberg, P. J. Lioy, G. B. Post, and J. M. Waldman (1993) "Assessment of Carcinogenic Risk from Personal Exposure to Benzo(a)pyrene in the Total Human Environmental Exposure Study (THEES)," *Journal of the Air & Waste Management Association,* vol. 43, pp. 970–977.

Byard, J. L. (1989) "Hazard Assessment of 1,1,1-Trichloroethane in Groundwater," in D. J. Paustenbach (ed.), *The Risk Assessment of Environmental Hazards,* John Wiley & Sons, New York, pp. 331–344.

Davis, M. L. and S. J. Masten (2009) *Principles of Environmental Engineering and Science,* 2nd ed., McGraw-Hill, New York.

Finley, B., and D. Paustenbach (1994) "The Benefits of Probabilistic Exposure Assessment; Three Case Studies Involving Contaminated Air, Water, and Soil," *Risk Analysis,* vol. 14, no. 1, pp. 53–73.

Heron, M., D. L. Hoyert, S. Murphy, et al. (2009) "Deaths: Final Data for 2006," *National Vital Statistics Report,* National Vital Statistics System, Centers for Disease Control and Prevention, vol. 53, No. 14.

Hutt, P. B. (1978) "Legal Considerations in Risk Assessment," *Food, Drugs, Cosmetic Law J,* vol. 33, pp. 558–559.

Loomis, T. A. (1978) *Essentials of Toxicology,* Lea & Febiger, Philadelphia, p. 2.

McKone, T. E. (1987) "Human Exposure to Volatile Organic Compounds in Household Tap Water: The Indoor Inhalation Pathway," *Environmental Science & Technology,* vol. 21, no. 12, pp. 1194–1201.

National Center for Health Statistics (2004) http://www.cdc.gov/nchs/fastats/deaths

Nazaroff, W. W., and L. Alverez-Cohen (2001) *Environmental Engineering Science,* John Wiley & Sons, Inc., New York, pp. 570–571.

NHTSA (2005) "2004 Annual Assessment," National Highway Traffic Safety Administration, Washington, DC, http//www.nhtsa.gov.

NHTSA (2009) "2008 Annual Assessment," National Highway Traffic Safety Administration, Washington, DC, http//www.nhtsa.gov.

NRC (1983) *Risk Assessment in the Federal Government: Managing the Process,* National Research Council, National Academy Press, Washington, DC, pp. 18–19.

Rodricks, I., and M. R. Taylor (1983) "Application of Risk Assessment to Good Safety Decision Making," *Regulatory Toxicology and Pharmacology,* vol. 3, pp. 275–284.

Silva, E., N. Rajapakse, and A. Kortenkamp (2002) "Something from "Nothing"— Eight Weak Estrogenic Chemicals Combined at Concentrations below NOECs Produce Significant Mixture Effects," *Environmental Science & Technology,* vol. 36, pp. 1751–1756.

Slovic, P., B. Fischoff, and S. Lichtenstein (1979) "Rating Risk," *Environment,* vol. 21, no. 3, pp. 1–20, 36–39.

Starr, C. (1969) "Social Benefit versus Technological Risk," *Science,* vol. 165, pp. 1232–1238.

U.S. EPA (1986) *Superfund Public Health Evaluation Manual,* Office of Emergency and Remedial Response, U.S. Environmental Protection Agency, Washington, DC.

U.S. EPA (1989) *Risk Assessment Guidance for Superfund, Volume I: Human Health Evaluation Manual (Part A),* U.S. Environmental Protection Agency Publication EPA/540/1-89/002, Washington, DC.

U.S. EPA (1994) *Annual Health Effects Assessment Summary Tables* (HEAST), U.S. Environmental Protection Agency Publication No. EPA 510-R-04-001, Washington, DC.

U.S. EPA (1996) National Center for Environmental Assessment—Provisional Value, http//www.epa.gov/ncea

U.S. EPA (1997) *Exposure Factor Handbook,* U.S. Environmental Protection Agency National Center for Environmental Assessment, Washington, DC.

U.S. EPA (2004) *Risk Assessment Guidance Manual for Superfund, Volume I: Human Health Evaluation Manual,* U.S. Environmental Protection Agency Publication EPA/540/R/99/005, Washington, DC.

U.S. EPA (2005) IRIS database, U.S. Environmental Protection Agency, Washington, DC.

Wilson, R. (1979) "Analyzing the Daily Risks of Life," *Technology Review,* vol. 81, no. 4, pp. 41–46.

CHAPTER

4

WATER RESOURCES ENGINEERING

4-1 INTRODUCTION

Clean, abundant water provides the basis for agriculture, energy production, commerce, industry, transportation, and recreation. Many disciplines are involved in the management of our water resources. Examples include agriculture, civil and environmental engineering, ecology, economics, law, and the social sciences. The rubric that encompasses the activities of civil and environmental engineers is water resources engineering. *Water resources engineering* may be divided into two major categories: hydrology and hydraulics. For civil and environmental engineers, hydrologic methods are used to answer the question *How much water can be expected?* Hydraulic methods are used to design the structures that withdraw, store, and convey water. This chapter is focused on an introduction to descriptive and quantitative hydrology. Hydraulic methods are the subject of other courses in traditional civil and environmental engineering curricula.

Applications

This chapter will provide you with elementary tools to estimate the following:

- Intensity, duration, and frequency of rainfall

- Amount of rainfall that will runoff a parking lot or a small parcel of land

- Risk of floods or droughts

- Volume of storage required for a retention pond or small reservoir

- How much water can be supplied by wells

4-2 FUNDAMENTALS

The Hydrologic Cycle

The global system that supplies and removes water from the earth's surface is known as the hydrologic cycle (Figure 4-1). Water is transferred to the earth's atmosphere through two processes: (1) *evaporation* and (2) *transpiration.** As moist air rises, it cools. Eventually enough moisture accumulates and the mass cools sufficiently to nucleate (form small crystals) on microscopic particles. Sufficient growth causes the droplets or snowflakes to become heavy enough to fall as precipitation. As they fall on the earth's surface, the droplets either run over the ground into streams and rivers (*surface runoff,* or just *runoff*) or percolate into the ground to form groundwater.

Surface Water Hydrology

Precipitation. Surface water hydrology begins before the precipitate hits the ground. The form the precipitate takes (rain, sleet, hail, or snow) is important. For example, about 100 mm of light fluffy snow is equivalent to about 10 mm of rain while 100 mm of heavy wet snow is equivalent to about 50 mm of rain. Other factors of importance are the size of the area over which the precipitation falls, the intensity of the precipitation, and its duration.

---

*Transpiration is the process whereby plants give off water vapor through the pores of their leaves. The moisture comes from the roots through capillary action.

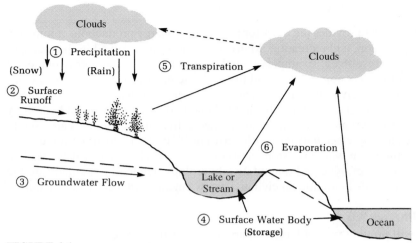

FIGURE 4-1
The hydrologic cycle.

Once the precipitation hits the ground, a number of things can happen. It can evaporate promptly. This is especially true if the surface is hot and impervious. If the soil is dry and/or porous, the precipitate may *infiltrate* into the ground or it may only wet the surface. This process and the process of wetting leaves and blades of grass is called *interception.* The precipitate may be trapped in small depressions or puddles. It may remain there until it evaporates or until the depressions fill and overflow. And last but not least, it may run off directly to the nearest stream or lake to become surface water. The four factors (evaporation, infiltration, interception, and trapping) that reduce the amount of direct runoff are called *abstractions.*

Streamflow.   The water that makes up our streams and rivers is derived from two sources: *direct runoff* and groundwater *exfiltration,* or *base flow,* as it is more commonly called. Direct runoff is a consequence of precipitation. Base flow is the dry weather flow that results from the seepage of groundwater out of stream banks.

The amount of water that reaches a stream is a function of the abstractions mentioned above and the catchment area or watershed that feeds the stream. The watershed, or *basin,* is defined by the surrounding topography (Figure 4-2). The perimeter of the watershed is called a *divide.* It is the highest elevation surrounding the watershed. All of the water that falls on the inside of the divide has the potential to be shed into the streams of the basin encompassed by the divide. Water falling outside of the divide is shed to another basin.

Groundwater Hydrology

Water Table (Unconfined) Aquifer.   As we mentioned earlier, part of the precipitation that falls on the soil may infiltrate. This water replenishes the soil moisture or is used by growing plants and returned to the atmosphere by transpiration. Water that drains downward below the root zone finally reaches a level at which all of the openings or voids in the earth's materials are filled with water. This zone is known as the *zone of saturation.*

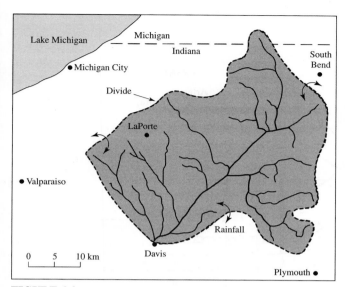

FIGURE 4-2
The Kankakee River Basin above Davis, IN. *Note:* Arrows indicate that precipitation falling inside the dashed line is in the Davis watershed, while that falling outside is in another watershed. The dashed line then "divides" the watersheds.

Water in the zone of saturation is referred to as groundwater. The geologic formation that bears the water is called an *aquifer.* The upper surface of the zone of saturation, if not confined by impermeable material, is called the *water table* (Figure 4-3). The aquifer is called a *water table aquifer* or an *unconfined aquifer.* Water will rise to the level of the water table in an unpumped water table well.

The smaller void spaces in the porous material just above the water table may contain water as a result of capillarity. This zone is referred to as the *capillary fringe* (Figure 4-4). It is not a source of supply since the water held will not drain freely by gravity. The region from the saturated zone to the surface is also called the *vadose zone.*

Springs. Because of the irregularities in underground deposits and in surface topography, the water table occasionally intersects the surface of the ground or the bed of a stream, lake, or ocean. At these points of intersection, groundwater moves out of the aquifer. The place where the water table breaks the ground surface is called a *gravity* or *seepage spring* (Figure 4-3).

Perched Water Table. A perched water table is a lens of water held above the surrounding water table by an impervious layer. It may cover an area from a few hundred square meters to several square kilometers.

Artesian (Confined) Aquifer. As water percolates into an aquifer and flows downhill, the lower layers come under pressure. This pressure is the result of the mass of water in the upper layers pressing on the water in the lower layers, much as deep sea divers are under greater and greater pressure as they go deeper and deeper into the sea. The system

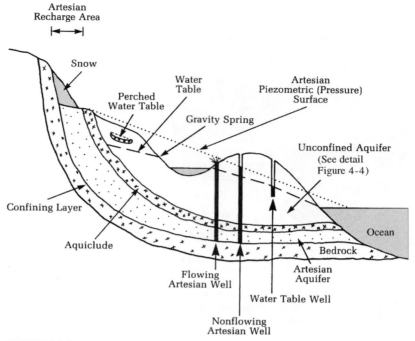

FIGURE 4-3
Schematic of groundwater aquifers.

is analogous to a manometer (Figure 4-5). When there is no constriction in the manometer, the water level in each leg rises to the same height. If the left leg is raised, the increased water pressure in that leg pushes the water up in the right leg until the levels are equal again. If the right leg is clamped shut then, of course, the water will not rise to the same level. However, at the point where the clamp is placed, the water pressure will increase. This pressure is the result of the height of water in the left leg.

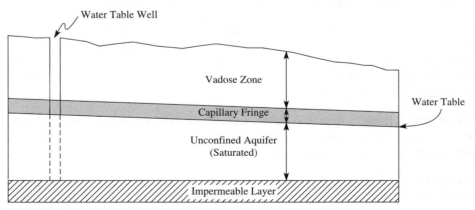

FIGURE 4-4
Detail of the unconfined aquifer.

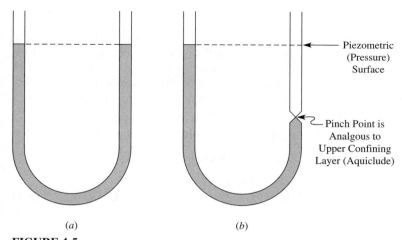

(a)   *(b)*

FIGURE 4-5
Manometer analogy to water in an aquifer. Manometer "a" is analogous to an unconfined aquifer. Manometer "b" is analogous to a confined aquifer.

A special type of groundwater system occurs when an overlying impermeable formation and an underlying impermeable formation restrict the water, much as the walls of a manometer. The impermeable layers are called *confining layers*. Other names given to these layers are *aquicludes* if they are essentially impermeable, or *aquitards* if they are less permeable than the aquifer but not truly impermeable. An aquifer between impermeable layers is called a *confined aquifer*. If the water in the aquifer is under pressure, it is called an *artesian aquifer* (Figure 4-3). The name "artesian" comes from the French province of Artois (*Artesium* in Latin) where, in the days of the Romans, water flowed to the surface of the ground from a well.

Water enters an artesian aquifer at some location where the confining layers intersect the ground surface. This is usually in an area of geological uplift. The exposed surface of the aquifer is called the *recharge area*. The artesian aquifer is under pressure for the same reason that the pinched manometer is under pressure, that is, because the recharge area is higher than the bottom of the top aquiclude and, thus, the height of the water above the aquiclude causes pressure in the aquifer. The greater the vertical distance between the recharge area and the bottom of the top aquiclude, the higher the height of the water, and the higher the pressure.

Piezometric Surfaces. If we place small tubes (*piezometers*) into an artesian aquifer along its length, the water pressure will cause water to rise in the tubes much as the water in the legs of a manometer rises to a point of equilibrium. The height of the water above the bottom of the aquifer is a measure of the pressure in the aquifer. An imaginary plane drawn through the points of equilibrium is called a *piezometric surface*. In an unconfined aquifer, the piezometric surface is the water table.

If the piezometric surface of a confined aquifer lies above the ground surface, a well penetrating into the aquifer will flow naturally without pumping. If the piezometric surface is below the ground surface, the well will not flow without pumping.

Hydrologic Mass Balance

Hydrologic problems of interest to civil and environmental engineers, such as the sizing of retention ponds and reservoirs and estimating the size of sewers for parking lots, streets, and airports, may be solved by the application of mass balance equations. The system shown in Figure 4-6a is an example of a small hydrologic system. A simplified mass balance diagram of this system is shown in Figure 4-6b. It may be described by a form of the mass balance equation given in Chapter 2 (Equation 2-3):

$$\text{Mass rate of accumulation} = \text{Mass rate of input} - \text{Mass rate of output}$$

Noting that mass rate = volumetric rate × density, we may write:

$$Q_S(\rho) = Q_P(\rho) + Q_{Q_{in}}(\rho) + Q_{I_{in}}(\rho)$$
$$- Q_{Q_{out}}(\rho) - Q_{I_{out}}(\rho) - Q_R(\rho) - Q_E(\rho) - Q_T(\rho) \tag{4-1}$$

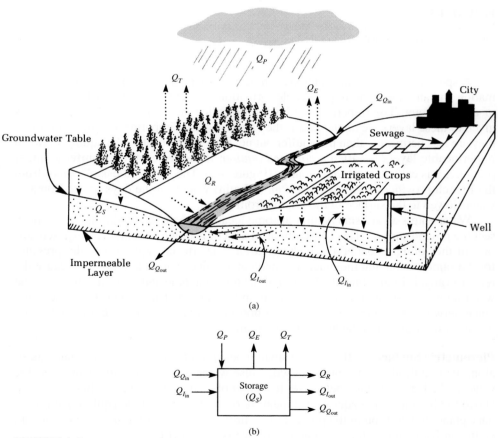

(a)

(b)

FIGURE 4-6

(a) Schematic diagram of a hydrologic subsystem; (b) mass balance diagram of a hydrologic subsystem.

where Q refers to the volume per unit of time (m³/s), ρ is the density of water (kg/m³), and the subscripts are defined as follows:

S = storage
P = precipitation
Q = river flow (in and out)
I = groundwater infiltration/exfiltration (in and out)
R = runoff
E = evaporation
T = transpiration

We often assume that the density of the water is constant throughout the system. Thus, we divide both sides of the equation by the density to yield an equation for the volumetric rate of accumulation. Equation 4-1 may then be written as:

$$Q_S = Q_P + Q_{Q_{in}} + Q_{I_{in}} - Q_{Q_{out}} - Q_{I_{out}} - Q_R - Q_E - Q_T \tag{4-2}$$

In many hydrology texts, the equation is further simplified by writing the expression in terms of the subscripts:

$$S = P + Q_{in} + I_{in} - Q_{out} - I_{out} - R - E - T \tag{4-3}$$

The common units of expression for the measurement of these terms are not consistent with one another. For example, the common unit of measure for precipitation, infiltration, evaporation, and transpiration is mm/h while the common unit of measure for storage, river flow, and runoff is m³/s. Because it is assumed that precipitation, infiltration, evaporation, and transpiration occur over the entire surface of the hydrologic system, we approximate the volumetric rate by multiplying the measurement (in units of length per unit time) by the surface area.

The terms of the mass balance equation for the hydrologic equation may be expanded to show their functional relationship to other physical phenomena. For example, the amount of runoff is a function of the characteristics of the surface (paved, cultivated, flat, steep-sloped). The amount of storage, for example, is a function of the type of soil or geological formation. These two aspects of the hydrologic equation are discussed in Sections 4-5 and 4-7. In the following paragraphs, we elaborate on the behavior of the other terms.

Infiltration. A typical infiltration curve is shown in Figure 4-7. This is the type of curve that results when the rainfall rate exceeds the infiltration rate. Beginning in 1911, numerous researchers (for example, Green and Ampt, 1911; Horton, 1935; and Holton, 1961) developed empirical relationships for infiltration capacity based on this type of curve. Of the numerous equations developed to describe infiltration, Horton's equation is useful to examine because it characterizes three phenomena of interest. Horton expressed the infiltration rate as (Horton, 1935):

$$f = f_c + (f_o - f_c)e^{-kt} \tag{4-4}$$

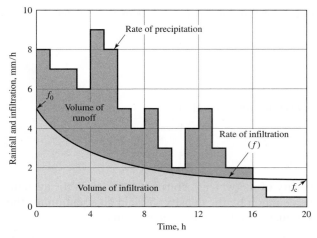

FIGURE 4-7
Typical infiltration curve.
(*Source:* Davis and Masten, 2009.)

where f = infiltration rate, mm/h
f_c = equilibrium or final infiltration rate, mm/h
f_o = initial infiltration rate, mm/h
k = empirical constant, h^{-1}
t = time, h

This expression assumes that the rate of precipitation is greater than the rate of infiltration. The parameters f_o and k have no physical basis. They cannot be determined from soil water properties. They must be determined by experiment.

The infiltration parameters are a function of the properties of the soil; thus, the values for f_o, f_c, and k are, as you might expect, a function of the soil type. Some examples are (in mm/h and h^{-1}):

	f_o	f_c	k
Dothan loamy sand	88	67	1.4
Fuquay pebbly loamy sand	159	61	4.7

Soil moisture content, vegetative cover, organic matter, and season affect these values.

The second property of interest is that the infiltration rate is an inverse exponential function of time. If the rate of precipitation exceeds the rate of infiltration, a plot of infiltration rate versus time will reveal that as rainfall continues, the rate at which the ground soaks it up decreases because the pore spaces in the soil fill up with water. Because typical values for f_o and f_c are greater than prevailing rainfall intensity, this may lead to calculated decreases in infiltration even though there is capacity to accept precipitation at higher rates.

The third property, which is directly related to hydrologic balances, is that the area under the infiltration curve represents the volume of water that infiltrates. Integration of Horton's equation yields the volume:

$$\Psi = f_c t + \frac{f_o - f_c}{k} (1 - e^{-kt}) \qquad (4\text{-}5)$$

Because of the extensive experimental data required to implement the infiltration models, civil and environmental engineers use lumped parameters to estimate direct runoff rather than attempt to estimate infiltration and subtract it from rainfall. The estimation of runoff is described in Sections 4-4 and 4-5.

Evaporation. The estimation of loss of water from lakes and reservoirs is an important component of water management activities to increase sustainability through drought mitigation.

The loss of water from the surface of a lake or reservoir is a function of solar radiation, air and water temperature, wind speed, and the difference in vapor pressures at the water surface and in the overlying air. As with estimates of infiltration rate, there are numerous methods for estimating evaporation. Dalton first expressed the fundamental relationship in the form (Dalton, 1802):

$$E = (e_s - e_a)(a + bu) \qquad (4\text{-}6)$$

where E = evaporation rate, mm/d
e_s = saturation vapor pressure, kPa
e_a = vapor pressure in overlying air, kPa
a, b = empirical constants
u = wind speed, m/s

Water vapor pressures at various air temperatures are listed in Table 4-1.

Empirical studies at Lake Hefner, Oklahoma, yielded a similar relationship:

$$E = 1.22(e_s - e_a)u \qquad (4\text{-}7)$$

TABLE 4-1
Water vapor pressures at various air temperatures

Temperature, °C	Vapor pressure, kPa
0	0.611
5	0.872
10	1.227
15	1.704
20	2.337
25	3.167
30	4.243
35	5.624
40	7.378
50	12.34

From these expressions, it is apparent that high wind speeds and low humidities (vapor pressure in the overlying air) result in large evaporation rates. You may note that the units for these expressions do not make much sense. This is because these are empirical expressions developed from field data. The constants have implied conversion factors in them. In applying empirical expressions, care must be taken to use the same units as those used by the author of the expression.

Because the cost to obtain data for the mass transfer estimation method is quite substantial, the most common method for estimating evaporation from a free water body is by using an *evaporation pan*. In the United States, the Standard National Weather Service *Class A Pan* is used. It is 122 cm (4 ft) in diameter by 254 mm (10 in) in depth. It is filled with water to a depth of 203 mm (8 in). It is made of unpainted galvanized iron. To circulate air under the pan, it is mounted on a wooden platform 305 mm (1 ft) above the ground.

In operation, the depth of water is measured with a hook gage in a stilling well. The evaporation is computed as the difference between measured levels with an adjustment for precipitation between observations.

Because it has been observed that pan evaporation occurs more rapidly than from larger bodies of water, a *pan coefficient* is used to correct the reading. The pan coefficient is the ratio of the evaporation from a water body and the pan. The pan coefficient ranges from 0.6 to 0.8, with an average of 0.7.

Estimation of reservoir evaporation for periods less than one year may result in serious errors. For lakes and reservoirs subject to significant transport of energy by water flowing into the water body (called *advected* energy), local pan-lake relationships are essential. The pan-to-lake coefficient is a major source of error in the pan method. Errors in the range of 10 to 15 percent of annual estimates and up to 50 percent for monthly estimates have been reported (Gupta, 2008).

Example 4-1.   Silk's Lake has a surface area of 70.8 ha. For the month of April the inflow was 1.5 m^3/s. The dam regulated the outflow (discharge) from Silk's Lake to be 1.25 m^3/s. If the precipitation recorded for the month was 7.62 cm and the storage volume increased by an estimated 650,000 m^3, what is the estimated evaporation in m^3 and cm? Assume that no water infiltrates out of the bottom of Silk's Lake and that the density of water is constant.

Solution.   Begin by drawing the mass balance diagram:

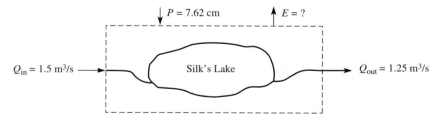

The mass balance equation is:

$$\text{Mass accumulation} = \text{Mass input} - \text{Mass output}$$

Using the assumption that the density of water is constant, we may write the mass balance equation as

$$\text{Volume accumulation} = \text{Volume in} - \text{Volume out}$$

The accumulation is given as 650,000 m³. The input consists of the inflow and the precipitation. The product of the precipitation depth and the area on which it fell (70.8 ha) will yield a volume. The output consists of outflow plus evaporation.

$$\Delta S = [(Q_{in})(t) + (P)(\text{area})]_{input} - [(Q_{out})(t) + E]_{output}$$

where

$$\Delta S = \text{Change in storage (a volume)}$$
$$(Q)(t) = (\text{flow rate})(\text{time}) = \text{volume}$$

Noting that April has 30 days and making the appropriate units conversions:

$$650,000 \text{ m}^3 = (1.5 \text{ m}^3/\text{s})(30 \text{ d})(86,400 \text{ s/d})$$
$$+ (7.62 \text{ cm})(70.8 \text{ ha})(10^4 \text{ m}^2/\text{ha})(1\text{m}/100 \text{ cm})$$
$$- (1.25 \text{ m}^3/\text{s})(30 \text{ d})(86,400 \text{ s/d}) - E$$

Solving for E:

$$E = 3.89 \times 10^6 \text{ m}^3 + 5.39 \times 10^4 \text{ m}^3 - 3.24 \times 10^6 \text{ m}^3 - 6.50 \times 10^5 \text{ m}^3$$
$$E = 5.39 \times 10^4 \text{ m}^3$$

For an area of 70.8 ha, the evaporation depth is:

$$E = \frac{5.39 \times 10^4 \text{ m}^3}{(70.8 \text{ ha})(10^4 \text{ m}^2/\text{ha})} = 0.076 \text{ m or 7.6 cm}$$

Comment: In an average year approximately 1,900 mm (6.25 ft) of water is estimated to be evaporated from the Elephant Butte Reservoir in New Mexico. This reservoir supplies irrigation water to southern New Mexico. The average surface area of the reservoir pool is about 5,570 ha (13,760 acres), so approximately 1.06×10^8 m³ (86,000 acre-ft) of water are evaporated annually. For a normal irrigation season in southern New Mexico, approximately 3 acre-ft per acre of cultivated land is required. The reservoir evaporation loss is approximately equal to the loss of 29,000 acres of irrigation. In 1988, New York City's water supply was about 1 billion gallons per day. The annual evaporated water loss from the Elephant Butte Reservoir could supply New York for about a month (Viesmann, et al., 1989).

Example 4-2. During April, the wind speed over Silk's Lake was estimated to be 4.0 m/s. The air temperature averaged 20°C and the relative humidity was 30 percent. The water temperature averaged 10°C. Estimate the evaporation rate using the empirical relationship in Equation 4-7.

Solution.   From the water temperature and Table 4-1, the saturation vapor pressure is estimated as $e_s = 1.227$ kPa. The vapor pressure in the air may be estimated as the product of the relative humidity and the saturation vapor pressure at the air temperature:

$$e_a = (2.337 \text{ kPa})(0.30) = 0.70 \text{ kPa}$$

The daily evaporation rate is then estimated to be:

$$E = 1.22(1.227 - 0.70)(4.0 \text{ m/s}) = 2.57 \text{ mm/d}$$

The monthly evaporation would then be estimated to be:

$$E = (2.56 \text{ mm/d})(30 \text{ d}) = 76.8 \text{ mm or } 7.7 \text{ cm}$$

Evapotranspiration (ET).   Water loss from plants (transpiration) is difficult to separate from losses from the soil surface or root zone. For mass balance calculations, these are often lumped together under the term evapotranspiration. The rate of evapotranspiration is a function of soil moisture, soil type, plant type, wind speed, net radiation, and temperature. Plant types may affect evapotranspiration rates dramatically. For Example see Table 4-2.

The *Penman–Monteith method* for estimating evapotranspiration is currently the most widely used relationship (Gupta, 2008). It considers the aerodynamic resistance to water vapor transfer and the movement of water vapor from inside the leaves to the ambient air. Another frequently used method is that developed by Blaney and Criddle (1945, 1962). It was developed for conditions in the arid western regions of the United States. With this method monthly evapotranspiration is estimated for a specific crop. Unlike the Penman–Monteith method, the Blaney–Criddle method does not separate climate and crop parameters. The literature is replete with proposed alternative methods and evaluations. Field measurements reveal the need for calibration to account for local conditions because the

TABLE 4-2
Some examples of evapotranspiration*

Crop	Evapotranspiration, mm	Time interval	Place
Alfalfa	841	Season	Wyoming
Alfalfa	587 – 621	4 mo	Wyoming
Corn	347	Season	China
Fescue grass	~480	2 mo in summer	Texas
	~120	2 mo in winter	Texas
Pasture	626	Season	Wyoming
Soybeans	~440	Season	Mississippi
Tomatoes	648	Season	California
Wheat	457	Season	China

* Measured experimental field data

Sources: Al-Kaisi, 2000; Hanson and May, 2005; Howell, et al., 1998; Thomas and Blaine, 2009; University of Wyoming, 1987; and Yang, et al., 2007

estimates may range from 40 percent under accounting to 20 percent over accounting. In regions where evapotranspiration losses exceed annual precipitation, these methods cannot be used without significant modification.

Another, and perhaps more practical, approach is to use local pan evaporation data corrected for a specific crop. This is particularly true when irrigation is practiced. Figure 4-8 illustrates the ratio of ET to pan evaporation for corn and soybeans in Iowa.

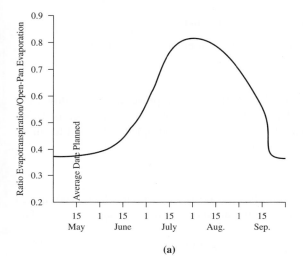

(a)

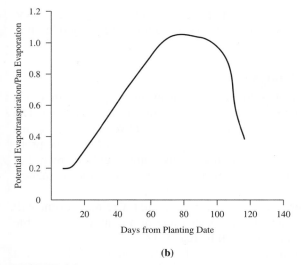

(b)

FIGURE 4-8

(*a*) Variation of corn ET factor over the growing season. For example, on June 7 ET from a cornfield would be 40 percent of measured pan evaporation. On August 1 it would be 82 percent. (*b*) Variation of soybean ET factor over the growing season. (*Source:* Iowa State University Integrated Crop Management, 2000)

4-3 RAINFALL ANALYSIS

Of the many variables of rainfall that might be of interest, we are concerned primarily with four:

1. Space: the average rainfall over the area

2. Intensity: how hard it rains

3. Duration: how long it rains at any given intensity

4. Frequency: how often it rains at any given intensity and duration

Point Precipitation Analysis

Data from a single nearby rain gage are often sufficiently representative to allow their use in the design of small projects. The analysis of data from a single gage is called point precipitation analysis. Spatial analysis is much more complex and is left for more advanced courses.

Intensity and Duration

Intensity is a measure of how hard it rains. The units of measure are mm/h. It is computed from rainfall records by determining the accumulation of rain for a selected time interval:

$$\frac{\Delta p}{\Delta t} = \frac{p_2 - p_1}{t_2 - t_1} \text{ (conversion factor)} \tag{4-8}$$

where p_1 and p_2 are the accumulated precipitation in mm at times t_1 and t_2. Because the time interval is typically less than an hour, the conversion factor is typically 60 min/h. The *duration* of precipitation is equal to $t_2 - t_1$.

Frequency from Probability Analysis

As we discussed in Chapter 3, the relative frequency of an event such as a coin toss is a probability. A rainfall of a given intensity for a given duration is such an *event*. The probability of a single rainfall event, say E_1, is defined as the relative number of occurrences of the event in a long period of record of rainfall events. Thus, $P(E_1)$, the probability of rainfall event E_1, is n_1/N for n_1 occurrences of the same event in a record of N events if N is sufficiently large. The number of occurrences of n_1 is the frequency, and n_1/N is the relative frequency.

$P(E_1) = 0.10$ implies a 10 percent chance each year that a rainfall event will "occur." Because the probability of any single, exact value of a continuous variable is zero, "occur" also means the rainfall event will be reached or exceeded.

Table 4-3 is the compilation of a partial series of rainfall events. Rather than a record of all rainfalls, it is a record of rainfall intensities above some practical minimum. It gives the *frequency* or number of times that a rainfall of given intensity and duration will be equaled or exceeded for the period of record. For example, looking at the first row in Table 4-3, one would expect seven rainfall events with an intensity of 160.0 mm/h or more and a duration of five minutes to occur in any 45-year period (1999 − 1954 = 45 years) in the Dismal Swamp.

TABLE 4-3
Rainfall record for the Dismal Swamp (1 Oct. 1954–30 Sep. 1999)

Duration (min)	Number of storms of stated intensity or more — Intensity (mm/h)										
	20.0	30.0	40.0	60.0	80.0	100.0	120.0	140.0	160.0	180.0	200.0
5						245	49	16	7	3	2
10					256	64	15	7	4	1	
15				241	94	18	6	3	2		
20		240	80	36	10	4	2	1			
30	202	44	17	9	2	2	1				
40	76	31	8	1							
50	30	12	3								
60	9	2									

You should note two other facts about the table. First, the numbers in the table are also ranks. If the rainfall events for 5-minute duration storms are arranged in descending order of intensity, then the seventh storm in the sequence or "the seventh-ranked" storm has an intensity of 160.0 mm/h or more. We assume that the ranks are spaced evenly between the recorded ranks, that is, that intensity and rank are linear. Thus, for the fifth-ranked storm of 5-minute duration, by interpolation, we can estimate that it would have an intensity of 170.0 mm/h or more.

The second fact that you should note is the ranks may be used to infer the probability that a given intensity storm will be equaled or exceeded. Again, using the seventh-ranked storm, we may infer that rainfall intensities of 160.0 mm/h or greater will occur with a frequency of seven times in 45 years. An annual average probability of occurrence would be $\frac{7}{45} = 0.16$, or 16 percent. Hydrologists and engineers often use the reciprocal of annual average probability because it has some temporal significance. The reciprocal is called the average *return period* or average *recurrence interval* (T):

$$T = \frac{1}{\text{Annual average probability}} \qquad (4\text{-}9)$$

For the case of our seventh-ranked storm, the average return period of a 160.0 mm/h, 5-minute storm is 6.25 years. This means that we would expect a storm of 160.0 mm/h or greater once every 6.25 years on the average.

Because the amount of reliable data available is limited,* it is customary to use Weibull's formula for calculating return period (Weibull, 1939):

$$T = \frac{n+1}{m} \qquad (4\text{-}10)$$

---

*Systematic measurement of precipitation was begun by the Surgeon General of the Army in 1819, while streamflow data collection did not begin until 1888.

where T = average return period in years
n = number of years of record
m = rank of storm, with most intense storm given a rank of 1

Weibull's formula allows for a small correction when the number of years of record is small. At larger values of n it closely approximates $T = n/m$.

From our previous discussion, $P(E_1)$, the probability of rainfall event E_1, is n_1/N and a $P(E_1) = 0.10$ implies a 10 percent chance each year that a rainfall event will "occur." That is, the probability of rainfall event E_1 being exceeded is

$$P(E_1) = \frac{1}{10}$$

In a sufficiently long run of data, the rainfall event would be equaled or exceeded on the average once in 10 years. This is the same as the definition of return period. From Chapter 3, we also note that this is also a "definition" of risk. Thus, return period, whether for rainfall events, floods, or droughts, is used as a convenient way to explain the "risk" of a hydrologic event to the public. Unfortunately, it also has led to serious misunderstanding by both the public and, in some cases, design engineers. For emphasis, we remind you of the following:

- A storm (or flood or drought) with a 20-year return period that occurred last year may occur next year (or it may not).
- A storm (or flood or drought) with a 20-year return period that occurred last year may not occur for another 100 years or more.
- A storm (or flood or drought) with a 20-year return period may, on the average, occur 5 times in 100 years.

Using the return period definition for T, the following general probability relations hold where E is the event (storm, flood, drought):

1. The probability that E will be equaled or exceeded in any year is

$$P(E) = \frac{1}{T} \qquad (4\text{-}11)$$

2. The probability that E will not be exceeded in any year is

$$P(\overline{E}) = 1 - P(E) = 1 - \frac{1}{T} \qquad (4\text{-}12)$$

3. The probability that E will not be equaled or exceeded in any of n successive years is

$$P(\overline{E})^n = \left(1 - \frac{1}{T}\right)^n \qquad (4\text{-}13)$$

4. The risk that E will be equaled or exceeded at least once in n successive years is

$$R = 1 - \left(1 - \frac{1}{T}\right)^n \qquad (4\text{-}14)$$

Intensity-Duration-Frequency Curves (IDF)

The family of curves that depicts the relationship between intensity, duration, and frequency of precipitation at a point is used for the design of storm sewers and retention ponds. Typically, civil and environmental engineers do not collect and analyze basic rainfall data in the United States. However, data collected by the U.S. Weather Bureau, the Department of Agriculture, and similar governmental agencies are used to develop IDF curves. Although some agencies such as state departments of transportation have prepared regional IDF curves, circumstances may require the development of curves more representative of local conditions. The methodology for developing a local IDF curve is explained in Example 4-3.

Example 4-3. Prepare a table of plotting points for an IDF curve for a 5-year storm at the Dismal Swamp. Compute points for each duration given in Table 4-3.

Solution. Because Table 4-3 is a table of ranks, we need to determine the rank of the 5-year storm. First, we rearrange Weibull's formula:

$$m = \frac{n + 1}{T}$$

where

$$n = 1999 - 1954 = 45 \text{ y}$$
$$T = 5 \text{ y}$$

thus

$$m = \frac{46}{5} = 9.2$$

Starting with the 5-minute duration, we note that the 9.2-ranked storm lies between the 16th- and seventh-ranked storm; that is,

Intensity (mm/h)		
140.0		160.0
16	9.2	7

We also note that the ranks increase from right to left while the intensities increase from left to right. Keeping this in mind, and recalling that we assume a linear relationship between intensity and rank, we may interpolate by simple proportions:

$$\frac{9.2 - 7}{16 - 7} (160.0 - 140.0) = 4.89$$

Thus, the 9.2-ranked storm is 4.89 mm/h less than 160.0 mm/h:

$$160.0 - 4.89 = 155.11 \text{ or } 155.1 \text{ mm/h}$$

The completed table would appear as follows:

**Intensity and duration values for a
5-year storm at the Dismal Swamp**

Duration (min)	Intensity (mm/h)
5	155.1
10	134.5
15	114.7
20	82.7
30	59.5
40	39.5
50	33.1
60	—

Note that a similar table could be constructed for each intensity given in Table 4-3. This would give us twice as many points to use for fitting the curve.

The IDF curve for Example 4-3 and the curve for a return period of 20 years are plotted in Figure 4-9. You should note that the frequency curves join occurrences that are not necessarily from the same storm. They represent the average intensity expected for a given duration. They do not represent a sequence of intensities during a single storm. Example IDF curves for four U.S. cities are shown in Figure 4-10.

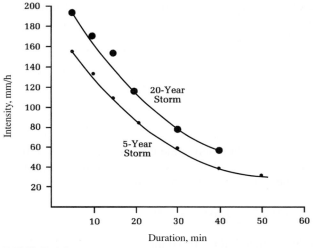

FIGURE 4-9
Intensity-duration-frequency curves for the Dismal Swamp.

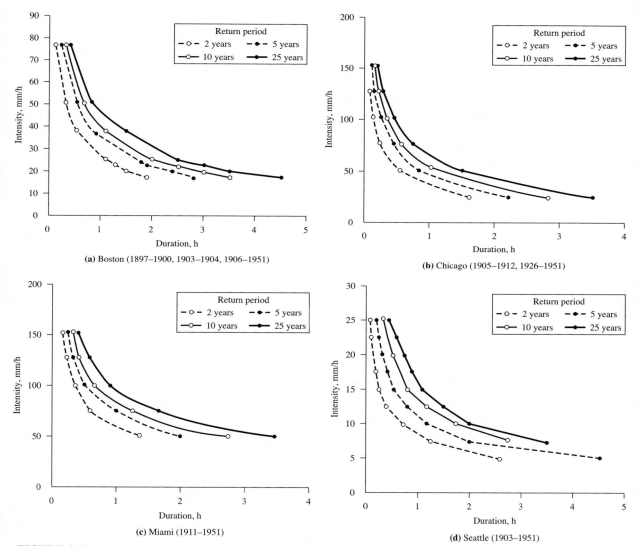

FIGURE 4-10

IDF Curves for (*a*) Boston, (*b*) Chicago, (*c*) Miami, and (*d*) Seattle. (*Source*: Gilman, 1964)

4-4 RUNOFF FROM SNOWMELT

Rainfall is not the only source of water for streamflow. In parts of the western states of the United States, melting snow produces a major portion of the annual runoff. The computation of snowmelt runoff is a very complex problem that we can only address in an abbreviated form in this introductory text.

Physics of Snowmelt

The melting of the snowpack is a heat transfer phenomenon. The sources of energy are (1) absorbed solar radiation, (2) net long-wave radiation, (3) condensation of atmospheric water vapor, (4) convective heat transfer by the wind, and (5) conductive heat transfer from rain. The effectiveness of radiation in melting snow is dependent on its reflectivity or *albedo*. Fresh, clean snow will reflect between 75 and 95 percent of the radiation. Old snow may have an albedo between 40 and 70 percent. Convective heat transfer from air is only effective when strong winds bring large quantities of warm air in contact with the snow. If the vapor pressure of the air is greater than that of snow at 0°C, turbulence brings moisture that can condense on the snow surface. Because snowmelt is essentially a process of converting ice into water, the enthalpy of condensation (2.49 MJ/kg) melts ice (enthalpy of fusion = 333 kJ/kg). The condensation of 1 kg of moisture on the snow surface results in the melting of approximately 7.5 kg of snow water.

Because rainfall is above 0°C, it also brings heat to snow. The depth of snowmelt water may be estimated from calorimetry as

$$D_{\text{snowmelt}} = \frac{PT_{\text{w}}}{80}$$ (4-15)

Where D_{snowmelt} = depth of snowmelt water, mm
P = rainfall depth, mm
T_{w} = wet bulb temperature, °C

Snowmelt Estimation

Snowmelt quantities cannot be measured directly. As a result, estimation techniques have been developed for use in estimating runoff.

The most comprehensive snowmelt estimation technique was developed by the U.S Army Corps of Engineers (1960). Their empirical equations were developed to estimate the depth of snowmelt in the western United States. The equations incorporate the results of extensive studies and heat transfer theory. These equations express snowmelt as a function of radiation, air temperature, vapor pressure, wind speed, rain, and forest cover. Because the air temperature drops about 5.5°C per 1,000 m of elevation, a number of zones are selected for evaluation. The melt in each zone is computed and the results are combined for the basin area.

A simpler technique for estimating snowmelt is called the *degree-day method*. A degree-day is defined as a departure of 1° in mean daily temperature above 0°C. Thus, a day with a mean temperature of 4°C is said to have 4 melting degree days. A degree-day factor is used to calculate the depth of water from melting snow. The factor is determined by dividing the volume of streamflow produced by melting snow within a given time period by the total degree days for the period. For basins with little range in elevation, the degree-day factor will range between 2 and 7 mm/degree-day. Because snow melting is also a function of humidity, wind, and solar radiation, the degree-day factor will vary from day to day.

There is no satisfactory simple method for areas with a wide range in elevation. The snowpack is not uniform in depth. It is shallower at lower elevations. The line of zero snow is known as the *snow line*. The snow line moves with the season so that no single station can provide measurements of the melting snow.

For basins with a wide range in elevation, a systematic method of observing the portion of the basin covered by snow and its depth is required. Satellite imaging is a powerful tool in providing these estimates. Data are derived from several regions of the electromagnetic spectrum. The extent of snowpack can be determined from the visible band. The infrared band provides estimates of the water content. Alternatively, microwave remote sensing provides information on the extent of the snowpack and the water equivalent (Gupta, 2008).

Rain and snowmelt.   Warm spring rain on a winter's snow accumulation is often an event that results in flooding. While neither the rainfall event nor the depth of accumulated snow may be significant in terms of event return periods, the combination, coupled with saturated soil, often produces flooding equivalent to that of significant return periods. Likewise, a mid-winter thaw can produce a significant event. Ice jams on the rivers will compound the problem.

4-5   RUNOFF ANALYSIS

Three runoff questions are of interest:

1. How much of the precipitation that falls on a watershed reaches the stream or storm sewer draining it?

2. How long does it take for the runoff to reach the stream or storm sewer?

3. How often does the runoff cause a flood?

Stream Gages

Streamflow measurements are made by recording the height of the surface of the water above a reference datum at a location with a fixed geometry (known as a *control section*). The control section allows the use of hydraulic equations to estimate the flowrate based on the depth of water above the reference datum. Several examples of control sections are illustrated at the text website: www.mhhe.com/davis.

The elevation (*stage*) readings are calibrated in terms of streamflow (*discharge*). At manual stations, readings are made from a graduated rod (*staff gage*) placed in the stream.

At automatic recording stations, three components are required: a device to sense the water stage, a method of recording the stage, and a method of storing the data. One of the oldest systems that is still in use is the float and cable system (Figure 4-11). A stilling well (Figure 4-12) is used to minimize the effects of wave action and to protect the float from floating logs and other debris. Newer installations may use a bubble-gage sensor. This is a pressure actuated system in which an orifice at the end of a piece of tubing is located below the water surface at the gage datum. A gas, typically nitrogen, is bubbled freely into the river. The gas pressure is equal to the height of water above the orifice. The pressure is recorded electronically. Another alternative is to place a pressure measuring device (*pressure transducer*) directly at the reference datum and record the reading directly.

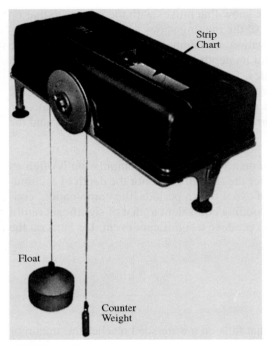

FIGURE 4-11
Float system and strip chart recorder for continuous stage
measurement. (Courtesy of Stevens Water Monitoring
Systems, Inc.)

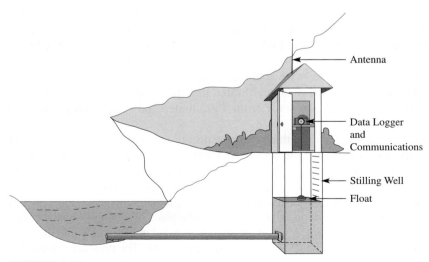

FIGURE 4-12
Stilling well. (Courtesy of Stevens Water Monitoring Systems, Inc.)

In each method of automatic recording, the data may be logged directly into a dedicated computer or it may be transmitted to a central computer. The transmission to a central site is particularly favored for remote gauging stations and large integrated water management systems such as that operated by California.

Estimation of Amount of Runoff

Measurement of streamflow with a stream-gauging network is the main and best source of surface water flow data. However, there is no national data collection program in the world that collects sufficient data to satisfy all the design and decision-making needs for every watershed. The approaches currently in use to estimate runoff fall into four categories (Gupta, 2008):

- Hydrograph analysis: This is a comparison of precipitation and the resulting streamflow record.

- Correlation with meteorological data: Statistical techniques such as cross-correlation, regression analysis, and frequency analysis are used in this approach.

- Correlation with hydrological data at another site: Stream gage data at one site may have transferable value at a nearby site on the same stream. This may involve extrapolation or interpolation of information gathered at two stations.

- Sequential data generation: Synthetic data are generated based on a time series that includes a random component in this technique.

For this introductory discussion we have selected two elementary ways of estimating runoff from rainfall that fall into the first category. They are called the *rational method* and the *unit hydrograph method*. The rational method is used to determine the diameter of a storm sewer to carry the runoff from a small watershed. The unit hydrograph method provides the starting point for the design of a retention pond.

Rational Method.   The starting point for this method reduces the hydrologic mass balance equation (Equation 4-2) to a simplified form:

$$Q_R = kQ_P \qquad\qquad (4\text{-}16)$$

Where Q_R = runoff
k = coefficient
Q_P = precipitation

The coefficient must account for all the other terms in the hydrologic mass balance: antecedent moisture, ground slope, ground cover, depression storage, shape of the drainage area, and so forth. This simplification is a major cause for criticism of the rational method. Despite the criticism, it is widely used for small urban projects.

A good example for us to look at is a paved parking lot (Figure 4-13). If the rainfall on the lot continues for a long enough period at a constant intensity, at some time the system will reach *steady state*. At steady state each drop of water that falls on the watershed conceptually displaces a drop through the storm sewer. Thus, further rainfall at the same intensity does not increase the discharge at the storm sewer. The

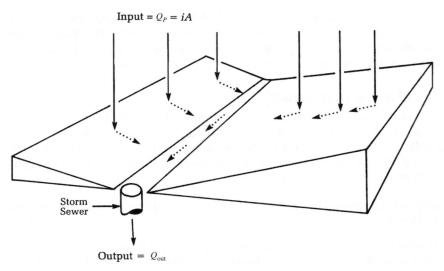

Input = Q_P = iA

Storm
Sewer →

Output = Q_{out}

FIGURE 4-13
The application of the hydrologic equation to a parking lot having area = A.

hyetograph (time versus rainfall) and corresponding direct runoff hydrograph for this situation are shown in Figure 4-14. The time that it takes for steady state to be achieved is called the *time of concentration* (t_c). The time of concentration is primarily a function of the basin geometry, surface conditions, and slope. A method for estimating t_c is given later in this section at Equation 4-21).

The input (Q_P) is a volume-per-unit time that may easily be shown to be equal to the product of the rainfall intensity (i) and the area of the watershed (A):

$$Q_P = iA \tag{4-17}$$

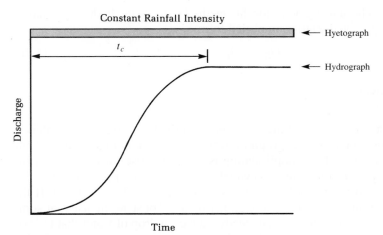

Constant Rainfall Intensity

← Hyetograph

t_c

← Hydrograph

Discharge

Time

FIGURE 4-14
Hyetograph and hydrograph for a parking lot.

The output volume per unit time (Q_R) is direct runoff. It is equal to the discharge (Q_{out} in Figure 4-13). Mulvaney (1851) and others recognized that for a constant rainfall intensity the ratio of the volume of water that is discharged past a gauging station at the outlet of a watershed to the volume of water that falls on the watershed as precipitation is approximately a constant. This ratio is called the *runoff coefficient*. It may be expressed as

$$C = \frac{Q_R}{Q_P} \tag{4-18}$$

The substitution of Equation 4-17 for Q_P and C for k in Equation 4-16 yields the *rational formula* in U.S. customary units as presented by Kuichling (1889):

$$Q = CiA \tag{4-19}$$

Where Q = peak runoff rate, ft^3/s
 C = runoff coefficient
 i = rainfall intensity, in/h
 A = area, acres

In SI units the equation is

$$Q = 0.0028\ CiA \tag{4-20}$$

Where Q = peak runoff rate, m^3/s
 C = runoff coefficient
 i = rainfall intensity, mm/h
 A = area, ha
 0.0028 = conversion factor, m^3 · h/mm · ha · s

Note that Q is **the peak discharge** and that the runoff coefficient is specific to a watershed for the assumption of steady state.

The original derivation of the rational formula was in what we now call U.S Customary units. In this system the use of intensity in inches/h and area in acres yields a runoff in ft^3/s without any conversion factor. Hence the name "rational" because the units work out rationally! This is demonstrated in Example 4-4.

Example 4-4.   Demonstrate that the units of the rational formula in U.S. Customary units yields an answer in ft^3/s (cfs) without a conversion factor.

Solution.

$$(1\ \text{in/h})(1\ \text{acre})\left(\frac{1\ \text{ft}}{12\ \text{in}}\right)\left(\frac{1\ \text{h}}{3{,}600\ \text{s}}\right)\left(\frac{43{,}560\ \text{ft}^2}{\text{acre}}\right) = 1.0083\ \text{or}\ 1\ \text{cfs}$$

Comment: For the typical design problem, only one or two significant figures may be justified. The rational formula units do yield a "rational" answer.

Although the basic principles of the rational method are applicable to large watersheds, an upper limit of 13 square kilometers is recommended (ASCE, 1969). A selected

TABLE 4-4
Selected runoff coefficients

Description of area or character of surface	Runoff coefficient	Description of area or character of surface	Runoff coefficient
Business		Railroad yard	0.20 to 0.35
Downtown	0.70 to 0.95	Unimproved	0.10 to 0.30
Neighborhood	0.50 to 0.70	Pavement	
Residential		Asphaltic and concrete	0.70 to 0.95
Single-family	0.30 to 0.50	Brick	0.70 to 0.85
Multi-units, detached	0.40 to 0.60	Roofs	0.75 to 0.95
Multi-units, attached	0.60 to 0.75	Lawns, sandy soil	
Residential (suburban)	0.25 to 0.40	Flat, 2 percent	0.05 to 0.10
Apartment	0.50 to 0.70	Average, 2 to 7 percent	0.10 to 0.15
Industrial		Steep, 7 percent	0.15 to 0.20
Light	0.50 to 0.80	Lawns, heavy soil	
Heavy	0.60 to 0.90	Flat, 2 percent	0.13 to 0.17
Parks, cemeteries	0.10 to 0.25	Average, 2 to 7 percent	0.18 to 0.22
Playgrounds	0.20 to 0.35	Steep, 7 percent	0.25 to 0.35

Extracted from ASCE, 1969.

list of runoff coefficients is given in Table 4-4. These are restricted to urban settings. The coefficients in Table 4-4 are applicable for storms with 5- to 10-year return periods. Less frequent, higher intensity storms will require the use of higher coefficients because infiltration and other losses have a proportionally smaller effect on runoff. The coefficients are based on the assumption that the design storm does not occur when the ground is frozen.

Example 4-5. What is the peak discharge from the grounds of the Beauregard Long Ashby High School during a 5-year storm? The school grounds encompass a 16.2 ha plot that is 1.3 km east of the Dismal Swamp rain gage. Assume that the time of concentration of the grounds is 41 minutes. (Note: The method for calculating the time of concentration is illustrated in Examples 4-8 and 4-9.) The composition of the grounds is as follows:

Character of surface	Area (m^2)	Runoff coefficient
Building	10,800	0.75
Parking lot, asphaltic	11,150	0.85
Lawns, heavy soil		
2.0% slope	35,000	0.17
6.0% slope	105,050	0.20
	$\Sigma = 162{,}000$	

Solution. We begin by computing the weighted runoff coefficient, that is, the product of the fraction of the area and its runoff coefficient.

$$AC = (10,800)(0.75) + (11,150)(0.85) + (35,000)(0.17) + (105,050)(0.20)$$
$$= 44,537.5 \text{ m}^2, \text{ or } 4.45 \text{ ha}$$

Because the Dismal Swamp rain gage is only 1.3 km away, we shall use the IDF curve obtained in Example 4-3 to determine the intensity. By definition, the peak discharge for a watershed occurs when the duration of the storm equals the time of concentration. Thus, we select a duration of 41 minutes and read a value of 38 mm/h at the 5-year storm curve in Figure 4-9.

The peak runoff is then

$$Q = (0.0028)(4.45)(38) = 0.47 \text{ m}^3/\text{s}$$

Thus, a storm sewer large enough to handle 0.47 m³/s of flow is required to carry storm water away from the BLAHS grounds.

Hydrographs. A graphical representation of the discharge of a stream at a single gaging station is called a *hydrograph* (Figure 4-15). As we mentioned earlier, during the period between storms the base flow is a result of exfiltration of groundwater from the banks of the stream. Discharge from precipitation excess, that is, that which remains after abstractions, causes a hump in the hydrograph. This hump is called the *direct runoff hydrograph* (DRH).

Obviously, any precipitation excess that occurs at the extremities of a watershed will not be recorded at the basin outlet until some time lapse has occurred. As precipitation continues, enough time elapses for the more distant areas to add to the discharge at the gaging station. The lag time of the peak and the shape of the DRH depend on the precipitation pattern and the characteristics of the basin (size, slope, shape, and storage capacity).

The area under the DRH (volume of water discharged) will be larger for a watershed with a large surface area than for one with a small area. Because the water will flow more quickly from a steeply sloped watershed, the lag time will be smaller than

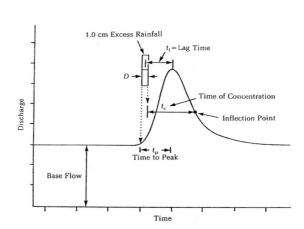

FIGURE 4-15

An idealized hydrograph showing a uniform base flow and a superimposed direct runoff hydrograph resulting from 1.0 cm of rainfall excess.

for a flat watershed having the same area. It will take longer for the water to reach the gaging station from a long, narrow watershed than from a short, broad watershed of the same area and slope. These effects are illustrated in Figure 4-16. The storage capacity of the watershed is dependent on a number of factors including, but not limited to, permeability of the soil, type of vegetative cover, time of year (frozen or not), and degree of development (urbanization). The effect of development on the discharge hydrograph is illustrated in Figure 4-17.

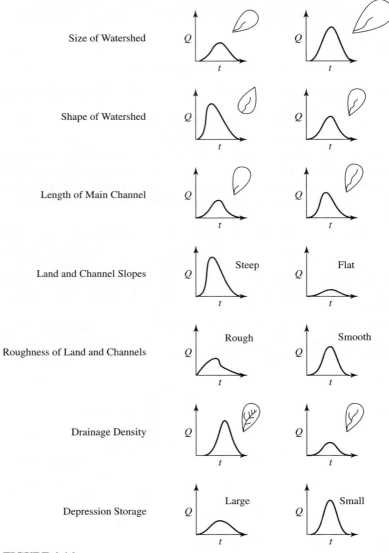

FIGURE 4-16
Factors that influence the characteristics of a hydrograph.
(*Source:* Gumaji, 1986.)

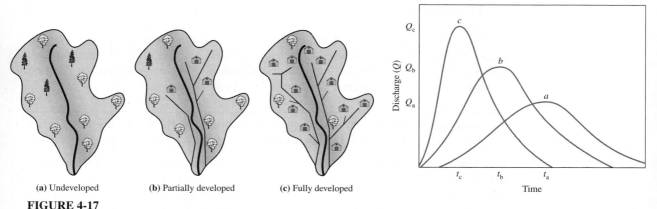

(a) Undeveloped   **(b)** Partially developed   **(c)** Fully developed

FIGURE 4-17
Effect of watershed development on a hydrograph. Note that $Q_c > Q_b > Q_a$ and that $t_c < t_a < t_a$.
(*Source:* Davis and Masten, 2009.)

Unit Hydrograph Method.   While the rational formula provides an estimate of the peak discharge, it does not provide an estimate of the total volume of water discharged from a storm. The unit hydrograph method provides a technique for estimating the volume of water discharged. A unit hydrograph (UH) is a DRH that results from a unit of precipitation excess over a watershed for a unit period of time. Although any unit depth may be selected (fathoms, furlongs, feet, hands, or cubits all would do), we have selected an excess of 1.0 cm after abstractions as a workable unit depth.* The presumption is that if you can determine an average UH, then you can approximate the DRH for any other rainfall excess over the same unit time by multiplying the UH ordinates by the amount of the rainfall excess (Sherman, 1932). For example, a 2.0-cm rainfall excess would yield a DRH with ordinates twice as large as a 1.0-cm UH. The method is limited to watersheds between 3,000- and 4,000-square kilometers in area (Viessman, et al., 1989). Because the intent of the UH is to portray discharge caused by direct runoff so we can use it to predict the DRH for other storms, the first step in constructing a UH is to remove the groundwater contribution. This step is called hydrograph separation. There are a number of graphical procedures for hydrograph separation.

The second step in the construction of the UH is to estimate the total volume of water that occurs as direct runoff. Because the hydrograph is a plot of discharge versus time, the area under the DRH is equal to the volume of direct runoff. The volume is computed by numerical integration of the area under the curve. This is simply a summation of the products of an arbitrary unit of time (Δt) and the height of the DRH ordinate at the center of the selected time interval.

The third step is to convert the volume of direct runoff to a storm depth of runoff. This is done by dividing the volume of direct runoff by the area of the watershed in square meters and then multiplying by a conversion factor of 100 cm/m.

---

*In the original development of the UH by Sherman, a unit depth was defined as 1.0 inch of rainfall excess.

The fourth step is to divide the ordinates of the DRH by the storm depth computed in step three. The quotients are the ordinates of the UH. They have units of $m^3/s \cdot cm$.

The unit duration of the UH is determined from the *hyetograph* (time-rainfall graph) of the storm that was used to develop the UH. Because all of the precipitation does not result in direct runoff, an effective duration of excess precipitation must be estimated. The effective duration becomes the UH *unit duration*.

Example 4-6.   Determine the unit hydrograph ordinates for the Triangle River hydrograph shown in Figure 4-18. The area of the watershed is 16.2 square kilometers.

Solution.   The first step is to determine the depth of the storm precipitation spread over the watershed. The depth is equivalent to the volume of water divided by the area. The volume is equal to the area under the hydrograph. Because of the rather symmetrical shape of this particular hydrograph, it would be easy to find the area from the principles of geometry. However, in the interest of developing a technique that will also be applicable to more customary hydrographs, we will numerically integrate the area under the curve. We do this by taking a convenient slice or Δt and multiplying it by the height of the direct runoff (DRH) ordinate. The direct runoff ordinate is simply the difference between the total ordinate and the base ordinate. In this particular instance the base ordinate is, by observation, 2.0 m^3/s for all time periods. Using a convenient time interval of 1 hour, the following tabular computations are used to numerically integrate the area under the curve:

Time interval (h)	Total ordinate (m^3/s)	Base ordinate (m^3/s)	DRH ordinate (m^3/s)	Volume increment (m^3)
10–11	2.5	2.0	0.5	1,800
11–12	3.5	2.0	1.5	5,400
12–13	4.5	2.0	2.5	9,000
13–14	4.5	2.0	2.5	9,000
14–15	3.5	2.0	1.5	5,400
15–16	2.5	2.0	0.5	1,800
				$\Sigma = 32,400$

The volume increment is calculated as follows: First, the difference between the total ordinate and the base ordinate is found for the time increment selected. In the first row, for the time period from 10 AM to 11 AM (1000 to 1100 hours) the total ordinate is read from the hydrograph (Figure 4-18) as 2.5 m^3/s:

$$\text{Total ordinate} - \text{Base ordinate} = \text{DRH ordinate}$$
$$2.5 \text{ m}^3/\text{s} \quad - \quad 2.0 \text{ m}^3/\text{s} \quad = 0.5 \text{ m}^3/\text{s}$$

To find the area (volume) represented by this slice, the flow rate is multiplied by the time interval selected (1 h) with appropriate units conversions:

$$(0.5 \text{ m}^3/\text{s})(1 \text{ h})(3,600 \text{ s/h}) = 1,800 \text{ m}^3$$

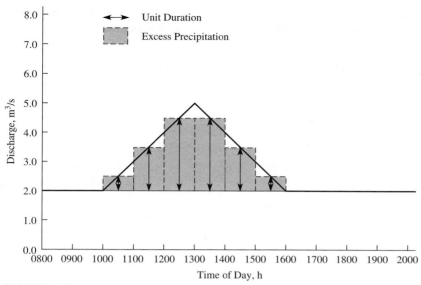

FIGURE 4-18
Triangle River hydrograph.

This process is continued for all the slices shown in Figure 4-18. The total volume (area under the curve) is estimated as 32,400 m^3. We can verify this by using the geometry of the triangle:

$$1/2(\text{base})(\text{height}) = (0.5)(6\text{ h})(5.0\text{ m}^3/\text{s} - 2.0\text{ m}^3/\text{s})(3{,}600\text{ s/h}) = 32{,}400\text{ m}^3$$

Because we wish to construct a *unit hydrograph,* we need to determine whether or not this storm produced 1.0 cm of rainfall excess over the watershed. If it did, then we may use the ordinates directly. If not, then we must adjust the ordinates so that they would be equivalent to that produced by a 1.0 cm rainfall excess. We can determine whether or not this storm produced 1.0 cm by dividing the volume of rainfall by the area of the watershed (given as 16.2 km^2):

$$\text{Storm depth} = \frac{32{,}400\text{ m}^3}{(16.2\text{ km}^2)(1 \times 10^6\text{ m}^2/\text{km}^2)} \times 100\text{ cm/m} = 0.20\text{ cm}$$

It is obvious that the storm is too small and, hence, the ordinates are too small. By dividing the ordinates by the storm depth, we can synthesize ordinates for a unit hydrograph. For example, for the first DRH ordinate:

$$\frac{\text{DRH ordinate}}{\text{Storm depth}} = \frac{0.5\text{ m}^3/\text{s}}{0.2\text{ cm}} = 2.5\text{ m}^3/\text{s} \cdot \text{cm}$$

This ordinate would be located at the center of the slice that was used to establish it, i.e., halfway between 1000 and 1100 hours (see the arrows in Figure 4-18), i.e., 1030.

For a generic hydrograph starting at a time equal to zero, the plotting point would be 0.5 h. The remaining unit hydrograph ordinates are tabulated below.

Triangle River plotting time (h)	Generic plotting time (h)	UH ordinate (m³/s · cm)
1030	0.5	2.5
1130	1.5	7.5
1230	2.5	12.5
1330	3.5	12.5
1430	4.5	7.5
1530	5.5	2.5

The unit "m³/s · cm" is read as

$$\frac{m^3}{(s)(cm)}$$

This means if we multiply a UH ordinate by the cm of excess rainfall, we will get units of m³/s for the ordinate.

We can check our logic by calculating the area under a similar triangle using these new ordinates.

Time interval (h)	DRH ordinate (m³/s)	Volume increment (m³)
10–11	2.5	9,000
11–12	7.5	27,000
12–13	12.5	45,000
13–14	12.5	45,000
14–15	7.5	27,000
15–16	2.5	9,000
		$\Sigma = 162{,}000$

Recalculating our storm depth:

$$\frac{162{,}000 \ m^3}{(16.2 \ km^2)(1 \times 10^6 \ m^2/km^2)} \times 100 \ cm/m = 1.00 \ cm$$

Comments: The unit hydrograph may be applied to a sequence of storms that have the same unit duration. There are two fundamental assumptions in the technique. The first is that storms of the same unit duration have ordinates that are in proportion to the unit hydrograph ordinates. Thus, simple ratios can account for differences in run-off excess. The second assumption is that a sequence of storms may be approximated by superimposing one hydrograph over another (with appropriate time lag) and adding the ordinates together. This is illustrated in the next example.

Example 4-7. Using the hyetograph in Figure 4-19, and the unit hydrograph ordinates from Example 4-6, determine the DRH ordinates and compound runoff.

Solution. The tabular computations are shown below. The explanation follows the table.

Time interval	Time (h)	Rainfall excess (cm)	DRH ordinates 1	DRH ordinates 2	DRH ordinates 3	Compound runoff (m^3/s)
1	0–1	0.5	1.25	N/A	N/A	1.25
2	1–2	2.0	3.75	5.0	N/A	8.75
3	2–3	1.0	6.25	15.0	2.5	23.75
4	3–4	0.0	6.25	25.0	7.5	38.75
5	4–5	0.0	3.75	25.0	12.5	41.25
6	5–6	0.0	1.25	15.0	12.5	28.75
7	6–7	0.0	0.0	5.0	7.5	12.5
8	7–8	0.0	0.0	0.0	2.5	2.5

The time interval is simply an enumeration of the segments. For the first hour, from the hyetograph in Figure 4-19, the rainfall excess is 0.5 cm. For the second and third hours, the rainfall excesses are 2.0 and 1.0 cm, respectively. No rain falls after the end of the third hour. The column labeled DRH 1 refers to the ordinates that are generated from the rainfall excess (0.5 cm) occurring in the first hour. Likewise, the DRH 2 refers to the ordinates resulting from the 2.0-cm rainfall excess in the second hour.

The first set of ordinates is obtained by multiplying the rainfall excess by each of the UH ordinates, that is:

$$(\text{Rainfall excess})(\text{UH ordinate}) = \text{DRH ordinate}$$

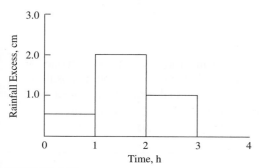

FIGURE 4-19
Hyetograph for Triangle River basin.

Using the UH ordinates from Example 3-5:

$(0.5 \text{ cm})(2.5 \text{ m}^3/\text{s} \cdot \text{cm}) = 1.25 \text{ m}^3/\text{s}$

$(0.5 \text{ cm})(7.5 \text{ m}^3/\text{s} \cdot \text{cm}) = 3.75 \text{ m}^3/\text{s}$

$(0.5 \text{ cm})(12.5 \text{ m}^3/\text{s} \cdot \text{cm}) = 6.25 \text{ m}^3/\text{s}$

$(0.5 \text{ cm})(12.5 \text{ m}^3/\text{s} \cdot \text{cm}) = 6.25 \text{ m}^3/\text{s}$

$(0.5 \text{ cm})(7.5 \text{ m}^3/\text{s} \cdot \text{cm}) = 3.75 \text{ m}^3/\text{s}$

$(0.5 \text{ cm})(2.5 \text{ m}^3/\text{s} \cdot \text{cm}) = 1.25 \text{ m}^3/\text{s}$

The values for the second DRH start an hour later. Thus, under the column DRH 2, the first row is not applicable (N/A) since the rain that falls in the second hour (time interval 2) cannot reach the stream in the first hour. Likewise, under the column DRH 3, the first and second rows are N/A because rain that falls in the third hour cannot reach the stream in the first or second hour.

The DRH ordinates for the second hour of rainfall excess are obtained in the same fashion as those for the first, that is by multiplying the rainfall excess by each of the UH ordinates:

$(2.0 \text{ cm})(2.5 \text{ m}^3/\text{s} \cdot \text{cm}) = 5.0 \text{ m}^3/\text{s}$

$(2.0 \text{ cm})(7.5 \text{ m}^3/\text{s} \cdot \text{cm}) = 15.0 \text{ m}^3/\text{s}$

$(2.0 \text{ cm})(12.5 \text{ m}^3/\text{s} \cdot \text{cm}) = 25.0 \text{ m}^3/\text{s}$

$(2.0 \text{ cm})(12.5 \text{ m}^3/\text{s} \cdot \text{cm}) = 25.0 \text{ m}^3/\text{s}$

$(2.0 \text{ cm})(7.5 \text{ m}^3/\text{s} \cdot \text{cm}) = 15.0 \text{ m}^3/\text{s}$

$(2.0 \text{ cm})(2.5 \text{ m}^3/\text{s} \cdot \text{cm}) = 5.0 \text{ m}^3/\text{s}$

You should note that the table is carried beyond the last rainfall period in the hyetograph until all of the ordinates are used because it takes some finite length of time for the last drop of rainfall excess to reach the stream.

The compound runoff is the sum of the DRH ordinates for each of the time intervals. For example:

$1.25 + \text{N/A} + \text{N/A} = 1.25$

$3.75 + 5.0 + \text{N/A} = 8.75$

$6.25 + 15.0 + 2.5 = 23.75$

To plot the compound runoff hydrograph, the compound runoff ordinates are plotted at 1.0-h intervals, starting 0.5 h from time zero in accordance with the plotting position of the UH ordinates specified earlier. A plot of the individual hydrographs for each of the storms, their superposition, and the resulting compound hydrograph are shown in Figure 4-20. These computations and the plot are easily executed on a spreadsheet.

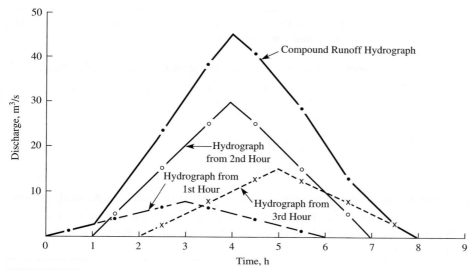

FIGURE 4-20
Compound runoff hydrograph for Triangle River. *Note:* Base flow is not shown.

Estimation of Time of Arrival

In addition to the quantity of discharge, it is often desirable to know when the peak flow will arrive at the watershed outlet or at some point along the discharge channel. This is particularly important when analyzing a series of watersheds that contribute to a river or sewer at various distances downstream from the headwater. The coincident arrival of two peaks would influence the design dramatically.

Lag Time. The time of arrival of the peak discharge is determined inherently in the UH method of estimating runoff. The lag time is the time from the midpoint of excess rainfall to the peak discharge as shown in Figure 4-15.

Time of Concentration. The time of concentration (t_c) is the time required for direct runoff to flow from the hydraulically most remote part of the drainage area to the watershed outlet. One of the major assumptions of the rational method is that the average rainfall intensity used in Equations 4-19 and 4-20 has continued for a period long enough to establish direct runoff and that rainfall has continued long enough to equal or exceed t_c. Thus, it is impossible to use the rational formula without being able to estimate t_c.

Although there are several methods for estimating t_c, the Federal Aviation Agency formula appears to be the easiest to use (FAA, 1970):

$$t_c = \frac{1.8(1.1 - C)\sqrt{3.28D}}{\sqrt[3]{S}} \tag{4-21}$$

where t_c = time of concentration, min
  C = runoff coefficient
  D = overland flow distance, m
  S = slope, %

Example 4-8.   Estimate t_c for the BLAHS 6-percent-slope lawn in Example 4-5. Assume that the overland flow distance was 300.0 m.

Solution.   From Example 4-5 we use the same value of C, namely 0.20. Thus,

$$t_c = \frac{1.8(1.1 - 0.20)\sqrt{(3.28)(300.0)}}{\sqrt[3]{6.0}}$$

$$t_c = \frac{50.82}{1.82} = 27.97, \text{ or } 28 \text{ min}$$

Comments:

1. Note that the equation requires the slope in percent rather than as a decimal fraction.
2. Because the overland flow distance selected for this problem was for the lawn with a 6.0 percent slope and not all of the school grounds, t_c is not the same as that used in Example 4-5.

For several areas that drain to a common outlet such as a drainage ditch or a storm sewer, the time of concentration is equal to the largest combination of the time of concentration for runoff to flow from the surface to the drainage inlet (t_c) plus the time of flow through the drainage ditch or storm sewer. Example 4-9 illustrates the computations.

Example 4-9.   Estimate the time of concentration for BLAHS. The table below shows the estimated values for t_c for each of the areas. The building and lawns drain to a common storm sewer inlet. The storm sewer flows to another manhole inlet where the parking lot runoff is collected. The total flow from the four areas is then carried to a municipal storm sewer. The arrangement of the areas and the storm sewers is shown in the sketch below. The storm sewer on the BLAHS grounds flows at a speed of 0.6 m/s.

Character of surface	t_c (min)
Building	8
Parking lot	10
Lawns	
2.0% slope	38
6.0% slope	28

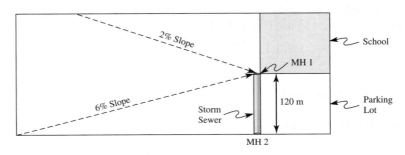

Solution. The time of concentration to be used for estimating the peak discharge is the largest combination of t_c and flow time.

For the distance marked on the sketch, the travel time of the flow from manhole 1 (MH 1) to manhole 2 (MH 2) on the school grounds is

$$\text{Travel time} = \frac{120 \text{ m}}{0.6 \text{ m/s}} = 200 \text{ s, or } 3.33 \text{ min}$$

The total time of concentration (T_c) for the school building is

$$T_c = t_c + \text{travel time} = 8 + 3.33 = 11.33, \text{ or } 11 \text{ min}$$

The time of concentration for each element is summarized as follows:

Character of surface	t_c (min)	Total T_c (min)
Building	8	11
Parking lot	10	10
Lawns		
2.0% slope	38	41
6.0% slope	28	31

Comments:

1. Note that the parking lot drains directly into MH 2. Thus, $T_c = t_c$.
2. According to this evaluation, the time of concentration to be used in entering the IDF curve is the maximum T_c. In this case the maximum T_c is 41 minutes. The estimate of the peak discharge using this time of concentration was shown in Example 4-5.

Estimation of Probability of Occurrence

The frequency of occurrence ($1/T$) used in the design of water supply or storm water control projects should be a function of the cost of the project and the benefits to be obtained from it. That is, the benefit-cost ratio should be greater than 1.0 to justify the project on economic grounds. While the cost of construction can be estimated in a straightforward manner, the environmental cost may be impossible to estimate. Likewise, the benefit beyond a cheaper supply of water or a reduced amount of flood damage is difficult to quantify.

The following paragraphs provide some guidance on the selection of a design frequency. They were obtained from the ASCE sewer design manual with their permission (ASCE, 1969).

In practice, benefit-cost studies usually are not conducted for the ordinary urban storm drainage project. Judgment supported by records of performance in other similar areas is usually the basis of selecting the design frequency.

The range of frequencies used in engineering offices is as follows:

1. For storm sewers in residential areas, 2 to 15 years, with 5 years most commonly reported.

2. For storm sewers in commercial and high-value districts, 10 to 50 years, depending on economic justification.

3. For flood protection works and reservoir design, economics usually dictate a 50-year return period.

Other factors that may affect choice of design frequency include:

1. Use of greater return periods for design of those parts of the system not economically susceptible to future relief.

2. Use of greater recurrence intervals for design of special structures, such as expressway drainage pumping systems, where runoff exceeding capacity would seriously disrupt an important facility. Design frequencies of 50 years or more may be justified in such cases, particularly in small drainage areas, even though the project may be located in a district justifying only 5-year frequency for normal drainage.

3. Adoption of shorter return periods than normal but commensurate with available funds so that some degree of protection can be provided.

The cost of storm sewers is not directly proportional to design frequency. Studies of effects of various factors on sewer cost show that sewer systems designed for 10-year storms may cost only about 6 to 11 percent more than systems designed for 5-year storms, depending on the sewer slope (Rousculp, 1939). The lesser increase applies to steeper sewers.

If the peak flow is estimated by the rational method, it is assumed that the return period (inverse probability) of the peak flow is the same as that of the rainfall used to obtain it. If you use the UH method to determine the discharge, you must resort to some other method of estimating the probability of occurrence.

Annual Series.   Extreme-value analysis is a probability analysis of the largest or smallest values in a data set. Each of the extreme values is selected from an equal time interval. For example, if the largest value in each year of record is used, the extreme-value analysis is called an *annual maxima* series. If the smallest value is used, it is called an *annual minima* series.

Because of the climatic effects on most hydrologic phenomena, *water year* or hydrologic year is adopted instead of a calendar year. The U.S. Geological Survey (U.S.G.S.) has adopted the 12-month period from October 1 to September 30 as the hydrologic year for the United States. This period was chosen for two reasons (Boyer, 1964). "(1) to break the record during the low-water period near the end of the summer season, and (2) to avoid breaking the record during the winter, so as to eliminate computation difficulties during the ice period."

The procedure for an annual maxima or minima analysis is as follows:

1. Select the minimum or maximum value in each 12-month interval (October to September) over the period of record.

2. Rank each value starting with the highest (for annual maxima) or lowest (for annual minima) as rank number one.

3. Compute a return period using Equation 4-10.

4. Plot the annual maxima series on a special probability paper known as *Gumbel paper* or flood data paper. Although the same paper may be used for annual minima series, Gumbel recommends a log extremal probability paper (axis of ordinates is log scale) for droughts (Gumbel, 1954).

From the Gumbel plot, the return period for a flood or drought of any magnitude may be determined. Conversely, for any magnitude of flood or drought, you may determine how frequently it will occur.

In statistical parlance a Gumbel plot is a linearization of a Type I probability distribution. The logarithmically transformed version of the Type I distribution is called a log-Pearson Type III distribution. The return period of the mean (X) of the Type I distribution occurs at $T = 2.33$ years. Thus, the U.S.G.S. takes the return period of the mean annual flood to be 2.33 years. This is marked by a vertical dashed line on Gumbel paper (Figure 4-21).

The data in Table 4-5 were used to plot the annual maxima line in Figure 4-21. The computations are explained in Example 4-10.

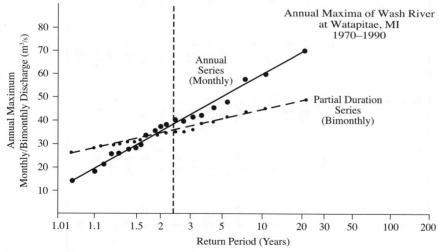

FIGURE 4-21
Gumbel plot of annual maxima of Wash River at Watapitae, MI.

TABLE 4-5
Average monthly discharge of the Wash River at Watapitae, MI (discharge in m^3/s)

Year	J	F	M	A	M	J	J	A	S	O	N	D
1969	2.92	5.10	1.95	4.42	3.31	2.24	1.05	0.74	1.02	1.08	3.09	7.62
1970	24.3	16.7	11.5	17.2	12.6	7.28	7.53	3.03	10.2	10.9	17.6	16.7
1971	15.3	13.3	14.2	36.3	13.5	3.62	1.93	1.83	1.93	3.29	5.98	12.7
1972	11.5	4.81	8.61	27.0	4.19	2.07	1.15	2.04	2.04	2.10	3.12	2.97
1973	11.1	7.90	41.1	6.77	8.27	4.76	2.78	1.70	1.46	1.44	4.02	4.45
1974	2.92	5.10	28.7	12.2	7.22	1.98	0.91	0.67	1.33	2.38	2.69	3.03
1975	7.14	10.7	9.63	21.1	10.2	5.13	3.03	10.9	3.12	2.61	3.00	3.82
1976	7.36	47.4	29.4	14.0	14.2	4.96	2.29	1.70	1.56	1.56	2.04	2.35
1977	2.89	9.57	17.7	16.4	6.83	3.74	1.60	1.13	1.13	1.42	1.98	2.12
1978	1.78	1.95	7.25	24.7	6.26	8.92	3.57	1.98	1.95	3.09	3.94	12.7
1979	13.8	6.91	12.9	11.3	3.74	1.98	1.33	1.16	0.85	2.63	6.49	5.52
1980	4.56	8.47	59.8	9.80	6.06	5.32	2.14	1.98	2.17	3.40	8.44	11.5
1981	13.8	29.6	38.8	13.5	37.2	22.8	6.94	3.94	2.92	2.89	6.74	3.09
1982	2.51	13.1	27.9	22.9	16.1	9.77	2.44	1.42	1.56	1.83	2.58	2.27
1983	1.61	4.08	14.0	12.8	33.2	22.8	5.49	4.25	5.98	19.6	8.5	6.09
1984	21.8	8.21	45.1	6.43	6.15	10.5	3.91	1.64	1.64	1.90	3.14	3.65
1985	8.92	5.24	19.1	69.1	26.8	31.9	7.05	3.82	8.86	5.89	5.55	12.6
1986	6.20	19.1	56.6	19.5	20.8	7.73	5.75	2.95	1.49	1.69	4.45	4.22
1987	15.7	38.4	14.2	19.4	6.26	3.43	3.99	2.79	1.79	2.35	2.86	10.9
1988	21.7	19.9	40.0	40.8	11.7	13.2	4.28	3.31	9.46	7.28	14.9	26.5
1989	31.4	37.5	29.6	30.8	11.9	5.98	2.71	2.15	2.38	6.03	14.2	11.5
1990	29.2	20.5	34.9	35.3	13.5	5.47	3.29	3.14	3.20	2.11	5.98	7.62

Example 4-10.   Perform an annual maxima extreme-value analysis on the data in Table 4-5. Determine the recurrence interval of monthly flows equal to or greater than 58.0 m³/s. Also determine the discharge of the mean monthly annual flood.

Solution.   To begin we select the maximum discharge in each hydrologic year. The first nine months of 1969 and the last three months of 1990 cannot be used because they are not complete hydrologic years. After selecting the maximum value in each year, we rank the data and compute the return period. The 1970 water year begins in October 1969.

The computations are summarized in Table 4-6. The return period and flows are plotted as the solid line in Figure 4-21. From Figure 4-21 we find that the return period for a flood of 58.0 m³/s is about 9.2 years. The mean annual flood is about 37 m³/s.

Partial-Duration Series.   It often happens that the second largest or second smallest flow in a water year is larger or smaller than the maxima or minima from a different

TABLE 4-6
Tabulated computations of annual maxima for the
Wash River at Watapitae, MI

Year	Discharge (m³/s)	Rank	$T = \dfrac{n + 1}{m}$
1970	24.3	18	1.22
1971	36.3	11	2.00
1972	27.0	16	1.38
1973	41.1	6	3.67
1974	28.7	14	1.57
1975	21.1	19	1.16
1976	47.4	4	5.50
1977	17.7	20	1.10
1978	24.7	17	1.29
1979	13.8	21	1.05
1980	59.8	2	11.00
1981	38.8	8	2.75
1982	27.9	15	1.47
1983	33.2	13	1.69
1984	45.1	5	4.40
1985	69.1	1	22.00
1986	56.6	3	7.33
1987	38.4	9	2.44
1988	40.8	7	3.14
1989	37.5	10	2.20
1990	35.3	12	1.83

TABLE 4-7
Theoretical relationship between partial series and annual series return periods

Partial series	Annual series
0.5	1.18
1.0	1.58
1.45	2.08
2.0	2.54
5.0	5.52
10.0	10.5
50.0	50.5
100.0	100.5

Source: Langbein, 1949.

water year. To take these events into consideration, a partial series of the data is examined. The theoretical relationship between an annual series and partial series is shown in Table 4-7. The partial series is approximately equal to the annual series for return periods greater than ten years (Langbein, 1949).

If the time period over which the event occurs is also taken into account, the analysis is termed a partial-duration series. While it is fairly easy to define a flood as "any" flow that exceeds the capacity of the drainage system, in order to properly define a drought we must specify the low flow and its duration. For example, if a roadway is covered with water for ten minutes, we can say that it is flooded. In contrast, if the flow in a river is below our demand for ten minutes, we certainly would not declare it a drought! Thus, a partial-duration series is particularly relevant for low-flow conditions.

From an environmental engineering point of view, three low-flow durations are of particular interest. The 10-year return period of seven days of low flow has been selected by many states as the critical flow for water pollution control. Wastewater treatment plants must be designed to provide sufficient treatment to allow effluent discharge without driving the quality of the receiving stream below the standard when the dilution capacity of the stream is at a 10-year low.

A longer duration low flow and longer return period are selected for water supply. In the Midwest, durations of 1 to 5 years and return periods of 25 to 50 years are used in the design of water-supply reservoirs. Where water supply is by direct draft (withdrawal) from a river, the duration selected may be on the order of 30 to 90 days with a 10-year return period.

The procedure for performing a partial-duration series analysis is very similar to that used for an annual series.

Complete Series. All of the observed data are used in a complete series analysis. This analysis is usually presented in one of two forms: as a duration curve (Figure 4-22), or as a cumulative probability distribution function (CDF) (Figure 4-23). In either

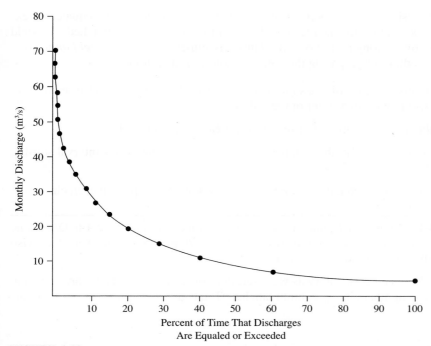

FIGURE 4-22
Duration curve for Wash River at Watapitae, MI.

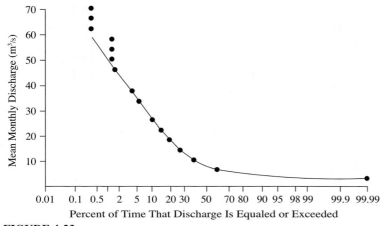

FIGURE 4-23
Cumulative probability distribution for Wash River at Watapitae, MI.

form the analysis shows the percent of time that a given flow will be equaled or exceeded. The percent of time is interpreted as the probability that a watershed will yield a given flow over a long period of time. Thus it is sometimes called a *yield analysis*.

The procedure for preparing the data for plotting is as follows:

1. Establish a series of class intervals that cover the range of discharge observations in increasing order of magnitude.

2. Tabulate the number of observations in each class interval.

3. Cumulatively sum the number of observations in each class interval starting with the highest flow.

4. Compute the percentage of all the observations that appear in each class.

Example 4-11. Perform a complete series analysis on the data in Table 4-6. Determine the percent of time that a monthly flow of 28.0 m^3/s is equaled or exceeded. Also determine the mean monthly discharge.

Solution. Eighteen class intervals were selected for the analysis. These are shown in Table 4-8. The numbers in the column labeled "Total" are the number of observations

TABLE 4-8
Tabulation of computations for yield analysis

Class mean (m^3/s)	Class interval[a] (m^3/s)	Total	Accumulated	Percent
2	0–4	106	264	100.0
6	4–8	54	158	59.9
10	8–12	30	104	39.4
14	12–16	23	74	28.0
18	16–20	13	51	19.3
22	20–24	8	38	14.4
26	24–28	6	30	11.4
30	28–32	8	24	9.1
34	32–36	4	16	6.1
38	36–40	5	12	4.6
42	40–44	2	7	2.6
46	44–48	1	5	1.9
50	48–52	0	4	1.5
54	52–56	0	4	1.5
58	56–60	3	4	1.5
62	60–64	0	1	0.4
66	64–68	0	1	0.4
70	68–72	1	1	0.4

[a]Class intervals were actually 0.00–4.00, 4.01–8.00, 8.01–12.00, and so on. Zeros were eliminated to save space.

in the class interval. The cumulative sum starting at the bottom of the column is in the next column. Thus, "Accumulated" means sum of observations equal to or greater than the class interval. "Percent" is the percent of readings equal to or greater than the class interval, computed by dividing the cumulative sum for each class interval by the total number of observations and multiplying by 100 percent.

The mean of each class interval is plotted against "Percent" in Figures 4-22 and 4-23. From these figures we find that a monthly flow of 28.0 m³/s is equaled or exceeded about 10 percent of the time. The mean monthly discharge is the flow that is equaled or exceeded 50 percent of the time. From either Figure 4-22 or 4-23 we find that the mean monthly discharge is about 8 m³/s.

When to Use Which Series.   You should not expect that the probability of occurrence $(1/T)$ computed from an annual series will be the same as that found from a complete series. There are many reasons for this difference: Among the most obvious is the fact that in an annual series we treat $1/12$ of the data as if it were all of the data when, in fact, it is not even a representative sample. It is only the extreme end of the possible range of values.

The following guidelines can be used to decide when to use which analysis:

1. Use an annual series to predict the size of flood that a storm sewer or drainage channel must handle.

2. Use a partial series to predict low-flow conditions for wastewater dilution and water supply.

3. Use a complete series to determine the long-time reliability (*safe yield*) for water supply or power generation.

In practice the complete-series analysis can be performed to decide whether or not it is worth doing a partial series for water supply. If the complete series indicates that the mean monthly flow will not supply the demand, then computation of a partial series to determine the storage requirement is not worth the trouble, since it would be impossible to store enough water.

4-6   STORAGE OF RESERVOIRS

Classification of Reservoirs

For our purpose we can classify reservoirs either by size or by use. The size of the reservoir is used to establish the degree of safety to be incorporated into the design of the dam and spillway. The use or uses of the reservoir are a basis for evaluating the benefit-cost ratio.

Major dams (reservoir capacity greater than 6×10^7 m³) are designed to withstand the maximum probable flood. Intermediate-sized dams (1×10^6 to 6×10^7 m³) are designed to handle the discharge from the most severe storm considered to be reasonably characteristic of the watershed. For minor reservoirs (less than 1×10^6 m³) the dams are designed to handle floods with return periods of 50 to 100 years.

Some of the benefits derived from reservoirs include the following: (1) flood control; (2) hydroelectric power; (3) irrigation; (4) water supply; (5) navigation; (6) preservation of aquatic life; and (7) recreation. The multipurpose or multiuse reservoir is the rule rather than the exception. Very seldom is it possible to justify the cost of a major reservoir on the basis of a single use.

Volume of Reservoirs

Mass Diagram.   The techniques for determining the storage volume required for a reservoir are dependent both on the size and use of the reservoir. We shall discuss the simplest procedure, which is quite satisfactory for small water-supply impoundments, storm-water retention ponds, and wastewater equalization basins. It is called the mass diagram or *Rippl method* (Rippl, 1883). The main disadvantage of the Rippl method is that it assumes that the sequence of events leading to a drought or flood will be the same in the future as it was in the past. More sophisticated techniques have been developed to overcome this disadvantage, but these techniques are left for more advanced classes.

The Rippl procedure for determining the storage volume is an application of the mass balance approach (Equation 2-4). In this case it is assumed that the only input is the flow into the reservoir (Q_{in}) and that the only output is the flow out of the reservoir (Q_{out}). Therefore,

$$\frac{dS}{dt} = \frac{d(\text{In})}{dt} - \frac{d(\text{Out})}{dt}$$

becomes

$$\frac{dS}{dt} = Q_{in} - Q_{out} \tag{4-22}$$

with the assumption that the density term cancels out because the change in density across the reservoir is negligible.

If we multiply both sides of the equation by dt, the inflow and outflow become volumes (flow rate $\times$ time = volume), that is,

$$dS = (Q_{in})(dt) - (Q_{out})(dt) \tag{4-23}$$

By substituting finite time increments (Δt), the change in storage is then

$$(Q_{in})(\Delta t) - (Q_{out})(\Delta t) = \Delta S \tag{4-24}$$

By cumulatively summing the storage terms, we can estimate the size of the reservoir. If the reservoir design is for water supply, then Q_{out} is the demand, and zero or positive values of storage (ΔS) indicate there is enough water to meet the demand. If the storage is negative, then the reservoir must have a capacity equal to the absolute value of cumulative storage to meet the demand. If the reservoir design is for flood protection, then Q_{out} is the capacity of the downstream river to hold water, and zero or negative values indicate the river is below flood stage. If the storage is positive, then the reservoir must have capacity equal to the cumulative storage to prevent flooding.

Example 4-12. Using the data in Table 4-5, determine the storage required to meet a demand of 2.0 m³/s for the period from August 1976 through December 1978.

Solution. The computations are summarized in the table below.

Month	Q_{in} (m³/s)	$Q_{in}(\Delta t)$ (10^6 m³)	Q_{out} (m³/s)	$Q_{out}(\Delta t)$ (10^6 m³)	ΔS (10^6 m³)	$\Sigma(\Delta S)$ (10^6 m³)
1976						
Aug	1.70	4.553	2.0	5.357	−0.8035	−0.8035
Sep	1.56	4.043	2.0	5.184	−1.140	−1.944
Oct	1.56	4.178	2.0	5.357	−1.178	−3.122
Nov	2.04	5.287	2.0	5.184	0.1036	−3.019
Dec	2.35	6.294	2.0	5.357	0.9374	−2.081
1977						
Jan	2.89	7.741	2.0	5.357	2.384	
Feb	9.57					
Mar	17.7					
Apr	16.4					
May	6.83					
Jun	3.74					
Jul	1.60	4.285	2.0	5.357	−1.071	−1.071
Aug	1.13	3.027	2.0	5.357	−2.330	−3.402
Sep	1.13	2.929	2.0	5.184	−2.255	−5.657
Oct	1.42	3.803	2.0	5.357	−1.553	−7.210
Nov	1.98	5.132	2.0	5.184	−0.052	−7.262
Dec	2.12	5.678	2.0	5.357	0.3214	−6.941
1978						
Jan	1.78	4.768	2.0	5.357	−0.5892	−7.530
Feb	1.95	4.717	2.0	4.838	−0.121	−7.651
Mar	7.25	19.418	2.0	5.357	14.061	
Apr	24.7					
May	6.26					
Jun	8.92					
Jul	3.57					
Aug	1.98	5.303	2.0	5.357	−0.0536	−0.0537
Sep	1.95	5.054	2.0	5.184	−0.1296	−0.1832
Oct	3.09	8.276	2.0	5.357	2.919	
Nov	3.94					
Dec	12.7					

The data in the first and second columns of the table were extracted from Table 4-5.

The third column is the product of the second column and the time interval for the month. For example, for August (31 d) and September (30 d), 1976:

$$(1.70 \text{ m}^3/\text{s})(31 \text{ d})(86,400 \text{ s/d}) = 4,553,280 \text{ m}^3$$
$$(1.56 \text{ m}^3/\text{s})(30 \text{ d})(86,400 \text{ s/d}) = 4,043,520 \text{ m}^3$$

The fourth column is the demand given in the problem statement.

The fifth column is the product of the demand and the time interval for the month. For example, for August and September 1976:

$$(2.0 \text{ m}^3/\text{s})(31 \text{ d})(86,400 \text{ s/d}) = 5,356,800 \text{ m}^3$$
$$(2.0 \text{ m}^3/\text{s})(30 \text{ d})(86,400 \text{ s/d}) = 5,184,000 \text{ m}^3$$

The sixth column (ΔS) is the difference between the third and fifth columns. For example, for August and September 1976:

$$4,553,280 \text{ m}^3 - 5,356,800 \text{ m}^3 = -803,520 \text{ m}^3$$
$$4,043,520 \text{ m}^3 - 5,184,000 \text{ m}^3 = -1,140,480 \text{ m}^3$$

The last column ($\Sigma(\Delta S)$) is the sum of the last value in that column and the value in the sixth column. For August 1976, it is $-803,520 \text{ m}^3$ since this is the first value. For September 1976, it is

$$(-803,520 \text{ m}^3) + (-1,140,480 \text{ m}^3) = -1,944,000 \text{ m}^3$$

The following logic is used in interpreting the table. From August through December 1976, the demand exceeds the flow, and storage must be provided. The maximum storage required for this interval is $3.122 \times 10^6 \text{ m}^3$. In January 1977, the storage (ΔS) exceeds the deficit ($\Sigma(\Delta S)$) from December 1976. If we view the deficit as the volume of water in a virtual reservoir with a total capacity of $3.122 \times 10^6 \text{ m}^3$, then in December 1976, the volume of water in the reservoir is $1.041 \times 10^6 \text{ m}^3$ ($3.122 \times 10^6 - 2.081 \times 10^6$). The January 1977 inflow exceeds the demand and fills the reservoir deficit of $2.081 \times 10^6 \text{ m}^3$.

Because the inflow (Q_{in}) exceeds the demand (2.0 m^3/s) for the months of February through June 1977, no storage is required during this period. Hence, no computations were performed.

From July 1977 through February 1978, the demand exceeds the inflow, and storage is required. The maximum storage required is $7.651 \times 10^6 \text{ m}^3$. Note that the computations for storage did not stop in December 1977, even though the inflow exceeded the demand. This is because the storage was not sufficient to fill the reservoir deficit. The storage was sufficient to fill the reservoir deficit in March 1978.

Comments:

1. You should note that these tabulations are particularly well suited to spreadsheets.

2. Rippl used a graphical technique to present his method of estimating the required storage volume. This is method of solving the problem is presented at the text website: www.mhhe.com/davis.

The storage volume determined by the Rippl method must be increased to account for water lost through evaporation and volume lost through the accumulation of sediment.

Application of the Mass Diagram. When existing storm sewers cannot handle run-off, consideration is given to providing a *retention basin* (storage pond) as an alternative to replacing the existing storm sewer. The retention basin serves two functions. First, it delays the coincidence of peak flows, and second, it can be designed to release storm water at a rate the storm sewer can handle. The procedure for applying the mass diagram to retention basins is basically the same as it is for water supply. The DRH for the design storm is substituted for the monthly discharge readings. The allowable discharge to the existing sewer is substituted for Q_{out}.

Wastewater treatment plants function best when the flow into the plant is constant. The general daily fluctuation in water use in most municipalities and industries results in an alternating pattern of high and low wastewater discharge. The peaks and valleys of flow can be evened out with an *equalization basin*. The volume of the equalization basin is chosen to shave off the peaks and fill in the valleys of the inflow mass diagram.

4-7 GROUNDWATER AND WELLS

Although the portion of the population of the United States supplied by surface water is 58 percent greater than that supplied by groundwater, the number of communities supplied by groundwater is almost 12 times that supplied by surface water (Figure 4-24). The reason for this pattern is that larger cities are supplied by surface water while many small communities use groundwater.

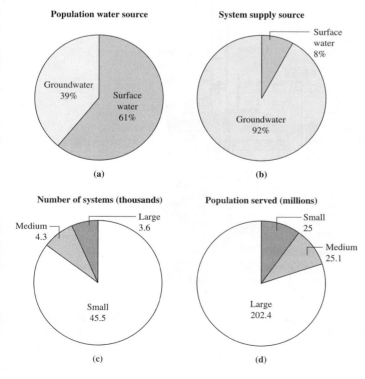

FIGURE 4-24

(*a*) Percentage of the population served by drinking-water system source. (*b*) Percentage of drinking-water systems by supply source. (*c*) Number of drinking-water systems (in thousands) by size. (*d*) Population served (in millions of people) by drinking-water system size. (*Source:* 1997 National Public Water Systems Compliance Report. U.S. EPA, Office of Water. Washington, D.C. 20460. (EPA-305-R-99-002).) (*Note:* Small systems serve 25–3,300 people; medium systems serve 3,301–10,000 people; large systems serve 10,000+ people.)

Groundwater has several characteristics that make it desirable as a water supply source. First, the groundwater system provides natural storage, which eliminates the cost of impoundment works. Second, since the supply frequently is available at the point of demand, the cost of transmission is reduced significantly. Third, because groundwater is filtered by the natural geologic strata, groundwater is clearer to the eye than surface water.

Groundwater is not without its drawbacks. It dissolves naturally occurring minerals, which may give the water undesirable characteristics such as hardness, red color from iron oxides, and toxic contaminants such as arsenic.

Construction of Wells

Modern wells consist of more than a simple hole in the ground (Figure 4-25). A steel pipe called a *casing* is placed in the well hole to maintain the integrity of the hole. The casing is sealed to the surrounding soil with a cement grout, and a screen is placed at the bottom of the casing to allow water in and to keep soil material out. Two types of pump may be used. In the diagram shown, the pump motor is at the ground surface and the pump itself is placed down in the well above the well screen. The alternative is a

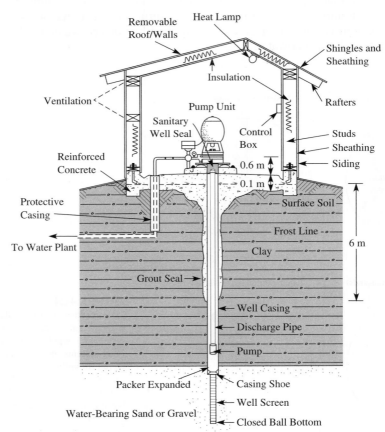

FIGURE 4-25
Pumphouse. (U.S. EPA, 1973)

submersible pump. In this case both the pump and the motor are lowered into the casing; water is pumped out of the well through a discharge pipe or drop pipe.

Sanitary Considerations.   The penetration of a water-bearing formation by a well provides a direct route for possible contamination of the groundwater. Although there are different types of wells and well construction, there are basic sanitary aspects that must be considered and followed (refer to Figure 4-26):

1. The annular space outside the casing should be filled with a watertight cement grout or puddled clay from the surface to the depth necessary to prevent entry of contaminated water. A minimum of 6 m is recommended.

2. For artesian aquifers, the casing should be sealed into the overlying impermeable formations so as to retain the artesian pressure.

3. When a water-bearing formation containing water of poor quality is penetrated, the formation should be sealed off to prevent the infiltration of water into the well and aquifer.

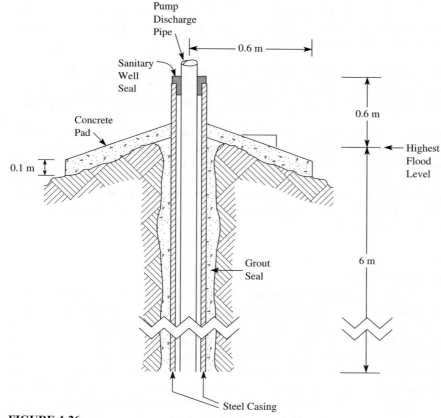

FIGURE 4-26
Sanitary considerations in well construction.

4. A sanitary well-seal with an approved vent should be installed at the top of the well casing to prevent the entrance of contaminated water or other objectionable material (U.S. EPA, 1973).

Well Covers and Seals. Every well should be provided with an overlapping, tight-fitting cover at the top of the casing or pipe sleeve to prevent contaminated water or other material from entering the well.

The seal in a well that is exposed to possible flooding should be elevated at least 0.6 m above the highest known flood level. When this is not possible, the seal should be watertight and equipped with a vent line whose opening to the atmosphere is at least 0.6 m above the highest known flood level.

Well covers and pump platforms should be elevated above the adjacent finished ground level. Pumproom floors should be constructed of reinforced, watertight concrete sloped away from the well so that surface and wastewater cannot stand near the well. The minimum thickness of such a slab or floor should be 0.1 m. Concrete slabs or floors should be poured separately from the grout formation seal and, where the threat of freezing exists, insulated from it and the well casing by a plastic or mastic coating or sleeve to prevent bonding of the concrete to either.

All water wells should be readily accessible at the top for inspection, servicing, and testing. This requires that any structure over the well be easily removable to provide full, unobstructed access for well-servicing equipment.

Disinfection of Wells. All newly constructed wells should be disinfected to neutralize contamination from equipment, material, or surface drainage introduced during construction. Every well should be disinfected promptly after construction or repair.

Pumphousing. A pumphouse installed above the surface of the ground should be used. It should be unnecessary to use an underground discharge connection if an insulated, heated pumphouse is provided. For individual installations in rural areas, two 60-watt light bulbs, a thermostatically controlled electric heater, or a heating cable will generally provide adequate protection when the pumphouse is properly insulated. Because power failures may occur, an emergency gasoline-driven power supply or pump should be considered.

Cone of Depression

When a well is pumped, the level of the piezometric surface in the vicinity of the well will be lowered (Figure 4-27). This lowering, or drawdown, causes the piezometric surface to take the shape of an inverted cone called a cone of depression. Because the water level in a pumped well is lower than that in the aquifer surrounding it, the water flows from the aquifer into the well. At increasing distances from the well, the drawdown decreases until the slope of the cone merges with the static water table. The distance from the well at which this occurs is called the radius of influence. The radius of influence is not constant but tends to expand with continued pumping. At a given pumping rate, the shape of the cone of depression depends on the characteristics of the water-bearing formation. Shallow and wide cones will form in aquifers composed of

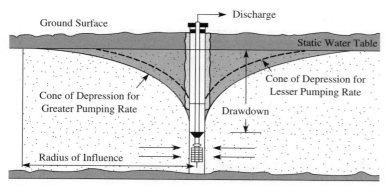

FIGURE 4-27
Effect of aquifer material on cone of depression. (U.S. EPA, 1973)

coarse sands or gravel. Deeper and narrower cones will form in fine sand or sandy clay (Figure 4-28). As the pumping rate increases, the drawdown increases. Consequently the slope of the cone steepens. When other conditions are equal for two wells, it may be expected that pumping costs will be higher for the well surrounded by the finer material because of greater drawdown.

When the cones of depression overlap, the local water table will be lowered (Figure 4-29). This requires additional pumping lifts to obtain water from the interior portion of the group of wells. A wider distribution of the wells over the groundwater basin will reduce the cost of pumping and will allow the development of a larger quantity of water. One rule of thumb is that two wells should be placed no closer together than two times the thickness of the water-bearing strata. For more than two wells, they should be spaced at least 75 meters apart.

Definition of Terms

The aquifer parameters identified and defined in this section are those relevant to determining the available volume of water and the ease of its withdrawal.

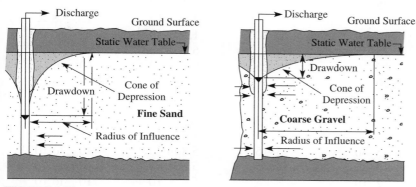

FIGURE 4-28
Effect of aquifer material on cone of depression. (U.S. EPA, 1973)

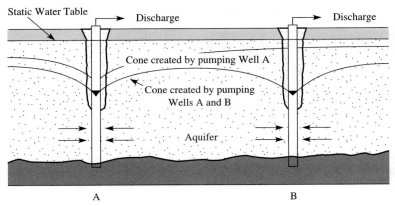

FIGURE 4-29
Effect of overlapping cones of depression. (U.S. EPA, 1973)

Porosity. The ratio of the volume of voids (open spaces) in the soil to the total volume is called *porosity*. It is a measure of the amount of water that can be stored in the spaces between soil particles. It does not indicate how much of this water is available for development. Some typical values are shown in Table 4-9.

TABLE 4-9
Values of aquifer parameters

Aquifer material	Typical porosity (%)	Range of porosities (%)	Range of specific yield (%)	Typical hydraulic conductivity (m/s)	Range of hydraulic conductivities (m/s)
Unconsolidated					
Clay	55	50–60	1–10	1.2×10^{-6}	$0.1–2.3 \times 10^{-6}$
Loam	35	25–45		6.4×10^{-6}	10^{-6} to 10^{-5}
Fine sand	45	40–50		3.5×10^{-5}	$1.1–5.8 \times 10^{-5}$
Medium sand	37	35–40	10–30	1.5×10^{-4}	10^{-5} to 10^{-4}
Coarse sand	30	25–35		6.9×10^{-4}	10^{-4} to 10^{-3}
Sand and gravel	20	10–30	15–25	6.1×10^{-4}	10^{-5} to 10^{-3}
Gravel	25	20–30		6.4×10^{-3}	10^{-3} to 10^{-2}
Consolidated					
Shale	<5		0.5–5	1.2×10^{-12}	
Granite	<1			1.2×10^{-10}	
Sandstone	15	5–30	5–15	5.8×10^{-7}	10^{-8} to 10^{-5}
Limestone	15	10–20	0.5–5	5.8×10^{-6}	10^{-7} to 10^{-5}
Fractured rock	5	2–10		5.8×10^{-5}	10^{-8} to 10^{-4}

Adapted from Bouwer, 1978; Linsley et al., 1975; and Walton, 1970.

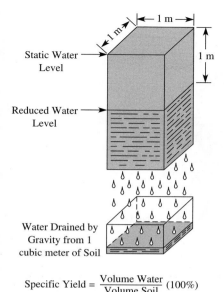

Static Water Level

Reduced Water Level

Water Drained by Gravity from 1 cubic meter of Soil

$$\text{Specific Yield} = \frac{\text{Volume Water}}{\text{Volume Soil}} \ (100\%)$$

FIGURE 4-30
Specific yield. (*Source:* Johnson Screens. A Weatherford company. Reprinted by permission.)

Specific Yield. The percentage of water that is free to drain from the aquifer under the influence of gravity is defined as *specific yield* (Figure 4-30). Specific yield is not equal to porosity because the molecular and surface tension forces in the pore spaces keep some of the water in the voids. Specific yield reflects the amount of water available for development. Some values are shown in Table 4-9.

Storage Coefficient (S). This parameter is akin to specific yield. The *storage coefficient* is the volume of available water resulting from a unit decline in the piezometric surface over a unit horizontal cross-sectional area. It has units of m^3 of water/m^3 of aquifer or, in essence, no units at all! Storage coefficient and specific yield may be used interchangeably for unconfined aquifers. Values of S for unconfined aquifers range from 0.01 to 0.35. For confined aquifers the values of S vary from 1×10^{-3} to 1×10^{-5}.

Hydraulic Gradient and Head. The slope of the piezometric surface is called the *hydraulic gradient.* It is measured in the direction of the steepest slope of the piezometric surface. Groundwater flows in the direction of the hydraulic gradient and at a rate proportional to the slope.

The vertical distance from a reference plane (zero datum) to the bottom of a well or piezometer (Figure 4-31) is called the *elevation head.* The height to which water will rise in the piezometer is called the *pressure head.* The sum of the elevation head and the pressure head is called the *total head.*

Using Figure 4-31 and the assumption that the groundwater is flowing in the plane of the page, we can define the gradient to be:

$$\text{Hydraulic gradient} = \frac{\text{Change in head}}{\text{Horizontal distance}} \tag{4-25}$$

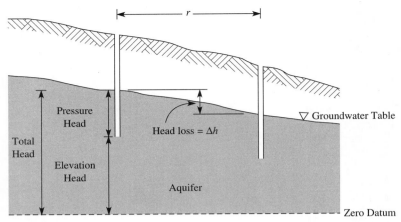

FIGURE 4-31
Geometry for definition of head and hydraulic gradient.

or in the differential sense:

$$\text{Hydraulic gradient} = \frac{\Delta h}{r} = \frac{dh}{dr} \tag{4-26}$$

We generally require three wells, one of which is out of plane with the other two, to define the hydraulic gradient. Heath suggests the following graphical procedure for finding the hydraulic gradient between three closely spaced wells (Heath, 1983):

1. Draw a line between the two wells with the highest and lowest head and divide it into equal intervals. By interpolation, find the place on the line between these two wells where the head is equal to the head of the third well.

2. Draw a line from the third well to the point on the line between the first two wells where the head is the same as that in the third well. This line is called an *equipotential line.* This means that the head anywhere along the line should be constant. Groundwater will flow in a direction perpendicular to this line.

3. Draw a line perpendicular to the equipotential line through the well with the lowest or highest head. The groundwater flow is in a direction parallel to this line. It is called a *flow line.*

4. Calculate the gradient as the difference in head between the head on the equipotential line and the head at the lowest or highest well divided by the distance from the equipotential line to that well.

Example 4-13. For the wells shown in plan view below, determine the direction of flow and the hydraulic gradient. The total head is given for each well as follows:

Well A = 10.4 m

Well B = 9.9 m

Well C = 10.0 m

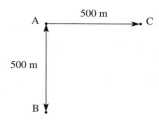

Solution. The stepwise graphical solution procedure is shown below.

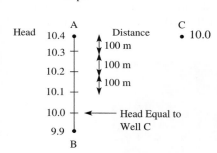

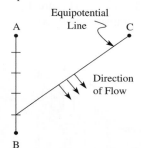

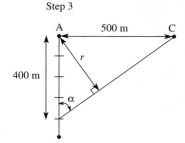

The distance r must be determined in order to calculate the hydraulic gradient. From the plan view, we may note that the wells form a right triangle with legs of 400 m and 500 m. The angle α may be computed as:

$$\tan^{-1}(\alpha) = \frac{500}{400}$$

and $\alpha = 51.34°$.

The distance r is

$$r = (400)\sin\alpha = 400\sin 51.34 = 312.35\text{ m}$$

The hydraulic gradient is then:

$$\text{Hydraulic gradient} = \frac{10.4\text{ m} - 10.0\text{ m}}{312.35\text{ m}} = 0.00128$$

Comment: Note that the hydraulic gradient has no units.

Hydraulic Conductivity (K).   The property of an aquifer that is a measure of its ability to transmit water under a sloping piezometric surface is called *hydraulic conductivity*. It is defined as the discharge that occurs through a unit cross section of aquifer (Figure 4-32) under a hydraulic gradient of 1.00. It has units of speed (m/s). Typical values are given in Table 4-9.

Transmissivity (T).   The coefficient of transmissivity (T) is a measure of the rate at which water will flow through a unit width vertical strip of aquifer extending through its full saturated thickness (Figure 4-32) under a unit hydraulic gradient. It has units of m^2/s. Values of the transmissivity coefficient range from 1.0×10^{-4} to 1.5×10^{-1} m^2/s.

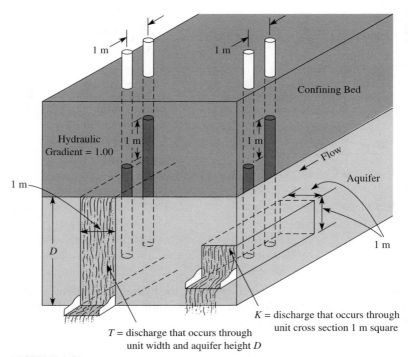

FIGURE 4-32
Illustration of definition of hydraulic conductivity (K) and transmissivity (T). (*Source:* Johnson Screens. A Weatherford company. Reprinted by permission.)

Well Hydraulics

Equations for calculating the discharge that results when the piezometric surface is lowered are based on the work of the French hydrologist Henri Darcy. In 1856 he discovered that the velocity of water flow in a porous aquifer was proportional to the hydraulic gradient. He proposed the following equation, which is now called Darcy's law (Darcy, 1856):

$$v = K\frac{dh}{dr}$$

(4-27)

where v = velocity, m/s
    K = hydraulic conductivity, m/s
    dh/dr = slope of the hydraulic gradient, m/m

The *Darcy velocity* is not a real velocity because it assumes that the full cross-sectional area of the aquifer is available for water to flow through. However, much of the cross-sectional area is soil material; thus, the actual area through which the water flows is much less. As a result, the actual linear velocity of the groundwater is considerably faster than the Darcy velocity.

We may determine the actual average linear velocity by taking into account the fraction of the cross-sectional area filled with soil. Let us define the flow (Q) to be the product of the total cross-sectional area and the Darcy velocity, that is $Q = Av$. The actual average linear velocity (v') through the voids times the area of the voids (A') is also equal to the flow (Q). Thus, the actual velocity is related to the Darcy velocity:

$$A'v' = Av = Q$$

(4-28)

If we solve this expression for the average linear velocity (also called the *seepage velocity*) v', we find:

$$v' = \frac{Av}{A'}$$

(4-29)

If we multiply the top and bottom by a unit length (L), then we have

$$v' = \frac{ALv}{A'L}$$

(4-30)

The product AL is the total volume of the soil. The product $A'L$ is the void volume. The ratio of the void volume to the total volume is the definition of porosity (η). The actual linear velocity may then be defined in terms of the porosity as

$$v' = \frac{\text{Darcy velocity}}{\text{Porosity}} = \frac{v}{\eta}$$

(4-31)

or in terms of the hydraulic gradient, the seepage velocity is

$$v' = \frac{K(dh/dr)}{\eta}$$

(4-32)

The gross discharge is the product of the velocity of flow and the area (A) through which it flows.

$$Q = vA = KA\frac{dh}{dr}$$ (4-33)

This equation has been solved for *steady state* and nonsteady or *transient flow*. Steady state is a condition under which no changes occur with time. It will seldom, if ever, occur in practice, but may be approached after very long periods of pumping. Transient-flow equations include a factor of time. The derivation of these equations is based on the following assumptions:

1. The well is pumped at a constant rate.

2. Flow towards the well is radial and uniform.

3. Initially the piezometric surface is horizontal.

4. The well fully penetrates the aquifer and is open for the entire height of the aquifer.

5. The aquifer is homogeneous in all directions and is of infinite horizontal extent.

6. Water is released from the aquifer in immediate response to a drop in the piezometric surface.

Steady Flow in a Confined Aquifer.   The equation describing steady, confined aquifer flow was first presented by Dupuit in 1863 (Dupuit, 1863) and subsequently extended by Thiem (1906). It may be written as follows (refer to Figure 4-33 for an explanation of the notation):

$$Q = \frac{2\pi T(h_2 - h_1)}{\ln(r_2/r_1)}$$ (4-34)

where
Q = well pumping flow rate, m^3/s
$T = KD$ = transmissivity, m^2/s
D = thickness of artesian aquifer, m
h_1, h_2 = height of piezometric surface above confining layer, m
r_1, r_2 = radius from pumping well, m
$\ln$ = logarithm to base e

Example 4-14.   An artesian aquifer 10.0 m thick with a piezometric surface 40.0 m above the bottom confining layer is being pumped by a fully penetrating well. The aquifer is a medium sand with a hydraulic conductivity of 1.50×10^{-4} m/s. Steady state drawdowns of 5.00 m and 1.00 m are observed at two nonpumping wells located 20.0 m and 200.0 m, respectively, from the pumped well. Determine the discharge at the pumped well.

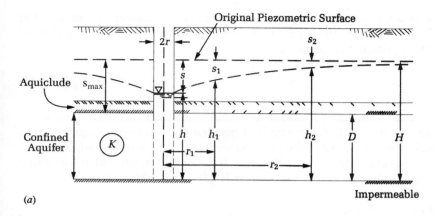

(a)

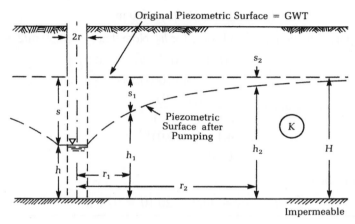

(b)

FIGURE 4-33
Geometry and symbols for a pumped well in (a) confined aquifer and (b) unconfined aquifer.
(Adapted from Bouwer, 1978.)

Solution. First we determine h_1 and h_2:

$$h_1 = 40.0 - 5.00 = 35.0 \text{ m}$$
$$h_2 = 40.0 - 1.00 = 39.0 \text{ m}$$

so

$$Q = \frac{(2\pi)(1.50 \times 10^{-4})(10.0)(39.0 - 35.0)}{\ln(200/20)}$$
$$Q = 0.0164 \text{ or } 0.016 \text{ m}^3/\text{s}$$

Comment: The maximum allowable drawdown is noted by s_{max} in Figure 4-33. Drawdown below the aquiclude may cause ground settling and structural failure of the well.

Steady Flow in an Unconfined Aquifer. For unconfined aquifers the factor D in Equation 4-34 is replaced by the height of the water table above the lower boundary of the aquifer. The equation then becomes

$$Q = \frac{\pi K(h_2^2 - h_1^2)}{\ln(r_2/r_1)} \qquad \text{[FE]} \quad (4\text{-}35)$$

Example 4-15. A 0.50 m diameter well fully penetrates an unconfined aquifer which is 30.0 m thick. The drawdown at the pumped well is 10.0 m and the hydraulic conductivity of the gravel aquifer is 6.4×10^{-3} m/s. If the flow is steady and the discharge is 0.014 m³/s, determine the drawdown at a site 100.0 m from the well.

Solution. First we calculate h_1

$$h_1 = 30.0 - 10.0 = 20.0 \text{ m}$$

Then we apply Equation 4-35 and solve for h_2. Note that $r_1 = 0.50$ m/2 $= 0.25$ m.

$$0.014 = \frac{\pi(6.4 \times 10^{-3})(h_2^2 - (20.0)^2)}{\ln(100/0.25)}$$

$$h_2^2 - 400.0 = \frac{(0.014)(5.99)}{\pi(6.4 \times 10^{-3})}$$

$$h_2 = (4.17 + 400.0)^{1/2}$$

$$h_2 = 20.10 \text{ m}$$

The drawdown is then

$$s_2 = H - h_2 = 30.0 - 20.10 = 9.90 \text{ m}$$

Unsteady Flow in a Confined Aquifer. A solution for the transient-flow problem was developed by Theis in 1935 (Theis, 1935). Using heat-flow theory as an analogy, he found the following for an infinitesimally small diameter well with radial flow:

$$s = \frac{Q}{4\pi T} \int_u^\infty \left(\frac{e^{-u}}{u} \right) du \qquad (4\text{-}36)$$

where s = drawdown $(H - h)$, m

$$u = \frac{r^2 S}{4Tt}$$

r = distance between pumping well and observation well, or radius of pumping well, m

S = storage coefficient

T = transmissivity, m²/s

t = time since pumping began, s

Some explanations of the terms may be of use here. The lower case s refers to the drawdown at some time, t, after the start of pumping. The time does not appear explicitly in Equation 4-36 but is used to compute the value of u to be used in the integration. The transmissivity and storage coefficient also are used to calculate u. The transmissivity may be determined from the hydraulic conductivity and the thickness of the aquifer as it was for steady-state flow. Field pumping tests may also be used to define T. You should note that the r term used to calculate the value of u may take on values ranging upward from the radius of the well. Thus, you could if you wished, calculate every point on the cone of depression (i.e., value of s) by iterating the calculation with values of r from the well radius to infinity. If you wish to calculate the drawdown at a specific distance from the pumping well, then, of course, you must use that distance for r. The integral in Equation 4-36 is called the well function of u and is evaluated by the following series expansion:

$$W(u) = -0.577216 - \ln u + u - \frac{u^2}{2 \cdot 2!} + \frac{u^3}{3 \cdot 3!} - \cdots \tag{4-37}$$

A table of values of $W(u)$ was prepared by Ferris, *et al.* (Ferris et al., 1962). It is reproduced in Table 4-10.

Example 4-16. If the storage coefficient is 2.74×10^{-4} and the transmissivity is 2.63×10^{-3} m^2/s, calculate the drawdown that will result at the end of 100 days of pumping a 0.61-m-diameter well at a rate of 2.21×10^{-2} m^3/s.

Solution. Begin by computing u. The radius is

$$r = \frac{0.61 \text{ m}}{2} = 0.305 \text{ m}$$

and

$$u = \frac{(0.305 \text{ m})^2 (2.74 \times 10^{-4})}{4(2.63 \times 10^{-3} \text{ m}^2/\text{s})(100 \text{ d})(86,400 \text{ s/d})} = 2.80 \times 10^{-10}$$

The factor of 86,400 is to convert days to seconds.

From the table of $W(u)$ versus u find that at 2.8×10^{-10}, $W(u) = 21.4190$. Compute s:

$$s = \frac{2.21 \times 10^{-2} \text{ m}^3/\text{s}}{4(3.14)2.63 \times 10^{-3} \text{ m}^3/\text{s}} (21.4190)$$

$$= 14.33 \text{ or } 14 \text{ m}$$

Unsteady Flow in an Unconfined Aquifer. There is no exact solution to the transient-flow problem for unconfined aquifers because T changes with time and r as the water table is lowered. Furthermore, vertical-flow components near the well invalidate the assumption of radial flow that is required to obtain an analytical solution. If the unconfined aquifer is very deep in comparison to the drawdown, the transient-flow solution for a confined aquifer may be used for an approximate solution. In general, however, numerical methods yield more satisfactory solutions.

TABLE 4-10
Values of $W(u)$

$N \backslash u$	$N \times 10^{-15}$	$N \times 10^{-14}$	$N \times 10^{-13}$	$N \times 10^{-12}$	$N \times 10^{-11}$	$N \times 10^{-10}$	$N \times 10^{-9}$	$N \times 10^{-8}$	$N \times 10^{-7}$	$N \times 10^{-6}$	$N \times 10^{-5}$	$N \times 10^{-4}$	$N \times 10^{-3}$	$N \times 10^{-2}$	$N \times 10^{-1}$	N
1.0	33.9616	31.6590	29.3564	27.0538	24.7512	22.4486	20.1460	17.8435	15.5409	13.2383	10.9357	8.6332	6.3315	4.0379	1.8229	0.2194
1.1	33.8662	31.5637	29.2611	26.9585	24.6559	22.3533	20.0507	17.7482	15.4456	13.1430	10.8404	8.5379	6.2363	3.9436	1.7371	0.1860
1.2	33.7792	31.4767	29.1741	26.8715	24.5689	22.2663	19.9637	17.6611	15.3586	13.0560	10.7534	8.4509	6.1494	3.8576	1.6595	0.1584
1.3	33.6992	31.3966	29.0940	26.7914	24.4889	22.1863	19.8837	17.5811	15.2785	12.9759	10.6734	8.3709	6.0695	3.7785	1.5889	0.1355
1.4	33.6251	31.3225	29.0199	26.7173	24.4147	22.1122	19.8096	17.5070	15.2044	12.9018	10.5993	8.2968	5.9955	3.7054	1.5241	0.1162
1.5	33.5561	31.2535	28.9509	26.6483	24.3458	22.0432	19.7406	17.4380	15.1354	12.8328	10.5303	8.2278	5.9266	3.6374	1.4645	0.1000
1.6	33.4916	31.1890	28.8864	26.5838	24.2812	21.9786	19.6760	17.3735	15.0709	12.7683	10.4657	8.1634	5.8621	3.5739	1.4092	0.08631
1.7	33.4309	31.1283	28.8258	26.5232	24.2206	21.9180	19.6154	17.3128	15.0103	12.7077	10.4051	8.1027	5.8016	3.5143	1.3578	0.07465
1.8	33.3738	31.0712	28.7686	26.4660	24.1634	21.8608	19.5583	17.2557	14.9531	12.6505	10.3479	8.0455	5.7446	3.4581	1.3089	0.06471
1.9	33.3197	31.0171	28.7145	26.4119	24.1094	21.8068	19.5042	17.2016	14.8990	12.5964	10.2939	7.9915	5.6906	3.4050	1.2649	0.05620
2.0	33.2684	30.9658	28.6632	26.3607	24.0581	21.7555	19.4529	17.1503	14.8477	12.5451	10.2426	7.9402	5.6394	3.3547	1.2227	0.04890
2.1	33.2196	30.9170	28.6145	26.3119	24.0093	21.7067	19.4041	17.1015	14.7969	12.4964	10.1938	7.8914	5.5907	3.3069	1.1829	0.04261
2.2	33.1731	30.8705	28.5679	26.2653	23.9628	21.6602	19.3576	17.0550	14.7524	12.4498	10.1473	7.8449	5.5443	3.2614	1.1454	0.03719
2.3	33.1286	30.8261	28.5235	26.2209	23.9183	21.6157	19.3131	17.0106	14.7080	12.4054	10.1028	7.8004	5.4999	3.2179	1.1099	0.03250
2.4	33.0861	30.7835	28.4809	26.1783	23.8758	21.5732	19.2706	16.9680	14.6654	12.3628	10.0603	7.7579	5.4575	3.1763	1.0762	0.02844
2.5	33.0453	30.7427	28.4401	26.1375	23.8349	21.5323	19.2298	16.9272	14.6246	12.3220	10.0194	7.7172	5.4167	3.1365	1.0443	0.02491
2.6	33.0060	30.7035	28.4009	26.0983	23.7957	21.4931	19.1905	16.8880	14.5854	12.2828	9.9802	7.6779	5.3776	3.0983	1.0139	0.02185
2.7	32.9683	30.6657	28.3631	26.0606	23.7580	21.4554	19.1528	16.8502	14.5476	12.2450	9.9425	7.6401	5.3400	3.0615	0.9849	0.01918
2.8	32.9319	30.6294	28.3268	26.0242	23.7216	21.4190	19.1164	16.8138	14.5113	12.2087	9.9061	7.6038	5.3037	3.0261	0.9573	0.01686
2.9	32.8968	30.5943	28.2917	25.9891	23.6865	21.3839	19.0813	16.7788	14.4762	12.1736	9.8710	7.5687	5.2687	2.9920	0.9309	0.01482
3.0	32.8629	30.5604	28.2578	25.9552	23.6526	21.3500	19.0474	16.7449	14.4423	12.1397	9.8371	7.5348	5.2349	2.9591	0.9057	0.01305
3.1	32.8302	30.5276	28.2250	25.9224	23.6198	21.3172	19.0146	16.7121	14.4095	12.1069	9.8043	7.5020	5.2022	2.9273	0.8815	0.01149
3.2	32.7984	30.4958	28.1932	25.8907	23.5880	21.2855	18.9829	16.6803	14.3777	12.0751	9.7726	7.4703	5.1706	2.8965	0.8583	0.01013
3.3	32.7676	30.4651	28.1625	25.8599	23.5573	21.2547	18.9521	16.6495	14.3470	12.0444	9.7418	7.4395	5.1399	2.8668	0.8361	0.008939
3.4	32.7378	30.4352	28.1326	25.8300	23.5274	21.2249	18.9223	16.6197	14.3171	12.0145	9.7120	7.4097	5.1102	2.8379	0.8147	0.007891
3.5	32.7088	30.4062	28.1036	25.8010	23.4985	21.1959	18.8933	16.5907	14.2881	11.9855	9.6830	7.3807	5.0813	2.8099	0.7942	0.006970
3.6	32.6806	30.3780	28.0755	25.7729	23.4703	21.1677	18.8651	16.5625	14.2599	11.9574	9.6548	7.3526	5.0532	2.7827	0.7745	0.006160
3.7	32.6532	30.3506	28.0481	25.7455	23.4429	21.1403	18.8377	16.5351	14.2325	11.9300	9.6274	7.3252	5.0259	2.7563	0.7554	0.005448
3.8	32.6266	30.3240	28.0214	25.7188	23.4162	21.1136	18.8110	16.5085	14.2059	11.9033	9.6007	7.2985	4.9993	2.7306	0.7371	0.004820
3.9	32.6006	30.2980	27.9954	25.6928	23.3902	21.0877	18.7851	16.4825	14.1799	11.8773	9.5748	7.2725	4.9735	2.7056	0.7194	0.004267
4.0	32.5753	30.2727	27.9701	25.6675	23.3649	21.0623	18.7598	16.4572	14.1546	11.8520	9.5495	7.2472	4.9482	2.6813	0.7024	0.003779
4.1	32.5506	30.2480	27.9454	25.6428	23.3402	21.0376	18.7351	16.4325	14.1299	11.8273	9.5248	7.2225	4.9236	2.6576	0.6859	0.003349
4.2	32.5265	30.2239	27.9213	25.6187	23.3161	21.0136	18.7110	16.4084	14.1058	11.8032	9.5007	7.1985	4.8997	2.6344	0.6700	0.002969
4.3	32.5029	30.2004	27.8978	25.5952	23.2926	20.9900	18.6874	16.3848	14.0823	11.7797	9.4771	7.1749	4.8762	2.6119	0.6546	0.002633
4.4	32.4800	30.1774	27.8748	25.5722	23.2696	20.9670	18.6644	16.3619	14.0593	11.7567	9.4541	7.1520	4.8533	2.5899	0.6397	0.002336
4.5	32.4575	30.1549	27.8523	25.5497	23.2471	20.9446	18.6420	16.3394	14.0368	11.7342	9.4317	7.1295	4.8310	2.5684	0.6253	0.002073
4.6	32.4355	30.1329	27.8303	25.5277	23.2252	20.9226	18.6200	16.3174	14.0148	11.7122	9.4097	7.1075	4.8091	2.5474	0.6114	0.001841
4.7	32.4140	30.1114	27.8088	25.5062	23.2037	20.9011	18.5985	16.2959	13.9933	11.6907	9.3882	7.0860	4.7877	2.5268	0.5979	0.001635
4.8	32.3929	30.0904	27.7878	25.4852	23.1826	20.8800	18.5774	16.2748	13.9723	11.6697	9.3671	7.0650	4.7667	2.5068	0.5848	0.001453
4.9	32.3723	30.0697	27.7672	25.4646	23.1620	20.8594	18.5568	16.2542	13.9516	11.6491	9.3465	7.0444	4.7462	2.4871	0.5721	0.001291
5.0	32.3521	30.0495	27.7470	25.4444	23.1418	20.8392	18.5366	16.2340	13.9314	11.6289	9.3263	7.0242	4.7261	2.4679	0.5598	0.001148
5.1	32.3323	30.0297	27.7271	25.4246	23.1220	20.8194	18.5168	16.2142	13.9116	11.6091	9.3065	7.0044	4.7064	2.4491	0.5478	0.001021
5.2	32.3129	30.0103	27.7077	25.4051	23.1026	20.8000	18.4974	16.1948	13.8922	11.5896	9.2871	6.9850	4.6871	2.4306	0.5362	0.0009086
5.3	32.2939	29.9913	27.6887	25.3861	23.0835	20.7809	18.4783	16.1758	13.8732	11.5706	9.2681	6.9659	4.6681	2.4126	0.5250	0.0008086
5.4	32.2752	29.9726	27.6700	25.3674	23.0648	20.7622	18.4596	16.1571	13.8545	11.5519	9.2494	6.9473	4.6495	2.3948	0.5140	0.0007198

TABLE 4-10
Values of W(u) (continued)

N	N × 10⁻¹⁵	N × 10⁻¹⁴	N × 10⁻¹³	N × 10⁻¹²	N × 10⁻¹¹	N × 10⁻¹⁰	N × 10⁻⁹	N × 10⁻⁸	N × 10⁻⁷	N × 10⁻⁶	N × 10⁻⁵	N × 10⁻⁴	N × 10⁻³	N × 10⁻²	N × 10⁻¹	N
5.5	32.2568	29.9542	27.6516	25.3491	23.0465	20.7439	18.4413	16.1387	13.8361	11.5336	9.2310	6.9289	4.6313	2.3775	0.5034	0.0006409
5.6	32.2388	29.9362	27.6336	25.3310	23.0285	20.7259	18.4233	16.1207	13.8181	11.5155	9.2130	6.9109	4.6134	2.3604	0.4930	0.0005708
5.7	32.2211	29.9185	27.6159	25.3133	23.0108	20.7082	18.4056	16.1030	13.8004	11.4978	9.1953	6.8932	4.5958	2.3437	0.4830	0.0005085
5.8	32.2037	29.9011	27.5985	25.2959	22.9934	20.6908	18.3882	16.0856	13.7830	11.4804	9.1779	6.8758	4.5785	2.3273	0.4732	0.0004532
5.9	32.1866	29.8840	27.5814	25.2789	22.9763	20.6737	18.3711	16.0685	13.7659	11.4633	9.1608	6.8588	4.5615	2.3111	0.4637	0.0004039
6.0	32.1698	29.8672	27.5646	25.2620	22.9595	20.6569	18.3543	16.0517	13.7491	11.4465	9.1440	6.8420	4.5448	2.2953	0.4544	0.0003601
6.1	32.1533	29.8507	27.5481	25.2455	22.9429	20.6403	18.3378	16.0352	13.7326	11.4300	9.1275	6.8254	4.5283	2.2797	0.4454	0.0003211
6.2	32.1370	29.8344	27.5318	25.2293	22.9267	20.6241	18.3215	16.0189	13.7163	11.4138	9.1112	6.8092	4.5122	2.2645	0.4366	0.0002864
6.3	32.1210	29.8184	27.5158	25.2133	22.9107	20.6081	18.3055	16.0029	13.7003	11.3978	9.0952	6.7932	4.4963	2.2494	0.4280	0.0002555
6.4	32.1053	29.8027	27.5001	25.1975	22.8949	20.5923	18.2898	15.9872	13.6846	11.3820	9.0795	6.7775	4.4806	2.2346	0.4197	0.0002279
6.5	32.0898	29.7872	27.4846	25.1820	22.8794	20.5768	18.2742	15.9717	13.6691	11.3665	9.0640	6.7620	4.4652	2.2201	0.4115	0.0002034
6.6	32.0745	29.7719	27.4693	25.1667	22.8641	20.5616	18.2590	15.9564	13.6538	11.3512	9.0487	6.7467	4.4501	2.2058	0.4036	0.0001816
6.7	32.0595	29.7569	27.4543	25.1517	22.8491	20.5465	18.2439	15.9414	13.6388	11.3362	9.0337	6.7317	4.4351	2.1917	0.3959	0.0001621
6.8	32.0446	29.7421	27.4395	25.1369	22.8343	20.5317	18.2291	15.9265	13.6240	11.3214	9.0189	6.7169	4.4204	2.1779	0.3883	0.0001448
6.9	32.0300	29.7275	27.4249	25.1223	22.8197	20.5171	18.2145	15.9119	13.6094	11.3068	9.0043	6.7023	4.4059	2.1643	0.3810	0.0001293
7.0	32.0156	29.7131	27.4105	25.1079	22.8053	20.5027	18.2001	15.8976	13.5950	11.2924	8.9899	6.6879	4.3916	2.1508	0.3738	0.0001155
7.1	32.0015	29.6989	27.3963	25.0937	22.7911	20.4885	18.1860	15.8834	13.5808	11.2782	8.9757	6.6737	4.3775	2.1376	0.3668	0.0001032
7.2	31.9875	29.6849	27.3823	25.0797	22.7771	20.4746	18.1720	15.8694	13.5668	11.2642	8.9617	6.6598	4.3636	2.1246	0.3599	0.00009219
7.3	31.9737	29.6711	27.3685	25.0659	22.7633	20.4608	18.1582	15.8556	13.5530	11.2504	8.9479	6.6460	4.3500	2.1118	0.3532	0.00008239
7.4	31.9601	29.6575	27.3549	25.0523	22.7497	20.4472	18.1446	15.8420	13.5394	11.2368	8.9343	6.6324	4.3364	2.0991	0.3467	0.00007364
7.5	31.9467	29.6441	27.3415	25.0389	22.7363	20.4337	18.1311	15.8286	13.5260	11.2234	8.9209	6.6190	4.3231	2.0867	0.3403	0.00006583
7.6	31.9334	29.6308	27.3282	25.0257	22.7231	20.4205	18.1179	15.8153	13.5127	11.2102	8.9076	6.6057	4.3100	2.0744	0.3341	0.00005886
7.7	31.9203	29.6178	27.3152	25.0126	22.7100	20.4074	18.1048	15.8022	13.4997	11.1971	8.8946	6.5927	4.2970	2.0623	0.3280	0.00005263
7.8	31.9074	29.6048	27.3023	24.9997	22.6971	20.3945	18.0919	15.7893	13.4868	11.1842	8.8817	6.5798	4.2842	2.0503	0.3221	0.00004707
7.9	31.8947	29.5921	27.2895	24.9869	22.6844	20.3818	18.0792	15.7766	13.4740	11.1714	8.8689	6.5671	4.2716	2.0386	0.3163	0.00004210
8.0	31.8821	29.5795	27.2769	24.9744	22.6718	20.3692	18.0666	15.7640	13.4614	11.1589	8.8563	6.5545	4.2591	2.0269	0.3106	0.00003767
8.1	31.8697	29.5671	27.2645	24.9619	22.6594	20.3568	18.0542	15.7516	13.4490	11.1464	8.8439	6.5421	4.2468	2.0155	0.3050	0.00003370
8.2	31.8574	29.5548	27.2523	24.9497	22.6471	20.3445	18.0419	15.7393	13.4367	11.1342	8.8317	6.5298	4.2346	2.0042	0.2996	0.00003015
8.3	31.8453	29.5427	27.2401	24.9375	22.6350	20.3324	18.0298	15.7272	13.4246	11.1220	8.8195	6.5177	4.2226	1.9930	0.2943	0.00002699
8.4	31.8333	29.5307	27.2282	24.9256	22.6230	20.3204	18.0178	15.7152	13.4126	11.1101	8.8076	6.5057	4.2107	1.9820	0.2891	0.00002415
8.5	31.8215	29.5189	27.2163	24.9137	22.6112	20.3086	18.0060	15.7034	13.4008	11.0982	8.7957	6.4939	4.1990	1.9711	0.2840	0.00002162
8.6	31.8098	29.5072	27.2046	24.9020	22.5995	20.2969	17.9943	15.6917	13.3891	11.0865	8.7840	6.4822	4.1874	1.9604	0.2790	0.00001936
8.7	31.7982	29.4957	27.1931	24.8905	22.5879	20.2853	17.9827	15.6801	13.3776	11.0750	8.7725	6.4707	4.1759	1.9498	0.2742	0.00001733
8.8	31.7868	29.4842	27.1816	24.8790	22.5765	20.2739	17.9713	15.6687	13.3661	11.0635	8.7610	6.4592	4.1646	1.9393	0.2694	0.00001552
8.9	31.7755	29.4729	27.1703	24.8678	22.5652	20.2626	17.9600	15.6574	13.3548	11.0523	8.7497	6.4480	4.1534	1.9290	0.2647	0.00001390
9.0	31.7643	29.4618	27.1592	24.8566	22.5540	20.2514	17.9488	15.6462	13.3437	11.0411	8.7386	6.4368	4.1423	1.9187	0.2602	0.00001245
9.1	31.7533	29.4507	27.1481	24.8455	22.5429	20.2404	17.9378	15.6352	13.3326	11.0300	8.7275	6.4258	4.1313	1.9087	0.2557	0.00001115
9.2	31.7424	29.4398	27.1372	24.8346	22.5320	20.2294	17.9268	15.6243	13.3217	11.0191	8.7166	6.4148	4.1205	1.8987	0.2513	0.000009988
9.3	31.7315	29.4290	27.1264	24.8238	22.5212	20.2186	17.9160	15.6135	13.3109	11.0083	8.7058	6.4040	4.1098	1.8888	0.2470	0.000008948
9.4	31.7208	29.4183	27.1157	24.8131	22.5105	20.2079	17.9053	15.6028	13.3002	10.9976	8.6951	6.3934	4.0992	1.8791	0.2429	0.000008018
9.5	31.7103	29.4077	27.1051	24.8025	22.4999	20.1973	17.8948	15.5922	13.2896	10.9870	8.6845	6.3828	4.0887	1.8695	0.2387	0.000007185
9.6	31.6998	29.3972	27.0946	24.7920	22.4895	20.1869	17.8843	15.5817	13.2791	10.9765	8.6740	6.3723	4.0784	1.8599	0.2347	0.000006439
9.7	31.6894	29.3868	27.0843	24.7817	22.4791	20.1765	17.8739	15.5713	13.2688	10.9662	8.6637	6.3620	4.0681	1.8505	0.2308	0.000005771
9.8	31.6792	29.3766	27.0740	24.7714	22.4688	20.1663	17.8637	15.5611	13.2585	10.9559	8.6534	6.3517	4.0579	1.8412	0.2269	0.000005173
9.9	31.6690	29.3664	27.0639	24.7613	22.4587	20.1561	17.8535	15.5509	13.2483	10.9458	8.6433	6.3416	4.0479	1.8320	0.2231	0.000004637

Source: Ferris et al., 1962.

Determining the Hydraulic Properties of a Confined Aquifer. The estimation of the transmissivity and storage coefficient of an aquifer is based on the results of a pumping test. The preferred situation is one in which one or more observation wells located at a distance from the pumping well are used to gather the data.

The transmissivity may be determined in the steady-state condition by using Equation 4-34. If we define drawdown as $s = H - h$, then the rearrangement of Equation 4-34 yields

$$T = \frac{Q\ln(r_2/r_1)}{2\pi(s_1 - s_2)} \tag{4-38}$$

where s_1 = drawdown at radius r_1, m
s_2 = drawdown at radius r_2, m

In the transient state we cannot solve for T and S directly. We have selected the Cooper and Jacob method from the several indirect methods that are available (Cooper and Jacob, 1946). For values of u less than 0.01, they found that Equation 4-36 could be rewritten as follows:

$$s = \frac{Q}{4\pi T} \ln \frac{2.25Tt}{r^2 S} \tag{4-39}$$

A semilogarithmic plot of s versus t (log scale) from the results of a pumping test (Figure 4-34) enables a direct calculation of T from the slope of the line. From Equation 4-39, the difference in drawdown at two points in time may be shown to be

$$s_2 - s_1 = \frac{Q}{4\pi T} \ln\frac{t_2}{t_1} \tag{4-40}$$

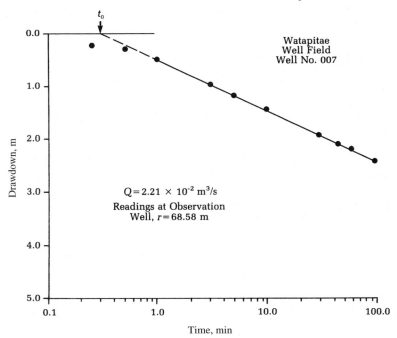

Watapitae
Well Field
Well No. 007

$Q = 2.21 \times 10^{-2}$ m^3/s

Readings at Observation
Well, $r = 68.58$ m

FIGURE 4-34
Pumping test results.

Solving for T we find

$$T = \frac{Q}{4\pi(s_2 - s_1)}\ln\frac{t_2}{t_1} \tag{4-41}$$

Cooper and Jacob showed that an extrapolation of the straight-line portion of the plot to the point where $s = 0$ yields a "virtual" (imaginary) starting time (t_0). At this virtual time, Equation 4-39 may be solved for the storage coefficient, S, as follows:

$$S = \frac{2.25Tt_0}{r^2} \tag{4-42}$$

The calculus implies that the distance to the observation well (r) may be as little as the radius of the pumping well itself. This means that drawdown measured in the pumping well may be used as a source of data.

Example 4-17.   Determine the transmissivity and storage coefficient for the Watapitae Wells based on the pumping test data plotted in Figure 4-34.

Solution.   Using Figure 4-33, we find $s_1 = 0.49$ m at $t_1 = 1.0$ min and at $t_2 = 100$ min we find $s_2 = 2.37$ m. Thus

$$T = \frac{2.21 \times 10^{-2}}{4(\pi)(2.37 - 0.49)}\ln\frac{100.0}{1.0}$$
$$= (9.35 \times 10^{-4})(4.61) = 4.31 \times 10^{-3}\ \text{m}^2/s$$

From Figure 4-34 we find that the extrapolation of the straight portion of the graph yields $t_0 = 0.30$ min. Using the distance between the pumping well and the observation ($r = 68.58$ m) we find

$$S = \frac{(2.25)(4.31 \times 10^{-3})(0.30)(60)}{(68.58)^2}$$
$$= 3.7 \times 10^{-5}$$

The factor of 60 is to convert minutes into seconds. Now we should check to see if our implicit assumption that u is less than 0.01 was true. We use $t = 100.0$ min for the check.

$$u = \frac{(68.58)^2(3.7 \times 10^{-5})}{4(4.31 \times 10^{-3})(10.0)(60)} = 0.0017$$

Comment: This meets the criterion that u be less than 0.01 for Equations 4-41 and 4-42 to be applicable.

Determining the Hydraulic Properties of an Unconfined Aquifer.   Transmissivity may be determined under steady-state conditions in a fashion similar to that for confined aquifers using Equation 4-38.

Providing that the basic assumptions can be met, the transient-state equations for a confined aquifer may be applied to an unconfined aquifer. Because the unconfined layer often does not release water immediately after a drop in the piezometric surface, the transient equations often are not valid. Thus, great care should be taken in using them.

Pumping Test Results from Nonhomogeneous Aquifers.   Of course, our assumption that an aquifer is homogeneous seldom applies in real life. Under a few circumstances the inhomogeneities even out to yield average aquifer parameters. Under many other circumstances more complex techniques are required.

Two special cases of inhomogeneity are of particular interest. The first case is where the cone of depression intercepts a barrier. This is shown in Figure 4-35. The net result is that a greater pumping lift is required to achieve the same well yield as would have been obtained if the barrier had not existed.

The second case is where the cone of depression intersects a recharge area (Figure 4-36) such as a lake, stream, or underground stream. The net result is to reduce the lift required to maintain the desired yield of the well. Of course, it is possible to dry up the lake or stream if the pumping rate is large enough. This is shown by the increase in slope after the plateau.

Calculating Interference.   As we mentioned earlier, the cones of depression of wells located close together will overlap; this interference will reduce the potential yield of both wells. In severe circumstances, well interference can cause drawdowns that leave shallow wells dry.

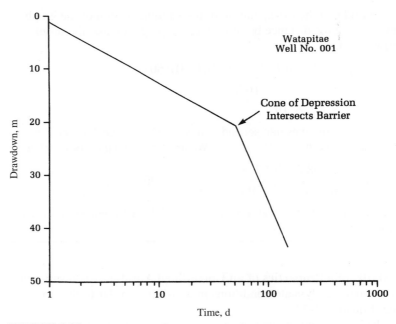

FIGURE 4-35
Pumping test curve showing effect of barrier at day 50.

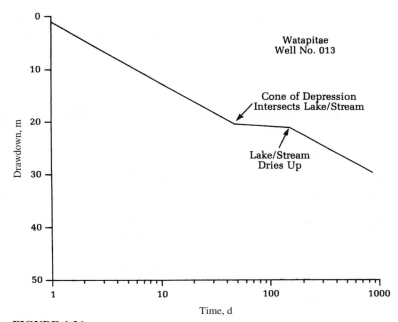

FIGURE 4-36
Pumping test curve showing effect of recharge from day 50 to day 150.

A solution to the well interference problem can be achieved by the method of superposition. This method assumes that the drawdown at a particular location is equal to the sum of the drawdowns from all of the influencing wells. Mathematically this can be represented as follows:

$$s_r = \sum_{i=1}^{n} s_i \tag{4-43}$$

where s_i = individual drawdown caused by well i at location r.

Example 4-18. Three wells are located at 75-m intervals along a straight line. Each well is 0.50 m in diameter. The coefficient of transmissivity is 2.63×10^{-3} m^2/s and the storage coefficient is 2.74×10^{-4}. Determine the drawdown at each well if each well is pumped at 4.42×10^{-2} m^3/s for 10 days.

Solution. The drawdown at each well will be the sum of the drawdown of each well pumping by itself plus the interference from each of the other two wells. Because each well is the same diameter and pumps at the same rate, we may compute one value of the term $Q/(4\pi T)$ and apply it to each well.

$$\frac{Q}{4\pi T} = \frac{4.42 \times 10^{-2}}{4(3.14)(2.63 \times 10^{-3})} = 1.34$$

In addition, because each well is identical, the individual drawdowns of the wells pumping by themselves will be equal. Thus, we may compute one value of u and apply it to each well.

$$u = \frac{(0.25)^2(2.74 \times 10^{-4})}{4(2.63 \times 10^{-3})(10)(86,400)} = 1.88 \times 10^{-9}$$

Using $u = 1.9 \times 10^{-9}$ and referring to Table 4-10 we find $W(u) = 19.5042$. The drawdown of each individual well is then

$$s = (1.34)(19.5042) = 26.14 \text{ m}$$

Before we begin calculating interference, we should label the wells so that we can keep track of them. Let us call the two outside wells A and C and the inside well B. Let us now calculate interference of well A on well B, that is, the increase in drawdown at well B as a result of pumping well A. Because we have pumped only for 10 days, we must use the transient-flow equations and calculate u at 75 m.

$$u_{75} = \frac{(75)^2(2.74 \times 10^{-4})}{4(2.63 \times 10^{-3})(10)(86,400)} = 1.70 \times 10^{-4}$$

From Table 4-10, $W(u) = 8.1027$. The interference of well A on B is then

$$s_{\text{A on B}} = (1.34)(8.1027) = 10.86 \text{ m}$$

In a similar fashion we calculate the interference of well A on well C.

$$u_{150} = (150)^2(3.0145 \times 10^{-8}) = 6.78 \times 10^{-4}$$

and $W(u) = 6.7169$

$$s_{\text{A on C}} = (1.34)(6.7169) = 9.00 \text{ m}$$

Since our well arrangement is symmetrical, the following equalities may be used:

$$s_{\text{A on B}} = s_{\text{B on A}} = s_{\text{B on C}} = s_{\text{C on B}}$$

and

$$s_{\text{A on C}} = s_{\text{C on A}}$$

The total drawdown at each well is computed as follows:

$s_{\text{A}} = s + s_{\text{B on A}} + s_{\text{C on A}}$

$s_{\text{A}} = 26.14 + 10.86 + 9.00 = 46.00 \text{ m}$

$s_{\text{B}} = s + s_{\text{A on B}} + s_{\text{C on B}}$

$s_{\text{B}} = 26.14 + 10.86 + 10.86 = 47.86 \text{ m}$

$s_{\text{C}} = s_{\text{A}} = 46.00 \text{ m}$

Drawdowns are measured from the undisturbed piezometric surface. Note that if the wells are pumped at different rates, the symmetry would be destroyed and the value $Q/(4\pi T)$ would have to be calculated separately for each case. Likewise, if the distances were not symmetric, then separate u values would be required. The results of these calculations have been plotted in Figure 4-37.

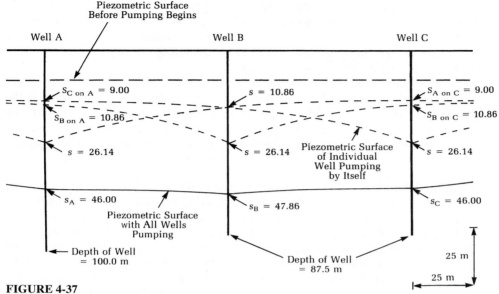

FIGURE 4-37
Interference drawdown of three wells.

Safe Yield for Groundwater

Unlike surface water supplies, groundwater is less subject to seasonal fluctuations and long-term droughts. The design basis is the long term or "safe" yield. The safe yield of a ground water basin is the amount of water which can be withdrawn from it annually without producing an undesired result. A yield analysis of the aquifer is performed because of the potential for over pumping the well with consequent failure to yield an adequate supply as well as the potential to cause dramatic ground surface settlement, detrimental dewatering of nearby ponds or streams or, in wells near the ocean, to cause salt water intrusion.

Confined Aquifer.   The components of the evaluation of the aquifer as a water supply are: (1) depth to the bottom of the aquiclude; (2) elevation of the existing piezometric surface; (3) drawdown for sustained pumping at the design rate of demand; and (4) recharge and drought implications.

The depth to the bottom of the aquiclude sets the limit of drawdown of the piezometric surface. If the piezometric surface drops below the bottom of the aquiclude, ground settlement will begin to occur and, in addition to structural failure of the well, structural damage will occur to buildings and roadways. In populated areas of the United States, regulatory agencies gather hydrogeologic data reported by well drillers and others that may be used to estimate the depth to the aquiclude. In less densely populated areas, exploration and evaluation by a professional hydrogeologist is required.

The existing piezometric surface sets the upper bound of the range of drawdown. That is, the difference between the existing piezometric surface and the bottom of the aquiclude (s_{max} in Figure 4-33a) is the maximum allowable drawdown for a safe yield. As noted above, in populated areas, regulatory agencies will have a database that includes this information. Otherwise, a hydrogeologic exploration will be required.

The maximum sustainable pumping rate is found by setting h_1 equal to the height of the aquifer (D in Figure 4-33a) and h_2 equal to the height of the piezometric surface before pumping (H in Figure 4-33a). If the required Q cannot be achieved using one well for the design flow, multiple wells may be required. Except for very small demands, this is the rule rather than the exception.

Multiple wells may be used to take advantage of the fact that wells will "recover" their original piezometric surface when pumping ends if there is adequate water in the aquifer. Thus, if the cones of depression of multiple wells do not interfere with one another, the wells can be operated on a schedule that allows them to recover. Theoretically, if the nonpumping time equals the pumping time, the recovery will be complete (Brown, 1963).

Unconfined Aquifer.   The components of the evaluation of the aquifer as a water supply are: (1) depth of the aquifer; (2) annual precipitation and resultant aquifer recharge; and (3) drawdown for sustained pumping at the design rate of demand.

The depth of the aquifer for an unconfined aquifer is measured from the static, unpumped, water level to the underlying impermeable layer (Figure 4-33b). In theory, the depth of the aquifer sets one dimension of the maximum extent of pumping. When the water level is lowered to the impermeable layer, the well "drys up." In actuality, this depth cannot be achieved because of other constraints. In populated areas, regulatory agencies have data that permit estimation of the depth of the aquifer. In less densely populated areas, exploration and evaluation by a professional hydrogeologist is required.

Aquifer Recharge.   A hydrologic mass balance is used to estimate the potential volume of water that recharges the aquifer. An annual time increment rather than the shorter monthly periods used in surface water analysis may be used for estimation purposes because the aquifer behaves as a large storage reservoir. Under steady-state conditions, the storage volume compensates for dry seasons with wet seasons. Thus, like the analysis of reservoirs, a partial duration series analysis for drought durations of 1 to 5 years with return periods of 25 to 50 years is used in evaluation of an unconfined aquifer as a water source.

Even though vast quantities of water may have accumulated in the aquifer over geologic time periods, the rate of pumping may exceed the rate of replenishment. Even with very deep aquifers where the well does not dry up, the removal of water results in removal of subsurface support. This, in turn, results in loss of surface elevation or land subsidence. Although this occurs in nearly every state in the United States, the San Joaquin Valley in California serves as a classic example. Figure 4-38 is a dramatic photograph showing the land surface as it was in 1977 in relation to its location in 1925. The distance between the 1925 sign and the 1977 sign is approximately 9 m.

Groundwater Contamination

The well hydraulics equations may be used to design purge or extraction wells to remove contaminated groundwater. The lowering of the piezometric surface will cause contaminants to flow to the well where they may be removed to the surface for treatment. In some cases, a well field may be required to intercept the contaminant plume. In these cases, the well interference effects may be put to good use in designing the distribution of the wells. The use of extraction wells is discussed in Chapter 12.

FIGURE 4-38
Land subsidence in the San
Joaquin Valley, 16 km south-
west of Mendota, CA. (*Source:*
U.S. Geological Survey.)

General Comments on Modeling Equations

We wish to emphasize that equations developed to model physical, chemical or biological behavior are simplifications of real-world behavior of the phenomenon being
modeled. When the authors of the equations present them in professional journals they
are careful to state the assumptions used in their development. Often the assumptions
cannot be completely met in actual situations. The models must be recognized as approximations to real-world behavior.

In some applications of the models unrealistic or even impossible results may be
obtained. For example, the well drawdown in Equation 4-34 or any of the other well
equations may be calculated to be below the bottom of the well! An understanding of
the physical situation must be employed to evaluate the results of the calculations. This
is called *engineering judgment*. Prudence in the application of models is always warranted. This is true not only for the well equations but for all engineering models.

4-8 CHAPTER REVIEW

When you have completed studying this chapter you should be able to do the following without the aid of your textbook or notes:

1. Sketch and explain the hydrologic cycle, labeling the parts as in Figure 4-1.

2. List and explain the four factors that reduce the amount of direct runoff.

3. Explain the difference between streamflow that results from direct runoff and that which results from baseflow.

4. Define evaporation, transpiration, runoff, baseflow, watershed, basin, and divide.

5. Sketch the groundwater hydrologic system, labeling the parts as in Figure 4-3.

6. Explain why water in an artesian aquifer is under pressure and why water may rise above the surface in some instances and not others.

7. Explain why infiltration rates decrease with time.

8. Explain return period or recurrence interval in terms of probability that an event will take place.

9. Define the rational formula and identify the units as used in this text.

10. Explain why the rational method may yield inadequate results when applied by inexperienced engineers, while experienced engineers may get adequate results.

11. Explain what a unit hydrograph is.

12. Explain the purpose of the unit hydrograph and explain how it might be applied in the analysis of a storm sewer design or a stream flood-control project.

13. Using a sketch, show how the groundwater contribution to a hydrograph is identified.

14. Define time of concentration and explain how it is used in conjunction with the rational method.

15. Sketch a subsurface cross-section from the results of a well boring log and identify pertinent hydrogeologic features.

16. Sketch a well and label the major sanitary protection features according to this text.

17. Sketch a piezometric profile for a single well pumping at a high rate, and sketch a profile for the same well pumping at a low rate.

18. Sketch a piezometric profile for two or more wells located close enough together to interfere with one another.

19. Sketch a well-pumping test curve which shows (*a*) the interception of a barrier, and (*b*) the interception of a recharge area.

20. Give two examples of methods to minimize water use.

With the aid of this text, you should be able to do the following:

21. Compute mass balances for open and closed hydrologic systems.

22. Compute infiltration rates by Horton's method and estimate the volume of infiltration.

23. Estimate the volume of water loss through transpiration given the air and water temperature, wind speed, and relative humidity.

24. Use Horton's equation and/or some form of Dalton's equation to solve complex mass balance problems.

25. Construct an intensity-duration-frequency curve from a compilation of intense rainfall occurrences.

26. Construct a unit hydrograph for a given stream-gaging station if you are provided with a rainfall and total flow at the gaging station.

27. Apply a given unit hydrograph to construct a compound hydrograph if you are provided with an observed rainfall and an estimate of the rainfall losses due to infiltration and evaporation.

28. Determine the peak flow (Q in the rational formula) and time of arrival (t_c) resulting from a rainfall of specified intensity and duration in a well-defined watershed.

29. Using the mass balance method, determine the volume of a reservoir or retention basin for a given demand or flood control given appropriate discharge data.

30. Calculate the drawdown at a pumped well or observation well if you are given the proper input data.

31. Calculate the transmissivity and storage coefficient for an aquifer if you are provided with the results of a pumping test.

32. Calculate the interference effects of two or more wells.

4-9 PROBLEMS

4-1. Lake Pleasant, AZ, has a surface area of approximately 2,000 ha. The inflow for the month of August is zero. The lake is dammed so that no water leaves the impoundment downstream of the dam. There is no rain in August. The evaporation is estimated to be 6.8 mm/d and the seepage is estimated to be 0.01 mm/d. Assuming that the lake shore is vertical, estimate the distance the lake elevation drops during the month of August. What distance will the shoreline recede if the lake shore slope is 5°?

> *Answers:* Vertical drop = 211.11 or 210 mm or 21 cm
>
> Contraction = 2,422 or 2,400 mm or 240 cm or 2.4 m

4-2. Lake Kickapoo, TX, is approximately 12 km in length by 2.5 km in width. The inflow for the month of March is 3.26 m^3/s and the outflow is 2.93 m^3/s. The total monthly precipitation is 15.2 cm and the evaporation is 10.2 cm. The seepage is estimated to be 2.5 cm. Estimate the change in storage (in m^3) during the month of March.

4-3. A 4,000-km^2 watershed receives 102 cm of precipitation in one year. The average flow of the river draining the watershed is 34.2 m^3/s. Infiltration is estimated to be 5.5×10^{-7} cm/s and evapotranspiration is estimated to be 40 cm/y. Determine the change in storage in the watershed over one year. The ratio of runoff to precipitation (both in cm) is termed the runoff coefficient. Compute the runoff coefficient for this watershed.

4-4. Using the values of f_o, f_c, and k for a Fuquay pebbly loamy sand (see Section 4-2), find the infiltration rate at times of 12, 30, 60, and 120 minutes. Compute the total volume of infiltration over 120 minutes.

4-5. Infiltration data from an experiment yield an initial infiltration rate of 4.70 cm/h and a final equilibrium infiltration rate of 0.70 cm/h after 60 minutes of steady precipitation. The value of k was estimated to be 0.1085 h^{-1}. Determine the total volume of infiltration for the following storm sequence: 30 mm/h for 30 minutes, 53 mm/h for 30 minutes, 23 mm/h for 30 minutes.

4-6. Using the empirical equation developed for Lake Hefner, estimate the evaporation from a lake on a day that the air temperature is 30°C, the water temperature is 15°C, the wind speed is 9 m/s, and the relative humidity is 30 percent.

Answer: 4.73 or 4.7 mm/d

4-7. Using the empirical equation developed for Lake Hefner, estimate the evaporation from a lake on a day that the air temperature is 40°C, the water temperature is 25°C, the wind speed is 2.0 m/s, and the relative humidity is 5 percent.

4-8. The Dalton-type evaporation equation implies that there is a limiting relative humidity above which evaporation will be nil regardless of the wind speed. Using the Lake Hefner empirical equation, estimate the relative humidity at which evaporation will be nil if the water temperature is 10°C and the air temperature is 25°C.

4-9. Prepare an IDF curve for a 2-year storm at Dismal Swamp using the data in Table 4-3. *Hint:* Curve should intersect 98.7 mm/h at 15 minutes duration.

4-10. Prepare an IDF curve for a 10-year storm at Dismal Swamp using the data in Table 4-3.

4-11. Prepare an intensity-duration-frequency curve for a 5-year storm with the data shown in the following table.

Annual maximum intensity (mm/h)

Year	30-minute duration	60-minute duration	90-minute duration	120-minute duration
1960	122.3	100.3	81.1	55.3
1961	104.6	82.9	64.5	39.7
1962	81.0	60.8	41.5	16.5
1963	145.1	123.7	104.7	81.7
1964	83.0	61.5	40.7	16.0
1965	70.1	51.1	30.1	11.3
1966	94.7	71.0	51.7	26.7
1967	63.7	41.5	21.8	10.1
1968	57.9	35.7	17.1	9.7
1969	71.5	50.0	31.3	15.9

Hint: Curve should intersect 96.8 mm/h at 60 minutes duration.

4-12. Prepare an intensity-duration-frequency curve for a 2-year storm with the data from Problem 4-11.

4-13. A large shopping mall parking lot may be arranged in either of the two configurations shown in Figure P-4-13. The drain commissioner prefers the longest time of concentration to allow for the passage of other peak flows in the drain. Using the highest runoff coefficient for asphaltic pavement, estimate the time of concentration for each configuration and recommend the one to submit to the drain commissioner. Note: area of (a) is same as area of (b).

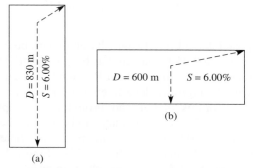

FIGURE P-4-13
Alternative parking lot configurations.

4-14. Mechanicsville has obtained a grant under the Federal Program for Urban Development of Greenspace and the Environment to build a condominium complex for the retired. The total land set aside is 74,010 m². Of this area 15,831 m² will be used for Swiss chalet condominiums with slate roofs. Public streets and drives will occupy 18,886 m². The remainder will be lawns. The area is flat and has a sandy soil. Using the most conservative estimates of C, find the peak discharge from the development for a 2-year storm. Assume an overland flow distance of 272 m and use the IDF curves given in Figure P-4-14.

Answer: $Q = 0.6097$ or 0.61 m³/s

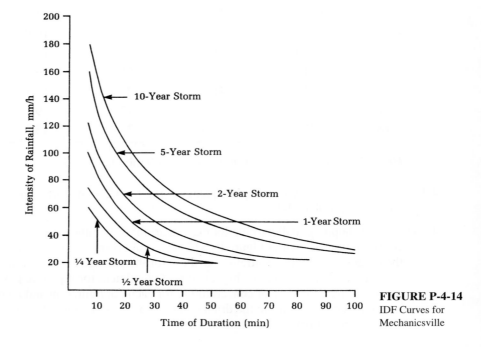

FIGURE P-4-14
IDF Curves for Mechanicsville

4-15. Repeat Problem 4-14 using the IDF curves for Miami, FL, shown in Figure 4-9c.

4-16. Paula Revere has proposed that a Little League baseball park be built near Boxborough, MA. A site on a 9.94-ha pasture has been selected. Drainage from the pasture passes through a culvert under Route 62. The capacity of the culvert is not known. The highway department uses a 5-year return period for design of culverts for this type of road. From the topographic data of the site, the overland flow distance is estimated to be 450 m, the slope of the pasture is estimated to be 2.00 percent, and the runoff coefficient is estimated to be 0.20. Using the IDF curves for Boston, MA (Figure 4-9a), estimate the peak flow capacity of the existing culvert in m³/s.

Answer: $Q = 0.2115$ or 0.21 m³/s

4-17. The proposed parking lot for Paula's baseball park (Problem 4-16) will occupy 2.64 ha of the site. It will be graded to a slope of 1.80 percent. The overland flow distance will be 200.0 m and the runoff coefficient is estimated to be 0.70. Ignoring the runoff from the pasture, determine whether the existing culvert has the capacity to carry the peak flow from the parking lot from a storm with a 5-year return period. Use the IDF curves for Boston, MA (Figure 4-9*a*).

4-18. A storm sewer is to be designed for a parking lot in Holland, MI, that has an area of 4.8 ha. The parking lot slope will be 1.00 percent and the estimated runoff coefficient is 0.85. The overland flow distance is 219.0 m. Using the rational method, estimate the peak discharge from the parking lot for a 5-year storm. The IDF curve for a 5-year storm at Holland, MI, can be described by the following equation:

$$i = \frac{1,193.80}{(t_c)^{0.8} + 7}$$

where i = intensity, mm/h, and t_c = time of concentration, min.

4-19. John Snow is planning to build a shopping mall and parking lot on the north side of Hindry Road as shown in Figure P-4-19. The existing culvert near the southwest corner of the proposed parking lot was designed for a 5-year storm. Determine the frequency that the capacity of the culvert will be exceeded if it is not enlarged when the mall is constructed. Also

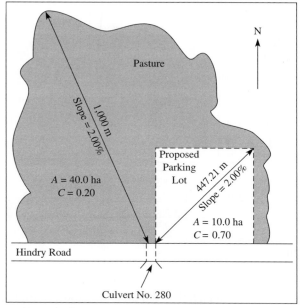

FIGURE P-4-19
Sketch of proposed shopping mall site.

determine the peak discharge that must he handled if the design criteria are changed to specify a 10-year storm. Use the IDF curves in Figure P-4-14.

Answers: $Q_{\text{design}} = 0.74$ m³/s; I that will cause flood $= 37.7$ mm/h; frequency of flooding ≈ 4 times/y; $Q_{10y} = 2.0$ m³/s

4-20. Dr. William Gorgas is planning to build a clinic on the east side of Okemos Road as shown in Figure P-4-20. The existing culvert (no. 481) was designed for a 5-year storm. Determine if the capacity of the culvert will be exceeded if it is not enlarged when the clinic is built. Also determine the peak discharge that must be handled if the design criteria are changed to specify a 10-year storm. Use the IDF curves in Figure P-4-14.

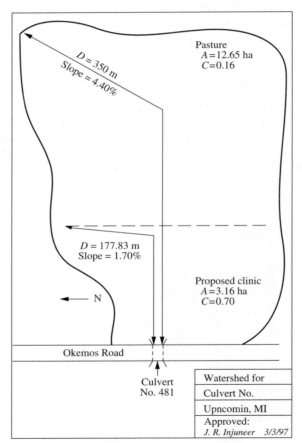

FIGURE P-4-20
Sketch of proposed clinic site.

4-21. Two adjacent parcels of land adjoin each other as shown in Figure P-4-21. Assume that the speed of the water moving to the east along the drainage ditch from point P is 0.60 m/s. Determine the peak runoff from a 5-year storm that must be carried by the drainage ditch using the IDF curves for Seattle, WA (Figure 4-9*d*). The repetitive calculations required in this problem make it a good candidate for a spreadsheet solution.

Answer: $Q = 0.11$ m^3/s

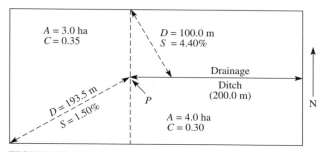

FIGURE P-4-21
Drainage from two adjacent parcels.

4-22. A parking lot is configured in three sections as shown in Figure P-4-22. There are storm water inlet manholes for each section. Storm water flows in the storm sewer from east to west. The storm water moves at a speed of 0.90 m/s once it reaches the storm sewer. Determine the peak runoff from a 5-year storm that must be carried by the storm sewer carrying storm water away from manhole no. 3. Use the IDF curves for Miami, FL (Figure 4-9*c*). The repetitive calculations required in this problem make it a good candidate for a spreadsheet solution.

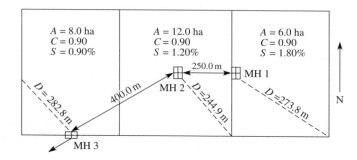

FIGURE P-4-22
Storm sewer capacity for three part parking lot.

4-23. Determine the unit hydrograph ordinates for the Isosceles River with the stream flow data shown in Figure P-4-23 on page 192. The basin area is 14.40 km^2, and the unit duration of the storm is 1 hour. For ease of computation, locate ordinates at hourly intervals, that is, 1500, 1600, and 1700 hours. Note that the time is military time; that is, 1500 h = 3 PM, 2000 h = 8 PM, etc.

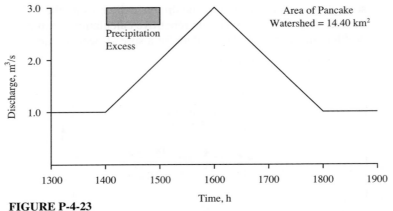

FIGURE P-4-23
Storm hydrograph for Isosceles River.

4-24. Determine the unit hydrograph ordinates for the Convex River flow data shown below. For ease of computation the total ordinates are tabulated below the hydrograph. Note that the base flow may be determined by simple graphical extrapolation. The area of the watershed is 100 ha. Note that the time is military time, that is, 1500 h = 3 PM, 2000 h = 8 PM, etc.

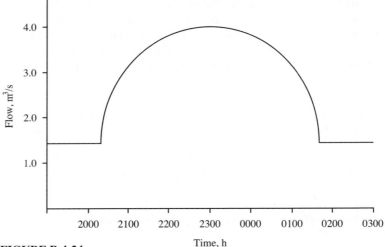

FIGURE P-4-24
Storm hydrograph for the Convex River.

Time (h)	Total stream flow (m³/s)
2100	3.0
2200	3.8
2300	4.0
0000	3.8
0100	3.0

4-25. Determine the unit hydrograph ordinates for the Verde River with the stream flow data shown in the table below that resulted from a 5-hour storm of uniform intensity. The basin area is 64.0 km^2.

Time (h)	Flow (m^3/s)	Time (h)	Flow (m^3/s)	Time (h)	Flow (m^3/s)
0	0.55	35	5.77	65	1.64
5	0.50	40	5.02	70	1.10
10	0.45	45	4.29	75	0.79
15	1.98	50	3.50	80	0.47
20	4.82	55	2.72	85	0.25
25	6.24	60	2.19	90	0.25
30	6.86				

Answer: The plotting points of the ordinates are at midpoints of the time intervals, i.e., 2.5 h, 7.5 h, 15.0 h, 25 h, 35 h, 45 h, 55 h, 65 h. The UH ordinates in m^3/s · cm are 0.51, 2.544, 4.95, 4.55, 3.29, 1.98, 1.07, 0.42.

4-26. Determine the unit hydrograph ordinates for the Crimson River with the stream flow data shown in the table below. The basin area is 626.0 km^2. The unit duration of the storm was five hours.

Time (h)	Flow (m^3/s)	Time (h)	Flow (m^3/s)	Time (h)	Flow (m^3/s)
0	1.60	35	21.55	70	5.15
5	1.73	40	18.13	75	3.46
10	1.57	45	15.77	80	2.48
15	1.41	50	13.48	85	1.48
20	6.22	55	11.03	90	0.79
25	15.14	60	8.55	95	0.77
30	19.60	65	6.88	100	0.77

4-27. Apply the unit hydrograph distribution shown below to the following observed rainfall. Compute the compound runoff.

Day	UH ord. (m^3/s · cm)
1	0.12
2	0.75
3	0.13

Day	Rain (cm)	Abstractions (cm)
1	0.50	0.20
2	0.30	0.10
3	0.0	0.0

4-28. Using the unit hydrograph for the Isosceles River developed in Problem 4-23, determine the stream flow hydrograph resulting from the storm shown in Figure P-4-28. Note that the time is military time; that is, 1500 h = 3 PM, 2000 h = 8 PM, etc.

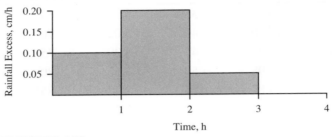

FIGURE P-4-28
Plot of rainfall on the Isosceles River watershed.

4-29. Using the unit hydrograph developed for the Verde River in Problem 4-25, determine the stream flow hydrograph resulting from the storm shown in Figure P-4-29 below.

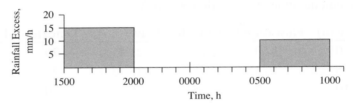

FIGURE P-4-29
Plot of rainfall on the Verde River watershed.

4-30. Using the unit hydrograph developed for the Crimson River in Problem 4-26, determine the stream flow hydrograph resulting from the storm shown in Figure P-4-30 below.

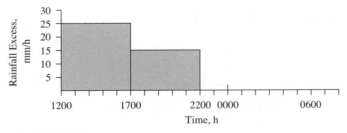

FIGURE P-4-30
Plot of rainfall on the Crimson River watershed.

4-31. A reservoir has been built to supply water during droughts. The design volume is $7.00 \times 10^6 \text{ m}^3$. If the average flow into the reservoir is 3.2 m³/s and the DNR requires that the discharge from the reservoir be at least 2.0 m³/s, how many days will it take to achieve the design volume?

4-32. The Town of Woebegone has requested that you estimate the size of water tower (volume in liters) it should provide, on the basis of the following estimate of their demand and the fact that the town has a well that pumps 36 L/s.

Time	Q_{out} (L/s)
Midnight–6 AM	0.0
6 AM–noon	54.0
Noon–6 PM	54.0
6 PM–midnight	36.0

4-33. The demand for a water supply from the Clear Fork Trinity River is 0.35 m³/s. Construct a mass balance for the period from July 1, 1951, through May 31, 1952, and determine the storage required (in m³) to meet this demand. Data are given in the table on page 196. Assume that the reservoir is full at the beginning of the analysis period and that flows greater than the demand are discharged downstream after the reservoir is filled. Is the reservoir full at the end of May 1952? Use a spreadsheet program you have written to solve this problem.

> *Answer:* $\Psi = 4{,}838{,}832$ or $4.8 \times 10^6 \text{ m}^3$; the reservoir is full at the end of May 1952.

4-34. Construct a mass balance for the Squannacook River to determine the storage (in cubic meters) required to meet a demand of 1.76 m³/s for the period from January 1, 1964 through December 31, 1967. Data are given in the table on page 197. Assume that the reservoir is full at the beginning of the analysis period and that flows greater than the demand are discharged downstream after the reservoir is filled. Is the reservoir full at the end of December 1967? Use a spreadsheet program you have written to solve this problem.

4-35. The demand for a water supply from the Hoko River is 0.325 m³/s. If the Department of Environmental Quality limits the withdrawal to 10 percent of the flow, what size reservoir (in cubic meters) is required to provide the required flow during the drought that occurs between May 1 and November 30, 1972? Data are given in the table on page 198. Assume that the reservoir is full at the beginning of the analysis period and that flows greater than the demand are discharged downstream after the reservoir is filled. Is the reservoir full at the end of November 1972? Work this problem by the mass balance method, using a spreadsheet program you have written.

Clear Fork Trinity River at Fort Worth, TX (Problem 4-33)
Mean monthly discharge (m^3/s)

Year	Jan	Feb	Mar	Apr	May	June	July	Aug	Sept	Oct	Nov	Dec
1940	—	—	—	—	—	—	—	—	—	0.00	5.63	15.4
1941	4.59	23.8	7.50	6.91	10.2	17.0	2.07	2.29	0.20	1.71	0.631	0.926
1942	0.697	0.595	0.614	58.93	24.1	9.09	0.844	0.714	1.21	10.0	2.38	1.87
1943	1.33	1.00	3.99	3.71	8.38	3.77	0.140	0.00	1.33	0.014	0.00	0.139
1944	0.311	4.93	2.83	2.25	13.3	1.68	0.210	0.609	4.11	0.985	0.515	1.47
1945	3.06	30.38	35.23	28.85	5.69	21.7	2.14	0.230	0.162	0.971	0.617	0.541
1946	1.88	5.75	3.54	1.89	6.57	5.86	0.153	1.45	4.02	0.906	12.2	10.3
1947	4.64	2.62	4.87	5.13	2.27	4.25	0.292	0.054	0.535	0.371	0.331	3.51
1948	3.99	16.9	9.06	1.91	2.64	1.11	1.22	0.00	0.00	0.00	0.00	0.003
1949	0.309	4.19	9.94	4.16	55.21	11.1	1.38	0.450	0.447	4.53	0.711	0.614
1950	3.28	14.7	3.26	12.7	15.1	2.50	3.60	2.44	10.6	1.12	0.711	0.801
1951	0.708	0.994	0.719	0.527	1.37	6.20	0.980	0.00	0.00	0.00	0.006	0.090
1952	0.175	0.413	0.297	1.93	3.65	0.210	0.003	0.029	0.007	0.00	0.368	0.167
1953	0.099	0.080	0.134	0.671	0.934	0.008	0.286	0.249	0.041	0.546	0.182	0.066
1954	0.108	0.092	0.114	0.088	0.278	0.017	0.021	0.015	0.008	0.047	0.024	0.063
1955	0.091	0.153	0.317	0.145	0.464	0.640	0.049	0.050	0.119	0.104	0.055	0.058
1956	0.069	0.218	0.026	0.306	1.35	0.30	0.026	0.019	0.029	0.266	0.030	0.170
1957	0.065	0.300	0.385	12.8	23.6	59.9	6.97	1.36	0.501	0.476	0.855	1.55
1958	1.65	1.61	4.59	5.69	28.1	0.589	0.524	0.456	0.549	0.572	0.490	0.566
1959	0.759	0.776	0.120	0.261	0.097	0.685	0.379	0.668	0.473	9.03	1.64	3.65
1960	11.8	4.45	3.26	1.42	0.631	0.379	0.660	0.566	0.467	0.498	0.241	0.648
1961	2.05	1.92	3.40	1.02	0.306	2.34	0.821	0.816	1.08	0.824	0.297	0.504
1962	0.311	2.03	0.467	0.759	0.459	0.236	0.745	1.41	6.94	1.31	0.405	0.767
1963	0.345	0.268	0.379	1.74	3.79	1.48	0.527	0.586	0.331	0.277	0.249	0.266
1964	0.416	0.266	1.16	0.813	1.02	0.374	0.535	0.963	3.96	0.351	1.47	0.886
1965	2.13	14.6	4.16	2.28	20.5	2.45	1.22	0.821	0.776	0.394	0.476	0.213
1966	0.169	0.354	0.462	6.40	23.1	18.5	5.32	0.951	0.294	1.37	0.15	0.134
1967	0.244	0.244	0.198	0.688	1.04	3.65	0.354	0.068	0.697	0.473	0.394	0.558
1968	1.64	3.85	23.2	1.89	15.7	6.12	0.583	0.144	0.220	0.419	0.396	0.206
1969	0.259	0.555	2.66	12.1	21.2	0.745	0.674	0.30	1.56	0.917	0.459	1.94
1970	5.78	6.37	27.0	3.31	15.0	1.03	0.521	0.697	1.23	—	—	—

Squannacook River near West Groton, MA (Problem 4-34)
Mean monthly discharge (m³/s)

Year	Jan	Feb	Mar	Apr	May	June	July	Aug	Sept	Oct	Nov	Dec
1951	3.48	8.18	6.63	6.63	3.20	2.38	2.40	1.49	1.06	1.82	6.60	4.47
1952	6.68	5.07	6.51	9.94	5.44	3.71	.87	1.01	.69	.45	1.05	3.37
1953	4.79	6.77	11.44	9.80	6.34	1.21	.52	.42	.29	.51	1.34	3.65
1954	2.06	3.20	4.67	4.53	9.71	2.75	1.21	1.05	6.94	2.27	6.26	7.16
1955	2.92	3.06	5.41	6.17	2.77	1.44	.46	1.63	.61	8.38	8.61	2.17
1956	9.15	3.29	3.82	14.56	5.21	2.50	.77	.40	.50	.54	1.14	2.33
1957	2.92	2.63	4.22	3.99	2.65	.87	.37	.22	.22	.29	.97	3.91
1958	5.89	3.48	6.60	12.40	5.35	1.29	.81	.49	.45	.62	1.13	1.49
1959	2.07	2.05	5.41	8.67	2.37	1.22	1.17	.59	.82	2.55	4.08	5.55
1960	3.51	3.96	3.03	14.73	5.52	2.41	1.09	1.21	2.71	2.18	3.34	2.49
1961	1.57	3.09	7.28	11.10	4.67	2.31	1.03	.80	1.23	.99	2.06	1.73
1962	3.14	1.80	5.47	10.93	3.71	1.25	.56	.69	.50	2.95	4.73	4.30
1963	2.19	1.76	6.83	7.53	2.66	.77	.38	.24	.25	.35	1.52	1.98
1964	3.77	2.57	7.33	6.57	1.85	.59	.38	.25	.21	.27	.36	.79
1965	.65	1.33	2.38	3.79	1.47	.59	.23	.20	.19	.27	.45	.64
1966	.61	1.96	5.55	2.92	2.46	.80	.26	.18	.27	.52	1.75	1.35
1967	1.68	1.53	2.64	10.62	6.29	3.17	2.22	.72	.47	.60	1.07	3.03
1968	2.02	2.14	9.60	3.79	3.82	4.79	1.92	.61	.48	.46	1.88	4.33
1969	2.21	2.17	5.81	10.70	2.80	1.01	.58	1.03	.93	.52	5.24	5.83

Hoko River near Sekiu, WA (Problem 4-35)
Mean monthly discharge (m³/s) 1963–1973

Year	Jan	Feb	Mar	Apr	May	June	July	Aug	Sept	Oct	Nov	Dec
1963	12.1	15.0	8.55	9.09	5.78	1.28	2.59	1.11	.810	13.3	26.1	20.3
1964	27.3	12.2	18.0	8.21	4.08	3.62	4.53	2.44	4.28	7.67	13.3	14.7
1965	27.4	27.3	5.01	5.61	6.68	1.38	.705	.830	.810	7.31	16.8	19.6
1966	29.8	11.3	17.6	5.18	2.67	2.10	1.85	.986	1.54	10.3	17.0	39.0
1967	35.6	26.8	18.5	6.51	3.43	1.46	.623	.413	.937	25.7	14.2	27.8
1968	34.2	22.4	15.7	9.20	3.68	2.65	1.72	1.55	9.12	16.8	16.5	25.2
1969	17.2	18.5	12.9	12.8	3.74	2.23	1.19	.810	6.15	7.84	9.15	15.9
1970	17.3	12.1	8.50	17.7	3.85	1.32	.932	.708	4.22	7.96	13.9	25.4
1971	32.7	21.0	21.1	8.13	3.43	2.83	1.83	.932	2.22	10.7	22.7	22.0
1972	27.4	26.9	25.4	14.6	3.00	1.00	5.32	.841	2.00	1.14	11.8	37.8
1973	28.0	9.23	11.3	4.13	5.30	4.93	1.63	.736	.810	13.1	29.8	31.5

4-36. A co-worker has been working on the design of a retention pond for the Bar Nunn Mall. Your boss has just informed you that your co-worker has come down with chicken pox and you must finish the calculations to determine the volume of the retention basin. Your boss gives you the work shown below. Determine the volume of the retention basin in cubic meters. *Note:* each interval is 1-hour long. Use a spreadsheet program you have written to solve this problem.

Interval	Inflow (L/s)	Volume in (m³)	Outflow (L/s)	Volume out (m³)
1	10.0	36.0	10.0	36.0
2	20.0	72.0	10.0	36.0
3	30.0		10.0	
4	20.0		10.0	
5	15.0		10.0	
6	5.0			

4-37. The Menominee River floods when the flow rate is greater than 100 m³/s. Using the data on page 200, determine the storage required (in cubic meters) to prevent a flood during the period from January 1, 1959, through December 31, 1960. Assume the reservoir is empty at the start of the analysis period. Also assume that, after the reservoir is full, the flow out of the reservoir is at a rate equal to Q_{out} until the reservoir is empty. Is the reservoir empty at the end of December 1960? Use a spreadsheet program you have written to solve this problem.

> *Answer:* $\forall = 1{,}226{,}525{,}760$ or 1.23×10^9 m³; the reservoir is not empty.

4-38. The Spokane River floods when the flow rate is greater than 250.0 m³/s. Using the data in the table on page 201, determine the storage required (in cubic meters) to prevent a flood during the period from January 1, 1957, through December 31, 1958. Assume the reservoir is empty at the start of the analysis period. Also assume that, after the reservoir is full, the flow out of the reservoir is at a rate equal to Q_{out} until the reservoir is empty. Is the reservoir empty at the end of December 1958? Use a spreadsheet program you have written to solve this problem.

4-39. The Rappahannock River floods if the flow rate exceeds 5.80 m³/s. Using the data in the table on page 202, determine the storage required (in cubic meters) to prevent a flood during the period from January 1, 1960, through December 31, 1962. Assume the reservoir is empty at the start of the analysis period. Also assume that, after the reservoir is full, the flow out of the reservoir is at a rate equal to Q_{out} until the reservoir is empty. Is the reservoir empty at the end of December 1962? Use a spreadsheet program you have written to solve this problem.

Menominee River below Koss, MI (Problem 4-37)
Mean monthly discharge (m³/s)

Year	Jan	Feb	Mar	Apr	May	June	July	Aug	Sept	Oct	Nov	Dec
1946	76.3	69.8	149.0	106.0	85.3	132.0	89.5	62.2	58.5	56.6	76.7	56.8
1947	54.0	53.0	60.1	157.0	164.0	103.0	72.1	55.5	52.7	51.0	59.3	47.1
1948	49.2	35.4	72.5	103.0	82.3	48.5	42.2	45.8	39.6	34.8	56.1	49.2
1949	45.9	47.9	57.4	87.7	84.7	62.0	102.0	51.7	57.2	57.0	57.8	58.6
1950	59.1	57.6	61.1	199.0	243.0	101.0	69.2	65.8	49.0	43.4	48.9	48.4
1951	50.5	44.5	68.0	267.0	171.0	143.0	159.0	93.3	122.0	147.0	117.0	87.0
1952	76.9	75.6	66.5	219.0	94.0	86.7	153.0	97.0	58.6	45.6	50.2	51.9
1953	59.4	62.5	103.0	167.0	133.0	166.0	174.0	87.5	67.0	56.1	53.4	63.6
1954	57.0	66.7	68.5	182.0	184.0	138.0	73.2	61.0	91.1	123.0	84.6	68.5
1955	66.2	61.0	70.5	254.0	108.0	106.0	48.7	56.1	38.0	59.7	63.0	57.4
1956	59.0	54.8	49.7	270.0	108.0	88.4	116.0	83.4	61.5	48.8	53.1	53.6
1957	49.4	46.1	69.7	130.0	93.0	65.0	41.4	36.3	52.3	52.7	73.3	59.8
1958	54.0	51.3	61.8	123.0	65.9	62.0	132.0	47.0	58.5	47.7	68.9	48.4
1959	46.7	43.1	55.0	110.0	105.0	56.7	48.3	78.0	142.0	155.0	122.0	78.2
1960	82.3	71.0	62.4	242.0	373.0	135.0	83.4	72.1	80.8	60.5	102.0	68.2
1961	52.6	49.2	77.0	158.0	186.0	82.5	60.8	53.8	48.9	57.9	67.6	63.1
1962	53.9	52.4	69.2	168.0	168.0	107.0	59.3	52.1	73.8	65.3	54.7	51.4
1963	46.8	43.8	56.1	87.7	120.0	99.1	43.6	40.6	38.0	34.2	35.6	35.7
1964	37.2	33.8	41.8	70.2	131.0	63.2	42.1	56.9	65.1	55.6	69.8	54.2
1965	49.3	44.1	50.9	173.0	361.0	83.6	51.7	45.6	56.2	67.5	76.6	87.2
1966	78.5	66.6	131.0	157.0	117.0	113.0	44.6	63.0	43.5	62.6	63.7	59.2
1967	59.8	64.0	65.2	295.0	133.0	135.0	100.0	66.3	51.6	86.0	108.0	63.3
1968	50.5	58.0	72.6	134.0	108.0	168.0	141.0	78.1	155.0	93.8	90.2	82.7
1969	89.9	90.0	84.4	229.0	157.0	118.0	87.1	51.1	42.6	65.3	68.9	58.2
1970	62.7	51.0	58.8	114.0	109.0	157.0	56.9	46.6	49.1	64.1	122.0	97.5
1971	72.4	64.7	89.4	284.0	155.0	93.3	67.5	51.0	47.0	92.4	88.2	80.2
1972	65.9	56.7	68.5	194.0	254.0	88.6	68.2	108.0	91.0	134.0	138.0	77.2
1973	85.3	73.8	226.0	240.0	290.0	111.0	72.6	80.3	69.0	66.9	82.5	66.3
1974	62.8	65.4	73.4	142.0	107.0	111.0	61.6	86.4	77.8	58.8	106.0	75.5
1975	66.3	66.0	68.4	193.0	209.0	116.0	54.8	41.5	63.2	43.4	69.2	86.3
1976	67.5	69.2	96.9	298.0	147.0	79.1	41.2	36.9	30.3	—	—	—

Spokane River near Otis Orchards, WA (Problem 4-38)
Mean monthly discharge (m³/s)

Year	Jan	Feb	Mar	Apr	May	June	July	Aug	Sept	Oct	Nov	Dec
1950	—	—	—	—	—	—	—	—	—	59.9	123.0	257.0
1951	217.0	492.0	200.0	422	460	156.0	36.3	4.25	8.44	75.6	98.7	145.0
1952	115.0	140.0	107.0	491	624	165.0	52.4	9.49	33.3	36.1	44.7	50.5
1953	169.0	311.0	163.0	246	525	358.0	32.3	12.8	29.8	52.2	53.0	97.8
1954	116.0	190.0	254.0	398	657	414.0	68.0	28.13	28.3	82.1	106.0	93.1
1955	60.0	76.3	62.4	257	544	468.0	82.3	19.5	26.0	86.0	166.0	354.0
1956	292.0	141.0	195.0	685	809	391.0	47.8	22.0	27.9	64.7	78.9	101.0
1957	99.0	61.0	278.0	461	792	329.0	33.6	12.5	15.7	55.5	66.9	73.0
1958	80.0	245.0	234.0	408	548	152.0	29.5	4.50	24.4	36.8	153.0	240.0
1959	356.0	233.0	192.0	465	351	410.0	35.5	11.6	45.2	84.4	224.0	172.0
1960	117.0	154.0	202.0	600	470	266.0	35.4	16.0	45.0	47.8	57.6	88.5
1961	93.0	454.0	406.0	389	559	352.0	8.86	4.96	19.1	24.5	55.9	50.7
1962	107.0	142.0	124.0	534	535	232.0	39.9	11.0	13.3	50.3	96.6	186.0
1963	152.0	268.0	208.0	353	303	92.1	26.9	5.32	10.1	21.2	68.0	71.2
1964	76.5	97.6	81.6	328	594	619.0	66.5	34.0	92.7	43.0	88.6	408.0
1965	282.0	314.0	268.0	475	602	158.0	57.3	28.9	23.2	32.5	82.7	76.7
1966	105.0	63.7	175.0	451	388	126.0	38.3	10.3	12.0	45.6	60.1	122.0
1967	205.0	282.0	191.0	272	504	413.0	40.8	13.3	30.8	49.3	68.5	93.6
1968	104.0	276.0	348.0	226	232	155.0	42.3	22.3	40.5	92.9	166.8	198.0
1969	254.0	138.0	162.0	638	651	220.0	50.7	25.2	40.5	45.8	49.0	60.8
1970	86.8	225.0	214.0	260	512	382.0	51.3	32.5	35.6	—	—	—

Rappahannock River near Warrenton, VA (Problem 4-39)
Mean monthly discharge (m³/s)

Year	Jan	Feb	Mar	Apr	May	June	July	Aug	Sept	Oct	Nov	Dec
1943	7.08	9.23	10.5	8.55	6.12	3.31	1.18	0.294	0.269	0.057	1.38	1.06
1944	4.67	3.14	8.44	5.66	3.71	1.88	0.334	0.450	4.13	4.62	1.87	5.27
1945	4.11	5.44	4.81	3.62	3.79	2.94	4.22	9.68	13.0	3.77	4.05	8.27
1946	8.44	7.90	7.36	5.38	9.94	7.84	2.60	4.70	2.02	3.14	2.35	2.38
1947	6.94	3.79	6.77	3.57	4.84	3.74	2.92	2.39	1.40	1.04	6.60	2.74
1948	3.85	6.09	7.28	9.77	11.5	4.05	4.16	13.1	2.86	7.73	10.4	14.1
1949	13.8	11.0	8.95	11.9	13.2	5.10	4.39	3.79	1.67	2.03	2.37	2.94
1950	2.76	7.45	7.73	4.42	7.56	5.44	4.25	1.60	5.75	4.19	5.63	20.0
1951	5.72	12.8	11.0	12.1	5.44	9.40	2.89	1.19	0.447	0.453	1.93	5.13
1952	7.62	9.74	12.5	18.0	9.32	3.43	2.21	1.98	2.50	0.951	9.80	7.28
1953	11.3	7.25	12.6	8.24	10.6	4.39	1.36	0.685	0.343	0.357	0.773	2.25
1954	2.47	2.54	5.69	5.72	4.39	1.95	0.875	0.572	0.131	1.90	1.80	3.57
1955	2.42	4.11	8.21	5.07	3.85	4.16	0.801	20.3	3.54	2.47	2.19	1.62
1956	2.21	6.54	7.31	6.23	2.46	1.06	10.3	3.60	2.17	4.42	6.34	3.91
1957	4.33	8.07	8.95	8.72	3.31	2.15	0.402	0.008	0.391	1.02	1.66	6.43
1958	10.1	6.82	12.0	11.8	9.03	2.97	5.07	3.57	1.33	1.56	1.81	1.89
1959	3.45	3.06	4.76	7.08	3.28	5.04	0.804	0.513	0.759	2.68	2.05	3.88
1960	4.11	9.71	7.70	13.3	11.3	9.97	2.97	1.85	2.77	1.10	1.23	1.31
1961	3.31	15.4	9.85	15.5	11.1	6.82	3.23	2.24	1.70	1.16	1.77	4.25
1962	5.44	5.61	16.8	10.7	5.27	6.88	3.57	1.51	0.855	0.932	4.73	3.60
1963	6.51	4.19	13.6	4.45	2.55	3.20	0.496	0.136	0.160	0.121	2.28	2.21
1964	9.97	8.18	9.63	11.8	7.25	1.37	1.44	0.660	0.697	1.72	1.83	3.82
1965	6.51	13.8	15.0	6.31	3.74	1.34	5.27	0.365	3.09	0.459	0.379	0.450
1966	0.694	5.83	4.45	4.45	6.63	1.70	0.225	0.135	4.47	3.51	3.31	4.42
1967	5.63	5.86	13.6	3.82	4.53	1.57	1.04	7.42	2.27	3.82	2.57	10.0
1968	12.9	6.68	9.40	4.56	4.02	4.59	2.64	1.64	1.03	0.971	4.93	3.00
1969	4.47	4.84	5.07	3.51	1.93	1.70	1.54	1.93	2.12	1.80	3.37	6.12
1970	5.92	11.4	6.14	9.83	4.30	2.34	3.62	1.62	0.413	—	—	—

4-40. Four monitoring wells have been placed around a leaking underground storage tank. The wells are located at the corners of a 100-ha square. The total piezometric head in each of the wells is as follows: NE corner, 30.0 m; SE corner, 30.0 m; SW corner, 30.6 m: NW corner 30.6 m. Determine the magnitude and direction of the hydraulic gradient.

> *Answer:* Hydraulic gradient $= 6.0 \times 10^{-4}$; direction $=$ west to east

4-41. After a long wet spell, the water levels in the wells described in Problem 4-40 were measured and found to be the following distances from the ground surface: NE corner, 3.0 m; SE corner, 3.0 m; SW corner, 3.6 m; NW corner, 3.4 m. Assume that the ground surface is at the same elevation for each of the wells. Determine the magnitude and direction of the hydraulic gradient.

4-42. In preparation for a groundwater modeling study of the movement of a plume of contamination, three wells were installed to determine the hydraulic gradient. Three wells were installed in a rectangular grid at the following locations: well A at $x = 0.0$ m and $y = 0.0$ m; well B at $x = 280$ m and $y = 0.0$ m; well C at $x = 0.0$ m and $y = 500$ m. The ground surface elevation for each well is at 186.66 m above mean sea level. The depth to the groundwater table in each well is as follows: well A $= 5.85$ m; well B $= 5.63$ m; well C $= 5.52$ m. Determine the magnitude and direction of the hydraulic gradient.

4-43. A gravelly sand has a hydraulic conductivity of 6.9×10^{-4} m/s, a hydraulic gradient of 0.00141, and a porosity of 20 percent. Determine the Darcy velocity and the average linear velocity.

> *Answer:* $v = 9.73 \times 10^{-7}$ m/s; $v' = 4.86 \times 10^{-6}$ m/s

4-44. A fine sand has a hydraulic conductivity of 3.5×10^{-5} m/s, a hydraulic gradient of 0.00141, and a porosity of 45 percent. Determine the Darcy velocity and the average linear velocity.

4-45. The results of a tracer study at a beach yielded an estimated average linear velocity of the groundwater of 0.60 m/d. Laboratory studies were used to determine a porosity of 30 percent and a hydraulic conductivity of 4.75×10^{-4} m/s for the sand. Estimate the Darcy velocity (in m/s) and hydraulic gradient of the groundwater.

4-46. Two piezometers have been placed along the direction of flow in a confined aquifer that is 30.0 m thick. The piezometers are 280 m apart. The difference in piezometric head between the two is 1.4 m. The aquifer hydraulic conductivity is 50 m/d and the porosity is 20 percent. Estimate the travel time for water to flow between the two piezometers.

4-47. A fully penetrating well in a 28.0 m thick artesian aquifer pumps at a rate of 0.00380 m³/s for 1,941 days and causes a drawdown of 64.05 m at an observation well 48.00 m from the pumping well. How much drawdown will occur at an observation well 68.00 m away? The original piezometric surface was 94.05 m above the bottom confining layer. The aquifer material is sandstone. Report your answer to two decimal places.

Answer: $s_2 = 51.08$ m

4-48. If a fully penetrating well in a 99.99 m thick artesian aquifer pumping at a rate of 0.0020 m^3/s for 1,812 days causes a drawdown of 12.73 m at an observation well 280.00 m from the pumping well, how much drawdown will occur at an observation well 1,492.0 m away? The original piezometric surface was 170.89 m above the bottom confining layer. The aquifer material is sandstone. Report your answer to two decimal places.

4-49. A 42.43 m thick fractured rock artesian aquifer supplies water to a fully penetrating well. Before pumping began, the piezometric surface in the pumping well was observed to be 70.89 m above the bottom confining layer. The pumping rate of the well is 0.0255 m^3/s. After 1,776 days of pumping an observation well is drilled 272.70 m from the pumping well. The drawdown in this well was observed to be 5.04 m. Estimate the drawdown at a distance of 64.28 m from the pumping well.

4-50. A long-term pumping test was conducted to determine the hydraulic conductivity of a 82.0 m thick confined aquifer. The nonpumping piezommetric surface was 109.5 m above the confining layer. The pumping rate was 0.0280 m^3/s. The steady-state drawdown at an observation well 41.0 m away from the pumping well was 3.55 m. The drawdown at an observation well 63.5 m away from the pumping well was 1.35 m. Determine the hydraulic conductivity to three significant figures.

4-51. It is undesirable to lower the piezometric surface of a confined aquifer below the aquiclude because this will destroy the structural integrity of the aquifer formation. Determine the maximum rate of pumping that would be permissible for the case described in Example 4-13 if the following conditions prevail:

1. The observation well at 200.0 m maintained the same drawdown.

2. Another observation well 2.0 m from the pumped well was used to observe lowering of the piezometric surface to the bottom of the aquiclude.

Report your answer to two decimal places.

4-52. An artesian aquifer 5.0 m thick with a piezometric surface 65.0 m above the bottom confining layer is being pumped by a fully penetrating well. The aquifer is a mixture of sand and gravel. A steady-state drawdown of 7.0 m is observed at a nonpumping well located 10.0 m away. If the pumping rate is 0.020 m^3/s, how far away is a second nonpumping well that has an observed drawdown of 2.0 m? Report your answer to two decimal places.

4-53. A test well was drilled to the underlying impervious stratum in an unconfined aquifer. The depth of the well was 18.3 m. Observation wells were drilled at 20.0 m and 110.0 m from the pumping test well. The static water level in each well was 4.57 m below the ground surface. The test well was pumped at a rate of 0.0347 m³/s until steady-state conditions were achieved. The drawdown in the observation wells was 2.78 m and 0.73 m at 20.0 m and 110.0 m, respectively. Determine the hydraulic conductivity in m/s. Report your answer to three significant figures.

4-54. For an unconfined aquifer, there is a possibility that drawdown will lower the piezometric surface to the bottom of the well and that the water will stop flowing. For the case described in Example 4-15, determine the maximum pumping rate that can be sustained indefinitely if the drawdown at the observation well 100.0 m from the pumped well is 9.90 m and the drawdown at the pumped well is limited by the depth of the aquifer, that is, 30.0 m. Report your answer to two decimal places.

Answer: $Q = 1.36$ m³/s

4-55. A contractor is trying to estimate the distance to be expected of a drawdown of 4.81 m from a pumping well under the following conditions:

Pumping rate 0.0280 m³/s

Pumping time $= 1,066$ d

Drawdown in observation well $= 9.52$ m

Observation well is located 10.00 m from the pumping well

Aquifer material $=$ medium sand

Aquifer thickness $= 14.05$ m

Assume that the well is fully penetrating in an unconfined aquifer. Report your answer to two decimal places.

4-56. A fully penetrating well in a 30.0 m thick unconfined aquifer is to be used to dewater the area to be excavated for a new wastewater treatment plant. The piezometric surface must be lowered 5.25 m below the static water level 45.45 m from the pumping well. In addition, it is desired that the piezometric surface be lowered 2.50 m at a distance of 53.56 m from the pumping well. The aquifer material is loam. Assuming steady state, what pumping rate (in m³/s) must be used to achieve the desired drawdowns. Report your answer to three significant figures.

4-57. A well with a 0.25 m diameter fully penetrates an unconfined aquifer that is 20.0 m thick. The well has a discharge of 0.0150 m³/s and a drawdown of 8.0 m. If the flow is steady and the hydraulic conductivity is 1.5×10^{-4} m/s, what is the height of the piezometric surface above the confining layer at a site 80 m from the well?

4-58. Repeat Problem 4-57 using a well diameter of 0.50 m.

4-59. A 0.30 m diameter well fully penetrates a confined aquifer that is 28.0 m thick. The aquifer material is fractured rock. If the drawdown in the pumped well is 6.21 m after pumping for 48 hours at a rate of 0.0075 m³/s, what will the drawdown be at the end of 48 days of pumping at this rate?

4-60. An aquifer yields the following results from pumping a 0.61 m diameter well at 0.0303 m^3/s: $s = 0.98$ m in 8 min; $s = 3.87$ m in 24 h. Determine its transmissivity. Report your answer to three significant figures.

Answer: $T = 4.33 \times 10^{-3}$ m^2/s

4-61. Determine the transmissivity of a confined aquifer that yields the following results from a pumping test of a 0.46 m diameter well that fully penetrates the aquifer.

Pumping rate $= 0.0076$ m^3/s

$s = 3.00$ m in 0.10 min

$s = 34.0$ m in 1.00 min

4-62. An aquifer yields a drawdown of 1.04 m at an observation well 96.93 m from a well pumping at 0.0170 m^3/s after 80 min of pumping. The virtual time is 0.6 min and the transmissivity is 5.39×10^{-3} m^2/s. Determine the storage coefficient.

Answer: $S = 4.647 \times 10^{-5}$ or 4.6×10^{-5}

4-63. Using the data from Problem 4-62, find the drawdown at the observation well 80 days after pumping begins.

4-64. If the transmissivity is 2.51×10^{-3} m^2/s and the storage coefficient is 2.86×10^{-4}, calculate the drawdown that will result at the end of 2 days of pumping a 0.50 m diameter well at a rate of 0.0194 m^3/s.

4-65. Determine the storage coefficient for an artesian aquifer from the pumping test results shown in the table below. The measurements were made at an observation well 300.00 m away from the pumping well. The pumping rate was 0.0350 m^3/s.

Time (min)	Drawdown (m)
100.0	3.10
500.0	4.70
1,700.0	5.90

Answer: $S = 1.9 \times 10^{-5}$

4-66. Rework Problem 4-65, but assume that the data were obtained at an observation well 100.0 m away from the pumping well.

4-67. Determine the storage coefficient for an artesian aquifer from the pumping test results shown in the table below. The measurements were made at an observation well 100.00 m away from the pumping well. The pumping rate was 0.0221 m^3/s.

Time (min)	Drawdown (m)
10.0	1.35
100.0	3.65
1,440.0	6.30

4-68. Rework Problem 4-67, but assume that the data were obtained at an observation well 60.0 m away from the pumping well.

4-69. Determine the storage coefficient for an artesian aquifer from the following pumping test results on a 0.76 m diameter well that fully penetrates the aquifer. The pumping rate was 0.00350 m^3/s. The drawdowns were measured in the pumping well.

Time (min)	Drawdown (m)
0.20	2.00
1.80	3.70
10.0	5.00

4-70. Two wells located 106.68 m apart are both pumping at the same time. Well A pumps at 0.0379 m^3/s and well B pumps at 0.0252 m^3/s. The diameter of each well is 0.460 m. The transmissivity is 4.35 $\times$ 10^{-3} m^2/s and the storage coefficient is 4.1 $\times$ 10^{-5}. What is the interference of well A on well B after 365 days of pumping? Report your answer to two decimal places.

Answer: Interference of well A on B is 9.29 m.

4-71. Using the data from Problem 4-70, find the total drawdown in well B after 365 days of pumping. Report your answer to two decimal places.

4-72. If two wells, no. 12 and no. 13, located 100.0 m apart, are pumping at rates of 0.0250 m^3/s and 0.0300 m^3/s, respectively, what is the interference of well no. 12 on well no. 13 after 280 days of pumping? The diameter of each well is 0.500 m. The transmissivity is 1.766 $\times$ 10^{-3} m^2/s and the storage coefficient is 6.675 $\times$ 10^{-5}. Report your answer to two decimal places.

4-73. Using the data from Problem 4-72, find the total drawdown in well 13 after 280 days of pumping. Report your answer to two decimal places.

4-74. Wells X, Y, and Z are located equidistant at 100.0 m intervals. Their pumping rates are 0.0315 m^3/s, 0.0177 m^3/s and 0.0252 m^3/s, respectively. The diameter of each well is 0.300 m. The transmissivity is 1.77 $\times$ 10^{-3} m^2/s. The storage coefficient is 6.436 $\times$ 10^{-5}. What is the interference of well X on well Y and on well Z after 100 days of pumping? Report your answer to two decimal places.

4-75. Using the data in Problem 4-74, find the total drawdown in well X at the end of 100 days of pumping. Report your answer to two decimal places.

4-76. For the well field layout shown in Figure P-4-76, determine the effect of adding a sixth well. Is there any potential for adverse effects on the well or the aquifer? Assume all wells are pumped for 100 days and that each well is 0.300 m in diameter. Well data are given in the table below. Aquifer data are shown below the well data.

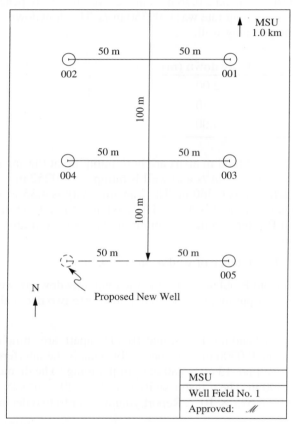

FIGURE P-4-76
Layout for MSU well field.

MSU Well Field No. 1

Well no.	Pumping rate (m³/s)	Depth of well (m)
1	0.0221	111.0
2	0.0315	112.0
3	0.0189	110.0
4	0.0177	111.0
5	0.0284	112.0
6 (proposed)	0.0252	111.0

The aquifer characteristics are as follows:

  Storage coefficient $= 6.418 \times 10^{-5}$

  Transmissivity $= 1.761 \times 10^{-3}$ m²/s

  Nonpumping water level $= 6.90$ m below grade

  Depth to top of artesian aquifer $= 87.0$ m

  Answer: Total drawdown for each well in numerical order: (1) 79.54 m, (2) 84.99 m, (3) 80.05 m, (4) 79.54 m, (5) 83.18 m, (6) 81.35 m. Drawdown is below the top of the aquiclude for wells 2, 5, and 6.

4-77. For the well field layout shown in Figure P-4-77, determine the effect of adding a sixth well. Is there any potential for adverse effects on the well or the aquifer? Assume all wells are pumped for 100 days and that each well is 0.300 m in diameter. Aquifer and well data are given below the figure.

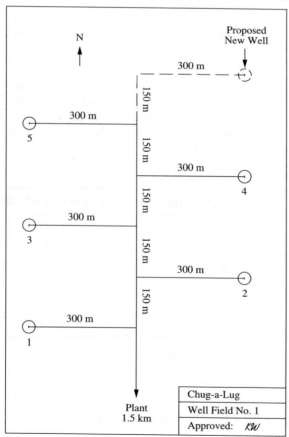

FIGURE P-4-77
Layout for brewery well field no. 1.

The aquifer characteristics are as follows:

  Storage coefficient $= 2.11 \times 10^{-6}$

  Transmissivity $= 4.02 \times 10^{-3}$ m²/s

Nonpumping water level = 9.50 m below grade

Depth to top of artesian aquifer = 50.1 m

Chug-a-Lug Brewery Well Field No. 1

Well no.	Pumping rate (m³/s)	Depth of well (m)
1	0.020	105.7
2	0.035	112.8
3	0.020	111.2
4	0.015	108.6
5	0.030	113.3
6 (proposed)	0.025	109.7

4-78. What pumping rate, pumping time, or combination thereof can be sustained by the new well in Problem 4-77 if all of the well diameters are enlarged to 1.50 m?

4-79. For the well field layout shown in Figure P-4-79, determine the effect of adding a sixth well. Is there any potential for adverse effects on the well or the aquifer? Assume all wells are pumped for 180 days and that each well is 0.914 m in diameter. Well data are given in the table below. Aquifer data are shown below the well field data.

Chug-a-Lug Brewery Well Field No. 2

Well No.	Pumping rate (m³/s)	Depth of well (m)
1	0.0426	169.0
2	0.0473	170.0
3	0.0426	170.0
4	0.0404	168.0
5	0.0457	170.0
6 (proposed)	0.0473	170.0

The aquifer characteristics are as follows:

Storage coefficient = 2.80×10^{-5}

Transmissivity = 1.79×10^{-3} m²/s

Nonpumping water level = 7.60 m below grade

Depth to top of artesian aquifer = 156.50 m

4-80. What pumping rate, pumping time, or combination thereof can be sustained by the new well in Problem 4-79 if all of the well diameters are enlarged to 1.80 m?

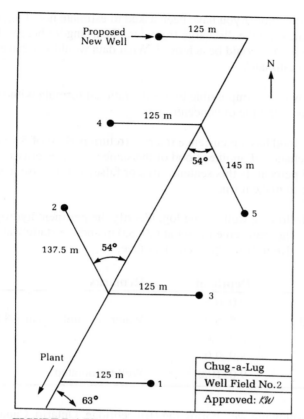

FIGURE P-4-79
Layout for brewery well field no. 2.

4-10 DISCUSSION QUESTIONS

4-1. An artesian aquifer is under pressure because of the weight of the overlying geologic strata. Is this sentence true or false? If it is false, rewrite the sentence to make it true.

4-2. Identify the base flow in the following hydrographs.

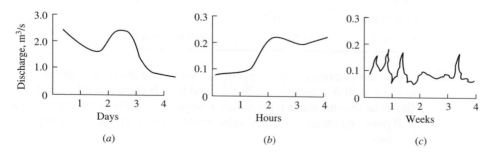

4-3. As a field engineer you have been asked to estimate how long you would have to measure the discharge from a mall parking lot before the maximum discharge would be achieved. What data would you have to gather to make the estimate?

4-4. Explain why it is impossible to use the rational formula without being able to estimate the time of concentration.

4-5. When a flood has a recurrence interval (return period) of 5 years, it means that the chance of another flood of the same or less severity occurring next year is 5 percent. Is this sentence true or false? If it is false, rewrite the sentence to make it true.

4-6. For the following well boring log, identify the pertinent hydrogeologic features. The well screen is set at 6.0–8.0 m and the static water level after drilling is 1.8 m from the ground surface.

Strata	Depth, m	Remarks
Top soil	0.0–0.5	
Sandy till	0.5–6.0	Water encountered at 1.8 m
Sand	6.0–8.0	
Clay	8.0–9.0	
Shale	9.0–10.0	Well terminated

4-7. For the following well boring log (Bracebridge, Ontario, Canada), identify the pertinent hydrogeologic features. The well screen is set at 48.0–51.8 m and the static water level after drilling is 10.2 m from the ground surface.

Strata	Depth, m	Remarks
Sand	0.0–6.1	
Gravelly clay	6.10–8.6	
Fine sand	8.6–13.7	
Clay	13.7–17.5	Casing sealed
Fine sand	17.5–51.8	
Bedrock	51.8	Well terminated

4-8. Sketch the piezometric profiles for two wells that interfere with one another. Well A pumps at 0.028 m^3/s and well B pumps at 0.052 m^3/s. Show the ground water table before pumping, the drawdown curve of each well pumping alone, and the resultant when both wells are operated together.

4-11 FE EXAM FORMATTED PROBLEMS

4-1. Lake Mead has a mean surface area of 124,200 acres and a mean annual evaporation of 849,000 acre-ft. If the New York City water supply requires 1.53×10^9 gallons per day, how many day's supply would the Lake Mead evaporation fulfill?

a. 365 d
c. 26 d

b. 181 d
d. 69 d

4-2. If a steady rain of 0.33 in/h falls on a parking lot with an area of 65,340 ft^2 and a runoff coefficient of 0.85, what is the estimated peak discharge?

a. 0.50 cfs
c. 5.1 cfs

b. 18,300 cfs
d. 0.42 cfs

4-3. Water is pumped at a rate of 33.42 ft^3/min from an unconfined aquifer that is 75 ft deep. Wells located at 100 and 150 ft from the pumping well have drawdowns of 9 and 7 ft, respectively. What is the hydraulic conductivity of the soil? (Note: The FE Manual uses permeability rather than hydraulic conductivity.)

a. 23 ft/d
c. 0.31 ft/d

b. 615 ft/d
d. 140 ft/d

4-4. What is the seepage velocity of water in an aquifer with a hydraulic conductivity of 6.9×10^{-4} m/s and porosity of 30 percent if the hydraulic gradient is 0.0014?

a. 0.28 m/d
c. 0.08 m/d

b. 3.22 m/d
d. 0.003 m/d

4-12 REFERENCES

Al-Kaisi, M. (2000) "Crop Water Use or Evapotranspiration," *Integrated Crop Management News*, Iowa State University Extension, http://www.ipm.iastate.edu/ipm/icm/2000/5-29-2000/wateruse.html

ASCE (1969) *Design and Construction of Sanitary and Storm Sewers,* Manual of Practice No. 37 (also Water Pollution Control Federation Manual of Practice No. 9), American Society of Civil Engineers, New York, pp. 43–46.

Blaney, H. F., and W. D. Criddle (1945) *Determining Water Requirements in Irrigated Areas from Climatological Data*, U.S. Depatment of Agriculture, Soil Conservation Service, Washington, DC.

Blaney, H. F., and W. D. Criddle (1962) *Determining Consumptive Use and Irrigation Water Requirements*, Technical Bulletin 1275, U.S. Department of Agriculture, Soil Conservation Service, Washington, DC.

Bouwer, H. (1978) *Groundwater Hydrology*, McGraw-Hill, New York, pp. 22, 38, 66, 68.

Boyer, M. C. (1964) "Streamflow Measurement," in V. T. Chow (ed.) *Handbook of Applied Hydrology*, McGraw-Hill, New York, pp. 15–41.

Brown, R. H. (1963) *Drawdowns Resulting from Cyclic Intervals of discharge,* in *Methods of Determining Permeability, Transmissibility, and Drawdowns,* U.S. Geologic Survey Water Supply Paper 1537-I, Washington, DC.

Cooper, H. H., and C. E. Jacob (1946) "A Generalized Graphical Method for Evaluating Formation Constants and Summarizing Well Field History," *Transactions American Geophysical Union,* vol. 27, pp. 526–534.

Dalton, J. (1802) "Experimental Essays on the Constitution of Mixed Gases; on the Force of Steam or Vapor from Waters and other Liquids, Both in a Torricellian Vacuum and in Air; on Evaporation; and on the Expansion of Gases by Heat," *Member Proceedings, Manchester Literary and Philosophical Society,* vol. 5, pp. 535–602.

Darcy, H. (1856) *Les Fontaines Publiques de la Ville de Dijon,* Victor Dalmont, Paris, pp. 570, 590–594.

Davis, M. L., and S. J. Masten (2004) *Principles of Environmental Engineering and Science,* McGraw-Hill, New York, p. 195.

Dupuit, J. (1863) *Etudes Théoriques et Practiques sur le Mouvement des Eaux dans Les Canaux Décoverts et à Travers les Terrains Perrméables*, Dunod, Paris.

FAA (1970) *Airport Drainage,* Advisory Circular A/C 150-5320-5B, Federal Aviation Agency, Department of Transportation, U.S. Government Printing Office, Washington, DC.

Ferris, J. G., D. B. Knowles, R. H. Brown, and R. W. Stallman (1962) *Theory of Aquifer Tests,* U.S. Geological Survey Water-Supply Paper 1536-E, pp. 69–174.

Gilman, C. S. (1964) "Rainfall," in V. T. Chow (ed.), *Handbook of Applied Hydrology,* McGraw-Hill, New York, p. 9–50.

Green, W. H. and G. Ampt (1911) "Studies of Soil Physics, Part I: The Flow of Air and Water Through Soils," *Journal of Agricultural Science*, vol. 4, pp. 1–24.

Gumaji, M. (1986) FEMA *Flood Mitigation Course Manual*, U.S. Federal Emergency Management Administration, Washington, DC.

Gumbel, E. J. (1954) "Statistical Theory of Droughts," *Proceedings of the American Society of Civil Engineers*, vol. 80, May, pp. 1–19.

Gupta, R. S. (2008) *Hydrology and Hydraulic Systems*, Waveland Press, Inc., Long Grove, IL, pp. 64, 74, 112, 343.

Hanson, B. R., and D. M. May (2005) "Crop Evapotranspiration of Processing Tomato in the San Joaquin Valley of California, USA," *Irrigation Science*, vol. 24, pp. 211–221.

Heath, R. C. (1983) *Basic Ground-Water Hydrology,* U.S. Geological Survey Water-Supply Paper 2220, U.S. Government Printing Office, Washington, DC.

Holton, R. E. (1961) "A Concept for Infiltration Estimates in Watershed Engineering," *Agricultural Research Service*, U.S. Department of Agriculture, pp. 41–51.

Horton, R. E. (1935) *Surface Runoff Phenomena: Part I, Analysis of the Hydrograph,* Horton Hydrologic Lab Publication 101, Edwards Bros., Ann Arbor, MI.

Howell, T. A., S. R. Evett, A. D. Schneider, R. W. Todd, and J. A. Tolk (1998) "Evoptranspiration of Irrigated Fescue Grass in a Semi-Arid Environment," presented at 1998 American Society of Agricultural Engineering International Meeting, Orlando, FL, July 12–16.

Iowa State University Integrated Crop Management (2000) http://www.ipm.iastate.edu/ipm/icm/2000/5-29-2000/wateruse.html

Johnson Screens. A Weatherford Company, 2005. *Ground Water and Wells*, St. Paul, MN, pp. 37, 102.

Kuichling, E. (1889) "The Relationship Between Rainfall and the Discharge of Sewers in Populous Districts," *Transactions of the American Society of Civil Engineers*, vol. 20, pp. 1–56.

Langbein, W. B. (1949) "Annual Floods and Partial Duration Series," *Transactions of the American Geophysical Union*, vol. 30, pp. 879–881.

Linsley, R. K., M. A. Kohler, and J. L. H. Paulhus (1975) *Hydrology for Engineers,* McGraw-Hill, New York, p. 200.

Mulvaney, T. J. (1851) "On the Use of Self-registering Rain and Flood Gages in Making Observations of the Relations of Rainfall and Flood Discharge in a Given Catchment, *Proceedings of the Institute of Civil Engineering* (Ireland), vol. 4, pp. 18–31.

Rippl, W. (1883) "The Capacity of Storage Reservoirs for Water Supply," *Proceedings of the Institution of Civil Engineers* (London), vol. 71, p. 270.

Rousculp, J. A. (1939) "Relation of Rainfall and Runoff to Cost of Sewers," *Transactions of the American Society of Civil Engineers,* vol. 104, p. 1473.

Sherman, L. K. (1932) "Stream-Flow from Rainfall by the Unit-Graph Method," *Engineering News Record,* vol. 108, pp. 501–505.

Theis, C. V. (1935) "The Relation Between Lowering of the Piezometric Surface and the Rate and Duration of Discharge of a Well Using Ground Water Storage," *Transactions of the American Geophysical Union,* vol. 16, pp. 519–524.

Thiem, G. (1906) *Hydrologische Methoden,* J. M. Gebhart, Leipzig, Germany.

Thomas, J. G., and A. Blaine (2009) *Soybean Irrigation,* Mississippi State University Extension Service, Publication 2185, http://msucares.com/pubs/publications/p2185.htm

University of Wyoming (1987) *Development of Evapotranspiration Crop Coefficients,* http://library.wrds.uwyo.edu/wrp/87-06/ch-04.html

U.S. Army Corps of Engineers (1960) *Runoff from Snowmelt,* Engineering Manual 1110-2-1406, Washington, DC.

U.S. EPA (1973) *Manual of Individual Water Supply Systems,* U.S. Environmental Protection Agency (EPA-430-9-73), Washington, DC, pp. 45–50, 107–109.

U.S. EPA (1997) *National Public Water Systems Compliance Report,* U.S. Environmental Protection Agency, Office of Water (EPA-305-R-99-002) Washington, DC.

Viesmann, W., G. L. Lewis, and J. W. Knapp (1989) *Introduction to Hydrology, 3rd Edition,* Harper & Row Publishers, New York, p. 84, 186.

Walton, W. C. (1970) *Groundwater Resource Evaluation,* McGraw-Hill, New York, p. 34.

Weibull, W. (1939) "Statistical Theory of the Strength of Materials," *Ing. Vetenskapsakad. Handl,* Stockholm, vol. 151, p. 15.

Yang, J., S. Wan, W. Deng, and G. Zhang (2007) " Water Fluxes at a Fluctuating Water Table and Groundwater Contributions to Wheat Water Use in the Lower Yellow River Flood Plain, China," *Hydrologic Processes,* vol. 21, no. 6, pp. 717–724.

CHAPTER
5

WATER CHEMISTRY

5-1 INTRODUCTION

A good working knowledge of water chemistry is critical to an engineer working in the environmental field. For most engineering students, studying chemistry is not what they had hoped to be doing during their engineering studies, but chemistry is the basis for many of the environmental problems that engineers must solve in this field. Mastering this topic will be a great help in understanding, and solving, environmental engineering challenges.

Water chemistry is important in the design of water and wastewater processes and in the remediation of environmental pollution, to give two examples. Of particular importance in water chemistry studies is the understanding of alkalinity chemistry, which, to a great extent, allows animals and plants to thrive. Natural alkalinity in the environment maintains the waters of the world—surface water, ground water, and oceans—at a near neutral pH thereby preventing them from becoming too acidic or too basic. When you jump in a lake to go swimming you don't even think about getting acid burns! It's the alkalinity balance of the environment that prevents that from happening. That same balance prevents many contaminants from dissolving into the water, which protects animal and plant life. In water environmental engineering we rely heavily on this phenomenon not only to protect water sources but also to treat water to render it potable or more aesthetically acceptable for consumption or to treat waste streams for discharge into the environment.

Another important aspect of water chemistry is that when we treat water we often have to add chemicals to react with impurities that are present and remove them or render them harmless. To do that, one needs to know how much of a chemical and what type of chemical should be added. Therefore a fundamental understanding of units for calculating doses is critical, as well as understanding basic reactions that determine how chemicals combine. Many contaminants are removed by adding a chemical that reacts with the contaminant to form a solid particle that can then be physically removed from the water. These are referred to as *precipitation reactions*, which are described in the coming sections. Some contaminants may be rendered harmless, acceptable to be left in the water, or will form a solid species if their oxidation state is changed. For example, iron in a plus 2 state (Fe^{+2}) is very soluble but when converted to a plus 3 state (Fe^{+3}) it tends to precipitate under natural conditions. Hence, chemistry associated with changing the oxidation state of compounds in water is important, and is referred to as *redox reactions*. Another important factor associated with using these precipitation reactions and redox reactions to treat water is a need to understand how long the reaction takes to occur so that if we are designing a tank for the reaction to take place we know how big to make it. The section on reaction kinetics addresses this concept.

Finally, some contaminants in water are volatile and can be removed from the water by transferring them to the air. Of course, this method of treating water is only acceptable if the contaminant in air does not pose a health threat. Generally, the amount of a volatile contaminant removed from water and dispersed into the air is at a low enough concentration to be acceptable or it can be removed from the air prior to discharge. The last section of the chapter deals with phase transfer from water to air. Some topics not addressed in this chapter but important in advanced environmental chemistry are further physical, analytical, and organic chemistry concepts.

Applications

This chapter will provide you with the tools to

- Convert general chemistry units of measure to those used in environmental engineering water chemistry.

- Estimate chemical doses for the design of chemical feed units for precipitation, neutralization, or oxidation of water contaminants.

- Estimate the mass of reaction products that must be handled by sludge disposal facilities.

- Estimate the amount of air required for biological wastewater treatment and air stripping of hazardous contaminants from drinking water.

5-2   BASIC WATER PROPERTIES AND UNITS

Physical Properties of Water

The basic physical properties of water relevant to water treatment are density and viscosity. Density is a measure of the concentration of matter and is expressed in three ways:

1. Mass density, ρ. *Mass density* is mass per unit volume and is measured in units of kg/m^3. Appendix A, Table A-1, shows the variation of density with temperature for pure water free from air. Dissolved impurities change the density in direct proportion to their concentration and their own density. In environmental engineering applications, it is common to ignore the density increase due to impurities in the water. However, environmental engineers do not ignore the density of the matter when dealing with high concentrations, such as thickened sludge or commercial liquid chemicals.

2. Specific weight, γ. *Specific weight* is weight (force) per unit volume, measured in units of kN/m^3. The specific weight of a fluid is related to its density by the acceleration of gravity, g, which is 9.81 m/s^2.

$$\gamma = \rho g \tag{5-1}$$

3. Specific gravity, S. Specific gravity is given by

$$S = \rho/\rho_0 = \gamma/\gamma_0 \tag{5-2}$$

where the subscript zero denotes the density of water at $3.98°C$, $1{,}000$ kg/m^3, and the specific weight of water, 9.81 kN/m^3.

For quick approximations, the density of water at normal temperature is taken as $1{,}000$ kg/m^3 (which is conveniently 1 kg/L) with a specific gravity $= 1.00$.

All substances, including liquids, exhibit a resistance to movement, an internal friction. The higher the friction, the harder it is to pump the liquid. A measure of the friction is viscosity. Viscosity is presented in one of two ways:

1. Dynamic viscosity, or absolute viscosity, μ, has dimensions of mass per unit length per time, with units of $Pa \cdot s$. Dynamic viscosity represents a measure of a fluid's internal resistance to flow. This is an important factor in the analysis of liquid flow, for example in designing pumps.

2. Kinematic viscosity, υ, is found by

$$\upsilon = \mu/\rho \tag{5-3}$$

and has dimensions of length squared per time with the corresponding units m^2/s.

This is an alternative numerical expression of the properties of a fluid. It is often used to evaluate the friction coefficient for flow in pipes.

States of Solution Impurities

From an environmental engineering point of view, substances can exist in water in one of three classifications—*suspended*, *colloidal*, or *dissolved*.

A dissolved substance is one that is truly in solution. The substance is homogeneously dispersed in the liquid. Dissolved substances can be simple atoms or complex molecular compounds. Dissolved substances are in the liquid, that is, there is only one phase present. The substance cannot be removed from the liquid without accomplishing a phase change such as distillation, precipitation, adsorption, extraction, or passage through "ionic" pore-sized membranes. In *distillation* either the liquid or the substance itself is changed from a liquid phase to a gas phase in order to achieve separation. In *precipitation* the substance in the liquid phase combines with another chemical to form a solid phase, thus achieving separation from the water. *Adsorption* also involves a phase change, wherein the dissolved substance reacts with a solid particle to form a solid particle-substance complex. *Liquid extraction* can separate a substance from water by extracting it into another liquid, hence a phase change from water to a different liquid. A membrane with pore sizes in the ionic-size range can separate dissolved substances from the solution by a high-pressure filtering process.

Suspended solids are large enough to settle out of solution or be removed by filtration. In this case there are two phases present, the liquid water phase and the suspended-particle solid phase. The lower size range of this class is 0.1 to $1.0 \mu m$, about the size of bacteria. In environmental engineering, suspended solids are defined as those solids that can be filtered by a glass fiber filter disc and are properly called filterable solids. Suspended solids can be removed from water by physical methods such as sedimentation, filtration, and centrifugation.

Colloidal particles are in the size range between dissolved substances and suspended particles. They are in a solid state and can be removed from the liquid by physical means such as very high-force centrifugation or filtration through membranes with very small pore spaces. However, the particles are too small to be removed by sedimentation or by normal filtration processes. Colloidal particles exhibit the Tyndall effect; that is, when light passes through a liquid containing colloidal particles, the light is reflected by the particles. The degree to which a colloidal suspension reflects light at a 90° angle to the entrance beam is measured by *turbidity*. Turbidity is a relative measure, and there are various standards against which a sample is compared. The most common standard is a nephelometric turbidity unit (NTU). For our purposes we will simply refer to the measure of turbidity as a turbidity unit (TU). For a given particle size, the higher the turbidity, the higher the concentration of colloidal particles.

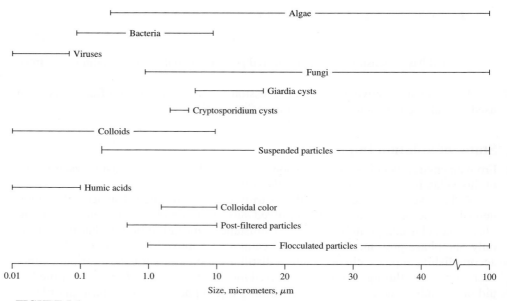

FIGURE 5-1

Particulates in water. (*Source:* McTigue and Cornwell, 1988.)

Another useful term in environmental engineering that is used to describe a solution state is *color*. Color is not separate from the above three categories, but rather is a combination of dissolved and colloidal materials. Color is widely used in environmental engineering because it, in itself, can be measured. However, it is very difficult to distinguish "dissolved color" from "colloidal color." Some color is caused by colloidal iron or manganese complexes, although the most common cause of color is from complex organic compounds that originate from the decomposition of organic matter. One common source of color is the degradation of soil humus, which produces humic acids. Humic acids impart reddish-brown color to the water. Humic acids have molecular weights between 800 and 50,000, the lower being dissolved and the greater, colloidal. Most color seems to be between 3.5 and 10μm, which is colloidal. Color is measured by the ability of the solution to absorb light. Color particles can be removed by the methods discussed for dissolved or colloidal particles, depending upon the state of the color. True color is a term used to describe the color after turbidity has been removed.

Figure 5-1 presents an overview by size of the types of particles that are often dealt with in water treatment. A technique that is being used in water treatment to help evaluate water quality is *particle counting*. A particle counter counts the number of particles in a water sample and reports the results by particle size, generally from 1 to 30μm. Figure 5-2 shows a sample count comparing the distribution of particles in a raw water to that of the finished water. While particle counting does not indicate anything about the kind of particle, it can be useful in assessing overall treatment efficiency as well as characterizing water sources.

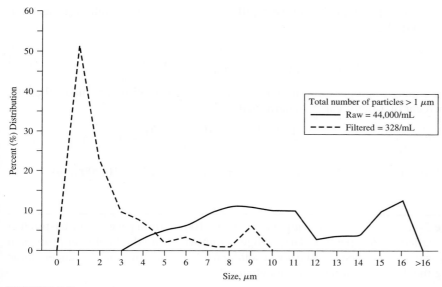

FIGURE 5-2
Particle distribution changes through treatment.

Chemical Units

Because solutes in solution are often analyzed by weight, the terms *weight percent* and *milligram per liter* are used. In order to perform stoichiometric calculations,* it is necessary to convert to common units, and the terms *molarity* and *normality* are used.

Weight percent, P, is sometimes employed to express approximate concentrations of commercial chemicals or of solid concentrations of sludges. The term specifies the grams of substance per 100 grams of solution and is mathematically expressed as

$$P = \frac{W}{W + W_0} \times 100\% \tag{5-4}$$

where P = percent of substance by weight
W = grams of substance
W_0 = grams of solution

Analysts generally give results directly in mass per volume (concentration), and the units are mg/L. In environmental engineering it is often assumed that the substance does not change the density of the water. This is generally untrue, but it does make for some useful conversions, and the assumption is not too inaccurate for dilute concentrations.

Stoichiometry is the part of chemistry concerned with measuring the proportions of elements involved in a reaction. Stoichiometric calculations are an application of the principle of conservation of mass to chemical reactions.

If such an assumption is made and we recall that 1 mL of water weighs 1 g (again an approximation), then

$$\frac{1 \text{ mg}}{L} = \frac{1 \text{ mg}}{1,000 \text{ g}} = \frac{1 \text{ mg}}{10^6 \text{ mg}} = 1 \text{ ppm} \qquad (5\text{-}5)$$

or 1 mg/L equals 1 part per million (ppm). If the same assumptions are made, then the weight percent of 1 mg/L can be determined:

$$P = \frac{W}{W + W_0} \times 100 = \frac{1 \text{ mg}(100)}{1 \text{ L}} = \frac{10^{-3} \text{ g}(100)}{10^3 \text{ g}} = 1 \times 10^{-4}\% \qquad (5\text{-}6)$$

or 1 mg/L equals $1 \times 10^{-4}\%$, which can be translated into $1\% = 10,000$ mg/L.

Example 5-1. A water treatment plant produces 100 kg of dry weight of by-product material, called *sludge,* per day. That sludge is removed from the treatment process in 10 m^3 of water. What is the actual percent solids and the approximate mg/L concentration assuming the density of water does not change?

Solution. Because 1 m^3 of water has a density at normal temperature of approximately 1,000 kg/m^3, the actual percent solids can be found by Equation 5-4.

$$P = \frac{100 \text{ kg}}{100 \text{ kg} + 10,000 \text{ kg}} \times 100\%$$

$$= 0.99\%$$

The approximate concentration in mg/L is found by dividing the weight of solids by the volume of water (1.0 m^3 = 1,000 L):

$$mg/L = \frac{100 \text{ kg} \times 10^6 \text{ mg/kg}}{10,000 \text{ L}}$$

$$mg/L = 10,000$$

Comment: Because P is approximately 1 percent, 10,000 mg/L is taken to be "1 percent." In a similar fashion, a 5.2 percent solution is taken to be 52,000 mg/L. Note that the error increases as the amount of dry solids increases. This is because the density of water increases as more solids are added and the approximation no longer holds. Typically, the density change must be accounted for when the solute concentration exceeds 10 percent.

In order to work with chemical reactions it is necessary to convert weight concentrations to molarity or normality. A *mole* is 6.02×10^{23} molecules of a substance. Chemical reactions are expressed in integral numbers of moles. A mole of a substance has a relative weight called its *molecular weight* (MW). Molecular weight is the sum of the atomic weights. A table of atomic weights is given inside the front cover of this book. *Molarity* is the number of moles in a liter of solution. A 1-molar ($1\,M$) solution has 1 mole of substance per liter of solution. Molarity is related to mg/L by

$$\text{mg/L} = \text{Molarity} \times \text{Molecular weight} \times 10^3 \qquad (5\text{-}7)$$
$$= (\text{moles/L})(\text{g/mole})(10^3\,\text{mg/g})$$

A second unit, *equivalent weight* (EW), is frequently used in softening and redox reactions. The equivalent weight is the molecular weight divided by the number (n) of electrons transferred in redox reactions or the number of protons transferred in acid/base reactions.

The value of n depends on how the molecule reacts. In this text we are concerned with molecules that react in acid/base reactions or precipitation reactions. **In an acid/base reaction, n is the number of hydrogen ions that the molecule transfers. That is, an acid gives up an EW of hydrogen ions, and a base accepts an EW of hydrogen ions. In a precipitation reaction, n is the valence of the element in question. For compounds, n is equal to the number of hydrogen ions that would be required to replace the cation; that is, for $CaCO_3$ it would take two hydrogen ions to replace the calcium, therefore, $n = 2$. In oxidation/reduction reactions, n is equal to the change in oxidation number that the compound undergoes in the reaction.** Obviously, it is difficult to recognize reaction capacity without the context of the reaction. Common valence states of elements found in water are listed in Appendix A.

Normality (N) is the number of equivalent weights per liter of solution and is related to molarity (M) by

$$N = Mn \qquad (5\text{-}8)$$

Example 5-2. Commercial sulfuric acid, H_2SO_4, is often purchased as a 93 weight percent solution. Find the mg/L of H_2SO_4 and the molarity and normality of the solution. Sulfuric acid has a specific gravity of 1.839.

Solution. Since 1 L of water weighs 1,000 g, 1 L of 100% H_2SO_4 weighs

$$1,000(1.839) = 1,839\,\text{g}$$

$(0.93)(1,839\,\text{g}) = 1,710\,\text{g}$ of H_2SO_4, or 1.7×10^6 mg/L of H_2SO_4 in a 93% solution. The molecular weight of H_2SO_4 is found by looking up the atomic weights on the inside cover of this book:

$$
\begin{aligned}
2H &= 2(1) &&= 2 \\
S &= &&\ 32 \\
4O &= 4(16) &&= \underline{64} \\
&&&\ 98\,\text{g/mole}
\end{aligned}
$$

The molarity is found by using Equation 5-7:

$$\frac{1,710\,\text{g/L}}{98\,\text{g/mole}} = 17.45\,\text{mole/L or } 17.45\,M$$

The normality is found from Equation 5-8, realizing that H_2SO_4 can give up two hydrogen ions and therefore $n = 2$ equivalents/mole:

$$N = 17.45\,\text{mole/L (2 equiv/mole)} = 34.9\,\text{equiv/L}$$

Example 5-3.   Find the weight of sodium bicarbonate, $NaHCO_3$, necessary to make a 1 M solution. Find the normality of the solution.

Solution.   The molecular weight of $NaHCO_3$ is 84; therefore by using Equation 5-7:

$$mg/L = (1 \text{ mole/L})(84 \text{ g/mole})(10^3 \text{ mg/g}) = 84{,}000$$

HCO_3^- is able to give or accept only one proton; therefore $n = 1$, and the normality is the same as the molarity.

Example 5-4.   Find the equivalent weight of each of the following: Ca^{2+}, CO_3^{2-}, $CaCO_3$.

Solution.   Equivalent weight was defined as

$$EW = \frac{\text{Atomic or molecular weight}}{n}$$

The units of EW are grams/equivalent (g/eq) or milligrams/milliequivalent (mg/meq).

For calcium, n is equal to the valence or oxidation state in water, so $n = 2$. From the table on the inside cover of the book, the atomic weight of Ca^{2+} is 40.08. The equivalent weight is then

$$EW = \frac{40.08}{2} = 20.04 \text{ g/eq or } 20.04 \text{ mg/meq}$$

For the carbonate ion (CO_3^{2-}) the oxidation state of 2^- is used for n since the base CO_3^{2-} can potentially accept 2 hydrogen ions (H^+). The molecular weight is

$$\begin{aligned} C = & & 12.01 \\ 3O = 3(16.00) = & & \underline{48.00} \\ & & 60.01 \end{aligned}$$

and the equivalent weight is

$$EW = \frac{60.01}{2} = 30.00 \text{ g/eq or } 30.00 \text{ mg/meq}$$

In $CaCO_3$, $n = 2$ since it would take two hydrogen ions to replace the cation (Ca^{2+}) to form carbonic acid, H_2CO_3. Its molecular weight is the sum of the atomic weights of Ca^{2+} and CO_3^{2-} and is, therefore, equal to $40.08 + 60.01 = 100.09$. Its equivalent weight is

$$EW = \frac{100.09}{2} = 50.04 \text{ g/eq or mg/meq}$$

5-3 CHEMICAL REACTIONS

There are four principal types of reactions of importance in environmental engineering: precipitation, acid/base, ion-association, and oxidation/reduction.

Dissolved ions can react with each other and form a solid compound. This phase-change reaction of dissolved to solid state is called a precipitation reaction. Typical of a precipitation reaction is the formation of calcium carbonate when a solution of calcium is mixed with a solution of carbonate:

$$Ca^{2+} + CO_3^{2-} \rightleftharpoons CaCO_3(s) \tag{5-9}$$

The (s) in the above reaction denotes that the $CaCO_3$ is in the solid state. When no symbol is used to designate state, it is assumed to be dissolved. The arrows in the reaction imply that the reaction is reversible and so could proceed to the right (that is, the ions are combining to form a solid) or to the left (that is, the solid is dissociating into the ions).

Often, out of convenience, we talk about compounds when in reality a compound does not exist in water. Take, for example, a water containing sodium chloride and calcium sulfate. We would say that the water has NaCl and $CaSO_4$ in it, but no implication is made regarding the association of Na and Cl or Ca and SO_4. The following reactions occur:

$$CaSO_4(s) \rightleftharpoons Ca^{2+} + SO_4^{2-} \tag{5-10}$$

and

$$NaCl(s) \rightleftharpoons Na^+ + Cl^- \tag{5-11}$$

such that the water consists of four unassociated ions: Na^+, Ca^{2+}, Cl^-, and SO_4^{2-}. Don't make the mistake of thinking that the sodium and chloride are together.

Acid/base reactions are a special type of ionization when a hydrogen ion is added to or removed from solution. An acid could be added to water to produce a hydrogen ion, as by the addition of hydrochloric acid to water with the reaction

$$HCl \rightleftharpoons H^+ + Cl^- \tag{5-12}$$

The above reaction is simplified in that it is assumed that water is present. The reaction is properly written

$$HCl + H_2O \rightleftharpoons H_3O^+ + Cl^- \tag{5-13}$$

A hydrogen ion could also be removed from water, as by the addition of a base:

$$NaOH + H_3O^+ \rightleftharpoons 2H_2O + Na^+ \tag{5-14}$$

In some cases, ions may exist in water complexed with other ions. Formation of dissolved complexes are ion-association reactions. In this case, the ions are "tied" together in the solution. The complex could be a neutral compound, such as soluble mercuric chloride:

$$Hg^{2+} + 2Cl^- \rightleftharpoons HgCl_2 \tag{5-15}$$

More often, the soluble complex has a charge and is itself an ion. Metal ion complexes are common examples:

$$Al^{3+} + OH^- \rightleftharpoons AlOH^{2+} \tag{5-16}$$

The $AlOH^{2+}$ is still soluble, but acts differently than did the individual species before complexation.

Oxidation/reduction reactions involve valence changes and the transfer of electrons. When iron metal corrodes, it releases electrons:

$$Fe^0 \rightleftharpoons Fe^{2+} + 2e^- \qquad (5\text{-}17)$$

If one element releases electrons, then another must be available to accept the electrons. In iron pipe corrosion, hydrogen gas is often produced:

$$2H^+ + 2e^- \rightleftharpoons H_2(g) \qquad (5\text{-}18)$$

where the symbol (g) indicates the hydrogen is in the gas phase.

Precipitation Reactions

All complexes are soluble in water to a certain extent. Likewise, all complexes are limited by how much can be dissolved in water. Some compounds, such as NaCl, are very soluble; other compounds, such as AgCl, are very insoluble—only a small amount will go into solution. Visualize a solid compound being placed in distilled water. Some of the compound will go into solution. At some time no more of the compound will dissolve, and equilibrium will be reached. The time to reach equilibrium may be seconds or centuries. The solubility reaction is written as follows:

$$A_aB_b(s) \rightleftharpoons aA^{b+} + bB^{a-} \qquad (5\text{-}19)$$

For example,

$$Ca_3(PO_4)_2(s) \rightleftharpoons 3Ca^{2+} + 2PO_4^{3-} \qquad (5\text{-}20)$$

Interestingly, the product of the activity of the ions (approximated by the molar concentration) is always a constant for a given compound at a given temperature. This constant is called the solubility constant, K_s. In the general form it is written as

$$K_s = [A]^a[B]^b \qquad \qquad (5\text{-}21)$$

where, in this text, [] denotes *molar* concentrations. **Do not use mg/L!** A table of constants is shown in Table 5-1 and in Appendix A. K_s values are often reported as pK_s, where

$$pK_s = -\log K_s \qquad (5\text{-}22)$$

TABLE 5-1
Selected solubility constants at 25°C

Substance	Equilibrium equation	pK_s	Application
Aluminum hydroxide	$Al(OH)_3(s) \rightleftharpoons Al^{3+} + 3OH^-$	32.9	Coagulation
Aluminum phosphate	$AlPO_4(s) \rightleftharpoons Al^{3+} + PO_4^{3-}$	20.0	Phosphate removal
Calcium carbonate	$CaCO_3(s) \rightleftharpoons Ca^{2+} + CO_3^{2-}$	8.305	Softening, corrosion control
Ferric hydroxide	$Fe(OH)_3(s) \rightleftharpoons Fe^{3+} + 3OH^-$	38.57	Coagulation, iron removal
Ferric phosphate	$FePO_4(s) \rightleftharpoons Fe^{3+} + PO_4^{3-}$	21.9	Phosphate removal
Magnesium hydroxide	$Mg(OH)_2(s) \rightleftharpoons Mg^{2+} + 2OH^-$	11.25	Softening

The constant works equally well whether we are dissolving a solid (reaction going to the right) or precipitating ions (reaction going to the left). If we place $A_aB_b(s)$ in water, for every a moles of A that dissolve, b moles of B will dissolve until equilibrium is reached. But kinetically* it might take years to happen.

When precipitating ions, it is possible to have a higher concentration of ions in solution than dictated by the solubility product. This is called a supersaturated solution.

Example 5-5. How many mg/L of PO_4^{3-} would be in solution at equilibrium with $AlPO_4(s)$?

Solution. The pertinent reaction is

$$AlPO_4(s) \rightleftharpoons Al^{3+} + PO_4^{3-}$$

The associated pK_s is found in Table 5-1 as 20.0 and calculated as

$$K_s = 10^{-20.0} = [Al][PO_4]$$

For every mole of $AlPO_4$ that dissolves, one mole of Al^{3+} and one mole of PO_4^{3-} are released into solution. At equilibrium, the molar concentration of Al^{3+} and PO_4^{3-} in solution will be equal, so we may say

$$[Al^{3+}] = [PO_4^{3-}] = X$$

Substituting X for each compound in the K_s expression,

$$10^{-20.0} = X^2$$

Solving for X (which is equal to PO_4^{3-}), we find $PO_4^{3-} = 10^{-10}$ moles per liter in solution. The molecular weight is 95 g/mole, so the concentration in mg/L is

$$(95 \text{ g/mole})(10^3 \text{ mg/g})(10^{-10} \text{ moles/L}) = 9.5 \times 10^{-6} \text{ mg/L}$$

Example 5-6. If 50.0 mg of CO_3^{2-} and 50.0 mg of Ca^{2+} are present in 1 L of water, what will be the final (equilibrium) concentration of Ca^{2+}?

Solution. The molecular weight of Ca^{2+} is 40.08 and that of CO_3^{2-} is 60.01, resulting in initial molar concentrations of 1.25×10^{-3} moles/L and 8.33×10^{-4} moles/L for Ca^{2+} and CO_3^{2-} respectively.

$$K_s = 10^{-pK_s} = 10^{-8.305} = [Ca^{2+}][CO_3^{2-}]$$

For every mole of Ca^{2+} that is removed from solution, one mole of CO_3^{2-} is removed from solution. If the amount removed is given by Z, then

---

Kinetics is the part of chemistry concerned with rates of reactions and factors that affect them.

$$10^{-8.305} = 4.95 \times 10^{-9} = [1.25 \times 10^{-3} - Z][8.33 \times 10^{-4} - Z]$$

$$1.04 \times 10^{-6} - (2.08 \times 10^{-3})Z + Z^2 = 0$$

$$Z = \frac{-b \pm \sqrt{b^2 - 4ac}}{2a}$$

$$= \frac{2.08 \times 10^{-3} \pm \sqrt{4.34 \times 10^{-6} - 4(1.04 \times 10^{-6})}}{2}$$

$$= 8.28 \times 10^{-4}$$

so that the final Ca^{2+} concentration is

$$[Ca^{2+}] = 1.25 \times 10^{-3} - 8.28 \times 10^{-4} = 4.22 \times 10^{-4} M$$

or

$$(4.22 \times 10^{-4} \text{ moles/L})(40 \text{ g/mole})(10^3 \text{ mg/g}) = 16.9 \text{ mg/L}$$

Acid/Base Reactions

For the purposes of this text, acids are defined as those compounds that release protons. Bases are those compounds that accept protons.

The simple reaction for the release of a proton is

$$HA \rightleftharpoons H^+ + A^- \tag{5-23}$$

In order for HA to release the proton (H^+), something must accept the proton. Often that something is water, that is,

$$H^+ + H_2O \rightleftharpoons H_3O^+ \tag{5-24}$$

resulting in the net reaction

$$HA + H_2O \rightleftharpoons H_3O^+ + A^- \tag{5-25}$$

It is understood that water is generally present. Hence Equation 5-23 is used in place of Equation 5-25. In the case of Equation 5-25, water is acting as the base; that is, it accepts the proton. If a base is added to water, the water can act as an acid.

$$B^- + H_2O \rightleftharpoons HB + OH^- \tag{5-26}$$

In the above reaction the base (B^-) accepts a proton from water. If a compound is a stronger acid than water, then water will act as a base. If a compound is a stronger base than water, then water will act as an acid.

You can quickly see that acid/base chemistry centers on water and that it is important to know how strong an acid water is. Water itself is ionized in water by the equation

$$H_2O \rightleftharpoons H^+ + OH^- \tag{5-27}$$

The degree of ionization of water is very small and can be measured by what is called the ion product of water, K_w. It is found by

$$K_w = [OH^-][H^+] \tag{5-28}$$

TABLE 5-2
Strong acids

Substance	Equilibrium equation	Significance
Hydrochloric acid	$HCl \rightarrow H^+ + Cl^-$	pH adjustment
Nitric acid	$HNO_3 \rightarrow H^+ + NO_3^-$	Analytical techniques
Sulfuric acid[a]	$H_2SO_4 \rightarrow 2H^+ + SO_4^{2-}$	pH adjustment, coagulation

[a]Dissociation of the second proton, $HSO_4^- \rightleftharpoons H^+ + SO_4^{2-}$, is actually a weak acid reaction with a pK_a of 1.92. As long as the pH of the solution is above 2.5, the release of both protons may be considered complete.

and has a value of 10^{-14} ($pK_w = 14$) at 25°C. A solution is said to be acidic if $[H^+]$ is greater than $[OH^-]$, neutral if equal, and basic if $[H^+]$ is less than $[OH^-]$. If the solution is neutral, then $[H^+] = [OH^-] = 10^{-7}$ M. If the solution is acidic, H^+ is greater than 10^{-7} M. A convenient expression for the hydrogen ion concentration is pH, given by

$$pH = -\log[H^+] \qquad (5\text{-}29)$$

Therefore, a neutral solution at 25°C has a pH of 7 (written pH 7), an acidic solution has a pH < 7, and a basic solution has a pH > 7.

Acids are classified as strong acids or weak acids. *Strong acids* have a tendency to donate their protons to water. For example,

$$HCl \rightarrow H^+ + Cl^- \qquad (5\text{-}30)$$

which we recall is the simplified form of

$$HCl + H_2O \rightarrow H_3O^+ + Cl^- \qquad (5\text{-}31)$$

A list of important strong acids is in Table 5-2. Note the use of the single arrow to signify that, for practical purposes, we may assume that the reaction proceeds completely to the right.

Example 5-7. If 100 mg of H_2SO_4 (MW = 98) is added to 1 L of water, what is the final pH?

Solution. Using the molecular weight of sulfuric acid we find

$$\left(\frac{100 \text{ mg}}{1 \text{ L } H_2O}\right)\left(\frac{1}{98 \text{ g/mole}}\right)\left(\frac{1}{10^3 \text{ mg/g}}\right) = 1.02 \times 10^{-3} \text{ mole/L}$$

The reaction is

$$H_2SO_4 \rightarrow 2H^+ + SO_4^{2-}$$

and therefore $2(1.02 \times 10^{-3})M$ H^+ is produced. The pH is

$$pH = -\log(2.04 \times 10^{-3}) = 2.69$$

TABLE 5-3
Selected weak acid dissociation constants at 25°C

Substance	Equilibrium equation	pK_a	Significance
Acetic acid	$CH_3COOH \rightleftharpoons H^+ + CH_3COO^-$	4.75	Anaerobic digestion
Carbonic acid	$H_2CO_3\ (CO_2 + H_2O) \rightleftharpoons H^+ + HCO_3^-$	6.35	Corrosion, coagulation,
	$HCO_3^- \rightleftharpoons H^+ + CO_3^{2-}$	10.33	softening, pH control
Hydrogen sulfide	$H_2S \rightleftharpoons H^+ + HS^-$	7.2	Aeration, odor control,
	$HS^- \rightleftharpoons H^+ + S^{2-}$	11.89	corrosion
Hypochlorous acid	$HOCl \rightleftharpoons H^+ + OCl^-$	7.54	Disinfection
Phosphoric acid	$H_3PO_4 \rightleftharpoons H^+ + H_2PO_4^-$	2.12	Phosphate removal
	$H_2PO_4^- \rightleftharpoons H^+ + HPO_4^{2-}$	7.20	plant nutrient,
	$HPO_4^{2-} \rightleftharpoons H^+ + PO_4^{3-}$	12.32	analytical

Weak acids are acids that do not completely dissociate in water. An equilibrium exists between the dissociated ions and undissociated compound. The reaction of a weak acid is

$$HW \rightleftharpoons H^+ + W^- \tag{5-32}$$

An equilibrium constant exists that relates the degree of dissociation:

$$K_a = \frac{[H^+][W^-]}{[HW]} \tag{5-33}$$

As with other K values,

$$pK_a = -\log K_a \tag{5-34}$$

A list of important weak acids in water and in wastewater treatment is in Table 5-3. By knowing the pH of a solution (which can be easily found with a pH meter) it is possible to get a rough idea of the degree of dissociation of the acid. For example, if the pH is equal to the pK_a (that is, $[H^+] = K_a$), then from Equation 5-33, $[HW] = [W^-]$ and the acid is 50 percent dissociated. If the $[H^+]$ is two orders of magnitude (100 times) less than the K_a, then $100[H^+] = K_a$ (or pH $\gg$ pK).

$$100[H^+] = \frac{[H^+][W^-]}{[HW]}$$

or $100\,[HW] = [W^-]$. We would conclude that essentially all the acid is dissociated ($W^- \gg HW$). Correspondingly, if pH $\ll$ pK then $[HW] \gg [W^-]$, and none of the acid is dissociated.*

Example 5-8.   If 15 mg/L of HOCl is added to a potable water for disinfection and the final measured pH is 7.0, what percent of the HOCl is not dissociated? Assume the temperature is 25°C.

---

*If $[H^+] < K_a$, then pH $>$ pK. The symbol $\gg$ means greater by two orders of magnitude.

Solution. The reaction is

$$HOCl \rightleftharpoons H^+ + OCl^-$$

From Table 5-3, we find the pK_a is 7.54 and

$$K_a = 10^{-7.54} = 2.88 \times 10^{-8}$$

Writing the equilibrium constant expression in the form of Equation 5-33

$$K_a = \frac{[H^+][OCl^-]}{[HOCl]}$$

and substituting the values for K_a and $[H^+]$

$$2.88 \times 10^{-8} = \frac{[10^{-7}][OCl^-]}{[HOCl]}$$

Solving for the HOCl concentration

$$[HOCl] = 3.47[OCl^-]$$

Since the fraction of HOCl that has not dissociated plus the OCl^- that was formed by the dissociation must, by the law of conservation of mass, equal 100 percent of the original HOCl added:

$$[HOCl] + [OCl^-] = 100\% \text{ (of the total HOCl added to the solution)}$$

then

$$3.47[OCl^-] + [OCl^-] = 100\%$$
$$4.47[OCl^-] = 100\%$$
$$[OCl^-] = \frac{100\%}{4.47} = 22.37\%$$

and

$$[HOCl] = 3.47(22.37\%) = 77.6\%$$

5-4 BUFFER SOLUTIONS

A solution that resists large changes in pH when an acid or base is added or when the solution is diluted is called a *buffer* solution. A solution containing a weak acid and its salt is an example of a buffer. Atmospheric carbon dioxide (CO_2) produces a natural buffer through the following reactions:

$$CO_2(g) \rightleftharpoons CO_2 + H_2O \rightleftharpoons H_2CO_3 \rightleftharpoons H^+ + HCO_3^- \rightleftharpoons 2H^+ + CO_3^{2-} \quad (5\text{-}35)$$

where H_2CO_3 = carbonic acid
HCO_3^- = bicarbonate ion
CO_3^{2-} = carbonate ion

This is perhaps the most important buffer system in water and wastewater treatment. We will be referring to it several times in this and subsequent chapters as the *carbonate buffer system.*

As depicted in Equation 5-35, the CO_2 in solution is in equilibrium with atmospheric $CO_2(g)$. Any change in the system components to the right of CO_2 causes the CO_2 either to be released from solution or to dissolve.

We can examine the character of the buffer system in resisting a change in pH by assuming the addition of an acid or a base and applying the law of mass action (Le Chatelier's principle). For example, if an acid is added to the system, it unbalances it by increasing the hydrogen ion concentration. Therefore, the carbonate combines with it to form bicarbonate. Bicarbonate reacts to form more carbonic acid, which in turn dissociates to CO_2 and water. The excess CO_2 can be released to the atmosphere in a thermodynamically open system. Alternatively, the addition of a base consumes hydrogen ions and the system moves to the right with the CO_2 being replenished from the atmosphere. When CO_2 is bubbled into the system or is removed by passing an inert gas such as nitrogen through the liquid (a process called *stripping*), the pH will change more dramatically because the atmosphere is no longer available as a source or sink for CO_2. Figure 5-3 summarizes the four general responses of the carbonate buffer system. The first two cases are common in natural settings when the reactions proceed over a relatively long period of time. In a water treatment plant, we can alter the reactions more quickly than the CO_2 can be replenished from the atmosphere. The second two cases are not common in natural settings. They are used in water treatment plants to adjust the pH.

In natural waters in equilibrium with atmospheric CO_2, the amount of CO_3^{2-} in solution is quite small in comparison to the HCO_3^- in solution. The presence of Ca^{2+} in the form of limestone rock or other naturally occurring sources of calcium results in the formation of calcium carbonate ($CaCO_3$), which is very insoluble. As a consequence, it precipitates from solution. The reaction of Ca^{2+} with CO_3^{2-} to form a precipitate is one of the fundamental reactions used to soften water.

Alkalinity

Alkalinity is defined as the sum of all titratable bases down to about pH 4.5. It is found by experimentally determining how much acid it takes to lower the pH of water to 4.5. In most waters the only significant contributions to alkalinity are the carbonate species and any free H^+ or OH^-. The total H^+ that can be taken up by a water containing primarily carbonate species is

$$\text{Alkalinity} = [HCO_3^-] + 2[CO_3^{2-}] + [OH^-] - [H^+] \tag{5-36}$$

where [] refers to concentrations in moles/L. In most natural water situations (pH 6 to 8), the OH^- and H^+ are negligible, such that

$$\text{Alkalinity} = [HCO_3^-] + 2[CO_3^{2-}] \tag{5-37}$$

Note that $[CO_3^{2-}]$ is multiplied by two because it can accept two protons. The pertinent acid/base reactions are

$$H_2CO_3 \rightleftharpoons H^+ + HCO_3^- \qquad pK_{a1} = 6.35 \text{ at } 25°C \tag{5-38}$$
$$HCO_3^- \rightleftharpoons H^+ + CO_3^{2-} \qquad pK_{a2} = 10.33 \text{ at } 25°C \tag{5-39}$$

Case I
Acid is added to carbonate buffer system[a]

Reaction shifts to the left as $H_2CO_3^*$ is formed when H^+ and HCO_3^- combine[b]

CO_2 is released to the atmosphere

pH is lowered slightly because the availability of free H^+ (amount depends on buffering capacity)

Case II
Base is added to carbonate buffer system

Reaction shifts to the right

CO_2 from the atmosphere dissolves into solution

pH is raised slightly because H^+ combines with OH^- (amount depends on buffering capacity)

Case III
CO_2 is bubbled into carbonate buffer system

Reaction shifts to the right because $H_2CO_3^*$ is formed when CO_2 and H_2O combine

CO_2 dissolves into solution

pH is lowered

Case IV
Carbonate buffer system is stripped of CO_2

Reaction shifts to the left to form more $H_2CO_3^*$ to replace that removed by stripping

CO_2 is removed from solution

pH is raised

[a]Refer to Equation 5-35
[b]The asterisk ∗ in the H_2CO_3 is used to signify the sum of CO_2 and H_2CO_3 in solution.

FIGURE 5-3
Behavior of the carbonate buffer system with the addition of acids and bases or the addition and removal of CO_2.

From the pK values, some useful relationships can be found. The more important ones are as follows:

1. Below pH of 4.5, essentially all of the carbonate species are present as H_2CO_3, and the alkalinity is negative (due to the H^+).

2. At a pH of 8.3 most of the carbonate species are present as HCO_3^-, and the alkalinity equals HCO_3^-.

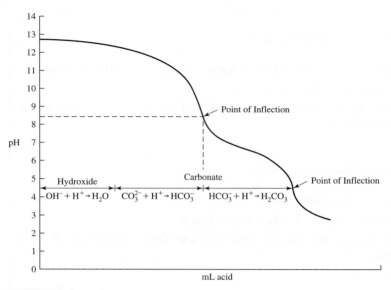

FIGURE 5-4
Titration curve for a hydroxide-carbonate mixture. (*Source:* Sawyer, McCarty, and Parkin, 1994.)

3. Above a pH of 12.3, essentially all of the carbonate species are present as CO_3^{2-}, and the alkalinity equals $2[CO_3^{2-}] + [OH^-]$. The $[OH^-]$ may not be insignificant at this pH.

Figure 5-4 schematically shows the change of species described above as the pH is lowered by the addition of acid to a water containing alkalinity. Note that the pH starts at above 12.3 and as acid is added the pH drops slowly as the first acid (H^+) addition is consumed by free hydroxide (OH^-), preventing a significant pH drop, and then the acid is consumed by carbonate (CO_3^{2-}) being converted to bicarbonate (HCO_3^-). At about pH 8.3 the carbonate is essentially all converted to bicarbonate, at which point there is another somewhat flat area where the acid is consumed by converting bicarbonate to carbonic acid.

From Equation 5-37 and our discussion of buffer solutions, it can be seen that alkalinity serves as a measure of buffering capacity. The greater the alkalinity, the greater the buffering capacity. In environmental engineering, then, we differentiate between alkaline water and water having high alkalinity. Alkaline water has a pH greater than 7, while a water with high alkalinity has a high buffering capacity. An alkaline water may or may not have a high buffering capacity. Likewise, a water with a high alkalinity may or may not have a high pH.

By convention, alkalinity is not expressed in molarity units as shown in the above equations, but rather in mg/L as $CaCO_3$. In order to convert species to mg/L as $CaCO_3$, multiply mg/L as the species by the ratio of the equivalent weight of $CaCO_3$ to the species equivalent weight:

$$\text{mg/L as } CaCO_3 = (\text{mg/L as species})\left(\frac{EW_{CaCO_3}}{EW_{species}}\right) \tag{5-40}$$

The alkalinity is then found by adding all the carbonate species and the hydroxide, and then subtracting the hydrogen ions. When using the units "mg/L as $CaCO_3$," the terms are added directly. The multiple of two for CO_3^{2-} has already been accounted for in the conversion.

Example 5-9.   A water contains 100.0 mg/L CO_3^{2-} and 75.0 mg/L HCO_3^- at a pH of 10. Calculate the alkalinity exactly at 25°C. Approximate the alkalinity by ignoring $[OH^-]$ and $[H^+]$.

Solution.   First, convert CO_3^{2-}, HCO_3^-, OH^-, and H^+ to mg/L as $CaCO_3$. The equivalent weights are

$$CO_3^{2-}: MW = 60, n = 2, EW = 30$$
$$HCO_3^-: MW = 61, n = 1, EW = 61$$
$$H^+: MW = 1, n = 1, EW = 1$$
$$OH^-: MW = 17, n = 1, EW = 17$$

and the concentration of H^+ and OH^- is calculated as follows: pH $= 10$; therefore $[H^+] = 10^{-10}\ M$. Using Equation 5-7,

$$mg/L = (10^{-10}\ moles/L)(1\ g/mole)(10^3\ mg/g) = 10^{-7}$$

Using Equation 5-28,

$$[OH^-] = \frac{K_w}{[H^+]} = \frac{10^{-14}}{10^{-10}} = 10^{-4}\ moles/L$$

and

$$mg/L = (10^{-4}\ moles/L)(17\ g/mole)(10^3\ mg/g) = 1.7$$

Now, the mg/L as $CaCO_3$ is found by using Equation 5-40 and taking the equivalent weight of $CaCO_3$ to be 50:

$$CO_3^{2-} = 100.0\left(\frac{50}{30}\right) = 167$$

$$HCO_3^- = 75.0\left(\frac{50}{61}\right) = 61$$

$$H^+ = 10^{-7}\left(\frac{50}{1}\right) = 5 \times 10^{-6}$$

$$OH^- = 1.7\left(\frac{50}{17}\right) = 5.0$$

The exact alkalinity (in mg/L) is found by

$$Alkalinity = 61 + 167 + 5.0 - (5 \times 10^{-6})$$
$$= 233\ mg/L\ as\ CaCO_3$$

It is approximated by $61 + 167 = 228$ mg/L as $CaCO_3$. This is a 2.2 percent error.

5-5   REACTION KINETICS

Many reactions that occur in the environment do not reach equilibrium quickly. Some examples include disinfection of water, gas transfer into and out of water, removal of organic matter from water, and radioactive decay. The study of how these reactions proceed is called *reaction kinetics*. The *rate of reaction, r*, is used to describe the rate of formation or disappearance of a compound. Reactions that take place in a single phase (that is, liquid, gas, or solid) are called *homogeneous* reactions. Those that occur at surfaces between phases are called *heterogeneous*. For each type of reaction, the rate may be defined as follows:

For homogeneous reactions

$$r = \frac{\text{Moles or milligrams}}{(\text{Unit volume})(\text{Unit time})} \tag{5-41}$$

For heterogeneous reactions

$$r = \frac{\text{Moles or milligrams}}{(\text{Unit surface})(\text{Unit time})} \tag{5-42}$$

Production of a compound results in a positive sign for the reaction rate $(+r)$, while disappearance of a substance yields a negative sign $(-r)$. Reaction rates are a function of temperature, pressure, and the concentration of reactants. For a stoichiometric reaction of the form:

$$a\text{A} + b\text{B} \rightarrow c\text{C}$$

where a, b, and c are the proportionality coefficients for the reactants A, B, and C, the change in concentration of compound A is equal to the reaction rate equation for compound A:

$$\frac{d[\text{A}]}{dt} = r_\text{A} = -k[\text{A}]^\alpha [\text{B}]^\beta = k[\text{C}]^\gamma \tag{5-43}$$

where [A], [B], and [C] are the concentrations of the reactants, and α, β, and γ are empirically determined exponents. The proportionality term, k, is called the *reaction rate constant*. It is often not a constant but, rather, is dependent on the temperature and pressure. Since A and B are disappearing, the sign of the reaction rate equation is negative. It is positive for C because C is being formed.

The *order of reaction* is defined as the sum of the exponents in the reaction rate equation. The exponents may be either integers or fractions. Some sample reaction orders are shown in Table 5-4.

TABLE 5-4
Example reaction orders

Reaction order	Rate Equation
Zero	$r_A = -k$
First	$r_A = -k[\text{A}]$
Second	$r_A = -k[\text{A}]^2$
Second	$r_A = -k[\text{A}][\text{B}]$

TABLE 5-5
**Plotting procedure to determine order of reaction by method of integration for
plug flow reactor and for a batch reactor**

Order	Rate equation	Integrated equation	Linear plot	Intercept	Slope
0	$\dfrac{d[A]}{dt} = -k$	$[A] - [A_0] = -kt$	t vs. $[A]$	$[A_0]$	$-k$
1	$\dfrac{d[A]}{dt} = -k[A]$	$\ln\dfrac{[A]}{[A_0]} = -kt$	t vs. $\ln[A]$	$\ln[A_0]$	$-k$
2	$\dfrac{d[A]}{dt} = -k[A]^2$	$\dfrac{1}{[A]} - \dfrac{1}{[A_0]} = kt$	t vs. $\dfrac{1}{[A]}$	$\dfrac{1}{[A_0]}$	k

For elementary reactions where the stoichiometric equation represents both the mass balance and the molecular scale process, the coefficients of proportionality (a, b, c) are equivalent to the exponents in the reaction rate equation:

$$r_A = -k[A]^a[B]^b \qquad (5\text{-}44)$$

The overall reaction rate, r, and the individual reaction rates are related:

$$r = \frac{r_A}{a} = \frac{r_B}{b} = \frac{r_C}{c} \qquad (5\text{-}45)$$

The reaction rate constant, k, may be determined experimentally by obtaining data on the concentrations of the reactants as a function of time and plotting on a suitable graph. The form of the graph is determined from the result of integration of the equations in Table 5-4. The integrated forms and the appropriate graphical forms are shown in Table 5-5.

5-6 GAS TRANSFER

An important example of time-dependent reactions is the mass transfer (dissolution or volatilization) of gas from water. In 1924 Lewis and Whitman postulated a two-film theory to describe the mass transfer of gases. According to their theory, the boundary between the gas phase and the liquid phase (also called the *interface*) is composed of two distinct films that serve as a barrier between the bulk phases (Figure 5-5). For a molecule of gas to go into solution, it must pass through the bulk of the gas, the gas film, the liquid film, and into the bulk of the liquid (Figure 5-5*a*). To leave the liquid, the gas molecule must follow the reverse course (Figure 5-5*b*). The driving force causing the gas to move, and hence the mass transfer, is the concentration gradient: $C_s - C$. C_s is the saturation concentration of the gas in the liquid, and C is the actual concentration. When C_s is greater than C, the gas will go into solution. When C is greater than C_s, the gas will desorb.

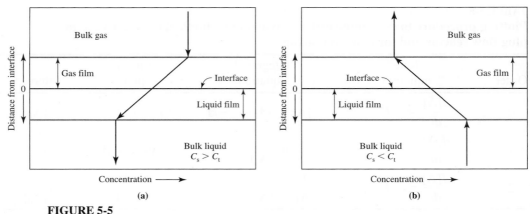

FIGURE 5-5

Two-film model of the interface between gas and liquid: (*a*) absorption mode and (*b*) desorption mode.

The relationship between the equilibrium concentration of gas dissolved in solution and the partial pressure of the gas is defined by *Henry's law*:

$$p = \frac{Hc}{P_T} \tag{5-46}$$

where c = the mole fraction of gas in water
  p = mole fraction of gas in air
  H = proportionality constant, known as Henry's constant (the slope of the straight-line portion of the distribution curve)
  P_T = total pressure atm

For water treatment, P_T is usually 1 atmosphere (atm). Various units are used by different investigators for the concentrations in the two phases and, therefore, the units for Henry's constant vary. Because of units, care must be taken in using the relationship, especially when obtaining constants from different sources. Below is a discussion of the primary methods of reporting Henry's law.

Probably the most common method of expressing Henry's law is with units of c and p as mole fractions:

$$p = \frac{Hc}{P_T} \tag{5-47}$$

where p = mol gas/mol air
  c = mol gas/mol water
  H = atm, actually = $\dfrac{\text{atm (mol gas/mol air)}}{\text{mol gas/mol water}}$
  P_T = atm, usually = 1

Recall that according to Dalton's law, a mole of gas per mole of air is the same as the partial pressure of the gas or is also the same as the volume of gas per volume of air. A useful conversion factor when calculating c is that 1 L of water contains 55.6 mol of water:

$$\frac{1000 \text{ g/L}}{18 \text{ g/mol}} = 55.6 \text{ mol/L}$$

Another method of reporting Henry's law is to utilize concentration units. In this case, the total pressure P_T is commonly defined as 1, and hence it is left off of the equation and atm is dropped from the units of H. In this case, any set of mass per volume or mole per volume units can be used (as long as p and c are the same), and hence it is often referred to as the dimensionless or unitless Henry's law constant:

$$p = H_u c \tag{5-48}$$

where p = concentration units, e.g., kg/m^3, mol/L, mg/L
$\quad H_u$ = unitless
$\quad c$ = same concentration units used for p

At 1 atm pressure and 0°C, 22.412 L of air is 1 mol of air. At other temperatures, 1 mol of air is $0.082T$ L [where T = temperature in kelvin (K)] of air. The following conversion between H and H_u can be made:

$$H_u = \left[H \frac{\text{atm (mol gas/mol air)}}{\text{mol gas/mol water}} \right] \left(\frac{\text{mol air}}{0.082T \text{ L air}} \right) \left(\frac{\text{L water}}{55.6 \text{ mol}} \right)$$

$$= \frac{H}{4.56T} \text{ or } H_u = H \times 7.49 \times 10^{-4} \text{ at } 20°C \tag{5-49}$$

Another method for reporting Henry's constant is to use mixed units for p and c. This is very common because units of partial pressure in the air phase and concentration units in the water phase tend to be used. Different variations are available. Two are shown below:

$$p = \frac{H_m c}{P_T} \tag{5-50}$$

where p = mol gas/mol air (partial pressure)
$\quad c$ = mol gas/m^3 water
$\quad H_m$ = atm $\times$ m^3 water/mol gas

$$= \left[H \frac{\text{atm(mol gas/mol air)}}{\text{mol gas/mol water}} \right] \left(\frac{m^3 \text{ water}}{55,600 \text{ mol}} \right)$$

$$= \frac{H}{55,600}$$

Finally, milligram per liter units for c may be used. This is very useful in water treatment:

$$p = \frac{H_D c}{P_T} \tag{5-51}$$

where p = mol gas/mol air (partial pressure)

c = mg/L

H_D = (atm)(L)/mg

$$H_D = \frac{H_m}{MW} = \frac{H}{55,600 \ MW}$$

MW = molecular weight of gas of interest

The Henry's law coefficient varies both with the temperature and the concentration of other dissolved substances. Henry's law constants are given in Appendix A.

Example 5-10. A water contains 6.84×10^{-8} moles of trichloroethylene (TCE, MW = 131.4) gas per mole of water. What is the concentration of TCE in the water as μg/L?

Solution. The number of grams of TCE present can be found by the molar concentration times the MW:

$$\text{grams TCE} = (6.84 \times 10^{-8} \text{ moles})(131.4)$$
$$= 9 \times 10^{-6} \text{ g} = 9 \ \mu g$$

This 9 μg of TCE is in one mole of water. Because a liter of water contains 55.6 moles, the TCE concentration is

$$\text{TCE} = \frac{9 \ \mu g}{\text{Mole water}} \times \frac{55.6 \text{ moles}}{\text{L water}}$$
$$= 500 \ \mu g/L$$

Example 5-11. If the Henry's constant for TCE is 560 at 20°C in units of atm, what will the water concentration be in the presence of 0.01 moles of air?

Solution. Using Equation 5-48

$$560 = H = p/c = \left(\frac{\text{moles TCE}}{\text{mole air}}\right) \Big/ \left(\frac{\text{moles TCE}}{\text{mole water}}\right)$$

For every mole of TCE that leaves the water (set at X), the same amount goes into 0.01 moles of air. Therefore, using 6.84×10^{-8} moles as the starting water concentration.

$$560 = \left(\frac{X \text{ moles of TCE}}{0.01 \text{ mole of air}}\right) \Big/ \left(\frac{6.84 \times 10^{-8} - X}{1.0 \text{ mole water}}\right)$$
$$X = 5.8 \times 10^{-8} \text{ moles TCE}$$

The equilibrium water concentration would be

$$6.84 \times 10^{-8} - 5.8 \times 10^{-8} = 1.04 \times 10^{-8}$$

By the same procedure as in Example 5–10

$$1.04 \times 10^{-8} \, (131.4)(10^6)(55.6) = 75.7 \, \mu g/L$$

Comment: In other words, if water containing 560 $\mu g/L$ TCE were brought into equilibrium with 0.01 moles of air, the TCE concentration in the water would reduce to 75.7 $\mu g/L$.

Example 5-12.   Very often to determine the amount of air needed to reduce a volatile gas in water, a unitless volumetric ratio called the air to water ratio (A:W) is used as a design value. Using Example 5-11, what is the unitless volume A:W ratio?

Solution.   We know that one mole of water is 1/55.6 L of water. This was exposed to 0.01 moles of air. Using the ideal gas law, the volume of 1.0 mole of air may be calculated as (R/P)(T) or (8.3143/101.325)(T):

$$0.082 \, (273 + 20) = 24 \, L$$

Therefore, the volumetric A:W ratio is

$$A{:}W = \frac{(0.01 \text{ moles air}) \left(\dfrac{24 \text{ L air}}{\text{mole air}} \right)}{(1 \text{ mole water}) \left(\dfrac{1}{55.6 \text{ moles/L water}} \right)} = 13{:}1 \text{ vol/vol}$$

The rate of mass transfer can be described by the following equation:

$$\frac{dC}{dt} = k_a(C_s - C) \tag{5-52}$$

where k_a = rate constant or mass transfer coefficient, s^{-1}.
   The difference between the saturation concentration and the actual concentration $(C_s - C)$ is called the *deficit*. Since the saturation concentration is a constant for a constant temperature and pressure, this is a first-order reaction.

Example 5-13.   A falling raindrop initially has no dissolved oxygen. The saturation concentration for the drop is 9.20 mg/L. If, after falling for two seconds, the droplet has an oxygen concentration of 3.20 mg/L, how long must the droplet fall (from the start of the fall) to achieve a concentration of 8.20 mg/L?

Solution.   We begin by calculating the deficit after two seconds, and that at a concentration of 8.20 mg/L:

$$\text{Deficit at 2 sec} = 9.20 - 3.20 = 6.00 \text{ mg/L}$$
$$\text{Deficit at } t \text{ sec} = 9.20 - 8.20 = 1.00 \text{ mg/L}$$

Now using the integrated form of the first-order rate equation from Table 5-5, noting that the rate of change is proportional to deficit and, hence, $[A] = (C_s - C)$ and that $[A_0] = (9.20 - 0.00)$,

$$\ln \frac{6.00}{9.20} = -k(2.00 \text{ s})$$

$$k = 0.2137 \text{ s}^{-1}$$

With this value of k, we can calculate a value for t:

$$\ln \frac{(9.20 - 8.20)}{9.20} = -(0.2137)(t)$$

$$t = 10.4 \text{ s}$$

5-7   CHAPTER REVIEW

When you have completed studying this chapter, you should be able to do the following without the aid of your textbook or notes:

1. Distinguish between dissolved substances, suspended solids, and colloidal substances based on their size and the mechanism by which they can be removed from water.

2. Define and calculate quantities of a given substance in water in percent by weight, parts per million (ppm), and milligrams per liter (mg/L), and convert from one unit of measure to the others.

3. Define buffer.

4. Define alkalinity in terms of all the chemical species found, that is, Equation 5-36.

5. Explain the effect of various chemical additions to the carbonate buffer system. Your explanation should include the effect on the displacement of the reaction (left or right), effect on CO_2 (into or out of solution), and effect on pH (increase, decrease, or no change).

With the aid of this text, you should be able to do the following:

6. Calculate the gram equivalent weight of a chemical or compound.

7. Calculate the molarity, normality, and concentration of a given chemical compound in milligrams per liter (mg/L) and convert from one unit of measure to the others.

8. Calculate the equilibrium concentration of a compound when it is in equilibrium with its precipitate.

9. Calculate the pH of a solution containing a strong or weak acid alone (neglecting the dissociation of water).

10. Convert a concentration of a compound to calcium carbonate equivalents.

11. Calculate the reaction rate constant (k) from a set of experimental data.

5-8 PROBLEMS

5-1. Show that a density of 1 g/mL is the same as a density of 1,000 kg/m³. (*Hint:* Some useful conversions are listed inside the back cover of this book.)

5-2. Show that a 4.50 percent by weight mixture contains 45.0 kg of substance in a cubic meter of water (that is, $4.50\% = 45.0$ kg/m³). Assume the density of water $= 1,000$ kg/m³.

5-3. What is the concentration of NH_3 (in mg/L) of household ammonia that contains 3.00 percent by weight of NH_3? Assume the density of water $= 1,000$ kg/m³.

> *Answer:* 30,000 mg/L

5-4. What is the concentration of chlorine (in mg/L) of household bleach that contains 5.25 percent by weight of Cl_2? Assume the density of water $= 1,000$ kg/m³.

5-5. Show that 1 mg/L $= 1$ g/m³.

5-6. In 2001 the U.S. Environmental Protection Agency promulgated a new standard MCL for arsenic in drinking water. The standard is now 10 parts per billion (ppb). What is the concentration in mg/L?

5-7. In a now antiquated and, we hope, soon to be forgotten system of measurements, it was common to consider water flows in terms of millions of gallons per day (MGD). Determine the number of MGD equivalent to the following flows in m³/s: 0.0438; 0.05; 0.438; 0.5; 4.38; and 5. Record both your calculated answer and the answer rounded to include only significant figures.

5-8. Calculate the molarity and normality of the following:

  a. 200.0 mg/L HCl
  b. 150.0 mg/L H_2SO_4
  c. 100.0 mg/L $Ca(HCO_3)_2$
  d. 70.0 mg/L H_3PO_4

> *Answers:*
>
> | Molarity (M) | Normality (N) |
> | --- | --- |
> | a. 0.005485 | 0.005485 |
> | b. 0.001529 | 0.003059 |
> | c. 0.0006168 | 0.001234 |
> | d. 0.000714 | 0.00214 |

5-9. Calculate the molarity and normality of the following:

    a. 80 μg/L HNO_3
    b. 135 μg/L $CaCO_3$
    c. 10 μg/L $Cr(OH)_3$
    d. 1000 μg/L $Ca(OH)_2$

5-10. Calculate the molarity and normality of the following:

    a. 0.05 mg/L As^{3+}
    b. 0.005 mg/L Cd^{2+}
    c. 0.002 mg/L Hg^{2+}
    d. 0.10 mg/L Ni^{2+}

5-11. Calculate the mg/L of the following:

    a. 0.01000 N Ca^{2+}
    b. 1.000 M HCO_3^-
    c. 0.02000 N H_2SO_4
    d. 0.02000 M SO_4^{2-}

        Answers: Ca^{2+} = 200.4 mg/L, HCO_3^- = 61.02 mg/L,
        H_2SO_4 = 980.7 mg/L, SO_4^{2-} = 1,921 mg/L

5-12. Calculate the μg/L of the following:

    a. 0.0500 N H_2CO_3
    b. 0.0010 M $CHCl_3$
    c. 0.0300 N $Ca(OH)_2$
    d. 0.0080 M CO_3^{2-}

5-13. Calculate the mg/L of the following:

    a. 0.2500 M NaOH
    b. 0.0704 M Na_2SO_4
    c. 0.0340 M $K_2Cr_2O_7$
    d. 0.1342 M KCl

5-14. How many mg/L of magnesium ion will remain in solution in water that is 0.001000 M in hydroxyl ion and at 25°C?

    Answer: 0.1367 mg/L

5-15. The Pherric, New Mexico, groundwater contains 1.800 mg/L of iron as Fe^{3+}. What pH is required to precipitate all but 0.300 mg/L of the iron at 25°C?

5-16. Determine the concentration, in moles per liter, to which the hydroxide concentration must be raised to produce a concentration of 0.200 mg/L of copper if the starting concentration is 2.00 mg/L. Estimate the resultant pH.

5-17. Given a saturated solution of calcium carbonate, how many moles of calcium ion will remain in solution after the addition of 3.16×10^{-4} M of Na_2CO_3 at 25°C? (Assume that the pH does not change.)

5-18. The solubility product of calcium fluoride (CaF_2) is 3.45×10^{-11} at 25°C. Will a fluoride concentration of 1.0 mg/L be soluble in a water containing 200 mg/L of calcium?

5-19. In preparation for a laboratory experiment, a technician makes up a saturated solution of $CaSO_4$. Because the container is unlabeled, a 5.00×10^{-3} M solution of Na_2SO_4 is accidently added to the container. What are the concentrations of calcium and sulfate after equilibrium is reached? Assume the pK_s of $CaSO_4$ is 4.31 at 25°C. Assume both solutions are at 25°C.

5-20. What amount of NaOH (a strong base), in mg, would be required to neutralize the acid in Example 5-7?

> *Answer:* 81.568 or 81.6 mg

5-21. The pH of a finished water from a softening process is 10.74. What amount of 0.02000 N sulfuric acid, in milliliters, is required to neutralize 1.000 L of the finished water? Assume the buffering capacity is zero.

5-22. How many milliliters of 0.02000 N hydrochloric acid would be required to perform the neutralization in Problem 5-21?

5-23. Using a computer spreadsheet program you have written, plot a titration curve of the pH of a 50.0 mL solution of 0.0200 N NaOH (a strong base) being titrated with 0.0200 N HCl (a strong acid) to a pH of 7.00.

5-24. Calculate the pH of a water at 25°C that contains 0.6580 mg/L of carbonic acid. Assume that $[H^+] = [HCO_3^-]$ at equilibrium and neglect the dissociation of water.

> *Answer:* pH = 5.66

5-25. If the pH in Problem 5-24 is adjusted to 4.50, what is the HCO_3^- concentration in moles/L?

5-26. What is the pH of a water at 25°C that contains 0.5000 mg/L of hypochlorous acid? Assume equilibrium has been achieved. Neglect the dissociation of water. Although it may not be justified by the data available to you, report the answer to two decimal places.

5-27. If the pH in Problem 5-26 is adjusted to 7.00, what would the OCl^- concentration in mg/L be at 25°C?

5-28. Convert the following from mg/L as the ion to mg/L as $CaCO_3$:

    a. 83.00 mg/L Ca^{2+}
    b. 27.00 mg/L Mg^{2+}
    c. 48.00 mg/L CO_2 (*Hint:* See footnote on p. 283)
    d. 220.00 mg/L HCO_3
    e. 15.00 mg/L CO_3^{2-}

Answers:

Ca^{2+} = 207.25 or 207.3 mg/L as $CaCO_3$
Mg^{2+} = 111.20 or l11.2 mg/L as $CaCO_3$
CO_2 = 109.18 or 109.2 mg/L as $CaCO_3$
HCO_3^- = 180.41 or 180.4 mg/L as $CaCO_3$
CO_3^{2-} = 25.02 or 25.0 mg/L $CaCO_3$

5-29. Convert the following from mg/L as the ion or compound to mg/L as $CaCO_3$:

    a. 200.00 mg/L HCl
    b. 280.00 mg/L CaO
    c. 123.45 mg/L Na_2CO_3
    d. 85.05 mg/L $Ca(HCO_3)_2$
    e. 19.90 mg/L Na^+

5-30. Convert the following from mg/L as $CaCO_3$ to mg/L as the ion or compound:

    a. 100.00 mg/L SO_4^{2-}
    b. 30.00 mg/L HCO_3^-
    c. 150.00 mg/L Ca^{2+}
    d. 10.00 mg/L H_2CO_3
    e. 150.00 mg/L Na^+

Answers:

SO_4^{2-} = 95.98, or 96.0 mg/L
HCO_3^- = 36.58, or 36.6 mg/L
Ca^{2+} = 60.07, or 60.1 mg/L
H_2CO_3 = 6.198, or 6.20 mg/L
Na^+ = 68.91 mg/L

5-31. Convert the following from mg/L as $CaCO_3$ to mg/L as the ion or compound:

    a. 10.00 mg/L CO_2
    b. 13.50 mg/L $Ca(OH)_2$
    c. 481.00 mg/L H_3PO_4
    d. 81.00 mg/L H_2PO_4
    e. 40.00 mg/L Cl^-

5-32. Convert 0.0100 N Ca^{2+} to mg/L as $CaCO_3$.

5-33. What is the "exact" alkalinity of a water that contains 0.6580 mg/L of bicarbonate, as the ion, at a pH of 5.66? No carbonate is present.

Answer: 0.4302 mg/L as $CaCO_3$

5-34. Calculate the "approximate" alkalinity (in mg/L as $CaCO_3$) of a water containing 120.0 mg/L of bicarbonate ion and 15.00 mg/L of carbonate ion.

5-35. Calculate the "exact" alkalinity of the water in Problem 5-34 if the pH is 9.43.

5-36. Calculate the "approximate" alkalinity (in mg/L as $CaCO_3$) of a water containing 15.00 mg/L of bicarbonate ion and 120.0 mg/L of carbonate ion.

5-37. Using Equations 5-28, 5-36, 5-38, and 5-39, derive two equations that allow calculation of the bicarbonate and carbonate alkalinities in mg/L as $CaCO_3$ from measurements of the total alkalinity (A) and the pH.

 Answers: (in mg/L as $CaCO_3$)

$$HCO_3^- = \frac{50{,}000\left\{\left(\dfrac{A}{50{,}000}\right) + [H^+] - \left(\dfrac{K_W}{[H^+]}\right)\right\}}{1 + \left(\dfrac{2K_2}{[H^+]}\right)}$$

$$CO_3^{2-} = \left(\frac{2K_2}{[H^+]}\right)(HCO_3^-)$$

where

A = total alkalinity, mg/L as $CaCO_3$
K_2 = second dissociation constant of carbonic acid
  = 4.68×10^{-11} at 25°C
K_W = ionization constant of water
  = 1×10^{-14} at 25°C
HCO_3^- = bicarbonate alkalinity in mg/L as $CaCO_3$
CO_3^{2-} = carbonate alkalinity in mg/L as $CaCO_3$

5-38. Using the solution to Problem 5-37, calculate the bicarbonate and carbonate alkalinities, in mg/L as $CaCO_3$, of a water having a total alkalinity of 233.0 mg/L as $CaCO_3$ and a pH of 10.47.

5-39. Using the solution to Problem 5-37, calculate the bicarbonate and carbonate alkalinities, in mg/L as $CaCO_3$, for a water sample taken from Well No. 1 at the Eastwood Manor Subdivision near McHenry, Illnois (Woller and Sanderson, 1976).

Well No. 1, Lab No. 02694, November 9, 1971

Iron	0.2	Silica (SiO_2)	20.0
Manganese	0.0	Fluoride	0.35
Ammonium	0.5	Boron	0.1
Sodium	4.7	Nitrate	0.0
Potassium	0.9	Chloride	4.5
Calcium	67.2	Sulfate	29.0
Magnesium	40.0	Alkalinity	284.0 as $CaCO_3$
Barium	0.5	pH	7.6 units

Note: All reported as "mg/L as the ion" unless stated otherwise.

5-40. If a water has a carbonate alkalinity of 120.00 mg/L as the ion and a pH of 10.30, what is the bicarbonate alkalinity in mg/L as the ion?

Answer: $HCO_3^- = 130.686$ or 130.7 mg/L

5-41. What is the pH of a water that contains 120.00 mg/L of bicarbonate ion and 15.00 mg/L of carbonate ion?

5-42. Calculate the alkalinity of the water in Problem 5-40, using the equations in Problem 5-37 and the "exact" method of Example 5-9.

5-43. The following data were obtained for an irreversible elementary reaction. Plot the data, determine the order of the reaction (zero, first, or second) and the rate constant k. Use a spreadsheet to plot the data and fit a curve.

Time, min	Reactant A Concentration, mmoles/L
0	2.80
1	2.43
2	2.12
5	1.39
10	0.69
20	0.17

5-44. Repeat Problem 5-43 for the following data.

Time, min	Reactant A Concentration, mmoles/L
0	48.0
1	6.22
2	3.32
3	2.27
5	1.39
10	0.704

5-9   DISCUSSION QUESTIONS

5-1. Would you expect a carbonated beverage to have a pH above, below, or equal to 7.0? Explain why.

5-2. Explain the word *turbidity* in terms that the mayor of a community could understand.

5-10   FE EXAM FORMATTED PROBLEMS

5-1.   The pH of a finished water from an excess lime softening process is 11.24. What volume of 0.0200 N sulfuric acid, in milliliters, is required to neutralize 1.00 L of the finished water? Assume the buffering capacity is zero.

a. 0.09 mL

b. 174 mL

c. 87 mL

d. 8.7 mL

5-2.   What pH is required to precipitate all but 0.20 mg/L of the iron from a raw water with an Fe^{3+} concentration of 2.1 mg/L? Assume the temperature is 25°C. The reaction is

$$Fe^{3+} + 3OH^- \rightleftharpoons Fe(OH)_3 \downarrow \text{ and } pK_s = 38.57.$$

a. 11.04

b. 1.96

c. 9.09

d. 2.96

5-3.   Estimate the pH of water that contains 0.6200 mg/L of carbonic acid. Assume that $[H^+] = [HCO_3^-]$ at equilibrium and that the dissociation of water may be neglected. The water temperature is 25°C. $pK_{a1} = 6.35$ and $pK_{a2} = 10.33$. GMW of $H_2CO_3 = 62$ g/mole

a. 2.67

b. 1.35

c. 5.67

d. 6.35

5-4.   Estimate the approximate alkalinity, in mg/L as $CaCO_3$, of water with a carbonate ion concentration of 17.0 mg/L and a bicarbonate ion concentration of 111.0 mg/L.

a. 119 mg/L as $CaCO_3$

b. 128 mg/L as $CaCO_3$

c. 148 mg/L as $CaCO_3$

d. 146 mg/L as $CaCO_3$

5-11   REFERENCES

Lewis, W. K., and W. G. Whitman (1924) "Principles of Gas Absorption," *Industrial Engineering Chemistry,* vol. 16, p. 1,215.

McTigue, N., and D. Cornwell, "The Use of Particle Counting for the Evaluation of Filter Performance," AWWA Seminar Proceedings, Filtration: Meeting New Standards, AWWA Conference, 1988.

Sawyer, C. N., P. L. McCarty, and G. F. Parkin (1994) *Chemistry for Environmental Engineering,* 4th ed., McGraw-Hill, Boston, p. 473.

Woller, D. M., and E. W. Sanderson (1976) *Public Water Supplies in McHenry County,* Illinois State Water Survey, Publication No. 60-19, Urbana, IL.

CHAPTER

6

WATER TREATMENT

6-1   INTRODUCTION

Approximately 80 percent of the United States population turn their taps on every day to take a drink of publicly supplied water. They all assume that when they take a drink it is safe. They probably never even think of safety.

There are approximately 170,000 public water systems in the United States. EPA classifies these water systems according to the number of people they serve, the source of their water, and whether they serve the same customers year-round or on an occasional basis. The following statistics are based on information in the *Safe Drinking Water Information System* (SDWIS), for the fiscal year ended September 2000, as reported to EPA by the states (U.S. EPA, 2000).

Public water systems provide water for human consumption through pipes or other constructed conveyances to at least 15 service connections or serve an average of at least 25 people for at least 60 days a year. EPA has defined three types of public water systems:

- Community Water System (CWS): A public water system that supplies water to the same population year-round.

- Non-Transient Non-Community Water System (NTNCWS): A public water system that regularly supplies water to at least 25 of the same people at least six months per year, but not year-round. Some examples are schools, factories, office buildings, and hospitals that have their own water systems.

- Transient Non-Community Water System (TNCWS): A public water system that provides water in a place such as a gas station or campground where people do not remain for long periods of time.

EPA also classifies water systems according to the number of people they serve:

- Very small water systems serve 25–500 people

- Small water systems serve 501–3,300 people

- Medium water systems serve 3,301–10,000 people

- Large water systems serve 10,001–100,000 people

- Very large water systems serve 100,000+ people

The following is the number of systems and population served in the year 2009 (*Note:* populations are not summed because some people are served by multiple systems and counted more than once):

- 51,651 CWS served 294.3 million people

- 18,395 NTNCWS served 6.2 million people

- 83,484 TNCWS served 13.3 million people

In developing nations, clean water is the exception rather than the rule.

In 2005 every 15 seconds a child under the age of 5 died from a water-related illness. Seventeen percent of the earth's population (1.2 billion people) do not have

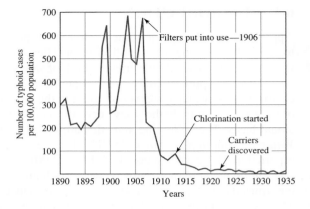

FIGURE 6-1

Typhoid fever cases per 100,000 population from 1890 to 1935, Philadelphia.

reliable drinking water and 40 percent of the population do not have access to adequate sanitation (www.waterforpeople.com).

The fact that the United States and developed countries have an outstanding water supply record is no accident. Since the early 1900s environmental engineers have been working in the United States to reduce waterborne disease. Figure 6-1 shows the incidence of typhoid cases in Philadelphia from 1890 to 1935. This is typical of the decrease in waterborne disease as public water supply treatment has increased. Philadelphia received its water from rivers and distributed the water untreated until about 1906, when slow sand filters were put into use. An immediate reduction in typhoid fever was realized. Disinfection of the water by the addition of chlorine further decreased the number of typhoid cases. A still greater decrease was accomplished after 1920 by careful control over infected persons who had become carriers. Since 1952 the death rate from typhoid fever in the United States has been less than 1 per 1,000,000. Many countries and organizations have committed their technological and financial resources to developing nations in recognition of the principle that reasonable access to safe and adequate drinking water is a fundamental right of all people.

Advances in public health tracking in the United States, and the more acute awareness of waterborne illnesses from a variety of organisms, have contributed to a better understanding of the link between water quality and illness. While many of the organisms associated with deaths or serious disease (e.g., typhoid, polio virus) have been eliminated in the United States, organisms with the potential to cause sickness and occasionally death are still being found. Between 1971 and 2006, 780 outbreaks of waterborne disease were reported in the United States (WQ&T, 2011) as shown in Figure 6-2a. Many of these illnesses are associated with gastrointestinal symptoms (diarrhea, fatigue, cramps). Table 6-1 shows the number of illnesses caused by various agents. Of those 418,819 cases, 403,000 cases are attributed to one outbreak of cryptosporidiosis in Milwaukee, Wisconsin, in 1993. It was estimated that 403,000 people became sick and 50 deaths occurred among severely immunocompromised individuals during this outbreak. At the time, the city complied with all water quality regulations. In 2006, EPA passed new regulations to address cryptosporidiosis.

Figure 6-2a graphically shows the number of outbreaks associated with drinking water by type of water system from 1987 to 2006 (Craun et al., 2010). The number of

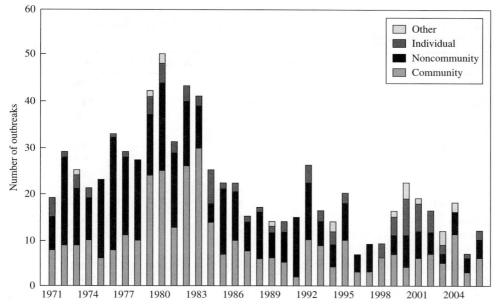

FIGURE 6-2a
Number of outbreaks associated with drinking water by water system type and year ($n = 780$), 1971 to 2006. "Other" includes outbreaks associated with bottled water ($n = 11$), mixed water system types ($n = 3$), unknown systems ($n = 3$), and bulk water purchase ($n = 1$).

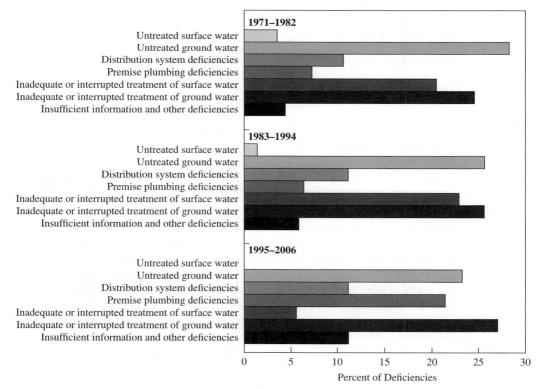

FIGURE 6-2b
Percentages of outbreak deficiencies ($n = 671$) in public water systems ($n = 656$) by time period, 1971 to 2006. (*Data Source:* Craun et al., 2010.)

TABLE 6-1
Agents causing waterborne disease outbreaks in the United States and territories, 1993–2006[a,b]

Agent	Number of outbreaks	Number of cases of illness
Unknown	**50**	**2761**
Bacteria	**63**	**3423**
Legionella spp.[c]	26	156
E. coli O157:H7	11	267
Campylobacter spp.	9	447
Shigella spp.	5	439
Salmonella spp.	4	928
Vibrio cholerae	2	114
Mixed bacteria	6	1072
Viruses	**17**	**3636**
Norovirus	15	3612
Hepatitis A	2	24
Parasites	**35**	**407001**
Giardia intestinalis (*G. lamblia*)	22	1908
Cryptosporidium parvum	10	404728
Entamoeba histolytica	1	59
Naegleria fowleri	1	2
Mixed parasite	1	304
Mixed Microbe Groups	**4**	**1698**
Chemicals	**34**	**300**
Copper	11	167
Nitrate/nitrite	6	21
Sodium hydroxide	4	38
Lead	3	3
Fluoride	2	43
Soap/cleaning product	2	15
Other chemical	6	13
Total	**203**	**418819**

(*Data Source:* WQ&T, 2011.)

outbreaks has decreased considerably since the early 1980s, although they continue to occur. The number of outbreaks associated with public water systems has decreased but those associated with individual water systems (private home or farm) has increased.

Figure 6-2*b* shows an evaluation of the outbreaks by type of deficiency that caused the outbreak. During the 36-year period, a trend analysis found a statistically significant decrease in the annual proportion of reported deficiencies in public water systems associated with contaminated, untreated surface water. In contrast, there was no statistically significant change in the annual proportion of reported deficiencies in public water system outbreaks associated with contaminated, untreated groundwater (Craun et al., 2010).

Craun reported that the annual proportion of treatment deficiencies involving surface water in public systems decreases over time. These deficiencies included inadequate disinfection of unfiltered surface water (e.g., contamination overwhelmed the chlorine dosage), interrupted disinfection (e.g, breakdown in equipment), inadequate filtration and/or pretreatment, and inadequate control of the addition of chemicals other than disinfectants (e.g., fluoride or sodium hydroxide). In contrast, no statistically significant change over time was observed in the annual proportion of treatment deficiencies involving groundwater in public systems. Inadequate or interrupted disinfection was the primary treatment deficiency for groundwater systems.

A water that can be consumed in any desired amount without concern for adverse health effects is termed a *potable* water. Potable does not necessarily mean that the water tastes good. This is in contrast to a *palatable* water, which is one that is pleasing to drink, but not necessarily safe. We have learned that we must provide a water that is both potable and palatable, for if it is not palatable people will turn to untreated water that may not be potable. The widespread availability of potable water in the United States does not mean that there are no operational or control deficiencies in water systems, especially in smaller systems. The scientific community is continually making advances in identifying contaminants and discovering potential long-term health effects of constituents that had not been previously identified.

Water Quality

Precipitation in the form of rain, hail, or sleet contains very few impurities. It may contain trace amounts of mineral matter, gases, and other substances as it forms and falls through the earth's atmosphere. The precipitation, however, has virtually no bacterial content (U.S. PHS, 1962).

Once precipitation reaches the earth's surface, many opportunities are presented for the introduction of mineral and organic substances, microorganisms, and other forms of pollution (contamination).* When water runs over or through the ground surface, it may pick up particles of soil. This is noticeable in the water as cloudiness or turbidity. It also picks up particles of organic matter and bacteria. As surface water seeps downward into the soil and through the underlying material to the water table, most of the suspended particles are filtered out. This natural filtration may be partially effective in removing bacteria and other particulate materials. However, the chemical characteristics of the water may change and vary widely when it comes in contact with mineral deposits. As surface water seeps down to the water table, it dissolves some of the minerals contained in the soil and rocks. Groundwater, therefore, often contains more dissolved minerals than surface water.

The following four categories are used to describe drinking water quality:

1. Physical: Physical characteristics relate to the quality of water for domestic use and are usually associated with the appearance of water, its color or turbidity, temperature, and, in particular, taste and odor.

2. Chemical: Chemical characteristics of waters are sometimes evidenced by their observed reactions, such as the comparative performance of hard and soft waters in laundering. Most often, differences are not visible.

---

*Pollution as used in this text means the presence in water of any foreign substances (organic, inorganic, radiological, or biological) that tend to lower its quality to such a point that it constitutes a health hazard or impairs the usefulness of the water.

3. **Microbiological:** Microbiological agents are very important in their relation to public health and may also be significant in modifying the physical and chemical characteristics of water.

4. **Radiological:** Radiological factors must be considered in areas where there is a possibility that the water may have come in contact with radioactive substances. The radioactivity of the water is of public health concern in these cases.

Consequently, in the development of a water supply system, it is necessary to examine carefully all the factors that might adversely affect the intended use of a water supply source.

Physical Characteristics

Turbidity.   The presence of suspended material such as clay, silt, finely divided organic material, plankton, and other particulate material in water is known as *turbidity*. The unit of measure is a Turbidity Unit (TU) or Nephlometric Turbidity Unit (NTU). It is determined by reference to a chemical mixture that produces a reproducible refraction of light. Turbidities in excess of 5 TU are easily detectable in a glass of water and are usually objectionable for aesthetic reasons.

Clay or other inert suspended particles in drinking water may not adversely affect health, but water containing such particles may require treatment to make it suitable for its intended use. Following a rainfall, variations in the groundwater turbidity may be considered an indication of surface or other introduced pollution.

Color.   Dissolved organic material from decaying vegetation and certain inorganic matter cause color in water. Occasionally, excessive blooms of algae or the growth of aquatic microorganisms may also impart color. While color itself is not usually objectionable from the standpoint of health, its presence is aesthetically objectionable and suggests that the water needs appropriate treatment.

Taste and Odor.   Taste and odor in water can be caused by foreign matter such as organic compounds, inorganic salts, or dissolved gases. These materials may come from domestic, agricultural, or natural sources. Drinking water should be free from any objectionable taste or odor at point of use.

Temperature.   The most desirable drinking waters are consistently cool and do not have temperature fluctuations of more than a few degrees. Groundwater and surface water from mountainous areas generally meet these criteria. Most individuals find that water having a temperature between $10° - 15°C$ is most palatable.

Chemical Characteristics

Chloride.   Most waters contain some chloride. The amount present can be caused by the leaching of marine sedimentary deposits or by pollution from sea water, brine, or industrial or domestic wastes. Chloride concentrations in excess of about 250 mg/L usually produce a noticeable taste in drinking water. Domestic water should contain less than 100 mg/L of chloride. In some areas, it may be necessary to use water with a chloride content in excess of 100 mg/L. In these cases, all of the other criteria for water purity must be met.

Fluoride. In some areas, water sources contain natural fluoride. Where the concentrations approach optimum levels, beneficial health effects have been observed. In such areas, the incidence of dental caries has been found to be below the levels observed in areas without natural fluoride. The optimum fluoride level for a given area depends upon air temperature, since temperature greatly influences the amount of water people drink. Excessive fluoride in drinking water supplies may produce fluorosis (mottling) of teeth, which increases as the optimum fluoride level is exceeded.* State or local health departments should be consulted for their recommendations, but acceptable levels are generally between 0.8 and 1.3 mg/L fluoride.

Iron. Small amounts of iron frequently are present in water because of the large amount of iron in the geologic materials. The presence of iron in water is considered objectionable because it imparts a brownish color to laundered goods and affects the taste of beverages such as tea and coffee.

Lead. Exposure of the body to lead, however brief, can be seriously damaging to health. Prolonged exposure to relatively small quantities may result in serious illness or death. Lead taken into the body in quantities in excess of certain relatively low "normal" limits is a cumulative poison.

Manganese. Manganese imparts a brownish color to water and to cloth that is washed in it. It flavors coffee and tea with a medicinal taste.

Sodium. The presence of sodium in water can affect persons suffering from heart, kidney, or circulatory ailments. When a strict sodium-free diet is recommended, any water should be regarded with suspicion. Home water softeners may be of particular concern because they add large quantities of sodium to the water. (See Section 6-3 for an explanation of the chemistry and operation of softeners).

Sulfate. Waters containing high concentrations of sulfate, caused by the leaching of natural deposits of magnesium sulfate (Epsom salts) or sodium sulfate (Glauber's salt), may be undesirable because of their laxative effects.

Zinc. Zinc is found in some natural waters, particularly in areas where zinc ore deposits have been mined. Zinc is not considered detrimental to health, but it will impart an undesirable taste to drinking water.

Arsenic. Arsenic occurs naturally in the environment and is also widely used in timber treatment, agricultural chemicals (pesticides), and manufacturing of gallium arsenide wafers, glass, and alloys. Arsenic in drinking water is associated with lung and urinary bladder cancer.

---

*Mottled teeth are characterized by black spots or streaks and may become brittle when exposed to large amounts of fluoride.

Toxic Inorganic Substances. *Nitrates* (NO_3), *cyanides* (CN), and *heavy metals* constitute the major classes of inorganic substances of health concern. Methemoglobinemia (infant cyanosis or "blue baby syndrome") has occurred in infants who have been given water or fed formula prepared with water having high concentrations of nitrate. CN ties up the hemoglobin sites that bind oxygen to red blood cells. This results in oxygen deprivation. A characteristic symptom is that the patient has a blue skin color. This condition is called cyanosis. CN causes chronic effects on the thyroid and central nervous system. The toxic heavy metals include arsenic (As), barium (Ba), cadmium (Cd), chromium (Cr), lead (Pb), mercury (Hg), selenium (Se), and silver (Ag). The heavy metals have a wide range of effects. They may be acute poisons (As and Cr^{6+} for example), or they may produce chronic disease (Pb, Cd, and Hg for example).

Toxic Organic Substances. There are over 120 toxic organic compounds listed on the U.S. Environmental Protection Agency's Priority Pollutant List. These include pesticides, insecticides, and solvents. Like the inorganic substances, their effects may be acute or chronic.

Microbiological Characteristics

Water for drinking and cooking purposes must be made free from disease-producing organisms (*pathogens*). These organisms include viruses, bacteria, protozoa, and helminths (worms).

Some organisms which cause disease in people originate with the fecal discharges of infected individuals. Others are from the fecal discharge of animals.

Unfortunately, the specific disease-producing organisms present in water are not easily identified. The techniques for comprehensive bacteriological examination are complex and time-consuming. It has been necessary to develop tests that indicate the relative degree of contamination in terms of an easily defined quantity. The most widely used test estimates the number of microorganisms of the *coliform group*. This grouping includes two genera: *Escherichia coli* and *Aerobacter aerogenes.* The name of the group is derived from the word *colon.* While *E. coli* are common inhabitants of the intestinal tract, *Aerobacter* are common in the soil, on leaves, and on grain; on occasion they cause urinary tract infections. The test for these microorganisms, called the *Total Coliform Test,* was selected for the following reasons:

1. The coliform group of organisms normally inhabits the intestinal tracts of humans and other mammals. Thus, the presence of coliforms is an indication of fecal contamination of the water.

2. Even in acutely ill individuals, the number of coliform organisms excreted in the feces outnumber the disease-producing organisms by several orders of magnitude. The large numbers of coliforms make them easier to culture than disease-producing organisms.

3. The coliform group of organisms survives in natural waters for relatively long periods of time, but does not reproduce effectively in this environment. Thus, the presence of coliforms in water implies fecal contamination rather than growth of the organism because of favorable environmental conditions. These

organisms also survive better in water than most of the bacterial pathogens. This means that the absence of coliforms is a reasonably safe indicator that pathogens are not present.

4. The coliform group of organisms is relatively easy to culture. Thus, laboratory technicians can perform the test without expensive equipment.

Current research indicates that testing for *Escherichia coli* specifically may be warranted. Some agencies prefer the examination for *E. coli* as a better indicator of biological contamination than total coliforms.

The two protozoa of most concern are *Giardia* cysts and *Cryptosporidium* oocysts. Both pathogens are carried by animals in the wild and on farms and make their way into the environment and water sources. Both are associated with gastrointestinal illness.

Radiological Characteristics

The development and use of atomic energy as a power source and the mining of radioactive materials, as well as naturally occurring radioactive materials, have made it necessary to establish limiting concentrations for the intake into the body of radioactive substances, including drinking water.

The effects of human exposure to radiation or radioactive materials are harmful, and any unnecessary exposure should be avoided. Humans have always been exposed to natural radiation from water, food, and air. The amount of radiation to which the individual is normally exposed varies with the amount of background radioactivity. Water with high radioactivity is not normal and is confined in great degree to areas where nuclear industries are situated.

Water Quality Standards

President Ford signed the National Safe Drinking Water Act (SDWA) into law on December 16, 1974. The Environmental Protection Agency (EPA) was directed to establish *maximum contaminant levels* (MCLs) for public water systems to prevent the occurrence of any known or anticipated adverse health effects with an adequate margin of safety. EPA defined a public water system to be any system that provides piped water for human consumption, if such a system has at least 15 service connections or regularly serves an average of at least 25 individuals daily at least 60 days out of the year. This definition includes private businesses, such as service stations, restaurants, motels, and others that serve more than 25 persons per day for greater than 60 days out of the year.

From 1975 through 1985, the EPA regulated 23 contaminants in drinking water supplied by public water systems. These regulations are known as interim primary drinking water regulations (IPDWRs). In June of 1986, the SDWA was amended. The amendments required EPA to set *maximum contaminant level goals* (MCLGs) and MCLs for 83 specific substances. This list included 22 of the IPDWRs (all except trihalomethanes). The amendments also required EPA to regulate 25 additional contaminants every three years beginning in January 1991 and continuing for an indefinite period of time. Had EPA met these statutory requirements, the number of regulated contaminants would have exceeded 250 in 2007. Figure 6-3 shows the timetable of the number of contaminants that have been regulated as of 2006.

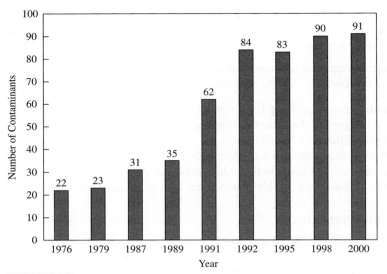

FIGURE 6-3
Number of regulated contaminants from 1976 through 2000. (*Data Source:* United States Environmental Protection Agency.)

The SDWA was amended again in 1996. These amendments contained four basic elements: (1) a new standard-setting process with language on how to select contaminants for regulation; (2) specific deadlines to regulate certain contaminants such as arsenic, sulfate, radon, and disinfection by-products; (3) new state programs for source water assessments, capacity development, operator certification, and a revolving loan fund; and (4) new public information programs.

Table 6-2 lists each regulated contaminant and summarizes its adverse health effects. Some of these contaminant levels are being considered for revision. The notation "TT" in the table means that a treatment technique is specified rather than a contaminant level. The treatment techniques are specific processes that are used to treat the water. Some examples include coagulation and filtration, lime softening, and ion exchange. These will be discussed in the following sections.

Lead and Copper. In 1991, EPA issued regulations for lead and copper, and established a monitoring program and a treatment technique for both. The MCL action levels are 0.015 mg/L for lead and 1.3 mg/L for copper.

Compliance with the regulations is also based on the quality of the water at the consumer's tap. Monitoring is required by means of collection of first-draw samples at residences. If the action level is exceeded, steps must be taken under this treatment technique to reduce the levels.

The SDWA amendments forbid the use of pipe, solder, or flux that is not lead-free in the installation or repair of any public water system or in any plumbing system providing water for human consumption. In 2007, EPA made short-term changes to the rule. The lead and copper rule is currently (2011) being revised.

TABLE 6-2
National primary drinking water standards

Contaminant	MCLG, mg/L	MCL, mg/L	Potential health effects	Sources of drinking water contamination
Fluoride Rule				
Fluoride	4.0	4.0	Skeletal and dental fluorosis	Natural deposits; fertilizer, aluminum industries; drinking water additive
Phase I Volatile Organics				
Benzene	zero	0.005	Cancer	Some foods; gas, drugs, pesticide, paint, plastic industries
Carbon tetrachloride	zero	0.005	Cancer	Solvents and their degradation products
p-Dichlorobenzene	0.075	0.075	Cancer	Room and water deodorants and mothballs
1,2-Dichloroethane	zero	0.005	Cancer	Leaded gas, fumigants, paints
1,1-Dichloro-ethylene	0.007	0.007	Cancer, liver, kidney effects	Plastics, dyes, perfumes, paints
Trichloro-ethylene	zero	0.005	Cancer	Textiles, adhesives, metal degreasers
1,1,1-Tri-chloroethane	0.2	0.2	Liver, nervous system effects	Adhesives, aerosols, textiles, paints, inks, metal degreasers
Vinyl chloride	zero	0.002	Cancer	May leach from PVC pipe; formed by solvent breakdown
Surface Water Treatment Rule and Total Coliform Rule				
Giardia lamblia	zero	TT	Gastroenteric disease	Human and animal fecal wastes
Legionella	zero	TT	Legionnaire's disease	Natural waters; can grow in water heating systems
Heterotrophic plate count	N/A	TT	Indicates water quality, effectiveness of treatment	
Total coliform	zero	< 5%+	Indicates gastroenteric pathogens	Human and animal fecal wastes
Turbidity	N/A	TT	Interferes with disinfection	Soil runoff
Viruses	zero	TT	Gastroenteric disease	Human and animal fecal wastes
Phase II Rule Inorganics				
Asbestos (>10 μm)	7 MFL	7 MFL	Cancer	Natural deposits; asbestos cement in water systems
Barium	2	2	Circulatory system effects	Natural deposits; pigments, epoxy sealants, spent coal
Cadmium	0.005	0.005	Kidney effects	Galvanized pipe corrosion; natural deposits; batteries, paints
Chromium (total)	0.1	0.1	Liver, kidney, circulatory disorders	Natural deposits; mining, electroplating, pigments
Mercury (inorganic)	0.002	0.002	Kidney, nervous system disorders	Crop runoff; natural deposits; batteries, electrical switches
Nitrate	10	10	Methemoglobinemia	Animal waste, fertilizer, natural deposits, septic tanks, sewage
Nitrite	1	1	Methemoglobinemia	Same as nitrate; rapidly converted to nitrate
Nitrate + nitrite	10	10		
Selenium	0.05	0.05	Liver damage	Natural deposits; mining, smelting, coal/oil combustion

TABLE 6-2
National primary drinking water standards (*continued*)

Contaminant	MCLG, mg/L	MCL, mg/L	Potential health effects	Sources of drinking water contamination
Phase II Rule Organics				
Acrylamide	zero	TT	Cancer, nervous system effects	Polymers used in sewage/waste-water treatment
Alachlor	zero	0.002	Cancer	Runoff from herbicide on corn, soybeans, other crops
Aldicarb	delayed	delayed	Nervous system effects	Insecticide on cotton, potatoes, other crops; widely restricted
Aldicarb sulfone	delayed	delayed	Nervous system effects	Biodegradation of aldicarb
Aldicarb sulfoxide	delayed	delayed	Nervous system effects	Biodegradation of aldicarb
Atrazine	0.003	0.003	Mammary gland tumors	Runoff from use as herbicide on corn and noncropland
Carbofuran	0.04	0.04	Nervous, reproductive system effects	Soil fumigant on corn and cotton; restricted in some areas
Chlordane	zero	0.002	Cancer	Leaching from soil treatment for termites
Chlorobenzene	0.1	0.1	Nervous system, liver effects	Waste solvent from metal degreasing processes
2,4-D	0.07	0.07	Liver and kidney damage	Runoff from herbicide on wheat, corn, rangelands, lawns
o-Dichloroethylene	0.1	0.1	Liver, kidney, blood cell damage	Paints, engine cleaning compounds, dyes, chemical
cis-1,2-Dichloroethylene	0.07	0.07	Liver, kidney, nervous, circulatory system effects	Waste industrial extraction solvents
trans-1,2-Dichloroethylene	0.1	0.1	Liver, kidney, nervous, circulatory system effects	Waste industrial extraction solvents
1,2-Dibromo-3-chloropropane	zero	0.0002	Cancer	Soil fumigant on soybeans, cotton, pineapple, orchards
1,2-Dichloropropane	zero	0.005	Liver, kidney effects; cancer	Soil fumigant; waste industrial solvents
Epichlorohydrin	zero	TT	Cancer	Water treatment chemicals; waste epoxy resins, coatings
Ethylbenzene	0.7	0.7	Liver, kidney, nervous system effects	Gasoline; insecticides; chemical manufacturing wastes
Ethylene dibromide	zero	0.00005	Cancer	Leaded gas additives; leaching of soil fumigant
Heptachlor	zero	0.0004	Cancer	Leaching of insecticide for termites, very few crops
Heptachlor epoxide	zero	0.0002	Cancer	Biodegradation of heptachlor
Lindane	0.0002	0.0002	Liver, kidney, nervous system, immune system, circulatory system effects	Insecticide on cattle, lumber, gardens; restricted in 1983
Methoxychlor	0.04	0.04	Growth; liver, kidney, nervous, system effects	Insecticide for fruits, vegetables, alfalfa, livestock, pets
Pentachlorophenol	zero	0.001	Cancer; liver, kidney effects	Wood preservatives, herbicide, cooling tower wastes
PCBs	zero	0.0005	Cancer	Coolant oils from electrical transformers; plasticizers
Styrene	1	1	Liver, nervous system effects	Plastics, rubber, resin, drug damage industries; leachate from city landfills
Tetrachloroethylene	zero	0.005	Cancer	Improper disposal of dry cleaning and other solvents
Toluene	1	1	Liver, kidney, nervous, system, circulatory system effects	Gasoline additive; manufacturing and solvent operations
Toxaphene	zero	0.003	Cancer	Insecticide on cattle, cotton, soybeans; cancelled in 1982
2,4,5-TP (Silvex)	0.05	0.05	Liver, kidney damage	Herbicide on crops, rights-of-way, golf courses; cancelled in 1983
Xylenes (total)	10	10	Liver, kidney, nervous system effects	By-product of gasoline refining; paints, inks, detergents

TABLE 6-2
National primary drinking water standards (*continued*)

Contaminant	MCLG, mg/L	MCL, mg/L	Potential health effects	Sources of drinking water contamination
Lead and Copper Rule				
Lead	zero	TT#	Kidney, nervous system effects	Natural/industrial deposits; plumbing solder, brass alloy faucets
Copper	1.3	TT#	Gastrointestinal irritation	Natural/industrial deposits; wood preservatives, plumbing
Phase V Inorganics				
Antimony	0.006	0.006	Cancer	Fire retardents, ceramics, electronics, fireworks
Beryllium	0.004	0.004	Bone, lung damage	Electrical, aerospace, defense industries
Cyanide	0.2	0.2	Thyroid, nervous system damage	Electroplating, steel, plastics, mining, fertilizer
Nickel	0.1	0.1	Heart, liver damage	Metal alloys, electroplating, batteries, chemical production
Thallium	0.0005	0.002	Kidney, liver, brain, intestinal effects	Electronics, drugs, alloys, glass
Phase V Organics				
Adipate, (di(2-ethylhexyl))	0.4	0.4	Decreased body weight	Synthetic rubber, food packaging, cosmetics
Benzo(a)pyrene (PAHs)	zero	0.0002	Cancer	Coal tar coatings; burning organic matter; volcanoes, fossil fuels
Dalapon	0.2	0.2	Liver, kidney effects	Herbicide on orchards, beans, coffee, lawns, roads, railways
Di(2-ethylhexyl) phathalate	zero	0.006	Cancer	PVC and other plastics
Dichloromethane	zero	0.005	Cancer	Paint stripper, metal degreaser, propellant, extractant
Di(2-ethylhexyl) phathalate	zero	0.006	Cancer	PVC and other plastics
Dinoseb	0.007	0.007	Thyroid, reproductive organ damage	Runoff of herbicide from crop and noncrop applications
Diquat	0.02	0.02	Liver, kidney, eye effects	Runoff of herbicide on land and aquatic weeds
Dioxin	zero	3×10^{-8}	Cancer	Chemical production by-product; impurity in herbicides
Endothall	0.1	0.1	Liver, kidney, gastrointestinal effects	Herbicide on crops, land/aquatic weeds; rapidly degraded
Glyphosate	0.7	0.7	Liver, kidney damage	Herbicide on grasses, weeds, brush
Hexachlorobenzene	zero	0.001	Cancer	Pesticide production waste by-product
Hexachlorocyclopentadiene	0.05	0.05	Kidney, stomach damage	Pesticide production intermediate
Oxamyl (Vydate)	0.2	0.2	Kidney damage	Insecticide on apples, potatoes, tomatoes
Picloram	0.5	0.5	Kidney, liver damage	Herbicide on broadleaf and woody plants
Simazine	0.004	0.004	Cancer	Herbicide on grass sod, some crops, aquatic algae
1,2,4-Trichlorobenzene	0.07	0.07	Kidney, liver damage	Herbicide production; dye carrier
1,1,2-Trichloroethane	0.003	0.003	Kidney, liver, nervous system damage	Solvent in rubber, other organic products; chemical production wastes

TABLE 6-2
National primary drinking water standards (*continued*)

Contaminant	MCLG, mg/L	MCL, mg/L	Potential health effects	Sources of drinking water contamination
Disinfectants and Disinfection By-Products				
Bromate	zero	0.010	Cancer	Ozonation by-product
Bromodichloromethane	zero	see TTHMs	Cancer, kidney, liver, reproductive effects	Chlorination by-product
Bromoform	zero	see TTHMs	Cancer, kidney, liver, nervous system effects	Chlorination by-product
Chlorine	4 (as Cl$_2$) (MRDLG)	4.0 (as Cl$_2$) (MRDL)		
Chloramies	4 (as Cl$_2$) (MRDLG)	4.0 (as Cl$_2$) (MRDL)		
Chlorine dioxide	0.8 (as ClO$_2$) (MRDLG)	0.8 (as ClO$_2$) (MRDL)		
Chloral hydrate	0.04	TT	Liver effects	Chlorination by-product
Chlorite	0.08	1.0	Developmental neurotoxicity	Chlorine dioxide by-product
Chloroform	zero	see TTHMs	Cancer, kidney, liver reproductive effects	Chlorination by-product
Dibromochloromethane	0.06	see TTHMs	Nervous system, kidney, liver, reproductive effects	Chlorination by-product
Dichloroacetic acid	zero	see HAA5	Cancer, reproductive, developmental effects	Chlorination by-product
Disinfectants and Disinfection By-Products				
Haloacetic acids (HAA5)[†]	zero	0.060	Cancer and other effects	Chlorination by-product
Trichloroacetic acid	0.3	see HAA5	Liver, kidney, spleen, mental effects	Chlorination by-product
Total trihalomethanes (TTHMs)	zero	0.010	Cancer and other effects	Chlorination by-product
Radionuclides				
Beta/photon emitters	zero	4 mrem/yr	Cancer	Natural and man-made deposits
Alpha particles	zero	15 pCi/L	Cancer	Natural deposits
Radium 226 and 228	zero	5 pCi/L	Bone cancer	Natural deposits
Uranium	zero	0.03	Cancer	Natural deposits
Arsenic				
Arsenic	zero	0.010	Skin, nervous system toxicity; cancer	Natural deposits; smelters; glass, electronics wastes; orchards

Notes:

\# Action level = 0.015 mg/L \#\# Action level = 1.3 mg/L

TT = treatment technique requirement MFL = million fibers per liter

*For water systems analyzing at least 40 samples per month, no more than 5.0 percent of the monthly samples may be positive for total coliforms. For systems analyzing fewer than 40 samples per month, no more than one sample per month may be positive for total coliforms.

[†]Sum of the concentrations of mono-, di-, and trichloroacetic acids and mono- and dibromoacetic acids.

Alternatives allowing public water systems the flexibility to select compliance options appropriate to protect the population served were proposed.
(*Source:* WQ&T, 2011.)

Disinfectants and Disinfectant By-Products (D-DBPs). The disinfectants used to destroy pathogens in water and the by-products of the reaction of these disinfectants with organic materials in the water are of potential health concern. One class of DBPs has been regulated since 1979. This class is known as trihalomethanes (THMs). THMs are formed when a water containing an organic precursor is chlorinated (*precursor* means forerunner). In this case it means an organic compound capable of reacting to produce a THM. The precursors are natural organic substances formed from the decay of vegetative matter, such as leaves, and aquatic organisms. THMs are of concern because they are potential carcinogens and may cause reproductive effects. The four THMs that were regulated in the 1979 rules are: chloroform ($CHCl_3$), bromo-dichloromethane ($CHBrCl_2$), dibromochloromethane ($CHBr_2Cl$), and bromoform ($CHBr_3$). Of these four, chloroform appears most frequently and is found in the highest concentrations; however, the brominated compounds may have a higher health risk.

The D-DBP rule was developed through a negotiated rule-making process, in which individuals representing major interest groups concerned with the rule (for example, public-water-system owners, state and local government officials, and environmental groups) publicly work with the EPA representatives to reach a consensus on the contents of the proposed rule. The D-DBP rule was implemented in stages. Stage 1 of the rule was promulgated in November 1998. Stage 2 was promulgated in 2006.

Maximum residual disinfectant level goals (MRDLGs) and maximum residual disinfectant levels (MRDLs) were established for chlorine, chloramine, and chlorine dioxide (Table 6-3). Because ozone reacts too quickly to be detected in the distribution system, no limits on ozone were set.

The MCLGs and MCLs for the disinfection by-products are listed in Table 6-4. In addition to regulating individual compounds, the D-DBP rule set levels for two groups of compounds: HAA5 and TTHMs. These groupings were made to recognize the potential cumulative effect of several compounds. HAA5 is the sum of five haloacetic acids (monochloroacetic acid, dichloroacetic acid, trichloroacetic acid, monobromo-acetic acid, and dibromoacetic acid). TTHMs (total trihalomethanes) is the sum of the concentrations of chloroform ($CHCl_3$), bromodichloromethane ($CHBrCl_2$), dibro-mochloromethane ($CHBr_2Cl$), and bromoform ($CHBr_3$).

The D-DBP rule is quite complex. In addition to the regulatory levels shown in the tables, levels are established for precursor removal. The amount of precursor required to be removed is a function of the alkalinity of the water and the amount of *total organic carbon* (TOC) present.

TABLE 6-3

Maximum residual disinfectant goals (MRDLGs) and maximum residual disinfectant levels (MRDLs)

Disinfectant residual	MRDLGs mg/L	MRDL mg/L
Chlorine (free)	4	4.0
Chloramines (as total chlorine)	4	4.0
Chlorine dioxide	0.8	0.8

TABLE 6-4

Maximum contaminant level goals (MCLGs) and maximum contaminant levels (MCLs) for disinfection by-products (DBPs)

Contaminant	MCLG mg/L	Stage 1 MCL mg/L	Stage 2 MCL mg/L
Bromate	Zero	0.010	
Bromodichloromethane	Zero		
Bromoform	Zero		
Chloral hydrate	0.005		
Chlorite	0.3	1.0	
Chloroform	0.07		
Dibromochloromethane	0.06		
Dichloroacetic acid	Zero		
Monochloroacetic acid	0.03		
Trichloroacetic acid	0.02		
HAA5		0.060	0.060*
TTHMs		0.080	0.080*

*Calculated differently in Stage 2.

When chlorine is added to a water that contains TOC, the chlorine and TOC slowly react to form THMs and HAA5. Therefore, THM and HAA5 are continuously increasing until the point of maximum formation is reached. In order to better determine the level of DBPs consumed at the tap, compliance with the regulation is based on collecting samples in the distribution system. Although the number of samples can vary, it is in the range of four distribution samples collected quarterly for each treatment plant. In the Stage 1 rule, the sample points (say, four) are averaged over four quarters of data (so 16 data points are averaged) to determine compliance with the MCLs of Table 6-4. This method of determining compliance is called a running annual average (RAA). For the Stage 2 rule, several changes were made to the number and locations of the samples, but also the method to calculate compliance was changed. In Stage 2, the four quarters from an individual sample site are averaged and each site must be below the MCLs. This is referred to as a locational running annual average (LRAA). Although the MCLs in Stage 1 and Stage 2 are the same, because of the use of the LRAA, compliance is more difficult in the Stage 2 rule. Also, the number of compliance sites is determined by the system population and is not set at four as in Stage 1.

Surface Water Treatment Rule (SWTR). The Surface Water Treatment Rule (SWTR) and its companion rules, the Interim Enhanced Surface Water Treatment Rule (IESWTR) and the Long-Term Enhanced Surface Water Treatment Rules (LT1ESWTR and LT2ESWTR), set forth primary drinking water regulations requiring treatment of surface water supplies or groundwater supplies under the direct influence of surface water. The regulations require a specific treatment technique—filtration and/or disinfection—in lieu of establishing maximum contaminant levels (MCLs) for turbidity, *Cryptosporidium*,

Giardia, viruses, *Legionella,* and heterotrophic bacteria, as well as many other pathogenic organisms that are removed by these treatment techniques. The regulations also establish a maximum contaminant level goal (MCLG) of zero for *Giardia, Cryptosporidium,* viruses, and *Legionella.* No MCLG is established for heterotrophic plate count or turbidity.

Turbidity Limits. Treatment by conventional or direct filtration must achieve a turbidity level of less than 0.3 NTU in at least 95 percent of the samples taken each month. Those systems using slow sand filtration must achieve a turbidity level of less than 5 NTU at all times and not more than 1 NTU in more than 5 percent of the samples taken each month. The 1 NTU limit may be increased by the state up to 5 NTU if it determines that there is no significant interference with disinfection. Other filtration technologies may be used if they meet the turbidity requirements set for slow sand filtration, provided they achieve the disinfection requirements and are approved by the state.

Turbidity measurements must be performed on representative samples of the system's filtered water every four hours or by continuous monitoring. For any system using slow sand filtration or a filtration treatment other than conventional treatment, direct filtration, or diatomaceous earth filtration, the state may reduce the monitoring requirements to once per day.

Disinfection Requirements. Filtered water supplies must achieve the same disinfection as required for unfiltered systems (that is, 99.9 or 99.99 percent removal, also known as 3-log and 6-log removal or inactivation, for *Giardia* and viruses respectively) through a combination of filtration and application of a disinfectant.

Giardia and viruses are both fairly well inactivated by chlorine, and hence with proper physical treatment and chlorination both can be controlled. *Cryptosporidium,* however, is resistant to chlorination. Depending on the source water concentration, EPA established levels of treatment, both additional physical barriers and disinfection techniques, that must be used to reduce the rush of illness due to *Cryptosporidium.* Ozone and ultraviolet light are effective disinfectants for *Cryptosporidium.*

Total Coliform. On June 19, 1989, the EPA promulgated the revised National Primary Drinking Water Regulations for total coliforms, including fecal coliforms and *E. coli.* These regulations apply to all public water systems.

The regulations establish a maximum contaminant level (MCL) for coliforms based on the presence or absence of coliforms. Larger systems that are required to collect at least 40 samples per month cannot obtain coliform-positive results in more than 5 percent of the samples collected each month to stay in compliance with the MCL. Smaller systems that collect fewer than 40 samples per month cannot have coliform-positive results in more than one sample per month.

The EPA will accept any one of the five analytical methods noted below for the determination of total coliforms:

Multiple-tube fermentation technique (MTF)
Membrane filter technique (MF)
Minimal media ONPG-MUG test (colilert system) (MMO-MUG)
Presence-absence coliform test (P-A)
Colisure technique

Regardless of the method used, the standard sample volume required for total coliform testing is 100 mL.

A public water system must report a violation of the total coliform regulations to the state no later than the end of the next business day. In addition to this, the system must make public notification according to the general public notification requirements of the Safe Drinking Water Act, but with special wording prescribed by the total coliform regulations.

Proposed changes to the total coliform rule (TCR) were made by EPA in 2010 and are expected to be finalized in 2012.

Secondary Maximum Contaminant Levels (SMCLs).   The National Safe Drinking Water Act also provided for the establishment of an additional set of standards to prescribe maximum limits for those contaminants that tend to make water disagreeable to use, but that do not have any particular adverse public health effect. These secondary maximum contaminant levels are the advisable maximum level of a contaminant in any public water supply system. The levels are shown in Table 6-5. SMCLs are recommended. They are not enforced.

TABLE 6-5
National secondary drinking water standards

Contaminant	Effect(s)	SMCL, mg/L
Aluminum	Colored water	0.05–0.2
Chloride	Salty taste	250
Color	Visible tint	15 color units
Copper	Metallic taste; blue-green stain	1.0
Corrosivity	Metallic taste; corrosion; fixture staining	Noncorrosive
Fluoride	Tooth discoloration	2
Foaming agents	Frothy, cloudy; bitter taste; odor	0.5
Iron	Rusty color; sediment; metallic taste; reddish or orange staining	0.3
Manganese	Black to brown color; black staining; bittermetallic taste	0.05
Odor	"Rotten egg," musty, or chemical smell	3 TON
pH	Low pH: bitter metallic taste, corrsosion high pH: slippery feel, soda taste, deposits	6.5–8.5
Silver	Skin discoloration; greying of the white part of the eye	0.10
Sulfate	Salty taste	250
Total dissolved solids (TDS)	Hardness; deposits; colored water; staining; salty taste	500
Zinc	Metallic taste	5

(*Source:* WQ&T, 2011.)

TABLE 6-6
General characteristics of groundwater and surface water

Ground	Surface
Constant composition	Varying composition
High mineralization	Low mineralization
Little turbidity	High turbidity
Low or no color	Color
Bacteriologically safe	Microorganisms present
No dissolved oxygen	Dissolved oxygen
High hardness	Low hardness
H_2S, Fe, Mn	Tastes and odors
	Possible chemical toxicity

Water Classification and Treatment Systems

Water Classification by Source.   Potable water is most conveniently classified as to its source, that is, groundwater or surface water. Generally, groundwater is uncontaminated but may contain aesthetically or economically undesirable impurities. However, groundwaters can be influenced by man-made spills and become contaminated. Surface water must be considered to be contaminated with bacteria, viruses, or inorganic substances that could present a health hazard. Surface water may also have aesthetically unpleasing characteristics for a potable water. Table 6-6 shows a comparison between groundwater and surface water.

Groundwater is further classified as to its source—deep or shallow wells. Municipal water quality factors of safety, temperature, appearance, taste and odor, and chemical balance are most easily satisfied by a deep well source. High concentrations of calcium, iron, manganese, and magnesium typify well waters. Some supplies contain hydrogen sulfide, while others may have excessive concentrations of chloride, sulfate, or carbonate.

Shallow wells are recharged by a nearby surface watercourse. They may have qualities similar to the deep wells, or they may take on the characteristics of the surface recharge water. A sand aquifer between the shallow well supply and the surface watercourse may act as an effective filter for removal of organic matter and as a heat exchanger for buffering temperature changes. To predict water quality from shallow wells, careful studies of the aquifer and nature of recharge water are necessary. EPA classifies some shallow wells as surface water.

Surface water supplies are classified as to whether they come from a lake, reservoir, or river. A comparison of lakes and rivers is shown in Table 6-7. Generally, a river has the lowest water quality and a lake the highest. Water quality in rivers depends upon the character of the watershed. River quality is largely influenced by pollution (or lack thereof) from municipalities, industries, and agricultural practices. The characteristics of a river can be highly variable. During rains or periods of runoff, turbidity may increase substantially. Many rivers will show an increase in color and taste and in odor-producing compounds. In warm months, algal blooms frequently cause taste and odor problems.

TABLE 6-7
Raw water quality as a function of water source

Source	Average turbidity (TU)	Average alum dose (mg/L)
Lake	16	20
River	26	23

Source: EE&T, 2008.

Reservoir and lake sources have much less day-to-day variation than rivers. Additionally, the quiescent conditions will reduce both the turbidity and, on occasion, the color. As in rivers, summer algal blooms can create taste and odor problems in lakes and reservoirs. While the averages are similar for rivers and lakes, rivers can have much higher extremes, with turbidities going into the hundreds.

Treatment Systems.   Treatment plants can be classified as simple disinfection, filter plants, or softening plants. Plants employing simple chlorination are usually groundwater sources that have a high water quality and chlorinate to ensure that the water reaching customers contains safe bacteria levels.

A groundwater source can contain naturally occurring contaminants such as arsenic, fluoride, radionuclides, iron, or manganese. In this case the water may be treated by adsorption, ion-exchange or precipitation processes to remove the contaminant. Groundwaters can also be contaminated by spills caused by humans. Spills from gasoline storage, degreasers, and dry cleaners are the most common. These contaminants are generally removed by aeration, because many of the contaminants are volatile, or by adsorption. Generally, a filtration plant is used to treat surface water and when necessary a softening plant is used to treat groundwater.

In a filtration plant, rapid mixing, flocculation, sedimentation, filtration, and disinfection are employed to remove color, turbidity, taste and odors, organic matter, and bacteria. Additional operations may include bar racks or coarse screens if floating debris and fish are a problem. Figure 6-4 shows a typical flow diagram of a filtration plant. The *raw* (untreated) *surface water* enters the plant via low-lift pumps or gravity. Usually screening has taken place prior to pumping. During mixing, chemicals called coagulants are added and rapidly dispersed through the water. The chemical reacts with the desired impurities and forms precipitates (*flocs*) that are slowly brought into contact

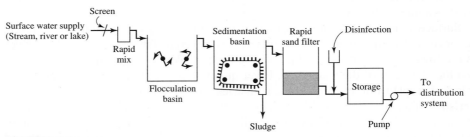

FIGURE 6-4
Flow diagram of a conventional surface water treatment plant ("filtration plant").

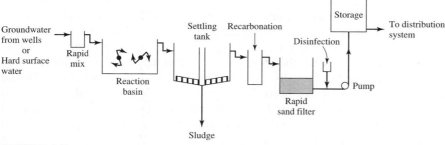

FIGURE 6-5
Flow diagram of water softening plant.

with one another during flocculation. The objective of flocculation is to allow the flocs to collide and "grow" to a settleable size. The particles are removed by gravity (sedimentation). This is done to minimize the amount of solids that are applied to the filters. For treatment works with a high-quality raw water, it may be possible to omit sedimentation and perhaps flocculation. This modification is called *direct filtration*. Filtration is the final polishing (removal) of particles. During filtration the water is passed through sand or similar media to screen out the fine particles that will not settle. Disinfection is the addition of chemicals (usually chlorine) to kill or reduce the number of pathogenic organisms. Storage may be provided at the plant or located within the community to meet peak demands and to allow the plant to operate on a uniform schedule. The high-lift pumps provide sufficient pressure to convey the water to its ultimate destination. The precipitated chemicals, original turbidity, and suspended material are removed from the sedimentation basins and from the filters. These residuals must be disposed of properly.

Softening plants utilize the same unit operations as filtration plants, but use different chemicals. The primary function of a softening plant is to remove hardness (calcium and magnesium). In a softening plant (a typical flow diagram is shown in Figure 6-5), the design considerations of the various facilities are different than those in filtration. Also the chemical doses are much higher in softening, and the corresponding sludge production is greater.

During rapid mix, chemicals are added to react with and precipitate the hardness. Precipitation occurs in the reaction basin. The other unit operations are the same as in a filtration plant except for the additional recarbonation step employed in softening to adjust the final pH. Softening plants sometimes have two reaction and settling tanks in series.

In the next two sections of this chapter, we discuss coagulation chemistry and softening chemistry, respectively. The subsequent sections describe the physical processes themselves. These are applicable to both filtration and softening plants.

6-2 COAGULATION

Surface waters must be treated to remove turbidity, color, and bacteria. When sand filtration was developed around 1885, it became immediately apparent that filtration alone would not produce a clear water. Experience has demonstrated that filtration alone is largely ineffective in removing bacteria, viruses, soil particles, and color except at very low loading rates, which is impractical for all but very small utilities.

The object of coagulation (and subsequently flocculation) is to turn the small particles of color, turbidity, and bacteria into larger flocs, either as precipitates or suspended particles. These flocs are then conditioned so that they will be readily removed in subsequent processes. Technically, coagulation applies to the removal of colloidal particles. However, the term has been applied more loosely to removal of dissolved ions, which is actually precipitation. Coagulation in this chapter will refer to colloid removal only. We define *coagulation* as a method to alter the colloids so that they will be able to approach and adhere to each other to form larger floc particles.

Colloid Stability

Before discussing colloid removal, we should understand why the colloids are suspended in solution and can't be removed by sedimentation or filtration. Very simply, the particles in the colloid range are too small to settle in a reasonable time period, and too small to be trapped in the pores of a filter. For colloids to remain stable they must remain small. Most colloids are stable because they possess a negative charge that repels other colloidal particles before they collide with one another.* The colloids are continually involved in *Brownian movement,* which is merely random movement. Charges on colloids are measured by placing DC electrodes in a colloidal dispersion. The particles migrate to the pole of opposite charge at a rate proportional to the potential gradient. Generally, the larger the surface charge, the more stable the suspension.

Colloid Destabilization

Colloids are stable because of their surface charge. In order to destabilize the particles, we must neutralize this charge. Such neutralization can take place by the addition of an ion of opposite charge to the colloid. Since most colloids found in water are negatively charged, the addition of sodium ions (Na^+) should reduce the charge. Figure 6-6 shows such an effect. The plot shows surface charge as a function of distance from the colloid for no-salt (NaCl) addition, low-salt addition, and high-salt addition. As we would have predicted, the higher the concentration of sodium we add, the lower the charge, and therefore the lower the repelling forces around the colloid. If, instead of adding a monovalent ion such as sodium, we add a divalent or trivalent ion, the charge is reduced even faster, as shown in Figure 6-7. In fact, it was found by Schulze and Hardy (Schulze, 1882, 1883; Hardy, 1900a, 1900b) that one mole of a trivalent ion can reduce the charge as much as 30 to 50 moles of a divalent ion and as much as 1,500 to 2,500 moles of a monovalent ion (often referred to as the *Schulze-Hardy rule*).

Coagulation

The purpose of coagulation is to alter the colloids so that they can adhere to each other. During coagulation a positive ion is added to water to reduce the surface charge to the point where the colloids are not repelled from each other. A *coagulant* is the substance

---

*Some colloids are stabilized by their affinity for water, but these types of colloids are of less importance.

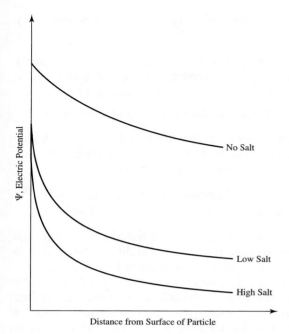

FIGURE 6-6
Effect of salt on electric potential.

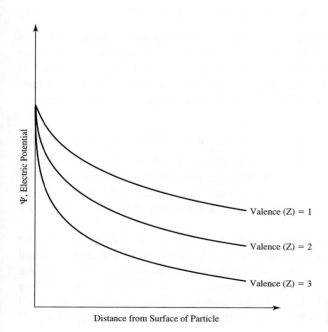

FIGURE 6-7
Effect of valence on electric potential.

(chemical) that is added to the water to accomplish coagulation. There are three key properties of a coagulant:

1. **Trivalent cation.** As indicated in the last section, the colloids most commonly found in natural waters are negatively charged, hence a cation is required to neutralize the charge. A trivalent cation is the most efficient cation.

2. **Nontoxic.** This requirement is obvious for the production of a safe water.

3. **Insoluble in the neutral pH range.** The coagulant that is added must precipitate out of solution so that high concentrations of the ion are not left in the water. Such precipitation greatly assists the colloid removal process.

The two most commonly used coagulants are aluminum (Al^{3+}) and ferric iron (Fe^{3+}). Both meet the above three requirements, and their reactions are outlined here.

Aluminum.   Aluminum can be purchased as either dry or liquid alum [$Al_2(SO_4)_3 \cdot 14H_2O$]. Commercial alum has an average molecular weight of 594, which is actually an average of 14.3 waters of hydration. Liquid alum is sold as approximately 48.8 percent alum (8.3% Al_2O_3) and 51.2 percent water. If it is sold as a more concentrated solution, there can be problems with crystallization of the alum during shipment and storage. A 48.8 percent alum solution has a crystallization point of $-15.6°C$. A 50.7 percent alum solution will crystallize at $+18.3°C$. Dry alum costs about 50 percent more than an equivalent amount of liquid alum so that only users of very small amounts of alum purchase dry alum.

When alum is added to a water containing alkalinity, the following reaction occurs:

$$Al_2(SO_4)_3 \cdot 14H_2O + 6HCO_3^- \rightleftharpoons 2Al(OH)_3 \cdot 3H_2O(s) + 6CO_2 + 8H_2O + 3SO_4^{2-} \tag{6-1}$$

such that each mole of alum added uses six moles of alkalinity and produces six moles of carbon dioxide. The above reaction shifts the carbonate equilibrium and decreases the pH. However, as long as sufficient alkalinity is present and $CO_2(g)$ is allowed to evolve, the pH is not drastically reduced and is generally not an operational problem. When sufficient alkalinity is not present to neutralize the sulfuric acid production, the pH may be greatly reduced:

$$Al_2(SO_4)_3 \cdot 14H_2O \rightleftharpoons 2Al(OH)_3 \cdot 3H_2O(s) + 3H_2SO_4 + 2H_2O \tag{6-2}$$

If the second reaction occurs, lime or sodium carbonate may be added to neutralize the acid.

Two important factors in coagulant addition are pH and dose. The optimum dose and pH must be determined from laboratory tests. The optimal pH range for alum is approximately 5.5 to 6.5 with adequate coagulation possible between pH 5 to pH 8 under some conditions.

An important aspect of coagulation is that the aluminum ion does not really exist as Al^{3+} and the final product is more complex than $Al(OH)_3$. When the alum is added to the water, it immediately dissociates, resulting in the release of an aluminum ion surrounded by six water molecules. The aluminum ion immediately starts reacting with the water, forming large $Al \cdot OH \cdot H_2O$ complexes. Some have suggested that it forms $[Al_8(OH)_{20} \cdot 28H_2O]^{4+}$ as the product that actually coagulates. Regardless of the actual species produced, the complex is a very large precipitate that removes many of the colloids by enmeshment as it falls through the water. This precipitate is referred to as a *floc*. Floc formation is one of the important properties of a coagulant for efficient colloid removal. The final product after coagulation has three water molecules associated with it in the solid form as indicated in the equations.

Example 6-1. One of the most common methods to evaluate coagulation efficiency is to conduct jar tests. Jar tests are performed in an apparatus such as shown in Figure 6-8. Six beakers are filled with water and then each is mixed and flocculated uniformly by

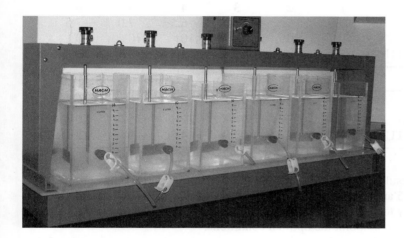

(*a*)

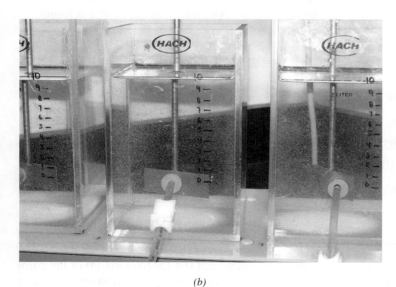

(*b*)

FIGURE 6-8
Jar tests: (*a*) Phipps and Bird jar tester; (*b*) close up of floc in Gator jar. (David Cornwell)

a gang stirrer. The jars are square in order to prevent swirling and are called gator jars. A test is often conducted by first dosing each jar with the same alum dose and varying the pH in each jar. The test can then be repeated in a second set of jars by holding the pH constant and varying the coagulant dose.

Two sets of such jar tests were conducted on a raw water containing 15 TU and an HCO_3^- alkalinity concentration of 50 mg/L expressed as $CaCO_3$. Given the data below, find the optimal pH, coagulant dose, and the theoretical amount of alkalinity that would be consumed at the optimal dose.

Jar test I

	1	2	3	4	5	6
pH	5.0	5.5	6.0	6.5	7.0	7.5
Alum dose (mg/L)	10	10	10	10	10	10
Settled turbidity (TU)	11	7	5.5	5.7	8	13

Jar test II

	1	2	3	4	5	6
pH	6.0	6.0	6.0	6.0	6.0	6.0
Alum dose (mg/L)	5	7	10	12	15	20
Settled turbidity (TU)	14	9.5	5	4.5	6	13

Solution. The results of the two jar tests are plotted in Figure 6-9. The optimal pH was chosen as 6.25 and the optimal alum dose was about 12.5 mg/L. The experimenter would probably try to repeat the test using a pH of 6.25 and varying the alum dose between 10 and 15 to pinpoint the optimal conditions.

The amount of alkalinity that will be consumed is found by using Equation 6-1, which shows us that one mole of alum consumes six moles of HCO_3^-. With the molecular weight of alum equal to 594, the moles of alum added per liter is found by using Equation 5-7:

$$\frac{12.5 \times 10^{-3} \text{ g/L}}{594 \text{ g/mole}} = 2.1 \times 10^{-5} \text{ moles/L}$$

which will consume

$$6(2.1 \times 10^{-5}) = 1.26 \times 10^{-4} \, M \, HCO_3^-$$

The molecular weight of HCO_3^- is 61, so

$$(1.26 \times 10^{-4} \text{ moles/L})(61 \text{ g/mole})(10^3 \text{ mg/g}) = 7.7 \text{ mg/L } HCO_3$$

are consumed, which can be expressed as $CaCO_3$ by using Equation 5-40:

$$7.7 \frac{50}{61} = 6.31 \text{ mg/L } HCO_3^- \text{ as } CaCO_3$$

Iron. Iron can be purchased as either the sulfate salt ($Fe_2(SO_4)_3 \cdot xH_2O$) or the chloride salt ($FeCl_3 \cdot xH_2O$). It is available in various forms, and the individual supplier

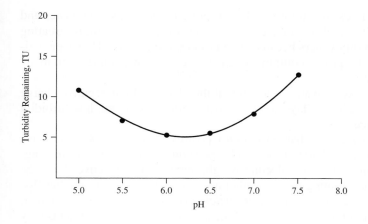

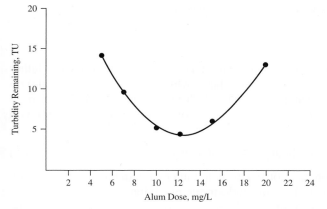

FIGURE 6-9
Results of jar test.

should be consulted for the specifics of the product. Dry and liquid forms are available. The properties of iron with respect to forming large complexes, dose, and pH curves are similar to those of alum. An example of the reaction of $FeCl_3$ in the presence of alkalinity is

$$FeCl_3 + 3HCO_3^- + 3H_2O \rightleftharpoons Fe(OH)_3 \cdot 3H_2O(s) + 3CO_2 + 3Cl^- \quad (6\text{-}3)$$

and without alkalinity

$$FeCl_3 + 6H_2O \rightleftharpoons Fe(OH)_3 \cdot 3H_2O(s) + 3HCL \quad (6\text{-}4)$$

forming hydrochloric acid which in turn lowers the pH. Ferric salts generally have a wider pH range for effective coagulation than aluminum, that is, pH ranges from 4 to 9.

Coagulant Aids.   The four basic types of coagulant aids are pH adjusters, activated silica, clay, and polymers. Acids and alkalies are both used to adjust the pH of the water into the optimal range for coagulation. The acid most commonly used for lowering the pH is sulfuric acid. Either lime [$Ca(OH)_2$] or soda ash (Na_2CO_3) are used to raise the pH.

When activated silica is added to water, it produces a stable solution that has a negative surface charge. The activated silica can unite with the positively charged aluminum or with iron flocs, resulting in a larger, denser floc that settles faster and enhances enmeshment. The addition of activated silica is especially useful for treating highly colored, low-turbidity waters because it adds weight to the floc. However, activation of silica does require proper equipment and close operational control, and many plants are hesitant to use it.

Clays can act much like activated silica in that they have a slight negative charge and can add weight to the flocs. Clays are also most useful for colored, low-turbidity waters, but are rarely used.

Polymers can have a negative charge (*anionic*), positive charge (*cationic*), positive and negative charge (*polyamphotype*), or no charge (*nonionic*). Polymers are long-chained carbon compounds of high molecular weight that have many active sites. The active sites adhere to flocs, joining them together and producing a larger, tougher floc that settles better. This process is called *interparticle bridging*. The type of polymer, dose, and point of addition must be determined for each water, and requirements may change within a plant on a seasonal, or even daily, basis.

6-3   SOFTENING

Hardness.   The term hardness is used to characterize a water that does not lather well, causes a scum in the bath tub, and leaves hard, white, crusty deposits (scale) on coffee pots, tea kettles, and hot water heaters. The failure to lather well and the formation of scum on bath tubs is the result of the reactions of calcium and magnesium with the soap. For example:

$$Ca^{2+} + (Soap)^- \rightleftharpoons Ca(Soap)_2(s) \tag{6-5}$$

As a result of this complexation reaction, soap cannot interact with the dirt on clothing, and the calcium-soap complex itself forms undesirable precipitates.

Hardness is defined as the sum of all polyvalent cations (in consistent units). The common units of expression are mg/L as $CaCO_3$ or meq/L. Qualitative terms used to describe hardness are listed in Table 6-8. Because many people object to water containing hardness greater than 150 mg/L as $CaCO_3$ suppliers of public water have considered it a benefit to *soften* the water, that is, to remove some of the hardness. A common water treatment goal is to provide water with a hardness in the range of 75 to 120 mg/L as $CaCO_3$.

Although all polyvalent cations contribute to hardness, the predominant contributors are calcium and magnesium. Thus, our focus for the remainder of this discussion will be on calcium and magnesium.

TABLE 6-8
Hard water classification

Hardness range (mg/L $CaCO_3$)	Description
0–75	Soft
75–100	Moderately hard
100–300	Hard
>300	Very hard

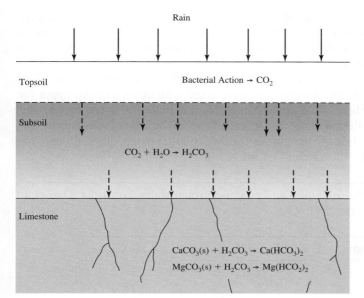

FIGURE 6-10
Natural process by which
water is made hard.

The natural process by which water becomes hard is shown schematically in Figure 6-10. As rainwater enters the topsoil, the respiration of microorganisms increases the CO_2 content of the water. As shown in Equation 5-35, the CO_2 reacts with the water to form H_2CO_3. Limestone, which is made up of solid $CaCO_3$ and $MgCO_3$ reacts with the carbonic acid to form calcium bicarbonate [$Ca(HCO_3)_2$] and magnesium bicarbonate [$Mg(HCO_3)_2$]. While $CaCO_3$ and $MgCO_3$ are both insoluble in water, the bicarbonates are quite soluble. Gypsum ($CaSO_4$) and $MgSO_4$ may also go into solution to contribute to the hardness.

Because calcium and magnesium predominate, it is often convenient in performing softening calculations to define the *total hardness* (TH) of a water as the sum of these elements

$$TH = Ca^{2+} + Mg^{2+} \qquad (6\text{-}6)$$

where the concentrations of each element are in consistent units (mg/L as $CaCO_3$ or meq/L). Total hardness is often broken down into two components: (1) that associated with the HCO_3^- anion (called *carbonate hardness* and abbreviated CH), and (2) that associated with other anions (called *noncarbonate hardness* and abbreviated NCH).* Total hardness, then, may also be defined as

$$TH = CH + NCH \qquad (6\text{-}7)$$

Carbonate hardness is defined as the amount of hardness equal to the total hardness or the total alkalinity, whichever is less. Carbonate hardness is often called temporary hardness because heating the water removes it. When the pH is less than 8.3, HCO_3^- is the dominant form of alkalinity, and total alkalinity is nominally taken to be equal to the concentration of HCO_3^-.

---

*Note that this does not imply that the compounds exist as compounds in solution. They are dissociated.

Noncarbonate hardness is defined as the total hardness in excess of the alkalinity. If the alkalinity is equal to or greater than the total hardness, then there is no noncarbonate hardness. Noncarbonate hardness is called permanent hardness because it is not removed when water is heated.

Bar charts of water composition are often useful in understanding the process of softening. By convention, the bar chart is constructed with cations in the upper bar and anions in the lower bar. In the upper bar, calcium is placed first and magnesium second. Other cations follow without any specified order. The lower bar is constructed with bicarbonate placed first. Other anions follow without any specified order. Construction of a bar chart is illustrated in Example 6-2.

Example 6-2.   Given the following analysis of a groundwater, construct a bar chart of the constituents, expressed as $CaCO_3$.

Ion	mg/L as ion	EW $CaCO_3$/EW ion	mg/L as $CaCO_3$
Ca^{2+}	103	2.50	258
Mg^{2+}	5.5	4.12	23
Na^+	16	2.18	35
HCO_3^-	255	0.82	209
SO_4^{2-}	49	1.04	51
Cl^-	37	1.41	52

Solution.   The concentrations of the ions have been converted to $CaCO_3$ equivalents. The results are plotted in Figure 6-11.

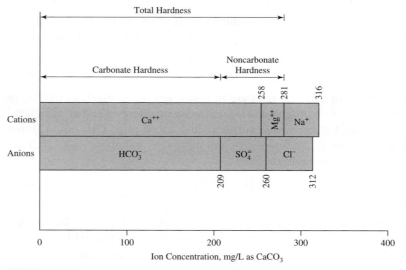

FIGURE 6-11
Bar graph of groundwater constituents.

The cations total 316 mg/L as $CaCO_3$, of which 281 mg/L as $CaCO_3$ is hardness. The anions total 312 mg/L as $CaCO_3$, of which the carbonate hardness is 209 mg/L as $CaCO_3$. There is a discrepancy between the cation and anion totals because there are other ions that were not analyzed. If a complete analysis were conducted, and no analytical error occurred, the equivalents of cations would equal exactly the equivalents of anions. Typically, a complete analysis may vary $\pm 5\%$ because of analytical errors.

The relationships between total hardness, carbonate hardness, and noncarbonate hardness are illustrated in Figure 6-12. In Figure 6-12a, the total hardness is 250 mg/L as $CaCO_3$, the carbonate hardness is equal to the alkalinity ($HCO_3^- = 200$ mg/L as $CaCO_3$), and the noncarbonate hardness is equal to the difference between the total hardness and the carbonate hardness (NCH = TH − CH = 250 − 200 = 50 mg/L as $CaCO_3$). In Figure 6-12b, the total hardness is again 250 mg/L as $CaCO_3$. However, since the alkalinity (HCO_3^-) is greater than the total hardness, and since the carbonate hardness cannot be greater than the total hardness (see Equation 6-7), the carbonate hardness is equal to the total hardness, that is, 250 mg/L as $CaCO_3$. With the carbonate hardness equal to the total hardness, then all of the hardness is carbonate hardness and there is no noncarbonate hardness. Note that in both cases it may be assumed that the pH is less than 8.3 because HCO_3^- is the only form of alkalinity present.

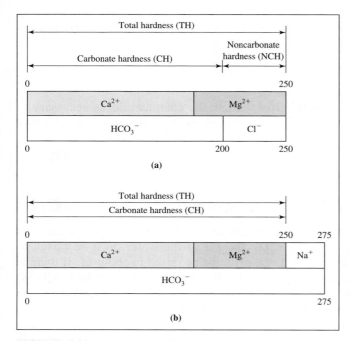

FIGURE 6-12
Relationships between total hardness, carbonate hardness, and noncarbonate hardness.

Example 6-3.   A water has an alkalinity of 200 mg/L as $CaCO_3$. The Ca^{2+} concentration is 160 mg/L as the ion, and the Mg^{2+} concentration is 40 mg/L as the ion. The pH is 8.1. Find the total, carbonate, and noncarbonate hardness.

Solution.   The molecular weights of calcium and magnesium are 40 and 24, respectively. Since each has a valence of 2^+, the corresponding equivalent weights are 20 and 12. Using Equation 5-40 to convert mg/L as the ion to mg/L as $CaCO_3$ and adding the two ions as shown in Equation 6-6, the total hardness is

$$TH = 160 \text{ mg/L}\left(\frac{50 \text{ mg/meq}}{20 \text{ mg/meq}}\right) + 40 \text{ mg/L}\left(\frac{50 \text{ mg/meq}}{12 \text{ mg/meq}}\right) = 567 \text{ mg/L as } CaCO_3$$

where 50 is the equivalent weight of $CaCO_3$.

By definition, the carbonate hardness is the lesser of the total hardness or the alkalinity. Since, in this case, the alkalinity is less than the total hardness, the carbonate hardness (CH) is equal to 200 mg/L as $CaCO_3$. The noncarbonate hardness is equal to the difference

$$NCH = TH - CH = 567 - 200 = 367 \text{ mg/L as } CaCO_3$$

Note that we can only add and subtract concentrations of Ca^{2+} and Mg^{2+} if they are in equivalent units, for example, moles/L or milliequivalents/L or mg/L as $CaCO_3$.

Softening can be accomplished by either the lime-soda process or by ion exchange. Both methods are discussed in the following sections.

Lime-Soda Softening

In lime-soda softening it is possible to calculate the chemical doses necessary to remove hardness. Hardness precipitation is based on the following two solubility reactions:

$$Ca^{2+} + CO_3^{2-} \rightleftharpoons CaCO_3(s) \qquad (6-8)$$

and

$$Mg^{2+} + 2OH^- \rightleftharpoons Mg(OH)_2(s) \qquad (6-9)$$

The objective is to precipitate the calcium as $CaCO_3$ and the magnesium as $Mg(OH)_2$. In order to precipitate $CaCO_3$, the pH of the water must be raised to about 10.3. To precipitate magnesium, the pH must be raised to about 11. If there is not sufficient naturally occurring bicarbonate alkalinity (HCO_3^-) for the $CaCO_3(s)$ precipitate to form (that is, there is noncarbonate hardness), we must add CO_3^{2-} to the water. Magnesium is more expensive to remove than calcium, so we leave as much Mg^{2+} in the water as possible. It is more expensive to remove noncarbonate hardness than carbonate hardness because we must add another chemical to provide the CO_3^{2-}. Therefore, we leave as much noncarbonate hardness in the water as possible.

Softening Chemistry. The chemical processes used to soften water are a direct application of the law of mass action. We increase the concentration of CO_3^{2-} and/or OH^- by the addition of chemicals, and drive the reactions given in Equations 6-8 and 6-9 to the right. Insofar as possible, we convert the naturally occurring bicarbonate alkalinity (HCO_3^-) to carbonate (CO_3^{2-}) by the addition of hydroxyl ions (OH^-). Hydroxyl ions cause the carbonate buffer system (Equation 5-35) to shift to the right and, thus, provide the carbonate for the precipitation reaction (Equation 6-8).

The common source of hydroxyl ions is calcium hydroxide [$Ca(OH)_2$]. Many water treatment plants find it more economical to buy quicklime (CaO), commonly called lime, than hydrated lime [$Ca(OH)_2$]. The quicklime is converted to hydrated lime at the water treatment plant by mixing CaO and water to produce a slurry of $Ca(OH)_2$, which is fed to the water for softening. The conversion process is called *slaking:*

$$CaO + H_2O \rightleftharpoons Ca(OH)_2 + heat \qquad \text{(6-10)}$$

The reaction is exothermic. It yields almost 1 MJ per gram mole of lime. Because of this high heat release, the reaction must be controlled carefully. All safety precautions for handling a strong base should be observed. Because the chemical is purchased as lime, it is common to speak of chemical additions as addition of "lime," when in fact we mean calcium hydroxide. When carbonate ions must be supplied, the most common chemical chosen is sodium carbonate (Na_2CO_3). Sodium carbonate is commonly referred to as *soda ash* or *soda.*

Softening Reactions. The softening reactions are regulated by controlling the pH. First, any free acids are neutralized. Then pH is raised to precipitate the $CaCO_3$; if necessary, the pH is raised further to remove $Mg(OH)_2$. Finally, if necessary, CO_3^{2-} is added to precipitate the noncarbonate hardness.

Six important softening reactions are discussed below. In each case, the chemical that has been added to the water is printed in bold type. Remember that (s) designates the solid form, and hence indicates that the substance has been removed from the water. The following reactions are presented sequentially, although in reality they occur simultaneously.

1. Neutralization of carbonic acid (H_2CO_3).

    In order to raise the pH, we must first neutralize any free acids that may be present in the water. CO_2 is the principal acid present in unpolluted, naturally occurring water.* You should note that no hardness is removed in this step.

$$CO_2 + \textbf{Ca(OH)}_2 \rightleftharpoons CaCO_3(s) + H_2O \qquad \text{(6-11)}$$

---

*CO_2 and H_2CO_3 in water are essentially the same:

$$CO_2 + H_2O \rightleftharpoons H_2CO_3$$

Thus, the number of reaction units (n) for CO_2 is two.

2. Precipitation of carbonate hardness due to calcium.

   As we mentioned previously, we must raise the pH to about 10.3 to precipitate calcium carbonate. To achieve this pH we must convert all of the bicarbonate to carbonate. The carbonate then serves as the common ion for the precipitation reaction.

$$Ca^{2+} + 2HCO_3^- + \mathbf{Ca(OH)_2} \rightleftharpoons 2CaCO_3(s) + 2H_2O \qquad \text{(6-12)}$$

3. Precipitation of carbonate hardness due to magnesium.

   If we need to remove carbonate hardness that results from the presence of magnesium, we must add more lime to achieve a pH of about 11. The reaction may be considered to occur in two stages. The first stage occurs when we convert all of the bicarbonate in step 2 above.

$$Mg^{2+} + 2HCO_3^- + \mathbf{Ca(OH)_2} \rightleftharpoons MgCO_3 + CaCO_3(s) + 2H_2O \qquad \text{(6-13)}$$

Note that the hardness of the water did not change because $MgCO_3$ is soluble. With the addition of more lime the hardness due to magnesium is removed.

$$Mg^{2+} + CO_3^{2-} + \mathbf{Ca(OH)_2} \rightleftharpoons Mg(OH)_2(s) + CaCO_3(s) \qquad \text{(6-14)}$$

4. Removal of noncarbonate hardness due to calcium.

   If we need to remove noncarbonate hardness due to calcium, no further increase in pH is required. Instead we must provide additional carbonate in the form of soda ash.

$$Ca^{2+} + \mathbf{Na_2CO_3} \rightleftharpoons CaCO_3(s) + 2Na^+ \qquad \text{(6-15)}$$

5. Removal of noncarbonate hardness due to magnesium.

   If we need to remove noncarbonate hardness due to magnesium, we will have to add both lime and soda. The lime provides the hydroxyl ion for precipitation of the magnesium.

$$Mg^{2+} + \mathbf{Ca(OH)_2} \rightleftharpoons Mg(OH)_2(s) + Ca^{2+} \qquad \text{(6-16)}$$

Note that although the magnesium is removed, there is no change in the hardness because the calcium is still in solution. To remove the calcium we must add soda.

$$Ca^{2+} + \mathbf{Na_2CO_3} \rightleftharpoons CaCO_3(s) + 2Na^+ \qquad \text{(6-17)}$$

Note that this is the same reaction as the one to remove noncarbonate hardness due to calcium.

These reactions are summarized in Figure 6-13.

Process Limitations and Empirical Considerations.   Lime-soda softening cannot produce a water completely free of hardness because of the solubility of $CaCO_3$ and $Mg(OH)_2$, the physical limitations of mixing and contact, and the lack of sufficient time for the reactions to go to completion. Thus, the minimum calcium hardness that can be achieved is about 30 mg/L as $CaCO_3$, and the minimum magnesium hardness is about 10 mg/L as $CaCO_3$. Because of the slimy condition that results when soap is

Neutralization of Carbonic Acid

$$CO_2 + \mathbf{Ca(OH)_2} \rightleftharpoons CaCO_3(s) + H_2O$$

Precipitation of Carbonate Hardness

$$Ca^{2+} + 2HCO_3^- + \mathbf{Ca(OH)_2} \rightleftharpoons 2CaCO_3(s) + 2H_2O$$

$$Mg^{2+} + 2HCO_3^- + \mathbf{Ca(OH)_2} \rightleftharpoons MgCO_3 + CaCO_3(s) + 2H_2O$$

$$MgCO_3 + \mathbf{Ca(OH)_2} = Mg(OH)_2(s) + CaCO_3(s)$$

Precipitation of Noncarbonate Hardness Due to Calcium

$$Ca^{2+} + \mathbf{Na_2CO_3} \rightleftharpoons CaCO_3(s) + 2Na^+$$

Precipitation of Noncarbonate Hardness Due to Magnesium

$$Mg^{2+} + \mathbf{Ca(OH)_2} \rightleftharpoons Mg(OH)_2(s) + Ca^{2+}$$

$$Ca^{2+} + \mathbf{Na_2CO_3} \rightleftharpoons CaCO_3(s) + 2Na^+$$

FIGURE 6-13

Summary of softening reactions. (*Note:* The chemical added is printed in bold type. The precipitate is designated by (s). The arrow indicates where a compound formed in one reaction is used in another reaction.)

used with a water that is too soft, we have traditionally set a goal for final total hardness of 75 to 120 mg/L as $CaCO_3$. Because of economic constraints, many utilities will operate at total hardness goals of up to 140 to 150 mg/L.

In order to achieve reasonable removal of hardness in a reasonable time period, an excess of $Ca(OH)_2$ beyond the stoichiometric amount usually is provided. Based on our empirical experience, a minimum excess of 20 mg/L of $Ca(OH)_2$ expressed as $CaCO_3$ must be provided.

Magnesium in excess of about 40 mg/L as $CaCO_3$ forms scales on heat exchange elements in hot water heaters. Because of the expense of removing magnesium, we normally remove only that magnesium which is in excess of 40 mg/L as $CaCO_3$. For magnesium removals less than 20 mg/L as $CaCO_3$, the basic excess of lime mentioned above is sufficient to ensure good results. For magnesium removals between 20 and 40 mg/L as $CaCO_3$, we must add an excess of lime equal to the magnesium to be removed. For magnesium removals greater than 40 mg/L as $CaCO_3$, the excess lime we need to add is 40 mg/L as $CaCO_3$. Addition of excess lime in amounts greater than 40 mg/L as $CaCO_3$ does not appreciably improve the reaction kinetics.

The chemical additions (as $CaCO_3$) to soften water may be summarized as follows:

Step	Chemical addition[a]	Reason
Carbonate hardness		
1.	Lime $= CO_2$	Destroy H_2CO_3
2.	Lime $= HCO_3^-$	Raise pH; convert HCO_3^- to CO_3^{2-}
3.	Lime $= Mg^{2+}$ to be removed	Raise pH; precipitate $Mg(OH)_2$
4.	Lime $=$ required excess	Drive reaction
Noncarbonate hardness		
5.	Soda $=$ noncarbonate hardness to be removed	Provide CO_3^{2-}

[a]The terms "Lime $=$" and "Soda $=$" refer to mg/L of $Ca(OH)_2$ and Na_2CO_3 as $CaCO_3$ equal to mg/L of ion (or gas in the case of CO_2) as $CaCO_3$.

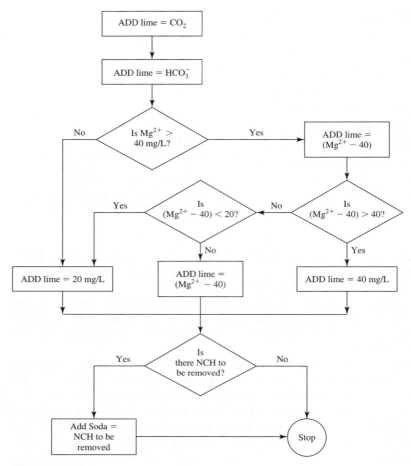

FIGURE 6-14
Flow diagram for solving softening problems. (*Note:* All additions are "as $CaCO_3$." NCH means noncarbonate hardness.)

These steps are diagrammed in the flow chart shown in Figure 6-14. The next three examples illustrate the technique.

Example 6-4. From the water analysis presented below, determine the amount of lime and soda (in mg/L as $CaCO_3$) necessary to soften the water to 80.00 mg/L hardness as $CaCO_3$.

Water Composition (mg/L)

Ca^{2+}: 95.20	CO_2: 19.36	HCO_3^-: 241.46
Mg^{2+}: 13.44		SO_4^{2-}: 53.77
Na^+: 25.76		Cl^-: 67.81

Solution. We begin by converting the elements and compounds to $CaCO_3$ equivalents.

Ion	mg/L as ion	EW $CaCO_3$/EW ion	mg/L as $CaCO_3$
Ca^{2+}	95.20	2.50	238.00
Mg^{2+}	13.44	4.12	55.37
Na^+	25.76	2.18	56.16
HCO_3^-	241.46	0.820	198.00
SO_4^{2-}	53.77	1.04	55.92
Cl^-	67.81	1.41	95.61
CO_2	19.36	2.28	44.14

The resulting bar chart would appear as shown below.

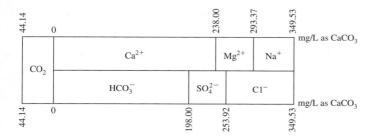

From the bar chart we note the following: CO_2 does not contribute to the hardness; the total hardness (TH) = 293.37 mg/L as $CaCO_3$; the carbonate hardness (CH) = 198.00 mg/L as $CaCO_3$; and, finally, the noncarbonate hardness (NCH) is equal to TH − CH = 95.37 mg/L as $CaCO_3$.

Using Figure 6-14 to guide our logic, we determine the lime dose as follows:

Step	Dose (mg/L as $CaCO_3$)
Lime = CO_2	44.14
Lime = HCO_3^-	198.00
Lime = Mg^{2+} − 40 = 55.37 − 40	15.37
Lime = excess	20.00
	277.51

The amount of lime to add is 277.51 mg/L as $CaCO_3$. The excess chosen was the minimum since (Mg^{2+} − 40) was less than 20.

Now we must determine if any NCH need be removed. The amount of NCH that can be left (NCH_f) is equal to the final hardness desired (80.00 mg/L) minus the CH left due to solubility and other factors (40.00 mg/L):

$$NCH_f = 80.00 - 40.00 = 40.00 \text{ mg/L}$$

Thus, 40.00 mg/L may be left. The NCH that must be removed (NCH_R) is the initial NCH_i (95.37 mg/L) minus the NCH_f:

$$NCH_R = NCH_i - NCH_f$$
$$NCH_R = 95.37 - 40.00 = 55.37 \text{ mg/L}$$

Thus, the amount of soda to be added is 55.37 mg/L as $CaCO_3$. The fact that this number is equal to the Mg^{2+} concentration is a coincidence of the numbers chosen.

Example 6-5. From the water analysis presented below, determine the amount of lime and soda (in mg/L as $CaCO_3$) necessary to soften the water to 90.00 mg/L as $CaCO_3$.

Water Composition (mg/L as CaCO₃)

Ca^{2+}: 149.2	CO_2: 29.3	HCO_3^-: 185.0	
Mg^{2+}: 65.8		SO_4^{2-}: 29.8	
Na^+: 17.4		Cl^-: 17.6	

Solution. The bar chart for this water may be plotted directly as shown below.

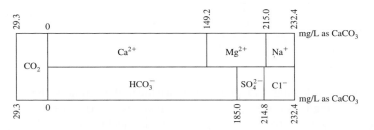

From the bar chart we note the following:

$$TH = 215.0 \text{ mg/L as CaCO}_3$$
$$CH = 185.0 \text{ mg/L as CaCO}_3$$
$$NCH = 30.0 \text{ mg/L as CaCO}_3$$

Following the logic of Figure 6-13, we calculate the lime dose as follows:

Step	Dose (mg/L as CaCO₃)
Lime = CO_2	29.3
Lime = HCO_3^-	185.0
Lime = $Mg^{2+} - 40 = 65.8 - 40 =$	25.8
Lime = excess	25.8
	265.9

The amount of lime to add is 265.9 mg/L as $CaCO_3$. The excess chosen was equal to the difference between the Mg^{2+} concentration and 40 since that difference was between 20 and 40, that is, $Mg^{2+} - 40 = 25.8$.

The amount of NCH_R is calculated as in Example 6-4:

$$NCH_f = 90.00 - 40.00 = 50.00$$
$$NCH_R = 30.00 - 50.00 = -20.00$$

Since NCH_R is a negative number, there is no need to remove NCH and, therefore, *no soda ash is required.*

Example 6-6.   Given the following water, determine the amount (mg/L) of 90 percent purity CaO and 97 percent purity Na_2CO_3 that must be purchased to treat the water to a final hardness of 85 mg/L; 120 mg/L.

Ion	(mg/L as $CaCO_3$)
CO_2	21
HCO_3^-	209
Ca^{2+}	183
Mg^{2+}	97

Solution.   First find the total hardness (TH), carbonate hardness (CH), and noncarbonate hardness (NCH):

$$TH = Ca^{2+} + Mg^{2+} = 183 + 97 = 280 \text{ mg/L}$$
$$CH = HCO_3^- = 209 \text{ mg/L}$$
$$NCH = TH - CH = 71 \text{ mg/L}$$

Use Figure 6-13 to find the lime dose as $CaCO_3$, assuming that we will leave 40 mg/L Mg^{2+} in the water:

$$Ca(OH)_2 = [21 + 209 + (97 - 40) + 40] = 327 \text{ mg/L as } CaCO_3$$

Since one mole of CaO equals one mole of $Ca(OH)_2$, we find 327 mg/L of CaO as $CaCO_3$. The molecular weight of CaO is 56 (equivalent weight = 28), and correcting for 90 percent purity:

$$CaO = 327\left(\frac{28}{50}\right)\left(\frac{1}{.9}\right) = 203 \text{ mg/L as CaO}$$

The amount of NCH that can be left in solution is equal to the final hardness desired (85 mg/L) minus the CH left behind due to solubility, inefficient mixing, etc. (40 mg/L), and is equal to $85 - 40 = 45$ mg/L.

Therefore, the NCH_R is the initial NCH (71 mg/L) minus the NCH which can be left (45 mg/L) and is $71 - 45 = 26$ mg/L. From Figure 6-14:

$$Na_2CO_3 = 26 \text{ mg/L as } CaCO_3$$

The equivalent weight of soda ash is 53 and the purity 97 percent:

$$Na_2CO_3 = 26\left(\frac{53}{50}\right)\left(\frac{1}{.97}\right) = 28 \text{ mg/L as } Na_2CO_3$$

If the final hardness desired is 120 mg/L, then the allowable final NCH is $120 - 40 = 80$ mg/L, which is greater than the initial NCH of 71, so no soda ash is necessary. The final hardness would be about 40 mg/L (carbonate hardness solubility) plus 71 mg/L (noncarbonate hardness) or 111 mg/L.

More Advanced Concepts in Lime-Soda Softening

Estimating CO_2 Concentration. CO_2 is of importance in two instances in softening. In the first instance, it consumes lime that otherwise could be used to remove Ca and Mg. When the concentration of CO_2 exceeds 10 mg/L as CO_2 (22.7 mg/L as $CaCO_3$ or 0.45 meq/L), the economics of removal of CO_2 by aeration (stripping) are favored over removal by lime neutralization. In the second instance, CO_2 is used to neutralize the high pH of the effluent from the softening process. These reactions are an application of the concepts of the carbonate buffer system discussed in Chapter 5.

The concentration of CO_2 may be estimated by using the equilibrium expressions for the dissociation of water and carbonic acid with the definition of alkalinity (Equation 5-36). The pH and alkalinity of the water must be determined to make the estimate. Example 6-7 illustrates a simple case where one of the forms of alkalinity predominates.

Example 6-7.   What is the estimated CO_2 concentration of a water with a pH of 7.65 and a total alkalinity of 310 mg/L as $CaCO_3$?

Solution.   When the raw water pH is less than 8.5, we can assume that the alkalinity is predominately HCO_3^-. Thus, we can ignore the dissociation of bicarbonate to form carbonate.

With this assumption, the procedure to solve the problem is

  a. Calculate the $[H^+]$ from the pH.

  b. Calculate the $[HCO_3^-]$ from the alkalinity.

  c. Solve the first equilibrium expression of the carbonic acid dissociation for $[H_2CO_3]$.

  d. Use the assumption that $[CO_2] = [H_2CO_3]$ to estimate the CO_2 concentration.

Following this approach, the $[H^+]$ concentration is

$$[H^+] = 10^{-7.65} = 2.24 \times 10^{-8} \text{ moles/L}$$

The $[HCO_3^-]$ concentration is

$$[HCO_3^-] = 310 \text{ mg/L} \left(\frac{61 \text{ mg/meq}}{50 \text{ mg/meq}} \right) \left(\frac{1}{(61 \text{ g/mole})(10^3 \text{ mg/g})} \right)$$
$$= 6.20 \times 10^{-3} \text{ moles/L}$$

Since the alkalinity is reported as mg/L as CaCO3, it must be converted to mg/L as the species using Equation 5-40 before the molar concentration may be calculated. The ratio 61/50 is the ratio of the equivalent weight of HCO_3^- to the equivalent weight of $CaCO_3$.

The equilibrium expression for the dissociation of carbonic acid is written in the form of Equation 5-33 using the reaction and pK_a given in Table 5-3.

$$K_a = \frac{[H^+][HCO_3^-]}{[H_2CO_3]}$$

where $K_a = 10^{-6.35} = 4.47 \times 10^{-7}$

Solving for $[H_2CO_3]$ gives

$$[H_2CO_3] = \frac{(2.24 \times 10^{-8} \text{ moles/L})(6.20 \times 10^{-3} \text{ moles/L})}{4.47 \times 10^{-7}}$$

$$[H_2CO_3] = 3.11 \times 10^{-4} \text{ moles/L}$$

We may assume that all the CO_2 in water forms carbonic acid. Thus, the estimated CO_2 concentration is

$$[CO_2] = 3.11 \times 10^{-4} \text{moles/L}$$

In other units for comparison and calculation:

$$CO_2 = (3.11 \times 10^{-4} \text{ moles/L})(44 \times 10^3 \text{ mg/mole}) = 13.7 \text{ mg/L as } CO_2$$

$$CO_2 = (13.7 \text{ mg/L as } CO_2)\left(\frac{50 \text{ mg/meq}}{22 \text{ mg/meq}}\right) = 31.14 \text{ or } 31.1 \text{ mg/L as } CaCO_3$$

The equivalent weight of CO_2 is taken as 22 because it effectively behaves as carbonic acid (H_2CO_3) and thus $n = 2$.

Softening to Practical Limits. In Example 6-5, the NCH_R was found to be a negative number. This implies that the water will be softened to a hardness less than the desired value. One way to overcome this difficulty is to treat a portion of the water to the practical limits and then blend the treated water with the raw water to achieve the desired hardness. Unlike the flowchart method used in Examples 6-4 through 6-6, calculations of chemical additions to soften to the practical limits of softening (that is 0.60 meq/L or 30 mg/L as $CaCO_3$ of Ca and 0.20 meq/L or 10 mg/L as $CaCO_3$ of $Mg(OH)_2$), do not take into account the desired final hardness. Stoichiometric amounts of lime and soda are added to remove all of the Ca and Mg. Example 6-8 illustrates the technique using both mg/L as $CaCO_3$ and milliequivalents/L as units of measure.

Example 6-8. Determine the chemical dosages for softening the following water to the practical solubility limits.

Constituent	mg/L	EW	EW CaCO₃/EW ion	mg/L as CaCO₃	meq/L
CO_2	9.6	22.0	2.28	21.9	0.44
Ca^{2+}	95.2	20.0	2.50	238.0	4.76
Mg^{2+}	13.5	12.2	4.12	55.6	1.11
Na^+	25.8	23.0	2.18	56.2	1.12
Alkalinity				198	3.96
Cl^-	67.8	35.5	1.41	95.6	1.91
SO_4^{2-}	76.0	48.0	1.04	76.0	1.58

Bar chart of raw water in mg/L as $CaCO_3$:

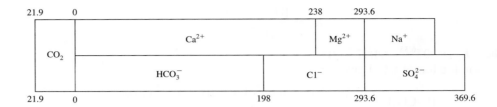

Solution. To soften to the practical solubility limits, lime and soda must be added as shown below.

Addition = to:	Lime mg/L as $CaCO_3$	Lime meq/L	Soda mg/L as $CaCO_3$	Soda meq/L
CO_2	21.9	0.44		
HCO_3^-	198.0	3.96		
$Ca-HCO_3^-$			40	0.80
Mg^{2+}	55.6	1.11	55.6	1.11
	275.5	5.51	95.6	1.91

Since the difference $Mg - 40 = 15.6$ mg/L as $CaCO_3$, the minimum excess lime of 20 mg/L as $CaCO_3$ is selected. The total lime addition is 295.5 mg/L as $CaCO_3$ or 165.5 mg/L as CaO. The soda addition is 95.6 mg/L as $CaCO_3$ or

$$95.6 \text{ mg/L as } CaCO_3 (53/50) = 101.3 \text{ mg/L as } Na_2CO_3$$

Note that (53/50) is the equivalent weight of Na_2CO_3/equivalent weight of $CaCO_3$. Reaction with CO_2:

$$CO_2 + Ca(OH)_2 \rightarrow CaCO_3(s) + H_2O$$

Bar chart after removal of CO_2:

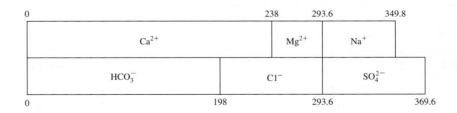

Reaction with HCO_3^-

$$Ca^{2+} + 2HCO_3^- + Ca(OH)_2 \rightarrow 2CaCO_3(s) + 2H_2O$$

Bar chart after reaction with HCO_3^-.

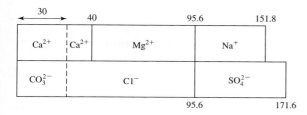

The 30 mg/L as $CaCO_3$ to the left of the dashed line is the calcium carbonate that remains because of solubility product limitations.

Reaction with calcium and soda:

$$Ca^{2+} + Na_2CO_3 \rightarrow CaCO_3(s) + 2Na^+$$

Bar chart after reaction with calcium and soda:

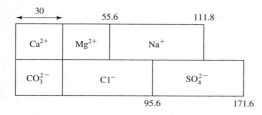

Reactions with magnesium, lime, and soda:

$$Mg^{2+} + Ca(OH)_2 \rightarrow Mg(OH)_2(s) + Ca^{2+}$$
$$Ca^{2+} + Na_2CO_3 \rightarrow CaCO_3(s) + 2Na^+$$

Bar chart of finished water:

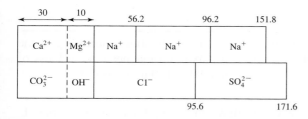

Total hardness of finished water is 30 mg/L as $CaCO_3$ + 10 mg/L as $CaCO_3$ = 40 mg/L as $CaCO_3$.

Split Treatment. As shown in Figure 6-15, in split treatment a portion of the raw water is by-passed around the softening reaction tank and the settling tank. This serves several functions. First, it allows the water to be tailored to yield a product water that has 0.80 meq/L or 40 mg/L as $CaCO_3$ of magnesium (or any other value above the

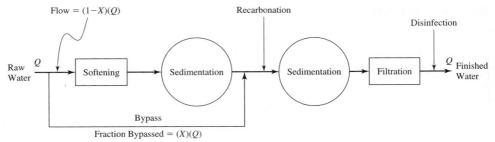

FIGURE 6-15
Split-flow treatment scheme.

solubility limit). Second, it allows for reduction in capital cost of tankage because the entire flow does not need to be treated. Third, it minimizes operating costs for chemicals by treating only a fraction of the flow. Fourth, it uses the natural alkalinity of the water to lower the pH of the product water and assist in stabilization. In many cases a second sedimentation basin is added after recarbonation and prior to filtration to reduce the solids loading onto the filters.

The fractional amount of the split is calculated as

$$X = \frac{Mg_f - Mg_i}{Mg_r - Mg_i} \qquad \text{(6-18)}$$

where Mg_f = final magnesium concentration, mg/L as $CaCO_3$
Mg_i = magnesium concentration from first stage, mg/L as $CaCO_3$
Mg_r = raw water magnesium concentration, mg/L as $CaCO_3$

The first stage is operated to soften the water to the practical limits of softening. Thus, the value for Mg_i is commonly taken to be 10 mg/L as $CaCO_3$. Since the desired concentration of Mg is nominally set at 40 mg/L as $CaCO_3$ as noted previously, Mg_f is commonly taken as 40 mg/L as $CaCO_3$.

Example 6-9. Determine the chemical dosages for split treatment softening of the following water. The finished water criteria is a maximum magnesium hardness of 40 mg/L as $CaCO_3$ and a total hardness in the range 80 to 120 mg/L as $CaCO_3$.

Constituent	mg/L	EW	EW $CaCO_3$/EW ion	mg/L as $CaCO_3$	meq/L
CO_2	11.0	22.0	2.28	25.0	0.50
Ca^{2+}	95.2	20.0	2.50	238	4.76
Mg^{2+}	22.0	12.2	4.12	90.6	1.80
Na^+	25.8	23.0	2.18	56.2	1.12
Alkalinity				198	3.96
Cl^-	67.8	35.5	1.41	95.6	1.91
SO_4^{2-}	76.0	48.0	1.04	76.0	1.58

Solution. In the first stage the water is softened to the practical solubility limits; lime and soda must be added as shown below.

Addition = to:	Lime mg/L as $CaCO_3$	Lime meq/L	Soda mg/L as $CaCO_3$	Soda meq/L
CO_2	25.0	0.50		
HCO_3^-	198.0	3.96		
Ca-HCO_3^-			40	0.80
Mg^{2+}	90.6	1.80	90.6	1.80
	313.6	6.26	130.6	2.60

The split is calculated in terms of mg/L as $CaCO_3$:

$$X = (40 - 10)/(90.6 - 10) = 0.372$$

The fraction of water passing through the first stage is then $1 - 0.372 = 0.628$. The total hardness of the water after passing through the first stage is the practical solubility limit, that is, 40 mg/L as $CaCO_3$. Since the total hardness in the raw water is $238 + 90.6 = 328.6$ mg/L as $CaCO_3$, the mixture of the treated and bypass water has a hardness of:

$$(0.372)(328.6) + (0.628)(40) = 147.4 \text{ mg/L as } CaCO_3$$

This is above the specified finished water criteria range of 80–120 mg/L as $CaCO_3$, so further treatment is required. Since the split is designed to yield the required 40 mg/L as $CaCO_3$ of magnesium, more calcium must be removed. Removal of the calcium equivalent to the bicarbonate will leave 40 mg/L as $CaCO_3$ of calcium hardness plus the 40 mg/L as $CaCO_3$ of magnesium hardness for a total of 80 mg/L as $CaCO_3$. The additions are as follows.

Constituent	Lime mg/L as $CaCO_3$	Lime meq/L
CO_2	25.0	0.50
HCO_3^-	198.0	3.96
	223.0	4.46

Addition of lime = to CO_2 and HCO_3^- (even in second stage) is necessary to get pH high enough.

The total chemical additions are in proportion to the flows:

Lime = 0.628(313.6) + 0.372(223) = 280 mg/L as $CaCO_3$

Soda = 0.628(130.6) + 0.372(0.0) = 82 mg/L as $CaCO_3$

Cases. The selection of chemicals and their dosage depends on the raw water composition and the desired final water composition. If we use a Mg concentration of

40 mg/L as $CaCO_3$ as a product water criterion, then six cases illustrate the dosage schemes. Three of the cases occur when the Mg concentration is less than 40 mg/L as $CaCO_3$ (Figure 6-16a, b, and c) and three cases occur when Mg is greater than 40 mg/L as $CaCO_3$ (Figure 6-17a, b, and c). In the cases illustrated in Figure 6-16, no split treatment is required. The cases illustrated in Figure 6-17 are for the first stage of a split

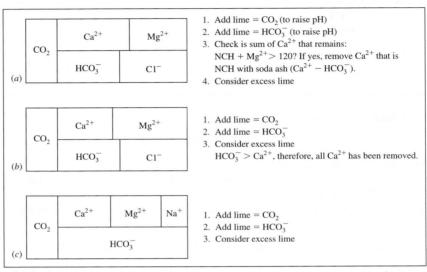

FIGURE 6-16
Dosage schemes when Mg^{2+} concentration is less than 40 mg/L as $CaCO_3$ and no split treatment is required. Note that no Mg^{2+} is removed and that reactions deal with CO_2 and Ca^{2+} only.

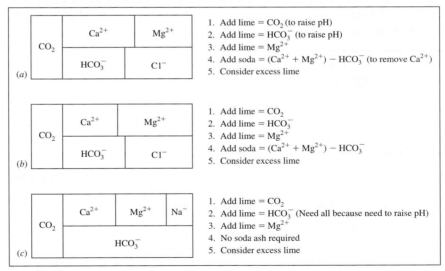

FIGURE 6-17
Cases when Mg^{2+} concentration is greater than 40 mg/L as $CaCO_3$ and split treatment is required. Note that these cases illustrate softening to the practical limits in the first stage of the split-flow scheme.

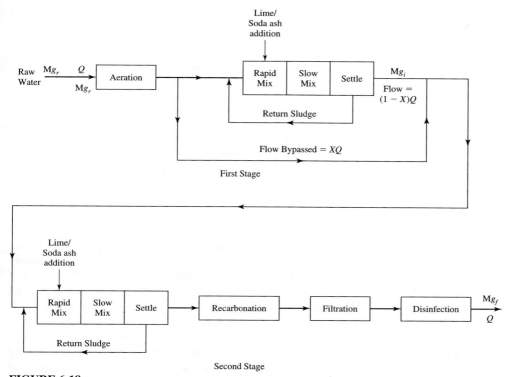

FIGURE 6-18
Flow diagram for a two stage split-treatment lime–soda ash softening plant. (Adapted from J. L. Cleasby and J. H. Dillingham, 1966.)

treatment flow scheme (softening to the practical limits). In the cases illustrated in Figure 6-17, the hardness of the mixture of the treated and raw water must be checked to see if an acceptable hardness has been achieved. If the hardness after blending is above the desired concentration, then further softening in a second stage is required (Figure 6-18). Since the design of the split is to achieve a desired Mg concentration of 40 mg/L as $CaCO_3$, no further Mg removal is required. Only treatment of the Ca is required.

Ion-Exchange Softening

Ion exchange can be defined as the reversible interchange of ions between a solid and a liquid phase in which there is no permanent change in the structure of the solid. Typically, in water softening by ion exchange, the water containing the hardness is passed through a column containing the ion-exchange material. The hardness in the water exchanges with an ion from the ion-exchange material. Generally, the ion exchanged with the hardness is sodium, as illustrated in Equation 6-19:

$$Ca(HCO_3)_2 + 2NaR \rightleftharpoons CaR_2 + 2NaHCO_3 \qquad (6\text{-}19)$$

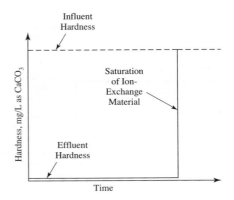

FIGURE 6-19

Hardness removal in ion-exchange column.

where R represents the solid ion-exchange material. By the above reaction, calcium (or magnesium) has been removed from the water and replaced by an equivalent amount of sodium, that is, two sodium ions for each cation. The alkalinity remains unchanged. The exchange results in essentially 100 percent removal of the hardness from the water until the exchange capacity of the ion-exchange material is reached, as shown in Figure 6-19. When the ion-exchange material becomes saturated, no hardness will be removed. At this point *breakthrough* is said to have occurred because the hardness passes through the bed. At this point the column is taken out of service, and the ion-exchange material is regenerated. That is, the hardness is removed from the material by passing water containing a large amount of Na^+ through the column. The mass action of having so much Na^+ in the water causes the hardness of the ion-exchange material to enter the water and exchange with the sodium:

$$CaR_2 + 2NaCl \rightleftharpoons 2NaR + CaCl_2 \tag{6-20}$$

The ion-exchange material can now be used to remove more hardness. The $CaCl_2$ is a waste stream that must be disposed of.

There are some large water treatment plants that utilize ion-exchange softening, but the most common application is for residential water softeners. The ion-exchange material can either be naturally occurring clays, called *zeolites*, or synthetically made resins. There are several manufacturers of synthetic resins. The resins or zeolites are characterized by the amount of hardness that they will remove per volume of resin material and by the amount of salt required to regenerate the resin. The synthetically produced resins have a much higher exchange capacity and require less salt for regeneration. However, they also cost much more. People who work in the water softening industry often work in units of grains of hardness per gallon of water (gr/gal). It is useful to remember that 1 gr/gal equals 17.1 mg/L.

Since the resin removes virtually 100 percent of the hardness, it is necessary to bypass a portion of the water and then blend in order to obtain the desired final hardness.

$$\%\text{Bypass} = (100) \frac{\text{Hardness}_{\text{desired}}}{\text{Hardness}_{\text{initial}}} \tag{6-21}$$

Example 6-10. A home water softener has 0.1 m^3 of ion-exchange resin with an exchange capacity of 57 kg/m^3. The occupants use 2,000 L of water per day. If the water contains 280.0 mg/L of hardness as $CaCO_3$ and it is desired to soften it to 85 mg/L as $CaCO_3$, how much should be bypassed? What is the time between regeneration cycles?

Solution. The percentage of water to be bypassed is found using Equation 6-21:

$$\%\text{Bypass} = (100)\frac{85}{280} = 30.36 \text{ or } 30\%$$

The length of time between regeneration cycles is determined from the exchange capacity of the ion-exchange material (media). This is also called the "time to break-through," that is, the time to saturate the exchange material. If 30 percent of the water is being bypassed, then 70 percent of the water is being treated and the hardness loading rate is

$$\text{Loading rate} = (0.7)(280 \text{ mg/L})(2,000 \text{ L/d}) = 392,000 \text{ mg/d}$$

Since the bed capacity is 57 kg/m^3 and the bed contains 0.1 m^3 of ion-exchange media, the breakthrough time is approximately

$$\text{Breakthrough time} = \frac{(57 \text{ kg/m}^3)(0.1 \text{ m}^3)}{(392,000 \text{ mg/d})(10^{-6} \text{ kg/mg})} = 14.5 \text{ d}$$

6-4   MIXING AND FLOCCULATION

Clearly, if the chemical reactions in coagulating and softening a water are going to take place, the chemical must be mixed with the water. In this section we will begin to look at the physical methods necessary to accomplish the chemical processes of co-agulation and softening.

Mixing, or *rapid mixing* as it is called, is the process whereby the chemicals are quickly and uniformly dispersed in the water. Ideally, the chemicals would be instantaneously dispersed throughout the water. During coagulation and softening, the chemical reactions that take place in rapid mixing form precipitates. Either aluminum hydroxide or iron hydroxide form during coagulation, while calcium carbonate and magnesium hydroxide form during softening. The precipitates formed in these processes must be brought into contact with one another so that they can agglomerate and form larger particles, called *flocs*. This contacting process is called *flocculation* and is accomplished by slow, gentle mixing.

In the treatment of water and wastewater, the degree of mixing is measured by the velocity gradient, G. The velocity gradient is best thought of as the amount of shear taking place; that is, the higher the G value, the more violent the mixing. The velocity gradient is a function of the power input into a unit volume of water. It is given by

$$G = \sqrt{\frac{P}{\mu\forall}} \qquad \qquad \text{(6-22)}$$

where G = velocity gradient, s^{-1}
P = power input, W
V = volume of water in mixing tank, m^3
μ = dynamic viscosity, Pa $\cdot$ s

From literature, experience, laboratory, or pilot plant work, it is possible to select a G value for a particular application. The total number of particle collisions is proportional to Gt, where t is the detention time in the basin as given by Equation 2-27.

Rapid Mix

Rapid mixing is probably the most important physical operation affecting coagulant dose efficiency. The chemical reaction in coagulation is completed in less than 0.1 s; therefore, it is imperative that mixing be as instantaneous and complete as possible. Rapid mixing can be accomplished within a tank utilizing a vertical shaft mixer (Figure 6-20), within a pipe using an in-line blender (Figure 6-21), or in a pipe using a static mixer (Figure 6-22). Other methods, such as Parshall flumes, hydraulic jumps, baffled channels, or air mixing may also be used.

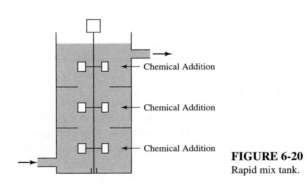

Chemical Addition

Chemical Addition

Chemical Addition

FIGURE 6-20
Rapid mix tank.

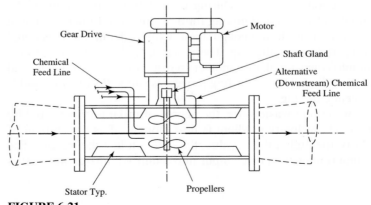

Gear Drive

Motor

Chemical
Feed Line

Shaft Gland

Alternative
(Downstream) Chemical
Feed Line

Stator Typ.

Propellers

FIGURE 6-21
Typical in-line blender. (*Source:* AWWA, 1998.)

FIGURE 6-22
STATIC MIXER: A succession of reversing, flow-twisting, and flow-splitting elements provides positive dispersion proportional to number of elements. (*Source: Chemical Engineering,* March 22, 1971.)

The selection of G and Gt values for coagulation is dependent on the mixing device, the chemicals selected, and the anticipated reactions. Coagulation occurs predominately by two mechanisms: adsorption of the soluble hydrolysis species on the colloid and destabilization or sweep coagulation where the colloid is trapped in the hydroxide precipitate. The reactions in adsorption-destabilization are extremely fast and occur within 1 second. Sweep coagulation is slower and occurs in the range of 1 to 7 seconds (Amirtharajah, 1978). Jar test data may be used to identify whether adsorption-destabilization or sweep coagulation is predominant. If charge reversal is apparent from the dose-turbidity curve (see, for example, Figure 6-9), then adsorption-destabilization is the predominant mechanism. If the dose-turbidity curve does not show charge reversal (that is, the curve is relatively flat at higher doses), then the predominant mechanism is sweep coagulation. G values in the range of 3,000 to 5,000 s^{-1} and detention times on the order of 0.5 s are recommended for adsorption-desorption reactions. These values are most commonly achieved in an in-line blender. For sweep coagulation, detention times of 1 to 10 s and G values in the range of 600 to 1,000 s^{-1} are recommended (Amirtharajah, 1978). An exception to these time requirements is for color or natural organic matter removal. It is found that rapid mix detention times of 2 to 5 minutes at low rapid mix G values of 300 to 700 s^{-1} work best.

Softening. For dissolution of $CaO/Ca(OH)_2$ mixtures for softening, detention times on the order of 5 to 10 minutes may be required. G values to disperse and maintain particles in suspension may be on the order of 700 s^{-1}. In-line blenders are not used to mix softening reagents.

Rapid-Mix Tanks. The volume of a rapid-mix tank seldom exceeds 8 m^3 because of mixing equipment and geometry constraints. The mixing equipment consists of an electric motor, gear-type speed reducer, and either a radial-flow or axial-flow impeller as shown in Figure 6-23. The radial-flow impeller provides more turbulence and is preferred for rapid mixing. The tanks should be horizontally baffled into at least two and preferably three compartments in order to provide sufficient residence time. They are also baffled vertically to minimize vortexing. Chemicals should be added below the impeller, the point of the most mixing. Some unitless geometric ratios for both rapid mixing and flocculation are shown in Table 6-9. These values can be used to select the proper basin depth and surface area and the impeller diameter. For rapid

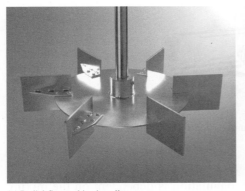

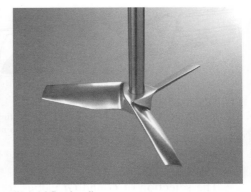

(*a*) Radial-flow turbine impeller

(*b*) Axial-flow impeller

FIGURE 6-23
Basic impeller styles. (*Source:* Courtesy of SPX Process Equipment.)

TABLE 6-9
Tank and impeller geometries for mixing

Geometric ratio	Allowed range
D/T (radial)	0.14–0.5
D/T (axial)	0.17–0.4
H/D (either)	2–4
H/T (axial)	0.34–1.6
H/T (radial)	0.28–2
B/D (either)	0.7–1.6

D = impeller diameter
T = equivalent tank diameter
H = water depth
B = water depth below impeller

mixing, in order to construct a reasonably sized basin, often more depth is required than allowed by the ratios in Table 6-9. In this case the tank is made deeper by using two impellers on the shaft. Rapid mixing is generally accomplished with a radial-flow-type turbine like that of Figure 6-23. When dual impellers are used, the top impeller is axial flow, while the bottom impeller is radial flow. Figure 6-24 shows the flow patterns of radial-and axial-flow impellers. When dual impellers are employed on gear-driven mixers, they are spaced approximately two impeller diameters apart. We normally assume an efficiency of transfer of motor power to water power of 0.8 for a single impeller.

Flocculation

While rapid mix is the most important physical factor affecting coagulant efficiency, flocculation is the most important factor affecting particle-removal efficiency. The

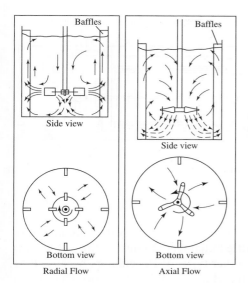

FIGURE 6-24
Basic flow patterns created by impellers. (*Source:* Cornwell and Bishop, 1983.)

objective of flocculation is to bring the particles into contact so that they will collide, stick together, and grow to a size that will readily settle. Enough mixing must be provided to bring the floc into contact and to keep the floc from settling in the flocculation basin. Too much mixing will shear the floc particles so that the floc is small and finely dispersed. Therefore, the velocity gradient must be controlled within a relatively narrow range. Flexibility should also be built into the flocculator so that the plant operator can vary the G value by a factor of two to three. The heavier the floc and the higher the suspended solids concentration, the more mixing is required to keep the floc in suspension. This is reflected in Table 6-10. Softening floc is heavier than coagulation floc and therefore requires a higher G value. An increase in the floc concentration (as measured by the suspended solids concentration) also increases the required G. With water temperatures of approximately 20°C, modern plants provide about 20 minutes of flocculation time (t) at plant capacity. With lower temperatures, the detention time is increased. At 15°C the detention time is increased by 7 percent, at 10°C it is increased 15 percent, and at 5°C it is increased 25 percent.

TABLE 6-10
Gt values for flocculation

Type	G (s^{-1})	Gt_o (unitless)
Low-turbidity, color removal coagulation	20–70	60,000 to 200,000
High-turbidity, solids removal coagulation	30–80	36,000 to 96,000
Softening, 10% solids	130–200	200,000 to 250,000
Softening, 39% solids	150–300	390,000 to 400,000

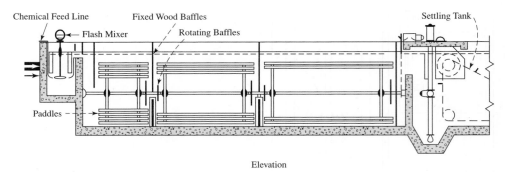

FIGURE 6-25
Paddle flocculator. (Courtesy of Siemens Water Technologies.)

Flocculation is normally accomplished with an axial-flow impeller (Figure 6-23), a paddle flocculator (Figure 6-25), or a baffled chamber (Figure 6-26). Axial-flow impellers have been recommended over the other types of flocculators because they impart a nearly constant G throughout the tank (Hudson, 1981). The flocculator basin should be divided into at least three compartments. The velocity gradient is tapered so that the G values decrease from the first compartment to the last, similar to that shown in Table 6-10.

Power Requirements

In the design of mixing equipment for rapid-mix and flocculation tanks, the power imparted to the liquid in a baffled tank by an impeller may be described by the following equation for fully turbulent flow developed by Rushton (1952).

$$P = N_p(n)^3(D_i)^5\rho \qquad (6-23)$$

where P = power, W
  N_p = impeller constant (also called power number)
  n = rotational speed, revolutions/s
  D_i = impeller diameter, m
  ρ = density of liquid, kg/m^3

The impeller constant of a specific impeller can be obtained from the manufacturer. For the radial-flow impeller of Figure 6-23, the impeller constant is 5.7 and for the axial-flow impeller, it is 0.31. It is recommended that for flocculation, the tangential velocity (tip speed) be limited to 2.7 m/s.

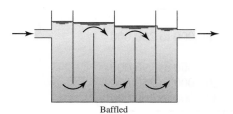

Baffled

FIGURE 6-26
Baffled chamber flocculator.

The power imparted by a paddle mixer (Figure 6-25) is a function of the drag force on the paddles

$$P = \frac{C_D A \rho (v_p)^3}{2}$$
(6-24)

where P = power imparted, W
C_D = coefficient of drag of paddle
ρ = density of fluid, kg/m^3
A = cross-sectional area of paddles, m^2
v_p = relative velocity of paddles with respect to fluid, m/s

It has been found that the peripheral velocity of the paddle blades should range from 0.1 to 1.0 m/s and that the relative velocity of the paddles to the fluid should be 0.6 to 0.75 times the paddle-tip speed. The drag coefficient (C_D) varies with the length-to-width ratio (for example: for L:W of 5, C_D = 1.20, for L:W of 20, C_D = 1.50, and for L:W of infinity, C_D = 1.90). It is also recommended that the total paddle-blade area on a horizontal shaft not exceed 15 to 20 percent of the total basin cross-sectional area to avoid excessive rotational flow.

For pneumatic mixing, the power imparted is given by

$$P = K Q_a \ln\left(\frac{h + 10.33}{10.33}\right)$$
(6-25)

where P = power imparted, W
K = constant = 1.689
Q_a = air flow rate at atmospheric pressure, m^3/min
h = air pressure at the point of discharge, m

The power imparted by static-mixing devices may be computed as

$$P = \gamma Q h$$
(6-26)

where P = power imparted, kW
γ = specific weight of fluid, kN/m^3
Q = flow rate, m^3/s
h = head loss through the mixer, m

The specific weight of water is equal to the product of the density and the acceleration due to gravity ($\gamma = \rho g$). At normal temperatures, the specific weight of water is taken to be 9.81 kN/m^3.

Upflow Solids-Contact.   Mixing, flocculation, and clarification may be conducted in a single tank such as that in Figure 6-27. The influent raw water and chemicals are mixed in the center cone-like structure. The solids flow down under the cone (sometimes called a "skirt"). As the water flows upward, the solids settle to form a *sludge blanket*. This design is called an upflow solids-contact basin. The main advantage of this unit is its reduced size. The units are best suited to treat a feed water that has a relatively constant quality. It is often favored for softening because the water quality from wells is relatively constant and the sludge blanket provides a further opportunity to drive the precipitation reactions to completion.

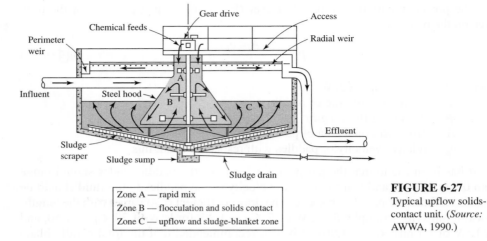

FIGURE 6-27
Typical upflow solids-contact unit. (*Source:* AWWA, 1990.)

Zone A — rapid mix	
Zone B — flocculation and solids contact	
Zone C — upflow and sludge-blanket zone	

Example 6-11. A city is planning the installation of a new water treatment plant to supply a growing population. There will be only one rapid mix basin and then the flow will be evenly split between two flocculator trains, each with three basins equal in volume. The required water depth of all basins is 4.0 m. Determine correct basin volumes, basin dimensions, tank equivalent diameter, required input power, impeller diameter from table below, and rotational speed using the following parameters:

$Q = 11.5 \times 10^3$ m³/d

Rapid mix $t_o = 2$ min.

Based on pilot studies the rapid mix $t_o = 2$ min. to remove color and natural organic matter

Rapid mix $G = 600$ s^{-1}

Total flocculation $t_o = 30$ min

Flocculators $G = 70, 50, 30$ s^{-1}

Water temperature = 5°C

Place impeller at one-third the water depth

Impeller type	Impeller diameters (m)			Power number (N_p)
Radial	0.8	1.1	1.4	5.7
Axial	0.8	1.4	2.0	0.31

Solution:

a. **Rapid-mix design**

Convert 11.5×10^3 m³/d to m³/min:

$$\frac{11.5 \times 10^3 \text{ m}^3}{d} \times \frac{d}{24 \text{ hr}} \times \frac{h}{60 \text{ min}} = 7.986 \text{ or } 8.0 \text{ m}^3/\text{min}$$

Now with Equation 2-27 we can determine the volume of the rapid mix basin using the flow rate and detention time.

$$V = (Q)(t_o) = (8.0 \text{ m}^3/\text{min})(2 \text{ min}) = 16.0 \text{ m}^3$$

Based on the required water depth of the basin, the corresponding area is:

$$\text{Area} = \frac{16 \text{ m}^3}{4.0 \text{ m}} = 4.0 \text{ m}^2$$

It is common practice to make the length and width of the mixing basins equal. Therefore, the square root of the required area will yield the side dimension.

$$\text{Length and width} = \sqrt{4.0 \text{ m}^2} = 2.0 \text{ m}$$

Calculating the equivalent tank diameter is useful at this point because it will be needed when calculating the geometric ratios for the impeller.

$$T_E = \sqrt{\frac{4 \times \text{Area}}{\pi}} = \sqrt{\frac{4 \times 4.0 \text{ m}^2}{\pi}} = 2.26 \text{ m}$$

The required input power can be calculated by using Equation 6-22. Using Table A-1 in Appendix A and the temperature of the water, find the dynamic viscosity of 1.52×10^{-3} Pa · s. As noted in the table footnote, the values in the table are multiplied by 10^{-3}.

$$P = G^2\mu V = (600 \text{ s}^{-1})^2(1.52 \times 10^{-3} \text{ Pa} \cdot \text{s})(16.0 \text{ m}^3) = 8,755 \text{ W}$$

By using Table 6-9, we can evaluate different size radial impellers using the geometric ratios. Below is a comparison of the ratios for the available sizes of radial impellers and the rapid mix basin.

Geometric ratio	Allowable range	Radial impeller diameter		
		0.8 m	1.1 m	1.40 m
D/T	0.14–0.5	0.35	0.49	**0.62**
H/D	2.0–4.0	**5.00**	3.64	2.85
H/T	0.28–2.0	1.77	1.77	1.77
B/D	0.7–1.6	**1.67**	1.21	0.95

The shaded areas indicate values that are not within the range, which leaves the 1.1-m radial impeller as best suited for this application. Finally, the required rotational speed for the impeller is calculated by using Equation 6-23.

$$n = \left[\frac{P}{N_p\rho(D_i)^5}\right]^{1/3} = \left[\frac{8,755 \text{ W}}{(5.7)(1,000 \text{ kg/m}^3)(1.1 \text{ m})^5}\right]^{1/3} = 0.98 \text{ rps} = 59 \text{ rpm}$$

b. Flocculator design

Because the flow is evenly split between two flocculation trains, the flow rate from the rapid mix calculations should be divided by 2.

$$Q_{\text{per train}} = \frac{8.0 \text{ m}^3/\text{min}}{2} = 4.0 \text{ m}^3/\text{min}$$

Using Equation 2-27 again, we can find the needed basin volume for each flocculation train.

$$\mathcal{V}_T = (Q)(t_o) = (4.0 \text{ m}^3/\text{min})(30 \text{ min}) = 120 \text{ m}^3$$

Because we need three equal flocculator tanks, the total volume per train is divided by 3.

$$\mathcal{V}_{\text{per basin}} = \frac{\mathcal{V}_T}{3} = \frac{120 \text{ m}^3}{3} = 40 \text{ m}^3$$

Using the same criteria as before we solve for the area, length, width, and equivalent tank diameter of the basins.

$$\text{Area} = \frac{40.0 \text{ m}^3}{4.0 \text{ m}} = 10.0 \text{ m}^2$$

$$\text{Length and width} = \sqrt{10.0 \text{ m}^2} = 3.16 \text{ m}$$

$$T_E = \sqrt{\frac{4 \times \text{Area}}{\pi}} = \sqrt{\frac{4 \times 10.0 \text{ m}^2}{\pi}} = 3.57 \text{ m}$$

The input power needed is going to be calculated three times because each flocculator has a unique velocity gradient.

$$P_{G=70} = G^2 \mu \mathcal{V} = (70 \text{ s}^{-1})^2 (1.52 \times 10^{-3} \text{ Pa} \cdot \text{s})(40.0 \text{ m}^3) = 298 \text{ W}$$

$$P_{G=50} = G^2 \mu \mathcal{V} = (50 \text{ s}^{-1})^2 (1.52 \times 10^{-3} \text{ Pa} \cdot \text{s})(40.0 \text{ m}^3) = 152 \text{ W}$$

$$P_{G=30} = G^2 \mu \mathcal{V} = (30 \text{ s}^{-1})^2 (1.52 \times 10^{-3} \text{ Pa} \cdot \text{s})(40.0 \text{ m}^3) = 54.7 \text{ W}$$

Select an impeller that will work with the flocculator basins by using the geometric ratios for an axial impeller.

Geometric ratio	Allowable range	Axial impeller diameter		
		0.8 m	1.4 m	2.0 m
D/T	0.17–0.4	0.22	0.39	**0.56**
H/D	2.0–4.0	**5.00**	2.86	2.00
H/T	0.34–1.6	1.12	1.12	1.12
B/D	0.7–1.6	**1.66**	0.95	**0.67**

Notice the gray areas again; this table shows that a 1.6-m impeller would best fit the tank sizes.

$$n_{G=70} = \left[\frac{P}{N_p\,\rho(D_i)^5}\right]^{1/3} = \left[\frac{298\ \text{W}}{(0.31)(1{,}000\ \text{kg/m}^3)(1.4\ \text{m})^5}\right]^{1/3} = 0.56\ \text{rps} = 34\ \text{rpm}$$

For flocculation, the tangential velocity (tip speed) must not exceed 2.7 m/s. If this is true for the flocculator with a $G = 70\ \text{s}^{-1}$, then it is true for the others since their rotational speed is lower.

$$\text{Tip speed} = (\text{rps})(\pi \times D_i) = (0.56)(\pi \times 1.4) = 2.46\ \text{or}\ 2.5\ \text{m/s}$$

6-5   SEDIMENTATION

Overview

Particles that will settle within a reasonable period of time can be removed in a sedimentation basin (also called a clarifier). Sedimentation basins are usually rectangular or circular with either a radial or upward water flow pattern. Regardless of the type of basin, the design can be divided into four zones: inlet, settling, outlet, and sludge storage. While our intent is to present the concepts of sedimentation and to design a sedimentation tank, a brief discussion of all four zones is helpful in understanding the sizing of the settling zone. A schematic showing the four zones is shown in Figure 6-28.

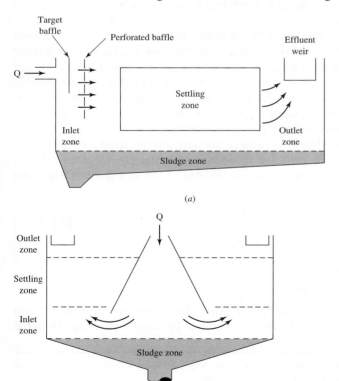

FIGURE 6-28

Zones of sedimentation: (*a*) horizontal flow clarifier; (*b*) upflow clarifier.

The purpose of the inlet zone is to evenly distribute the flow and suspended particles across the cross section of the settling zone.* The inlet zone consists of a series of inlet pipes and baffles placed about 1 m into the tank and extending the full depth of the tank. Following the baffle system, the flow takes on a flow pattern determined by the inlet structure. At some point the flow pattern is evenly distributed, and the water velocity slowed to the design velocity of the sedimentation zone. At that point the inlet zone ends and the settling zone begins. With a well-designed inlet baffle system, the inlet zone extends approximately 1.5 m down the length of the tank. Proper inlet zone design may well be the most important aspect of removal efficiency.

With improper design, the inlet velocities may never subside to the settling-zone design velocity. Typical design numbers are usually conservative enough that an inlet zone length does not have to be added to the length calculated for the settling zone. In an accurate design, the inlet and settling zones are each designed separately and their lengths added together.

The configuration and depth of the sludge storage zone depends upon the method of cleaning, the frequency of cleaning, and the quantity of sludge estimated to be produced. All these variables can be evaluated and a sludge storage zone designed. In lieu of these design details, some general guidelines can be presented. With a well-flocculated solid and good inlet design, over 75 percent of the solids may settle in the first fifth of the tank. For coagulant floc, Hudson recommends a sludge storage depth of about 0.3 m near the outlet and 2 m or more near the inlet (Hudson, 1981).

If the tank is long enough, storage depth can be provided by bottom slope; if not, a sludge hopper is necessary at the inlet end or the overall tank is made deeper. Mechanically-cleaned basins may be equipped with a bottom scraper, such as shown in Figure 6-29 for a circular tank. The sludge is continuously scraped to a hopper where it is pumped out. For mechanically cleaned basins, a 1 percent slope toward the sludge withdrawal point is used. A sludge hopper is designed with sides sloping with a vertical to horizontal ratio of 1.2:1 to 2:1. For horizontal tanks a bottom collection device is used such as shown in Figure 6-30. These devices travel the length of the basin and remove sludge via pump or siphon action.

The outlet zone is designed so as to remove the settled water from the basin without carrying away any of the floc particles. A fundamental property is that the velocity of flowing water is proportional to the flow rate divided by the area through which the water flows, that is,

$$v = \frac{Q}{A_c}$$ 🏴 (6-27)

where v = water velocity, m/s
Q = water flow, m^3/s
A_c = cross-sectional area, m^2

Within the sedimentation tank, the flow is going through a very large area (basin depth times width); consequently, the velocity is slow. To remove the water from the

---

*The cross section is the area through which the flow moves. For example, in Figure 6-28a the cross section is the settling zone width × depth and in 6-28b it is the bottom circular area.

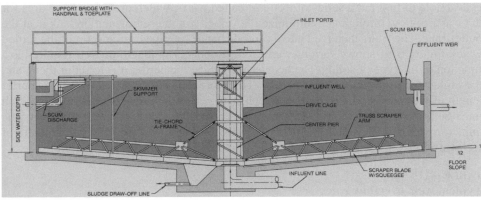

FIGURE 6-29
Photograph and schematic diagram of circular sludge scraper. (*Source:* David Cornwell/Courtesy of Siemens Water Technologies.)

basin quickly, it is desirable to direct the water into a pipe or small channel for easy transport, which will produce a significantly higher velocity. If a pipe were to be placed at the end of the sedimentation basin, all the water would "rush" to the pipe. This rushing water would create high velocity profiles in the basin, which would tend to raise the settled floc from the basin and into the effluent water. This phenomenon of washing out the floc is called *scouring,* and one way to create scouring is with an

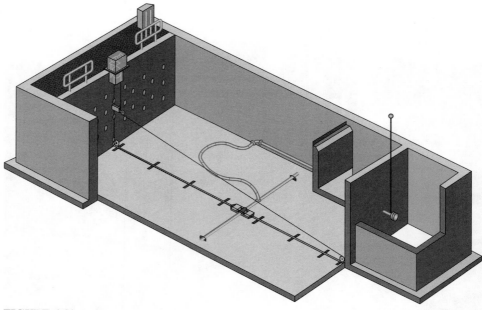

FIGURE 6-30
Vacuum sludge collection.

improper outlet design. Rather than put a pipe at the end of the sedimentation basin, it is desirable to first put a series of troughs, called *weirs,* which provide a large area for the water to flow through and minimize the velocity in the sedimentation tank near the outlet zone. The weirs then feed into a central channel or pipe for transport of the settled water. Figure 6-31 shows various weir arrangements. The length of weir required is a function of the type of solids. The heavier the solids, the harder it is to scour them, and the higher the allowable outlet velocity. Therefore, heavier particles require a shorter length of weir than do light particles. Each state generally has a set of standards which must be followed, but Table 6-11 shows typical design values for weir loadings. The units for weir overflow rates are $m^3/d \cdot m$, which is water flow (m^3/d) per unit length of weir (m). In U. S. Customary System (USCS) units, this is expressed as gpm/ft.

TABLE 6-11
Typical weir overflow rates

Type of floc	Weir overflow rate ($m^3/d \cdot m$)
Light alum floc (low-turbidity water)	143–179
Heavier alum floc (higher-turbidity water)	179–268
Heavy floc from lime softening	268–322

Source: Walker Process Equipment, 1973.

(*a*)

(*b*)

FIGURE 6-31
Weir arrangements: (*a*) rectangular; (*b*) circular. (*Source:* David Cornwell/ Mackenzie L. Davis.)

In a rectangular basin, the weirs often cover at least one-third of the basin length. Spacing may be as large as 5 to 6 m on-centers but is frequently on the order of one-half this distance.

Example 6-12.   The town of Urbana has a low-turbidity raw water and is designing its overflow weir at a loading rate of 150 m³/d · m. If its plant flow rate is 0.5 m³/s, how many linear meters of weir are required?

Solution

$$\frac{(0.5 \text{ m}^3/\text{s})(86{,}400 \text{ s/d})}{150 \text{ m}^3/\text{d} \cdot \text{m}} = 288 \text{ m}$$

Sedimentation Concepts

There are two important terms to understand in sedimentation zone design. The first is the particle (floc) *settling velocity, v_s.* The second is the velocity at which the tank is designed to operate, called the *overflow rate, v_o.* The easiest way to understand these two concepts is to view an upward-flow sedimentation tank as shown in Figure 6-32. In this design, the particles fall downward and the water rises vertically. The rate at which the particle is settling downward is the particle-settling velocity, and the velocity of the liquid rising is the overflow rate. Obviously, if a particle is to be removed from the bottom of the clarifier and not go out in the settled water, then the particle-settling velocity must be greater than the liquid-rise velocity or overflow rate ($v_s > v_o$). If v_s is greater than v_o, one would expect 100 percent particle removal, and if v_s is less than v_o, one would expect 0 percent removal. In design, the procedure would be to determine the particle-settling velocity and set the overflow rate at some lower value. Often v_o is set at 50 to 70 percent of v_s for an *upflow clarifier.*

Let us now consider why the liquid-rise velocity is called an overflow rate and what its units are. The term *overflow rate* is used since the water is flowing over the top of the tank into the weir system. It is sometimes referred to as the *surface loading rate* because it has units of m³/d · m². The units are flow of water (m³/d) being applied

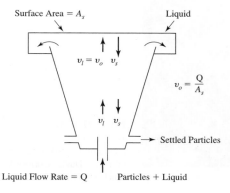

FIGURE 6-32
Settling in an upflow clarifier. (Legend: v_l = velocity of liquid; v_s = terminal settling velocity of particle.)

to a m^2 of surface area. This can be thought of as the amount of water that goes through each m^2 of tank surface area per day, which is similar to a loading rate. Recall from Equation 6-27 that the velocity of flow is equal to the flow rate divided by the area through which it flows. Hence an overflow rate is the same as a liquid velocity:

$$v_o = \frac{\text{Volume/Time}}{\text{Surface area}} = \frac{(\text{Depth})(\text{Surface area})}{(\text{Time})(\text{Surface area})} = \frac{\text{Depth}}{\text{Time}} = \text{Liquid velocity} \quad (6\text{-}28)$$

$$v_o = \frac{\cancel{V}/t_o}{A_s} = \frac{(h)(A_s)}{(t_o)(A_s)} = \frac{h}{t_o}$$

It can be seen from the above discussion that particle removal is independent of the depth of the sedimentation tank. As long as v_s is greater than v_o, the particles will settle downward and be removed from the bottom of the tank regardless of the depth. Sedimentation zones vary from a depth of a few centimeters to a depth of 6 m or greater.*

We can show that particle removal in a horizontal sedimentation tank is likewise dependent only upon the overflow rate. An ideal horizontal sedimentation tank is based upon three assumptions (Hazen, 1904, and Camp, 1946):

1. Particles and velocity vectors are evenly distributed across the tank cross section. This is the function of the inlet zone.

2. The liquid moves as an ideal slug down the length of the tank.

3. Any particle hitting the bottom of the tank is removed.

Using Figure 6-33 to illustrate the concept, let us consider a particle which is released at point A. In order to be removed from the water it must have a settling velocity great enough so that it reaches the bottom of the tank during the detention time (t_o) of the water in the tank. We may say its settling velocity must equal the depth of the tank divided by the detention time, that is,

$$v_s = \frac{h}{t_o} \quad (6\text{-}29)$$

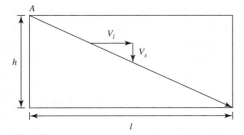

FIGURE 6-33
Ideal horizontal sedimentation tank.

---

*Tube settlers are designed with a very shallow settling zone. Their use is beyond the presentation of this text.

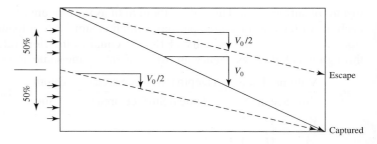

FIGURE 6-34
Partial solids removal in ideal sedimentation tank.

Now we can also show that the settling velocity of the particle must be equal to or greater than the overflow rate of the tank in order to be removed. Using the definition of detention time from Equation 2-27 and substituting into Equation 6-29:

$$v_s = \frac{h}{(\forall/Q)} = \frac{hQ}{\forall} \tag{6-30}$$

But we know that the tank volume is described by the product of the height, length, and width so we can rewrite this as:

$$v_s = \frac{hQ}{l \times w \times h} = \frac{Q}{l \times w} \tag{6-31}$$

And the product $(l \times w)$ is the surface area (A_s) so

$$v_s = \frac{Q}{A_s} \tag{6-32}$$

which is the overflow rate (v_o). This implies that the removal of a horizontal clarifier is independent of depth! This is indeed strange and runs counter to our intuitive feeling of how the sedimentation tank should work. The tank depth is important because particles rarely settle discretely, and the removal efficiency for flocculating particles is dependent on depth.

In a horizontal sedimentation tank, unlike an upflow clarifier, some percentage of the particles with a v_s less than v_o will be removed. For example, consider particles having a settling velocity of 0.5 v_o entering uniformly into the settling zone. Figure 6-34 shows that 50 percent of these particles (those below half the depth of the tank) will be removed. That is, they will hit the bottom of the tank before being carried out because they only have to settle one-half the tank depth. Likewise, one-fourth of the particles having a settling velocity of 0.25 v_o will be removed. The percentage of particles removed, P, with a settling velocity of v_s in a sedimentation tank designed with an overflow rate of v_o is

$$P = 100\frac{v_s}{v_o} \qquad \text{FE} \tag{6-33}$$

Example 6-13. The town of San Jose has an existing horizontal-flow sedimentation tank with an overflow rate of 17 $m^3/d \cdot m^2$, and it wishes to remove particles that have settling velocities of 0.1 mm/s, 0.2 mm/s, and 1 mm/s. What percentage of removal should be expected for each particle in an ideal sedimentation tank?

Solution:

a. $v_s = 0.1$ mm/s

First we need to convert the overflow rate to compatible units:

$$17 \frac{m^3}{d \cdot m^2} = 17 \frac{m}{d} \frac{(1,000 \text{ mm/m})}{(86,400 \text{ s/d})} = 0.2 \text{ mm/s}$$

Since $v_s < v_o$ for a v_s of 0.1 mm/s, some fraction of the particles will be removed, as given by Equation 6-33.

$$P = 100 \frac{(0.1)}{(0.2)} = 50\%$$

b. $v_s = 0.2$ mm/s

These particles have $v_s = v_o$, and ideally will be 100 percent removed.

c. $v_s = 1$ mm/s

These particles have $v_s > v_o$, and 100 percent of the particles should be easily removed.

In the United States, design engineers have to work, or at least report, in USCS units. State regulations for design parameters such as overflow rate are in USCS units. Overflow rate, or any hydraulic loading rate such as a filter loading rate, is reported as gpm/ft^2 or gpd/ft^2 (volume flow per surface area). Metric overflow rates or loading rates are generally reported as m/h. Converting m/h to gpm/ft^2 can be confusing unless we remember the simple trick that a gallon is easily changed to a ft^3. The following example illustrates this conversion.

Example 6.14. A sedimentation basin has an overflow rate of 1.25 m/h. What is the loading rate in gpm/ft^2.

Solution. There are several ways to make the conversion, but the easiest is to first convert m/h to ft/h

$$1.25 \frac{m}{h} \times \frac{3.28 \text{ ft}}{m} = 4.1 \frac{\text{ft}}{h}$$

A ft is the same as $\frac{\text{ft}^3}{\text{ft}^2}$ and a cubic foot (ft^3) of water is equal to 7.48 gallons:

$$4.1 \frac{\text{ft}^3}{h \cdot \text{ft}^2} \times \frac{7.48 \text{ gal}}{\text{ft}^3} \times \frac{h}{60 \text{ min}}$$
$$= 0.51 \text{ gpm/ft}^2$$

Determination of v_s

In design of an ideal sedimentation tank, one first determines the settling velocity (v_s) of the particle to be removed and then sets the overflow rate (v_o) at some value less than or equal to v_s.

Determination of the particle-settling velocity is different for different types of particles. Settling properties of particles are often categorized into one of three classes:

Type I Sedimentation.   Type I sedimentation is characterized by particles that settle discretely at a constant settling velocity. They settle as individual particles and do not flocculate or stick to other particles during settling. Examples of these particles are sand and grit material. Generally speaking, the only application of Type I settling is during presedimentation for sand removal prior to coagulation in a potable water plant, in settling of sand particles during cleaning of rapid sand filters, and in grit chambers.

Type II Sedimentation.   Type II sedimentation is characterized by particles that flocculate during sedimentation. Because they flocculate, their size is constantly changing; therefore, the settling velocity is changing. Generally the settling velocity is increasing. These types of particles occur in alum or iron coagulation, in primary sedimentation, and in settling tanks in trickling filtration.

Type III or Zone Sedimentation.   In zone sedimentation the particles are at a high concentration (greater than 1,000 mg/L) such that the particles tend to settle as a mass, and a distinct clear zone and sludge zone are present. Zone settling occurs in lime-softening sedimentation, activated-sludge sedimentation, and sludge thickeners.

Determination of v_o

There are different ways to determine effective particle-settling velocities and consequently to determine overflow rates.

Discrete Settling.   In the case of Type I sedimentation, the particle-settling velocity can be calculated and the basin designed to remove a specific size particle. In 1687, Sir Isaac Newton showed that a particle falling in a quiescent fluid accelerates until the frictional resistance, or drag, on the particle is equal to the gravitational force of the particle (Figure 6-35) (Newton, 1687). The three forces are defined as follows:

$$F_G = (\rho_s)g V_p \tag{6-34}$$

$$F_B = (\rho)g V_p \tag{6-35}$$

$$F_D = C_D A_P(\rho)\frac{v^2}{2} \tag{6-36}$$

where F_G = gravitational force
$\quad F_B$ = buoyancy force
$\quad F_D$ = drag force
$\quad \rho_s$ = density of particle, kg/m^3
$\quad \rho$ = density of fluid, kg/m^3
$\quad g$ = acceleration due to gravity, m/s^2
$\quad V_p$ = volume of particle, m^3
$\quad C_D$ = drag coefficient
$\quad A_p$ = cross sectional area of particle, m^2
$\quad v$ = velocity of particle, m/s

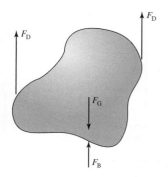

FIGURE 6-35
Forces acting on a free-falling particle in a fluid
(F_D = drag force; F_G = gravitational force; F_B = buoyancy force).

The driving force for acceleration of the particle is the difference between the gravitational and buoyant force:

$$F_G - F_B = (\rho_s - \rho)g\,\mathcal{V}_p \qquad (6\text{-}37)$$

When the drag force is equal to the driving force, the particle velocity reaches a constant value called the *terminal settling velocity* (v_s).

$$F_G - F_B = F_D \qquad (6\text{-}38)$$

$$(\rho_s - \rho)g\mathcal{V}_p = C_D A_p(\rho)\,\frac{v_s^2}{2} \qquad (6\text{-}39)$$

For spherical particles with a diameter = d,

$$\frac{\mathcal{V}_p}{A_p} = \frac{4/3\,(\pi)(d/2)^3}{(\pi)(d/2)^2} = \frac{2}{3}\,d \qquad (6\text{-}40)$$

Using Equation 6-39 and 6-40 to solve for the terminal settling velocity:

$$v_s = \left[\frac{4\,g(\rho_s - \rho)\,d}{3\,C_D\rho}\right]^{1/2} \qquad (6\text{-}41)$$

The drag coefficient takes on different values depending on the flow regime surrounding the particle. The flow regime may be characterized qualitatively as laminar, turbulent, or transitional. In laminar flow, the fluid moves in layers, or laminas, one layer gliding smoothly over adjacent layers with only molecular interchange of momentum. In turbulent flow, the fluid motion is very erratic with violent transverse interchange of momentum. Osborne Reynolds (1883) developed a quantitative means of describing the different flow regimes using a dimensionless ratio that is called the *Reynolds number*. For spheres moving through a liquid this number is defined as

$$\mathbf{R} = \frac{(d)\,v_s}{v} \qquad (6\text{-}42)$$

where $\mathbf{R}$ = Reynolds number
d = diameter of sphere, m
v_s = velocity of sphere, m/s
v = kinematic viscosity, m²/s = μ/ρ
ρ = density of fluid, kg/m³
μ = dynamic viscosity, Pa · s

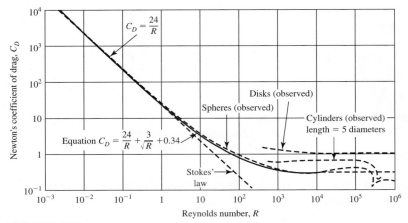

FIGURE 6-36
Newton's coefficient of drag as a function of Reynolds number. (*Source:* T. R. Camp, 1946.)

Thomas Camp (1946) developed empirical data relating the drag coefficient to Reynolds number (Figure 6-36). For eddying resistance for spheres at high Reynolds numbers ($\mathbf{R} > 10^4$), C_D has a value of about 0.4. For viscous resistance at low Reynolds numbers ($\mathbf{R} < 0.5$) for spheres:

$$C_D = \frac{24}{\mathbf{R}} \tag{6-43}$$

For the transition region of $\mathbf{R}$ between 0.5 and 10^4, the drag coefficient for spheres may be approximated by the following:

$$C_D = \frac{24}{\mathbf{R}} + \frac{3}{\mathbf{R}^{1/2}} + 0.34 \tag{6-44}$$

Sir George Gabriel Stokes showed that, for spherical particles falling under laminar (quiescent) conditions, Equation 6-41 reduces to the following (Stokes, 1845):

$$v_s = \frac{g(\rho_s - \rho)d^2}{18\mu} \tag{6-45}$$

where μ = dynamic viscosity, Pa · s
$\quad$ 18 = a constant
$\quad \rho_s$ = density of particle, kg/m^3
$\quad \rho$ = density of fluid, kg/m^3

Equation 6-45 is called *Stokes' law* (Stokes, 1845). Dynamic viscosity (also called absolute viscosity) is a function of the water temperature. A table of dynamic viscosities is given in Appendix A. Fair, Geyer, and Okun (1968) recommend that v_o be set at 0.33 to 0.7 times v_s depending upon the efficiency desired.

Flocculant Sedimentation. There is no adequate mathematical relationship that can be used to describe Type II settling. The Stokes equation cannot be used because the flocculating particles are continually changing in size and shape, and when water is entrapped in the floc, in specific gravity. Jar tests are used to develop design data.

Jar testing (Example 6-15) can also be used to determine flocculated particles settling velocity and hence set the overflow rate. Note in the pictures of the jar test apparatus that there is a sample port. In a properly designed jar this port is set at 10 cm below the water surface. After flocculation is complete the mixer can be turned off and the floc allowed to settle. Samples are taken and measured for turbidity from the sample tube at various times. The settling velocity corresponding to a given turbidity at time t is given by

$$v_s = \frac{10 \text{ cm}}{t, \text{ min}} \tag{6-46}$$

Typical overflow rates are 0.67 to 1.67 m/h or 1 to 3 cm/min. Therefore, sample collections need to be between about 2 and 10 minutes and often are done between about 2 and 20 minutes.

Example 6-15. A jar test was performed using alum as the coagulant and turbidity samples were taken between 2 and 20 minutes of settling as shown in Figure 6-37. If a settled turbidity of 1 NTU is desired, what is the effective floc settling velocity (cm/min), and the corresponding design overflow rate (m/h and gpm/ft^2).

Solution. At 1 NTU the settling time was about 3.6 min, therefore, the effective particle settling velocity was

$$v_s = \frac{10}{3.6} = 2.8 \text{ cm/min}$$

If the overflow rate were set equal to the settling velocity, the overflow rate would be

$$v_o = 2.8 \frac{\text{cm}}{\text{min}} \times \frac{60 \text{ min}}{\text{h}} \times \frac{\text{m}}{100 \text{ cm}}$$
$$v_o = 1.68 \text{ m/h or } 40.3 \text{ m/d}$$

and,

$$v_o = \frac{1.68 \text{ m}}{\text{h}} \times \frac{3.28 \text{ ft}}{m} \times \frac{\text{h}}{60 \text{ min}} \times \frac{7.48 \text{ gal}}{\text{ft}^3}$$
$$= 0.69 \text{ gpm/ft}^2$$

Zone Sedimentation Lab Data. For zone sedimentation, values can also be obtained from the lab, but it is a fairly involved procedure and data obtained through pilot studies are more useful. The design overflow is again set at about 0.5 to 0.7 times the lab value.

Experience. Typical design numbers exist for all types of sedimentation basins. These numbers can be used in lieu of laboratory or pilot work. The applicability of the typical numbers to particles in different situations is unknown. For this reason, the

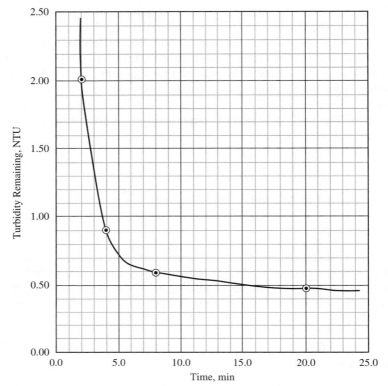

FIGURE 6-37
Jar test settling curve.

typical numbers are often quite conservative. These conservative numbers also correct for ineffective inlet or outlet zone design. A design engineer with sufficient time and funds can generally save his or her client capital costs by designing a good inlet system and conducting lab tests for proper sedimentation zone design. However, clients are not always willing to expend such funds and the engineer has little choice other than to select conservative design numbers. Table 6-12 shows some overflow rates for potable water treatment.

Typical detention times for waters coagulated with alum or iron salts are on the order of 2 to 8 hours. In lime-soda softening plants, the detention times range from 4 to 8 hours (Reynolds and Richards, 1996).

Example 6-16. Determine the surface area of a settling tank for the city of Urbana's 0.5 m^3/s design flow using the design overflow rate found in Example 6-15. Compare this surface area with that which results from assuming a typical overflow rate of 20 m^3/d $\cdot$ m^2. Find the depth of the clarifier for the overflow rate found in Example 6-15 at a detention time of 95 min.

TABLE 6-12
Typical sedimentation tank overflow rates

Application	Long rectangular and circular $m^3/d \cdot m^2$	Upflow solids-contact $m^3/d \cdot m^2$
Alum or iron coagulation		
Turbidity removal	40	50
Color removal	30	35
High algae	20	
Lime softening		
Low magnesium	70	130
High magnesium	57	105

Source: AWWA, 1990.

Solution

a. Find the surface area.

First change the flow rate to compatible units:

$$\left(\frac{0.5 \text{ m}^3}{\text{s}}\right)\left(\frac{86,400 \text{ s}}{\text{d}}\right) = 43,200 \text{ m}^3/\text{d}$$

Using the overflow rate from Example 6-15, the surface area is

$$A_s = \frac{43,200 \text{ m}^3/\text{d}}{40.3 \text{ m}^3/\text{d} \cdot \text{m}^2} = 1,071.96 \text{ or } 1,072 \text{ m}^2$$

Using the conservative value

$$A_s = \frac{43,200 \text{ m}^3/\text{d}}{20 \text{ m}^3/\text{d} \cdot \text{m}^2} = 2,160 \text{ m}^2$$

Obviously, the use of conservative data would, in this case, result in a 100 percent overdesign of the tank area.

Common length-to-width ratios for settling are between 2:1 and 5:1, and lengths seldom exceed 100 m. A minimum of two tanks is always provided.

Assuming two tanks, each with a width of 12 m, a total surface area of 1,072 m² would imply a tank length of

$$\text{Length} = \frac{1,072 \text{ m}^2}{(2 \text{ tanks})(12 \text{ m wide})} = 44.7 \text{ or } 45 \text{ m}$$

This meets our length-to-width ratio criteria.

b. Find the tank depth.

First find the total tank volume from Equation 2-27 using a detention time of 95 minutes:

$$V = (0.5 \text{ m}^3/\text{s})(95 \text{ min})(60 \text{ s/min}) = 2,850 \text{ m}^3$$

This would be divided into two tanks as noted above. The depth is found as the total tank volume divided by the total surface area:

$$\text{Depth} = \frac{2,850 \text{ m}^3}{1,072 \text{ m}^2} = 2.66$$

This depth would not include the sludge storage zone.

Comment: The final design would then be two tanks, each having the following dimensions: 12 m wide $\times$ 45 m long $\times$ 2.7 m deep plus sludge storage depth.

6-6 FILTRATION

The water leaving the sedimentation tank still contains floc particles. The settled water turbidity is generally in the range from 1 to 10 TU with a typical value being 2 TU. In order to reduce this turbidity to less than 0.3 TU, a filtration process is normally used. Water filtration is a process for separating suspended or colloidal impurities from water by passage through a porous medium, usually a bed of sand or other medium. Water fills the pores (open spaces) between the sand particles, and the impurities are left behind, either clogged in the open spaces or attached to the sand itself.

There are several methods of classifying filters. One way is to classify them according to the type of medium used, such as sand, coal (called anthracite), dual media (coal plus sand), or mixed media (coal, sand, and garnet). Another common way to classify the filters is by allowable loading rate. *Loading rate* is the flow rate of water applied per unit area of the filter. It is the velocity of the water approaching the face of the filter:

$$v_a = \frac{Q}{A_s}$$

🏴 (6-47)

where v_a = face velocity, m/d
   = loading rate, m^3/d $\cdot$ m^2
Q = flow rate onto filter surface, m^3/d
A_s = surface area of filter, m^2

Based on loading rate, the filters are described as being slow sand filters, rapid sand filters, or high-rate sand filters.

Slow sand filters were first introduced in the 1800s. The water is applied to the sand at a loading rate of 2.9 to 7.6 m^3/d $\cdot$ m^2 (0.05 to 0.13 gpm/ft^2). As the suspended or colloidal material is applied to the sand, the particles begin to collect in the top 75 mm and to clog the pore spaces. As the pores become clogged, water will no longer pass through the sand. At this point the top layer of sand is scraped off, cleaned, and replaced. Slow sand filters require large areas of land and are operator intensive.

In the early 1900s there was a need to install filtration systems in large numbers in order to prevent epidemics. Rapid sand filters were developed to meet this need. These filters have graded (layered) sand within the bed. The sand grain size distribution is selected to optimize the passage of water while minimizing the passage of particulate matter.

Rapid sand filters are cleaned in place by forcing water backwards through the sand. This operation is called *backwashing*. The washwater flow rate is such that the sand is expanded and the filtered particles are removed from the bed. After backwashing, the sand settles back into place. The largest particles settle first, resulting in a fine sand layer on top and a coarse sand layer on the bottom. Rapid sand filters are the most common type of filter used in water treatment today.

Traditionally, rapid sand filters have been designed to operate at a loading rate of 120 m³/d · m² (2 gpm/ft²). Experiments conducted at the Chicago water treatment plant have demonstrated that satisfactory water quality can be obtained with rates as high as 235 m³/d · m² (4 gpm/ft²) (AWWA, 1971). Filters now operate successfully at even higher loading rates through the use of proper media selection and improved pretreatment. Normally, a minimum of two filters are constructed to ensure redundancy. For larger plants (>0.5 m³/s), a minimum of four filters is suggested (Montgomery, 1985). The surface area of the filter tank (often called a filter box) is generally restricted in size to about 100 m², except for very large plants.

In the wartime era of the early 1940s, dual-media filters were developed. They are designed to utilize more of the filter depth for particle removal. In a rapid sand filter, the finest sand is on the top; hence, the smallest pore spaces are also on the top. Therefore, most of the particles will clog in the top layer of the filter. In order to use more of the filter depth for particle removal, it is necessary to have the large particles on top of the small particles. This was accomplished by placing a layer of coarse coal on top of a layer of fine sand. Coal has a lower specific gravity than sand, so, after backwash, it settles slower than the sand and ends up on top. Dual-media filters are operated up to loading rates of 350 m³/d · m² (6 gpm/ft²) or higher.

In the mid 1980s, deep-bed monomedia filters came into use. The filters are designed to achieve higher loading rates while at the same time producing lower finished water turbidities. The filters typically consist of 1.0 mm to 1.5 mm diameter anthracite about 1.5 m to 2.5 m deep. They operate at loading rates up to 700 m³/d · m² (12 gpm/f²).

Example 6-17. As part of their proposed new treatment plant, Urbana is going to install rapid sand filters after their sedimentation tanks. The design loading rate to the filter is 200 m³/d · m². How much filter surface area should be provided for their design flow rate of 0.5 m³/s? How large should the filter be if four filters are used?

Solution. The surface area required is the flow rate divided by the loading rate:

$$A_s = \frac{Q}{V_a} = \frac{(0.5 \text{ m}^3/\text{s})(86,400 \text{ s/d})}{200 \text{ m}^3/\text{d} \cdot \text{m}^2} = 216 \text{ m}^2$$

If four filters are used, the surface area per filter is

$$\text{Surface area per filter} = \frac{216 \text{ m}^2}{4} = 54 \text{ m}^2$$

However, many states require that the design loading rate be met with one filter out of service. In this case we would need each filter to be

$$\text{Surface area per filter} = \frac{216 \text{ m}^2}{3} = 72 \text{ m}^2$$

if we were to have four total.

Because we generally do not design an odd number of filters, the alternative would be to use six filters with one out of service

$$\text{Surface area per filter} = \frac{216 \text{ m}^2}{5} = 43.2 \text{ m}^2$$

Because filters can be 100 m² for even small plants, we would use the four-filter design.

Filter boxes are often built square, but cost minimization data suggests that filters should have a length to width ratio of 3 to 6 (Cleasby, 1972). If we use a ratio of 3

$$3 \text{ w} \times \text{w} = 54 \text{ m}^2$$

$$\text{w} = 4.24 \text{ m}$$

$$\text{and } 1 = 12.73 \text{ m}$$

Comment: If we had some loading rate leeway we would make each filter 13 m × 4 m (52 m²), or if not we could use around 13 m × 4.2 m or we would go up in size to 14 m × 4 m.

Figure 6-38 shows a cutaway drawing of a rapid sand filter. The bottom of the filter consists of a support media and collection system. The support media is designed to keep the sand in the filter and prevent it from leaving with the filtered water. Layers of graded gravel (large on bottom, small on top) traditionally have been used for the support. The underdrain blocks collect the filtered water.

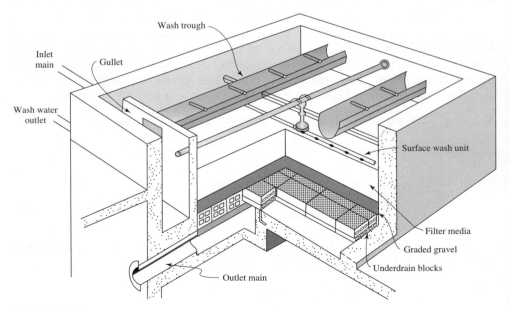

FIGURE 6-38
Typical gravity filter box. (*Source: F. B. Leopold Co.*)

On top of the support media is a layer of graded sand. The sand depth varies between 0.5 and 0.75 m. If a dual media filter is used, the sand is about 0.3 m thick and the coal about 0.45 m thick. Approximately 0.7 m to 1 m above the top of the sand are the washwater troughs. The washwater troughs collect the backwash water used to clean the filter. The troughs are placed high enough above the sand layer so that sand will not be carried out with the backwash water. Generally a total depth of 1.8 m to 3 m is allowed above the sand layer for water to build up above the filter. This depth of water provides sufficient pressure to force the water through the sand during filtration.

Figure 6-39 shows a slightly simplified version of a rapid sand filter. Water from the settling basins enters the filter and seeps through the sand and gravel bed, through a false floor, and out into a clear well that acts as a storage tank for finished water. During filtration, valves A and C are open.

During filtration the filter bed will become more and more clogged. As the filter clogs, the water level will rise above the sand as it becomes harder to force water through the bed. Eventually, the water level will rise to the point that the filter bed must be cleaned. This point is called *terminal head loss*. Very often valve "C" is a "rate of flow controller". This valve will open as head loss builds up. In this case the water level in the filter will remain fairly constant. When terminal head loss occurs, the operator turns off valves A and C. This stops the supply of water from the sedimentation tank and prevents any more water from entering the clear well. The operator then opens

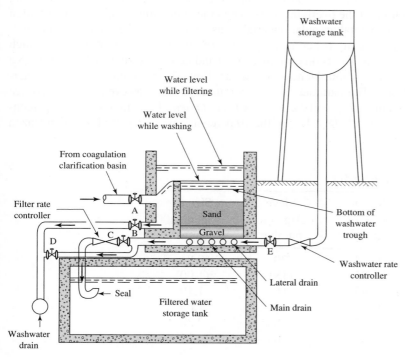

FIGURE 6-39
Operation of a rapid sand filter. (*Source:* Steel and McGhee, 1979).

valves E and B. This allows a large flow of washwater (clean water stored in an elevated tank or pumped from a clear well) to enter below the filter bed. This rush of water forces the sand bed to expand and sets individual sand particles in motion. By rubbing against each other, the light colloidal particles that were trapped in the pore spaces are released and escape with the washwater. The washwater is a waste stream that must be treated. After a few minutes the washwater is shut off and filtration resumed.

Grain Size Characteristics

The size distribution or variation of a sample of granular material is determined by sieving the sample through a series of standard sieves (screens). One such standard series is the U.S. Standard Sieve Series. The U.S. Standard Sieve Series (Table 6-13) is based on a sieve opening of 1 mm. Sieves in the "fine series" stand successively in the ratio of $(2)^{1/4}$ to one another, the largest opening in this series being 5.66 mm and the smallest 0.037 mm. All material that passes through the smallest sieve opening in the series is caught in a pan that acts as the terminus of the series (Fair and Geyer, 1954).

The grain size analysis begins by placing the sieve screens in ascending order with the largest opening on top and the smallest opening on the bottom. A sand sample is placed on the top sieve and the stack is shaken for a prescribed amount of time. At the end of the shaking period, the mass of material retained on each sieve is determined. The cumulative mass is recorded and converted into percentages by mass equal to or less than the size of separation of the overlying sieve. Then the cumulative frequency distribution is plotted. For many natural granular materials, this curve approaches geometric normality. Logarithmic-probability paper, therefore, assures an almost straight-line plot which facilitates interpolation. The geometric mean (X_g) and geometric standard deviation (S_g) are useful parameters of central tendency and variation. Their magnitudes may be determined from the plot. The parameters most commonly used, however, are the *effective size, E*, or 10 percentile, P_{10}, and the *uniformity coefficient, U*, or ratio of the 60 percentile to the 10 percentile, P_{60}/P_{10}. Use of the 10 percentile was suggested by Allen Hazen

TABLE 6-13
U.S. Standard Sieve Series

Sieve designation number	Size of opening (mm)	Sieve designation number	Size of opening (mm)
200	0.074	20	0.84
140	0.105	(18)	(1.00)
100	0.149	16	1.19
70	0.210	12	1.68
50	0.297	8	2.38
40	0.42	6	3.36
30	0.59	4	4.76

Source: Fair and Geyer, 1954.

because he had observed that resistance to the passage of water offered by a bed of sand within which the grains are distributed homogeneously remains almost the same, irrespective of size variation (up to a uniformity coefficient of about 5.0), provided that the 10 percentile remains unchanged (Hazen, 1892). Use of the ratio of the 60 percentile to the 10 percentile as a measure of uniformity was suggested by Hazen because this ratio covered the range in size of half the sand.* On the basis of logarithmic normality, the probability integral establishes the following relations between the effective size and uniformity coefficient and the geometric mean size and geometric standard deviation:

$$E = P_{10} = (X_g)(S_g)^{-1.282} \qquad \text{FE (6-48)}$$

$$U = P_{60}/P_{10} = (S_g)^{1.535} \qquad \text{FE (6-49)}$$

Experience has suggested that, for silica sand, the effective size should be in the range of 0.35 to 0.55 mm with a maximum of about 1.0 mm. The uniformity coefficient should range between 1.3 and 1.7. Smaller effective sizes result in a product water that is lower in turbidity, but they also result in higher pressure losses in the filter and shorter operating cycles between cleanings. Diameters corresponding to P_{10} and P_{60} are referred to as the d_{10} and d_{60} sizes. Another useful diameter size is that corresponding to the 90 percentile point, d_{90}.

Example 6-18. For the size frequencies by weight and by count of the sample of sand listed below, find the effective size, E, and uniformity coefficient, U.

U. S. Standard Sieve No.	Analysis of stock sand (Cumulative mass % passing)
140	0.2
100	0.9
70	4.0
50	9.9
40	21.8
30	39.4
20	59.8
16	74.4
12	91.5
8	96.8
6	99.0

Solution. First we must plot the data on log-probability paper as shown in Figure 6-40. From this plot we then find the effective size:

$$E = P_{10} = 0.30 \text{ mm}$$

---

*It would be more logical to speak of this ratio as a coefficient of nonuniformity because the coefficient increases in magnitude as the nonuniformity increases.

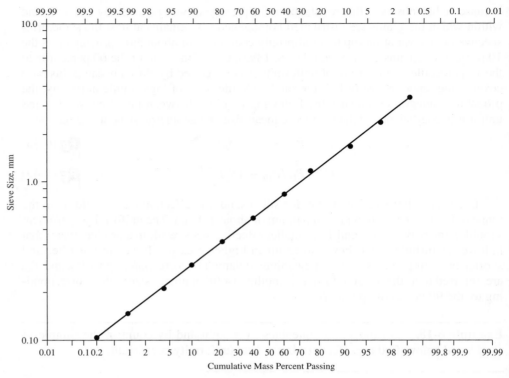

FIGURE 6-40
Grain size analysis of run-of-bank sand.

The uniformity coefficient is then

$$U = \frac{P_{60}}{P_{10}} = \frac{0.85}{0.30} = 2.8$$

Sand excavated from a natural deposit is called *run-of-bank* sand. Run-of-bank sand may be too coarse, too fine, or too nonuniform for use in filters. Within economical limits, proper sizing and uniformity are obtained by screening out coarse components and washing out fine components. In rapid sand filters, the removal of "fines" may be accomplished by stratifying the bed through backwashing and then scraping off the layer that includes the unwanted sand.

Filter Hydraulics

The head loss (i.e., pressure drop) that occurs when clean water flows through a bed of clean filter media is an important design value to determine the overall head required to operate a filter. The flow through a clean filter of ordinary grain size (i.e., 0.5–1.0 mm) at ordinary filtration velocities (5 to 12 m/h or 2 to 5 gpm/ft^2) would be in the laminar range of flow depicted by the Blake-Kozeny equation* (Blake, 1922; Kozeny, 1927) that is dimensionally homogeneous (i.e., any consistent units may be used that are dimensionally homogeneous):

$$\frac{h}{D} = \frac{(k_z)(\mu)(1 - \epsilon)^2}{(\rho)(g)(\epsilon)^3}\left(\frac{a}{V}\right)^2 v \qquad \text{(6-50)}$$

where h = head loss in depth of bed D, m
g = acceleration of gravity, m/s^2
a/V = grain surface area per unit of grain volume (i.e., specific surface $S_v = 6/d$ for spheres and $6/\Psi d_{eq}$ for irregular grains), d_{eq} is the grain diameter of a sphere of equal volume, m^2/m^3 or m^{-1}
ψ = sphericity
μ = dynamic viscosity of the fluid, Pa · s
ρ = mass density of the fluid, kg/m^3
k_z = dimensionless Kozeny constant, commonly found to be close to 5 under most filtration conditions (Fair et al., 1968)
v = applied water velocity, or loading rate, m^3/s · m^2 or m/s

Properties of grain density (ρ_s) bed porosity (ϵ), and sphericity (ψ) sand and anthracite are found in Table 6-14.

As the filter clogs, the head loss will increase so the calculated results are the minimum expected head losses. Initial head losses in excess of 0.6 m to 1.0 m imply either that the loading rate is too high or that the sand has too large a proportion of fine grain sizes. The design of the filter must account for the additional losses that occur as the filter runs. Thus, the filter box must be at least as deep as the highest design head loss. As mentioned previously, this is about 3 m maximum.

Example 6-19. Calculate the head loss for a 0.3-m deep bed for the filter sand shown in Example 6-18 at a filter loading rate of 9.8 m/h and a temperature of 20°C using the Blake-Kozeny equation. From Table 6-14 use a porosity of 0.42 and a sphericity of 0.75.

TABLE 6-14
Typical properties of common filter media for granular-bed filters

	Silica sand	Anthracite coal
Grain density (ρ_s), kg/m^3	2650	1450–1730
Loose-bed porosity (ϵ)	0.42–0.47	0.50–0.60
Sphericity (ψ)	0.7–0.8	0.46–0.60

---

*Fair et al., (1968) suggest that "credit for this formulation should go to Blake as well as Kozeny."

Solution. The Blake-Kozeny equation will be used to find the total bed head loss. Assuming a specific surface for spheres and correcting for irregular grains

$$\frac{a}{V} = \frac{6}{\psi d_{eq}} = \frac{6}{0.75\,d_{eq}}$$

$$\frac{\mu}{\rho} = 1.004 \times 10^{-6}\,\text{m}^2/\text{s at } 20°C$$

Note: μ/ρ = kinematic viscosity, Appendix A. Assuming a Kozeny constant of 5 and converting the loading rate to compatible units

$$v = 9.8\,\text{m/h} = 2.72 \times 10^{-3}\,\text{m/s}$$

The head loss as a function of diameter is

$$\frac{h}{D} = \frac{5(1.004 \times 10^{-6})}{9.81}\frac{(1 - 0.42)^2}{0.42^3}\left(\frac{6}{0.75\,d_{eq}}\right)^2 2.72 \times 10^{-3}$$

$$\frac{h}{D} = \frac{4.05 \times 10^{-7}}{d_{eq}^2}$$

From Figure 6-40 we can divide the sand into five equal depths by selecting the diameters at d_{10}, d_{30}, d_{50}, d_{70}, and d_{90} and calculate the head loss for each layer as shown in the following table.

Size	Diameter (mm)	h/D	Layer depth (m)	h (m)
d_{10}	0.3	4.5	0.06	0.27
d_{30}	0.5	1.62	0.06	0.10
d_{50}	0.7	0.83	0.06	0.05
d_{70}	1.0	0.41	0.06	0.02
d_{90}	1.7	0.14	0.06	0.01
Total head loss				0.45 m

The calculation of minimum fluidization velocity is important in determining minimum backwash flow-rate requirements. Minimum fluidization velocity is the velocity necessary to overcome the weight of the media and raise it into the water. By achieving fluidization velocity the media is cleaned of accumulated turbidity and allowed to re-stratify when the backwash is turned off. If the backwash velocity is not high enough to fluidize the large particles then they will tend to sink to the bottom of the filter. The calculation of the minimum fluidization velocity is based on the fixed bed head loss being equal to the constant head loss of the fluidized bed at the point of incipient fluidization. An equation called the Ergun equation can be equated to the constant-head loss equation and solved for the minimum fluidization velocity, that is, v_{mf}. The accuracy of the result depends on using realistic values for sphericity ψ and fixed bed porosity ϵ. By using an approximate relation between sphericity and porosity, Wen and Yu (1966) developed a simplified equation for the calculation of v_{mf}. The equation shown is based on the v_{mf} required to fluidize the d_{90} size particle which is representative of the largest particle designed to be fluidized. The resulting equation is

$$v_{mf} = \frac{\mu}{\rho d_{90}} (33.7^2 + 0.0408 Ga)^{0.5} - \frac{33.7\,\mu}{\rho d_{90}} \tag{6-51}$$

where Ga = Galileo number

$$Ga = d_{90}^3 \frac{\rho(\rho_s - \rho)g}{\mu^2} \tag{6-52}$$

where ρ, μ, g = as defined above

ρ_s = density of the grain (sand or anthracite), kg/m^3

For a bed containing a gradation in particle sizes, the minimum fluidization velocity is not the same for all particles. Smaller grains become fluidized at a lower superficial velocity than larger grains. Therefore, a gradual change from the fixed bed to the totally fluidized state occurs. In applying the equation to a filter bed with grains graded in size, calculating v_{mf} for the coarser grains in the bed is necessary to ensure that the entire bed is fluidized. The d_{90} sieve size would be practical diameter choice in this calculation. The minimum backwash rate selected is set higher than v_{mf} for the d_{90} sieve size to allow free movement of these coarse grains during backwashing. A backwash rate equal to 1.3 v_{mf} of the d_{90} particle has been suggested to ensure adequate free movement of the grains (Cleasby and Fan, 1981). The fluidization velocity is strongly affected by the water temperature and hence it is necessary to seasonally adjust backwash rates as the water temperature changes.

Example 6-20.   Find the required backwash rate to fluidize the sand in Example 6-18 if the water temperature is 20°C.

Solution.   Calculate the v_{mf} for the d_{90} particle size. The d_{90} particle size was found in the last example to be 1.7 mm (1.7×10^{-3} m).

The following physical properties of water are from Appendix A: $\mu = 1.002 \times 10^{-3}$ Pa $\cdot$ s; $\rho = 998.21$ kg/m^3. The acceleration due to gravity is $g = 9.81$ m^2/s

$$Ga = (1.7 \times 10^{-3})^3 \frac{998(2,650 - 998)9.81}{(1.02 \times 10^{-3})^2}$$

$Ga = 79,145$ (dimensionless)

$$v_{mf} = \frac{1.002 \times 10^{-3}}{998(1.7 \times 10^{-3})} [33.7^2 + 0.0408(79,145)]^{0.5} - \frac{33.7(1.002 \times 10^{-3})}{998(1.7 \times 10^{-3})}$$

$v_{mf} = 0.019$ m/s

The recommended backwash rate would be 1.3 v_{mf}

$$\text{Backwash rate} = 1.3(0.019) = 0.025 \text{ m/s}$$
$$= 90 \text{ m/h} = 36.9 \text{ gpm/ft}^2$$

Comment: Note that this is a high backwash rate and is a result of this sand having a high uniformity coefficient resulting in a high d_{90}. In order to bring the backwash rate down, a lower uniformity coefficient sand would be selected. For example, at a uniformity coefficient of 2.0 the backwash rate would be reduced to about 37 m/h.

The *head loss* through a clean stratified-sand filter with uniform porosity may also be calculated using the Rose equation (Rose, 1945)*:

$$h_L = \frac{1.067\,(v_a)^2(D)}{(\phi)(g)(\epsilon)^4} \sum_{i=1}^{n} \frac{(C_D)(f)}{d}$$ FE (6-53)

where h_L = frictional head loss through the filter, m
   v_a = approach velocity, m/s
   D = depth of filter sand, m
   C_D = drag coefficient
   f = mass fraction of sand particles of diameter d
   d = diameter of sand grains, m
   ϕ = shape factor
   g = acceleration due to gravity, m/s^2
   ϵ = porosity

The drag coefficient is defined in Equations 6-43 and 6-44. The Reynolds number used to calculate the drag coefficient is multiplied by the shape factor to account for nonspherical sand grains. The summation term may be calculated using the size distribution of the sand particles found from a sieve analysis. Like the Kozeny equation, the Rose equation is limited to clean filter beds. Thus, the filter box must be at least as deep as the highest design head loss. As mentioned previously, this is about 3 m maximum.

The hydraulic head loss that occurs during backwashing is calculated to determine the placement of the backwash troughs above the filter bed. The trough bottom should be at least 0.15 m above the expanded bed to prevent loss of filter material. Fair and Geyer (1954) developed the following relationship to predict the depth of the expanded bed:

$$D_e = (1 - \epsilon)(D) \sum_{i=1}^{n} \frac{f}{(1 - \epsilon_e)}$$ FE (6-54)

where D_e = depth of the expanded bed, m
   D = depth of bed before expansion, m
   ϵ = porosity of the bed
   ϵ_e = porosity of expanded bed
   f = mass fraction of sand with expanded porosity

The porosity of the expanded bed may be calculated for a given particle by

$$\epsilon_e = \left(\frac{v_b}{v_s}\right)^{0.22}$$ (6-55)

where v_b = velocity of backwash, m/s
   v_s = settling velocity, m/s

---

*Other headloss equations include those by Carmen-Kozeny, Fair-Hatch, and Hazen. [These equations are summarized by Cleasby and Logsdon (WQ&T, 1999) and by Metcalf & Eddy, Inc. (2003).]

Strictly speaking, this form of the expanded bed porosity is applicable only for laminar conditions. Since the conditions during backwash are turbulent, a more representative model equation is that given by Richardson and Zaki (1954):

$$\epsilon_e = \left(\frac{v_b}{v_s}\right)^{0.2247\mathbf{R}^{0.1}}$$

(6-56)

where the Reynolds number is defined as

$$\mathbf{R} = \frac{v_s\, d_{60\%}}{v}$$

where $d_{60\%}$ = 60 percentile diameter, m. A more sophisticated model developed by Dharmarajah and Cleasby (1986) is also available.

The determination of D_e is not straightforward. From Equation 6-55, it is obvious that the expanded bed porosity is a function of the settling velocity. The particle settling velocity is determined by Equation 6-41. To solve Equation 6-41 you must calculate the drag coefficient (C_D). The drag coefficient is a function of the Reynolds number which, in turn, is a function of the settling velocity. Thus, you need the settling velocity to find the settling velocity! To resolve this dilemma, you must begin with an estimated settling velocity.

6-7   DISINFECTION

Disinfection is used in water treatment to reduce pathogens (disease-producing microorganisms) to an acceptable level. Disinfection is not the same as sterilization. Sterilization implies the destruction of all living organisms. Drinking water need not be sterile.

Three categories of human enteric pathogens are normally of consequence: bacteria, viruses, and amebic cysts. Purposeful disinfection must be capable of destroying all three.

To be of practical service, such water disinfectants must possess the following properties:

1. They must destroy the kinds and numbers of pathogens that may be introduced into water within a practicable period of time over an expected range in water temperature.

2. They must meet possible fluctuations in composition, concentration, and condition of the waters or wastewaters to be treated.

3. They must be neither toxic to humans and domestic animals nor unpalatable or otherwise objectionable in required concentrations.

4. They must be dispensable at reasonable cost and safe and easy to store, transport, handle, and apply.

5. Their strength or concentration in the treated water must be determined easily, quickly, and (preferably) automatically.

6. They must persist within disinfected water in a sufficient concentration to provide reasonable residual protection against its possible recontamination before use, or—because this is not a normally attainable property—the disappearance of residuals must be a warning that recontamination may have taken place.

Disinfection Kinetics

Under ideal conditions, when an exposed microorganism contains a single site vulnerable to a single unit of disinfectant, the rate of die-off follows *Chick's law,* which states that the number of organisms destroyed in a unit time is proportional to the number of organisms remaining (Chick, 1908):

$$-\frac{dN}{dt} = kN \tag{6-57}$$

This is a first-order reaction. Under real conditions the rate of kill may depart significantly from Chick's law. Increased rates of kill may occur because of a time lag in the disinfectant reaching vital centers in the cell. Decreased rates of kill may occur because of declining concentrations of disinfectant in solution or poor distribution of organisms and disinfectant.

Chlorine Reactions in Water

Chlorine is the most common disinfecting chemical used. The term *chlorination* is often used synonymously with disinfection. Chlorine may be used as the element (Cl_2), as sodium hypochlorite (NaOCl), or as calcium hypochlorite [$Ca(OCl)_2$].

When chlorine is added to water, a mixture of hypochlorous acid (HOCl) and hydrochloric acid (HCl) is formed:

$$Cl_2(g) + H_2O \rightleftharpoons HOCl + H^+ + Cl^- \tag{6-58}$$

This reaction is pH dependent and essentially complete within a very few milliseconds. In dilute solution and at pH levels above 1.0, the equilibrium is displaced to the right and very little Cl_2 exists in solution. Hypochlorous acid is a weak acid and dissociates poorly at levels of pH below about 6. Between pH 6.0 and 8.5 there occurs a very sharp change from undissociated HOCl to almost complete dissociation:

$$HOCl \rightleftharpoons H^+ + OCl^- \tag{6-59}$$

$$pK = 7.537 \text{ at } 25°C$$

Thus, chlorine exists predominantly as HOCl at pH levels between 4.0 and 6.0. Below pH 1.0, depending on the chloride concentration, the HOCl reverts back to Cl_2 as shown in Equation 6-58. At 20°C, above about pH 7.5, and at 0°C, above about pH 7.8, hypochlorite ions (OCl^-) predominate. Hypochlorite ions exist almost exclusively at levels of pH around 9 and above. Chlorine existing in the form of HOCl and/or OCl^- is defined as free available chlorine.

Hypochlorite salts dissociate in water to yield hypochlorite ions:

$$NaOCl \rightleftharpoons Na^+ + OCl^- \tag{6-60}$$

$$Ca(OCl)_2 \rightleftharpoons Ca^{2+} + 2OCl^- \tag{6-61}$$

The hypochlorite ions establish equilibrium with hydrogen ions (in accord with Equation 6-59), again depending on pH. Thus, the same active chlorine species and equilibrium are established in water regardless of whether elemental chlorine or hypochlorites are used. The significant difference is in the resultant pH and its influence on the relative amounts of HOCl and OCl^- existing at equilibrium. Elemental chlorine tends to decrease pH; each mg/L of chlorine added reduces the alkalinity by up to 1.4 mg/L as $CaCO_3$. Hypochlorites, on the other hand, always contain excess alkali to enhance their stability and tend to raise the pH somewhat. We seek to maintain the design pH within a range of 6.5 to 7.5 to optimize disinfecting action.

Chlorine-Disinfecting Action

Chlorine disinfection involves a very complex series of events and is influenced by the kind and extent of reactions with chlorine-reactive materials (including nitrogen), temperature, pH, the viability of test organisms, and numerous other factors. Such factors greatly complicate attempts to determine the precise mode of action of chlorine on bacteria and other microorganisms. Over the years, several theories have been advanced. One early theory held that chlorine reacts directly with water to produce nascent oxygen; another held that the action of chlorine is due to complete oxidative destruction of organisms. These theories were nullified because small concentrations of hypochlorous acid were observed to destroy bacteria whereas other oxidants (such as hydrogen peroxide or potassium permanganate) failed to do likewise. A later theory suggested that chlorine reacts with protein and amino acids of cells to alter and ultimately destroy cell protoplasm. Currently it is considered that the bactericidal action of chlorine is physiochemical, but among yet-unanswered questions are those pertaining to phenomena such as variations in resistance of bacteria, spores, cysts, and viruses, and the appearance of mutants.

Microorganism kill by disinfectants is assumed to follow the CT concept, that is, the product of disinfectant concentration (C) and time (T) yields a constant. CT is widely used in the Surface Water Treatment Rule (SWTR) as a criteria for cyst and virus disinfection. CT is an empirical expression for defining the nature of biological inactivation where:

$$CT = 0.9847C^{0.1758}pH^{2.7519}temp^{-0.1467} \qquad (6\text{-}62)$$

where C = disinfectant concentration
  T = contact time between the microorganism and the disinfectant
  pH = $-\log [H^+]$
  temp = temperature, °C

The relationship shown in Equation 6-62 means that the combination of concentration and time (CT) required to produce a 3-log reduction in *Giardia* cysts by free chlorine can be estimated if the free chlorine concentration, pH, and water temperature are known.

Table 6-15 shows examples of the U.S. EPA–required CT times for free chlorine under the SWTR. EPA used empirical data and a safety factor to develop the table. Therefore, the numbers in the table do not exactly match Equation 6-62. Generally, for a conventional coagulation plant, 0.5-log inactivation of *Giardia* is required and for an untreated surface water, 3-log inactivation is required.

TABLE 6-15
CT values (in mg/L · min) for inactivation of *Giardia* cysts by free chlorine at 10°C

Chlorine concentration (mg/L)	pH = 6.0 Log inactivations						pH = 7.0 Log inactivations						pH = 8.0 Log inactivations						pH = 9.0 Log inactivations					
	0.5	1.0	1.5	2.0	2.5	3.0	0.5	1.0	1.5	2.0	2.5	3.0	0.5	1.0	1.5	2.0	2.5	3.0	0.5	1.0	1.5	2.0	2.5	3.0
≤0.4	12	24	37	49	61	73	17	35	52	69	87	104	25	50	75	99	124	149	35	70	105	139	174	209
0.6	13	25	38	50	63	75	18	36	54	71	89	107	26	51	77	102	128	153	36	73	109	145	182	218
0.8	13	26	39	52	65	78	18	37	55	73	92	110	26	53	79	105	132	158	38	75	113	151	188	226
1	13	26	40	53	66	79	19	37	56	75	93	112	27	54	81	108	135	162	39	78	117	156	195	234
1.2	13	27	40	53	67	80	19	38	57	76	95	114	28	55	83	111	138	166	40	80	120	160	200	240
1.4	14	27	41	55	68	82	19	39	58	77	97	116	28	57	85	113	142	170	41	82	124	165	206	247
1.6	14	28	42	55	69	83	20	40	60	79	99	119	29	58	87	116	145	174	42	84	127	169	211	253
1.8	14	29	43	57	72	86	20	41	61	81	102	122	30	60	90	119	149	179	43	86	130	173	216	259
2	15	29	44	58	73	87	21	41	62	83	103	124	30	61	91	121	152	182	44	88	133	177	221	265
2.2	15	30	45	59	74	89	21	42	64	85	106	127	31	62	93	124	155	186	45	90	136	181	226	271
2.4	15	30	45	60	75	90	22	43	65	86	108	129	32	63	95	127	158	190	46	92	138	184	230	276
2.6	15	31	46	61	77	92	22	44	66	87	109	131	32	65	97	129	162	194	47	94	141	187	234	281
2.8	16	31	47	62	78	93	22	45	67	89	112	134	33	66	99	131	164	197	48	96	144	191	239	287
3	16	32	48	63	79	95	23	46	69	91	114	137	34	67	101	134	168	201	49	97	146	195	243	292

Source: U.S. EPA, 1991.

Another pathogen that may require inactivation is *Cryptosporidium. Cryptosporidium* is not inactivated by chlorine, and either ozone or ultraviolet light (UV) is required. Those processes are discussed in later sections.

It has become common in the water industry to express the inactivation credit or to express the physical removal achieved in a plant as *log removal*. This term does not imply that the removal of physical particles is a logarithmic process, but rather that the percent removal found at a point in time can be mathematically represented by a log removal function. Log removal (LR) can be found as:

$$LR = \log\left(\frac{\text{influent concentrations}}{\text{effluent concentrations}}\right) \qquad (6\text{-}63)$$

If the log removal for a series of data is to be determined, then the averages for the influent and effluent concentrations can be used. The percent removal (or inactivation) is related to the log removal or inactivation by

$$\% \text{ removal} = 100 - \frac{100}{10^{LR}} \qquad (6\text{-}64)$$

Example 6-21. A city measured the concentration of aerobic spores in its raw and finished water as an indicator of plant performance. Spores are often plentiful in water supplies and are conservative indicators of how well a plant is able to remove *cryptosporidium*. The city data are as follows:

Day of Week	(spores/L)	
	Raw	Finished
Sunday	200,000	16
Monday	145,000	4
Tuesday	170,000	2
Wednesday	150,000	8
Thursday	170,000	10
Friday	180,000	2
Saturday	180,000	3

Calculate the log removal and convert that to percent removal.

Solution. The log removal is found by finding the average raw and finished water concentrations and then using Equation 6-63. The average raw concentration is 170,714 spores/L and the finished average concentration is 6.43. Therefore,

$$LR = \log\left(\frac{170,714}{6.43}\right) = 4.42$$

And the percent removal is

$$\% = 100 - \frac{100}{10^{4.42}} = 99.996$$

Chlorine/Ammonia Reactions

The reactions of chlorine with ammonia are of great significance in water chlorination processes. When chlorine is added to water that contains natural or added ammonia (ammonium ion exists in equilibrium with ammonia and hydrogen ions), the ammonia reacts with HOCl to form various *chloramines* which, like HOCl, retain the oxidizing power of the chlorine. The reactions between chlorine and ammonia may be represented as follows (AWWA, 1971):

$$NH_3 + HOCl \rightleftharpoons NH_2Cl + H_2O \tag{6-65}$$
$$\text{Monochloramine}$$

$$NH_2Cl + HOCl \rightleftharpoons NHCl_2 + H_2O \tag{6-66}$$
$$\text{Dichloramine}$$

$$NHCl_2 + HOCl \rightleftharpoons NCl_3 + H_2O$$
$$\text{Trichloramine or nitrogen trichloride} \tag{6-67}$$

The distribution of the reaction products is governed by the rates of formation of monochloramine and dichloramine, which are dependent upon pH, temperature, time, and initial $Cl_2:NH_3$ ratio. In general, high $Cl_2:NH_3$ ratios, low temperatures, and low pH levels favor dichloramine formation.

Chlorine also reacts with organic nitrogenous materials, such as proteins and amino acids, to form organic chloramine complexes. Chlorine that exists in water in chemical combination with ammonia or organic nitrogen compounds is defined as *combined available chlorine.*

The oxidizing capacity of free chlorine solutions varies with pH because of variations in the resultant $HOCl:OCl^-$ ratios. This applies also in the case of chloramine solutions as a result of varying $NHCl_2:NH_2Cl$ ratios, and where monochloramine predominates at high pH levels. The disinfecting ability of the chloramines is much lower than that of free available chlorine, indicating that the ammonia chloramines also are less reactive than free available chlorine.

Practices of Water Chlorination

Evolution.   Early water chlorination practices (variously termed "plain chlorination," "simple chlorination," and "marginal chlorination") were applied for the purpose of disinfection. Chlorine-ammonia treatment was soon thereafter introduced to limit the development of objectionable tastes and odors often associated with marginal chlorine disinfection. Subsequently, "super-chlorination" was developed for the additional purpose of destroying objectionable taste- and odor-producing substances often associated with chlorine-containing organic materials. The introduction of "breakpoint chlorination" and the recognition that chlorine residuals can exist in two distinct forms established contemporary water chlorination as one of two types: combined residual chlorination or free residual chlorination.

Combined Residual Chlorination.   Combined residual chlorination practice involves the application of chlorine to water in order to produce, with natural or added ammonia, a combined available chlorine residual, and to maintain that residual through part or all of a water-treatment plant or distribution system. Combined available chlorine forms have lower oxidation potentials than free available chlorine forms and, therefore, are less effective as oxidants. Moreover, they are also less effective disinfectants. In fact, about 25 times as much combined available residual chlorine as free available residual chlorine is necessary to obtain equivalent bacterial kills (*S. typhosa*) under the same conditions of pH, temperature, and contact time. And about 100 times longer contact is required to obtain equivalent bacterial kills under the same conditions and for equal amounts of combined and free available chlorine residuals.

When a combined available chlorine residual is desired, the characteristics of the water will determine how it can be accomplished:

1. If the water contains sufficient ammonia to produce with added chlorine a combined available chlorine residual of the desired magnitude, the application of chlorine alone suffices.

2. If the water contains too little or no ammonia, the addition of both chlorine and ammonia is required.

3. If the water has an existing free available chlorine residual, the addition of ammonia will convert the residual to combined available residual chlorine. A combined available chlorine residual should contain little or no free available chlorine.

The practice of combined residual chlorination is especially applicable after filtration (post treatment) for controlling certain algae and bacterial growths and for providing and maintaining a stable residual throughout the system to the point of consumer use.

Although combined chlorine residual is not as good a disinfectant as free chlorine, it has an advantage over free chlorine residual in that it is reduced more slowly and, therefore, persists for a longer time in the distribution system. Thus, it is useful as an indicator of major contamination. Water plant personnel routinely monitor the chlorine level in the distribution system. The presence of available chlorine (either combined or free) indicates that no major contamination has occurred. If major contamination does take place, the combined chlorine residual will be depleted, albeit at a slow rate. This depletion serves as a warning that contamination may have taken place.

A chloramine residual also has the advantage of minimizing the formation of disinfection by-products (DBPs), as chloramines are not a strong enough oxidant to react with the organic matter in the water and form chlorinated hydrocarbons. An increasingly common disinfection strategy is to add chlorine in order to achieve *Giardia* and virus *CT* and then add ammonia to form chloramines and stop or reduce the DBP formation.

Because of its relatively poor disinfecting power, combined residual chlorination is often preceded by free residual chlorination to ensure the production of potable water.

Free Residual Chlorination. Free residual chlorination practice involves the application of chlorine to water to produce, either directly or through the destruction of ammonia, a free available chlorine residual and to maintain that residual through part or all of a water treatment plant or distribution system. Free available chlorine forms have higher oxidation potentials than combined available chlorine forms and therefore are more effective as oxidants. Moreover, as already noted, they are also the most effective disinfectants.

When free available chlorine residual is desired, the characteristics of the water will determine how it can be accomplished:

1. If the water contains no ammonia (or other nitrogenous materials), the application of chlorine will yield a free residual.

2. If the water does contain ammonia that results in the formation of a combined available chlorine residual, it must be destroyed by applying an excess of chlorine.

Figure 6-41 shows what is referred to as the breakpoint curve and depicts the reactions between chlorine and ammonia. The reactions associated with breakpoint chlorination are not well understood even to this day; but are generally described by equations 6-65 to 6-66. In the first zone chlorine reacts with the ammonia to form primarily monochloramine (NH_2Cl) in accordance with Equation 6-65. Some amount of dichloramine ($NHCl_2$) will also form depending on the pH and temperature. The combined chlorine residual reaches a maximum at 1:1 Cl_2:NH_3 molar ratio, which is a 5:1 Cl_2:NH_3-N weight ratio. Further chlorine addition moves the reaction into zone 2 where dichloramine forms according to Equation 6-66. However, also in this zone reduction of chlorine and oxidation of ammonia takes place such that the chlorine and ammonia concentrations are both lowered (nitrogen leaves solution as a gas and chlorine is reduced to chloride). When operating with a combined chlorine residual, it is desirable to operate in zone 1 (Cl_2:NH_3-N weight ratio <5:1) to maximize the residual. Also,

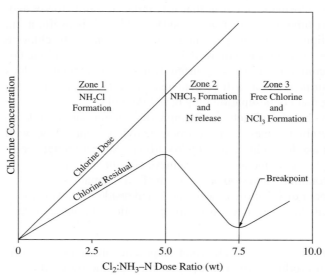

FIGURE 6-41
Breakpoint chlorination.

dichlorimane has objectionable taste and odors. In zone 3 primarily free chlorine results with possibly some minor amounts of NCl_3 according to Equation 6-67.

Chlorine Dioxide

Another very strong oxidant is chlorine dioxide. Chlorine dioxide (ClO_2) is formed on-site by combining chlorine and sodium chlorite. Chlorine dioxide is often used as a primary disinfectant, inactivating the bacteria and cysts, followed by the use of chloramine as a distribution system disinfectant. Chlorine dioxide does not maintain a residual long enough to be useful as a distribution-system disinfectant. The advantage of chlorine dioxide over chlorine is that chlorine dioxide does not react with precursors to form DBPs.

When chlorine dioxide reacts in water, it forms two by-products, chlorite and, to some extent, chlorate. These compounds may be associated with a human health risk, and therefore many state regulatory agencies limit the dose of chlorine dioxide to 1.0 mg/L. In many cases this may not be a sufficient dose to provide good disinfection. Chlorine dioxide use has also been associated with tastes and odors in some communities. The health concerns, taste and odors, and relatively high cost have tended to limit the use of chlorine dioxide. However, many utilities have successfully used chlorine dioxide as a primary disinfectant.

Ozonation

Ozone is a pungent-smelling, unstable gas. It is a form of oxygen in which three atoms of oxygen are combined to form the molecule O_3. Because of its instability, it is generated at the point of use. The ozone-generating apparatus commonly is a discharge electrode. To reduce corrosion of the generating apparatus, air is passed through a drying process and then into the ozone generator. The generator consists of two plates or a wire and tube with an electric potential of 15,000 to 20,000 volts. The oxygen in the air is dissociated by the impact of electrons from the discharge electrode. The atomic oxygen then combines with atmospheric oxygen to form ozone in the following reaction:

$$O + O_2 \rightarrow O_3 \qquad (6\text{-}68)$$

Approximately 0.5 to 1.0 percent by volume of the air exiting from the apparatus will be ozone. The resulting ozone-air mixture is then diffused into the water that is to be disinfected.

Ozone is widely used in drinking water treatment in Europe and is continuing to gain popularity in the United States. It is a powerful oxidant, more powerful even than hypochlorous acid. It has been reported to be more effective than chlorine in destroying viruses and cysts. Table 6-16 shows the *CT* values for ozone, chlorine dioxide, and chloramine for 3-log inactivation (99.90 percent removal) of *Giardia* cysts and for 2-log inactivation of *Cryptosporidium* oocysts. Table 6-16 can be compared to the values for chlorine in Table 6-15 to note how strong an oxidant ozone is. (Also the weak disinfection ability of chloramines is obvious.)

In addition to being a strong oxidant, ozone has the advantage of not forming THMs or any of the chlorinated DBPs. As with chlorine dioxide, ozone will not persist in the water, decaying back to oxygen in minutes. Therefore, a typical flow schematic

TABLE 6-16
CT values for Giardia and Cryptosporidium inactivation

Temperature, (°C)	Chlorine dioxide	Ozone	Chloramine
Giardia* at 3-log inactivation			
0.5	63	2.9	3,800
5	26	1.9	2,288
10	23	1.43	1,850
15	19	0.95	1,500
20	15	0.72	1,100
25	11	0.48	750
Cryptosporidium[†] at 2-log inactivation			
0.5	1,275	48	N/A
5	858	32	N/A
10	553	20	N/A
15	357	12	N/A
20	232	7.8	N/A
25	150	4.9	N/A

Sources: *U.S EPA, 1991.
[†]U.S. Environmental Protection Agency, National Primary Drinking Water Regulations: Long Term 2 Enhanced Surface Water Treatment Rule.

would be to add ozone either to the raw water or between the sedimentation basins and filter for primary disinfection, followed by chloramine addition after the filters as the distribution system disinfectant.

Ultraviolet Radiation

Light is important to almost all life forms. We "see" only a very small fraction of the colors of light. The ultraviolet range of light falls beyond the violet end of the rainbow. Table 6-17 outlines the spectral ranges of interest in photochemistry.

TABLE 6-17
Spectral ranges of interest in photochemistry

Range name	Wavelength range (nm)
Near infrared	700–1,000
Visible	400–700
Ultraviolet	
UVA	315–400
UVB	280–315
UVC	200–280
Vacuum ultraviolet (VUV)	100–200

Light photons with wavelengths longer than 1000 nanometers (nm) have a photon energy too small to cause chemical change when absorbed, and photons with wavelengths shorter than 100 nm have so much energy that ionization and molecular disruptions characteristic of radiation chemistry prevail.

Little photochemistry occurs in the near infrared range except in some photosynthetic bacteria. The visible range is completely active for photosynthesis in green plants and algae. Most studies in photochemistry involve the ultraviolet range. The ultraviolet range is divided into three categories connected with the human skin's sensitivity to ultraviolet light. The UVA range causes changes to the skin that lead to tanning. The UVB range can cause skin burning and is prone to induce skin cancer. The UVC range is extremely dangerous since it is absorbed by proteins and can lead to cell mutations or cell death.

Disinfection of water and wastewater using ultraviolet light has been practiced extensively in Europe and is becoming more common in the United States. Ultraviolet light disinfects water by rendering pathogenic organisms incapable of reproducing. This is accomplished by disrupting the genetic material in cells. The genetic material in cells, namely DNA, will absorb light in the ultraviolet range—primarily between 200 nm and 300 nm wavelengths (UV light is most strongly absorbed by DNA at 253.7 nm). If the DNA absorbs too much UV light, it will be damage and will be unable to replicate. It has been found that the energy required to damage the DNA is much less than that required to actually destroy the organism. The effect is the same, however, since if a microorganism cannot reproduce, it cannot cause an infection.

The inactivation of microorganisms by UV is directly related to UV dose, a concept similar to *CT* used for other common disinfections including chlorine and ozone. The average UV dose is calculated as follows:

$$D = It \qquad (6\text{-}69)$$

where D = UV dose
I = average intensity, mW/cm^2
t = average exposure time, s

The survival fraction is calculated as follows:

$$\text{Survival fraction} = \log\left(\frac{N}{N_o}\right) \qquad (6\text{-}70)$$

where N = organism concentration after inactivation
N_o = organism concentration before inactivation

The equation for UV dose indicates that dose is directly proportional to exposure time and thus inversely proportional to system flow rate. UV intensity (I) is a function of water UV transmittance and UV reactor geometry as well as lamp age and fouling. UV intensity can be estimated by mathematical modeling and verified by bioassay. Exposure time is estimated from the UV reactor specific hydraulic characteristics and flow patterns.

The major factor affecting the performance of UV disinfection systems is the influent water quality. Particles, turbidity, and suspended solids can shield pathogens from UV light or scatter UV light to prevent it from reaching the target microorganism, thus reducing its effectiveness as a disinfectant. Some organic compounds and inorganic compounds (such as iron and permanganate) can reduce UV transmittance by absorbing

UV energy, requiring higher levels of UV to achieve the same dose. Therefore, it is recommended that UV systems be installed downstream of the filters so that removal of particles and organic and inorganic compounds is maximized upstream of UV.

Water turbidity and UV transmittance are commonly used as process controls at UV facilities. The UV percent transmittance of a water sample is measured by a UV-range spectrophotometer set at a wavelength of 253.7 nm using a 1-cm-thick layer of water. The water UV transmittance is related to UV absorbance (A) at the same wavelength by the equation:

$$\text{Percent transmittance} = 100 \times 10^{-A} \tag{6-71}$$

For example, a water UV absorbance of 0.022 per cm corresponds to a water percent transmittance of 95 (i.e., at 1 cm from the UV lamp, 95 percent of lamp output is left). Similarly, a UV absorbance of 0.046 per cm is equivalent to 90 percent UV transmittance.

UV has been found to be very effective for the disinfection of *Cryptosporidium, Giardia,* and viruses. U.S. EPA has established UV dose requirements. Table 6-18 shows the U.S. EPA requirements.

Although the *Cryptosporidium* log credit requirements for say 2-log inactivation are 5.8 mJ/cm^2, many design engineers will use values of 20 to 40 mJ/cm^2 for additional safety.

UV electromagnetic energy is typically generated by the flow of electrons from an electrical source through ionized mercury vapor in a lamp. Several manufactures have developed systems to align UV lamps in vessels or channels to provide UV light in the germicidal range for inactivation of bacteria, viruses, and other microorganisms. The UV lamps are similar to household fluorescent lamps, except that fluorescent lamps are coated with phosphorus, which converts the UV light to visible light. Ballasts (i.e., transformers) that control the power of the UV lamps are either electronic or electromagnetic. Electronic ballasts offer several potential advantages including lower lamp operating temperatures, higher efficiencies, and longer ballast life.

Both low pressure and medium-pressure lamps are available for disinfection applications. Low-pressure lamps emit their maximum energy output at a wavelength of 253.7 nm, while medium-pressure lamps emit energy with wavelengths ranging from 180 to 1370 nm. The intensity of medium-pressure lamps is much greater than

TABLE 6-18
UV dose requirements for *Cryptosporidium, Giardia lamblia,* and virus

Log credit	*Cryptosporidium* UV dose (mJ/cm^2)	*Giardia lamblia* UV dose (mJ/cm^2)	Virus UV dose (mJ/cm^2)
0.5	1.6	1.5	39
1.0	2.5	2.1	58
1.5	3.9	3.0	79
2.0	5.8	5.2	100
2.5	8.5	7.7	121
3.0	12	11	143
3.5	15	15	163
4.0	22	22	186

low-pressure lamps. Thus, fewer medium-pressure lamps are required for an equivalent dosage. For small systems, the medium-pressure system may consist of a single lamp. Although both types of lamps work equally well for inactivation of organisms, low-pressure UV lamps are recommended for small systems because of the reliability associated with multiple low-pressure lamps as opposed to a single medium-pressure lamp, and for adequate operation during cleaning cycles.

Most conventional UV reactors are available in two types: namely, closed vessel and open channel. For drinking water applications, the closed vessel is generally the preferred UV reactor for the following reasons (U.S. EPA, 1996):

- Smaller footprint
- Minimized pollution from airborne material
- Minimal personnel exposure to UV
- Modular design for installation simplicity

Figure 6-42 shows a conventional closed-vessel UV reactor.

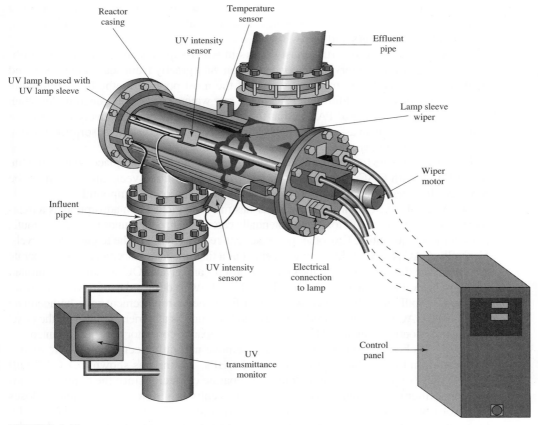

FIGURE 6-42
UV disinfection system schematic. (*Source:* Aquionics.)

Advanced Oxidation Processes (AOPs)

AOPs are combinations of disinfectants designed to produce hydroxyl radicals (OH·). Hydroxyl radicals are highly reactive nonselective oxidants able to decompose many organic compounds. Most noteworthy of the AOP processes is ozone plus hydrogen peroxide.

6-8 ADSORPTION

Adsorption is a mass transfer process wherein a substance is transferred from the liquid phase to the surface of a solid where it is bound by chemical or physical forces.

Generally, in water treatment, the adsorbent (solid) is activated carbon, either granular (GAC) or powdered (PAC). PAC is fed to the raw water in a slurry and is generally used to remove taste- and odor-causing substances or to provide some removal of synthetic organic chemicals (SOCs). GAC is added to the existing filter system by replacing the anthracite with GAC, or an additional contactor is built and is placed in the flow scheme after primary filtration. The design of the GAC contactor is very similar to a filter box, although deeper.

At present, the applications of adsorption in water treatment in the United States are predominately for taste and odor removal. However, adsorption is increasingly being considered for removal of SOCs, VOCs, and naturally occurring organic matter, such as THM precursors and DBPs.

Biologically derived earthy-musty odors in water supplies are a widespread problem. Their occurrence interval and concentration vary greatly from season to season and is often unpredictable. As mentioned, one of the most popular methods for removing these compounds is the addition of PAC to the raw water. The dose is generally less than 10 mg/L. The advantage of PAC is that the capital equipment is relatively inexpensive and it can be used on an as-needed basis. The disadvantage is that the adsorption is often incomplete. Sometimes even doses of 50 mg/L are not sufficient.

As an alternative for taste and odor control, many plants have replaced the anthracite in the filters with GAC. The GAC will last from one to three years and then must be replaced. It is very effective in removing many taste and odor compounds.

Concern about SOCs in drinking water has motivated interest in adsorption as a treatment process for removal of toxic and potentially carcinogenic compounds present in minute, but significant, quantities. Few other processes can remove SOCs to the required low levels. Generally, GAC is used for SOC removal either as a filter media replacement or as a separate contactor. The data for how long the GAC will last for any given SOC are somewhat limited and must be evaluated on a case-by-case basis. If the GAC is designed to remove SOCs from a periodic "spill" into the source water, then filter media replacement may be adequate because the GAC is not being used every day and is acting as a barrier. However, if the GAC is to be used continuously for SOC removal, then a separate contactor may be warranted.

GAC has been proposed to be used to remove naturally occurring organic matter that would, in turn, reduce the formation of DBPs. Testing has shown that GAC will remove these organics. It must operate in a separate contactor since the depth of a conventional filter is inadequate. The GAC will typically last 90 to 120 days until it loses its adsorptive capacity. Because of its short life, the GAC needs to be regenerated by burning in a high-temperature furnace. This can be done on-site or can be done by returning the GAC to the manufacture.

GAC has also been considered for removal of THMs. However, the capacity is very low and the carbon may only last up to 30 days. GAC is not considered practical for THM removal.

6-9 MEMBRANES

A membrane is a thin layer of material that is capable of separating materials as a function of their physical and chemical properties when a driving force is applied across the membrane. In the case of water treatment, the driving force is supplied by using a high pressure pump and the membrane type is selected based on the constituents to be removed.

In the membrane process, the feed stream is divided into two streams, the concentrate or reject stream, and the permeate or product stream as shown in Figure 6-43. The membrane is at the heart of every membrane process and can be considered a barrier between the feed and product water that does not allow certain contaminants to pass, as represented by Figure 6-44. There is a pressure drop across the membrane, called the transmembrane pressure (TMP) that must be overcome by the water feed pump. The performance or efficiency of a given membrane is determined by its selectivity and the applied flow. The efficiency is called the flux and is defined as the volume flowing through the membrane per unit area and time, $m^3/m^2 \cdot s$ (gpm/ft^2 or gpd/ft^2). The selectivity of a membrane toward a mixture is expressed by its retention, R, and is given by

$$R = \frac{c_p - c_f}{c_p} \times 100\% \tag{6-72}$$

where c_f = contaminant concentration in feed
c_P = contaminant concentration in permeate

The value of R varies between 100 percent (complete retention) and 0 percent (no retention).

Membrane technology is becoming increasingly popular as an alternative treatment technology for drinking water. The anticipation of more stringent water quality regulations, a decrease in availability of adequate water resources, and an emphasis on water for reuse has made membrane processes more viable as a water treatment process. As advances are being made in membrane technology, capital and operation and maintenance costs continue to decline, further endorsing the use of membrane treatment techniques.

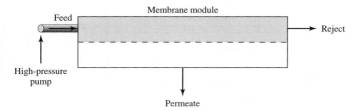

FIGURE 6-43
Schematic representation of a membrane process.

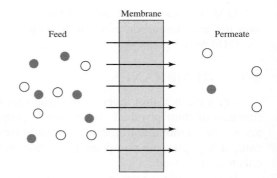

FIGURE 6-44
Schematic representation of contaminants separated by a membrane.

Membranes can be described by a variety of criteria including: (Jacangelo et al., 1997)

- membrane pore size
- molecular weight cutoff (MWCO)
- membrane material and geometry
- targeted materials to be removed
- type of water quality to be treated, and/or
- treated water quality

Along with these criteria, membrane processes can also be categorized broadly into pressure-driven and electrically driven processes. This discussion is limited to pressure-driven membrane processes. Figure 6-45 summarizes the various pressure-

Operating pressures (kPa)	700	200		70	35		
Molecular range (daltons*)	200 1,000	100,000					
Size (μm)	0.001 0.01	0.1	1.0	10	100	1,000	
Relative size of various materials in water	Aqueous salts / Humic acids / Metal ions	Viruses	Clays	Bacteria / Algae / Cysts / Silt / Asbestos fibers	Sand		
Separation processes	Reverse osmosis / Nano-filtration / Ultrafiltration			Microfiltration / Conventional filtration processes			

*Dalton is psi × 7 = kPa a unit of mass equal to 1/16 the mass of the lightest and most abundant isotope of oxygen.

FIGURE 6-45
Schematic comparison of selected separation processes. (*Source:* Jacangelo et. al., 1997.)

driven membrane processes and selected materials removed by each. A brief description of each follows.

- *Reverse osmosis (RO)*. This process has traditionally been employed for the removal of salts from brackish water and seawater. It depends upon applying high pressures across the membrane (in the range of 1,000 to 8,000 kPa) in order to overcome the osmotic pressure differential between the saline feed and product waters.

- *Nano-filtration (NF)*. This process, also called membrane softening, lies between RO and ultrafiltration. This membrane process employs 500 to 1,000 kPa for operation. While it provides removal of ions contributing to hardness (i.e., calcium and magnesium), the technology is also very effective for removal of color and DBP precursors.

- *Ultrafiltration (UF)*. UF membranes cover a wide range of MWCOs and pore sizes. Operational pressures range from 70 to 700 kPa, depending on the application. "Tight" UF membranes (MWCO = 1,000 daltons) may be employed for removal of some organic materials from freshwater, while the objective of "loose" membranes (MWCO > 50,000 daltons, 70 to 200 kPa) is primarily for liquid/solid separation, i.e., particle and microbial removal.

- *Microfiltration (MF)*. A major difference between MF and loose UF is membrane pore size; the pores of MF (= 0.1 μm or greater) are approximately an order of magnitude greater than those of UF. The primary application for this membrane process is particulate and microbial removal.

6-10 WATER PLANT RESIDUALS MANAGEMENT

The precipitated chemicals and other materials removed from raw water to make it potable and palatable are termed *residuals*. Satisfactory treatment and disposal of water treatment plant residuals can be the single most complex and costly operation in the plant.

Residuals withdrawn from coagulation and softening plants are composed largely of water, and the residuals are often referred to a sludge. As much as 98 percent of the sludge mass may be water. Thus, for example, 20 kg of solid chemical precipitate is accompanied by 980 kg of water. Assuming equal densities for the precipitate and water (a bad assumption at best), approximately 1 m^3 of sludge is produced for each 20 kg of chemicals added to the water. For even a small plant (say 0.05 m^3/s) this might mean up to 800 m^3/y of sludge—a substantial volume to say the least!

Water treatment plants and the residuals they produce can be broadly divided into four general categories. First are those treatment plants that coagulate, filter, and oxidize a surface water for removal of turbidity, color, bacteria, algae, some organic compounds, and often iron and/or manganese. These plants generally use alum or iron salts for coagulation and produce two waste streams. The majority of the waste produced from these plants is sedimentation basin (or clarifier) sludge and spent filter backwash water (SFBW). The second type of treatment plants are those that practice softening for the removal of calcium and magnesium by the addition of lime, sodium hydroxide, and/or soda ash. These plants produce clarifier basin sludges and SFBW. On occasion, plants practice both coagulation and softening. Softening plant wastes can also contain trace inorganics such as radium that could affect their proper handling. The third type of plants are those that are designed to specifically remove trace inorganics such as nitrate, fluoride, radium, arsenic, and so forth. These plants use processes such as ion exchange, reverse osmosis, or adsorption. They produce liquid residuals or solid residuals such as spent adsorption material. The fourth category of treatment plants are those that produce air-phase residuals during the stripping of volatile compounds. The major types of treatment plant residuals produced are shown in Table 6-19. Because 95 percent of the residuals produced are coagulants or softening sludge, they will be stressed in this section.

Hydrolyzing metal salts or synthetic organic polymers are added in the water treatment process to coagulate suspended and dissolved contaminants and yield relatively clean water suitable for filtration. Most of these coagulants and the impurities they remove settle to the bottom of the settling basin where they become part of the sludge. These sludges are referred to as alum, iron, or polymeric sludge according to which primary coagulant is used. These wastes account for approximately 70 percent of the water plant waste generated. The sludges produced in treatment plants where water softening is practiced using lime or lime and soda ash account for an additional 25 percent of the industry's waste production. It is therefore apparent that most of the waste generation involves water treatment plants using coagulation or softening processes.

TABLE 6-19
Major water treatment plant residuals

Solid/Liquid Residuals
1. Alum sludges
2. Iron sludges
3. Polymeric sludges
4. Softening sludges
5. SFBW
6. Spent GAC or discharge from carbon systems
7. Slow sand filter cleanings
8. Residuals from iron and manganese removal plants
9. Spent precoat filter media

Liquid-Phase Residuals
10. Ion-exchange regenerant brine
11. Pregenerant from activated alumina
12. Reverse osmosis reject

Gas-Phase Residuals
13. Air stripping off-gases

The most logical sludge management program attempts to use the following approach in disposing of the sludge:

1. Minimization of sludge generation

2. Chemical recovery of precipitates

3. Sludge treatment to reduce volume

4. Ultimate disposal in an environmentally safe manner

With a short digression to identify the sources of water plant sludges and their production rates, we have organized the following discussion along these lines.

Sludge Production and Characteristics

In water treatment plants, sludge is most commonly produced in the following treatment processes: presedimentation, sedimentation, and filtration (filter backwash).

Presedimentation. When surface waters are withdrawn from watercourses that contain a large quantity of suspended materials, presedimentation prior to coagulation may be practiced. The purpose of this is to reduce the accumulation of solids in subsequent units. The settled material generally consists of fine sand, silt, clays, and organic decomposition products.

Softening Sedimentation Basin. The residues from softening by precipitation with lime $[Ca(OH)_2]$ and soda ash (Na_2CO_3) will vary from a nearly pure chemical to a highly variable mixture. The softening process discussed earlier in this chapter produces a sludge containing primarily $CaCO_3$ and $Mg(OH)_2$.

Theoretically, each mg/L of calcium hardness removed produces 1 mg/L of $CaCO_3$ sludge; each mg/L of magnesium hardness removed produces 0.6 mg/L of sludge; and each mg/L of lime added produces 1 mg/L of sludge. The theoretical sludge production can be calculated as:

$$M_s = 86.4 \, Q(2 \, \text{CaCH} + 2.6 \, \text{MgCH} + \text{CaNCH} + 1.6 \, \text{MgNCH} + CO_2) \qquad (6\text{-}73)$$

where M_s = sludge production (kg/d)
   CaCH = calcium carbonate hardness removed as $CaCO_3$ (mg/L)
   MgCH = magnesium carbonate hardness removed as $CaCO_3$ (mg/L)
   Q = plant flow, m^3/s
   CaNCH = noncarbonate calcium hardness removed as $CaCO_3$ (mg/L)
   MgNCH = noncarbonate magnesium hardness removed as $CaCO_3$ (mg/L)
   CO_2 = carbon dioxide removed by lime addition, as $CaCO_3$ (mg/L)

When surface waters are softened, this equation is not valid. There will be additional sludge from coagulation of suspended materials and precipitation of metal coagulants. The solids content of lime-softening sludge in the sedimentation basin ranges between 2 and 15 percent. A nominal value of 10 percent solids is often used.

Coagulation Sedimentation Basin. Aluminum or iron salts are generally used to accomplish coagulation. (The chemistry of the two salts was discussed earlier in this chapter.) The pH range of 6 to 8 is where most water treatment plants effect the coagulation process. In this range the insoluble aluminum hydroxide complex of $Al(H_2O)_3(OH)_3$ probably predominates. This species results in the production of 0.44 kg of chemical sludge for each kg of alum added. Any suspended solids present in the water will produce an equal amount of sludge. The amount of sludge produced per turbidity unit is not as obvious; however, in many waters a correlation does exist. Carbon, polymers, and clay will produce about 1 kg of sludge per kg of chemical addition. The sludge production for alum coagulation may then be approximated by:

$$M_s = 86.40 \, Q(0.44A + SS + M) \tag{6-74}$$

where M_s = dry sludge produced, kg/d
$\quad Q$ = plant flow, m^3/s
$\quad A$ = alum dose, mg/L
$\quad SS$ = suspended solids in raw water, mg/L
$\quad M$ = miscellaneous chemical additions such as clay, polymer, and carbon, mg/L

Alum sludge leaving the sedimentation basin usually has a suspended solids content in the range of 0.5 to 2 percent. It is often less than 1 percent. Twenty to 40 percent of the solids are organic; the remainder are inorganic or silts. The pH of alum sludge is normally in the 5.5 to 7.5 range. Alum sludge from sedimentation basins may include large numbers of microorganisms, but it generally does not exhibit an unpleasant odor. The sludge flow rate is often in the range of 0.3 to 1 percent of the treatment plant flow.

Spent Filter Backwash Water. All water treatment plants that practice filtration produce a large volume of washwater containing a low suspended solids concentration. The volume of backwash water is usually 3 to 7 percent of the treatment plant flow. The solids in backwash water resemble those found in sedimentation units. Since filters can support biological growth, the spent filter backwash water may contain a larger fraction of organic solids than do the solids from the sedimentation basins.

Mass Balance Analysis. Clarifier sludge production can be estimated by a mass balance analysis of the sedimentation basin. Since there is no reaction taking place, the mass balance equation reduces to the form:

$$\text{Accumulation rate} = \text{Input rate} - \text{Output rate} \tag{6-75}$$

The input rate of solids may be estimated from Equation 6-73 or 6-74. To estimate the mass flow (output rate) of solids leaving the clarifier through the weir, you must have an estimate of the concentration of solids and the flow rate. The mass flow out through the weir is then

$$\text{Weir output rate} = (\text{Concentration, } g/m^3)(\text{Flow rate, } m^3/s) = g/s$$

Example 6-22. A coagulation treatment plant with a flow of 0.5 m^3/s is dosing alum at 23.0 mg/L. No other chemicals are being added. The raw-water suspended-solids concentration is 37.0 mg/L. The effluent suspended-solids concentration is measured at 12.0 mg/L. The sludge solids content is 1.00 percent and the specific gravity of the sludge solids is 3.01. What volume of sludge must be disposed of each day?

Solution. The mass-balance diagram for the sedimentation basin is

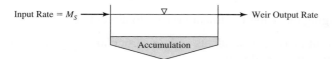

The mass of solids (sludge) flowing into the clarifier is estimated from Equation 6-74:

$$M_s = 86.40(0.50 \text{ m}^3/\text{s})(0.44(23.0 \text{ mg/L}) + 37.0 \text{ mg/L} + 0)$$
$$= 2{,}035.58 \text{ kg/d}$$

Recognizing that $\text{g/m}^3 = \text{mg/L}$, the mass of solids leaving the weir is

$$\text{Weir output rate} = (12.0 \text{ g/m}^3)(0.50 \text{ m}^3/\text{s})(86{,}400 \text{ s/d})(10^{-3} \text{ kg/g})$$
$$= 518.4 \text{ kg/d}$$

The accumulation is then

$$\text{Accumulation} = 2{,}035.58 - 518.4 = 1{,}517.18 \text{ or } 1{,}517 \text{ kg/d}$$

Because this is a dry mass and the sludge has only 1.00 percent solids, we must account for the volume of water in estimating the volume to be removed each day. Using Equation 5-4, we can solve for the mass of water (W_o):

$$1.00 = \frac{1{,}517}{1{,}517 + W_o}(100)$$
$$W_o = 150{,}183 \text{ kg/d}$$

Now we use the definition of density (mass per unit volume) to find the volume of sludge and water:

$$\text{Volume} = \frac{\text{Mass}}{\text{Density}}$$
$$V_T = \text{volume of solids} + \text{volume of water}$$
$$= \frac{1{,}517 \text{ kg/d}}{(3.01)(1{,}000 \text{ kg/m}^3)} + \frac{150{,}183 \text{ kg/d}}{1{,}000 \text{ kg/m}^3}$$
$$= 0.50 + 150.18 = 150.7 \text{ m}^3/\text{d or } 150 \text{ m}^3/\text{d}$$

The specific gravity of the solids is 3.01 and the density of water is 1,000 kg/m^3.

Comment: Obviously, the solids account for only a small fraction of the total volume. This is why sludge dewatering is an important part of the water treatment process.

Minimization of Sludge Generation for Sustainability

Minimizing sludge generation can have an advantageous effect on the requirements and economics of handling, treating, and disposing of water treatment plant sludges. Minimization also results in the conservation of raw materials, energy, and labor.

There are three methods to minimize the quantity of metal hydroxide precipitates in the sludge:

1. Changing the water treatment process to direct filtration

2. Substituting other coagulants and, in particular, using polymers that are more effective at lower dosage

3. Conserving chemicals by determining optimum dosage at frequent intervals as raw water characteristics change

Sludge Treatment

The treatment of solid/liquid wastes produced in water treatment processes involves the separation of the water from the solid constituents to the degree necessary for the selected disposal method. Therefore, the required degree of treatment is a direct function of the ultimate disposal method.

There are several sludge treatment methodologies which have been practiced in the water industry. Figure 6-46 shows the most common sludge-handling options available, listed by general categories of thickening, dewatering, and disposal. In choosing a combination of the possible treatment process trains, it is probably best to first identify the available disposal options and their requirements for a final cake solids concentration. Most landfill applications will require a "handleable" sludge and this may limit the type of dewatering devices that are acceptable. Methods and costs of transportation may affect the decision "how dry is dry enough." The criteria should not be to simply reach a given solids concentration but rather to reach a solids concentration of desired properties for the handling, transport, and disposal options available.

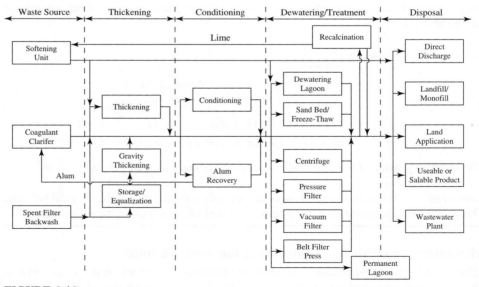

FIGURE 6-46
Sludge handling options.

TABLE 6-20
Range of cake solid concentrations obtainable

	Lime sludge, %	Coagulation sludge, %
Gravity thickening	15–30	3–4
Scroll centrifuge	35–70	20–25
Belt filter press	25–50	18–25
Vacuum filter	45–65	n/a
Pressure filter	55–70	30–45
Sand drying beds	50	20–25
Storage lagoons	50–60	7–15

Table 6-20 shows a generalized range of results that have been obtained for final solids concentrations from different dewatering devices for coagulant and lime sludges.

To give you an appreciation of these solids concentrations, a coagulant sludge cake with 25 percent solids would have the consistency of soft clay, while a 15 percent sludge would have a consistency much like rubber cement.

After removal of the sludge from the clarifier or sedimentation basin, the first treatment step is usually thickening. Thickening assists the performance of any subsequent treatment, gets rid of much water quickly, and helps to equalize flows to the subsequent treatment device. An approximation for determining sludge volume reduction via thickening is given by:

$$\frac{V_2}{V_1} = \frac{P_1}{P_2} \tag{6-76}$$

where V_1 = volume of sludge before thickening, m^3
V_2 = volume of sludge after thickening, m^3
P_1 = percent of solids concentration of sludge before thickening
P_2 = percent of solids concentration after thickening

Thickening is usually accomplished by using circular settling basins similar to a clarifier (Figure 6-47). Thickeners can be designed based on pilot evaluations or using data obtained from similar plants. Lime sludges are typically loaded at 100 to 200 kg/m$^2 \cdot$ d, (20 to 40 lb$_m$/ft$^2 \cdot$ d) and coagulant sludge loading rates are about 15 to 25 kg/m$^2 \cdot$ d.

Following thickening of the sludge, dewatering can take place by either mechanical or nonmechanical means. In nonmechanical devices, sludge is spread out with the free water draining and the remaining water evaporating. Sometimes the amount of free water available to drain is enhanced by natural freeze-thaw cycles. In mechanical dewatering, some type of device is used to force the water out of the sludge.

We begin our discussion with the nonmechanical methods and follow with the mechanical methods.

Lagoons. A lagoon is essentially a large hole that is dug out for the sludge to flow into. Lagoons can be constructed as either storage lagoons or dewatering lagoons. Storage lagoons are designed to store and collect the solids for some predetermined

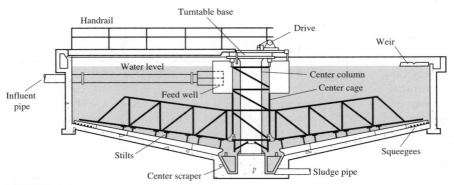

FIGURE 6-47
Continuous-flow gravity thickener. (Courtesy of Metcalf and Eddy, Inc. 2003.)

amount of time. They will generally have decant capabilities but no underdrain system. Storage lagoons should be equipped with sealed bottoms to protect the groundwater. Once the storage lagoon is full or decant can no longer meet discharge limitations, it must be abandoned or cleaned. To facilitate drying, the standing water may be removed by pumping, leaving a wet sludge. Coagulant sludges can only be expected to reach a 7 to 10 percent solids concentration in storage lagoons. The remaining solids must be either cleaned out wet or allowed to evaporate. Depending upon the depth of the wet solids, evaporation can take years. The top layers will often form a crust, preventing evaporation of the bottom layers of sludge.

The primary difference between a dewatering lagoon and a storage lagoon is that a dewatering lagoon has a sand and underdrain-bottom, similar to a drying bed. Dewatering lagoons can be designed to achieve a dewatered sludge cake. The advantage of a dewatering lagoon over a drying bed is that storage is built into the system to assist in meeting peak solids production or to assist in handling sludge during wet weather. The disadvantage of bottom sand layers compared to conventional drying beds is that the bottom sand layers can plug up or "blind" with multiple loadings, thereby increasing the required surface area. Polymer treatment can be useful in preventing this sand blinding.

Storage lagoons, which are generally earthen basins, have no size limitations but have been designed in areas from 2,000 to 60,000 m^2, ranging in depth from 2 to 10 or more meters. Storage and dewatering lagoons may be equipped with inlet structures designed to dissipate the velocity of the incoming sludge. This minimizes turbulence in the lagoons and helps prevent carryover of solids in the decant. The lagoon outlet structure is designed to skim the settled supernatant and is sometimes provided with flash boards to vary the draw-off depths. Any design of a storage lagoon must consider how the sludge will be ultimately removed unless the site is to be abandoned.

The basis for design of dewatering lagoons is essentially the same as that for sand drying beds. The difference is that the applied depth is high and the number of applications per year is greatly reduced.

Sand-Drying Beds.   Sand-drying beds operate on the simple principle of spreading the sludge out and letting it dry. As much water as possible is removed by drainage or

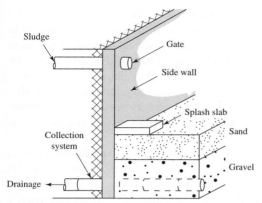

FIGURE 6-48
Typical sludge drying bed construction. (*Source:* U.S. Environmental Protection Agency, 1979.)

decant and the rest of the water must evaporate until the desired final solids concentration is reached. Sand-drying beds have been built simply by cleaning an area of land, dumping the sludge, and hoping something happens. At the other end of the spectrum sophisticated automated drying systems have been built.

Drying beds may be roughly categorized as follows:

1. Conventional rectangular beds with side walls and a layer of sand on gravel with underdrain piping to carry away the liquid. They are built either with or without provisions for mechanical removal of the dried sludge and with or without a roof or a greenhouse-type covering (see Figure 6-48).

2. Paved rectangular drying beds with a center sand drainage strip with or without heating pipes buried in the paved section and with or without covering to prevent incursion of rain.

3. "Wedge-water" drying beds that include a wedge wire septum incorporating provision for an initial flood with a thin layer of water, followed by introduction of liquid sludge on top of the water layer, controlled formation of cake, and provision for mechanical cleaning.

4. Rectangular vacuum-assisted drying beds with provision for application of vacuum to assist gravity drainage.

The dewatering of sludge on sand beds is accomplished by two major factors: drainage and evaporation. The removal of water from sludge by drainage is a two-step process. First, the water is drained from the sludge, into the sand, and out the underdrains. This process may last a few days until the sand is clogged with fine particles or until all the free water has drained away. Further drainage by decanting can occur once a supernatant layer has formed (if beds are provided with a means of removing surface water). Decanting for removal of rain water can also be particularly important with sludges that do not crack.

The water that does not drain or is not decanted must evaporate. Obviously climate plays a role here. Phoenix would be a more efficient area for a sand bed than Seattle! Much of the lower Midwest, for example, would have an annual evaporation rate of about 0.75 m, so typical annual loadings to a sand bed in that area would be 100 kg/m^2 · y. This may be applied and cleaned in about 10 cycles during the year.

The filtrate from the sand-drying beds can be either recycled, treated, or discharged to a watercourse depending on its quality. Laboratory testing of the filtrate should be performed in conjunction with sand-drying bed pilot testing before a decision is made as to what is to be done with it.

Current United States practice is to make drying beds rectangular with dimensions of 4 to 20 m by 15 to 50 m with vertical side walls. Usually 100 to 230 mm of sand is placed over 200 to 460 mm of graded gravel or stone. The sand is usually 0.3 to 1.2 mm in effective diameter and has a uniformity coefficient less than 5.0. Gravel is normally graded from 3 to 25 mm in effective diameter. Underdrain piping has normally been of vitrified clay, but plastic pipe is also becoming acceptable. The pipes should be no less than 100 mm in diameter, should be spaced 2.2 to 6 m apart, and should have a minimum slope of 1 percent. When the cost of labor is high, newly constructed beds are designed for mechanical sludge removal.

Freeze Treatment. Dewatering sludge by either of the nonmechanical methods may be enhanced by physical conditioning of the sludge through alternate natural freezing and thawing cycles. The freeze-thaw process dehydrates the sludge particles by freezing the water that is closely associated with the particles. The dewatering process takes place in two stages. The first stage reduces sludge volume by selectively freezing the water molecules. Next, the solids are dehydrated when they become frozen. When thawed, the solid mass forms granular-shaped particles. This coarse material readily settles and retains its new size and shape. This residue sludge dewaters rapidly and makes suitable landfill material.

The supernatant liquid from this process can be decanted, leaving the solids to dewater by natural drainage and evaporation. Pilot-scale systems can be utilized to evaluate this method's effectiveness and to establish design parameters. Elimination of rain and snow from the dewatering system by the provision of a roof will enhance the process considerably.

The potential advantages of a freeze-thaw system are

1. Insensitivity to variations in sludge quality

2. No conditioning required

3. Minimum operator attention

4. Natural process in cold climates (winter)

5. Solids cake more acceptable at landfills

6. Sludge easily worked with conventional equipment

Several natural freeze-thaw installations are located in New York state. At the alum coagulation plant of the Metropolitan Water Board of Oswego County, SFBW is discharged to lagoons that act as decant basins. Thickened sludge is pumped from the

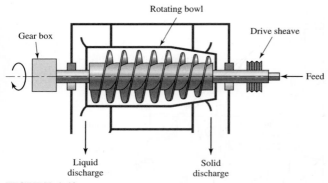

FIGURE 6-49
Solid bowl centrifuge. (Courtesy of Andritz Separation Inc.)

lagoons to special freeze-thaw basins in layers about 450 mm thick. The sludge has never been deeper than 300 mm during freezing because of additional water losses. The 300 mm sludge layer reduces to about 75 mm of dried material after freeze-thaw.

Centrifuging.   A centrifuge uses centrifugal force to speed up the separation of sludge particles from the liquid. In a typical unit (Figure 6-49), sludge is pumped into a horizontal, cylindrical bowl, rotating at 800 to 2,000 rpm. Polymers used for sludge conditioning also are injected into the centrifuge. The solids are spun to the outside of the bowl where they are scraped out by a screw conveyor. The liquid, or *centrate,* is returned to the treatment plant. Two types of centrifuges are currently used for sludge dewatering: the solid bowl and the basket bowl. For dewatering water-treatment-plant sludges, the solid bowl has proven to be more successful than the basket bowl. Centrifuges are very sensitive to changes in the concentration or composition of the sludge, as well as to the amount of polymer applied.

Because of its calcium carbonate content, lime softening-sludge dewaters with relative ease in a centrifuge. A cake dryness of 20 to 25 percent can be achieved for a centrifuged alum sludge. A solids content of 50 percent or higher can be achieved with a lime sludge.

Vacuum Filtration.   A vacuum filter consists of a cylindrical drum covered with a filtering material or fabric, which rotates partially submerged in a vat of conditioned sludge (Figure 6-50). A vacuum is applied inside the drum to extract water, leaving the solids, or *filter cake,* on the filter medium. As the drum completes its rotational cycles, a blade scrapes the filter cake from the filter and the cycle begins again. Two basic types of rotary-drum vacuum filters are used in water treatment: the *traveling medium* and the *precoat medium* filters. The traveling medium filter is made of fabric or stainless steel coils. This filter is continuously removed from the drum, allowing it to be washed from both sides without diluting the sludge in the sludge vat. The precoat medium filter is coated with 50 to 75 mm of inert material, which is shaved off in 0.1 mm increments as the drum moves.

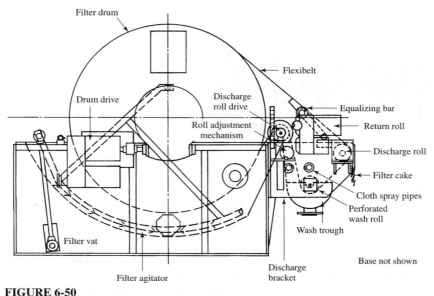

FIGURE 6-50
Vacuum filter. (Courtesy of Komline-Sanderson Engineering Corporation.)

Continuous Belt Filter Press (CBFP). The belt filter press operates on the principle that bending a sludge cake contained between two filter belts around a roll introduces shear and compressive forces in the cake, allowing water to work its way to the surface and out of the cake, thereby reducing the cake moisture content. The device employs double moving belts to continuously dewater sludges through one or more stages of dewatering (Figure 6-51). Typically the CBFP includes the following stages of treatment:

1. A reactor/conditioner to remove free-draining water

2. A low pressure zone of belts with the top belt being solid and the bottom belt being a sieve; here further water removal occurs and a sludge mat having significant dimensional stability is formed

3. A high pressure zone of belts with a serpentine or sinusoidal configuration to add shear to the pressure dewatering mechanisms

Plate Pressure Filters. The basic component of a plate filter is a series of recessed vertical plates. Each plate is covered with cloth to support and contain the sludge cake. The plates are mounted in a frame consisting of two head supports connected by two horizontal parallel bars. A schematic cross section is illustrated in Figure 6-52. Conditioned sludge is pumped into the pressure filter and passes through feed holes in the filter plates along the length of the filter and into the recessed chambers. As the sludge cake forms and builds up in the chamber, the pressure gradually increases to a point where further sludge injection would be counterproductive. At this time the injection ceases.

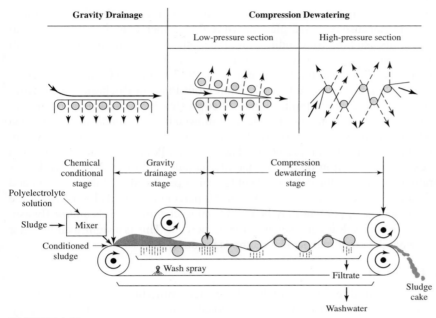

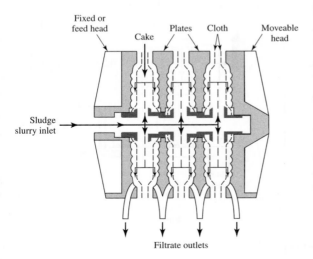

FIGURE 6-51

Continuous belt filter press. (*Source:* U.S. EPA, 1979.)

A typical pressure filtration cycle begins with the closing of the press to the position shown on Figure 6-52. Sludge is fed for a 20- to 30-minute period until the press is effectively full of cake. The pressure at this point is generally the designed maximum (700 to 1,700 kPa) and is maintained for 1 to 4 hours, during which more filtrate is removed and the desired cake solids content is achieved. The filter is then mechanically opened, and the dewatered cake is dropped from the chambers onto a conveyor

FIGURE 6-52

Schematic cross section of a fixed volume recessed plate filter assembly. (*Source:* U.S. EPA, 1979.)

belt for removal. Cake breakers are usually required to break up the rigid cake into conveyable form. Because recessed-plate pressure filters operate at high pressures and because many units use lime for conditioning, the cloths require routine washing with high-pressure water, as well as periodic washing with acid.

The Erie County Water Authority's Sturgeon Point Plant in Erie County, New York, and the Monroe County Water Authority's Shoremont Plant in Rochester, New York, serve as typical examples of the application of filter presses. Both plants include gravity settling and chemical conditioning of the sludge followed by mechanical dewatering via pressure plate filtration. A typical process flow diagram of the sludge treatment system used at both plants is shown in Figure 6-53. The Sturgeon Point Plant includes a 1.6-m diameter press, while the Monroe County facility includes a 2-m by 2-m press. Under actual operating conditions, the dewatered sludge cakes produced at Sturgeon Point and at the Shoremont Plant contain between 45 and 50 percent dry solids by mass. Of the dry solids, as much as 30 percent may be conditioning chemical solids and/or fly ash. Thus, the corrected dry solids achieved is about 35 percent. This is a significantly better product than what the other mechanical methods produce.

Ultimate Disposal

After all possible sludge treatment has been accomplished, a residual sludge remains, which must be sent to ultimate disposal or used in a beneficial manner.

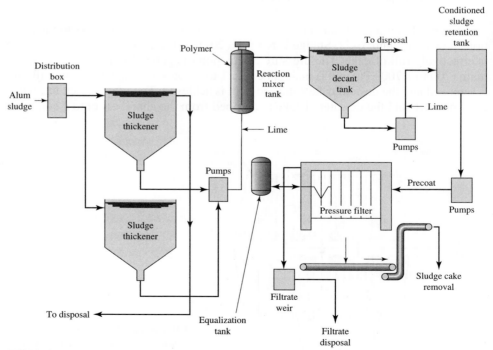

FIGURE 6-53
Pressure plate filter process for alum sludge treatment.

Of the many theoretical alternatives for ultimate disposal, only three are of practical interest:

1. Co-disposal with sewage sludge
2. Landfilling
3. Beneficial use

Due to the increasing costs and decreasing availability of landfills, beneficial use options are becoming more popular. Beneficial use options include land application to agricultural or forest lands, soil replacement for turf farms, additives to yard waste or biosolids compost, top soil additions, and brick or cement manufacturing. For detailed information on beneficial use options refer to: *Commercial Application and Marketing of Water Plant Residuals* by Cornwell et al. (2000)

6-11 CHAPTER REVIEW

When you have completed studying this chapter, you should be able to do the following without the aid of your textbook or notes:

1. List the four categories of water quality for drinking water.

2. List the four categories of physical characteristics.

3. Select the appropriate category of chemical standard for a given constituent, for example, zinc—esthetics, iron—esthetics/economics, nitrates—toxicity.

4. Define pathogen, SDWA, MCL, VOC, SOC, DBP, THM, and SWTR.

5. Identify the microorganism group used as an indicator of fecal contamination of water and explain why it was selected.

6. Sketch a water softening plant and a filtration plant, labeling all of the parts and explaining their functions.

7. Define the Schulze-Hardy rule and use it to explain the effectiveness of ions of differing valence in coagulation (Figures 6-6 and 6-7).

8. Explain the significance of alkalinity in coagulation.

9. Differentiate between coagulation and flocculation.

10. Write the reaction chemistry of alum and ferric chloride when alkalinity is present and when no alkalinity is present.

11. Explain the effect of pH on alum and ferric chloride solubility.

12. Explain how to conduct a jar test to obtain an optimum coagulant dose.

13. List the four basic types of coagulant aids; explain how each aid works and when it should be employed.

14. Define hardness in terms of the chemical constituents that cause it and in terms of the results as seen by the users of hard water.

15. Using diagrams and chemical reactions, explain how water becomes hard.

16. Given the total hardness and alkalinity, calculate the carbonate hardness and noncarbonate hardness.

17. Write the general equations for softening by ion exchange and by chemical precipitation.

18. Explain the significance of alkalinity in lime-soda softening.

19. Calculate the theoretical detention time or volume of tank if you are given the flow rate and the volume or detention time.

20. Explain how an upward-flow sedimentation tank (upflow clarifier) works, using a vector arrow diagram of a settling particle.

21. Define overflow rate in terms of liquid flow and settling-basin geometry, and state its units.

22. Using a vector arrow diagram, show how, in a horizontal-flow clarifier, a particle with a settling velocity that is less than the overflow rate may be captured (see Figure 6-34).

23. Calculate the percent of particles retained in a settling basin given the overflow rate, settling velocity, and basin flow scheme (that is, horizontal flow or upward flow).

24. Explain the difference between Type I, Type II, and Type III sedimentation.

25. Compare slow sand filters, rapid sand filters, and dual media filters with respect to operating procedures and loading rates.

26. Explain how a rapid sand filter is cleaned.

27. Sketch and label a rapid sand filter identifying the following pertinent features: inlet main, outlet main, washwater outlet, collection laterals, support media (graded gravel), graded filter sand, and backwash troughs.

28. Define effective size and uniformity coefficient and explain their use in designing a rapid sand filter.

29. Explain why a disinfectant that has a residual is preferable to one that does not.

30. Write the equations for the dissolution of chlorine gas in water and the subsequent dissociation of hypochlorous acid.

31. Explain the difference between free available chlorine and combined available chlorine, and state which is the more effective disinfectant.

32. Sketch a breakpoint chlorination curve and label the axes, breakpoint, and regions of predominantly combined and predominantly free residual.

33. Define the terms thickening, conditioning, and dewatering.

34. List and describe three methods of nonmechanical dewatering of sludge.

35. List and describe four mechanical methods of dewatering of sludge.

With the aid of this text, you should be able to do the following:

36. Calculate consumption of alkalinity for a given dose of alum or ferric chloride.

37. Calculate the production of CO_2, SO_4^{2-}, or acid for a given dose of alum or ferric chloride.

38. Calculate the amount of lime (as $CaCO_3$) that must be added if insufficient alkalinity is present in order to neutralize the acid produced by the addition of alum or ferric chloride to a given water.

39. Estimate the amount of lime and soda ash required to soften water of a stated composition.

40. Calculate the fraction of the "split" for a lime-soda softening system or an ion-exchange softening system.

41. Size a rapid-mix and flocculation basin for a given type of water treatment plant and determine the required power input.

42. Design a mixer system for either a rapid-mix or flocculation basin.

43. Size a sedimentation basin and estimate the required weir length.

44. Perform a grain size analysis and determine the effective size and uniformity coefficient.

45. Size a rapid sand filter and determine the clean-sand head loss and the depth of the expanded bed during backwash.

6-12   PROBLEMS

6-1. The following mineral analysis was reported for a water sample taken from Well No. 1 at the Eastwood Manor Subdivision near McHenry, Illnois (Woller and Sanderson, 1976a).

Well No. 1, Lab No. 02694, November 9, 1971

Iron	0.2	Silica (SiO_2)	20.0
Manganese	0.0	Fluoride	0.35
Ammonium	0.5	Boron	0.1
Sodium	4.7	Nitrate	0.0
Potassium	0.9	Chloride	4.5
Calcium	67.2	Sulfate	29.0
Magnesium	40.0	Alkalinity	284.0 as $CaCO_3$
Barium	0.5	pH	7.6 units

Note: All reported as "mg/L as the ion" unless stated otherwise.

Determine the total, carbonate, and noncarbonate hardness in mg/L as $CaCO_3$ using the predominant polyvalent cation definition in Section 6-3.

Answers: TH = 332.8 mg/L as $CaCO_3$
CH = 284.0 mg/L as $CaCO_3$
NCH = 48.8 mg/L as $CaCO_3$

6-2. Calculate the total, carbonate, and noncarbonate hardness in Problem 6-1 using all of the polyvalent cations. What is the percent error in using only the predominant cations?

6-3. The following mineral analysis was reported for a water sample taken from Well No. 1 at Magnolia, Illinois (Woller and Sanderson, 1976b). Determine the total, carbonate and noncarbonate hardness in mg/L as $CaCO_3$ using the predominant polyvalent cation definition of hardness.

Well No. 1, Lab No. B109535, April 23, 1973

Iron	0.42	Zinc	0.01
Manganese	0.04	Silica (SiO_2)	20.0
Ammonium	11.0	Fluoride	0.3
Sodium	78.0	Boron	0.3
Potassium	2.6	Nitrate	0.0
Calcium	78.0	Chloride	9.0
Magnesium	32.0	Sulfate	0.0
Barium	0.5	Alkalinity	494.0 as $CaCO_3$
Copper	0.01	pH	7.7 units

Note: All reported as "mg/L as the ion" unless stated otherwise.

6-4. The following mineral analysis was reported for Michigan State University well water (MDEQ, 1979). Determine the total, carbonate, and noncarbonate hardness in mg/L as $CaCO_3$ using the predominant polyvalent cation definition of hardness. *Note:* All units are mg/L as the ion.

Michigan State University Well Water

Fluoride	1.1	Silica (SiO_2)	3.4
Chloride	4.0	Bicarbonate	318.0
Nitrate	0.0	Sulfate	52.0
Sodium	14.0	Iron	0.5
Potassium	1.6	Manganese	0.07
Calcium	96.8	Zinc	0.27
Magnesium	30.4	Barium	0.2

6-5. An analysis of bottled water from the Kool Artesian Water Bottling Company is listed below. Determine the total, carbonate, and noncarbonate hardness in mg/L as $CaCO_3$ using the predominant polyvalent cation definition of hardness. *Note:* All units are mg/L as the ion unless otherwise stated. (*Hint:* use the solution to Problem 5-37 to find the bicarbonate concentration.)

Kool Artesian Water

Calcium	37.0	Silica	11.5
Magnesium	18.1	Sulfate	5.0
Sodium	2.1	Potassium	1.6
Fluoride	0.1	Zinc	0.02
Alkalinity	285.0 mg/L as CaCO₃		
pH	7.6 units		

6-6. Shown below are the results of water quality analyses of the Thames River in London. If the water is treated with 60.00 mg/L of alum to remove turbidity, how much alkalinity will remain? Ignore side reactions with phosphorus and assume all the alkalinity is HCO_3^-.

Thames River, London

Constituent	Expressed as	Milligrams per liter
Total hardness	$CaCO_3$	260.0
Calcium hardness	$CaCO_3$	235.0
Magnesium hardness	$CaCO_3$	25.0
Total iron	Fe	1.8
Copper	Cu	0.05
Chromium	Cr	0.01
Total alkalinity	$CaCO_3$	130.0
Chloride	Cl	52.0
Phosphate (total)	PO_4	1.0
Silica	SiO_2	14.0
Suspended solids		43.0
Total solids		495.0
pH[a]		7.4

[a]Not in mg/L.

Answer: Alkalinity remaining = 99.69 or 100 mg/L as CaCO₃

6-7. Shown below are the results of water quality analyses of the Mississippi River at Baton Rouge, Louisiana. If the water is treated with 30.00 mg/L of ferric chloride for turbidity coagulation, how much alkalinity will remain? Ignore the side reactions with phosphorus and assume all the alkalinity is HCO_3^-.

Mississippi River, Baton Rouge, Louisiana

Constituent	Expressed as	Milligrams per liter
Total hardness	$CaCO_3$	164.0
Calcium hardness	$CaCO_3$	108.0
Magnesium hardness	$CaCO_3$	56.0
Total iron	Fe	0.9
Copper	Cu	0.01
Chromium	Cr	0.03
Total alkalinity	$CaCO_3$	136.0
Chloride	Cl	32.0
Phosphate (total)	PO_4	3.0
Silica	SiO_2	10.0
Suspended solids		29.9
Turbidity[a]	NTU	12.0
pH[a]		7.6

[a]Not in mg/L.

6-8. Shown below are the results of water quality analyses of Crater Lake at Mount Mazama, Oregon. If the water is treated with 40.00 mg/L of alum for turbidity coagulation, how much alkalinity will remain? Assume all the alkalinity is HCO_3^-.

Crater Lake, Mount Mazama, Oregon

Constituent	Expressed as	Milligrams per liter
Total hardness	$CaCO_3$	28.0
Calcium hardness	$CaCO_3$	19.0
Magnesium hardness	$CaCO_3$	9.0
Total iron	Fe	0.02
Sodium	Na	11.0
Total alkalinity	$CaCO_3$	29.5
Chloride	Cl	12.0
Sulfate	SO_4	12.0
Silica	SiO_2	18.0
Total dissolved solids		83.0
pH[a]		7.2

[a]Not in mg/L.

6-9. Prepare a bar chart of the water described in Problem 6-1. (*Note:* Valences may be found in Appendix A.) Because all of the constituents were not analyzed, an ion balance is not achieved.

6-10. Prepare a bar chart of the water described in Problem 6-3. Because all of the constituents were not analyzed, an ion balance is not achieved.

6-11. Prepare a bar chart of the water described in Problem 6-4. Because all of the constituents were not analyzed, an ion balance is not achieved.

6-12. Prepare a bar chart of the Lake Michigan water analysis shown below. Because all of the constituents were not analyzed, an ion balance is not achieved. For the estimate of the CO_2 concentration, ignore the carbonate alkalinity.

Lake Michigan at Grand Rapids, MI, Intake

Constituent	Expressed as	Milligrams per liter
Total hardness	$CaCO_3$	143.0
Calcium	Ca	38.4
Magnesium	Mg	11.4
Total iron	Fe	0.10
Sodium	Na	5.8
Total alkalinity	$CaCO_3$	119
Bicarbonate alkalinity	$CaCO_3$	115
Chloride	Cl	14.0
Sulfate	SO_4	26.0
Silica	SiO_2	1.2
Total dissolved solids		180.0
Turbidity[a]	NTU	3.70
pH[a]		8.4

[a]Not in mg/L.

6-13. Determine the lime and soda ash dose, in mg/L as $CaCO_3$, to soften the following water to a final hardness of 80.0 mg/L as $CaCO_3$. The ion concentrations reported below are all mg/L as $CaCO_3$.

$$Ca^{2+} = 120.0$$
$$Mg^{2+} = 30.0$$
$$HCO_3^- = 70.0$$
$$CO_2 = 10.0$$

Answers: Total lime addition = 100 mg/L as $CaCO_3$
Total soda ash addition = 40 mg/L as $CaCO_3$

6-14. What amount of lime and/or soda ash, in mg/L as $CaCO_3$, is required to soften the Village of Lime Ridge's water to 80.0 mg/L hardness as $CaCO_3$.

Compound	Concentration, mg/L as $CaCO_3$
CO_2	4.6
Ca^{2+}	257.9
Mg^{2+}	22.2
HCO_3^-	248.0
SO_4^{2-}	32.1

6-15. Determine the lime and soda ash dose, in mg/L as CaO and Na_2CO_3, to soften the following water to a final hardness of 80.0 mg/L as $CaCO_3$. The ion concentrations reported below are all mg/L as $CaCO_3$. Assume the lime is 90 percent pure and the soda ash is 97 percent pure.

$$Ca^{2+} = 210.0$$
$$Mg^{2+} = 23.0$$
$$HCO_3^- = 165.0$$
$$CO_2 = 5.0$$

Answers: Total lime addition = 118 mg/L as CaO

Total soda ash addition = 31 mg/L as Na_2CO_3

6-16. Determine the lime and soda ash dose, in mg/L as $CaCO_3$ to soften the following water to a final hardness of 70.0 mg/L as $CaCO_3$. The ion concentrations reported below are all as $CaCO_3$.

$$Ca^{2+} = 220.0$$
$$Mg^{2+} = 75.0$$
$$HCO_3^- = 265.0$$
$$CO_2 = 17.0$$

Answers: Total lime addition = 352 mg/L as $CaCO_3$;

Add no soda ash

6-17. What amount of lime and/or soda ash, in mg/L as $CaCO_3$, is required to soften the Village of Sarepta's water to a hardness of 80.0 mg/L as $CaCO_3$.

Compound	Concentration, mg/L as $CaCO_3$
CO_2	39.8
Ca^{2+}	167.7
Mg^{2+}	76.3
HCO_3^-	257.9
SO_4^{2-}	109.5

6-18. Determine the lime and soda ash dose, in mg/L as CaO and Na_2CO_3, to soften the following water to a final hardness of 80.0 mg/L as $CaCO_3$. The ion concentrations reported below are all as $CaCO_3$. Assume the lime is 93 percent pure and the soda ash is 95 percent pure.

$$Ca^{2+} = 137.0$$
$$Mg^{2+} = 56.0$$
$$HCO_3^- = 128.0$$
$$CO_2 = 7.0$$

6-19. What amount of lime and/or soda ash, in mg/L as $CaCO_3$, is required to soften the Village of Zap's water to 80.0 mg/L hardness as $CaCO_3$?

Compound	Concentration, mg/L as $CaCO_3$
CO_2	44.2
Ca^{2+}	87.4
Mg^{2+}	96.3
HCO_3^-	204.6
SO_4^{2-}	73.8

6-20. Determine the lime and soda ash dose, in mg/L as $CaCO_3$, to soften the water described in Problem 6-1 to a final hardness of 100.0 mg/L as $CaCO_3$.

Answers: $CO_2 = 44.2$ mg/L as $CaCO_3$

Total lime addition $= 481$ mg/L as $CaCO_3$

Add no soda ash

6-21. Determine the lime and soda ash dose, in mg/L as $CaCO_3$, to soften the water described in Problem 6-6 to a final hardness of 90.0 mg/L as $CaCO_3$.

6-22. The Village of Galena wants to use a softening process to remove lead from their water. The water analysis is shown next page. Determine the lime and soda ash dose, in mg/L as CaO and Na_2CO_3, to soften the following water to a final hardness of 80.0 mg/L as $CaCO_3$. The ion concentrations reported below are all as the ion. Assume the lime is 93 percent pure and the soda ash is 95 percent pure.

Village of Galena

Constituent	Expressed as	Milligrams per liter
Calcium	Ca	177.8
Magnesium	Mg	16.2
Total iron	Fe	0.20
Lead[a]	Pb	20[a]
Sodium	Na	4.9
Carbonate alkalinity	$CaCO_3$	0.0
Bicarbonate alkalinity	$CaCO_3$	276.6
Chloride	Cl	0.0
Sulfate	SO_4	276.0
Silica	SiO_2	1.2
Total dissolved solids		667
pH[b]		8.2

[a]Parts per billion.

[b]Not in mg/L.

Will the softening process remove lead?

6-23. Determine the lime and soda ash dose, in mg/L as $CaCO_3$, to soften the following water to a final hardness of 80.0 mg/L as $CaCO_3$. If the price of lime, purchased as CaO, is $100.00 per megagram (Mg), and the price of soda ash, purchased as Na_2CO_3 is $200.00 per Mg, what is the annual chemical cost of treating 0.500 m^3/s of this water?

Assume the lime and soda ash are 100 percent pure. The ion concentrations reported below are all mg/L as $CaCO_3$.

$$Ca^{2+} = 200.0$$
$$Mg^{2+} = 100.0$$
$$HCO_3^- = 150.0$$
$$CO_2 = 22.0$$

Answers: Lime = 272.0 mg/L as $CaCO_3$

Soda = 110.0 mg/L as $CaCO_3$

Total annual cost = $607,703.25, or $608,000

6-24. Determine the lime and soda ash dose, in mg/L as $CaCO_3$, to soften the following water to a final hardness of 120.0 mg/L as $CaCO_3$. If the price of lime, purchased as CaO, is $61.70 per megagram (Mg), and the price of soda ash, purchased as Na_2CO_3, is $172.50 per Mg, what is the annual chemical cost of treating 1.35 m^3/s of this water? Assume the lime is 87 percent pure and the soda ash is 97 percent pure. The ion concentrations reported below are all mg/L as $CaCO_3$.

$$Ca^{2+} = 293.0$$
$$Mg^{2+} = 55.0$$
$$HCO_3^- = 301.0$$
$$CO_2 = 3.0$$

6-25. Determine the lime and soda ash dose, in mg/L as $CaCO_3$, to soften the Hardin, Illinois, water to a final hardness of 95.00 mg/L as $CaCO_3$ (Woller, 1975). Using the price and purity information supplied in Problem 6-23, determine the annual chemical cost of treating 0.150 m^3/s of this water.

Well No. 2, Hardin, IL

Iron	0.10	Zinc	0.13
Manganese	0.64	Silica (SiO_2)	21.6
Ammonium	0.38	Fluoride	0.3
Sodium	21.8	Boron	0.38
Potassium	3.0	Nitrate	8.4
Calcium	102.0	Chloride	32.0
Magnesium	45.2	Sulfate	65.0
Copper	0.01	Alkalinity	344.0 as $CaCO_3$
pH	7.2 units		

Note: All reported as "mg/L as the ion" unless stated otherwise.

6-26. Determine the lime and soda ash dose, in mg/L as $CaCO_3$, to soften the following water to a final hardness of 90.0 mg/L as $CaCO_3$. If the price of lime, purchased as CaO, is $61.70 per megagram (Mg), and the price of soda ash, purchased as Na_2CO_3, is $172.50 per Mg. What is the annual chemical cost of treating 0.050 m^3/s of this water? Assume the lime is 90 percent pure and the soda ash is 97 percent pure. The ion concentrations reported below are all mg/L as $CaCO_3$.

$$Ca^{2+} = 137.0$$
$$Mg^{2+} = 40.0$$
$$HCO_3^- = 197.0$$
$$CO_2 = 9.0$$

6-27. Design a split treatment softening process (flow scheme/split, chemical dose in mg/L as $CaCO_3$) for the following water. The final hardness must be ≤ 120 mg/L as $CaCO_3$. Compounds are given in mg/L as the ion stated unless otherwise specified:

CO_2	42.7	HCO_3^-	344.0 mg/L as $CaCO_3$
Ca^{2+}	102.0	SO_4^{2-}	65.0
Mg^{2+}	45.2	Cl^-	32.0
Na^+	21.8		

6-28. Given the following water (all in meq/L), design a process to soften the water (flow scheme/split; amount of lime and/or soda required in mg/L as CaO and Na_2CO_3 respectively) and find the final hardness. The final hardness must be ≤ 120 mg/L as $CaCO_3$.

CO_2	0.40	Mg^{2+}	1.12
Ca^{2+}	2.16	HCO_3^-	2.72

6-29. Design a softening process for the City of What Cheer to achieve a magnesium concentration of 40 mg/L as $CaCO_3$ and a final total hardness less than 120 mg/L as $CaCO_3$. Show the flow scheme, calculated split, amount of lime and/or soda in mg/L as $CaCO_3$, and the final hardness. The water analysis is shown below.

Compound	Concentration, mg/L as $CaCO_3$
CO_2	39.8
Ca^{2+}	167.7
Mg^{2+}	76.3
HCO_3^-	257.9
SO_4^{2-}	109.5

6-30. What is the volume required for a rapid-mix basin that is to be used to treat 0.05 m^3/s of water if the detention time is 10 seconds?

Answer: 0.5 m^3

6-31. Two parallel flocculation basins are to be used to treat a water flow of 0.150 m^3/s. If the design detention time is 20 minutes, what is the volume of each tank?

6-32. Two sedimentation tanks operate in parallel. The combined flow to the two tanks is 0.1000 m^3/s. The volume of each tank is 720 m^3. What is the detention time of each tank?

6-33. Determine the power input required for the tank designed in Problem 6-30 if the water temperature is 20°C and the velocity gradient is 700 s^{-1}.

Answer: 245.49 or 250 W or 0.25 kW

6-34. The flocculation tanks in Problem 6-31 were designed for an average velocity gradient of 36 s^{-1} and a water temperature of 17°C. What power input is required?

6-35. What power input is required for Problem 6-33 if the water temperature falls to 10°C?

6-36. What power input is required for Problem 6-34 if the water temperature falls to 8°C?

6-37. Complete Example 6-11 by computing the rotational speed of the impellers in compartments 2 and 3.

6-38. The town of Eau Gaullie has requested proposals for a new coagulation water treatment plant. The design flow for the plant is 0.1065 m^3/s. The average annual water temperature is 19°C. The following design assumptions for a rapid-mix tank have been made:

1. Number of tanks = 1 (with 1 backup spare)
2. Tank configuration: circular with liquid depth = 2 × diameter
3. Detention time = 10 s
4. Velocity gradient = 800 s^{-1}
5. Impeller type: turbine, 6 flat blades, N_p = 5.7
6. Available impeller diameters: 0.45, 0.60, and 1.2 m
7. Assume $B = \dfrac{1}{3}H$

Design the rapid-mix system by providing the following:

1. Water power input in kW
2. Tank dimensions in m
3. Diameter of the impeller in m
4. Rotational speed of impeller in rpm

> *Answers:* P = 0.700 kW
> Diameter = 0.88 m; depth = 1.76 m
> Impeller diameter = 0.45 m
> Rotational speed = 399 or 400 rpm

6-39. Laramie is planning for a new softening plant. The design flow is 0.168 m^3/s. The average water temperature is 5°C. The following design assumptions for a rapid-mix tank have been made:

1. Tank configuration: square plan with depth = width
2. Detention time = 5 s
3. Velocity gradient 700 s^{-1}
4. Impeller type: turbine, 6 flat blades, N_P = 5.7

5. Available impeller diameters: 0.45, 0.60, and 1.2 m

6. Assume $B = \frac{1}{3}H$

Design the rapid-mix system by providing the following:

1. Number of tanks
2. Water power input in kW
3. Tank dimensions in m
4. Diameter of the impeller in m
5. Rotational speed of impeller in rpm

6-40. Your boss has assigned you the job of designing a rapid-mix tank for the new water treatment plant for the town of Waffle. The design flow rate is 0.050 m³/s. The average water temperature is 8°C. The following design assumptions for a rapid-mix tank have been made:

1. Number of tanks = 1 (with 1 backup)
2. Tank configuration: circular with liquid depth = 1.0 m
3. Detention time = 5 s
4. Velocity gradient = 750 s^{-1}
5. Impeller type: turbine, 6 flat blades, $N_P = 3.6$
6. Available impeller diameters: 0.25, 0.50, and 1.0 m
7. Assume $B = \frac{1}{3}H$

Design the rapid-mix system by providing the following:

1. Water power input in kW
2. Tank dimensions in m
3. Diameter of the impeller in m
4. Rotational speed of impeller in rpm

6-41. Continuing the preparation of the proposal for the Eau Gaullie treatment plant (Problem 6-38), design the flocculation tank by providing the following for the first two compartments only:

1. Water power input in kW
2. Tank dimensions in m
3. Diameter of the impeller in m
4. Rotational speed of impeller in rpm

Use the following assumptions:

1. Number of tanks = two
2. Tapered G in three compartments: 90, 60, and 30 s^{-1}

3. $G\theta = 120,000$
4. Compartment length = width = depth
5. Impeller type: axial-flow impeller, three blades, $N_P = 0.31$
6. Available impeller diameters: 1.0, 1.5, and 2.0 m
7. Assume $B = \dfrac{1}{3}H$

   Answers for first compartment only:
   $P = 295.31$, or 295 W or 0.295 kW
   $L = W = D = 3.3$ m
   Impeller diameter = 1.5 m
   Rotational speed = 30 rpm

6-42. Continuing the preparation of the proposal for the Laramie treatment plant (Problem 6-39), design the flocculation tank by providing the following for the first two compartments only:

1. Water power input in kW
2. Tank dimensions in m
3. Diameter of the impeller in m
4. Rotational speed of impeller in rpm

Use the following assumptions:

1. Number of tanks = two
2. Tapered G in three compartments: 90, 60, and 30 s^{-1}
3. $Gt = 120,000$
4. Compartment length = width = depth
5. Impeller type: axial-flow impeller, three blades, $N_P = 0.40$
6. Available impeller diameters: 1.0, 1.8, and 2.4 m
7. Assume $B = \dfrac{1}{3}H$

6-43. Continuing the preparation of the proposal for the Waffle treatment plant (Problem 6-40), design the flocculation tank by providing the following for the first two compartments only:

1. Water power input in kW
2. Tank dimensions in m
3. Diameter of the impeller in m
4. Rotational speed of impeller in rpm

Use the following assumptions:

1. Number of tanks = 1 (with 1 backup)
2. Tapered G in three compartments: 60, 50, and 20 s^{-1}

3. Detention time = 30 min

4. Depth = 3.5 m

5. Impeller type: axial-flow impeller, three blades, $N_p = 0.43$

6. Available impeller diameters: 1.0, 1.5, and 2.0 m

7. Assume $B = \dfrac{1}{3}H$

6-44. If the settling velocity of a particle is 0.70 cm/s and the overflow rate of a horizontal flow clarifier is 0.80 cm/s, what percent of the particles are retained in the clarifier?

Answer: 88 percent

6-45. If the settling velocity of a particle is 2.80 mm/s and the overflow rate of an upflow clarifier is 0.560 cm/s, what percent of the particles are retained in the clarifier?

6-46. If the settling velocity of a particle is 0.30 cm/s and the overflow rate of a horizontal flow clarifier is 0.25 cm/s, what percent of the particles are retained in the clarifier?

6-47. If the flow rate of the original plant in Problem 6-44 is increased from 0.150 m³/s to 0.200 m³/s, what percent removal of particles would be expected?

6-48. If the flow rate of the original plant in Problem 6-45 is doubled, what percent removal of particles would be expected?

6-49. If the flow rate of the original plant in Problem 6-46 is doubled, what percent removal of particles would be expected?

6-50. If a 1.0-m³/s flow water treatment plant uses 10 sedimentation basins with an overflow rate of 15 m³/d · m², what should be the surface area (m²) of each tank?

Answer: 576.0 m²

6-51. Assuming a conservative value for an overflow rate, determine the surface area (in m²) of each of two sedimentation tanks that together must handle a flow of 0.05162 m³/s of lime softening floc.

6-52. Repeat Problem 6-51 for an alum or iron floc.

6-53. Two sedimentation tanks operate in parallel. The combined flow to the two tanks is 0.1000 m³/s. The depth of each tank is 2.00 m and each has a detention time of 4.00 h. What is the surface area of each tank and what is the overflow rate of each tank in m³/d · m²?

6-54. For a flow of 0.8 m³/s, how many rapid sand filter boxes of dimensions 10 m × 20 m are needed for a loading rate of 110 m³/d · m²?

Answer: 4 filters (rounding to nearest even number)

6-55. If a dual-media filter with a loading rate of 300 $m^3/d \cdot m^2$ were built instead of the standard filter in Problem 6-54, how many filter boxes would be required?

6-56. The water flow meter at the Troublesome Creek water plant is on the blink. The plant superintendent tells you the four dual-media filters (each 5.00 m × 10.0 m) are loaded at a velocity of 280 m/d. What is the flow rate through the filters in m^3/s?

6-57. A plant expansion is planned for Urbana (Example 6-17). The new design flow rate is 1.0 m^3/s. A deep-bed monomedia filter with a design loading rate of 600 $m^3/d \cdot m^2$ of filter is to be used. If each filter box is limited to 50 m^2 of surface area, how many filter boxes will be required? Check the design loading with one filter box out of service. Propose an alternative design if the design loading rate is exceeded with one filter box out of service.

6-58. The Orono Sand and Gravel Company has made a bid to supply sand for Eau Gaullie's new sand filter. The request for bids stipulated that the sand have an effective size in the range 0.35 to 0.55 mm and a uniformity coefficient in the range 1.3 to 1.7. Orono supplied the following sieve analysis as evidence that their sand will meet the specifications. Perform a grain size analysis (semi-log plot) and determine whether or not the sand meets the specifications. Use a spreadsheet program you have written to plot the data and fit a curve.

U.S. Standard Sieve No.	Mass percent retained
8	0.0
12	0.01
16	0.39
20	5.70
30	25.90
40	44.00
50	20.20
70	3.70
100	0.10

6-59. The Lexington Sand and Gravel Company has made a bid to supply sand for Laramie's new sand filter. The request for bids stipulated that the sand have an effective size in the range 0.35 to 0.55 mm and a uniformity coefficient in the range 1.3 to 1.7. Lexington supplied the following sieve analysis (sample size = 500.00 g) as evidence that its sand will meet the specifications. Perform a grain size analysis (log-log plot) and determine whether or not the sand meets the specifications. Use a spreadsheet program you have written to plot the data and fit a curve.

U.S. Standard Sieve No.	Mass retained, g
12	0.00
16	2.00
20	65.50
30	272.50
40	151.0
50	8.925
70	0.075

6-60. Rework Example 6-18 with the 70, 100, and 140 sieve fractions removed. Assume original sample contained 100 g.

6-61. The rapid sand filter being designed for Eau Gaullie has the characteristics and sieve analysis shown below. Determine the head loss for the clean filter bed in a stratified condition.

Depth = 0.60 m

Filter loading = 120 $m^3/d \cdot m^2$

Sand specific gravity = 2.50

Shape factor = 1.00

Stratified bed porosity = 0.42

Water temperature = 19°C

Sand Analysis

U.S. Standard Sieve No.	Mass percent retained
8–12	0.01
12–16	0.39
16–20	5.70
20–30	25.90
30–40	44.00
40–50	20.20
50–70	3.70
70–100	0.10

6-62. Determine the height of the expanded bed for the sand used in Problem 6-61 if the backwash rate is 1,000 m/d. Assume Equation 6-55 applies.

6-63. The rapid sand filter being designed for Laramie has the characteristics shown below. Determine the head loss for the clean filter bed in a stratified condition.

Depth $= 0.75$ m
Filter loading $= 230$ m^3/d $\cdot$ m^2
Sand specific gravity $= 2.80$
Shape factor $= 0.91$
Stratified bed porosity $= 0.50$
Water temperature $= 5°C$

Sand Analysis

U.S. Standard Sieve No.	Mass percent retained
8–12	0.00
12–16	0.40
16–20	13.10
20–30	54.50
30–40	30.20
40–50	1.785
50–70	0.015

6-64. Determine the minimum fluidization velocity and the height of the backwash troughs above the sand used in Problem 6-63. Assume Equation 6-55 applies.

6-65. What is the equivalent percent reduction for a 2.5 log reduction of *Giardia lambia*?

6-66. What is the log reduction of *Giardia lambia* that is equivalent to 99.96 percent reduction?

6-13 DISCUSSION QUESTIONS

6-1. Which of the chemicals added to treat a surface water aids in making the water palatable?

6-2. Microorganisms play a role in the formation of hardness in groundwater. True or false? Explain.

6-3. If there is no bicarbonate present in a well water that is to be softened to remove magnesium, which chemicals must you add?

6-4. Use a scale drawing to sketch a vector diagram of a horizontal-flow sedimentation tank that shows how 25 percent of the particles with a settling velocity one-quarter that of the overflow rate will be removed.

6-5. In the United States, chlorine is preferred as a disinfectant over ozone because it has a residual. Why is the presence of a residual important?

6-6. A new water softening plant is being designed for Lubbock, Texas. The climate is dry and land is readily available at a reasonable cost. What methods of sludge dewatering would be most appropriate? Explain your reasoning.

6-14   FE EXAM FORMATTED PROBLEMS

6-1. Three parallel flow flocculation basins are to be used to treat a flow rate of 3 MGD. What is the design volume, in cubic feet, of one of these tanks if the detention time is 30 minutes?

   a. $1,117 \text{ ft}^3$ b. $2,800 \text{ ft}^3$
   c. $8,375 \text{ ft}^3$ d. $173,700 \text{ ft}^3$

6-2. Determine the number of rapid sand filters to treat a flow rate of $75.7 \times 10^3 \text{ m}^3/\text{d}$ if the design loading rate is $300 \text{ m}^3/\text{d} \cdot \text{m}^2$. The maximum dimension is 7.5 m and the length to width ratio is 1.2:1.

   a. 4 filters   b. 3 filters
   c. 6 filters   d. 5 filters

6-3. Estimate the amount of lime, in tons/d, required to soften 5 MGD of water to the practical solubility limits. The constituent concentrations are as follows:

   $CO_2 = 0.44 \text{ meq/L}$ Alkalinity $= 3.96 \text{ meq/L}$
   $Ca^{2+} = 4.76 \text{ meq/L}$ $Cl^- = 1.91 \text{ meq/L}$
   $Mg^{2+} = 1.11 \text{ meq/L}$ $SO_4^{2-} = 1.58 \text{ meq/L}$

   a. 1.15 tons/d   b. 6.4 tons/d
   c. 2.92 tons/d   d. 3.2 tons/d

6-4. Calculate the theoretical percent removal of a particle having a settling velocity of 0.25 cm/s settling in a horizontal flow clarifier with an overflow rate of 0.05 cm/s.

   a. 100%   b. 20%
   c. 50%   d. 1.25%

6-15 REFERENCES

Amirtharajah, A. (1978) "Design of Raid Mix Units," in R. L. Sanks (ed.) *Water Treatment Plant Design for the Practicing Engineer,* Ann Arbor Science, Ann Arbor, MI, pp. 132, 141, 143.

AWWA (1990) *Water Treatment Plant Design,* 2nd ed., McGraw-Hill, New York.

AWWA (1971) *Water Quality and Treatment,* 3rd ed., American Water Works Association, McGraw-Hill, New York, pp. 188, 259.

AWWA (1998) *Water Treatment Plant Design,* 3rd ed., American Water Works Association, McGraw-Hill, New York.

Blake, C. F. (1922) "The Resistance of Packing to Fluid Flow," *Transactions of American Institute of Civil Engineers,* vol. 14, p. 415.

Camp, T. R. (1946) "Sedimentation and Design of Settling Tanks," *Transactions of the American Society of Civil Engineers,* vol. 111, p. 895.

Chick, H. (1908) "Investigation of the Law of Disinfection," *Journal of Hygiene,* vol. 8, p. 92.

Cleasby, J. L., and J. H. Dillingham (1966) "Rational Aspects of Split Treatment," *Proceedings American Society of Civil Engineers, Journal Sanitary Engineering Division,* vol. 92, SA2, pp. 1–7.

Cleasby, J. L. (1972) "Filtration," in W. J. Weber (ed.) *Physiochemical Processes for Water Quality Control,* Wiley-Interscience, New York.

Cleasby, J. L., and K. S. Fan (1981) "Predicting Fluidization and Expansion of Filter Media," *Journal of the Environmental Engineering Division*, ASCE 107 (3), pp. 455–471.

Cleasby, J. L., and G. S Logsdon, "Filtration," in Letterman, R. D. (Ed.) (1999) *Water Quality and Treatment,* 5th ed., American Water Works Association, McGraw-Hill, New York, pp. 8.11–8.16.

Cornwell, D. A., and M. Bishop (1983) "Determining Velocity Gradients in Laboratory and Full-Scale Systems," *Journal of American Water Works Association,* vol. 75, September, pp. 470–475.

Cornwell, D. A., R. M. Mutter, C. Vandermeyden, (2000) "Commercial Application and Marketing of Water Plant Residuals," AWWA Research Foundation and American Water Works Association, Denver, CO.

Craun, G. F. et al. (2010) "Causes of Outbreaks Associated with Drinking Water in the United States from 1971 to 2006," *Clinical Microbiology Reviews*, July, pp 507–528.

Dharmarajah, A. H., and J. L. Cleasby (1986) "Predicting the Expansion Behavior of Filter Media," *Journal of the American Water Works Association,* December, pp. 66–76.

EE & T (2008) *Costing Analysis to Support National Drinking Water Treatment Plant Residuals Management Regulatory Options*, Environmental Engineering & Technology, Inc., Newport News, Virginia.

Fair, G. M., and J. C. Geyer (1954) *Water Supply and Wastewater Disposal,* John Wiley & Sons, New York, pp. 664–671, 678, 27–15.

Fair, G. M., J. C. Geyer, and D. A. Okun (1968) *Water and Wastewater Engineering,* vol. 2, John Wiley & Sons, Inc., New York, pp. 25–14, 27–15.

Hardy, W. B. (1900a) *Proceedings of the Royal Society of London,* vol. 66, p. 110.

Hardy, W. B. (1900b) *Z. Physik Chem.,* vol. 33, p. 385.

Hazen, A. (1892) *Annual Report of the Massachusetts State Board of Health,* Boston.

Hazen, A. (1904) "On Sedimentation," *Transactions of the American Society of Civil Engineers,* vol. 53, p. 45.

Hudson, H. E. (1981) *Water Clarification Processes, Practical Design and Evaluation,* Van Nostrand Reinhold, New York, pp. 115–117.

Jacangelo, J. G., S. Adham, and J. Laine (1997) *Membrane Filtration for Microbial Removal,* AWWA Research Foundation and American Water Works Association.

Kozeny, J. (1927) "Ueber kapillare Leitung des Wassers im Boden." *Sitzungsber Akad. Wiss.,* Vienna, Abt IIIa, vol. 136, p. 276.

Lide, D. R. (2000) *CRC Handbook of Chemistry and Physics,* 81st ed., CRC Press, Boca Raton, FL, pp. 8-111–8-112.

MDEQ (1979) *Annual Data Summary,* Michigan Department of Public Health, Lansing, MI.

Metcalf & Eddy, Inc. (2003) *Wastewater Engineering: Treatment and Reuse,* revised by G. Tchobanoglous, F. L. Burton, and H. D. Stensel, McGraw-Hill, New York, p. 1,051.

Montgomery, J. M. (1985) *Water Treatment Principles and Design,* John Wiley & Sons, New York, p. 535.

Newton, I. (1687) *Philosophiae Naturalis Principia Mathematica.*

Reynolds, O. (1883) "An Experimental Investigation of the Circumstances Which Determine Whether the Motion of Water Shall Be Direct or Sinuous and the Laws of Resistance in Parallel Channels," *Transactions of the Royal Society of London,* vol. 174.

Reynolds, T. D., and P. A. Richards (1996) *Unit Operations and Processes in Environmental Engineering,* PWS-Kent, Boston, p. 256.

Richardson, J. F., and W. N. Zaki, (1954) "Sedimentation and Fluidization, Part I, *Transactions of the Institute of Chemical Engineers* (Brit.), vol. 32, pp. 35–53.

Rose, H. E. (1945) "On the Resistance Coefficient–Reynolds Number Relationship of Fluid Flow through a Bed of Granular Material," *Proceedings of the Institute of Mechanical Engineers,* vol. 153, p. 493.

Rushton, J. H. (1952) "Mixing of Liquids in Chemical Processing," *Industrial & Engineering Chemistry,* vol. 44, no. 12, p. 2,931.

Schulze, H. (1882) *J. Prakt. Chem.,* vol. 25, p. 431.

Schulze, H. (1883) *J. Prakt. Chem.,* vol. 27, p. 320.

Steel, E. W., and T. J. McGhee (1979) *Water Supply and Sewerage,* McGraw-Hill, New York.

Stokes, G. G. (1845) *Transactions of the Cambridge Philosophical Society,* vol. 8, p. 287.

U.S. EPA (1979) *Process Design Manual, Sludge Treatment and Disposal,* U. S. Environmental Protection Agency, Washington, DC.

U.S. EPA (1991) *Guidance Manual for Compliance with Filtration and Disinfection Requirements for Public Water Systems Using Surface Waters,* Compliance and Standards Division, Office of Drinking Water, U.S. Environmental Protection Agency, NTIS Pub. No. PB93-222933.

U.S. EPA (1996) *Ultraviolet Light Disinfection Technology in Drinking Water Application—An Overview,* U.S. Environmental Protection Agency Office of Ground Water and Drinking Water, Pub. No. 811-R-96-002.

U.S. EPA (2000) *Safe Drinking Water Information System,* U.S. Environmental Protection Agency, www.epa.gov/safewater/getdata.html.

U.S. PHS (1962) *Manual of Individual Water Supply Systems,* U.S. Public Health Service Publication No. 24, U.S. Department of Health, Education and Welfare, Washington, DC.

Walker Process Equipment (1973) *Walker Process Circular Clarifiers,* Bulletin 9-w-65, Aurora, IL.

Wen, C. Y., and Y. H. Yu (1966) "Mechanics of Fluidization," *Chemical Engineering Progress Symposium Series,* vol. 62, pp. 100–111.

Woller, D. M. (1975) *Public Groundwater Supplies in Calhoun County,* Illinois State Water Survey, Publication No. 60-16, Urbana, IL.

Woller, D. M., and E. W. Sanderson (1976a) *Public Water Supplies in McHenry County,* Illinois State Water Survey, Publication No. 60-19, Urbana, IL.

Woller, D. M., and E. W. Sanderson (1976b) *Public Groundwater Supplies in Putnam County,* Illinois State Water Survey, Publication No. 60-15, Urbana, IL.

WQ&T (1999) *Water Quality and Treatment: A Handbook of Community Water Supplies,* R. D. Letterrman (ed.), American Water Works Association, McGraw-Hill, New York, pp. 23, 24.

WQ&T (2011) *Water Quality and Treatment: A Handbook of Drinking Water,* J. K. Edzwald (ed.), American Waterworks Association, McGraw-Hill, New York.

CHAPTER

7

WATER POLLUTION

7-1 INTRODUCTION

The uses we make of water in lakes, rivers, ponds, and streams is greatly influenced by the quality of the water found in them. Activities such as fishing, swimming, boating, shipping, and waste disposal have very different requirements for water quality. Water of a particularly high quality is needed for potable water supplies. In many parts of the world, the introduction of pollutants from human activity has seriously degraded water quality even to the extent of turning pristine trout streams into foul open sewers with few life forms and fewer beneficial uses.

Water quality management is concerned with the control of pollution from human activity so that the water is not degraded to the point that it is no longer suitable for intended uses. The lone frontier family, settled on the banks of the Ohio River, did not significantly degrade water quality in that mighty river even though it threw all its wastes into the river. The city of Cincinnati, however, could not discharge its untreated wastes into the Ohio River without disastrous consequences. Thus, water quality management is also the science of knowing how much is too much for a particular water body.

To know how much waste can be tolerated (the technical term is *assimilated*) by a water body, you must know the type of pollutants discharged and the manner in which they affect water quality. You must also know how water quality is affected by natural factors such as the mineral heritage of the watershed, the geometry of the terrain, and the climate of the region. A small, tumbling mountain brook will have a very different assimilative capacity than a sluggish, meandering lowland river, and lakes are different from moving waters.

Originally, the intent of water quality management was to protect the intended uses of a water body while using water as an economic means of waste disposal within the constraints of its assimilative capacity. In 1972, the Congress of the United States established that it was in the national interest to "restore and maintain the chemical, physical, and biological integrity of the nation's waters." In addition to making the water safe to drink, the Congress also established a goal of "water quality which provides for the protection and propagation of fish, shellfish, and wildlife, and provides for recreation in and on the water" (FWPCA, 1972). By understanding the impact of pollutants on water quality, the environmental engineer can properly design the treatment facilities to remove these pollutants to acceptable levels.

This chapter deals first with the major types of pollutants and their sources. In the remainder of the chapter, pollution in rivers, lakes, and groundwater is discussed, placing the emphasis on the categories of pollutants found in domestic wastewaters. For both rivers and lakes, the natural factors affecting water quality will be discussed as the basis for understanding the impact of human activities on water quality.

Applications

This chapter has both qualitative and quantitative applications. The qualitative discussion of pollutants and their sources provides a basis for developing management

---

The first edition of this chapter was written by John A. Eastman, Ph.D., of Lockwood, Jones and Beals, Inc., Kettering, OH.

strategies to control pollution before it enters surface or ground water. The quantitative discussion provides methods for:

- Evaluating the strength of a waste
- Estimating the maximum amount of pollutant that a water body can receive and still meet water quality standards
- Estimating the pumping rate that will minimize salt water intrusion into a well

7-2   WATER POLLUTANTS AND THEIR SOURCES

Point Sources

The wide range of pollutants discharged to surface waters can be grouped into broad classes, as shown in Table 7-1. Domestic sewage and industrial wastes are called *point sources* because they are generally collected by a network of pipes or channels and conveyed to a single point of discharge into the receiving water. Domestic sewage consists of wastes from homes, schools, office buildings, and stores. The term municipal sewage is used to mean domestic sewage into which industrial wastes are also discharged.

Under the Federal Water Pollution Control Act of 1972 some *animal feeding operations* (AFOs) may be designated as a point source. AFOs are designated point sources if the feeding operation may be classified as a concentrated animal feeding operation (CAFO). The regulations define AFOs as facilities where animals are fed and confined for 45 days or more in any 12-consecutive-month period, and where crops, vegetation, forage growth, or postharvest residues are not grown or sustained in the feedlot or facility. The latter requirement is to distinguish feedlots from pastures. To qualify as a CAFO (and,

TABLE 7-1
Major pollutant categories and principal sources of pollutants

	Point sources		Non-point sources	
Pollutant category	Domestic sewage	Industrial wastes	Agricultural runoff	Urban runoff
Oxygen-demanding material	X	X	X	X
Nutrients	X	X	X	X
Pathogens	X	X	X	X
Suspended solids/sediments	X	X	X	X
Salts		X	X	X
Toxic metals		X		X
Toxic organic chemicals		X	X	
Endocrine-disrupting chemicals	X	X	X	
Pharmaceuticals	X	X	X	
Personal care products	X			
Heat		X		

thus, be a point source) an AFO must meet one of three criteria (DiMura, 2003): (1) all AFOs with 1,000 animal units* or more are CAFOs, (2) AFOs with between 300 and 1,000 animal units that discharge directly or indirectly to surface waters are CAFOs (AFOs with less than 300 animal units may not be classified as CAFOs), and (3) any AFO may be designated a CAFO if it is found to contribute significantly to pollution of surface waters.

In general, point source pollution can be reduced or eliminated through waste minimization and proper wastewater treatment prior to discharge to a natural water body.

Non-Point Sources

Urban and agricultural runoff are characterized by multiple discharge points. These are called non-point sources. Often the polluted water flows over the surface of the land or along natural drainage channels to the nearest water body. Even when urban or agricultural runoff waters are collected in pipes or channels, they are generally transported the shortest possible distance for discharge, so that wastewater treatment at each outlet is not economically feasible. Much of the non-point source pollution occurs during rainstorms or spring snowmelt resulting in large flow rates that make treatment even more difficult. Non-point pollution from urban storm water and, in particular, storm water collected in *combined sewers* that carry both storm water and municipal sewage requires major engineering work to correct. The original design of combined sewers provided a flow structure that diverted excess storm water mixed with raw sewage (above the design capacity of the wastewater treatment plant) directly to the nearest river or stream. The elimination of *combined sewer overflow* (CSO) may involve not only provision of separate storm and sanitary sewers but also the provision of storm water retention basins and expanded treatment facilities to treat the storm water. This is particularly complex and expensive because the combined sewers frequently occur in the oldest, most developed portions of the community. Thus, paved streets, utilities, and commercial activities will be disrupted. The installation of combined sewers is now prohibited in the United States.

Runoff from agricultural land is a significant non-point source. Fertilizer, whether in the form of manure or commercial fertilizer, contributes nutrients. Agricultural runoff carries nutrients and toxic organic compounds. Soil erosion contributes suspended solids. Implementation of *Best Management Practices* (BMP) to reduce excess application of fertilizer and pesticides along with erosion control programs conserves the farmers economic investment while protecting the river.

Oxygen-Demanding Material

Anything that can be oxidized in the receiving water with the consumption of dissolved molecular oxygen is termed oxygen-demanding material. This material is usually biodegradable organic matter but also includes certain inorganic compounds. The consumption of *dissolved oxygen,* DO (pronounced "dee oh"), poses a threat to fish and other higher forms of aquatic life that must have oxygen to live. The critical level of DO varies greatly among species. For example, brook trout may require about

---

*Generally, one animal unit is equal to 450 kg of live animal mass.

7.5 mg/L of DO, while carp may survive at 3 mg/L. As a rule, the most desirable commercial and game fish require high levels of dissolved oxygen. Oxygen-demanding materials in domestic sewage come primarily from human waste and food residue. Particularly noteworthy among the many industries that produce oxygen-demanding wastes are the food processing and paper industries. Almost any naturally occurring organic matter, such as animal droppings, crop residues, or leaves, that get into the water from non-point sources, contribute to the depletion of DO.

Nutrients

Nitrogen and phosphorus, two nutrients of primary concern, are considered pollutants because they are too much of a good thing. All living things require these nutrients for growth. Thus, they must be present in rivers and lakes to support the natural food chain.* Problems arise when nutrient levels become excessive and the food web is grossly disturbed, which causes some organisms to proliferate at the expense of others. As will be discussed in a later section, excessive nutrients often lead to large growths of algae, which in turn become oxygen-demanding material when they die and settle to the bottom. Some major sources of nutrients are phosphorus-based detergents, fertilizers, and food-processing wastes.

Pathogenic Organisms

Microorganisms found in wastewater include bacteria, viruses, and protozoa excreted by diseased persons or animals. When discharged into surface waters, they make the water unfit for drinking (that is, nonpotable). If the concentration of pathogens is sufficiently high, the water may also be unsafe for swimming and fishing. Certain shellfish can be toxic because they concentrate pathogenic organisms in their tissues, making the toxicity levels in the shellfish much greater than the levels in the surrounding water.

Cholera and typhoid are *endemic* diseases in the world with over 384,000 cases of cholera and 16 million cases of typhoid per year. These killers cause 20,000 and 600,000 deaths per year respectively (*Population Reports,* 1998). Yet, in the United States, they are unheard of outside of a historic context. The widespread disease-causing organisms in the United States are *Giardia lambia* and *Cryptosporidium parvum*. These are protozoan pathogens from both human and animal sources.

Bacteria that have developed immunity to antibiotics are now being found in natural waters (Ash et al., 1999; Sternes, 1999; Bennett and Kramer, 1999). The resistance of the bacteria appearing in natural waters is an ominous prelude to the future effectiveness of antibiotics.

Suspended Solids

Organic and inorganic particles that are carried by the wastewater into a receiving water are termed *suspended solids* (SS). When the speed of the water is reduced by flowing into a pool or a lake, many of these particles settle to the bottom as sediment.

---

*In simplistic terms, a food chain is the collection of interrelated organisms in which the lower levels are the "eatees" and the upper levels are the "eaters."

In common usage, the word sediment also includes eroded soil particles that are being carried by water even if they have not yet settled. Colloidal particles, that do not settle readily, cause the turbidity found in many surface waters. Organic suspended solids may also exert an oxygen demand. Inorganic suspended solids are discharged by some industries but result mostly from soil erosion, which is particularly bad in areas of logging, strip mining, and construction activity. As excessive sediment loads are deposited into lakes and reservoirs, the usefulness of the water is reduced. Even in rapidly moving mountain streams, sediment from mining and logging operations has destroyed many living places (*ecological habitats*) for aquatic organisms. For example, salmon eggs can only develop and hatch in stream beds of loose gravel. As the pores between the pebbles are filled with sediment, the eggs suffocate and the salmon population is reduced.

Salts

Although most people associate salty water with oceans and salt lakes, all water contains some salt. These salts are often measured by evaporation of a filtered water sample. The salts and other things that don't evaporate are called *total dissolved solids* (TDS). A problem arises when the salt concentration in normally fresh water increases to the point where the natural population of plants and animals is threatened or the water is no longer useful for public water supplies or irrigation. High concentrations of salts are discharged by many industries, and the use of salt on roads during the winter causes high salt levels in urban runoff, especially during the spring snowmelt. Of particular concern in arid regions, where water is used extensively for irrigation, is that the water picks up salts every time it passes through the soil on its way back to the river. In addition, evapotranspiration causes the salts to be further concentrated. Thus, the salt concentration continuously increases as the water moves downstream. If the concentration gets too high, crop damage or soil poisoning can result.

Pharmaceuticals and Personal Care Products (PPCPs)

These are a class of compounds that are applied externally or ingested by humans, pets, and other domesticated animals. They are released to the environment through the disposal of expired, unwanted, or excess medications to the sewage system by individuals, pharmacies, or physicians. Another source of PPCPs in the environment is through metabolic excretion—the excretion of the chemically unaltered parent compound and metabolized by-products in urine and feces. PCPs, such as deodorants and sunscreens, can be washed into our waterways during bathing, washing, and swimming. Some PCPs are also used as pest-control agents!

The presence of pharmaceuticals in raw and treated wastewater has been well documented with concentrations averaging from less than 10 μg/L in finished wastewater to greater than 100 μg/L in raw wastewater. Examples of pharmaceuticals that have been detected in wastewater are common prescription and veterinary drugs such as beta-blockers (e.g., metoprolol, propranolol), analgesics (e.g., ibuprofen, naproxen), and antibiotics (e.g., erythromycin, trimethoprim, ciprofloxacin, tetracylcline, clindomycin, sulfonamides, tetracycline, fluoroquinolone, macrolides, and trimethoprim) (Hullman, 2009).

CAFOs commonly administer antibiotics and hormones to livestock for disease control and added into feed at subtherapeutic levels to improve the feeding efficiency, growth rate, and animal health. About 11,400 megagrams (Mg) or 70 percent of antibiotics produced for medical and agricultural purposes in the United States are used for nontherapeutic treatment of livestock (Mellon et al., 2001). As much as 75 percent of the antibiotics used for animal growth are excreted back into manure (Campagnolo et al., 2002), of which the United States produces 266 gigagrams (Gg) of dry weight per year (Burkholder et al., 2007). The most common growth promoters of livestock are tetracycline, chlorotetracycline, and bacitracen (Jindal et al., 2006; Mackie et al., 2006). All of the antibiotics listed above are also used in human therapy. After a period of storage, manure is often land-applied as plant fertilizer, and as consequence the pharmaceuticals are discharged to surface and ground water via surface runoff and percolation.

Personal care products (PCPs) such as hand and dish soap, oral hygiene products, and air fresheners, may contain compounds such as triclosan (a registered pesticide), phthalates, and phenols. Seven carcinogenic compounds that are commonly found in PCPs are hydroquinone, ethylene dioxide, formaldehyde, nitrosamine, acrylamides, and PAHs. Hydroquinone was found in products used by 94 percent of women and 69 percent of men. All of these compounds have been found in raw and treated wastewater (Goodman, 2008).

Endocrine-Disrupting Chemicals

The class of chemicals known as *endocrine disrupters,* or EDCs, alters the normal physiological function of the endocrine system and can affect the synthesis of hormones. Some naturally occurring EDCs, such as 17 β-estradiol that is both a naturally occurring estrogen and a synthetic estrogen used as a method of birth control, are excreted into wastewater. Other EDCs include the polychlorinated biphenyls, commonly used pesticides such as atrazine and other triazine chemicals, and the phthalates. EDCs can mimic estrogens, androgens, or thyroid hormones, or their antagonists, although the structures of many EDCs bear little resemblance to that of natural hormones with which they interfere. They can also interfere with the regulation of reproductive and developmental processes in mammals, birds, reptiles and fish (Sadik and Witt, 1999; Harries et al., 2000). The chemicals can also alter the normal physiological function of the endocrine system and can affect the synthesis of hormones in the body. EDCs can also target tissues where the hormones exert their effects. Figure 7-1 shows the most frequently detected pharmaceuticals and EDCs in recent survey of U.S. waters (U.S.G.S., 2002).

Pesticides

The major classes of pesticides are herbicides, insecticides, and fungicides. Herbicides are used to kill weeds. The most commonly used herbicides are atrazine, metolachlor, and alachlor. Chloropyrifos, diazinon, and malathion are the typical insecticides used to kill insects that would otherwise destroy crops, gardens, or structures. Common fungicides are sulfur, chlorothalonil, and mancozeb.

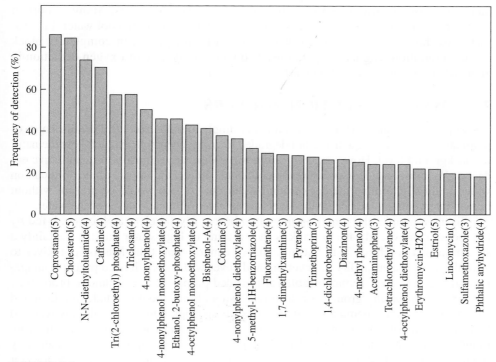

FIGURE 7-1
Most frequently detected pharmaceutical- and endocrine-disrupting compounds. (*Source:* U.S.G.S., 2002.)

The presence of pesticides in surface- and groundwater is ubiquitous in the United States. In a study conducted by the U.S. Geological Survey (U.S.G.S.) of 20 national water quality assessment program (NAWQA) study units across the country, it was found that more than 90 percent of water and fish samples from all streams contained one or, more often, several pesticides. The most commonly occurring pesticides found in water were primarily those currently in use, whereas those found in fish and sediment are the organochlorine insecticides, such as DDT, that were heavily used decades ago. About 50 percent of the wells sampled contained one or more pesticides, with the highest detection frequencies in shallow groundwater beneath agricultural and urban areas and the lowest frequencies in deeper aquifers (U.S.G.S., 1999).

Heat

Although heat is not often recognized as a pollutant, those in the electric power industry are well aware of the problems of disposing of waste heat. Also, waters released by many industrial processes are much hotter than the receiving waters. In some environments an increase of water temperature can be beneficial. For example, production of clams and oysters can be increased in some areas by warming the water. On the other hand, increases in water temperature can have negative impacts. Many important commercial and game fish, such as salmon and trout, live only in cool water. In some instances the discharge of heated water from a power plant can completely block salmon migration. Higher temperatures also increase the rate of oxygen depletion in areas where oxygen-demanding wastes are present.

7-3   WATER POLLUTION IN RIVERS

The objective of water quality management is simple to state: to control the discharge of pollutants so that water quality is not degraded to an unacceptable extent below the natural background level. However, controlling waste discharges must be a quantitative endeavor. We must be able to measure the pollutants, predict the impact of the pollutant on water quality, determine the background water quality that would be present without human intervention, and decide the levels acceptable for intended uses of the water.

To most people, the tumbling mountain brook, crystal clear and icy cold, fed by the melting snow, and safe to drink is the epitome of high water quality. Certainly a stream in that condition is a treasure, but we cannot expect the Mississippi River to have the same water quality. It never did and never will. Yet both need proper management if the water is to remain usable for intended purposes. The mountain brook may serve as the spawning ground for desirable fish and must be protected from heat and sediment as well as chemical pollution. The Mississippi, however, is already warmed from hundreds of kilometers of exposure to the sun and carries the sediment from thousands of square kilometers of land. But even the Mississippi can be damaged by organic matter and toxic chemicals. Fish do live there and the river is used as a water supply for millions of people.

The impact of pollution on a river depends both on the nature of the pollutant and the unique characteristics of the individual river.* Some of the most important characteristics include the volume and speed of water flowing in the river, the river's depth, the type of bottom, and the surrounding vegetation. Other factors include the climate of the region, the mineral heritage of the watershed, land use patterns, and the types of aquatic life in the river. Water quality management for a particular river must consider all these factors. Thus, some rivers are highly susceptible to pollutants such as sediment, salt, and heat, while other rivers can tolerate large inputs of these pollutants without much damage.

Total Maximum Daily Load (TMDL)

Under Section 303(d) of the 1972 Clean Water Act, states, territories, and authorized tribes are required to develop lists of *impaired waters*. Impaired waters are those that do not meet water quality standards that the states, territories, and authorized tribes have established for them. This assessment is made after assuming that point sources of pollution have installed minimum levels of pollution control technology. The law requires that these jurisdictions establish priority rankings for waters on the lists and develop *total maximum daily loads* (TMDL) for these waters. A TMDL specifies the maximum amount of pollutant that a water body can receive and still meet water quality

---

*Here we will use the word "river" to include streams, brooks, creeks, and any other channel of flowing, fresh water.

standards. In addition, the TMDL allocates pollutant *loadings* (that is, the mass of pollutant) that may be contributed among point and non-point sources. The TMDL is computed on a pollutant-by-pollutant basis for a list of pollutants similar to those in Table 7-1. Additional categories include acids/bases (measured as pH), pesticides, and mercury. The TMDL computation is defined as:

$$TMDL = \Sigma WLA + \Sigma LA + MOS \qquad (7\text{-}1)$$

where WLA = waste load allocations, that is, portions of the TMDL assigned to existing and future point sources

LA = load allocations, that is, portions of the TMDL assigned to existing and future non-point sources

MOS = margin of safety

The MOS is to account for uncertainty about the relationships between loads and water quality. A software system called *Better Assessment Science Integrating Point and Non-point Sources* (BASINS) that integrates a *geographic information system* (GIS), national watershed and meteorological data, and state-of-the-art environmental assessment and modeling tools is used to develop the TMDL (Ahmad, 2002; U.S EPA, 2005).

Example 7-1 illustrates the calculation of waste load using the concentration of the waste and the flow rate of the wastewater carrying it.

Example 7-1. Estimate the mass discharge of 30 mg/L of suspended solids in a wastewater flow rate of 28 MGD.

Solution. Assuming that 1 mg/L $\approx$ 1 ppm and using the conversion factor of 8.34 lb_m/gal of water from Appendix C:

$$\left(\frac{30 \text{ parts}}{10^6 \text{ parts}} \right) (28 \times 10^6 \text{ gal/d})(8.34 \text{ } 1b_m/\text{gal}) = 7{,}005.6 \text{ or } 7{,}000 \text{ } 1b_m/d$$

Comments:

1. The assumption that 1 mg/L $\approx$ 1 ppm is reasonable for water and wastewater with concentrations less than 10%. It is ***not*** valid for concentrations in air.

2. The estimate of mass flow rate of a compound in mg/L and volume in MG or volumetric flow rate in MGD frequently occurs in water and wastewater calculations using the U.S. Customary System of units.

Some pollutants, particularly oxygen-demanding wastes and nutrients, are so common and have such a profound impact on almost all types of rivers that they deserve special emphasis. This is not to say that they are always the most significant pollutants in any one river, but rather that no other pollutant category has as much overall effect on our nation's rivers. For these reasons, the next sections of this chapter will be devoted to a more detailed look at how oxygen-demanding material and nutrients affect water quality in rivers.

Effect of Oxygen-Demanding Wastes on Rivers

The introduction of oxygen-demanding material, either organic or inorganic, into a river causes depletion of the dissolved oxygen in the water. This poses a threat to fish and other higher forms of aquatic life if the concentration of oxygen falls below a critical point. To predict the extent of oxygen depletion, it is necessary to know how much waste is being discharged and how much oxygen will be required to degrade the waste. However, because oxygen is continuously being replenished from the atmosphere and from photosynthesis by algae and aquatic plants, as well as being consumed by organisms, the concentration of oxygen in the river is determined by the relative rates of these competing processes. Organic oxygen-demanding materials are commonly measured by determining the amount of oxygen consumed during degradation in a manner approximating degradation in natural waters. This section begins by considering the factors affecting oxygen consumption during the degradation of organic matter, then moves on to inorganic nitrogen oxidation. Finally, the equations for predicting dissolved oxygen concentrations in rivers from degradation of organic matter are developed and discussed.

Biochemical Oxygen Demand

The amount of oxygen required to oxidize a substance to carbon dioxide and water may be calculated by stoichiometry if the chemical composition of the substance is known. This amount of oxygen is known as the *theoretical oxygen demand* (ThOD). The calculation of ThOD is illustrated in Example 7-2.

Example 7-2.   Compute the ThOD of 108.75 mg/L of glucose ($C_6H_{12}O_6$).

Solution.   We begin by writing a balanced equation for the reaction.

$$C_6H_{12}O_6 + 6O_2 \rightleftharpoons 6CO_2 + 6H_2O$$

Next, compute the gram molecular weights of the reactants using the table on the inside cover of the book.

$$
\begin{array}{cc}
glucose & oxygen \\
6C = 72 & (6)(2)O = 192 \\
12H = 12 & \\
6O = \underline{96} & \\
180 & \\
\end{array}
$$

Thus, it takes 192 g of oxygen to oxidize 180 g of glucose to CO_2 and H_2O.
     The ThOD of 108.75 mg/L of glucose is

$$(108.75 \text{ mg/L of glucose})\left(\frac{192 \text{ g O}_2}{180 \text{ g glucose}}\right) = 116 \text{ mg/L O}_2$$

In contrast to the ThOD, the *chemical oxygen demand,* COD (pronounced "see oh dee"), is a measured quantity that does not depend on knowledge of the chemical composition of the substances in the water. In the COD test, a strong chemical oxidizing agent (chromic acid) is mixed with a water sample and then boiled. The difference between the amount of oxidizing agent at the beginning of the test and that remaining at the end of the test is used to calculate the COD.

If the oxidation of an organic compound is carried out by microorganisms using the organic matter as a food source, the oxygen consumed is known as *biochemical oxygen demand,* or BOD (pronounced "bee oh dee"). The actual BOD is less than the ThOD due to the incorporation of some of the carbon into new bacterial cells. The test is a bioassay that utilizes microorganisms in conditions similar to those in natural water to measure indirectly the amount of biodegradable organic matter present. *Bioassay* means to measure by biological means. A water sample is inoculated with bacteria that consume the biodegradable organic matter to obtain energy for their life processes. Because the organisms also utilize oxygen in the process of consuming the waste, the process is called *aerobic* decomposition. This oxygen consumption is easily measured. The greater the amount of organic matter present, the greater the amount of oxygen utilized. The BOD test is an indirect measurement of organic matter because we actually measure only the change in dissolved oxygen concentration caused by the microorganisms as they degrade the organic matter. Although not all organic matter is biodegradable and the actual test procedures lack precision, the BOD test is still the most widely used method of measuring organic matter because of the direct conceptual relationship between BOD and oxygen depletion in receiving waters.

Only under rare circumstances will the ThOD, COD, and BOD be equal. If the chemical composition of all of the substances in the water is known and they are capable of being completely oxidized both chemically and biologically, then the three measures of oxygen demand will be the same.

When a water sample containing degradable organic matter is placed in a closed container and inoculated with bacteria, the oxygen consumption typically follows the pattern shown in Figure 7-2. During the first few days the rate of oxygen depletion is rapid because of the high concentration of organic matter present. As the

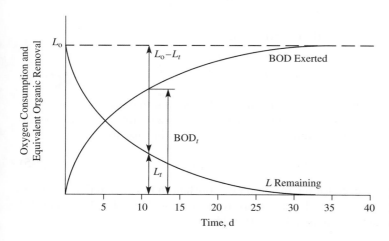

FIGURE 7-2
BOD and oxygen-equivalent relationships.

concentration of organic matter decreases, so does the rate of oxygen consumption. During the last part of the BOD curve, oxygen consumption is mostly associated with the decay of the bacteria that grew during the early part of the test. It is generally assumed that the rate at which oxygen is consumed is directly proportional to the concentration of degradable organic matter remaining at any time. As a result, the BOD curve in Figure 7-2 can be described mathematically as a first-order reaction. Using our definition of reaction rate and reaction order from Chapter 5, this may be expressed as:

$$\frac{dL_t}{dt} = -r_A \tag{7-2}$$

where L_t = oxygen equivalent of the organics remaining at time t, mg/L
$-r_A = -kL_t$
k = reaction rate constant, d^{-1}

Rearranging Equation 7-2 and integrating yields:

$$\frac{dL_t}{L_t} = -k\, dt$$

$$\int_{L_o}^{L} \frac{dL_t}{L_t} = -k \int_0^t dt$$

$$\ln \frac{L_t}{L_o} = -kt$$

or

$$L_t = L_o\, e^{-kt} \tag{7-3}$$

where

$$L_o = \text{oxygen equivalent of organic compounds at time } t = 0$$

Rather than L_t, our interest is in the amount of oxygen used in the consumption of the organics (BOD_t). From Figure 7-2, it is obvious that BOD_t is the difference between the initial value of L_o and L_t, so

$$\text{BOD}_t = L_o - L_t$$
$$= L_o - L_o\, e^{-kt}$$
$$= L_o(1 - e^{-kt}) \tag{7-4}$$

L_o is often referred to as the *ultimate BOD,* that is, the maximum oxygen consumption possible when the waste has been completely degraded. Equation 7-4 is called the *BOD rate equation* and in older literature it is often written in base 10:

$$\text{BOD}_t = L_o(1 - 10^{-Kt}) \tag{7-5}$$

Note that lower case k is used for the reaction rate constant in base e and that capital K is used for the constant in base 10. They are related: $k = 2.303(K)$. Example 7-3 illustrates the use of rate constants in base 10 and base e.

Example 7-3. If the BOD_3 of a waste is 75 mg/L and the K is 0.150 d^{-1}, what is the ultimate BOD?

Solution. Note that the rate constant is given in base 10 (K versus k), and substitute the given values into Equation 7-5 and solve for L_o:

$$75 = L_o(1 - 10^{-(.150)(3)}) = 0.645\, L_o$$

or

$$L_o = \frac{75}{0.645} = 116 \text{ or } 120 \text{ mg/L}$$

In base e,

$$k = 2.303(K) = 0.345, \text{ and}$$

$$75 = L_o(1 - e^{-(.345)(3)}) = 0.645\, L_o$$

so

$$L_o = 116 \text{ or } 120 \text{ mg/L}$$

Comment: Because two significant figures were used to report the 3-day BOD, the calculated ultimate BOD is reported to two significant figures.

You should note that the ultimate BOD (L_o) is defined as the maximum BOD exerted by the waste. It is denoted by the horizontal line in Figure 7-2. Because BOD_t approaches L_o asymptotically, it is difficult to assign an exact time to achieve ultimate BOD. Indeed, based on Equation 7-3, it is achieved only in the limit as t approaches infinity. However, from a practical point of view, we can observe that when the BOD curve is approximately horizontal, the ultimate BOD has been achieved. In Figure 7-2, we would take this to be at about 35 days. In computations, we use a rule of thumb that if BOD_t and L_o agree when rounded to three significant figures, then the time to reach ultimate BOD has been achieved. Given the vagaries of the BOD test, there are occasions when rounding to two significant figures would not be unrealistic.

While the ultimate BOD best expresses the concentration of degradable organic matter, it does not, by itself, indicate how rapidly oxygen will be depleted in a receiving water. Oxygen depletion is related to both the ultimate BOD and the BOD rate constant (k). While the ultimate BOD increases in direct proportion to the concentration of degradable organic matter, the numerical value of the rate constant is dependent on the following:

1. The nature of the waste

2. The ability of the organisms in the system to utilize the waste

3. The temperature

Nature of the Waste. There are literally thousands of naturally occurring organic compounds, not all of which can be degraded with equal ease. Simple sugars and starches are

TABLE 7-2
Typical values for the BOD rate constant

Sample	k (20°C) (day^{-1})
Raw sewage	0.35–0.70
Well-treated sewage	0.12–0.23
Polluted river water	0.12–0.23

rapidly degraded and will therefore have a very large BOD rate constant. Cellulose (for example, toilet paper) degrades much more slowly, and hair and fingernails are almost undegradable in the BOD test or in normal wastewater treatment. Other compounds are intermediate between these extremes. The BOD rate constant for a complex waste depends very much on the relative proportions of the various components. A summary of typical BOD rate constants is shown in Table 7-2. The lower rate constants for treated sewage compared to raw sewage result from the fact that easily degradable organic compounds are more completely removed than less readily degradable organic compounds during wastewater treatment.

Ability of Organisms to Utilize Waste. Any given microorganism is limited in its ability to utilize organic compounds. As a consequence, many organic compounds can be degraded by only a small group of microorganisms. In a natural environment receiving a continuous discharge of organic waste, that population of organisms which can most efficiently utilize this waste will predominate. However, the culture used to inoculate the BOD test may contain only a very small number of organisms that can degrade the particular organic compounds in the waste. This problem is especially common when analyzing industrial wastes. The result is that the BOD rate constant is lower in the laboratory test than in the natural water. This is an undesirable outcome. The BOD test should therefore be conducted with organisms that have been acclimated to the waste so that the rate constant determined in the laboratory can be compared to that in the river.*

Temperature. Most biological processes speed up as the temperature increases and slow down as the temperature drops. Because oxygen utilization is caused by the metabolism of microorganisms, the rate of utilization is similarly affected by temperature. Ideally, the BOD rate constant should be experimentally determined for the temperature of the receiving water. There are two difficulties with this ideal. Often the receiving-water temperature changes throughout the year, so a large number of tests would be required to define k. An additional difficulty is the task of comparing data from various locations having different temperatures. Laboratory testing is therefore done at a standard temperature of 20°C, and the BOD

---

*The word "acclimated" means that the organisms have had time to adapt their metabolisms to the waste or that organisms that can utilize the waste have been given the chance to predominate in the culture.

rate constant is adjusted to the receiving-water temperature using the following expression:

$$k_T = k_{20}(\theta)^{T-20} \qquad (7\text{-}6)$$

where T = temperature of interest, °C
k_T = BOD rate constant at the temperature of interest, d^{-1}
k_{20} = BOD rate constant determined at 20°C, d^{-1}
θ = temperature coefficient. This has a value of 1.135 for temperatures between 4 and 20°C and 1.056 for temperatures between 20 and 30°C (Schroepfer, et al., 1964).

Example 7-4. A waste is being discharged into a river that has a temperature of 10°C. What fraction of the maximum oxygen consumption has occurred in four days if the BOD rate constant determined in the laboratory under standard conditions is 0.115 d^{-1} (base e)?

Solution.   Determine the BOD rate constant for the waste at the river temperature using Equation 7-6:

$$k_{10°C} = 0.115(1.135)^{10-20}$$
$$= 0.032 \ d^{-1}$$

Use this value of k in Equation 7-4 to find the fraction of maximum oxygen consumption occurring in four days:

$$\frac{BOD_4}{L_o} = [1 - e^{-(.032)(4)}]$$
$$= 0.12$$

Graphical Determination of BOD Constants

A variety of methods may be used to determine k and L_o from an experimental set of BOD data. The simplest and least accurate method is to plot BOD versus time. This results in a hyperbolic first-order curve of the form shown in Figure 7-2. The ultimate BOD is estimated from the asymptote of the curve. The rate equation is used to solve for k. It is often difficult to fit an accurate hyperbola to data that are frequently scattered. Methods that linearize the data are preferred. The usual graphical methods for first-order reactions cannot be used because the semilog plot requires knowledge of the initial concentration which, in this case, is one of the constants we are trying to determine, that is, L_o! One simple method around this impasse is called Thomas' graphical method (Thomas, 1950). The method relies on the similarity of the series expansion of the following two functions:

$$F_1 = 1 - e^{-kt} \qquad (7\text{-}7)$$

and

$$F_2 = (kt)(1 + (1/6) \ kt)^{-3} \qquad (7\text{-}8)$$

The series expansion of these functions yields:

$$F_1 = (kt)\left[1 - 0.5(kt) + \frac{1}{6}(kt)^2 - \frac{1}{24}(kt)^3 + \cdots\right] \tag{7-9}$$

$$F_2 = (kt)\left[1 - 0.5(kt) + \frac{1}{6}(kt)^2 - \frac{1}{21.9}(kt)^3 + \cdots\right] \tag{7-10}$$

The first two terms are identical and the third differs only slightly. Replacing Equation 7-9 by 7-10 in the BOD rate equation results in the following approximate equation:

$$BOD_t = L_o(kt)[1 + (1/6)kt]^{-3} \tag{7-11}$$

By rearranging terms and taking the cube root of both sides, Equation 7-11 can be transformed to:

$$\left(\frac{t}{BOD_t}\right)^{1/3} = \frac{1}{(kL_o)^{1/3}} + \frac{(k)^{2/3}}{6(L_o)^{1/3}}(t) \tag{7-12}$$

A plot of $(t/BOD_t)^{1/3}$ versus t is linear (Figure 7-3). The intercept is defined as:

$$A = (kL_o)^{-1/3} \tag{7-13}$$

The slope is defined by:

$$B = \frac{(k)^{2/3}}{6(L_o)^{1/3}} \tag{7-14}$$

Solving for $L_o^{1/3}$ in Equation 7-13 substituting into Equation 7-14 and solving for k yields:

$$k = 6(B/A) \tag{7-15}$$

Likewise, substituting Equation 7-15 into 7-13 and solving for L yields:

$$L_o = \frac{1}{6(A)^2(B)} \tag{7-16}$$

The procedure for determining the BOD constants by this method is as follows:

1. From the experimental results of BOD for various values of t, calculate $(t/BOD_t)^{1/3}$ for each day.

2. Plot $(t/BOD_t)^{1/3}$ versus t on arithmetic graph paper and draw the line of best fit by eye.

3. Determine the intercept (A) and slope (B) from the plot.

4. Calculate k and L_o from Equations 7-15 and 7-16.

Example 7-5 illustrates the technique for estimating k.

Example 7-5. The following data were obtained from an experiment to determine the BOD rate constant and ultimate BOD for an untreated wastewater:

Day	0	1	2	4	6	8
BOD, mg/L	0	32	57	84	106	111

Solution. First calculate values of $(t/BOD_t)^{1/3}$

Day	0	1	2	4	6	8
$(t/BOD_t)^{1/3}$	—	0.315	0.327	0.362	0.384	0.416

The plot of $(t/BOD_t)^{1/3}$ versus time for these data is shown in Figure 7-3. From the figure, $A = 0.30$ and

$$B = \frac{(0.416 - 0.300)}{(8 - 0)} = 0.0145$$

Substituting in Equations 7-15 and 7-16:

$$k = \frac{6(0.0145)}{0.30} = 0.29 \text{ d}^{-1}$$

$$L_o = \frac{1}{6(0.30)^2(0.0145)} = 128 \text{ or } 130 \text{ mg/L}$$

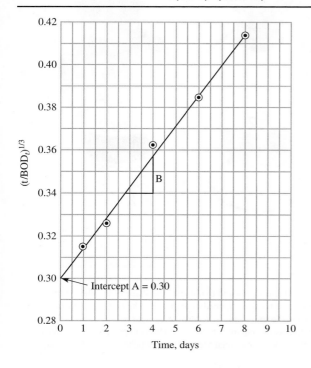

FIGURE 7-3
Plot of $(t/BOD_t)^{1/3}$ versus t for Thomas' graphical method.

Laboratory Measurement of Biochemical Oxygen Demand

In order to have as much consistency as possible, it is important to standardize testing procedures when measuring BOD. In the paragraphs that follow, the standard BOD test is outlined with emphasis placed on the reason for each step rather than the details. The detailed procedures can be found in *Standard Methods for the Examination of Water and Wastewater* (APHA, 2005) which is the authoritative reference of testing procedures in the water pollution control field.

Step 1.   A special 300 mL BOD bottle (Figure 7-4) is completely filled with a sample of water that has been appropriately diluted and inoculated with microorganisms. The bottle is then stoppered to exclude air bubbles. Samples require dilution because the only oxygen available to the organisms is dissolved in the water. The most oxygen that can dissolve is about 9 mg/L, so the BOD of the diluted sample should be between 2 and 6 mg/L. Samples are diluted with a special dilution water that contains all of the trace elements required for bacterial metabolism so that degradation of the organic matter is not limited by lack of bacterial growth. The dilution water also contains an inoculum of microorganisms so that all samples tested on a given day contain approximately the same type and number of microorganisms.

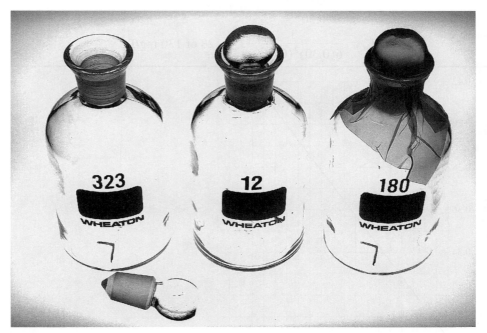

FIGURE 7-4
BOD bottles. The bottle on the left is shown with the cap removed to illustrate the shape of the glass stoppers. The point on the end of the stopper is to ensure that no air is trapped in the bottle. The bottle in the center is shown with the stopper in place. Water is placed in the small cup formed by the lip. This acts as a seal to further exclude air. The bottle on the right is shown with plastic wrap over the stopper. This is to prevent evaporation of the water seal. (Photo courtesy of Harley Seeley of the Instructional Media Center, Michigan State University.)

The ratio of undiluted to diluted sample is called the *sample size,* usually expressed as a percentage, while the inverse relationship is called the *dilution factor.* Mathematically, these are

$$\text{Sample size }(\%) = \frac{\text{vol. of undiluted sample}}{\text{vol. of diluted sample}} \times 100 \qquad (7\text{-}17)$$

$$\text{Dilution factor} = \frac{\text{vol. of diluted sample}}{\text{vol. of undiluted sample}} = \frac{100}{\text{sample size }(\%)} \qquad (7\text{-}18)$$

The appropriate sample size to use can be determined by dividing 4 mg/L (the midpoint of the desired range of diluted BOD) by the estimated BOD concentration in the sample being tested. A convenient volume of undiluted sample is then chosen to approximate to this sample size.

Example 7-6. The BOD of a wastewater sample is estimated to be 180 mg/L. What volume of undiluted sample should be added to a 300 mL bottle? Also, what are the sample size and dilution factor using this volume? Assume that 4 mg/L BOD can be consumed in the BOD bottle.

Solution. Estimate the sample size needed:

$$\text{Sample size} = \frac{4}{180} \times 100 = 2.22\%$$

Estimate the volume of undiluted sample needed since the volume of diluted sample is 300 mL:

$$\text{Vol. of undiluted sample} = 0.0222 \times 300 \text{ mL} = 6.66 \text{ mL}$$

Therefore a convenient sample volume would be 7.00 mL.

Compute the actual sample size and dilution factor:

$$\text{Sample size} = \frac{7.00 \text{ mL}}{300 \text{ mL}} \times 100 = 2.33\%$$

$$\text{Dilution factor} = \frac{300 \text{ mL}}{7.00 \text{ mL}} = 42.9$$

Step 2. Blank samples containing only the inoculated dilution water are also placed in BOD bottles and stoppered. Blanks are required to estimate the amount of oxygen consumed by the added inoculum of microorganisms (called *seed*) in the absence of the sample.

Step 3. The stoppered BOD bottles containing diluted samples and blanks are incubated in the dark at 20°C for the desired number of days. For most purposes, a standard time of five days is used. To determine the ultimate BOD and the BOD rate constant, additional times are used. The samples are incubated in the dark to prevent photosynthesis from adding oxygen to the water and invalidating the oxygen consumption results. As mentioned earlier, the BOD test is conducted at a standard temperature of 20°C so that the effect of temperature on the BOD rate constant is eliminated and results from different laboratories can be compared.

Step 4.   After the desired number of days has elapsed, the samples and blanks are removed from the incubator and the dissolved oxygen concentration in each bottle is measured. The BOD of the undiluted sample is then calculated using the following equation:

$$BOD_t = (DO_{b,t} - DO_{s,t}) \times \text{dilution factor} \qquad (7\text{-}19)$$

where $DO_{b,t}$ = dissolved oxygen concentration in blank after t days of incubation, mg/L
$DO_{s,t}$ = dissolved oxygen concentration in sample after t days of incubation, mg/L

Example 7-7.   What is the BOD_5 of the wastewater sample of Example 7-6 if the DO values for the blank and diluted sample after five days are 8.7 and 4.2 mg/L, respectively?

Solution.   Substitute the appropriate values into Equation 7-19:

$$BOD_5 = (8.7 - 4.2) \times 42.9 = 193 \text{ or } 190 \text{ mg/L}$$

Comments: Equation 7-19 is only appropriate if the dilution water is *not* seeded. If the dilution water is seeded then a correction factor for the oxygen depletion due the seed is required (APHA, 2005).

Additional Notes on Biochemical Oxygen Demand

Although the 5-day BOD has been chosen as the standard value for most wastewater analysis and for regulatory purposes, ultimate BOD is actually a better indicator of total waste strength. For any one type of waste having a defined BOD rate constant, the ratio between ultimate BOD and BOD_5 is constant so that BOD_5 indicates relative waste strength. For different types of wastes having the same BOD_5, the ultimate BOD is the same only if, by chance, the BOD rate constants are the same. This is illustrated in Figure 7-5 for a municipal wastewater having a $k = 0.340$ d^{-1} and an industrial

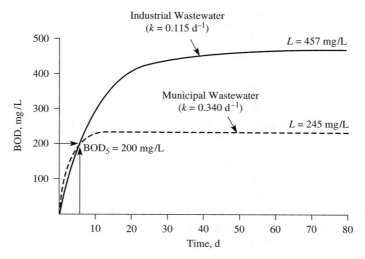

FIGURE 7-5
The effect of k on ultimate BOD for two wastewaters having the same BOD_5.

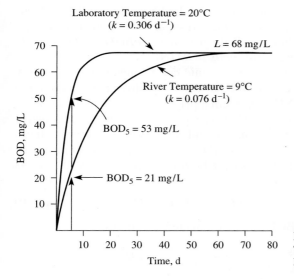

Laboratory Temperature = 20°C
($k = 0.306$ d^{-1})

$L = 68$ mg/L

River Temperature = 9°C
($k = 0.076$ d^{-1})

BOD$_5$ = 53 mg/L

BOD$_5$ = 21 mg/L

FIGURE 7-6
The effect of k on BOD$_5$, when the ultimate BOD is constant.

wastewater having a $k = 0.115$ d^{-1}. Although both wastewaters have a BOD$_5$ of 200 mg/L, the industrial wastewater has a much higher ultimate BOD and can be expected to have a greater impact on dissolved oxygen in a river. For the industrial wastewater, a smaller fraction of the BOD was exerted in the first five days due to the lower rate constant.

Proper interpretation of BOD$_5$ values can also be illustrated in another way. Consider a sample of polluted river water for which the following values were determined using standard laboratory techniques at a temperature of 20°C: BOD$_5$ = 53 mg/L, and $k = 0.306$ d^{-1}. The ultimate BOD calculated from Equation 7-4 is, therefore, 68 mg/L. However, because the river temperature is 9°C, the k value in the river is only 0.076 d^{-1}. As shown graphically in Figure 7-6, the laboratory value of BOD$_5$ seriously overestimates the actual oxygen consumption in the river. Again, for a constant ultimate BOD, a smaller fraction of the BOD is exerted in five days when the BOD rate constant is lower.

The 5-day BOD was chosen as the standard value for most purposes because the test was devised by environmental engineers in England, where rivers have travel times to the sea of less than 5 days, so there was no need to consider oxygen demand at longer times. Since there is no other time that is any more rational than 5 days, this value has become firmly entrenched.*

Nitrogen Oxidation

Up to this point an unstated assumption has been that only the carbon in organic matter is oxidized. Actually many organic compounds, such as proteins, also contain nitrogen that can be oxidized with the consumption of molecular oxygen. However, because the

---

*Numerous investigations of the kinetics of the BOD test have demonstrated that the selection of a 5-day incubation period may be justified on scientific grounds regardless of the travel time of streams. For example, one explanation is that after five days, the microbial system in the BOD bottle is in the autodigestive phase. That is, there is no carbon source outside of the microbial cells that they can consume and that any oxygen uptake that occurs beyond five days is the result of consumption of carbon that they have stored (Gaudy and Gaudy, 1980).

mechanisms and rates of nitrogen oxidation are distinctly different from those of carbon oxidation, the two processes must be considered separately. Logically, oxygen consumption due to oxidation of carbon is called *carbonaceous BOD* (CBOD), while that due to nitrogen oxidation is called *nitrogenous BOD* (NBOD).

The organisms that oxidize the carbon in organic compounds to obtain energy cannot oxidize the nitrogen in these compounds. Instead, the nitrogen is released into the surrounding water as ammonia (NH_3). At normal pH values, this ammonia is actually in the form of the ammonium cation (NH_4^+). The ammonia released from organic compounds, plus that from other sources such as industrial wastes and agricultural runoff (that is, fertilizers), is oxidized to nitrate (NO_3^-) by a special group of nitrifying bacteria as their source of energy in a process called *nitrification*. The overall reaction for ammonia oxidation is

$$NH_4^+ + 2O_2 \xrightleftharpoons{\text{microorganisms}} NO_3^- + H_2O + 2H^+ \tag{7-20}$$

From this reaction the theoretical NBOD can be calculated as follows:

$$NBOD = \frac{\text{grams of oxygen used}}{\text{grams of nitrogen oxidized}} = \frac{4 \times 16}{14} = 4.57 \text{ g } O_2/\text{g N}$$

The actual nitrogenous BOD is slightly less than the theoretical value due to the incorporation of some of the nitrogen into new bacterial cells, but the difference is only a few percent.

Example 7-8.   Compute the theoretical NBOD of a wastewater containing 30 mg/L of ammonia as nitrogen. (We often say "ammonia nitrogen" and write the expression as NH_3-N.) If the wastewater analysis was reported as 30 mg/L of ammonia (NH_3), what would the theoretical NBOD be?

Solution.   In the first part of the problem, the amount of ammonia was reported as NH_3-N. Therefore, we can use the theoretical relationship developed from Equation 7-20.

$$\text{Theo. NBOD} = (30 \text{ mg N/L})(4.57 \text{ mg } O_2/\text{mg N}) = 137 \text{ mg } O_2/\text{L}$$

To answer the second question, we must convert mg/L of ammonia to NH_3-N by multiplying by the ratio of gram molecular weights of N to NH_3.

$$(30 \text{ mg } NH_3/\text{L}) \left(\frac{14 \text{ g N}}{17 \text{ g } NH_3} \right) = 24.7 \text{ mg N/L}$$

Now we may use the relationship developed from Equation 7-20.

$$\text{Theo. NBOD} = (24.7 \text{ mg N/L})(4.57 \text{ mg } O_2/\text{mg N}) = 133 \text{ mg } O_2/\text{L}$$

The rate at which the NBOD is exerted depends heavily on the number of nitrifying organisms present. In untreated sewage, there are few of these organisms, while in a well-treated effluent, the concentration is high. When samples of untreated and

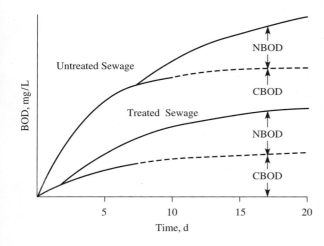

FIGURE 7-7
BOD curves showing both carbonaceous and nitrogenous BOD.

treated sewage are subjected to the BOD test, oxygen consumption follows the pattern shown in Figure 7-7. In the case of untreated sewage, the NBOD is exerted after much of the CBOD has been exerted. The lag is due to the time it takes for the nitrifying bacteria to reach a sufficient population for the amount of NBOD exertion to be significant compared with that of the CBOD. In the case of the treated sewage, a higher population of nitrifying organisms in the sample reduces the lag time. Once nitrification begins, however, the NBOD can be described by Equations 7-4 and 7-5 with a BOD rate constant comparable to that for the CBOD of a well-treated effluent ($k = 0.12$ to 0.23 d^{-1}). Because the lag before the nitrogenous BOD is highly variable, BOD$_5$ values are often difficult to interpret. When measurement of only carbonaceous BOD is desired, chemical inhibitors are added to stop the nitrification process. The rate constant for nitrification is also affected by temperature and can be adjusted using Equation 7-6.

Streeter-Phelps DO Sag Curve

The concentration of dissolved oxygen in a river is an indicator of the general health of the river. All rivers have some capacity for self-purification. As long as the discharge of oxygen-demanding wastes is well within the self-purification capacity, the DO level will remain high and a diverse population of plants and animals, including game fish, can be found. As the amount of waste increases, the self-purification capacity can be exceeded, causing detrimental changes in plant and animal life. The stream loses its ability to cleanse itself and the DO level decreases. When the DO drops below about 4 to 5 mg/L, most game fish will have been driven out. If the DO is completely removed, fish and other higher animals are killed or driven out and extremely noxious conditions result. The water becomes blackish and foul smelling as the sewage and dead animal life decompose under *anaerobic* conditions (that is, without oxygen).

One of the major tools of water quality management in rivers is the ability to assess the capability of a stream to absorb a waste load. This is done by determining the profile of DO concentration downstream from a waste discharge. This profile

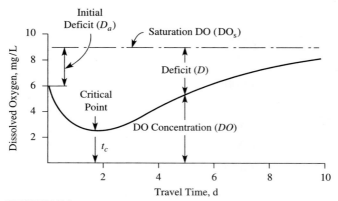

FIGURE 7-8
Typical Streeter-Phelps DO sag curve.

is called the DO sag curve (see Figure 7-8) because the DO concentration dips as oxygen-demanding materials are oxidized and then rises again further downstream as the oxygen is replenished from the atmosphere. As depicted in Figure 7-9, the biota of the stream are often a reflection of the dissolved oxygen conditions in the stream.

To develop a mathematical expression for the DO sag curve, the sources of oxygen and the factors affecting oxygen depletion must be identified and quantified. The only significant sources of oxygen are reaeration from the atmosphere and photosynthesis of aquatic plants. Oxygen depletion is caused by a larger range of factors, the most important being the BOD, both carbonaceous and nitrogenous, of the waste discharge, and the BOD already in the river upstream of the waste discharge. The second most important factor is that the DO in the waste discharge is usually less than that in the river. Thus, the DO at the river is lowered as soon as the waste is added, even before any BOD is exerted. Other factors affecting dissolved oxygen depletion include non-point source pollution, the respiration of organisms living in the sediments (*benthic demand*), and the respiration of aquatic plants. Following the classical approach, the DO sag equation (also known as the *Streeter-Phelps* equation in recognition of the engineers who developed it) will be developed by considering only initial DO reduction, carbonaceous BOD, and reaeration from the atmosphere (Streeter-Phelps, 1925). Subsequently, the equation will be expanded to include the nitrogenous BOD. Finally, the other factors affecting DO levels will be discussed qualitatively; a quantitative discussion is beyond the scope of this book.

Mass Balance Approach.   Simplified mass balances help us understand and solve the DO sag curve problem. Three conservative (those without chemical reaction) mass balances may be used to account for initial mixing of the waste stream and the river. DO, carbonaceous BOD, and temperature all change as the result of mixing of the waste stream and the river. Once these are accounted for, the DO sag curve may be viewed as a nonconservative mass balance, that is, one with reactions.

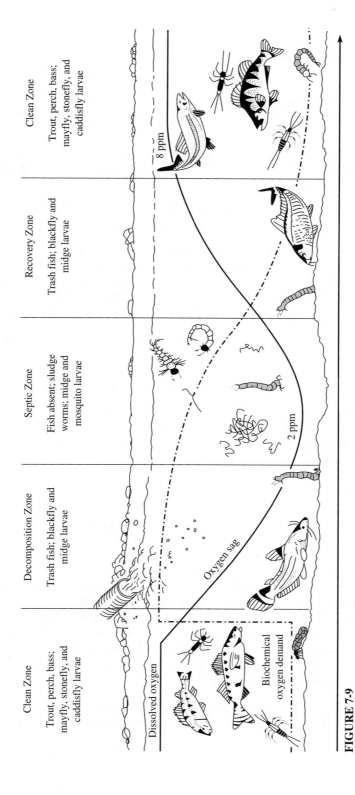

FIGURE 7-9

Oxygen sag downstream of an biodegradable organic chemical source. *Source:* U.S. EPA.

FIGURE 7-10
Conservative mass balance diagram for DO mixing.

The conservative mass balance diagram for oxygen (mixing only) is shown in Figure 7-10. The product of the water flow and the DO concentration yields a mass of oxygen per unit of time:

$$\text{Mass of DO in wastewater} = Q_w \text{DO}_w \qquad (7\text{-}21)$$

$$\text{Mass of DO in river} = Q_r \text{DO}_r \qquad (7\text{-}22)$$

where Q_w = volumetric flow rate of wastewater, m^3/s
   Q_r = volumetric flow rate of the river, m^3/s
   DO_w = dissolved oxygen concentration in the wastewater, g/m^3
   DO_r = dissolved oxygen concentration in the river, g/m^3

The mass of DO in the river after mixing equals the sum of the mass flows:

$$\text{Mass of DO after mixing} = Q_w \text{DO}_w + Q_r \, \text{DO}_r \qquad (7\text{-}23)$$

In a similar fashion for ultimate BOD:

$$\text{Mass of BOD after mixing} = Q_w L_w + Q_r L_r \qquad (7\text{-}24)$$

where L_w = ultimate BOD of the wastewater, mg/L
   L_r = ultimate BOD of the river, mg/L

The concentrations of DO and BOD after mixing are the respective masses per unit time divided by the total flow rate (that is, the sum of the wastewater and river flows):

$$\text{DO} = \frac{Q_w \text{DO}_w + Q_r \, \text{DO}_r}{Q_w + Q_r} \qquad (7\text{-}25)$$

$$L_a = \frac{Q_w L_w + Q_r L_r}{Q_w + Q_r} \qquad (7\text{-}26)$$

where L_a = initial ultimate BOD after mixing.

Example 7-9.   The town of State College discharges 17,360 m^3/d of treated wastewater into the Bald Eagle Creek. The treated wastewater has a BOD_5 of 12 mg/L and a k of 0.12 d^{-1} at 20°C. Bald Eagle Creek has a flow rate of 0.43 m^3/s and an ultimate BOD of 5.0 mg/L. The DO of the river is 6.5 mg/L and the DO of the wastewater is 1.0 mg/L. The stream temperature is 10°C and the wastewater temperature is 10°C. Compute the DO and initial ultimate BOD after mixing.

Solution.   The DO after mixing is given by Equation 7-25. To use this equation we must convert the wastewater flow to compatible units, that is, m^3/s.

$$Q_w = \frac{(17,360 \ m^3/d)}{(86,400 \ s/d)} = 0.20 \ m^3/s$$

The DO after mixing is then

$$DO = \frac{(0.20 \text{ m}^3/\text{s})(1.0 \text{ mg/L}) + (0.43 \text{ m}^3/\text{s})(6.5 \text{ mg/L})}{0.20 \text{ m}^3/\text{s} + 0.43 \text{ m}^3/\text{s}} = 4.75 \text{ mg/L}$$

Because the rate constant for the wastewater is given for a reference temperature of 20°C, we must first convert it to the temperature of the mixture of the wastewater and stream water. Because both have a temperature of 10°C, the mixture will be at 10°C. Using Equation 7-6 with $\theta = 1.135$:

$$k_{10} = (0.12 \text{ d}^{-1})(1.135)^{10-20} = 0.0338$$

Solving Equation 7-4 for L_o:

$$L_o = \frac{BOD_5}{(1 - e^{-kt})} = \frac{12 \text{ mg/L}}{(1 - e^{-(0.0338 \times 5)})} = \frac{12 \text{ mg/L}}{(1 - 0.844)} = 77.12 \text{ mg/L}$$

Note that we used the subscript of five days in BOD_5 to determine the value of t in the equation. Now setting $L_w = L_o$, we can determine the initial ultimate BOD after mixing using Equation 7-26:

$$L_a = \frac{(0.20 \text{ m}^3/\text{s})(77.12 \text{ mg/L}) + (0.43 \text{ m}^3/\text{s})(5.0 \text{ mg/L})}{0.20 \text{ m}^3/\text{s} + 0.43 \text{ m}^3/\text{s}} = 27.90, \text{ or } 28 \text{ mg/L}$$

For temperature, we must consider a heat balance rather than a mass balance. As noted in Chapter 2, this is an application of a fundamental principle of physics:

$$\text{Loss of heat by hot bodies} = \text{gain of heat by cold bodies} \qquad (7\text{-}27)$$

The change in *enthalpy* or "heat content" of a mass of a substance may be defined by the following equation:

$$H = mc_p \Delta T \qquad (7\text{-}28)$$

where H = change in enthalpy, J
$\quad\quad m$ = mass of substance, g
$\quad\quad c_p$ = specific heat at constant pressure, J/g · K
$\quad\quad \Delta T$ = change in temperature, K

The specific heat of water varies slightly with temperature. For natural waters, a value of 4.19 will be a satisfactory approximation. Using our fundamental heat loss = heat gain equation, we may write

$$(m_w)(4.19)\Delta T_w = (m_r)(4.19)\Delta T_r \qquad (7\text{-}29)$$

The temperature after mixing is found by solving this equation for the final temperature by recognizing that ΔT on each side of the equation is the difference between the final river temperature (T_f) and the starting temperature of the wastewater and the river water, respectively:

$$T_f = \frac{Q_w T_w + Q_r T_r}{Q_w + Q_r} \qquad (7\text{-}30)$$

Oxygen Deficit.   The DO sag equation has been developed using oxygen deficit rather than dissolved oxygen concentration, to make it easier to solve the integral equation that results from the mathematical description of the mass balance. The oxygen deficit is the amount by which the actual dissolved oxygen concentration is less than the saturation value with respect to oxygen in the air:

$$D = \text{DO}_s - \text{DO} \tag{7-31}$$

where D = oxygen deficit, mg/L
   DO_s = saturation concentration of dissolved oxygen at the temperature of the river after mixing, mg/L
   DO = actual concentration of dissolved oxygen, mg/L

The saturation value of dissolved oxygen is heavily dependent on water temperature—it decreases as the temperature increases. Values of DO_s for fresh water are given in Appendix A.

Initial Deficit.   The beginning of the DO sag curve is at the point where a waste discharge mixes with the river. The initial deficit is calculated as the difference between saturated DO and the concentration of the DO after mixing (Equation 7-25):

$$D_a = \text{DO}_s - \frac{Q_w\text{DO}_w + Q_r\text{DO}_r}{Q_w + Q_r} \tag{7-32}$$

where D_a = initial deficit after river and waste have mixed, mg/L.

Example 7-10.   Calculate the initial deficit of the Bald Eagle Creek after mixing with the wastewater from the town of State College (see Example 7-9 for data). The stream temperature is 10°C and the wastewater temperature is 10°C.

Solution.   With the stream temperature, the saturation value of dissolved oxygen (DO_s) can be determined from the table in Appendix A. At 10°C, DO_s = 11.33 mg/L. Since we calculated the concentration of DO after mixing as 4.75 mg/L in Example 7-9, the initial deficit after mixing is

$$D_a = 11.33 \text{ mg/L} - 4.75 \text{ mg/L} = 6.58 \text{ mg/L}$$

Because wastewater commonly has a higher temperature than river water, especially during the winter, the river temperature downstream of the discharge is usually higher than that upstream. Since we are interested in downstream conditions, it is important to use the downstream temperature when determining the saturation concentration of dissolved oxygen.

DO Sag Equation.   A mass balance diagram of DO in a small *reach* (stretch) of river is shown in Figure 7-11a. This is a comprehensive mass balance that accounts for all of the inputs and outputs. We are going to limit our development to the classical

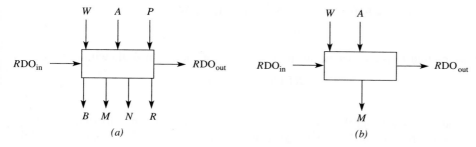

Legend

RDO_{in}, RDO_{out} = mass of DO flowing in and out of reach
W = mass of DO in wastewater flowing into reach
A = mass of DO entering from atmosphere
P = mass of DO entering from algae photosynthetic oxygen production
B = mass of DO consumed by benthic demand
M = mass of DO removed by microbial degradation of carbonaceous BOD
N = mass of DO removed by microbial degradation of nitrogenous BOD
R = mass of DO consumed by algal respiration

FIGURE 7-11
Mass balance diagram of DO in a small reach of river (*a*) and simplified mass balance for Streeter-Phelps model (*b*).

Streeter-Phelps model. The simplified mass balance diagram is shown in Figure 7-11*b*. The mass balance equation is then:

$$RDO_{in} + W + A - M - RDO_{out} = 0 \qquad (7\text{-}33)$$

where RDO_{in} = mass of DO in river flowing into reach
W = mass of DO in wastewater flowing into reach
A = mass of DO added from atmosphere
M = mass of DO removed by microbial degradation of carbonaceous BOD
RDO_{out} = mass of DO in river flowing out of reach

In Equation 7-33 we can account for $RDO_{in} + W$. Our goal is to find RDO_{out} in terms of mass per unit volume (mg/L). This leaves A and M to be accounted for before we can solve the mass balance equation.

The rate at which DO disappears from the stream as a result of microbial action (M) is exactly equal to rate of increase in the deficit. With the assumption that the saturation value for DO remains constant [$d(DO_s)/dt = 0$], differentiation of Equation 7-31 yields:

$$\frac{d(DO)}{dt} + \frac{dD}{dt} = 0$$

and

$$\frac{d(DO)}{dt} = -\frac{dD}{dt} \qquad (7\text{-}34)$$

The rate at which DO disappears coincides with the rate that BOD is degraded, so

$$\frac{d(DO)}{dt} = -\frac{dD}{dt} = -\frac{d(BOD)}{dt} \qquad (7\text{-}35)$$

Remembering that in Equation 7-4 BOD_t was defined as

$$BOD_t = L_o - L_t$$

and noting that L_o is a constant, we may say that the change in BOD with time is

$$\frac{d(\text{BOD})}{dt} = -\frac{dL_t}{dt} \qquad (7\text{-}36)$$

This leads us to see that the rate of change in deficit at time t due to BOD is a first-order reaction proportional to the oxygen equivalent of the organic compounds remaining:

$$\frac{dD}{dt} = kL_t \qquad (7\text{-}37)$$

The rate constant, k, is called the *deoxygenation rate constant* and is designated k_d.

The rate of oxygen mass transfer into solution from the air (A) has been shown to be a first-order reaction proportional to the difference between the saturation value and the actual concentration:

$$\frac{d(\text{DO})}{dt} = k(\text{DO}_s - \text{DO}) \qquad (7\text{-}38)$$

From Equations 7-31 and 7-34 we can see that

$$\frac{dD}{dt} = -kD \qquad (7\text{-}39)$$

The rate constant is called the *reaeration rate constant, k_r*.

From Equations 7-37 and 7-39 we can see that the oxygen deficit is a function of the competition between oxygen utilization and reaeration from the atmosphere:

$$\frac{dD}{dt} = k_d L - k_r D \qquad (7\text{-}40)$$

where $\dfrac{dD}{dt}$ = the change in oxygen deficit (D) per unit of time, mg/L · d

k_d = deoxygenation rate constant, d^{-1}
L = ultimate BOD of river water, mg/L
k_r = reaeration rate constant, d^{-1}
D = oxygen deficit in river water, mg/L

By integrating Equation 7-40 and using the initial conditions (at $t = 0$, $D = D_a$) we obtain the *Streeter-Phelps* DO sag equation:

$$D = \frac{k_d L_a}{k_r - k_d} (e^{-k_d t} - e^{-k_r t}) + D_a(e^{-k_r t}) \qquad \text{FE} \quad (7\text{-}41)$$

where D = oxygen deficit in river water after exertion of BOD for time, t, mg/L
L_a = initial ultimate BOD after river and wastewater have mixed (Equation 7-26), mg/L
k_d = deoxygenation rate constant, d^{-1}

k_r = reaeration rate constant, d^{-1}

t = time of travel of wastewater discharge downstream, d

D_a = initial deficit after river and wastewater have mixed (Equation 7-32), mg/L

When $k_r = k_d$, Equation 7-41 reduces to:

$$D = (k_d t L_a + D_a)(e^{-k_d t}) \tag{7-42}$$

where the terms are as previously defined.

Deoxygenation Rate Constant.   The deoxygenation rate constant differs from the BOD rate constant because there are physical and biological differences between a river and a BOD bottle. In general, BOD is exerted more rapidly in a river because of turbulent mixing, larger numbers of "seed" organisms, and BOD removal by organisms on the stream bed as well as by those suspended in the water. While k rarely has a value greater than 0.7 day^{-1}, k_d may be as large as 7-day^{-1} for shallow, rapidly flowing streams. However, for deep, slowly moving rivers, the value of k_d is very close to that for k.

Bosko has developed a method of estimating k_d from k using characteristics of the stream (Bosko, 1966).

$$k_d = k + \frac{v}{H}\eta \tag{7-43}$$

where k_d = deoxygenation rate constant at 20°C, d^{-1}

v = average speed of stream flow, m/s

k = BOD rate constant determined in laboratory at 20°C, d^{-1}

H = average depth of stream, m

η = bed-activity coefficient

The bed-activity coefficient may vary from 0.1 for stagnant or deep water to 0.6 or more for rapidly flowing streams. Note that the bed-activity coefficient includes a conversion factor to make the second term dimensionally correct. After determining k_d from Equation 7-43, it should be corrected for temperature using Equation 7-6 if the stream temperature is not 20°C.

Example 7-11.   Determine the deoxygenation rate constant for the reach of Bald Eagle Creek (Examples 7-8 and 7-9) below the wastewater outfall (discharge pipe). The average speed of the stream flow in the creek is 0.03 m/s. The depth is 5.0 m and the bed-activity coefficient is 0.35.

Solution.   From Example 7-9, the value of k is 0.12 d^{-1}. Using Equation 7-43, the deoxygenation rate constant at 20°C is

$$k_d = 0.12 \text{ d}^{-1} + \frac{0.03 \text{ m/s}}{5.0 \text{ m}}(0.35) = 0.1221 \text{ or } 0.12 \text{ d}^{-1}$$

Note that the units are not consistent. As we have noted before, empirical expressions, such as that in Equation 7-43, may have implicit conversion factors. Thus, you must be careful to use the same units as those used by the author of the equation.

We also note that the deoxygenation rate constant of 0.1221 d^{-1} is at 20°C. In Example 7-10, we noted that the stream temperature was 10°C. Thus, we must correct the estimated k_d using Equation 7-6.

$$k_d \text{ at } 10°C = (0.1221 \text{ d}^{-1})(1.135)^{10-20} = (0.1221)(0.2819)$$
$$= 0.03442 \text{ or } 0.034 \text{ d}^{-1}$$

Reaeration. The value of k_r depends on the degree of turbulent mixing, which is related to stream velocity, and on the amount of water surface exposed to the atmosphere compared to the volume of water in the river. A narrow, deep river will have a much lower k_r than a wide, shallow river. O'Connor and Dobbins (1958) developed a generalized empirical equation to estimate the reaeration constant based on the characteristics of the stream and the molecular diffusion of oxygen into water:

$$k_r = \frac{3.9 \, v^{0.5}}{H^{1.5}} \tag{7-44}$$

where k_r = reaeration rate constant at 20°C, day^{-1}
v = average stream velocity, m/s
H = average depth, m

Note that the factor of 3.9 includes a conversion factor to make the equation dimensionally correct. The reaeration rate constant is also affected by temperature and can be adjusted to the river temperature using Equation 7-6 **but with a temperature coefficient (θ) of 1.024.** For various streams, k_r can range from 0.05 to greater than 18 d^{-1}.

Travel Time. With the speed of the flow in the river and the distance downstream, the travel time may be estimated from the basic relationship between time, distance, and speed:

$$t = \frac{\text{distance}}{\text{speed}} \tag{7-45}$$

To relate travel time to a physical distance downstream, one must also know the average stream velocity. Once D has been found at any point downstream, the DO can be found from Equation 7-31.

Critical Point. The lowest point on the DO sag curve, which is called the *critical point,* is of major interest since it indicates the worst conditions in the river. The time to the critical point (t_c) can be found by differentiating Equation 7-41, setting it equal to zero, and solving for t using base e values for k_r and k_d:

$$t_c = \frac{1}{k_r - k_d} \ln \left[\frac{k_r}{k_d} \left(1 - D_a \frac{k_r - k_d}{k_d L_a} \right) \right] \tag{7-46}$$

or when $k_r = k_d$:

$$t_c = \frac{1}{k_d} \left(1 - \frac{D_a}{L_a} \right) \tag{7-47}$$

The critical deficit (D_c) is then found by using this critical time in Equation 7-41.

Interpreting Results.   As noted in Chapter 4, modeling equations may predict impossible results. This is true for the Streeter-Phelps model. For example, if the deficit calculated from Equation 7-41 is greater than the DO saturation, the result will be a negative number. Because it is physically impossible for the DO to be less than zero, this means that all the oxygen was depleted at some point upstream and the DO is zero. Assuming your calculations are correct, you must use your engineering judgment and **report it as zero because it cannot be less than zero!**

In some instances there may not be a sag in the DO downstream. The lowest DO may occur in the mixing zone. In these instances Equation 7-46 will not give a useful value.

Example 7-12 illustrates the calculation of the DO at a given point downstream and the calculation of the critical DO.

Example 7-12.   Determine the DO concentration at a point 5 km downstream from the State College discharge into the Bald Eagle Creek (Examples 7-9, 7-10, 7-11). Also determine the critical DO and the distance downstream at which it occurs.

Solution.   All of the appropriate data are provided in the three previous examples. With the exceptions of the travel time, t, and the reaeration rate, the values needed for Equations 7-41 and 7-46 have been computed in Examples 7-9, 7-10, and 7-11. The first step then is to calculate k_r.

$$k_r \text{ at } 20°C = \frac{(3.9)(0.03 \text{ m/s})^{0.5}}{(5.0 \text{ m})^{1.5}} = 0.0604 \text{ d}^{-1}$$

Because this is at 20°C and the stream temperature is at 10°C, Equation 7-6 must be used to correct for the temperature difference.

$$k_r \text{ at } 10°C = (0.0604 \text{ d}^{-1})(1.024)^{10-20} = (0.0604)(0.7889) = 0.04766 \text{ d}^{-1}$$

Note that the temperature coefficient is the one noted in the text above rather than the ones reported with Equation 7-6.

The travel time t is computed from the distance downstream and the speed of the stream (Equation 7-45):

$$t = \frac{(5 \text{ km})(1,000 \text{ m/km})}{(0.03 \text{ m/s})(86,400 \text{ s/d})} = 1.929 \text{ d}$$

Although it is not warranted by the significant figures in the computation, we have elected to keep four significant figures because of the computational effects of truncating the value.

The deficit is estimated using Equation 7-41.

$$D = \frac{(0.03442)(27.90)}{0.04766 - 0.03442}[e^{-(0.03442)(1.929)} - e^{-(0.04766)(1.929)}] + 6.58[e^{-(0.04766)(1.929)}]$$

$$D = (72.53)(0.9358 - 0.9122) + 6.58(0.9122) = 7.714 \text{ or } 7.71 \text{ mg/L}$$

and the dissolved oxygen is

$$DO = 11.33 - 7.71 = 3.62 \text{ mg/L}$$

The critical time is computed using Equation 7-46:

$$t_c = \frac{1}{0.04766 - 0.03442} \ln \left\{ \frac{0.04766}{0.03442} \left[1 - 6.58 \frac{0.04766 - 0.03442}{(0.03442)(27.90)} \right] \right\}$$

$$t_c = 17.40 \text{ d}$$

Using t_c for the time in Equation 7-41, calculate the critical deficit as

$$D_c = \frac{(0.03442)(27.90)}{0.04766 - 0.03442} \left[e^{-(0.03442)(17.40)} - e^{-(0.04766)(17.40)} \right] + 6.58 \left[e^{-(0.04766)(17.40)} \right]$$

$$D_c = 11.07 \text{ mg/L}$$

Solving Equation 7-31 for DO at the critical point (DO_c) and using $DO_c = 11.33 \text{ mg/L}$ from Table A-3 of Appendix A gives the critical DO:

$$DO_c = 11.33 - 11.07 = 0.26 \text{ mg/L}$$

The critical DO occurs downstream at a distance of

$$(6.45 \text{ d})(86,400 \text{ s/d})(0.03 \text{ m/s}) \left(\frac{1}{1,000 \text{ m/km}} \right) = 16.7 \text{ km}$$

from the wastewater discharge point. (Note that 0.03 m/s is the speed of the stream.)

Comments:

1. The DO at both the 5 km distance downstream and at the critical point are very low. The management strategy for improving the DO level is discussed in the following paragraph.

2. Because of the large number of calculations, we recommend that you set up a spreadsheet with all of the steps using Examples 7-8 through 7-11 to check the computations before attempting the end of chapter problems. This will make it easier to spot math errors. This is particularly true if the end of chapter problem solutions yield unusual results such as negative DO or no DO sag.

3. An example that demonstrates the effect of changes in the reaeration rate and the deoxygenation rate on the deficit and the location of the critical point is presented at the text website: www.mhhe.com/davis

---

TMDL Mass Loading. As noted earlier in this chapter, TMDL calculations are based on mass loading. To use the DO sag equation as presented here, the mass loading (kg/d of ultimate BOD) must be converted to a concentration. Following our general approach for calculating concentration from mass flow (Chapter 2); we divide the mass flow (kg/d) by the flow of the water carrying the waste (Q_w, Q_r, or the sum $Q_w + Q_r$):

$$\frac{\text{Mass flow of ultimate BOD (kg/d)}}{\text{Flow of water carrying waste (m}^3\text{/s)}} \tag{7-48}$$

The mass discharge units are then converted to mg/d and the water flow rate to L/d so that the days cancel.

$$\frac{(kg/d) \times (1 \times 10^6 \text{ mg/kg})}{(m^3/s) \times (86,400 \text{ s/d}) \times (1 \times 10^3 \text{ L/m}^3)} \tag{7-49}$$

Example 7-13 illustrates the calculation.

Example 7-13. Convert the discharge of 129.60 kg/d ultimate BOD in 0.0500 m^3/s of wastewater flow to a concentration in units of mg/L.

Solution. Using Equations 7-48 and 7-49

$$L_w = \frac{(129.60 \text{ kg/d})(1 \times 10^6 \text{ mg/kg})}{(0.0500 \text{ m}^3/\text{s})(86,400 \text{ s/d})(1 \times 10^3 \text{ L/m}^3)} = 30.00 \text{ mg/L}$$

Management Strategy. The beginning point for water quality management in rivers using the DO sag curve is to determine the minimum DO concentration that will protect the aquatic life in the stream. This value, called the DO standard, is generally set to protect the most sensitive species that exist or could exist in the particular river. For a known waste discharge and a known set of river characteristics, the DO sag equation can be solved to find the DO at the critical point. If this value is higher than the standard, the stream can adequately assimilate the waste. If the DO at the critical point is less than the standard, then additional waste treatment is needed. Usually, the environmental engineer has control over just two parameters, L_a and D_a. By increasing the efficiency of the existing treatment processes or by adding additional treatment steps, the ultimate BOD of the waste discharge can be reduced, thereby reducing L_a. Often a relatively inexpensive method for improving stream quality is to reduce D_a by adding oxygen to the wastewater to bring it close to saturation prior to discharge. To determine whether a proposed improvement will be adequate, the new values for L_a and D_a are used to determine whether the DO standard will be violated at the critical point. Under unusual conditions, the engineer may artificially aerate the river with mechanical systems to increase the DO.

When using the DO sag curve to determine the adequacy of wastewater treatment, it is important to use the river conditions that will cause the lowest DO concentration. Usually these conditions occur in the late summer when river flows are low and temperatures are high. A frequently used criterion is the "10-year, 7-day low flow," which is the recurrence interval return period of the average low flow for a 7-day period. Low river flows reduce the dilution of the waste entering the river, causing higher values for L_a and D_a. The value of k_r is usually reduced by low river flows because of reduced velocities. In addition, higher temperatures increase k_d more than k_r and also decrease DO saturation, thus making the critical point more severe.

Nitrogenous BOD. Up to this point, only carbonaceous BOD has been considered in the DO sag curve. However, in many cases nitrogenous BOD has at least as much impact on dissolved oxygen levels. Modern wastewater treatment plants can routinely produce effluents with $CBOD_5$ of less than 30 mg/L. A typical effluent also contains approximately 30 mg/L of nitrogen, which would mean an NBOD of about 137 mg/L if it were discharged as ammonia (see Example 7-8). Nitrogenous BOD can be incorporated into the DO sag curve by adding an additional term to Equation 7-41:

$$D = \frac{k_d L_a}{k_r - k_d} (e^{-k_d t} - e^{-k_r t}) + D_a(e^{-k_r t}) + \frac{k_n L_n}{k_r - k_n}(e^{-k_n t} - e^{-k_r t}) \quad (7\text{-}50)$$

where k_n = the nitrogenous deoxygenation coefficient, d^{-1}; L_n = ultimate nitrogenous BOD after waste and river have mixed, mg/L; and the other terms are as previously defined. It is important to note that with the additional term for NBOD, it is not possible to find the critical time using Equation 7-46. Instead, it must be found by a trial and error solution of Equation 7-50.

Other Factors Affecting DO Levels in Rivers.   The classical DO sag curve assumes that there is only one point-source discharge of waste into the river. In reality, this is rarely the case. Multiple point sources can be handled by dividing the river up into reaches with a point source at the head of each reach. A *reach* is a length of river specified by the engineer on the basis of its homogeneity, that is, channel shape, bottom composition, slope, and so forth. The oxygen deficit and residual BOD can be calculated at the end of each reach. These values are then used to determine new values of D_a and L_a at the beginning of the following reach. Non-point source pollution can also be handled this way if the reaches are made small enough. Non-point source pollution can also be incorporated directly into the DO sag equation for a more sophisticated analysis. Dividing the river into reaches is also necessary whenever the flow regime changes, since the reaeration coefficient would also change. In small rivers, rapids play a major role in maintaining high DO levels. Eliminating rapids by dredging or damming a river can have a severe impact on DO, although DO levels immediately downstream of dams are usually high because of the turbulence of the falling water.

Some rivers contain large deposits of organic matter in the sediments. These can be natural deposits of leaves and dead aquatic plants or can be sludge deposits from wastewaters receiving little or no treatment. In either case, decomposition of this organic matter places an additional burden on the stream's oxygen resources, since the oxygen demand must be supplied from the overlying water. When this benthic demand is significant, compared to the oxygen demand in the water column, it must be included quantitatively in the sag equation.

Aquatic plants can also have a substantial effect on DO levels. During the day, their photosynthetic activities produce oxygen that supplements the reaeration and can even cause oxygen supersaturation. However, plants also consume oxygen for respiration processes. Although there is a net overall production of oxygen, plant respiration can severely lower DO levels during the night. Plant growth is usually highest in the summer when flows are low and temperatures are high, so that large nighttime respiration requirements coincide with the worst cases of oxygen depletion from BOD exertion. In addition, when aquatic plants die and settle to the bottom, they increase the benthic demand. As a general rule, large growths of aquatic plants are detrimental to the maintenance of a consistently high DO level.

Effect of Nutrients on Water Quality in Rivers

Although oxygen-demanding wastes are the most important river pollutants on an overall basis, nutrients can also contribute to deteriorating water quality in rivers by causing excessive plant growth. Nutrients are those elements required by plants for their growth. They include, in order of abundance in plant tissue: carbon, nitrogen, phosphorus, and a variety of trace elements. When there are sufficient quantities of all nutrients available, plant growth is possible. By limiting the availability of any one nutrient, further plant growth is prevented.

Some plant growth is desirable, since plants form the base of the food chain and thus support the animal community. However, excessive plant growth can create a number of undesirable conditions such as thick slime layers on rocks and dense growths of aquatic weeds.

The availability of nutrients is not the only requirement for plant growth. In many rivers, the turbidity caused by eroded soil particles, bacteria, and other factors prevents light from penetrating far into the water, thereby limiting total plant growth in deep water. It is for this reason that slime growths on rocks usually occur in shallow water. Strong water currents also prevent rooted plants from taking hold, and thus limit their growth to quiet backwaters where the currents are weak and the water is shallow enough for light to penetrate.

Effects of Nitrogen.   There are three reasons why nitrogen is detrimental to a receiving body:

1. In high concentrations, NH_3-N is toxic to fish.

2. NH_3, in low concentrations, and NO_3^- serve as nutrients for excessive growth of algae.

3. The conversion of NH_4^+ to NO_3^- consumes large quantities of dissolved oxygen.

Effects of Phosphorus.   The major deleterious effect of phosphorus is that it serves as a vital nutrient for the growth of algae. If the phosphorus availability meets the growth demands of the algae, there is an excessive production of algae. When the algae die, they become an oxygen-demanding organic material as bacteria seek to degrade them. This oxygen demand frequently overtaxes the DO supply of the water body and, as a consequence, causes fish to die.

Management Strategy.   The strategy for managing water quality problems associated with excessive nutrients is based on the sources for each nutrient. Except under rare circumstances, there is plenty of carbon available for plant growth. Plants use carbon dioxide, which is available from the bicarbonate alkalinity of the water and from the bacterial decomposition of organic matter. As carbon dioxide is removed from the water, it is replenished from the atmosphere. Generally, the major source of trace elements is the natural weathering of rock minerals, a process over which the environmental engineer has little control. However, since the acid rain caused by air pollution accelerates the weathering process, air pollution control can help reduce the supply of trace elements. Even when substantial amounts of trace elements are found in wastewater, their removal is difficult. In addition, such small amounts are needed for plant growth that nitrogen or phosphorus is more likely to be the limiting nutrient. Therefore, the practical control of nutrient-caused water-quality problems in streams is based on removal of nitrogen and/or phosphorus from wastewaters before they are discharged and using *best management practices* (BMPs) to minimize nutrients in agricultural runoff.

7-4 WATER POLLUTION IN LAKES

Oxygen-demanding wastes can also be important lake pollutants, especially when the waste is discharged to a contained area such as a bay. Pathogens are of particular concern near bathing beaches. Again, as with rivers, there are special classes of lakes which are most seriously affected by other pollutants such as toxic chemicals from industrial discharges. However, phosphorus so dominates other pollutants in controlling water quality in the vast majority of lakes that we will give it special emphasis.

A knowledge of lake systems is essential to an understanding of the role of phosphorus in lake pollution. The study of lakes is called *limnology*. This section is essentially a short course in limnology as it relates to phosphorus pollution.

Stratification and Turnover

Nearly all lakes in the temperate climatic zone become stratified during the summer and overturn (*turnover*) in the fall due to changes in the water temperature that result from the annual cycle of air temperature changes. In addition, lakes in cold climates undergo winter stratification and spring overturn as well. These physical processes, which are described below, occur regardless of the water quality in the lake. Nonetheless, they do help determine the water quality.

During the summer, the surface water of a lake is heated both indirectly by contact with warm air and directly by sunlight. Warm water, being less dense than cool water, remains near the surface until mixed downward by turbulence from wind, waves, boats, and other forces. Because this turbulence extends only a limited distance below the water surface, the result is an upper layer of well-mixed, warm water (the *epilimnion*) floating on the lower water (the *hypolimnion*), which is poorly mixed and cool, as shown in Figure 7-12a. Because of good mixing the epilimnion will be *aerobic* (have DO). The hypolimnion will have a lower DO and may become *anaerobic* (devoid of oxygen). The boundary is called the *thermocline* because of the sharp temperature change (and therefore density change) that occurs within a relatively short distance. The thermocline may be defined as a change in temperature with depth that is greater than 1°C/m. You may have experienced the thermocline while swimming in a small lake. As long as you are swimming horizontally, the water is warm, but as soon as you tread water or dive, the water turns cold. You have penetrated the thermocline. The depth of the epilimnion is related to the size of the lake. It is as little as one meter in small lakes and as much as 20 meters or more in large lakes. The depth of the epilimnion is also related to storm activity in the spring when stratification is developing. A major storm at the right time will mix warmer water to a substantial depth and thus create a deeper than normal epilimnion. Once formed, lake stratification is very stable.

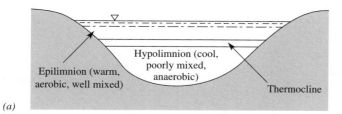

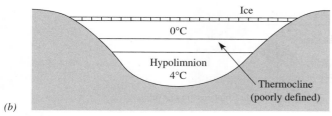

FIGURE 7-12
Stratification of a lake during (*a*) summer and (*b*) winter.

It can be broken only by exceedingly violent storms. In fact, as the summer progresses, the stability increases because the epilimnion continues to warm, while the hypolimnion remains at a fairly constant temperature.

In the fall, as temperatures drop, the epilimnion cools until it is more dense than the hypolimnion. The surface water then sinks, causing *overturning*. The water of the hypolimnion rises to the surface where it cools and again sinks. The lake thus becomes completely mixed. If the lake is in a cold climate, this process stops when the temperature reaches 4°C, since this is the temperature at which water is most dense. Further cooling or freezing of the surface water results in winter stratification, as shown in Figure 7-12*b*. As the water warms in the spring, it again overturns and becomes completely mixed. Thus, temperate climate lakes have at least one, if not two, cycles of stratification and turnover every year.

Biological Zones

Lakes contain several distinct zones of biological activity, largely determined by the availability of light and oxygen. The most important biological zones, shown in Figure 7-13, are the euphotic, littoral, and benthic zones.

Euphotic Zone. The upper layer of water through which sunlight can penetrate is called the *euphotic zone*. All plant growth occurs in this zone. In deep water, algae are the most important plants, while rooted plants grow in shallow water near the shore. The depth of the euphotic zone is determined by the amount of turbidity blocking sunlight penetration. In most lakes, the turbidity is due to algal growth, although color and suspended clays may substantially reduce sunlight penetration in some lakes. In the euphotic zone, plants produce more oxygen by photosynthesis than they remove by respiration. Below the euphotic zone lies the *profundal zone*. The transition between

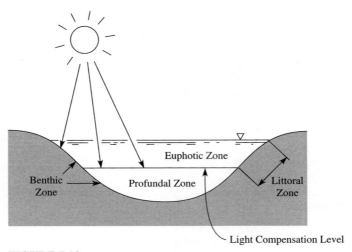

FIGURE 7-13
Biological zones in a lake.

the two zones is called the *light compensation level.* The light compensation level corresponds roughly to a depth at which the light intensity is about 1 percent of unattenuated sunlight. It is important to note that the bottom of the euphotic zone only rarely coincides with the thermocline.

Littoral Zone.   The shallow water near the shore in which rooted water plants can grow is called the *littoral zone.* The extent of the littoral zone depends on the slope of the lake bottom and the depth of the euphotic zone. The littoral zone cannot extend deeper than the euphotic zone.

Benthic Zone.   The bottom sediments comprise the *benthic zone.* As organisms living in the overlying water die, they settle to the bottom where they are decomposed by organisms living in the benthic zone. Bacteria are always present. The presence of higher life forms such as worms, insects, and crustaceans depends on the availability of oxygen.

Lake Productivity

The productivity of a lake is a measure of its ability to support a food web. Algae form the base of this food web, supplying food for the higher organisms. A lake's productivity may be determined by measuring the amount of algal growth that can be supported by the available nutrients. Although a more productive lake usually will have a higher fish population, the number of the most desirable fish may decline. In fact, increased productivity generally results in reduced water quality because of undesirable changes that occur as algal growth increases. Because of the important role productivity plays in determining water quality, it forms a basis for classifying lakes. Table 7-3 shows a lake classification based on productivity.

TABLE 7-3
Lake classification based on productivity

Lake classification		Chlorophyll *a* concentration, μg/L	Secchi depth, m	Total phosphorus concentration, μg/L
Oligotrophic	Average	1.7	9.9	8
	Range	0.3–4.5	5.4–28.3	3.0–17.7
Mesotrophic	Average	4.7	4.2	26.7
	Range	3–11	1.5–8.1	10.9–95.6
Eutrophic	Average	14.3	2.5	84.4
	Range	3–78	0.0–7.0	15–386

Source: Wetzel, 1983.

Oligotrophic Lakes.   Oligotrophic lakes have a low level of productivity due to a severely limited supply of nutrients to support algal growth. As a result, the water is clear enough that the bottom can be seen at considerable depths. In this case, the euphotic zone often extends into the hypolimnion, which is aerobic. Oligotrophic lakes, therefore, support cold water game fish. Lake Tahoe on the California-Nevada border is a classic example of an oligotrophic lake.

Eutrophic Lakes.   Eutrophic lakes have a high productivity because of an abundant supply of algal nutrients. The algae cause the water to be highly turbid, so the euphotic zone may extend only partially into the epilimnion. As the algae die, they settle to the lake bottom where they are decomposed by benthic organisms. In a eutrophic lake, this decomposition is sufficient to deplete the hypolimnion of oxygen during summer stratification. Because the hypolimnion is anaerobic during the summer, eutrophic lakes support only warm-water fish. In fact, most cold-water fish are driven out of the lake long before the hypolimnion becomes anaerobic because they generally require dissolved oxygen levels of at least 5 mg/L. Highly eutrophic lakes may also have large mats of floating algae that typically impart unpleasant tastes and odors to the water.

Mesotrophic Lakes.   Lakes which are intermediate between oligotrophic and eutrophic are called mesotrophic. Although substantial depletion of oxygen may have occurred in the hypolimnion, it remains aerobic.

Senescent Lakes.   These are very old, shallow lakes which have thick organic sediments and rooted water plants in great abundance. These lakes will eventually become marshes.

Eutrophication

Eutrophication is a natural process in which lakes gradually become shallower and more productive through the introduction and cycling of nutrients. Thus, oligotrophic lakes gradually pass through the mesotrophic, eutrophic, and senescent stages, eventually filling completely. The time for this process to occur depends on the original size of the lake and on the rate at which sediments and nutrients are introduced. In some lakes the eutrophication process is so slow that thousands of years may pass with little change in water quality. Other lakes may have been eutrophic from the day they were formed, if nutrient levels were high at that time.

Cultural eutrophication is caused when human activity speeds the processes naturally occurring by increasing the rate at which sediments and nutrients are added to the lake. Thus, lake pollution can be seen as the intensification of a natural process. This is not to say that eutrophic lakes are necessarily polluted, but that pollution contributes to eutrophication. Water quality management in lakes is primarily concerned with slowing eutrophication to at least the natural rate. To understand the factors involved in eutrophication, it is necessary to understand the factors contributing to algal growth.

Algal Growth Requirements

All algae require macronutrients, such as carbon, nitrogen, and phosphorus, and micronutrients, such as trace elements. For algae to grow, all nutrients must be available. Lack of any one nutrient will limit the total algal population. The availability of each nutrient and its natural cycle are summarized below.

Carbon. Algae obtain their carbon from carbon dioxide dissolved in the water. Because the carbon dioxide is in equilibrium with the bicarbonate buffer system (see Chapter 6), the immediately available carbon is determined by the alkalinity of the water. However, as carbon dioxide is removed from the water, it is replenished from the atmosphere. The atmosphere is, of course, a virtually inexhaustible source of this gas. When algae are either consumed by higher organisms or die and decompose, the organic carbon is oxidized back to carbon dioxide which returns either to the water or to the atmosphere to complete the carbon cycle.

Nitrogen. Nitrogen in lakes is usually in the form of nitrate (NO_3^-) and comes from external sources by way of inflowing streams or groundwater. When taken up for algal growth, the nitrogen is chemically reduced to amino-nitrogen (NH_2^-) and incorporated into organic compounds. When dead algae undergo decomposition, the organic nitrogen is released to the water as ammonia (NH_3). The ammonia is then oxidized back to nitrate by bacteria in the same nitrification process discussed earlier in river systems.

Nitrogen cycles from nitrate to organic nitrogen, to ammonia, and back to nitrate as long as the water remains aerobic. However, in anaerobic sediments, and in the hypolimnion of eutrophic lakes, when algal decomposition has depleted the oxygen supply, nitrate is reduced by anaerobic bacteria to nitrogen gas (N_2) and lost from the system in a process called *denitrification*. Denitrification reduces the average time nitrogen remains in the lake system. The denitrification reaction is

$$2NO_3^- + \text{organic carbon} \rightleftharpoons N_2 + CO_2 + H_2O \qquad (7\text{-}51)$$

Some photosynthetic microorganisms can also fix nitrogen gas from the atmosphere by converting it to organic nitrogen. In lakes the most important nitrogen-fixing microorganisms are photosynthetic bacteria called cyanobacteria, formerly known as blue-green algae because of the pigments they contain. Because of their nitrogen-fixing ability, cyanobacteria have a competitive advantage over green algae when nitrate and ammonium concentrations are low but other nutrients are sufficiently abundant. These cyanobacteria are generally undesirable because of their tendency to aggregate in unsightly floating mats and because they impart unpleasant odor and taste to the water. Cyanobacteria can also produce toxins which kill fish. Fortunately, these nuisance organisms are not prevalent unless the supply of soluble fixed nitrogen is reduced to low levels.

Phosphorus. Phosphorus in lakes originates from external sources and is taken up by algae in the inorganic form (PO_4^{3-}) and incorporated into organic compounds. During algal decomposition, phosphorus is returned to the inorganic form. The release of phosphorus from dead algal cells is so rapid that only a little of it leaves the epilimnion with the settling algal cells. However, little by little, phosphorus is transferred to the sediments, some of it in undecomposed organic matter; some of it in precipitates of iron, aluminum, and calcium; and some bound to clay particles. To a large extent, the permanent removal of phosphorus from the overlying waters to the sediments depends on the amount of iron, aluminum, calcium, and clay entering the lake along with phosphorus.

Trace Elements. The quantities of trace elements required to support algal growth are so small that most fresh waters have sufficient amounts for a substantial algal population.

The Limiting Nutrient

In 1840, Justin Liebig formulated the idea that "growth of a plant is dependent on the amount of foodstuff that is presented to it in minimum quantity." This is now known as *Liebig's law of the minimum*. As applied to algae, it means that algal growth will be limited by the nutrient that is least available. Of all the nutrients, only phosphorus is not readily available from the atmosphere or the natural water supply. For this reason, phosphorus is deemed the *limiting nutrient* in lakes. The amount of phosphorus controls the quantity of algal growth and therefore the productivity of lakes. This can be seen from Figure 7-14 in which the concentration of *chlorophyll a* is plotted against phosphorus concentration. Chlorophyll *a*, one of the green pigments involved in photosynthesis, is found in all algae, so it is used to distinguish the amount of algae in the water from other organic solids such as bacteria. It has been estimated that the phosphorus concentration should be below 0.010 to 0.015 mg/L to limit algae blooms (Vollenweider, 1975).

Control of Phosphorus in Lakes

Because phosphorus is usually the limiting nutrient, control of cultural eutrophication must be accomplished by reducing the input of phosphorus to the lake. Once the input is reduced, the phosphorus concentration will gradually fall as phosphorus is buried in the sediment or flushed from the lake. Other strategies for reversing or slowing the eutrophication process, such as precipitating phosphorus with additions of aluminum (alum) or removing phosphorus-rich sediments by dredging, have been proposed. However, if the input of phosphorus is not also curtailed, the eutrophication process will continue. Thus, dredging or precipitation alone can result only in

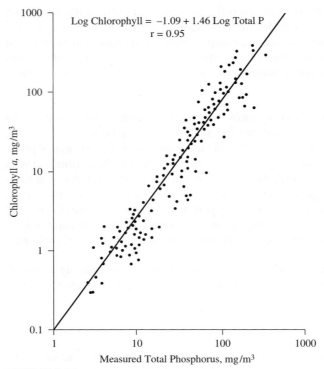

FIGURE 7-14

Relationship between summer levels of chlorophyll *a* and measured total phosphorus concentration for 143 lakes. (*Source:* Copyright © 1976 Water Environment Federation. Used with permission.)

temporary improvement in water quality. In conjunction with reduced phosphorus inputs, these measures can help speed up the removal of phosphorus already in the lake system. Of course, the need to speed the recovery process must be weighed against the potential damage from inundating shoreline areas with sludge and stirring up toxic compounds buried in the sediment.

To be able to reduce phosphorus inputs, it is necessary to know the sources of phosphorus and the potential for their reduction. The natural source of phosphorus is the weathering of rock. Phosphorus released from the rock can enter the water directly, but more commonly it is taken up by plants and enters the water in the form of dead plant matter. It is exceedingly difficult to reduce the natural inputs of phosphorus. If these sources are large, the lake is generally naturally eutrophic. For many lakes the principal sources of phosphorus are the result of human activity. The most important sources are municipal and industrial wastewaters, seepage from septic tanks, and agricultural runoff that carries phosphorus fertilizers into the water.

Municipal and Industrial Wastewaters. All municipal sewage contains phosphorus from human excrement. Many industrial wastes are high in this nutrient. In these cases, the only effective way of reducing phosphorus is through advanced waste treatment processes, which are discussed in Chapter 6. Municipal wastewaters also contain large quantities of phosphorus from detergents containing polyphosphate, which is a chain of phosphate ions (usually three) linked together. The polyphosphate binds with hardness in water to make the detergent a more effective cleaning agent. By the 1970s, phosphorus loading from detergents was approximately twice that from human excrement. Today's detergents do not contain phosphorus because the manufacturers have replaced it with other chemicals.

Septic Tank Seepage.   The shores of many lakes are dotted with homes and summer cottages, each with its own septic tank and tile field for waste disposal. As treated wastewater moves through the soil toward the lake, phosphorus is adsorbed by soil particles, especially clay. Thus, during the early life of the tile field, very little phosphorus gets to the lake. However, with time, the capacity of the soil to adsorb phosphorus is exceeded and any additional phosphorus will pass on into the lake, contributing to eutrophication. The time it takes for phosphorus to break through to the lake depends on the type of soil, the distance to the lake, the amount of wastewater generated, and the concentration of phosphorus in that wastewater. To prevent phosphorus from reaching the lake, it is necessary to put the tile field far enough from the lake that the adsorption capacity of the soil is not exceeded. If this is not possible, it may be necessary to replace the septic tanks and tile fields with a sewer to collect the wastewater and transport it to a treatment facility.

Agricultural Runoff.   Because phosphorus is a plant nutrient, it is an important ingredient in fertilizers. As rain water washes off fertilized fields, some of the phosphorus is carried into streams and then into lakes. Most of the phosphorus not taken up by growing plants is bound to soil particles. Bound phosphorus is carried into streams and lakes through soil erosion. Waste minimization can be applied to the control of phosphorus loading to lakes from agricultural fertilization by encouraging farmers to fertilize more often with smaller amounts and to take effective action to stop soil erosion.

Acidification of Lakes

Pure rainwater is slightly acid. As we discussed in Chapter 5, CO_2 dissolves in water to form carbonic acid (H_2CO_3). The equilibrium concentration of H_2CO_3 results in a rainwater pH of approximately 5.6. Thus, acid rain is usually defined to be precipitation with a pH less than 5.6. The northeastern U.S. and Canada frequently record rainwater pH values between 4 and 5 (Figure 7-15). These low pH values have been attributed to emissions of sulfur and nitrogen oxides from the combustion of fossil fuels (see Chapter 9).

   Fish, and in particular trout and Atlantic salmon, are very sensitive to low pH levels. Most are severely stressed if the pH drops below 5.5, and few are able to survive if the pH falls below 5.0. If the pH falls below 4.0, cricket frogs and spring peepers experience mortalities in excess of 85 percent.

   High aluminum concentrations are often the trigger that kills fish. Aluminum is abundant in soil but it is normally bound up in the soil minerals. At normal pH values aluminum rarely occurs in solution. Acidification of the water releases highly toxic Al^{3+} to the water.

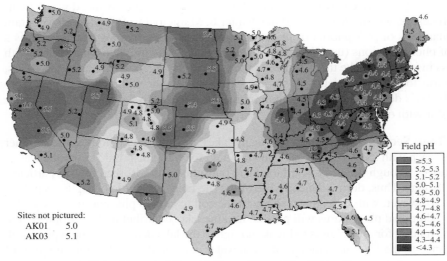

National Atmospheric Deposition Program/National Trends Network
http://nadp.sws.uiuc.edu

FIGURE 7-15

Hydrogen ion concentration as pH from measurements of rainwater. (*Source:* National Atmospheric Deposition Program (NRSP-3)/National Trends Network (2003). NADP Program Office, Illinois State Water Survey, 2204 Griffith Dr. Champaign, IL 61820. Data for 2001 now available. http://nadp.sws. uiuc.edu/isopleths/maps1997/phfield.gif.)

Most lakes are buffered by the carbonate buffer system (see Chapter 5). To the extent that the buffer capacity of the lake is not exceeded, the pH of the lake will not be appreciably affected by acid rain. If there is a source of carbonate to replace that consumed by the acid rain, the buffering capacity can be quite large. Calcareous soils are those containing large quantities of calcium carbonate ($CaCO_3$). As shown in Figure 6-10, carbonic acid releases bicarbonate into solution. H^+ from acid rain will also release bicarbonate. Thus, lakes formed in calcareous soils tend to be resistant to acidification.

Other factors that affect the susceptibility of a lake to acidification are the permeability and depth of the soil, the bedrock, the slope and size of the watershed, and the type of vegetation. Thin, impermeable soils provide little time for contact between the soil and the precipitation. This reduces the potential for the soil to buffer the acid precipitation. Likewise, small watersheds with steep slopes reduce the time for buffering to occur. Deciduous foliage tends to decrease acidity. Coniferous foliage tends to yield runoff that is more acid than the precipitation itself. Granite bedrock offers little potential to buffer acid rain. Galloway and Cowling (1978) used bedrock geology to predict areas where lakes are potentially most sensitive to acid rain (Figure 7-16). You may note that the predicted areas of sensitivity are also those subjected to very acid precipitation.

The control of lake acidification is related to the control of atmospheric emissions of sulfur and nitrogen oxides. The role of air pollution in acid deposition is discussed in more detail in Chapter 9.

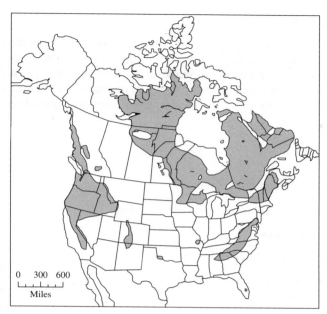

FIGURE 7-16
Regions in North America containing lakes sensitive to acidification
by acid precipitation. The shaded areas have igneous or metamorphic
bedrock geology; the unshaded areas have calcereous or sedimentary
bedrock geology. Regions having low alkalinity lakes are concurrent with
regions of igneous and metamorphic bedrock geology. (*Source*: Galloway
and Cowling, 1978.)

7-5 WATER POLLUTION IN ESTUARIES

An *estuary* is formed along the coastline where freshwater from rivers and streams
flows into the ocean. It is a place of transition, where freshwater mixes with saltwater.
Estuaries are influenced by tides. The Bay of Fundy, Boston Harbor, Chesapeake Bay,
Corpus Christi Bay, Florida Bay, Hudson River estuary, Mobile Bay, Puget Sound, and
San Francisco Bay are examples of estuaries.

The mix of saltwater and freshwater combined with the diurnal cycle of tidal
wetting and drying and changes in salinity make the estuary ecosystem extremely
complex. Numerous different habitats support an abundance of a wide variety of
different animals. Each of these is sensitive to minor changes in the quality of the
water.

Like other surface waters, estuaries are subject to a plethora of pollutants. Although the technologies and policies that serve to manage the water quality of
lakes and rivers also apply to estuaries, the management of the water quality is
complicated by the number of political jurisdictions that touch on the estuary as
well as the competing needs of organisms that live there. Competing with the environmental value is the economic value of the estuary to commerce (Costanza
and Voinov, 2000).

7-6   GROUNDWATER POLLUTION

Groundwater, by its very location, has a large measure of protection from contaminants that are found in surface waters. However, once groundwater becomes contaminated, its location and low rate of replacement with freshwater makes it difficult to return it to a pristine state. Two major source of contaminants are of concern: uncontrolled releases of biological and chemical contaminants and saltwater intrusion from over pumping of wells.

Uncontrolled Releases

Uncontrolled releases come from a variety of sources:

- Discharge from improperly operated or located septic systems

- Leaking underground storage tanks

- Improper disposal of hazardous or other chemical wastes

- Spills from pipelines or transportation accidents

- Recharge of groundwater with contaminated surface water

- Leaking landfills

- Leaking retention ponds or lagoons

The properties of the contaminant and the aquifer material govern their migration of the contaminants. Water-soluble chemicals are likely to move vertically down through the soil to the aquifer and migrate with the water as shown in Figure 7-17. Nitrate, methyl tertiary butyl ketone (a gasoline additive), and methamidophos (a pesticide) are examples of water soluble chemicals.

Chemicals that are only sparingly soluble migrate through the aquifer as a separate nonaqueous phase. These chemicals are known as *nonaqueous phase liquids* (NAPLs). The NAPLs are divided into two categories based on their density. The light NAPLs or (LNAPLs) are less dense than water and will tend to "float" on the water table as shown in Figure 7-18. Some of the chemical will dissolve into the groundwater and some of the chemical may volatilize into the pore spaces of the unsaturated zone. Examples of LNAPLs include gasoline and aviation fuel. The dense NAPLs (DNAPLs) have a density greater than that of water. They will tend to "sink" in the aquifer until they reach an impervious barrier. The behavior of DNAPLs is shown in Figure 7-19. Trichloroethylene, tetrachloroethylene, and PCB-laden oils are examples of DNAPLs.

Management techniques for contaminant control in groundwater are discussed in Chapter 12.

Saltwater Intrusion

Freshwater aquifers near oceans or above saline aquifers may become contaminated with saltwater when water is pumped from wells that draw water that is too close to the freshwater-saltwater interface.

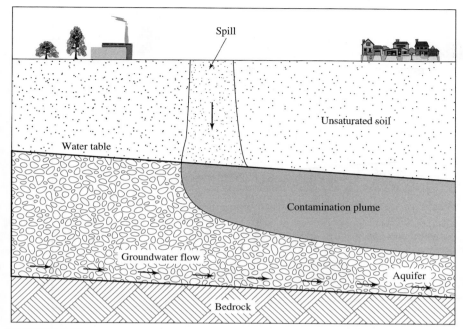

FIGURE 7-17
Dissolved contamination plume.

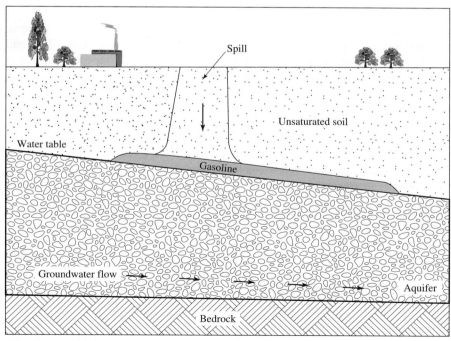

FIGURE 7-18
Immiscible plume less dense than water.

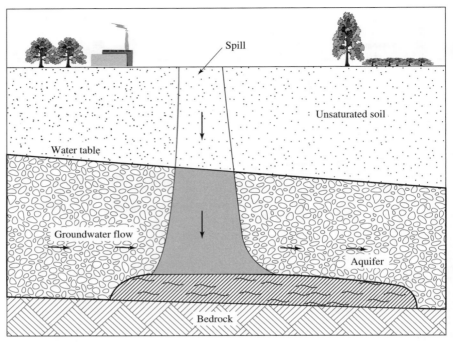

FIGURE 7-19
Immiscible plume denser than water.

Because seawater is heavier than freshwater, it will form a saltwater wedge in aquifers that drain into the ocean (Figure 7-20). Using the notation in Figure 7-20 and assuming that fresh water moves horizontally to the ocean and that the interface is abrupt and that it occurs at the shoreline, we can express the freshwater pressure at any point of the interface as

$$P_f = (h + z)\rho_f \tag{7-52}$$

where P_f = freshwater pressure, Pa
$\quad h$ = height of water table above sea level, m
$\quad z$ = distance of interface below sea level, m
$\quad \rho_f$ = density of fresh water, kg/m^3

This pressure must be the same as the saltwater pressure on the other side of the interface; that is, P_f is also equal to $z\rho_s$. Equating the two expressions and solving for z gives

$$z = \frac{\rho_f}{\rho_s - \rho_f} h \tag{7-53}$$

where ρ_s = density of saltwater, g/mL
$\quad \rho_f$ = density of freshwater, g/mL

Taking the density of fresh water as 1.000 g/mL and that of seawater as 1.025 g/mL, we can use this expression to estimate that, for coastal waters, $z \approx 40h$. Thus, for every

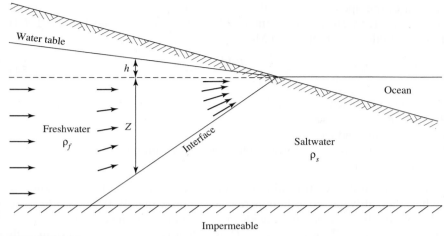

FIGURE 7-20
Freshwater-saltwater interface in coastal aquifer draining into ocean. (*Source:* Bouwer, 1978.)

meter that the water table is above sea level, there will be 40 m of freshwater before saltwater is encountered. Actual conditions are not as simple as those assumed for Figure 7-20 and a much more complex expression must be used for accurate estimates of the position of the interface (Bouwer, 1978).

When fresh groundwater is underlain by saline water, pumping a well in the freshwater will cause the interface to rise below the well as shown in Figure 7-21. This *upconing* is in response to the pressure reduction on the interface that results from the drawdown of the water table around the well. If the well screen is close to the saline water or the pumping rate is too high, saltwater may be drawn into the well.

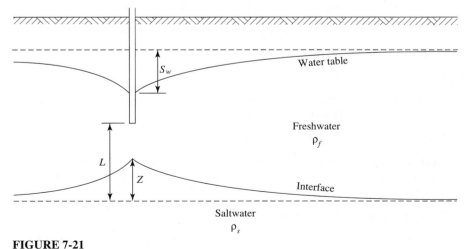

FIGURE 7-21
Geometry and symbols for upconing of saltwater beneath a pumped well (dashed lines represent static positions of water table and interface). (*Source:* Bouwer, 1978.)

The height of the upconing below the center of a well after an infinite pumping time at the freshwater well may be estimated by using the Bear and Dagan (1968) equation as modified by Schmorak and Mercado (1969):

$$z_\infty = \frac{\rho_f Q}{2\pi(\rho_s - \rho_f)KL} \tag{7-54}$$

where z_∞ = rise of the cone center at $t = \infty$, m
  Q = well discharge, m^3/d
  K = hydraulic conductivity, m/d
  L = depth of freshwater-saltwater interface below well bottom prior to pumping, m

A critical point in the relationship between the upconing and the drawdown is reached when the saltwater cone height reaches approximately $0.4L$ to $0.6L$ (see Figure 7-21). When the cone height exceeds this critical height, the cone may "jump" to the bottom of the well (Bouwer, 1978). Thus, when freshwater is underlain with saline water, prediction of the upconing is important to prevent *saltwater intrusion*.

7-7   CHAPTER REVIEW

When you have completed studying this chapter, you should be able to do the following without the aid of your textbook or notes:

1. List the major pollutant categories (there are four) that are produced by each of the four principal sources of wastewater.

2. Explain the difference between point sources and non-point sources of pollution.

3. Explain the difference between AFOs and CAFOs and determine whether or not an animal feeding operation is a CAFO on the basis of data you are given.

4. List the two nutrients of primary concern with respect to a receiving body of water.

5. Define TMDL and explain how it is calculated.

6. Define biochemical oxygen demand (BOD).

7. Explain the procedure for determining BOD and specify the nominal values of temperature and time used in the test.

8. List three reasons why the BOD rate constant may vary.

9. Sketch a graph showing the effect of varying rate constant on 5-day BOD if the ultimate BOD is the same, and the effect on ultimate BOD if the 5-day BOD is the same.

10. Utilizing Equation 7-20 in your answer, explain what causes nitrogenous BOD.

11. Sketch a series of curves that show the deoxygenation, reaeration, and DO sag in a river. Show the effect of a change in the deoxygenation or reaeration rate on the location of the critical point and the magnitude of the DO deficit.

12. List three reasons why ammonia nitrogen is detrimental to a receiving body of water and its inhabitants.

13. Sketch and compare the epilimnion and hypolimnion with respect to the following: location in a lake, temperature, and oxygen abundance (that is, DO).

14. Describe the process of stratification and turnover in lakes.

15. Explain what determines the euphotic zone of a lake and what significance this has for biological growth.

16. Given a description of a lake that includes the productivity, clarity, and oxygen levels, classify it as oligotrophic, mesotrophic, eutrophic, or senescent.

17. Explain the process of eutrophication.

18. State Liebig's law of the minimum.

19. Name the most common "limiting nutrient" in lakes and explain why it is a limiting nutrient.

20. List three sources of phosphorus that must be controlled to reduce cultural eutrophication of lakes.

21. Explain why the pH of pure rainwater is about 5.6.

22. Define acid rain.

23. Explain why acid rain is of concern.

24. Explain the role of calcareous soils in protecting lakes from acidification.

25. Other than rainwater pH, list six variables that determine the extent of lake acidification and explain how increasing or decreasing the value of each might be expected to change the extent of acidification.

26. Identify the two components that define an estuary.

27. Describe the process of saltwater intrusion into a freshwater well.

With the aid of this text, you should be able to do the following:

28. Calculate the ThOD of a compound given the balanced oxidation reaction(s).

29. Calculate the BOD_5, given the sample size and oxygen consumption, or calculate the sample size, given the allowable oxygen consumption and estimated BOD_5.

30. Calculate the ultimate BOD (L_o), given the BOD exerted (BOD_t) in time t and rate constant k, or calculate the rate constant k, given L_o and BOD_5.

31. Calculate a new k for a temperature other than 20°C, given a value at T °C.

32. Calculate the BOD rate constant k and ultimate BOD (L_o) from experimental data of BOD versus time.

33. Calculate the oxygen deficit D in a length of stream (reach), given the required input data.

34. Calculate the critical oxygen deficit DO at the DO sag point (minimum).

35. Estimate the height of upconing from a calculation of drawdown (Chapter 4).

7-8 PROBLEMS

7-1. Glutamic acid ($C_5H_9O_4N$) is used as one of the reagents for a standard to check the BOD test. Determine the theoretical oxygen demand of 63 mg/L of glutamic acid. Assume the following reactions apply:

$$C_5H_9O_4N + 4.5O_2 \rightleftharpoons 5CO_2 + 3H_2O + NH_3$$
$$NH_3 + 2O_2 \rightleftharpoons NO_3 + H^+ + H_2O$$

Answer: ThOD = 89.14 or 89 mg/L

7-2. Bacterial cells have been represented by the chemical formula $C_5H_7NO_2$. Compute the theoretical oxygen demand of 30 mg/L of bacterial cells, assuming the following reactions apply:

$$C_5H_7NO_2 + 5O_2 \rightleftharpoons 5CO_2 + 2H_2O + NH_3$$
$$NH_3 + 2O_2 \rightleftharpoons NO_3 + H^+ + H_2O$$

7-3. A step in the anaerobic decomposition of organic wastes produces acetic acid (CH_3COOH). Determine the theoretical oxygen demand of 300 mg/L of acetic acid. Assume the following reaction applies:

$$CH_3COOH + 2O_2 \rightleftharpoons 2CO_2 + H_2O$$

7-4. If the BOD_5 of a waste is 220.0 mg/L and the ultimate BOD is 320.0 mg/L, what is the rate constant? Assume the temperature is 20 °C.

Answer: $k = 0.233$ d^{-1}

7-5. If the BOD of a municipal wastewater at the end of 7 days is 60.0 mg/L and the ultimate BOD is 85.0 mg/L, what is the rate constant? Assume the temperature is 20°C.

7-6. If the BOD_6 of a municipal wastewater is 213 mg/L and the ultimate BOD is 318.4 mg/L, what is the rate constant (base e)? Assume the temperature is 20°C.

7-7. Assuming that the data in Problem 7-4 were taken at 20°C, compute the rate constant at a temperature of 15°C.

Answer: $k = 0.1235$ d^{-1}

7-8. Assuming that the data in Problem 7-5 were taken at 25°C, compute the rate constant at 16°C.

7-9. What is the BOD_5 of a waste that yields an oxygen consumption of 2.00 mg/L from a 1.00 percent sample?

Answer: $BOD_5 = 200$ mg/L

7-10. What sample size (in percent) is required for a BOD_5 of 350.0 mg/L if the oxygen consumed is to be limited to 4.00 mg/L?

7-11. If the BOD_5 of a waste is 327 mg/L, what sample size (in percent) should be selected to yield an oxygen consumption of 4.8 mg/L?

7-12. If the ultimate BOD of two wastes having k values of $0.0800\ d^{-1}$ and $0.120\ d^{-1}$ is 280.0 mg/L, what would be the 5-day BOD for each?

    Answer: For $k = 0.08\ d^{-1}$, $BOD_5 = 92$ mg/L; for $k = 0.12\ d^{-1}$, $BOD_5 = 126$ mg/L

7-13. If the BOD_5 of two wastes having k values of $0.0800\ d^{-1}$ and $0.120\ d^{-1}$ is 280.0 mg/L, what would be the ultimate BOD for each?

7-14. Using a computer spreadsheet program you have written, plot the BOD curves that would result for the data given in Problem 7-12. At approximately what day (± 5.0 d) does the ultimate BOD occur? Check your answer, using Equation 7-4.

7-15. Using a computer spreadsheet program you have written, plot the BOD curves that would result for the data given in Problem 7-13. At approximately what day (± 5.0 d) would the ultimate BOD occur for each waste? Check your answer, using Equation 7-4.

7-16. Using Thomas' graphical method and a computer spreadsheet program you have written, calculate the BOD rate constant in base e and the ultimate BOD from the following data:

Day	BOD, mg/L
2	70.0
5	102.4
7	111.00
8	114.0
10	118.8

    Answers: $k = 0.36\ d^{-1}$; $L_0 = 129.69$, or 130 mg/L

7-17. Using Thomas' graphical method and a computer spreadsheet program you have written, calculate the BOD rate constant in base e from the following data:

Day	BOD, mg/L
2	119
5	210
10	262
20	279
35	279.98

7-18. Using Thomas' graphical method and a computer spreadsheet program you have written, calculate the BOD rate constant in base e from the following data:

Day	BOD, mg/L
2	86
5	169
10	236
20	273
35	279.55

7-19. Using the data from Problem 7-1, calculate the theoretical NBOD of glutamic acid.

> *Answer:* Theo. NBOD = 27.42 or 27 mg/L

7-20. Using the data from Problem 7-2, calculate the theoretical NBOD of bacterial cells.

7-21. Calculate the NBOD of 200 mg/L of a casein ($C_8H_{12}O_3N_2$) dairy waste. Assume the following reactions:

$$C_8H_{12}O_3N_2 + 8O_2 \rightleftharpoons 8CO_2 + 3H_2O + 2NH_3$$
$$NH_3 + 2O_2 \rightleftharpoons NO_3 + H^+ + H_2O$$

7-22. Derive an expression for the final temperature T_s of the mixture of wastewater flow Q_w at temperature T_w and river flow Q_r at temperature T_r. Assume that the specific heat and density of the wastewater and the river are the same.

7-23. A tannery with a wastewater flow of 0.011 m³/s and a BOD$_5$ of 590 mg/L discharges into the Cattaraugus Creek (Nemerow, 1974). The creek has a 10-year, 7-day low flow of 1.7 m³/s. Upstream of the tannery, the BOD$_5$ of the creek is 0.6 mg/L. The BOD rate constants k are 0.115 d^{-1} for the tannery and 3.7 d^{-1} for the creek. The temperature of both the creek and the tannery wastewater is 20°C. Calculate the initial ultimate BOD after mixing.

> *Answers:* $L_{o\text{Tannery}}$ = 1,349.2 mg/L, $L_{o\text{Creek}}$ = 0.6 mg/L, L_a = 9.27 or 9 mg/L

7-24. The town of Pittsburgh discharges 0.126 m³/s of treated wastewater into Cherry Creek. The BOD$_5$ of the wastewater is 34 mg/L. Cherry Creek has a 10-year, 7-day low flow of 0.126 m³/s. Upstream of the wastewater outfall from Pittsburgh, the BOD$_5$ is 1.2 mg/L. The BOD rate constants k are 0.222 d^{-1} and 0.090 d^{-1} for the wastewater and creek respectively. The temperature of both the creek and the municipal wastewater is 20°C. Calculate the initial ultimate BOD after mixing.

7-25. A short distance downstream from the tannery in Problem 7-23, a glue factory and a municipal wastewater treatment plant also discharge into Cattaraugus Creek. The wastewater flows and ultimate BODs for these discharges are listed below. Determine the initial ultimate BOD after mixing of the creek and the three wastewater discharges.

Source	Flow, m³/s	Ultimate BOD, mg/L	Temperature, °C
Glue factory	0.13	255	20
Municipal WWTP	0.02	75	20

7-26. Cherry Creek is joined by Peach Tree Creek and Apple Creek to form the Ambrosia River. The flows and ultimate BODs for these creeks are listed below. Determine the initial ultimate BOD after mixing of the Ambrosia River.

Source	Flow, m³/s	Ultimate BOD, mg/L	Temperature, °C
Cherry Creek	0.252	27	20
Peach Tree Creek	0.13	8	20
Apple Creek	0.02	16	20

7-27. Compute the deoxygenation rate constant and reaeration rate constant (base e) for the following wastewater and stream conditions.

Source	k, d⁻¹	Temperature, °C	H, m	v, m/s	η
Wastewater	0.20	20			
Stream		20	1.0	0.5	0.4

7-28. As a result of snowmelt and spring flooding, the stream conditions in Problem 7-27 change as shown below. Determine the values of k_d and k_r for flood conditions.

Source	Q, m³/s	k, d⁻¹	Temperature, °C	H, m	v, m/s	η
Wastewater	0.126	0.20	20			
Stream	0.252		10	4.0	2.5	0.6

7-29. The initial ultimate BOD after mixing of the Noir River is 50 mg/L. The DO in the river after the wastewater and river have mixed is at saturation. The river temperature is 10°C. At 10°C, the deoxygenation rate constant k_r is 0.30 d⁻¹ and the reaeration rate constant k_r is 0.30 d⁻¹. Determine the critical point t_c and the critical DO.

7-30. Repeat Problem 7-29, assuming the river temperature rises to 15°C so that k_d and k_r change.

7-31. Churchill, Elmore, and Buckingham (1962) developed the following reaeration equation based on studies of Tennessee Valley rivers:

$$k_r = \frac{5.23v}{H^{1.67}}$$

Compare this equation with the O'Connor and Dobbins equation (Equation 7-44) using a computer spreadsheet program you have written. Plot the values of k_r as a function of the speed of the stream flow for each of these equations, using four speeds (0.05, 0.10, 0.20, and 0.40 m/s) and a depth of 1.0 m. Assume the stream temperature is 20°C. Discuss the possible reasons for the results you observe.

7-32. The discharge from a sugar beet plant causes the DO at the critical point to fall to 4.0 mg/L. The stream has a negligible BOD and the initial deficit after the river and wastewater have mixed is zero. What DO will result if the concentration of the waste (L_w) is reduced by 50 percent? Assume that the flows remain the same and that the saturation value of DO is 10.83 mg/L in both cases.

7-33. The Big Bear town council has asked that you determine whether the discharge of the town's wastewater into the Salmon River will reduce the DO below the state standard of 5.00 mg/L at Alittlebit, 5.79 km downstream, or at any other point downstream. The pertinent data are as follows:

Parameter	Big Bear wastewater	Salmon River
Flow, m³/s	0.280	0.877
Ultimate BOD at 28°C, mg/L	6.44	7.00
DO, mg/L	1.00	6.00
k_d at 28°C, d^{-1}	N/A	0.458
k_r at 28°C, d^{-1}	N/A	0.852
Speed, m/s	N/A	0.650
Temperature, °C	28°C	28°C

Answers: DO at Alittlebit = 4.79 mg/L
Critical DO = 4.74 mg/L at t_c = 0.3113 d

7-34. If the Salmon River temperature in Problem 7-33 decreases to 12°C, will the discharge from Big Bear reduce the DO in the river below 5.00 mg/L at a distance of 5.79 km downstream? *Note:* You must calculate a new temperature of the mixed river water and wastewater, then correct k_d and k_r of the river only.

7-35. Calculate the DO at a point 1.609 km downstream from a waste discharge point for the following conditions. Report your answer to two decimal places. Rate constants are already temperature-adjusted.

Parameter	Stream
k_d	1.911 d^{-1}
k_r	4.49 d^{-1}
Flow	2.4 m^3/s
Speed	0.100 m/s
D_a (after mixing)	0.00
Temperature	17.0°C
BOD$_L$ (after mixing)	1,100.00 kg/d

Answer: DO = 8.69 or 8.7 mg/L

7-36. Calculate the DO at a point 2.880 km downstream from a waste discharge point for the following conditions. Report your answer to two decimal places. Rate constants are already temperature adjusted.

Parameters	Stream
k_d	4.215 d^{-1}
k_r	4.675 d^{-1}
Flow	0.30 m^3/s
Speed	0.100 m/s
D_a (after mixing)	0.00
Temperature	18.0°C
BOD$_L$ (after mixing)	1,125.00 kg/d

Hint: Problems 7-37 through 7-47 require solution of a long series of sequential equations. Some of the problems require iterative solutions. Even when an iterative solution is not required, a computer spreadsheet solution that computes the results of each of the sequential equations is recommended. This will allow you to correct small mistakes in equations you might have made early in the sequence without time-consuming recomputation of the subsequent equations. Your time is valuable, make efficient use of it!

7-37. The town of Avepitaeonmi has filed a complaint with the state Department of Natural Resources (DNR) that the City of Watapitae is restricting its use of the Wash River because of the discharge of raw sewage. The DNR water quality criterion for the Wash River is 5.00 mg/L of DO. Avepitaeonmi is 15.55 km downstream from Watapitae. What is the DO at Avepitaeonmi? What is the critical DO and where (at what distance) downstream does it occur? Is the assimilative capacity of the river restricted? The following data pertain to the 7-year, 10-day low flow at Watapitae.

Parameter	Watapitae wastewater	Wash River
Flow, m^3/s	0.1507	1.08
BOD$_5$ at 16°C, mg/L	128.00	N/A
BOD$_u$ at 16°C, mg/L	N/A	11.40
DO, mg/L	1.00	7.95
Temperature, °C	16.0	16.0
k at 20°C, d^{-1}	0.4375	N/A
Speed, m/s	N/A	0.390
Depth, m	N/A	2.80
Bed-activity coefficient	N/A	0.200

7-38. Under the provisions of the Clean Water Act, the U.S. Environmental Protection Agency established a requirement that municipalities had to provide secondary treatment of their waste. This was defined to be treatment that results in an effluent BOD$_5$ that does not exceed 30 mg/L. The discharge from Watapitae (Problem 7-37) is clearly in violation of this standard. Given the data in Problem 7-37, rework the problem assuming that Watapitae provides treatment to lower the BOD$_5$ to 30.00 mg/L.

7-39. The Blue Ox Tannery has asked you to assist them in preparations to file for a NPDES permit to discharge wastewater into the Zmellsbad River. The DNR water quality criterion for the Zmellsbad River is 5.00 mg/L of DO. They have asked you the following questions: What is the critical DO and where (at what distance) downstream does it occur? They have provided you with the following data. They pertain to the 7-year, 10-day low flow at the tannery.

Parameter	Blue Ox wastewater	Zmellsbad River
Flow, m^3/s	1.148	7.222
BOD$_5$ at 15°C, mg/L	90.00	N/A
BOD$_u$ at 15°C, mg/L	N/A	7.66
DO, mg/L	1.00	6.00
Temperature, °C	15.0	15.0
k at 20°C, d^{-1}	0.3685	N/A
Speed, m/s	N/A	0.300
Depth, m	N/A	2.92
Bed-activity coefficient	N/A	0.100

7-40. Under the provisions of the Clean Water Act, the U.S. Environmental Protection Agency established a requirement that municipalities had to provide secondary treatment of their waste. This was defined to be treatment that results in an effluent BOD_5 that does not exceed 30 mg/L. The discharge from the Blue Ox Tannery (Problem 7-39) is clearly in violation of this standard. Given the data in Problem 7-39, rework the problem to determine the amount of ultimate BOD (in kg/d) that the tannery may discharge to keep the DO above the DEQ water quality criteria of 5.00 mg/L at the critical point.

Hint: This problem cannot be worked backward starting with DO = 5.00 mg/L because there are two unknowns in the equation: L_a and t. Two approaches may be used. One is to set up the equations on a computer spreadsheet and decrease L_w until the DO_c becomes greater than 5.00 mg/L.

7-41. If the population and water use of Watapitae (Problems 7-37 and 7-38) are growing at 5 percent per year with a corresponding increase in wastewater flow, how many years' growth may be sustained before secondary treatment becomes inadequate? Assume that the treatment plant continues to maintain an effluent BOD_5 of 30.00 mg/L.

7-42. When ice covers a river, it severely limits the reaeration. There is some compensation for the reduced aeration because of the reduced water temperature. The lower temperature reduces the biological activity and, thus, the deoxygenation rate and, at the same time, the DO saturation level increases. Assuming a winter condition, rework Problem 7-37 with the reaeration reduced to zero and the river water temperature at 2°C.

7-43. What combination of BOD reduction and/or wastewater DO increase is required so the Big Bear wastewater in Problem 7-33 does not reduce the DO below 5.00 mg/L anywhere along the Salmon River? Assume that the cost of BOD reduction is 3 to 5 times that of increasing the effluent DO. Because the cost of adding extra DO is high, limit the excess above the minimum amount such that the critical DO falls between 5.00 mg/L and 5.25 mg/L.

Answer: Raising the wastewater DO to 2.7 mg/L is the most cost-effective remedy.

7-44. What amount of ultimate BOD, in kg/d, may Watapitae (Problem 7-37) discharge and still allow Avepitaeonmi 1.50 mg/L of DO above the DNR water quality criteria for assimilation of its waste?

7-45. Assuming that the mixed oxygen deficit D_a is zero and that the ultimate BOD L_r of the Looking Glass River above the wastewater outfall from Carrollville is zero, calculate the amount of ultimate BOD, in kg/d, that can be discharged if the DO must be kept at 4.00 mg/L at a point 8.05 km downstream. The stream deoxygenation rate k_d is 4.14 d^{-1} at 12°C, and the reaeration rate k_r is 5.07 d^{-1} at 12°C. The river temperature is 12°C. The river flow is 5.95 m³/s with a speed of 0.300 m/s. The Carrollville wastewater flow is 0.0130 m³/s.

Answer: $Q_w L_w = 1.14 \times 10^4$ kg/d of ultimate BOD

7-46. Assume that the Carrollville wastewater (Problem 7-45) also contains 3.0 mg/L of ammonia nitrogen with a stream deoxygenation rate of 0.900 d^{-1} at 12°C. What is the amount of ultimate carbonaceous BOD, in kg/d, that Carrollville can discharge and still meet the DO level of 4.00 mg/L at a point 8.05 km downstream? Assume also that the theoretical amount of oxygen will ultimately be consumed in the nitrification process.

7-47. As part of a TMDL evaluation, you have been asked to determine the effect of the wastewater discharge from the μBrew Bottling Company on the dissolved oxygen of the Big Head River for the winter and summer conditions shown below. Use a spreadsheet to plot the DO sag curve and calculate the DO at the critical point.

Parameter	μBrew wastewater	Big Head River Winter	Big Head River Summer
Flow, m^3/s	0.200	0.483	0.241
BOD$_5$, mg/L	100	N/A	N/A
k at 20°C, d^{-1}	0.3685	N/A	N/A
BOD$_U$, mg/L	N/A	7.66	7.66
Temperature, °C	28	4	28
DO, mg/L	0.0	8.0	8.0
Speed, m/s	N/A	0.150	0.150
Depth, m	N/A	2.0	1.0
Bed Activity Coefficient	N/A	0.3	0.3

7-48. For the case of salt water intrusion, we made the statement that for coastal waters $z \approx 40h$, where z is the elevation of freshwater above sea level. Show by calculation that this is true.

7-49. A freshwater well is located in a coastal aquifer with a hydraulic conductivity of 4.63×10^{-5} m/s. The well screen is located 30 m above the freshwater-saltwater interface. To prevent saltwater from entering the well, upconing should be less than 10 m. What is the maximum permissible discharge of the well for an infinite pumping time? (After Bouwer, 1978.)

7-9   DISCUSSION QUESTIONS

7-1. Students in a graduate-level environmental engineering laboratory took samples of the influent (raw sewage) and effluent (treated sewage) of a municipal wastewater treatment plant. They used these samples to determine the BOD rate constant (k). Would you expect the rate constants to be the same or different? If different, which would be higher and why?

7-2. If it were your job to set standards for a water body and you had a choice of either BOD_5 or ultimate BOD, which would you choose and why?

7-3. A summer intern has turned in his log book for temperature measurements for a limnology survey. He was told to take the measurements in the air 1 m above the lake, 1 m deep in the lake, and at a depth of 10 m. He turned in the following results but did not record which temperatures were taken where. If the measurements were made at noon in July in Missouri, what is your best guess as to the location of the measurements (i.e. air, 1 m deep, 10-m deep)? The recorded values were: 33°C, 18°C, and 21°C.

7-4. If the critical point in a DO sag curve is found to be 18 km downstream from the discharge point of untreated wastewater, would you expect the critical point to move upstream (toward the discharge point), downstream, or remain in the same place, if the wastewater is treated?

7-5. You have been assigned to conduct an environmental study of a remote lake in Canada. Aerial photos and a ground-level survey reveal no anthropogenic waste sources are contributing to the lake. When you investigate the lake, you find a highly turbid lake with abundant mats of floating algae and a hypolimnion DO of 1.0 mg/L. What productivity class would you assign to this lake? Explain your reasoning.

7-6. The lakes in Illinois, Indiana, western Kentucky, the lower peninsula of Michigan, and Ohio do not appear to be subject to acidification even though the rainwater pH is 4.4. Based on your knowledge (or what you can discover by research) of the topography, vegetation, and bedrock, explain why the lakes in these areas are not acidic.

7-10 FE EXAM FORMATTED PROBLEMS

7-1. If the BOD_5 and ultimate BOD of a municipal wastewater are 200 mg/L and 457 mg/L respectively, what is the reaction rate constant?

    a. 0.265 d^{-1} b. 0.050 d^{-1}
    c. 0.115 d^{-1} d. 0.165 d^{-1}

7-2. Twenty-six million gallons per day of wastewater with a DO of 1.00 mg/L is discharged into a river with a DO of 6.00 mg/L. If the flowrate of the river is 165×10^6 gal/d and saturation value of dissolved oxygen is 9.17 mg/L, what is the oxygen deficit after compete mixing of the two flows?

    a. 3.9 mg/L   b. 6.0 mg/L
    c. 4.3 mg/L   d. 5.3 mg/L

7-3. What is the mass loading (in lb_m/d) of a 33 MGD wastewater discharge with an ultimate BOD of 30.0 mg/L?

    a. 73,000 lb_m/d   b. 2,200 lb_m/d
    c. 8,300 lb_m/d   d. 62,000 lb_m/d

7-4. A stream survey revealed that the reaeration constant is 0.05 d^{-1} and the deoxygenation constant is 0.030 d^{-1}. If the initial deficit is 5.00 mg/L and the ultimate BOD after mixing is 12.0 mg/L, how long will it take to reach the critical point in the DO sag?

a. 16.43 d b. 25.46 d
c. 17.05 d d. 9.27 d

7-11 REFERENCES

Ahmad, R. (2002) "Watershed Assessment Power for Your PC," *Water Environment & Technology,* April, pp. 25–29.

APHA (2005) "Part 5210B: 5-day Biochemical Oxygen Demand," *Standard Methods for the Examination of Water and Wastewater,* 21st ed., American Public Health Association, Washington, DC.

Ash, R., J., B. Mauck, M. Morgan, et al. (1999) "Antibiotic Resistant Bacteria in U.S. Rivers," (Abstract Q-383), in *Abstracts of the 99th General Meeting of the American Society for Microbiology,* Chicago, May 30–June 3, p. 607.

Bear, J., and G. Dagan (1968) "Solving the Problem of Local Interfaced Upconing in a Coastal Aquifer by the Method of Small Perturbations," *Journal of Hydraulic Research,* vol. 6, no. 1, pp. 16–44.

Bennett, J., and G. Kramer (1999) "Multidrug Resistant Strains of Bacteria in the Streams of Dubuque County, Iowa," (Abstract Q-86), in *Abstracts of the 99th General Meeting of the American Society for Microbiology,* Chicago, May 30–June 3, p. 464.

Bosko, K. (1966) "An Explanation of the Rate of BOD Progression Under Laboratory and Stream Conditions," *Advances in Water Pollution Research,* Proceedings of the Third International Conference, Munich, p. 43.

Bouwer, H. (1978) *Groundwater Hydrology,* McGraw-Hill, New York, pp. 402–406.

Burkholder J. B., B. Libra, P. Weyer, S. Heathcote, D. Kolpin, P. S. Thorne, and M. Wichman (2007) "Impacts of Waste from Concentrated Animal Feeding Operations on Water Quality," *Environmental Health Perspectives*, vol. 115, no. 2, pp. 308–312.

Campagnolo E. R., K. R. Johnson, A. Karpati, C. S. Rubin., D. Kolpin, M. T. Meyer, E. Esteban, R. W. Currier, K. Smith, K. M. Thu, and M. McGreehin (2002) "Antimicrobial Residues in Animal Waste and Water Resources Proximal to Large-Scale Swine and Poultry Feeding Operations," *Science of the Total Environment*, vol. 299, pp. 89–95.

Churchill, M. A., H. L. Elmore, and R. A. Buckingham (1962) "Prediction of Stream Reaeration Rates," *Journal of the Sanitary Engineering Division,* American Society of Civil Engineers, vol. 88, SA4, p. 1.

Costanza, R., and A. Voinov (2000) "Integrated Ecological Economic Regional Modeling," in J. E. Hobbie (ed.), *Estuarine Science: A Synthetic Approach to Research and Practices,* Island Press, Washington, DC, pp. 461–506.

DiMura, J. (2003) "Permitting Agricultural Sources of Water Pollution," *Water Environment & Technology,* May, pp. 50–53.

FWPCA (1972) Federal Water Pollution Control Act Amendments, PL 92-500.

Galloway, J. N., and E. B. Cowling (1978) "The Effects of Precipitation on Aquatic and Terrestrial Ecosystems: A Proposed Precipitation Chemistry Network," *Journal of the Air Pollution Control Association,* vol. 28, pp. 229–235.

Gaudy, A., and E. Gaudy (1980) *Microbiology for Environmental Scientists and Engineers,* McGraw-Hill, New York, pp. 209–210, 485–492.

Goodman, A. (2008) "Sources and Origins of Compounds of Microconstituents and Endocrine Disrupting Compounds (EDCs)," presented at the Michigan American Water Works Association Research and Technical Practices Seminar, May 20, 2008, Lansing, MI.

Harries, J. E., T. Ryunnalls, E. Hill, et al. (2000) "Development of a Reproductive Performance Test for Endocrine Disrupting Chemicals Using Pair-Breeding Fathead Minnows," *Environmental Science and Technology,* vol. 34, pp. 3003–3011.

Hullman, R. (2009) *Fate of Pharmaceuticals in Drinking Water Utilities*, MS Thesis, Michigan State University. East Lansing, MI, p.4.

Jindal A., S. Kocherginskaya, A. Mehboob, M. Robert, R. I. Mackie, L. Raskin, and J. L. Zilles (2006) "Antimicrobial Use and Resistance in Swine Waste Treatment Systems," *Applied and Environmental Microbiology*, vol. 72, no. 12, pp. 7813–7820.

Jones, J. R., and R. W. Bachmann (1976) "Prediction of Phosphorus and Chlorophyll Levels in Lakes," *Journal of the Water Pollution Control Federation,* vol. 48, p. 2176.

Mackie R. I., S. Koike, I. Krapac, J. Chee-Sanford, S. Maxwell, and R. I. Aminov (2006) "Tetracycline Residues and Tetracylcine Resistance Genes in Groundwater Impacted by Swine Production Facilities," *Animal Biotechnology*, vol. 17, pp. 157–176.

Mellon M., C. Benbrook, and K.L. Benbrook (2001) "Hogging It: Estimates of Antimicrobial Abuse in Livestock," Union of Concerned Scientists.

Nemerow, N. L. (1974) *Scientific Stream Pollution Analysis*, McGraw-Hill, New York, p. 272.

O'Connor, D. J., and W. E. Dobbins (1958) "Mechanisms of Reaeration in Natural Streams," *American Society of Civil Engineers Transactions,* vol. 153, p. 641.

Population Reports (1998) "Solutions for a Water-Short World," Population Information Program, Center for Communication Programs, The Johns Hopkins University School of Public Health, Baltimore, MD, vol. 26, no. 1, p. 14.

Sadik, O. A., and D. M. Witt (1999) "Monitoring Endocrine-Disrupting Chemicals," *Environmental Science and Technology,* vol. 33, pp. 368A–374A.

Schmorak, S., and A. Mercado (1969) "Upconing of Fresh Water–Sea Water Interface Below Pumping Wells, Field Study," *Water Resources Research,* vol. 5, pp. 1290–1311.

Schroepfer, G. J., M. L. Robins, and R. H. Susag (1964) "Research Program on the Mississippi River in the Vicinity of Minneapolis and St. Paul," *Advances in Water Pollution Research,* vol. 1, part 1, p. 145.

Sternes, K. L. (1999) "Presence of High-level Vancomycin Resistant Enterococci in the Upper Rio Grande," (Abstract Q-63) in *Abstracts of the 99th General Meeting of the American Society for Microbiology*, Chicago, May 30–June 3, p. 545.

Streeter, H. W., and E. B. Phelps (1925) *A Study of the Pollution and Natural Purification of the Ohio River,* U.S. Public Health Service Bulletin No. 146.

Thomas, H. A. (1950) "Graphical Determination of B.O.D. Curve Constants," *Water and Sewage Works,* pp. 123–124.

U.S. EPA (2005) http://www.epa.gov/owow/tmdl/intro.html and http://www.epa.gov/waterscience/models/allocation/def.html.

U.S.G.S. (1999) *The Quality of Our Nation's Waters—Nutrients and Pesticides,* U.S. Geological Survey Circular 1225, Reston, VA.

U.S.G.S. (2002) http://toxics.usgs.gov/pubs/OFR-02-94

Vollenweider, R. A. (1975) "Input-Output Models with Special Reference to the Phosphorus Loading Concept in Limnology," *Schweiz. Z. Hydro,* vol. 37, pp. 53–83.

Wetzel, R. G. (1983) *Limnology,* W. B. Saunders, Philadelphia, p. 767.

CHAPTER

8

WASTEWATER TREATMENT

8-1 INTRODUCTION

Wastewater has historically been considered a nuisance to be discarded in the cheapest, least offensive manner possible. This meant the use of on-site disposal systems such as the pit privy and direct discharge into our lakes and streams. Over the last century it has been recognized that this approach produces an undesirable impact on the environment. This led to a variety of treatment techniques that characterize the municipal treatment systems today that are the focal point of this chapter. As we look forward, it becomes obvious that in the interest of sustainability as well as fundamental economic efficiency, we must view the wastewater as a raw material to be conserved. Clean water is a scarce commodity; it should be treated as such and conserved and reused. The contents of wastewater are often viewed as pollutants. The abundance of nutrients such as phosphorus and nitrogen are, in some treatment schemes (as discussed in Section 8-10), recovered for crop growth. This approach must become more prevalent to achieve a sustainable future. The organic compounds in wastewater are a source of energy. Currently we utilize the processes described in Section 8-11 to recover some of this energy. Other efforts that focus on improving the efficiency of energy utilization in wastewater treatment are discussed in Chapter 13.

Applications

This chapter provides an overview of the characteristics of wastewater, treatment standards, and wastewater treatment processes. Treatment standards provide the basic rules for the water quality to be produced by a treatment system. The characteristics of the wastewater identify the constituents that must be addressed by the design engineer in selecting appropriate wastewater treatment processes. This chapter will enable you to make a preliminary selection and organization of appropriate processes for treatment of a municipal wastewater.

This chapter will provide you with tools to do the following:

- Estimate the volume of a basin to dampen flow and BOD variations

- Size sedimentation tanks

- Estimate the size of biological treatment processes

- Estimate the size of a blower for aeration of a biological treatment process

- Estimate the amount of sludge generated in a biological treatment process

- Select appropriate sludge treatment units to prepare sludge for disposal

- Select appropriate sludge disposal alternatives

8-2 CHARACTERISTICS OF WASTEWATER

Physical Characteristics of Domestic Wastewater

Fresh, aerobic, domestic wastewater has been said to have the odor of kerosene or freshly turned earth. Aged, septic sewage is considerably more offensive to the olfactory nerves. The characteristic rotten-egg odor of hydrogen sulfide and the mercaptans is indicative of septic sewage. Fresh sewage is typically gray in color. Septic sewage is black.

Wastewater temperatures normally range between 10 and 20°C. In general, the temperature of the wastewater will be higher than that of the water supply. This is because of the addition of warm water from households and heating within the plumbing system of the structure.

One cubic meter of wastewater weighs approximately 1,000,000 grams. It will contain about 500 grams of solids. One-half of the solids will be dissolved solids such as calcium, sodium, and soluble organic compounds. The remaining 250 grams will be insoluble. The insoluble fraction consists of about 125 grams of material that will settle out of the liquid fraction in 30 minutes under quiescent conditions. The remaining 125 grams will remain in suspension for a very long time. The result is that wastewater is highly turbid.

Chemical Characteristics of Domestic Wastewater

Because the number of chemical compounds found in wastewater is almost limitless, we normally restrict our consideration to a few general classes of compounds. These classes often are better known by the name of the test used to measure them than by what is included in the class. The biochemical oxygen demand (BOD_5) test, which we discussed in Chapter 7, is a case in point. Another closely related test is the *chemical oxygen demand* (COD) test.

The COD test is used to determine the oxygen equivalent of the organic matter that can be oxidized by a strong chemical oxidizing agent (potassium dichromate) in an acid medium. The COD of a waste, in general, will be greater than the BOD_5 because more compounds can be oxidized chemically than can be oxidized biologically, and because BOD_5 does not equal ultimate BOD.

The COD test can be conducted in about 3 hours. If it can be correlated with BOD_5, it can be used to aid in the operation and control of the wastewater treatment plant (WWTP).

Total Kjeldahl nitrogen (TKN) is a measure of the total organic and ammonia nitrogen in the wastewater.* TKN gives a measure of the availability of nitrogen for building cells, as well as the potential nitrogenous oxygen demand that will have to be satisfied to meet discharge standards that protect receiving bodies of water.

Phosphorus may appear in many forms in wastewater. Among the forms found are the orthophosphates, polyphosphates, and organic phosphate. For our purpose, we will lump all of these together under the heading "Total Phosphorus (as P)."

---

*Pronounced "kell dall" after J. Kjeldahl, who developed the test in 1883.

TABLE 8-1
Typical composition of untreated domestic wastewater

Constituent	Weak	Medium	Strong
	(all mg/L except settleable solids)		
Alkalinity (as $CaCO_3$)[a]	50	100	200
BOD_5 (as O_2)	100	200	300
Chloride[a]	30	50	100
COD (as O_2)	250	500	1,000
Suspended solids (SS)	100	200	350
Settleable solids, mL/L	5	10	20
Total dissolved solids (TDS)	200	500	1,000
Total Kjeldahl nitrogen (TKN) (as N)	20	40	80
Total organic carbon (TOC) (as C)	75	150	300
Total phosphorus (as P)	5	10	20

[a]To be added to amount in domestic water supply. Chloride is exclusive of contribution from water-softener backwash.

TABLE 8-2
EPA's conventional and nonconventional pollutant categories

Conventional	*Nonconventional*
Biochemical oxygen demand (BOD_5)	Ammonia (as N)
Total suspended solids (TSS)	Chromium VI (hexavalent)
Oil and grease	Chemical oxygen demand (COD)
Oil (animal, vegetable)	COD/BOD_7
Oil (mineral)	Fluoride
pH	Manganese
	Nitrate (as N)
	Organic nitrogen (as N)
	Pesticide active ingredients (PAI)
	Phenols, total
	Phosphorus, total (as P)
	Total organic carbon (TOC)

Source: CFR, 2005a.

Three typical compositions of untreated domestic wastewater are summarized in Table 8-1. The pH for all of these wastes will be in the range of 6.5 to 8.5, with a majority being slightly on the alkaline side of 7.0.

Characteristics of Industrial Wastewater

Industrial processes generate a wide variety of wastewater pollutants. The characteristics and levels of pollutants vary significantly from industry to industry. The Environmental Protection Agency (EPA) has grouped the pollutants into three categories: conventional pollutants, nonconventional pollutants, and priority pollutants. The conventional and nonconventional pollutants are listed in Table 8-2. The priority pollutants are listed in Table 8-3.

TABLE 8-3
EPA's priority pollutant list

1. Antimony	43. Trichloroethylene	86. Fluoranthene
2. Arsenic	44. Vinyl chloride	87. Fluorene
3. Beryllium	45. 2-Chlorophenol	88. Hexachlorobenzene
4. Cadmium	46. 2,4-Dichlorophenol	89. Hexachlorobutadiene
5a. Chromium (III)	47. 2,4-Dimethylphenol	90. Hexachlorocyclopentadiene
5b. Chromium (VI)	48. 2-Methyl-4-chlorophenol	91. Hexachloroethane
6. Copper	49. 2,4-Dinitrophenol	92. Indeno(1,2,3-cd)pyrene
7. Lead	50. 2-Nitrophenol	93. Isophorone
8. Mercury	51. 4-Nitrophenol	94. Naphthalene
9. Nickel	52. 3-Methyl-4-chlorophenol	95. Nitrobenzene
10. Selenium	53. Pentachlorophenol	96. N-Nitrosodimethylamine
11. Silver	54. Phenol	97. N-Nitrosodi-n-propylamine
12. Thallium	55. 2,4,6-Trichlorophenol	98. N-Nitrosodiphenylamine
13. Zinc	56. Acenaphthene	99. Phenanthrene
14. Cyanide	57. Acenaphthylene	100. Pyrene
15. Asbestos	58. Anthracene	101. 1,2,4-Trichlorobenzene
16. 2,3,7,8-TCDD (Dioxin)	59. Benzidine	102. Aldrin
17. Acrolein	60. Benzo(a)anthracene	103. alpha-BHC
18. Acrylonitrile	61. Benzo(a)pyrene	104. beta-BHC
19. Benzene	62. Benzo(a)fluoranthene	105. gamma-BHC
20. Bromoform	63. Benzo(ghi)perylene	106. delta-BHC
21. Carbon tetrachloride	64. Benzo(k)fluoranthene	107. Chlordane
22. Chlorobenzene	65. bis(2-Chloroethoxy)methane	108. 4,4'-DDT
23. Chlorodibromomethane	66. bis(2-Chloroethyl)ether	109. 4,4'-DDE
24. Chloroethane	67. bis(2-Chloroisopropyl)ether	110. 4,4'-DDD
25. 2-Chloroethylvinyl ether	68. bis(2-Ethylhexyl)phthalate	111. Dieldrin
26. Chloroform	69. 4-Bromophenyl phenyl ether	112. alpha-Endosulfan
27. Dichlorobromomethane	70. Butylbenzyl phthalate	113. beta-Endosulfan
28. 1,1-Dichloroethane	71. 2-Chloronaphthalene	114. Endosulfan sulfate
29. 1,2-Dichloroethane	72. 4-Chlorophenyl phenyl ether	115. Endrin
30. 1,1-Dichloroethylene	73. Chrysene	116. Endrin aldehyde
31. 1,2-Dichloropropane	74. Dibenzo(a,h)anthracene	117. Heptachlor
32. 1,3-Dichloropropylene	75. 1,2-Dichlorobenzene	118. Heptachlor epoxide
33. Ethylbenzene	76. 1,3-Dichlorobenzene	119. PCB-1242
34. Methyl bromide	77. 1,4-Dichlorobenzene	120. PCB-1254
35. Methyl chloride	78. 3,3-Dichlorobenzidine	121. PCB-1221
36. Methylene chloride	79. Diethyl phthalate	122. PCB-1232
37. 1,2,2,2-Tetrachloroethane	80. Dimethyl phthalate	123. PCB-1248
38. Tetrachloroethylene	81. Di-n-butyl phthalate	124. PCB-1260
39. Toluene	82. 2,4-Dinitrotoluene	125. PCB-1016
40. 1,2-trans-dichloroethylene	83. 2,6-Dinitrotoluene	126. Toxaphene
41. 1,1,1-Trichloroethane	84. Di-n-octyl phthalate	
42. 2,4 Dichlorophenol	85. 1,2-Diphenylhydrazine	

Source: Code of Federal Regulations, 40 CFR 131.36, July 1, 1993.

TABLE 8-4
Examples of industrial wastewater concentrations for BOD_5 and suspended solids

Industry	BOD_5, mg/L	Suspended solids, mg/L
Ammunition	50–300	70–1,700
Fermentation	4,500	10,000
Slaughterhouse (cattle)	400–2,500	400–1,000
Pulp and paper (kraft)	100–350	75–300
Tannery	700–7,000	4,000–20,000

TABLE 8-5
Examples of industrial wastewater concentrations for nonconventional pollutants

Industry	Pollutant	Concentration, mg/L
Coke by-product (steel mill)	Ammonia (as N)	200
	Organic nitrogen (as N)	100
	Phenol	2,000
Metal plating	Chromium VI	3–550
Nylon polymer	COD	23,000
	TOC	8,800
Plywood-plant glue waste	COD	2,000
	Phenol	200–2,000
	Phosphorus (as PO_4)	9–15

Because of the wide variety of industries and levels of pollutants, we can only present a snapshot view of the characteristics. A sampling of a few industries for two conventional pollutants is shown in Table 8-4.

A similar sampling for nonconventional pollutants is shown in Table 8-5.

8-3 WASTEWATER TREATMENT STANDARDS

In Public Law 92-500, the Congress required municipalities and industries to provide *secondary treatment* before discharging wastewater into natural water bodies. The U.S. Environmental Protection Agency (EPA) established a definition of secondary treatment based on three wastewater characteristics: BOD_5, suspended solids, and hydrogen-ion concentration (pH). The definition is summarized in Table 8-6.

PL 92-500 also directed that the EPA establish a permit system called the National Pollutant Discharge Elimination System (NPDES). Under the NPDES program, all facilities that discharge pollutants from any point source into waters of the United States are required to obtain a NPDES permit. Although some states elected to have EPA administer their permit system, most states administer their own program. Before a permit is granted, the administering agency will model the response of the receiving

TABLE 8-6
U.S. Environmental Protection Agency definition of secondary treatment[a,b]

Characteristic of discharge	Units	Average monthly concentration[c]	Average weekly concentration[c]
BOD$_5$	mg/L	30[d]	45
Suspended solids	mg/L	30[d]	45
Hydrogen-ion Concentration	pH units	Within the range 6.0–9.0 at all times[e]	
CBOD$_5$[f]	mg/L	25	40

[a]*Source:* CFR, 2005b.
[b]Present standards allow stabilization ponds and trickling filters to have higher 30-day average concentrations (45 mg/L) and 7-day average concentrations (65 mg/L) of BOD and suspended solids as long as the water quality of the receiving body of water is not adversely affected. Other exceptions are also permitted. The CFR and the *NPDES Permit Writers' Manual* (U.S. EPA, 1996) should be consulted for details on the exceptions.
[c]Not to be exceeded.
[d]Average removal shall not be less than 85 percent.
[e]Only enforced if caused by industrial wastewater or by in-plant inorganic chemical addition.
[f]May be substituted for BOD$_5$ at the option of the permitting authority.

body to the proposed discharge to determine if the receiving body is adversely affected (for an example of modeling, see Section 7-3). The permit may require lower concentrations than those specified in Table 8-6 to maintain the quality of the receiving body of water.

In addition, the states may impose additional conditions in the NPDES permit. For example, in Michigan, a limit of 1 mg/L of phosphorus is contained in permits for discharges to surface waters that do not have substantial problems with high levels of nutrients. More stringent limits are required for discharges to surface waters that are very sensitive to nutrients.

CBOD$_5$ limits are placed in the NPDES permits for all facilities that have the potential to contribute significant quantities of oxygen-consuming substances. The nitrogenous oxygen demand from ammonia nitrogen is typically the oxygen demand of concern from municipal discharges (see Section 7-3). It is computed separately from the CBOD$_5$ and then combined to establish a discharge limit. Ammonia is also evaluated for its potential toxicity to the stream's biota.

Bacterial effluent limits may also be included in the NPDES permit. For example, municipal wastewater treatment plants in Michigan must comply with limits of 200 fecal coliform bacteria (FC) per 100 mL of water as a monthly average and 400 FC/100 mL as a 7-day average. More stringent requirements are imposed to protect waters that are used for recreation. Total-body-contact recreation waters must meet limits of 130 *Escherichia coli* per 100 mL of water as a 30-day average and 300 *E. coli* per 100 mL at any time. Partial-body-contact recreation is permitted for water with less than 1,000 *E. coli* per 100 mL of water.

TABLE 8-7
NPDES Limits for the city of Hailey, Idaho[a,b]

Parameter	Average monthly limit	Average weekly limit	Instantaneous maximum limit
BOD$_5$	30 mg/L	45 mg/L	N/A
	43 kg/d	64 kg/d	
Suspended solids	30 mg/L	45 mg/L	N/A
	43 kg/d	64 kg/d	
E. coli bacteria	126/100 mL	N/A	406/100 mL
Fecal coliform bacteria	N/A	200 colonies/100 mL	N/A
Total ammonia as N	1.9 mg/L	2.9 mg/L	3.3 mg/L
	4.1 kg/d	6.4 kg/d	7.1 kg/d
Total phosphorus	6.8 kg/d	10.4 kg/d	N/A
Total Kjeldahl nitrogen	25 kg/d	35 kg/d	N/A

[a]*Source:* U.S. EPA, 2005.
[b]Renewal announcement, 7 February 2001.

For thermal discharges such as cooling water, temperature limits may be included in the permit. Michigan rules state that the Great Lakes and connecting waters and inland lakes shall not receive a heat load that increases the temperature of the receiving water more than 1.7°C above the existing natural water temperature after mixing. For rivers, streams, and impoundments the temperature limits are 1°C for cold-water fisheries and 2.8°C for warm-water fisheries. (See Section 2-4 for a discussion of energy balances and Problems 2-36, 2-37, and 2-40 for typical problems in thermal discharge analysis.)

An example of NPDES limits is shown in Table 8-7.* Note that in addition to concentration limits, mass discharge limits are also established. As shown in Example 8-1, specification of both the concentration and mass loading limits effectively limits the WWTP's allowable flow rate.

Example 8-1. Demonstrate that the specification of the BOD$_5$ limit of 30 mg/L and the mass discharge limit of 43 kg/d of BOD$_5$ fixes the allowable monthly flow rate.

Solution. The average monthly flow rate of wastewater (Q_{ww}) is

$$Q_{ww} = \frac{43 \text{ kg/d}}{(30 \text{ mg/L})(10^{-6} \text{ kg/mg})(10^3 \text{ L/m}^3)} \ 1{,}433.33 \text{ or } 1{,}400 \text{ m}^3/\text{d}$$

*This table outlines only the quantitative limits. The entire permit is 22 pages long.

Pretreatment of Industrial Wastes

Industrial wastewaters can pose serious hazards to municipal systems because the collection and treatment systems have not been designed to carry or treat them. The wastes can damage sewers and interfere with the operation of treatment plants. They may pass through the wastewater treatment plant (WWTP) untreated or they may concentrate in the sludge, rendering it a hazardous waste.

The Clean Water Act gives the EPA the authority to establish and enforce pretreatment standards for discharge of industrial wastewaters into municipal treatment systems. Specific objectives of the pretreatment program are:

- To prevent the introduction of pollutants into WWTPs that will interfere with their operation, including interference with their use or with disposal of municipal sludge.

- To prevent the introduction of pollutants to WWTPs that will pass through the treatment works or otherwise be incompatible with such works.

- To improve opportunities to recycle and reclaim municipal and industrial wastewaters and sludge.

EPA has established "prohibited discharge standards" (40 CFR 403.5) that apply to all nondomestic discharges to the WWTP and "categorical pretreatment standards" that are applicable to specific industries (40 CFR 405-471).* Congress assigned the primary responsibility for enforcing these standards to local WWTPs.

In the General Pretreatment Regulations, industrial users (IUs) are prohibited from introducing the following into a WWTP:

1. Pollutants that create a fire or explosion hazard in the municipal WWTP, including, but not limited to, waste streams with a closed-cup flash point of less than or equal to 60°C, using the test methods specified in 40 CFR 261.21.

2. Pollutants that will cause corrosive structural damage to the municipal WWTP (but in no case discharges with a pH lower than 5.0) unless the WWTP is specifically designed to accommodate such discharges.

3. Solid or viscous pollutants in amounts that will cause obstruction to the flow in the WWTP resulting in interference.

4. Any pollutant, including oxygen-demanding pollutants (such as BOD), released in a discharge at a flow rate and/or concentration that will cause interference with the WWTP.

5. Heat in amounts that will inhibit biological activity in the WWTP and result in interference, but in no case heat in such quantities that the temperature at the WWTP exceeds 40°C unless the approval authority, on request of the publicly owned treatment works (POTW), approves alternative temperature limits.

---

*CFR = Code of Federal Regulations.

6. Petroleum oil, nonbiodegradable cutting oil, or products of mineral oil origin in amounts that will cause interference or will pass through.

7. Pollutants that result in the presence of toxic gases, vapors, or fumes within the POTW in a quantity that may cause acute worker health and safety problems.

8. Any trucked or hauled pollutants, except at discharge points designated by the POTW.

8-4 MUNICIPAL WASTEWATER TREATMENT SYSTEMS

The alternatives for municipal wastewater treatment fall into three major categories (Figure 8-1): (1) primary treatment, (2) secondary treatment, and (3) tertiary treatment. It is commonly assumed that each of the "degrees of treatment" noted in Figure 8-1 includes the previous steps. For example, primary treatment is assumed to include the pretreatment processes: bar rack, grit chamber, and equalization basin. Likewise, secondary treatment is assumed to include all the processes of primary treatment: bar rack, grit chamber, equalization basin, and primary settling tank.

The purpose of pretreatment is to provide protection to the wastewater treatment plant (WWTP) equipment that follows. In some older municipal plants the equalization step may not be included.

The major goal of primary treatment is to remove from wastewater those pollutants that will either settle or float. Primary treatment will typically remove about 60 percent of the suspended solids in raw sewage and 35 percent of the BOD_5. Soluble pollutants are not removed. At one time, this was the only treatment used by many cities. Although primary treatment alone is no longer acceptable, it is still frequently used as the first treatment step in a secondary-treatment system. The major goal of secondary treatment is to remove the soluble BOD_5 that escapes the primary process and to provide added removal of suspended solids. Secondary treatment is typically achieved by using biological processes. These provide the same biological reactions that would occur in the receiving water if it had adequate capacity to assimilate the wastewater. The secondary-treatment processes are designed to speed up these natural processes so that the breakdown of the degradable organic pollutants can be achieved in relatively short time periods. Although secondary treatment may remove more than 85 percent of the BOD_5 and suspended solids, it does not remove significant amounts of nitrogen, phosphorus, or heavy metals, nor does it completely remove pathogenic bacteria and viruses.

In cases where secondary levels of treatment are not adequate, additional treatment processes are applied to the secondary effluent to provide tertiary wastewater treatment. These processes may involve chemical treatment and filtration of the wastewater.

Advanced Wastewater Treatment (AWT) processes produce a sparkling clean, colorless, odorless effluent indistinguishable in appearance from a high-quality drinking water. They are used to prepare tertiary treated wastewater for reuse in recharging groundwater aquifers, indirect potable reuse through surface water augmentation, and direct potable reuse (Metcalf & Eddy, 2007). The processes used are often the same as those used for treatment of drinking water.

Most of the impurities removed from the wastewater do not simply vanish. Some organic compounds are broken down into harmless carbon dioxide and water. Most of

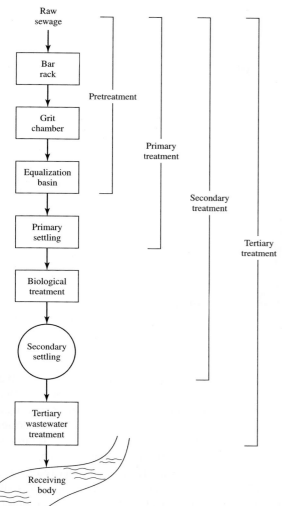

FIGURE 8-1
Degrees of treatment.

the impurities are removed from the wastewater as a solid, that is, sludge. Because most of the impurities removed from the wastewater are present in the sludge, sludge handling and disposal must be carried out carefully to achieve satisfactory pollution control.

8-5   UNIT OPERATIONS OF PRETREATMENT

Several devices and structures are placed upstream of the primary treatment operation to provide protection to the wastewater treatment plant (WWTP) equipment. These devices and structures are classified as pretreatment because they have little effect in reducing BOD_5. In industrial WWTPs where only soluble compounds are present, bar racks and grit chambers may be absent. Equalization is frequently required in industrial WWTPs.

FIGURE 8-2
Bar rack. (Courtesy of Siemens Water Technologies.)

Bar Racks

Typically, the first device encountered by the wastewater entering the plant is a bar rack (Figure 8-2). The primary purpose of the rack is to remove large objects that would damage or foul pumps, valves, and other mechanical equipment. Rags, logs, and other objects that find their way into the sewer are removed from the wastewater on the racks. In modern WWTPs, the racks are cleaned mechanically. The solid material is stored in a hopper and removed to a sanitary landfill at regular intervals.

Bar racks (or bar screens) may be categorized as trash racks, manually cleaned racks, and mechanically cleaned racks. Trash racks are those with large openings, 40 to 150 mm, that are designed to prevent very large objects such as logs from entering the plant. These are normally followed by racks with smaller openings. Manually cleaned racks have openings that range from 25 to 50 mm. Channel approach velocities are designed to be in the range of 0.3 to 0.6 m/s. As mentioned above, manually cleaned racks are not frequently employed. They do find application in bypass channels that are infrequently used. Mechanically cleaned racks have openings ranging from 15 to 75 mm. Maximum channel approach velocities range from 0.6 to 1.0 m/s. Minimum velocities of 0.3 to 0.5 m/s are necessary to prevent grit accumulation. Regardless of the type of rack, two channels with racks are provided to allow one to be taken out of service for cleaning and repair.

Grit Chambers

Inert dense material, such as sand, broken glass, silt, and pebbles, is called *grit*. If these materials are not removed from the wastewater, they abrade pumps and other

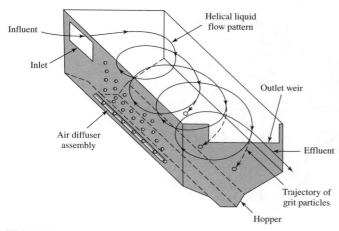

FIGURE 8-3
Aerated grit chamber.

mechanical devices, causing undue wear. In addition, they have a tendency to settle in corners and bends, reducing flow capacity and, ultimately, clogging pipes and channels.

There are two basic types of grit-removal devices: aerated and vortex.

Aerated Grit Chambers. The spiral roll of the aerated grit chamber liquid "drives" the grit into a hopper which is located under the air diffuser assembly (Figure 8-3). The shearing action of the air bubbles is supposed to strip the inert grit of much of the organic material that adheres to its surface.

Aerated grit chamber performance is a function of the roll velocity and detention time. The roll velocity is controlled by adjusting the air feed rate. Nominal air flow values are in the range of 0.1 to 0.75 cubic meters per minute of air per meter of tank length ($m^3/min \cdot m$). Liquid detention times are usually set to be about three minutes at maximum flow. Length-to-width ratios range from 2.5:1 to 5:1 with depths on the order of 2 to 5 m.

Grit accumulation in the chamber varies greatly, depending on whether the sewer system is a combined type or a separate type, and on the efficiency of the chamber. For combined systems, 90 m^3 of grit per million cubic meters of sewage ($m^3/10^6 m^3$) is not uncommon. In separate systems you might expect something less than 40 $m^3/10^6 m^3$. Normally the grit is buried in a sanitary landfill.

Vortex Grit Chambers. Wastewater is brought into the chamber tangentially (Figure 8-4). At the center of the chamber, a rotating turbine with adjustable-pitch blades along with the cone-shaped floor produces a spiraling, doughnut-shaped flow pattern. This pattern tends to lift the lighter organic particles and settle the grit into a grit sump. The effluent outlet has twice the width of the influent flume. This results in a lower exit velocity than the influent velocity and thus prevents grit from being drawn into the effluent flow. It should be noted that centrifugal acceleration does not play a significant role in removing the particles. The velocities are too low.

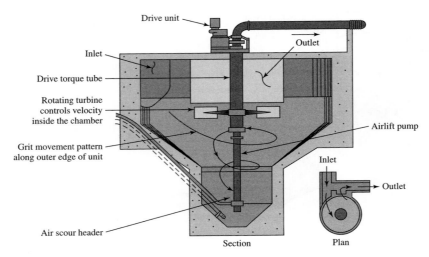

FIGURE 8-4
Vortex grit chamber.

Comminutors and Macerators

Devices that are used to macerate wastewater solids (rags, paper, plastic, and other materials) by revolving cutting bars are called *comminutors*. Comminutors are most commonly used in small WWTPs with flows less than 0.2 m³/s. These devices are placed downstream of the grit chambers to protect the cutting bars from abrasion. They are used as a replacement for the downstream bar rack but must be installed with a hand-cleaned rack in parallel in case they fail. Comminutors may create a string of material such as rags that collect on downstream equipment. Because of this and a high maintenance cost, newer installations use a screen or a *macerator* (Figure 8-5).

Macerators are slow-speed grinders. One type consists of counterrotating blades. The material is chopped as it passes between the blades. The chopping action reduces the potential for producing ropes of rags or plastic.

Equalization

Flow equalization is not a treatment process in itself, but a technique that can be used to improve the effectiveness of both secondary and tertiary wastewater treatment processes (U.S. EPA, 1979a). Wastewater does not flow into a municipal wastewater treatment plant at a constant rate (see Figure 1-4); the flow rate varies from hour to hour, reflecting the living habits of the area served. In most towns, the pattern of daily activities sets the pattern of sewage flow and strength. Above-average sewage flows and strength occur in midmorning. The constantly changing amount and strength of wastewater to be treated makes efficient process operation difficult. Also, many treatment units must be designed for the maximum flow conditions encountered, which actually results in their being oversized for average

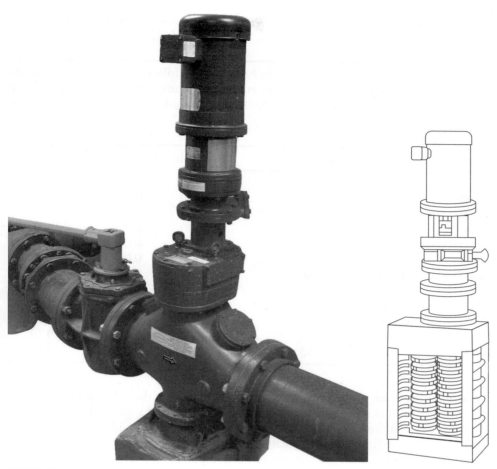

FIGURE 8-5
Photo and isometric drawing of a macerator. (*Source:* Mackenzie L. Davis.)

conditions. The purpose of flow equalization is to dampen these variations so that the wastewater can be treated at a nearly constant flow rate. Flow equalization can significantly improve the performance of an existing plant and increase its useful capacity. In new plants, flow equalization can reduce the size and cost of the treatment units.

Flow equalization is usually achieved by constructing large basins that collect and store the wastewater flow and from which the wastewater is pumped to the treatment plant at a constant rate. These basins are normally located near the head end of the treatment works, preferably downstream of pretreatment facilities such as bar screens, comminutors, and grit chambers. Adequate aeration and mixing must be provided to prevent odors and solids deposition. The required volume of an equalization basin is estimated from a mass balance of the flow into the treatment plant with the average flow the plant is designed to treat. The theoretical basis is the same as that used to size reservoirs (see Section 4-5).

Example 8-2.   Design an equalization basin for the following cyclic flow pattern. Provide a 25 percent excess capacity for equipment, unexpected flow variations, and solids accumulation. Evaluate the impact of equalization on the mass loading of BOD_5.

Time, h	Flow, m^3/s	BOD_5, mg/L	Time, h	Flow, m^3/s	BOD_5, mg/L
0000	0.0481	110	1200	0.0718	160
0100	0.0359	81	1300	0.0744	150
0200	0.0226	53	1400	0.0750	140
0300	0.0187	35	1500	0.0781	135
0400	0.0187	32	1600	0.0806	130
0500	0.0198	40	1700	0.0843	120
0600	0.0226	66	1800	0.0854	125
0700	0.0359	92	1900	0.0806	150
0800	0.0509	125	2000	0.0781	200
0900	0.0631	140	2100	0.0670	215
1000	0.0670	150	2200	0.0583	170
1100	0.0682	155	2300	0.0526	130

Solution.   Because of the repetitive and tabular nature of the calculations, a computer spreadsheet is ideal for this problem. The spreadsheet solution is easy to verify if the calculations are set up with judicious selection of the initial value. If the initial value is the first flow rate greater than the average after the sequence of nighttime low flows, then the last row of the computation should result in a storage value ($\Sigma \, dS$) of zero.

The first step is to calculate the average flow. In this case it is 0.05657 m³/s. Next, the flows are arranged in order beginning with the time and flow that first exceeds the average. In this case it is at 0900 h with a flow of 0.0631 m³/s. The tabular arrangement is shown on page 472. An explanation of the calculations for each column follows.

The third column converts the flows to volumes using the time interval between flow measurements:

$$V = (0.0631 \text{ m}^3/\text{s})(1 \text{ h})(3600 \text{ s/h}) = 227.16 \text{ m}^3$$

The fourth column is the average volume that leaves the equalization basin.

$$V = (0.05657 \text{ m}^3/\text{s})(1 \text{ h})(3600 \text{ s/h}) = 203.655 \text{ m}^3$$

The fifth column is the difference between the inflow volume and the outflow volume.

$$dS = V_{in} - V_{out} = 227.16 \text{ m}^3 - 203.655 \text{ m}^3 = 23.505 \text{ m}^3$$

The sixth column is the cumulative sum of the difference between the inflow and outflow. For the second time interval, it is

$$\text{Storage} = \sum dS = 37.55 \text{ m}^3 + 23.51 \text{ m}^3 = 61.06 \text{ m}^3$$

Note that the last value for the cumulative storage is 0.12 m³. It is not zero because of round-off truncation in the computations. At this point the equalization basin is empty and ready to begin the next day's cycle.

The required volume for the equalization basin is the maximum cumulative storage. With the requirement for 25 percent excess, the volume would then be

$$\text{Storage volume} = (863.74 \text{ m}^3)(1.25) = 1{,}079.68, \text{ or } 1{,}080 \text{ m}^3$$

The mass of BOD_5 into the equalization basin is the product of the inflow (Q), the concentration of BOD_5 (S_o), and the integration time (Δt):

$$M_{\text{BOD-in}} = (Q)(S_o)(\Delta t)$$

The mass of BOD_5 out of the equalization basin is the product of the average outflow (Q_{avg}), the average concentration (S_{avg}) in the basin, and the integration time (Δt):

$$M_{\text{BOD-out}} = (Q_{\text{avg}})(S_{\text{avg}})(\Delta t)$$

The average concentration is determined as

$$S_{\text{avg}} = \frac{(V_i)(S_o) + (V_s)(S_{\text{prev}})}{V_i + V_s}$$

where V_i = volume of inflow during time interval Δt, m³
  S_o = average BOD_5 concentration during time interval Δt, g/m³
  V_s = volume of wastewater in the basin at the end of the previous time interval, m³
  S_{prev} = concentration of BOD_5 in the basin at the end of the previous time interval
   = (previous S_{avg}), g/m³

Noting that 1 mg/L = 1 g/m³, we find that the first row (the 0900 h time) computations are

$$M_{\text{BOD-in}} = (0.0631 \text{ m}^3/\text{s})(140 \text{ g/m}^3)(1 \text{ h})(3{,}600 \text{ s/h})(10^{-3} \text{ kg/g})$$
$$= 31.8 \text{ kg}$$

$$S_{\text{avg}} = \frac{(227.16 \text{ m}^3)(140 \text{ g/m}^3) + 0}{227.16 \text{ m}^3 + 0}$$
$$= 140 \text{ mg/L}$$

$$M_{\text{BOD-out}} = (0.05657 \text{ m}^3/\text{s})(140 \text{ g/m}^3)(1 \text{ h})(3{,}600 \text{ s/h})(10^{-3} \text{ kg/g})$$
$$= 28.5 \text{ kg}$$

The flow rate of 0.05657 m³/s is the calculated average flow rate.

Time (1)	Flow, m³/s (2)	Vol$_{in}$, m³ (3)	Vol$_{out}$, m³ (4)	dS, m³ (5)	Σ dS, m³ (6)	BOD$_5$, mg/L (7)	M$_{BOD\text{-}in}$, kg (8)	S, mg/L (9)	M$_{BOD\text{-}out}$, kg (10)
0900	0.0631	227.16	203.65	23.51	23.51	140	31.80	140.00	28.51
1000	0.067	241.2	203.65	37.55	61.06	150	36.18	149.11	30.37
1100	0.0682	245.52	203.65	41.87	102.93	155	38.06	153.83	31.33
1200	0.0718	258.48	203.65	54.83	157.76	160	41.36	158.24	32.23
1300	0.0744	267.84	203.65	64.19	221.95	150	40.18	153.06	31.17
1400	0.075	270	203.65	66.35	288.3	140	37.80	145.89	29.71
1500	0.0781	281.16	203.65	77.51	365.81	135	37.96	140.51	28.62
1600	0.0806	290.16	203.65	86.51	452.32	130	37.72	135.86	27.67
1700	0.0843	303.48	203.65	99.83	552.15	120	36.42	129.49	26.37
1800	0.0854	307.44	203.65	103.79	655.94	125	38.43	127.89	26.04
1900	0.0806	290.16	203.65	86.51	742.45	150	43.52	134.67	27.43
2000	0.0781	281.16	203.65	77.51	819.96	200	56.23	152.61	31.08
2100	0.067	241.2	203.65	37.55	857.51	215	51.86	166.79	33.97
2200	0.0583	209.88	203.65	6.23	863.74	170	35.68	167.42	34.10
2300	0.0526	189.36	203.65	−14.29	849.45	130	24.62	160.69	32.73
0000	0.0481	173.16	203.65	−30.49	818.96	110	19.05	152.11	30.98
0100	0.0359	129.24	203.65	−74.41	744.55	81	10.47	142.42	29.00
0200	0.0226	81.36	203.65	−122.29	622.26	53	4.31	133.61	27.21
0300	0.0187	67.32	203.65	−136.33	485.93	35	2.36	123.98	25.25
0400	0.0187	67.32	203.65	−136.33	349.6	32	2.15	112.79	22.97
0500	0.0198	71.28	203.65	−132.37	217.23	40	2.85	100.46	20.46
0600	0.0226	81.36	203.65	−122.29	94.94	66	5.37	91.07	18.55
0700	0.0359	129.24	203.65	−74.41	20.53	92	11.89	91.61	18.66
0800	0.0509	183.24	203.65	−20.41	0.12	125	22.91	121.64	24.77

Note that the zero values in the computation of S_{avg} are valid only at startup of an empty basin. Also note that in this case M_{BOD-in} and $M_{BOD-out}$ differ only because of the difference in flow rates. For the second row (1000 h), the computations are

$$M_{BOD-in} = (0.0670 \text{ m}^3/\text{s})(150 \text{ g/m}^3)(1 \text{ h})(3,600 \text{ s/h})(10^{-3} \text{ kg/g})$$
$$= 36.2 \text{ kg}$$

$$S_{avg} = \frac{(241.20 \text{ m}^3)(150 \text{ g/m}^3) + (23.51 \text{ m}^3)(140 \text{ g/m}^3)}{241.20 \text{ m}^3 + 23.51 \text{ m}^3}$$
$$= 149.11 \text{ mg/L}$$

$$M_{BOD-out} = (0.05657 \text{ m}^3/\text{s})(149.11 \text{ g/m}^3)(1 \text{ h})(3,600 \text{ s/h})(10^{-3} \text{ kg/g})$$
$$= 30.37 \text{ kg}$$

Note that V_s is the volume of wastewater in the basin at the end of the previous time interval. Therefore, it equals the accumulated dS. The concentration of BOD_5 (S_{prev}) is the average concentration at the end of previous interval (S_{avg}) and *not* the influent concentration for the previous interval (S_o).

For the third row (1100 h), the concentration of BOD_5 is

$$S_{avg} = \frac{(245.52 \text{ m}^3)(155 \text{ g/m}^3) + (61.06 \text{ m}^3)(149.11 \text{ g/m}^3)}{245.52 \text{ m}^3 + 61.06 \text{ m}^3}$$
$$= 153.83 \text{ mg/L}$$

Comments:

1. In order to demonstrate the calculation procedure, a very large number of significant figures were used. The final results, $\Sigma \, dS$ and $M_{BOD-out}$, should only have two significant figures because, realistically, both the flow rate and BOD can only be measured accuately to two significant figures.

2. Theoretically, if the final $\Sigma \, dS$ at the end of a cycle is positive, the equalization basin will fill up. Likewise, a negative value for $\Sigma \, dS$ implies that there is no wastewater to pump to downstream processes. In reality, it is unlikely that the inflow will be perfectly sinusoidal. In addition, the provision of 25 percent extra volume and the ability to increase the pumping rate compensate for these eventualities.

8-6 PRIMARY TREATMENT

With the screening completed and the grit removed, the wastewater still contains light organic suspended solids, some of which can be removed from the sewage by gravity in a sedimentation tank. These tanks can be round or rectangular. The mass of settled solids is called *raw sludge*. The sludge is removed from the sedimentation tank by mechanical scrapers and pumps (Figure 8-6). Floating materials, such as grease and oil, rise to the surface of the sedimentation tank, where they are collected by a surface skimming system and removed from the tank for further processing.

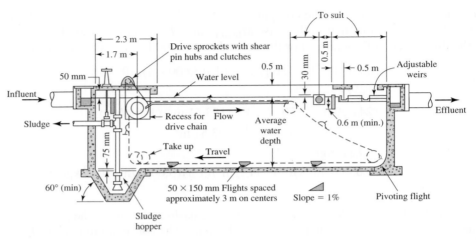

FIGURE 8-6
Primary settling tank.

Primary sedimentation basins (*primary tanks*) are characterized by Type II floc-culant settling. The Stokes equation cannot be used because the flocculating particles are continually changing in size, shape, and, when water is entrapped in the floc, spe-cific gravity. There is no adequate mathematical relationship that can be used to de-scribe Type II settling. Laboratory tests with settling columns are used to develop design data (see Chapter 6).

Rectangular tanks with common-wall construction are frequently chosen because they are advantageous for sites with space constraints. Typically, these tanks range from 30 to 100 m in length and 3 to 24 m in width. Common length-to-width ratios for the design of new facilities range from 3:1 to 5:1. Existing plants have length-to-width ratios ranging from 1.5:1 to 15:1. The width is often controlled by the availability of sludge collection equipment. Side water depths range from 3 to 5 m. Typically the depth is about 4 m.

Circular tanks have diameters from 3 to 100 m. Side water depths range from 3 to 5 m.

As in water treatment clarifier design, overflow rate is the controlling parame-ter for the design of primary settling tanks. At average flow, overflow rates typi-cally range from 30 to 50 $m^3/m^2 \cdot d$ (or 30 to 50 m/d). When waste-activated sludge is returned to the primary tank, a lower range of overflow rates is chosen (25 to 35 m/d). Under peak flow conditions, overflow rates may be in the range of 60 to 120 m/d.

Hydraulic detention time in the sedimentation basin ranges from 1.5 to 2.5 hours under average flow conditions. A 2.0-hour detention time is typical.

The Great Lakes–Upper Mississippi River Board of State Sanitary Engineers (GLUMRB) recommends that *weir loading* (hydraulic flow over the effluent weir) rates not exceed 250 m^3/d of flow per m of weir length ($m^3/d \cdot m$) for plants with aver-age flows less than 0.04 m^3/s. For larger flows, the recommended rate is 375 $m^3/d \cdot m$

(GLUMRB, 2004). If the side water depths exceed 3.5 m, the weir loading rates have little effect on performance.

Two different approaches have been used to place the weirs. Some designers believe in the "long" approach and place the weirs to cover 33 to 50 percent of the length of the tank. Those of the "short school" (for example, see Metcalf & Eddy, Inc., 2003) assume the weir length is less important and place it across the width of the end of the tank as shown in Figure 8-6. The spacing may vary from 2.5 to 6 m between weirs.

As mentioned previously, approximately 50 to 60 percent of the raw sewage suspended solids and as much as 30 to 35 percent of the raw sewage BOD_5 may be removed in the primary tank.

Example 8-3. Evaluate the following primary tank design with respect to detention time, overflow rate, and weir loading.

Design data:

Flow = 0.150 m³/s

Influent SS = 280 mg/L

Sludge concentration = 6.0%

Efficiency = 60%

Length = 40.0 m (effective)

Width = 10.0 m

Liquid depth = 2.0 m

Weir length = 75.0 m

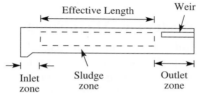

The detention time is simply the volume of the tank divided by the flow:

$$t_o = \frac{\Psi}{Q} = \frac{40.0 \text{ m} \times 10.0 \text{ m} \times 2.0 \text{ m}}{0.150 \text{ m}^3\text{/s}}$$

$$= 5333.33 \text{ s or } 1.5 \text{ h}$$

This is a reasonable detention time.

The overflow rate is the flow divided by the surface area:

$$v_o = \frac{0.150 \text{ m}^3\text{/s}}{40.0 \text{ m} \times 10.0 \text{ m}}$$

$$= 3.75 \times 10^{-4} \text{ m/s} \times 86,400 \text{ s/d} = 32 \text{ m/d}$$

This is an acceptable overflow rate.

The weir loading is calculated in the same fashion:

$$WL = \frac{0.150 \text{ m}^3\text{/s}}{75.0 \text{ m}}$$

$$= 0.0020 \text{ m}^3\text{/s} \cdot \text{m} \times 86,400 \text{ s/d} = 172.8 \text{ or } 173 \text{ m}^3\text{/d} \cdot \text{m}$$

This is an acceptable weir loading.

8-7   UNIT PROCESSES OF SECONDARY TREATMENT

Overview

The major purpose of secondary treatment is to remove the soluble BOD that escapes primary treatment and to provide further removal of suspended solids. The basic ingredients needed for conventional aerobic secondary biologic treatment are the availability of many microorganisms, good contact between these organisms and the organic material, the availability of oxygen, and the maintenance of other favorable environmental conditions (for example, favorable temperature and sufficient time for the organisms to work). A variety of approaches have been used in the past to meet these basic needs. The most common approaches are (1) activated sludge, (2) trickling filters, and (3) oxidation ponds (or lagoons). A process that does not fit precisely into either the trickling filter or the activated sludge category but does employ principles common to both is the *rotating biological contactor* (RBC).

Because the secondary treatment processes are fundamentally microbiological, we begin this discussion with an introduction to wastewater microbiology.

Wastewater Microbiology

Role of Microorganisms.   The stabilization of organic matter is accomplished biologically using a variety of microorganisms. The microorganisms convert the colloidal and dissolved carbonaceous organic matter into various gases and into protoplasm. Because protoplasm has a specific gravity slightly greater than that of water, it can be removed from the treated liquid by gravity settling.

It is important to note that unless the protoplasm produced from the organic matter is removed from the solution, complete treatment will not be accomplished because the protoplasm, which itself is organic, will be measured as BOD in the effluent. If the protoplasm is not removed, the only treatment that will be achieved is that associated with the bacterial conversion of a portion of the organic matter originally present to various gaseous end products (Metcalf & Eddy, Inc., 1979.)

Classification of Microorganisms by Kingdoms.   Microorganisms are organized into five broad groups based on their structural and functional differences. The groups are called *kingdoms*. The five kingdoms are *animals, plants, protista, fungi,* and *bacteria.* Representative examples and characteristics of differentiation are shown in Figure 8-7.

Classification by Energy and Carbon Source.   The relationship between the source of carbon and the source of energy for the microorganism is important. Carbon is the basic building block for cell synthesis. A source of energy must be obtained from outside the cell to enable synthesis to proceed. Our goal in wastewater treatment is to convert both the carbon and the energy in the wastewater into the cells of microorganisms, which we can remove from the water by settling. Therefore, we wish to encourage the growth of organisms that use organic material for both their carbon and energy source.

If the microorganism uses organic material as a supply of carbon, it is called *heterotrophic. Autotrophs* require only CO_2 to supply their carbon needs.

Organisms that rely only on the sun for energy are called *phototrophs. Chemotrophs* extract energy from organic or inorganic oxidation/reduction reactions.

KINGDOMS EXAMPLES

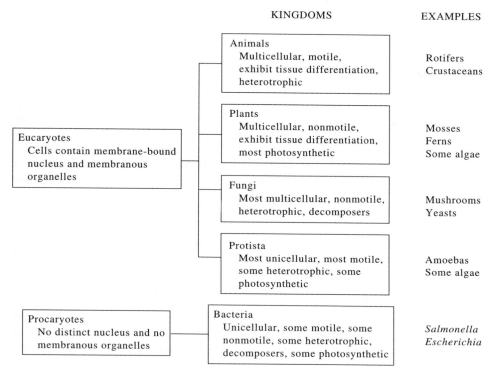

FIGURE 8-7
Classification of microorganisms by kingdom.

Organotrophs use organic materials, while *lithotrophs* oxidize inorganic compounds (Bailey and Ollis, 1977).

Classification by their Relationship to Oxygen. Bacteria also are classified by their ability or inability to utilize oxygen as a terminal electron acceptor* in oxidation/reduction reactions. *Obligate aerobes* are microorganisms that must have oxygen as the terminal electron acceptor. When wastewater contains oxygen and can support obligate aerobes, it is called *aerobic*.

---

*An organic substrate is not directly oxidized to carbon dioxide and water in a single chemical step because there is no energy-conserving mechanism that could trap so much energy. Thus, biological oxidation occurs in small steps. Oxidation requires the transfer of an electron from the substance being oxidized to some acceptor molecule that will subsequently be reduced. In most biological systems, each step in the oxidation process involves the removal of two electrons and the simultaneous loss of two protons (H^+). The combination of the two losses is equivalent to the molecule having lost two hydrogen atoms. The reaction is often referred to as *dehydrogenation*. The electrons and protons are not released into the cell, but are transferred to an acceptor molecule. The acceptor molecule will not accept the protons until it has accepted the electrons and thus it is referred to as an electron acceptor. Since the net result of accepting an electron and proton is the same as accepting a hydrogen atom, such acceptors are also called hydrogen acceptors. (Grady and Lim, 1980.)

Obligate anaerobes are microorganisms that cannot survive in the presence of oxygen. They cannot use oxygen as a terminal electron acceptor. Wastewater that is devoid of oxygen is called *anaerobic. Facultative anaerobes* can use oxygen as the terminal electron acceptor and, under certain conditions, they can also grow in the absence of oxygen.

Under *anoxic* conditions, a group of facultative anaerobes called *denitrifiers* utilizes nitrites (NO_2^-) and nitrates (NO_3^-) as the terminal electron acceptor. Nitrate nitrogen is converted to nitrogen gas in the absence of oxygen. This process is called *anoxic denitrification.*

Classification by their Preferred Temperature Regime. Each species of bacteria reproduces best within a limited range of temperatures. Four temperature ranges are used to classify bacteria. Those that grow best at temperatures below 20°C are called *psychrophiles. Mesophiles* grow best at temperatures between 25 and 40°C. Between 45 and 60°C, the *thermophiles* grow best. Above 60°C, *stenothermophiles* grow best. The growth range of *facultative thermophiles* extends from the thermophilic range into the mesophilic range. These ranges are qualitative and somewhat subjective. You will note the gaps between 20 and 25°C and between 40 and 45°C. Don't make the mistake of saying that an organism that grows well at 20.5°C is a mesophile. The rules just aren't that hard and fast. Bacteria will grow over a range of temperatures and will survive at a very large range of temperatures. For example, *Escherichia coli,* classified as mesophiles, will grow at temperatures between 20 and 50°C and will reproduce, albeit very slowly, at temperatures down to 0°C. If frozen rapidly, they and many other microorganisms can be stored for years with no significant death rate.

Bacteria. The highest population of microorganisms in a wastewater treatment plant will belong to the bacteria. They are single-celled organisms which use soluble BOD as food. Conditions in the treatment plant are adjusted so that chemoheterotrophs predominate. No particular species is selected as "the best."

Fungi. Fungi are multicellular, nonphotosynthetic, heterotrophic organisms. Fungi are obligate aerobes that reproduce by a variety of methods including fission, budding, and spore formation. Their cells require only half as much nitrogen as bacteria so that in a nitrogen-deficient wastewater, they predominate over the bacteria (McKinney, 1962).

Algae. This group of microorganisms are photoautotrophs and may be either unicellular or multicellular. Because of the chlorophyll contained in most species, they produce oxygen through photosynthesis. In the presence of sunlight, the photosynthetic production of oxygen is greater than the amount used in respiration. At night they use up oxygen in respiration. If the daylight hours exceed the night hours by a reasonable amount, there is a net production of oxygen.

Protozoa. Protozoa are single-celled organisms that can reproduce by *binary fission* (dividing in two). Most are aerobic chemoheterotrophs, and they often consume bacteria. They are desirable in wastewater effluents because they act as polishers in consuming the bacteria.

Rotifers and Crustaceans. Both rotifers and crustaceans are animals—aerobic, multicellular chemoheterotrophs. The rotifer derives its name from the apparent rotating

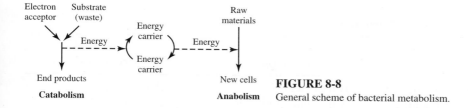

FIGURE 8-8

General scheme of bacterial metabolism.

motion of two sets of cilia on its head. The cilia provide mobility and a mechanism for catching food. Rotifers consume bacteria and small particles of organic matter.

Crustaceans, a group that includes shrimp, lobsters, and barnacles, are characterized by their shell structure. They are a source of food for fish and are not found in wastewater treatment systems to any extent except in underloaded lagoons. Their presence is indicative of a high level of dissolved oxygen and a very low level of organic matter.

Bacterial Biochemistry

The general term that describes all of the chemical activities performed by a cell is *metabolism*. This in turn is divided into two parts: catabolism and anabolism. *Catabolism* includes all the biochemical processes by which a substrate is degraded to end products with the release of energy.* In wastewater treatment, the substrate is oxidized. The oxidation process releases energy that is transferred to an energy carrier, which stores it for future use by the bacterium (Figure 8-8). Some chemical compounds released by catabolism are used by the bacterial cell for its life functions.

Anabolism includes all the biochemical processes by which the bacterium synthesizes new chemical compounds needed by the cells to live and reproduce. The synthesis process is driven by the energy that was stored in the energy carrier.

Decomposition of Waste.   The type of electron acceptor available for catabolism determines the type of decomposition (that is, aerobic, anoxic, or anaerobic) used by a mixed culture of microorganisms. Each type of decomposition has peculiar characteristics that affect its use in waste treatment.

Aerobic Decomposition.   From our discussion of bacterial metabolism you will recall that molecular oxygen (O_2) must be present as the terminal electron acceptor for decomposition to proceed by aerobic oxidation. As in natural water bodies, the oxygen is measured as DO. When oxygen is present, it is the only terminal electron acceptor used. Hence, the chemical end products of decomposition are primarily carbon dioxide, water, and new cell material (Table 8-8). Odiferous gaseous end products are kept to a minimum. In healthy natural water systems, aerobic decomposition is the principal means of self-purification.

---

*Substrate is food. For our application, "food" is the organic material from the human digestive tract and other biodegradable wastes.

TABLE 8-8
Representative end products

Substrates	Aerobic and anoxic decomposition	Anaerobic decomposition
Proteins and other nitrogen-containing compounds	Amino acids Ammonia → nitrites → nitrates*	Amino acids Ammonia Hydrogen sulfide Methane Carbon dioxide Alcohols Organic acids
Carbohydrates	Alcohols → CO_2 + H_2O Fatty acids	Carbon dioxide Fatty acids Methane
Fats and related substances	Fatty acids + glycerol Alcohols → CO_2 + H_2O Lower fatty acids	Fatty acids + glycerol Carbon dioxide Alcohols Lower fatty acids Methane

*Under anoxic conditions the nitrates are converted to nitrogen gas.
Source: After Pelczar, M. J., and R. D. Reid, 1958.

A wider spectrum of organic material can be oxidized aerobically than by any other type of decomposition. This fact, coupled with the fact that the final end products are oxidized to a very low energy level, results in a more stable end product (that is, one that can be disposed of without damage to the environment and without creating a nuisance condition) than can be achieved by the other oxidation systems.

Because of the large amount of energy released in aerobic oxidation, most aerobic organisms are capable of high growth rates. Consequently, there is a relatively large production of new cells in comparison with the other oxidation systems. This means that more biological sludge is generated in aerobic oxidation than in the other oxidation systems.

Aerobic decomposition is the method of choice for large quantities of dilute wastewater (BOD_5 less than 500 mg/L) because decomposition is rapid, efficient, and has a low odor potential. For high-strength wastewater (BOD_5 is greater than 1,000 mg/L), aerobic decomposition is not suitable because of the difficulty in supplying enough oxygen and because of the large amount of biological sludge produced. In small communities and in special industrial applications where aerated lagoons (see "Oxidation Ponds," below) are used, wastewaters with BOD_5 up to 3,000 mg/L may be treated satisfactorily by aerobic decomposition.

Anoxic Decomposition. Some microorganisms can use nitrate (NO_3^-) as the terminal electron acceptor in the absence of molecular oxygen. Oxidation by this route is called denitrification.

The end products from denitrification are nitrogen gas, carbon dioxide, water, and new cell material. The amount of energy made available to the cell during denitrification is about the same as that made available during aerobic decomposition. As a consequence, the rate of production of new cells, although not as high as in aerobic decomposition, is relatively high.

Denitrification is of importance in wastewater treatment where nitrogen must be removed to protect the receiving body. In this case, a special treatment step is added to the conventional process for removal of carbonaceous material. Denitrification will be discussed in detail later.

One other important aspect of denitrification is in relation to final clarification of the treated wastewater. If the environment of the final clarifier becomes anoxic, the formation of nitrogen gas will cause large globs of sludge to float to the surface and escape from the treatment plant into the receiving water. Thus, it is necessary to ensure that anoxic conditions do not develop in the final clarifier.

Anaerobic Decomposition.   In order to achieve anaerobic decomposition, molecular oxygen and nitrate must not be present as terminal electron acceptors. Sulfate (SO_4^{2-}), carbon dioxide, and organic compounds that can be reduced serve as terminal electron acceptors. The reduction of sulfate results in the production of hydrogen sulfide (H_2S) and a group of equally odiferous organic sulfur compounds called *mercaptans.*

The anaerobic decomposition (fermentation) of organic matter generally is considered to be a three-step process. In the first step, waste components are hydrolysed. In the second step, complex organic compounds are fermented to low-molecular-weight fatty acids (volatile acids). In the third step, the organic acids are converted to methane. Carbon dioxide serves as the electron acceptor.

Anaerobic decomposition yields carbon dioxide, methane, and water as the major end products. Additional end products include ammonia, hydrogen sulfide, and mercaptans. As a consequence of these last three compounds, anaerobic decomposition is characterized by an unbelievably horrid stench!

Because only small amounts of energy are released during anaerobic oxidation, the amount of cell production is low. Thus, sludge production is low. We make use of this fact in wastewater treatment by using anaerobic decomposition to stabilize sludges produced during aerobic and anoxic decomposition.

Direct anaerobic decomposition of wastewater generally is not feasible for dilute waste.* The optimum growth temperature for the anaerobic bacteria is at the upper end of the mesophilic range. Thus, to get reasonable biodegradation, we must elevate the temperature of the culture. For dilute wastewater, this is not practical. For concentrated wastes (BOD_5 greater than 1,000 mg/L), anaerobic digestion is quite appropriate.

Population Dynamics.   In the discussion of the behavior of bacterial cultures that follows, there is the inherent assumption that all the requirements for growth are initially present. Because these requirements are fairly extensive and stringent, it is worth

---

*Some researchers are exploring the use of anaerobic systems for treatment of dilute wastes, especially groundwater contaminated with hazardous waste.

taking a moment to recapitulate them. The following list summarizes the major requirements that must be satisfied:

1. A terminal electron acceptor

2. Macronutrients

   a. Carbon to build cells
   b. Nitrogen to build cells
   c. Phosphorus for ATP (energy carrier) and DNA

3. Micronutrients

   a. Trace metals
   b. Vitamins are required by some bacteria

4. Appropriate environment

   a. Moisture
   b. Temperature
   c. pH

As an illustration of growth in pure cultures, let us examine a hypothetical situation in which 1,400 bacteria of a single species are introduced into a synthetic liquid medium. Initially nothing appears to happen. The bacteria must adjust to their new environment and begin to synthesize new protoplasm. On a plot of bacterial growth versus time (Figure 8-9), this phase of growth is called the *lag phase.*

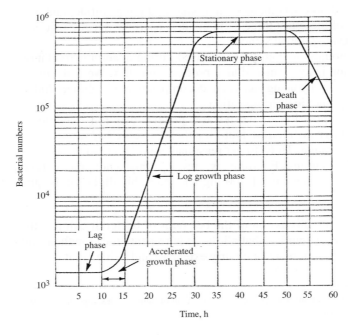

FIGURE 8-9

Bacterial growth in a pure culture: the "log-growth curve."

At the end of the lag phase the bacteria begin to divide. Because all of the organisms do not divide at the same time, there is a gradual increase in population. This phase is labeled *accelerated growth* on the growth plot.

At the end of the accelerated growth phase, the population of organisms is large enough and the differences in generation time are small enough that the cells appear to divide at a regular rate. Because reproduction is by binary fission (each cell divides producing two new cells), the increase in population follows in geometric progression: $1 \rightarrow 2 \rightarrow 4 \rightarrow 8 \rightarrow 16 \rightarrow 32$, and so forth. The population of bacteria (P) after the nth generation is given by the following expression:

$$P = P_0(2)^n \qquad \qquad (8\text{-}1)$$

where P_0 is the initial population at the end of the accelerated growth phase. If we take the log of both sides of Equation 8-1, we obtain the following:

$$\log P = \log P_0 + n \log 2 \qquad \qquad (8\text{-}2)$$

This means that if we plot bacterial population on a logarithmic scale, this phase of growth would plot as a straight line of slope n and intercept P_0 at t_0 equal to the end of the accelerated growth phase. Thus, this phase of growth is called the *log growth* or *exponential growth phase.*

The log growth phase tapers off as the substrate becomes exhausted or as toxic by-products build up. Thus, at some point the population becomes constant either as a result of cessation of fission or a balance in death and reproduction rates. This is depicted by the *stationary phase* on the growth curve.

Following the stationary phase, the bacteria begin to die faster than they reproduce. This death phase is due to a variety of causes that are basically an extension of those that lead to the stationary phase.

In wastewater treatment, as in nature, pure cultures of microorganisms do not exist. Rather, a mixture of species compete and survive within the limits set by the environment. *Population dynamics* is the term used to describe the time-varying success of the various species in competition. It is expressed quantitatively in terms of relative mass of microorganisms.*

The prime factor governing the dynamics of the various microbial populations is the competition for food. The second most important factor is the predator-prey relationship.

The relative success of a pair of species competing for the same substrate is a function of the ability of the species to metabolize the substrate. The more successful species will be the one that metabolizes the substrate more completely. In so doing, it will obtain more energy for synthesis and consequently will achieve a greater mass.

Because of their relatively smaller size and, thus, larger surface area per unit mass, which allows a more rapid uptake of substrate, bacteria will predominate over fungi. For the same reason, the fungi predominate over the protozoa.

---

*If each individual organism of species A has, on the average, twice the mass at maturity as each individual organism of species B, and both compete equally, we would expect that both would have the same total biomass, but that there would be twice as many of species B as there would be of A.

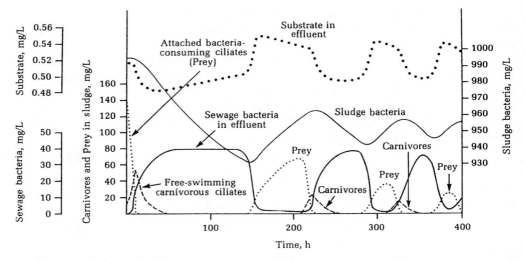

FIGURE 8-10
Population dynamics in a closed system. (*Source:* Curds, 1973.)

When the supply of soluble organic substrate becomes exhausted, the bacterial population is less successful in reproduction and the predator populations increase. In a closed system with an initial inoculum of mixed microorganisms and substrate, the populations will cycle as the bacteria give way to higher level organisms which in turn die for lack of food and are then decomposed by a different set of bacteria (Figure 8-10). In an open system, such as a wastewater treatment plant or a river, with a continuous inflow of new substrate, the predominant populations will change through the length of the plant (Figure 8-11). This condition is known as *dynamic equilibrium.* It is a highly sensitive state, and changes in influent characteristics must be regulated closely to maintain the proper balance of the various populations.

For the large numbers and mixed cultures of microorganisms found in waste treatment systems, it is convenient to measure biomass rather than numbers of organisms.* In the log-growth phase, the rate expression for biomass increase is

$$\frac{dX}{dt} = \mu X \tag{8-3}$$

where $\dfrac{dX}{dt}$ = growth rate of the biomass, mg/L · t

μ = growth rate constant, t^{-1}

X = concentration of biomass, mg/L

*Frequently, this is done by measuring suspended solids or volatile suspended solids (those that burn at 550 ± 50°C). When the wastewater contains only soluble organic matter, the volatile suspended solids test is reasonably representative. The presence of organic particles (which is often the case in municipal wastewater) confuses the issue completely.

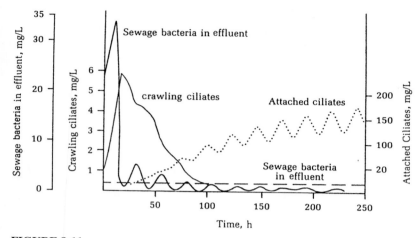

FIGURE 8-11
Population dynamics in an open system. (*Source:* Curds, 1973.)

Because of the difficulty of direct measurement of μ in mixed cultures, Monod (1949) developed a model equation that assumes that the rate of food utilization, and therefore the rate of biomass production, is limited by the rate of enzyme reactions involving the food compound that is in shortest supply relative to its need. The *Monod equation* is

$$\mu = \frac{\mu_m S}{K_s + S}$$
(8-4)

where μ_m = maximum growth rate constant, t^{-1}
S = concentration of limiting food in solution, mg/L
K_s = half saturation constant, mg/L
 = concentration of limiting food when $\mu = 0.5 \, \mu_m$

The growth rate of biomass follows a hyperbolic function as shown in Figure 8-12.

Two limiting cases are of interest in the application of Equation 8-4 to wastewater treatment systems. In those cases where there is an excess of the limiting food, then $S \gg K_s$ and the growth rate constant μ is approximately equal to μ_m. Equation 8-3 then becomes zero-order in substrate. At the other extreme, when $S \ll K_s$, the system is food-limited and the growth rate becomes first-order with respect to substrate.

Equation 8-4 assumes only growth of microorganisms and does not take into account natural die-off. It is generally assumed that the death or decay of the microbial mass is a first-order expression in biomass and hence Equations 8-3 and 8-4 are expanded to

$$\frac{dX}{dt} = \frac{\mu_m SX}{K_s + S} - k_d X$$
(8-5)

where k_d = endogenous decay rate constant, t^{-1}.

If all of the food in the system were converted to biomass, the rate of food utilization (dS/dt) would equal the rate of biomass production. Because of the inefficiency of

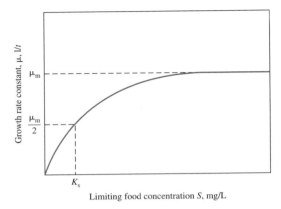

FIGURE 8-12

Monod growth rate constant as a function of limiting food concentration.

the conversion process, the rate of food utilization will be greater than the rate of biomass utilization, so

$$-\frac{dS}{dt} = \frac{1}{Y}\frac{dX}{dt} \tag{8-6}$$

where Y = decimal fraction of food mass converted to biomass

$$= \text{yield coefficient, } \frac{\text{mg/L biomass}}{\text{mg/L food utilized}}$$

Combining Equations 8-3, 8-4, and 8-6

$$-\frac{dS}{dt} = \frac{1}{Y}\frac{\mu_m SX}{K_s + S} \tag{8-7}$$

Equations 8-5 and 8-7 are a fundamental part of the development of the design equations for wastewater treatment processes.

Activated Sludge

The activated sludge process is a biological wastewater treatment technique in which a mixture of wastewater and biological sludge (microorganisms) is agitated and aerated. Because the microorganisms are suspended in the liquid wastewater, this process is known as a *suspended growth* process. The biological solids are subsequently separated from the treated wastewater and returned to the aeration process as needed.

The activated sludge process derives its name from the biological mass formed when air is continuously injected into the wastewater. In this process, microorganisms are mixed thoroughly with the organic compounds under conditions that stimulate their growth through use of the organic compounds as food. As the microorganisms grow and are mixed by the agitation of the air, the individual organisms clump together (flocculate) to form an active mass of microbes (biologic floc) called *activated sludge.*

In practice, wastewater flows continuously into an aeration tank (Figure 8-13) where air is injected to mix the activated sludge with the wastewater and to supply the oxygen needed for the organisms to break down the organic compounds. The mixture of activated sludge and wastewater in the aeration tank is called *mixed liquor.* The

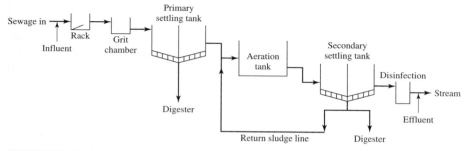

FIGURE 8-13
Conventional activated sludge plant.

mixed liquor flows from the aeration tank to a secondary clarifier where the activated sludge is settled out. Most of the settled sludge is returned to the aeration tank (and is called *return sludge*) to maintain the high population of microbes that permits rapid breakdown of the organic compounds. Because more activated sludge is produced than is desirable in the process, some of the return sludge is diverted or wasted to the sludge handling system for treatment and disposal. In conventional activated sludge systems, the wastewater is typically aerated for six to eight hours in long, rectangular aeration basins. About 8 m^3 of air is provided for each m^3 of wastewater treated. Sufficient air is provided to keep the sludge in suspension (Figure 8-14). The air is injected near the

FIGURE 8-14
Activated sludge aeration tank under air. (*Source:* E. Lansing WWTP, photo courtesy of Harley Seeley of the Instructional Media Center, Michigan State University.)

bottom of the aeration tank through a system of diffusers (Figure 8-15). The volume of sludge returned to the aeration basin is typically 20 to 30 percent of the wastewater flow.

The activated sludge process is controlled by wasting a portion of the microorganisms each day in order to maintain the proper amount of microorganisms to efficiently degrade the BOD_5. *Wasting* means that a portion of the microorganisms is discarded from the process. The discarded microorganisms are called *waste activated sludge* (WAS). A balance is then achieved between growth of new organisms and their removal by wasting. If too much sludge is wasted, the concentration of microorganisms in the mixed liquor will become too low for effective treatment. If too little sludge is wasted, a large concentration of microorganisms will accumulate and, ultimately, overflow the secondary tank and flow into the receiving stream.

(a)

(c)

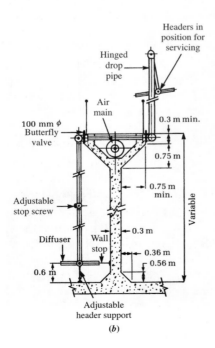

(b)

FIGURE 8-15

(a) Diffuser system for activated sludge tank; (b) cross section through tank illustrating hinged drop pipe; (c) close-up of a set of diffusers. (*Photo Source*: E. Lansing WWTP, courtesy of Harley Seeley of the Instructional Media Center, Michigan State University.)

The *mean cell residence time* θ_c, also called *solids retention time* (SRT) or *sludge age,* is defined as the average amount of time that microorganisms are kept in the system.

Many modifications of the conventional activated sludge process have been developed to address specific treatment problems. A brief description of these is given in Table 8-9. Process flow diagrams for some of these processes may be found at the text website: www.mhhe.com/davis. We have selected the completely mixed and conventional plug-flow processes for further discussion.

TABLE 8-9
Description of activated sludge processes and process modifications[a]

Process or process modification	Description
Conventional plug-flow	Settled wastewater and recycled activated sludge enter the head end of the aeration tank and are mixed by diffused-air or mechanical aeration. Air application is generally uniform throughout tank length. During the aeration period, adsorption, flocculation, and oxidation of organic matter occur. Activated sludge solids are separated in a secondary settling tank.
Complete-mix	Process is an application of the flow regime of a continuous-flow stirred-tank reactor. Settled wastewater and recycled activated sludge are introduced typically at several points in the aeration tank. The organic load on the aeration tank and the oxygen demand are uniform throughout the tank length.
Tapered aeration	Tapered aeration is a modification of the conventional plug-flow process. Varying aeration rates are applied over the tank length depending on the oxygen demand. Greater amounts of air are supplied to the head end of the aeration tank, and the amounts diminish as the mixed liquor approaches the effluent end. Tapered aeration is usually achieved by using different spacing of the air diffusers over the tank length.
Step-feed aeration	Step feed is a modification of the conventional plug-flow process in which the settled wastewater is introduced at several points in the aeration tank to equalize the food-to-microorganism (F/M) ratio, thus lowering peak oxygen demand. Generally three or more parallel channels are used. Flexibility of operation is one of the important features of this process.
Modified aeration	Modified aeration is similar to the conventional plug-flow process except that shorter aeration times and higher F/M ratios are used. BOD removal efficiency is lower than other activated sludge processes.
Contact stabilization	Contact stabilization uses two separate tanks or compartments for the treatment of the wastewater and stabilization of the activated sludge. The stabilized activated sludge is mixed with the influent (either raw or settled) wastewater in a contact tank. The mixed liquor is settled in a secondary settling tank and return sludge is aerated separately in a reaeration basin to stabilize the organic matter. Aeration volume requirements are typically 50 percent less than conventional plug-flow.

(continued)

TABLE 8-9
(continued)

Process or process modification	Description
Extended aeration	Extended aeration process is similar to the conventional plug-flow process except that it operates in the endogenous respiration phase of the growth curve, which requires a low organic loading and long aeration time. Process is used extensively for prefabricated package plants for small communities.
High-rate aeration	High-rate aeration is a process modification in which high MLSS concentrations are combined with high volumetric loadings. This combination allows high F/M ratios and long mean cell-residence times with relatively short hydraulic detention times. Adequate mixing is very important.
Kraus process	Kraus process is a variation of the step aeration process used to treat wastewater with low nitrogen levels. Digester supernatant is added as a nutrient source to a portion of the return sludge in a separate aeration tank designed to nitrify. The resulting mixed liquor is then added to the main plug-flow aeration system.
High-purity oxygen	High-purity oxygen is used instead of air in the activated sludge process. Oxygen is diffused into covered aeration tanks and is recirculated. A portion of the gas is wasted to reduce the concentration of carbon dioxide. pH adjustment may also be required. The amount of oxygen added is about four times greater than the amount that can be added by conventional aeration systems.
Oxidation ditch	The oxidation ditch consists of a ring- or oval-shaped channel and is equipped with mechanical aeration devices. Screened wastewater enters the ditch, is aerated, and circulates at about 0.25 to 0.35 m/s. Oxidation ditches typically operate in an extended aeration mode with long detention and solids retention times. Secondary sedimentation tanks are used for most applications.
Sequencing batch reactor	The sequencing batch reactor is a fill-and-draw type reactor system involving a single complete-mix reactor in which all steps of the activated sludge process occur. Mixed liquor remains in the reactor during all cycles, thereby eliminating the need for separate secondary sedimentation tanks.
Deep-shaft reactor	The deep vertical-shaft reactor is a form of the activated sludge process. A vertical shaft about 120 to 150 m deep replaces the primary clarifiers and aeration basin. The shaft is lined with a steel shell and fitted with a concentric pipe to form an annular reactor. Mixed liquor and air are forced down the center of the shaft and allowed to rise upward through the annulus.
Single-stage nitrification	In single-stage nitrification, both BOD and ammonia reduction occur in a single biological stage. Reactor configurations can be either a series of complete-mix reactors or plug flow.
Separate stage nitrification	In separate stage nitrification, a separate reactor is used for nitrification, operating on a feed waste from a preceding biological treatment unit. The advantage of this system is that operation can be optimized to conform to the nitrification needs.

[a]*Source:* Metcalf & Eddy, Inc., 1991.

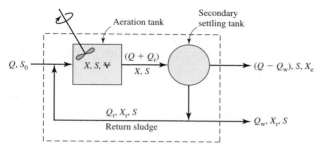

FIGURE 8-16
Completely mixed biological reactor with solids recycle.

Completely Mixed Activated Sludge Process.   The design formulas for the completely mixed activated sludge process are a mass balance application of the equations used to describe the kinetics of microbial growth. A mass balance diagram for the completely mixed system (CSTR) is shown in Figure 8-16. The mass balance equations are written for the system boundary shown by the dashed line. Two mass balances are required to define the design of the reactor: one for biomass and one for food (soluble BOD$_5$).

Under steady-state conditions, the mass balance for biomass may be written as:

$$\begin{matrix} \text{Biomass in} \\ \text{influent} \end{matrix} + \begin{matrix} \text{Net biomass} \\ \text{growth} \end{matrix} = \begin{matrix} \text{Biomass in} \\ \text{effluent} \end{matrix} + \begin{matrix} \text{Biomass} \\ \text{wasted} \end{matrix} \qquad (8\text{-}8)$$

The biomass in the influent is the product of the concentration of microorganisms in the influent (X_o) and the flow rate of wastewater (Q). The concentration of microorganisms in the influent (X_o) is measured as suspended solids (mg/L). The biomass that grows in the aeration tank is the product of the volume of the tank (V) and the Monod expression for growth of microbial mass (Equation 8-5)

$$(V)\left(\frac{\mu_m SX}{K_s + S} - k_d X\right) \qquad (8\text{-}9)$$

The biomass in the effluent is the product of flow rate of treated wastewater leaving the plant ($Q - Q_w$) and the concentration of microorganisms that does not settle in the secondary clarifier (X_e). The flow rate of wastewater leaving the plant does not equal the flow rate into the plant because some of the microorganisms must be wasted. The flow rate of wasting (Q_w) is deducted from the flow exiting the plant.

The biomass that is wasted is the product of concentration of microorganisms in the WAS flow (X_r) and the WAS flow rate (Q_w). The narrative mass balance equation may be rewritten as

$$QX_o + (V)\left(\frac{\mu_m SX}{K_s + S} - k_d X\right) = (Q - Q_w)X_e + Q_w X_r \qquad (8\text{-}10)$$

The variables are summarized as follows:

Q = wastewater flow rate into the aeration tank, m³/d

X_o = microorganism concentration (volatile suspended solids or VSS)* entering aeration tank, mg/L

V = volume of aeration tank, m³

μ_m = maximum growth rate constant, d⁻¹

S = soluble BOD_5 in aeration tank and effluent, mg/L

X = microorganism concentration (mixed-liquor volatile suspended solids or MLVSS)† in the aeration tank, mg/L

K_s = half velocity constant

= soluble BOD_5 concentration at one-half the maximum growth rate, mg/L

k_d = decay rate of microorganisms, d⁻¹

Q_w = flow rate of liquid containing microorganisms to be wasted, m³/d

X_e = microorganism concentration (VSS) in effluent from secondary settling tank, mg/L

X_r = microorganism concentration (VSS) in sludge being wasted, mg/L

At steady-state, the mass balance equation for food (soluble BOD_5) may be written

$$\frac{\text{Food in}}{\text{influent}} - \frac{\text{Food}}{\text{consumed}} = \frac{\text{Food in}}{\text{effluent}} + \frac{\text{Food in}}{\text{WAS}} \tag{8-11}$$

The food in the influent is the product of the concentration of soluble BOD_5 in the influent (S_o) and the flow rate of wastewater (Q). The food that is consumed in the aeration tank is the product of the volume of the tank (V) and the expression for rate of food utilization (Equation 8-7)

$$(V)\left(\frac{\mu_m S X}{Y(K_s + S)}\right) \tag{8-12}$$

The food in the effluent is the product of flow rate of treated wastewater leaving the plant ($Q - Q_w$) and the concentration of soluble BOD_5 in the effluent (S). The concentration of soluble BOD_5 in the effluent (S) is the same as that in the aeration tank because we have assumed that the aeration tank is completely mixed. Since the BOD_5 is soluble, the secondary settling tank will not change the concentration. Thus, the effluent concentration from the secondary settling tank is the same as the influent concentration.

---

*Suspended solids means that the material will be retained on a filter, unlike dissolved solids such as NaCl. The amount of the suspended solids that volatilizes at 500 ± 50°C is taken to be a measure of active biomass concentration. The presence of nonliving organic particles in the influent wastewater will cause some error (usually small) in the use of volatile suspended solids as a measure of biomass.

†Mixed-liquor volatile suspended solids is a measure of the active biological mass in the aeration tank. The term "mixed liquor" implies a mixture of activated sludge and wastewater. The phrase "volatile suspended solids" has the same meaning as in the definition of X_o.

The food in the waste activated sludge flow is the product of the concentration of soluble BOD_5 in the influent (S) and the WAS flow rate (Q_w). The narrative mass balance equation for steady-state conditions may be rewritten as

$$QS_o - (V)\left(\frac{\mu_m SX}{Y(K_s + S)}\right) = (Q - Q_w)S + Q_w S \tag{8-13}$$

where Y = yield coefficient (see Equation 8-6).

To develop working design equations we make the following assumptions:

1. The influent and effluent biomass concentrations are negligible compared to that in the reactor.

2. The influent food (S_o) is immediately diluted to the reactor concentration in accordance with the definition of a CSTR (see Chapter 2).

3. All reactions occur in the CSTR.

From the first assumption we may eliminate the following terms from Equation 8-10: QX_o, and $(Q - Q_w)X_e$ because X_o, and X_e, are negligible compared to X. Equation 8-10 may be simplified to

$$(V)\left(\frac{\mu_m SX}{K_s + S} - k_d X\right) = +Q_w X_r \tag{8-14}$$

For convenience, we may rearrange Equation 8-14 in terms of the Monod equation

$$\left(\frac{\mu_m S}{K_s + S}\right) = \frac{Q_w X_r}{VX} + k_d \tag{8-15}$$

Equation 8-13 may also be rearranged in terms of the Monod equation

$$\left(\frac{\mu_m S}{K_s + S}\right) = \frac{Q}{V}\frac{Y}{X}(S_o - S) \tag{8-16}$$

Noting that the left side of Equations 8-15 and 8-16 are the same, we set the right-hand side of these equations equal and rearrange to give:

$$\frac{Q_w X_r}{VX} = \frac{Q}{V}\frac{Y}{X}(S_o - S) - k_d \tag{8-17}$$

Two parts of this equation have physical significance in the design of a completely mixed activated sludge system. The inverse of Q/V is the *hydraulic detention time* (t_o) of the reactor:

$$\frac{V}{Q} = t_o \tag{8-18}$$

The inverse of the left side of Equation 8-17 defines the mean cell-residence time (θ_c):

$$\frac{VX}{Q_w X_r} = \theta_c \tag{8-19}$$

The mean cell-residence time expressed in Equation 8-19 must be modified if the effluent biomass concentration is not negligible. Equation 8-20 accounts for effluent losses of biomass in calculating θ_c.

$$\theta_c = \frac{\forall X}{Q_w X_r + (Q - Q_w)(X_e)} \qquad \text{(8-20)}$$

From Equation 8-15, it can be seen that once θ_c is selected, the concentration of soluble BOD_5 in the effluent (S) is fixed:

$$S = \frac{K_s(1 + k_d\theta_c)}{\theta_c(\mu_m - k_d) - 1} \qquad \text{(8-21)}$$

Typical values of the microbial growth constants are given in Table 8-10. Note that the concentration of soluble BOD_5 leaving the system (S) is affected only by the mean cell-residence time and not by the amount of BOD_5 entering the aeration tank or by the hydraulic detention time. It is also important to reemphasize that S is the soluble BOD_5 and not the total BOD_5. Some fraction of the suspended solids that do not settle in the secondary settling tank also contributes to the BOD_5 load to the receiving body. To achieve a desired effluent quality both the soluble and insoluble fractions of BOD_5 must be considered. Thus, to use Equation 8-21 to achieve a desired effluent quality (S) by solving for θ_c, some estimate of the BOD_5 of the suspended solids must be made first. This value is then subtracted from the total allowable BOD_5 in the effluent to find the allowable S:

$$S = \text{Total } BOD_5 \text{ allowed } - BOD_5 \text{ in suspended solids} \qquad \text{(8-22)}$$

From Equation 8-17, it is also evident that the concentration of microorganisms in the aeration tank is a function of the mean cell-residence time, hydraulic detention time, and difference between the influent and effluent concentrations:

$$X = \frac{\theta_c(Y)(S_o - S)}{t_o(1 + k_d\theta_c)} \qquad \text{(8-23)}$$

TABLE 8-10
Values of growth constants for domestic wastewater[a]

Parameter	Basis	Value[b]	
		Range	Typical
K_s	mg/L BOD_5	25–100	60
k_d	d^{-1}	0–0.30	0.10
μ_m	d^{-1}	1–8	3
Y	mg VSS/mg BOD_5	0.4–0.8	0.6

[a]*Sources:* Metcalf & Eddy, Inc., 2003, and Shahriari et al., 2006.
[b]Values are for 20°C.

Plug Flow with Recycle. A plug-flow reactor is one in which the fluid particles pass through the tank in sequence (see Section 2-3 for further discussion). Although it is difficult to achieve true plug flow, many long, narrow aeration tanks may be better approximated by plug flow than by completely mixed models. A kinetic model of a plug-flow system is difficult to develop from basic mass balance equations. With two simplifying assumptions, Lawrence and McCarty (1970) have developed a useful equation. The assumptions are:

1. The concentration of microorganisms in the influent to the aeration tank is approximately the same as that in the effluent from the aeration tank. This assumption applies if θ_c/t_o is greater than 5.

2. The rate of soluble BOD_5 utilization as the waste passes through the aeration tank is given by

$$r_u = -\frac{\mu_m S X_{avg}}{K_s + S} \tag{8-24}$$

where X_{avg} is the average concentration of microorganisms in the aeration tank. The design equation is

$$\frac{1}{\theta_c} = \frac{Y\mu_m (S_o - S)}{(S_o - S) + (1 + \alpha)K_s \ln(S_i/S)} - k_d \tag{8-25}$$

where α = recycle ratio, Q_r/Q
  ln = logarithm to base e
  S_i = influent concentration to aeration tank after dilution with recycle flow, mg/L

$$= \frac{S_o + \alpha S}{1 + \alpha}$$

Other terms are the same as those defined previously.

Example 8-4. The town of Gatesville has been directed to upgrade its primary WWTP to a secondary plant that can meet an effluent standard of 30.0 mg/L BOD_5 and 30.0 mg/L suspended solids (SS). They have selected a completely mixed activated sludge system.

Assuming that the BOD_5 of the SS may be estimated as equal to 63 percent of the SS concentration, estimate the required volume of the aeration tank. The following data are available from the existing primary plant.

Existing plant effluent characteristics
  Flow = 0.150 m³/s
  BOD_5 = 84.0 mg/L

Assume the following values for the growth constants: K_s = 100 mg/L BOD_5; μ_m = 2.5 d^{-1}; k_d = 0.050 d^{-1}; Y = 0.50 mg VSS/mg BOD_5 removed.

Solution.   Assuming that the secondary clarifier can produce an effluent with only 30.0 mg/L SS, we can estimate the allowable soluble BOD_5 in the effluent using the 63-percent assumption from above and Equation 8-22.

$$S = BOD_5 \text{ allowed } - BOD_5 \text{ in suspended solids}$$
$$S = 30.0 - (0.630)(30.0) = 11.1 \text{ mg/L}$$

The mean cell-residence time can be estimated with Equation 8-21 and the assumed values for the growth constants.

$$11.1 = \frac{100.0(1 + (0.050)\theta_c)}{\theta_c(2.5 - 0.050) - 1}$$

Solving for θ_c

$$(11.1)(2.45\theta_c - 1) = 100.0 + 5.00\theta_c$$
$$27.20\,\theta_c - 11.1 = 100.0 + 5.00\theta_c$$
$$\theta_c = \frac{111.1}{22.2} = 5.00 \text{ or } 5.0 \text{ d}$$

If we assume a value of 2,000 mg/L for the MLVSS, we can solve Equation 8-23 for the hydraulic detention time.

$$2,000 = \frac{5.00(0.50)(84.0 - 11.1)}{\theta(1 + (0.050)(5.00))}$$
$$t_o = \frac{2.50(72.9)}{2,000(1.25)}$$
$$t_o = 0.073 \text{ d or } 1.8 \text{ h}$$

The volume of the aeration tank is then estimated using Equation 8-18:

$$1.8 = \frac{\forall}{0.150 \times 3,600 \text{ s/h}}$$
$$\forall = 972 \text{ m}^3 \text{ or } 970 \text{ m}^3$$

Note: This answer results from rounding off the time. If the exact time is used, the volume us 945.63 m^3.

Another commonly used parameter in the activated sludge process is the food to microorganism ratio (F/M), which is defined as:

$$\frac{F}{M} = \frac{QS_o}{\forall X} \qquad \text{🏴 (8-26)}$$

The units of the F/M ratio are

$$\frac{\text{mg } BOD_5/d}{\text{mg MLVSS}} = \frac{\text{mg}}{\text{mg} \cdot \text{d}}$$

The F/M ratio is controlled by wasting part of the microbial mass, thereby reducing the MLVSS. A high rate of wasting causes a high F/M ratio. A high F/M yields organisms that are saturated with food. The result is that efficiency of treatment is poor. A low rate of wasting causes a low F/M ratio, which yields organisms that are starved. This results in more complete degradation of the waste.

A long θ_c (low F/M) is not always used, however, because of certain trade-offs that must be considered. A long θ_c means a larger and more costly aeration tank. It also means a higher requirement for oxygen and, thus, higher power costs. Problems with poor sludge "settleability" in the final clarifier may be encountered if θ_c is too long. However, because the waste is more completely degraded to final end products and less of the waste is converted into microbial cells when the microorganisms are starved at a low F/M, there is less sludge to handle.

Because both the F/M ratio and the cell-detention time are controlled by wasting of organisms, they are interrelated. A high F/M corresponds to a short θ_c and a low F/M corresponds to a long θ_c. F/M values typically range from 0.1 to 1.0 mg/mg · d for the various modifications of the activated sludge process.

Example 8-5. Two "fill and draw," batch-operated sludge tanks are operated at the "extreme" conditions described below. What is the effect on the operating parameters?

Tank A is settled once each day, and half the liquid is removed with care not to disturb the sludge that settles to the bottom. This liquid is replaced with fresh settled sewage. A plot of MLVSS concentration versus time takes the shape shown below.

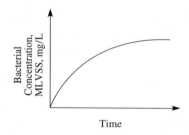

Tank B is not settled. Rather, once each day, half the mixed liquor is removed while the tank is being violently agitated. The liquid is replaced with fresh settled sewage. A plot of MLVSS concentration versus time is shown below.

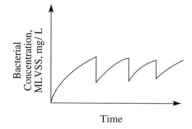

A comparison of the operating characteristics of the two systems is shown in the following table.

Parameter	Tank A	Tank B
F/M	Low	High
θ_c	Long	Short
Sludge	None	Much
Oxygen required	High	Low
Power	High	Low

The optimum choice is somewhere between these extremes. A balance must be struck between the cost of sludge disposal and the cost of power to provide oxygen (air).

Example 8-6. Compute the F/M ratio for the new activated-sludge plant at Gatesville (Example 8-4).

Solution. Using the data from Example 8-4 and Equation 8-26:

$$\frac{F}{M} = \frac{(0.150 \text{ m}^3/\text{s})(84.0 \text{ mg/L})(86,400 \text{ s/d})}{(970 \text{ m}^3)(2,000 \text{ mg/L})}$$

$$= 0.56 \text{ mg/mg} \cdot \text{d}$$

This is well within the typical range of F/M ratios.

Sludge Return. Among the major decisions in developing a suspended growth reactor design is the selection of the mixed liquor volatile suspended solids (MLVSS) concentration and the corresponding mixed liquor suspended solids (MLSS) concentration. This selection is not simple. It depends on the objective of the reactor, settling characteristics of the sludge, and the rate of recycle of sludge (called the *sludge return rate* or *sludge return*, or *return activated sludge* (RAS).

A high MLVSS concentration is desirable because it leads to a smaller reactor and lower construction costs. But this may lead to a larger settling tank to handle the sludge load. In addition, a higher MLVSS also requires a higher aeration rate to meet the increase in oxygen demand. Increasing MLVSS also requires increasing the rate at which sludge is returned from the settling tank. Finally, a higher MLVSS may lead to a higher effluent suspended solids and BOD in the effluent.

A mass balance around the settling tank in Figure 8-16 is the basis for selecting a return sludge rate. Assuming that the amount of sludge in the secondary settling tank remains constant (steady-state conditions) and that the effluent suspended solids (X_e) are negligible, the mass balance is

$$\text{Accumulation} = \text{Inflow} - \text{Outflow} \tag{8-27}$$

$$0 = (Q + Q_r)(X') - (Q_r X'_r + Q_w X'_r) \tag{8-28}$$

where
Q = wastewater flow rate, m³/d
Q_r = return sludge flow rate, m³/d
X' = mixed liquor suspended solids (MLSS), g/m³
X'_r = maximum return sludge concentration, g/m³
Q_w = sludge wasting flow rate, m³/d

Solving for the return sludge flow rate:

$$Q_r = \frac{QX' - Q_w X_r'}{X_r' - X'} \tag{8-29}$$

Using Equation 8-19, this may be rewritten as

$$Q_r = \frac{QX' - \left(\dfrac{\forall X'}{\theta_c}\right)}{X_r' - X'} \tag{8-30}$$

Frequently, the assumption that the effluent suspended solids are negligible is not valid. If the effluent suspended solids are significant, the mass balance may then be expressed as

$$0 = (Q + Q_r)(X') - (Q_r X_r' + Q_w X_r') + (Q - Q_w)X_e \tag{8-31}$$

Solving for the return sludge flow rate gives

$$Q_r = \frac{QX' - Q_w X_r' - (Q - Q_w)X_e}{X_r' - X'} \tag{8-32}$$

Note that X_r' and X' include both the volatile and inert fractions. Thus, they differ from X_r and X by a constant factor. It is generally assumed that VSS is 60 to 80 percent of MLVSS. Thus, MLSS may be estimated by dividing MLVSS by a factor of 0.6 to 0.8 (or multiplying by 1.25 to 1.67).

With the volume of the tank and the mean cell-residence time, the sludge wasting flow rate can be determined with Equation 8-19 if the maximum return sludge concentration (X_r') can be determined. Operating experience has demonstrated a range of values for X_r'. For a typical good settling sludge the maximum X_r' is in the range of 10,000 to 14,000 mg/L. With poorly settling sludges the maximum X_r' may be as low as 3,000 to 6,000 mg/L.

For operational flexibility the return sludge pumping rate must be adjustable. Typically, the maximum sludge return rate is set equal to the design flowrate. Most activated sludge plants are designed to permit variable sludge return from 10 to 100 percent of the raw wastewater flow. Example 8-7 illustrates the estimation of a maximum RAS flowrate.

Example 8-7. Determine the return sludge concentration (X_r') that results in the maximum return sludge flow rate for the proposed activated sludge upgrade at Gatesville (Example 8-4). Also estimate the mass flow rate of sludge wasting. Use the following assumptions:

MLVSS fraction of MLSS = 0.70

Solution. The Solver* program in a spreadsheet was used to perform the iterations for solution of this problem. The spreadsheet cells are shown in Figure 8-17. The cell locations used in the figure are identified by brackets [ ] in the discussion below.

---

* *Solver* is a "tool" in Excel®. Other spreadsheets may have a different name for this program. For those using Microsoft Excel®, the *Solver* program may be found under the "tools" pull-down on the tool bar. If this program has not been installed, it may be installed from the "Add-ins." Use the Microsoft Excel Help under the heading "Load or unload add-in programs" to find and install this program.

	A	B	C	D
3	**Input Data**			
4				
5	Q =	12960	m³/d	
6	MLVSS =	2	g/L	
7	MLVSS fraction =	0.7		
8	MLSS	2.86	g/L	This is X'
9	Volume of AT =	970	m³	
10	SRT =	5	d	
11	X'$_e$ =	0.03	g/L	
12				
13	**Calculation of Q$_r$ and X$_r$'**			
14				
15	X$_r$' =	5.64	g/L	
16				
17	Q$_w$ =	98.2	m³/d	
18				
19	Q$_r$ =	12960.0	m³/d	
20				
21	Q$_w$ X$_r$' =	554.3	kg/d	

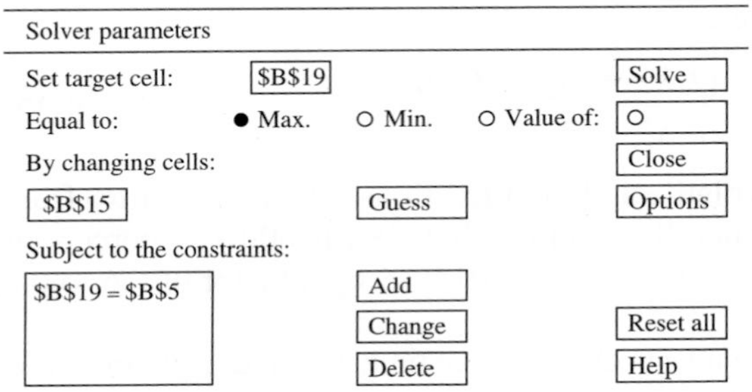

Solver parameters

Set target cell: [B19]   [Solve]

Equal to: ● Max.   ○ Min.   ○ Value of: [0]

By changing cells: [Close]

[B15]   [Guess]   [Options]

Subject to the constraints:

[B19 = B5]   [Add]
                       [Change]   [Reset all]
                       [Delete]   [Help]

FIGURE 8-17
Example 8-7 spreadsheet wih Solver dialog box.

Begin with the average design flow by setting the fixed parameters as follows:

[B5] $Q = 0.15 \text{ m}^3/\text{s} = 12,960 \text{ m}^3/\text{d}$
[B6] MLVSS = 2,000 mg/L = 2.0 g/L
[B7] MLVSS fraction = 0.70

In cell [B8] write an equation to convert MLVSS to MLSS

$$= \frac{[B6]}{[B7]} = \frac{2.0 \text{ g/L}}{0.7} = 2.86 \text{ g/L}$$

Continue setting the fixed parameters from Example 8-4:

[B9]   Volume of aeration tank = 970 m³
[B10] θ_c or SRT = 5.0 d
[B11] $X'_e = 0.030$ g/L

In cell [B15] insert a guess at X'_r between 3 and 20 g/L
In cell [B17] calculate Q_w using Equation 8-19

$$= \frac{([B9]*[B8])}{([B10]*[B15])}$$

In cell [B19] calculate Q_r using Equation 8-32

$$= \frac{((([B5]*[B8]) - ([B15]*[B17]) - (([B5] - [B17])*[B11]))}{([B15] - [B8])h}$$

In cell [B21] calculate the estimated mass flowrate $Q_w X_r'$

$$= [B17]*[B15]$$

Activate the dialog box for solver and designate the target cell as [B19], i.e., the one for the return sludge flow rate.

Set *Equal to* to "Max."

By changing the cell containing the return sludge MLSS (X_r') , i.e., [B15].

Add the following *constraint* in the dialog box: [B19] $\leq$ [B5]

Execute solve to find:

$$X_r' = 5.64 \text{ g/L}$$
$$Q_w = 98.2 \text{ m}^3/\text{d}$$
$$Q_r = 12,960 \text{ m}^3/\text{d}$$
$$Q_w X_r' = 554.3 \text{ kg/d}$$

Comments:

1. The constraint in the dialog box forces the maximum return sludge rate equal to the design flow rate in accordance with the typical design criterion. This criterion also reduces the problem to the solution of two simultaneous equations so that the answer may be obtained without the Solver function.

2. The spreadsheet can be used without the Solver tool to explore the other relationships, i.e., X_r', Q_w, Q_r, and $Q_w X_r'$.

Sludge Volume Index (SVI). SVI is determined from a standard laboratory test (APHA, 2005). The procedure involves measuring the MLSS and sludge settleability. A one-liter sample of mixed liquor is obtained from the aeration tank at the discharge end. The sludge settleability is measured by filling a standard one-liter graduated cylinder to the 1.0 liter mark, allowing undisturbed settling for 30 minutes, and then reading the volume occupied by the settled sludge. The MLSS is determined by filtering, drying, and weighing a second portion of the mixed liquor. SVI is defined as the volume in milliliters occupied by 1 g of activated sludge after the aerated liquor has settled 30 min. It is calculated as:

$$\text{SVI} = \frac{(\text{Settled volume of sludge, mL/L})(10^3 \text{ mg/g})}{\text{MLSS, mg/L}} = \frac{\text{mL}}{\text{g}} \qquad \text{FE} \quad (8\text{-}33)$$

An SVI of 100 mL/g or less is considered a good settling sludge. An SVI above 150 is typically associated with filamentous growth and poorly settling sludge (Parker, et al., 2001).

There are several alternative methods for conducting the test. The preferred technique is to use a vessel that is larger than a 1- or 2-liter cylinder. The vessel is equipped with a slow-speed stirring device.

A rearrangement of Equation 8-33 has been used as a design tool to predict the MLSS concentration in return sludge. The MLSS is estimated by assuming desired SVI:

$$X' = \frac{(1,000 \text{ mg/g})(1,000 \text{ mL/L})}{\text{SVI}} \tag{8-34}$$

where X' = MLSS, mg/L

Because it is impossible to relate SVI to MLSS by any fundamental principles, its use as a basis of design is problematic at best and may, in the worst case, lead to significant design errors. It assumes that the underflow concentration from the secondary settling tank is fixed by the concentration of the sludge and is totally independent of the design of the final settling tank or the manner in which it is operated (Dick, 1976).

The use of an assumed SVI for design is **not** recommended. It rationalizes a bold assumption. It is more realistic to make an assumption about the MLSS concentration within typical design ranges, provide a rigorous secondary clarifier design, and provide sufficient flexibility in the operation of the system to achieve good settling.

Nonetheless, the SVI may be the best available method for operators to evaluate performance of their clarifiers because its conservative characteristics compensate for unknowns in the behavior of the sedimentation basin (Dick, 1976).

The test is empirical. Because of this the results may lead to significant errors in interpretation. This is illustrated in Example 8-8.

Example 8-8.   Estimate the SVI for each of the following test results.

Test 1
Settled volume in 1 liter graduated cylinder = 250 mL
MLSS = 2,500 mg/L

Test 2
Settled volume in 1 liter graduated cylinder = 1,000 mL
MLSS = 10,000 mg/L

Solution.   For Test 1,

$$\text{SVI} = \frac{(250 \text{ mL/L})(10^3 \text{ mg/g})}{2,500 \text{ mg/L}} = 100 \text{ mL/g}$$

For Test 2,

$$\text{SVI} = \frac{(1,000 \text{ mL/L})(10^3 \text{ mg/g})}{10,000 \text{ mg/L}} = 100 \text{ mL/g}$$

Comments:

1. Obviously, the Test 2 SVI makes no sense as the sludge did not settle at all! This is another case where engineering judgment is essential in analyzing the results of a calculation.

2. Experienced operators can use the SVI results to monitor the settling process. The emphasis is on **experience**.

3. The preferred method for assessing secondary settling tank operating conditions is called *state point analysis*. A discussion of this technique may be found in *Water and Wastewater Engineering* (Davis, 2010).

Sludge Production.   The activated sludge process removes substrate, which exerts an oxygen demand by converting the food into new cell material and degrading this cell material while generating energy. This cell material ultimately becomes sludge that must be disposed of. Despite the problems in doing so, researchers have attempted to develop enough basic information on sludge production to permit a reliable design basis. Heukelekian and Sawyer both reported that a net yield of 0.5 kg MLVSS/kg BOD_5 removed could be expected for a completely soluble organic substrate (Heukelekian et al., 1951, and Sawyer, 1956). Most researchers agree that, depending on the inert solids in the system and the SRT, 0.40 to 0.60 kg MLVSS/kg BOD_5 removed will normally be observed.

The amount of sludge that must be wasted each day is the difference between the amount of increase in sludge mass and the suspended solids (SS) lost in the effluent:

$$\text{Mass to be wasted} = \text{Increase in MLSS} - \text{SS lost in effluent} \qquad (8\text{-}35)$$

The net activated sludge produced each day is determined by:

$$Y_{obs} = \frac{Y}{1 + k_d \theta_c} \qquad (8\text{-}36)$$

and

$$P_x = Y_{obs} Q (S_o - S)(10^{-3} \text{ kg/g}) \qquad (8\text{-}37)$$

where P_x = net waste activated sludge produced each day in terms of VSS, kg/d
Y_{obs} = observed yield, kg MLVSS/kg BOD_5 removed

Other terms are as defined previously.

The increase in MLSS may be estimated by assuming that VSS is some fraction of MLSS. It is generally assumed that VSS is 60 to 80 percent of MLVSS. Thus, the increase in MLSS in Equation 8-37 may be estimated by dividing P_x by a factor of 0.6 to 0.8 (or multiplying by 1.25 to 1.667). The mass of suspended solids lost in the effluent is the product of the flow rate $(Q - Q_w)$ and the suspended solids concentration (X_e).

Example 8-9.   Estimate the mass of sludge to be wasted each day from the new activated sludge plant at Gatesville (Examples 8-4 and 8-6).

Solution.   Using the data from Example 8-4, calculate Y_{obs}:

$$Y_{obs} = \frac{0.50 \text{ kg VSS/kg BOD}_5 \text{ removed}}{1 + [(0.050 \text{ d}^{-1})(5\,\text{d})]}$$

$$= 0.40 \text{ kg VSS/kg BOD}_5 \text{ removed}$$

The net waste activated sludge produced each day is

$$P_x = (0.40)(0.150 \text{ m}^3/\text{s})(84.0 \text{ g/m}^3 - 11.1 \text{ g/m}^3)(86{,}400 \text{ s/d})(10^{-3} \text{ kg/g})$$

$$= 377.9 \text{ kg/d of VSS}$$

The total mass produced includes inert materials. Using the relationship between MLSS and MLVSS in Example 8-6,

$$\text{Increase in MLSS} = (1.43)(377.9 \text{ kg/d}) = 540.4 \text{ kg/d}$$

The mass of solids (both volatile and inert) lost in the effluent is

$$(Q - Q_w)(X_e) = (0.150 \text{ m}^3/\text{s} - 0.001 \text{ m}^3/\text{s})(30 \text{ g/m}^3)(86{,}400 \text{ s/d})(10^{-3} \text{ kg/g})$$

$$= 385.9 \text{ kg/d}$$

The mass to be wasted is then

$$\text{Mass to be wasted} = 540.4 - 385.9 = 154.5 \text{ kg/d}$$

Note that this mass is calculated as dry solids. Because the sludge is mostly water, the actual mass will be considerably larger. This is discussed further in Section 8-11.

Oxygen Demand.   Oxygen is used in those reactions required to degrade the substrate to produce the high-energy compounds required for cell synthesis and for respiration. For long SRT systems, the oxygen needed for cell maintenance can be of the same order of magnitude as substrate metabolism. A minimum residual of 0.5 to 2 mg/L DO is usually maintained in the reactor basin to prevent oxygen deficiencies from limiting the rate of substrate removal.

An estimate of the oxygen requirements may be made from the BOD_5 of the waste and amount of activated sludge wasted each day. If we assume all of the BOD_5 is converted to end products, the total oxygen demand can be computed by converting BOD_5 to BOD_L. Because a portion of waste is converted to new cells that are wasted, the BOD_L of the wasted cells must be subtracted from the total oxygen demand. An approximation of the oxygen demand of the wasted cells may be made by assuming cell oxidation can be described by the following reaction:

$$\underbrace{C_5H_7NO_2}_{\text{cells}} + 5O_2 \rightleftharpoons 5CO_2 + 2H_2O + NH_3 + \text{energy} \qquad (8\text{-}38)$$

The ratio of gram molecular weights is

$$\frac{5(32)}{113} = 1.42$$

Thus the oxygen demand of the waste activated sludge may be estimated as 1.42 (P_x).
The mass of oxygen required may be estimated as:

$$M_{O_2} = \frac{Q(S_o - S)(10^{-3} \text{ kg/g})}{f} - 1.42(P_x) \tag{8-39}$$

where Q = wastewater flow rate into the aeration tank, m^3/d
S_o = influent soluble BOD$_5$, mg/L
S = effluent soluble BOD$_5$, mg/L
f = conversion factor for converting BOD$_5$ to ultimate BOD$_L$
P_x = waste activated sludge produced (see Equation 8-37)

The volume of air to be supplied must take into account the percent of air that is oxygen and the transfer efficiency of the dissolution of oxygen into the wastewater.

Example 8-10. Estimate the volume of air to be supplied (m^3/d) for the new activated sludge plant at Gatesville (Examples 8-4 and 8-9). Assume that BOD$_5$ is 68 percent of the ultimate BOD and that the oxygen transfer efficiency is 8 percent.

Solution. Using the data from Examples 8-4 and 8-9 gives

$$M_{O_2} = \frac{(0.150 \text{ m}^3/\text{s})(84.0 \text{ g/m}^3 - 11.1 \text{ g/m}^3)(86{,}400 \text{ s/d})(10^{-3} \text{ kg/g})}{0.68}$$

$$- 1.42(377.9 \text{ kg/d of VSS})$$

$$= 1{,}389.4 - 536.6 = 852.8 \text{ kg/d of oxygen}$$

From Table A-5 in Appendix A, air has a density of 1.185 kg/m^3 at standard conditions. By mass, air contains about 23.2 percent oxygen. At 100 percent transfer efficiency, the volume of air required is

$$\frac{852.8 \text{ kg/d}}{(1.185 \text{ kg/m}^3)(0.232)} = 3{,}101.99 \text{ or } 3{,}100 \text{ m}^3/\text{d}$$

At 8 percent transfer efficiency the volume of air required is

$$\frac{3{,}101.99 \text{ m}^3/\text{d}}{0.08} = 38{,}774.9 \text{ or } 38{,}000 \text{ m}^3/\text{d}$$

Process Design Considerations. The SRT (i.e., θ_c) selected for design is a function of the degree of treatment required. A high SRT (or older sludge age) results in a higher quantity of solids being carried in the system and a higher degree of treatment being obtained. A long SRT also results in the production of less waste sludge.

SRT values for design of carbonaceous BOD$_5$ removal are a function of the minimum temperature at which the reactor basin will be operated. They range from 5d at 20°C to 15d at 5°C. It is expected that the soluble BOD$_5$ in the effluent from the aeration system will be 4 to 8 mg/L.

If industrial wastes are discharged to the municipal system, several additional concerns must be addressed. Municipal wastewater generally contains sufficient nitrogen and phosphorus to support biological growth. The presence of large volumes of industrial wastewater that is deficient in either of these nutrients will result in poor removal efficiencies. Addition of supplemental nitrogen and phosphorus may be required. The ratio of nitrogen to BOD_5 should be 1:32. The ratio of phosphorus to BOD_5 should be 1:150.

Although toxic metals and organic compounds may be at low enough levels that they do not interfere with the operation of the plant, two other unwanted effects may result if they are not excluded in a pretreatment program. Volatile organic compounds may be stripped from solution into the atmosphere in the aeration tank. Thus, the WWTP may become a source of air pollution. The toxic metals may precipitate into the waste sludges. Thus, otherwise nonhazardous sludges may be rendered hazardous.

Oil and grease that pass through the primary treatment system will form grease balls on the surface of the aeration tank. The microorganisms cannot degrade this material because it is not in the water where they can physically come in contact with it. Special consideration should be given to the surface skimming equipment in the secondary clarifier to handle the grease balls.

Secondary Clarifier Design Considerations.   Although the secondary settling tank (Figure 8-18) is an integral part of both the trickling filter and the activated sludge process, environmental engineers have focused particular attention on the secondary clarifier used after the activated sludge process. A secondary clarifier is important

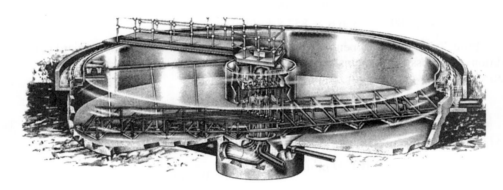

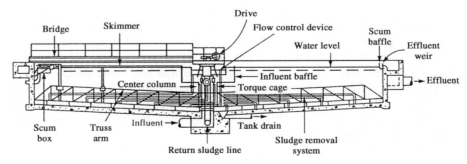

FIGURE 8-18
Rendering and cross-sectional diagram of secondary settling tank. (Courtesy of FMC Corporation.)

TABLE 8-11
Final settling basin side water depth

Tank diameter, m	Side water depth, m	
	Minimum	Recommended
<12	3.0	3.4
12 to 20	3.4	3.7
20 to 30	3.7	4.0
30 to 42	4.0	4.3
>42	4.3	4.6

Source: Adapted from WEF, 1998.

because of the high solids loading and fluffy nature of the activated sludge biological floc. Also, it is highly desirable that sludge recycle be well thickened.

Secondary settling tanks for activated sludge are generally characterized as having Type III settling. Some authors would argue that Types I and II also occur.

The following guidance has been excerpted from the Joint Task Force of the Water Pollution Control Federation and the American Society of Civil Engineers (WEF, 1998). The design factors discussed here are the result of the experiences of investigators, plant superintendents, and equipment manufacturers. The criteria primarily apply to circular (or square) center-fed basins, which comprise the majority of activated sludge secondary settling units designed in the last 40 years.

An overflow rate between 16 and 30 m/d for the average flow in a conventional process can be expected to result in good separation of liquid and SS. The design engineer also must check the peak hydraulic rates that will be imposed on the settling basin.

Suggested secondary settling tank *side water depths* (SWD) are shown the Table 8-11. Solids loading rates should be in the range 100 to 150 kg MLSS/d · m^2.

GLUMRB (2004) specifies that plants with a capacity less than 3,800 m^3/d should not have a hydraulic loading greater than 250 m^3/d · m of weir length at peak hourly flow. For plants larger than 3,800 m^3/d, the weir loading rate is specified as less than 350 m^3/d · m.

Current design consensus is that weir placement and settling tank configuration have more effect on clarifier performance than the hydraulic loading. Davis (2010) provides a concise discussion of *energy dissipating inlets* (EDIs), baffles, and weir configurations.

Example 8-11. The secondary clarifier for the Gatesville plant (Examples 8-4, 8-6, 8-7, 8-9, and 8-10) must be able to handle an MLSS load of 2,860 mg/L. The flow rate to the secondary clarifier is 0.300 m^3/s, of which one-half is contributed by return flow. Determine the tank diameter, side water depth, and weir length.

Solution. The overflow rate and solids loading rate are computed using the diameter of the tank. Thus, the solution may require iteration to obtain a diameter that satisfies

	A	B	C	D	E
4	**Input Data**				
5					
6	Flow rate	0.3	m³/s		
7	Q/2 =	0.15	m³/s		
8	v₀	17.5	m/d		
9	MLSS	2860	mg/L		
10					
11	Calculation of area, tank diameter, and loading rate				
12					
13	A_s =	741.312	m²		
14					
15	Diameter of tank =	30.7	m		
16					
17	Solids loading rate =	100.0	kg/d-m²		
18					
19					
20	**Solver parameters**				
21					
22	Set target cell =	B17			
23	Equal to:	min			
24	By changing cells: B8				
25	Subject to constraints:				
26		B8 <= 30			
27		B8 >= 16			
28		B17 <= 150			
29		B17 >= 100			
30					

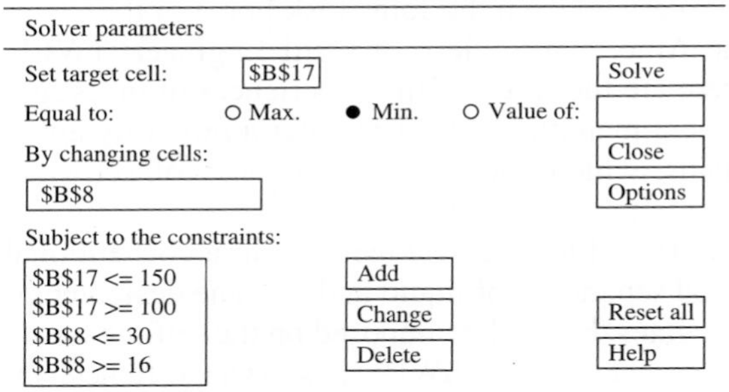

Solver parameters

Set target cell: B17   Solve

Equal to: ○ Max.   ● Min.   ○ Value of:

By changing cells: Close

B8   Options

Subject to the constraints:

B17 <= 150   Add
B17 >= 100   Change   Reset all
B8 <= 30   Delete   Help
B8 >= 16

FIGURE 8-19
Solver spreadsheet and dialog box for Example 8-11.

both the criteria for acceptable overflow rate and solids loading rate. The *Solver** program in a spreadsheet was used to perform the iterations for solution of this problem. The spreadsheet cells are shown in Figure 8-19. (Note that the figure shows the solution after the Solver program has been executed.) The cell locations used in the figure are identified by brackets [ ] in the discussion below.

a. Calculation of the overflow rate and solids loading rate.

Begin with the average design flow by setting the initial parameters as follows:

$$[B6] \quad Q + Q_r = 0.300 \text{ m}^3/\text{s}$$

---

* *Solver* is a "tool" in Excel®. Other spreadsheets may have a different name for this program. For those using Microsoft Excel®, the *Solver* program may be found under the "tools" pull-down on the tool bar. If this program has not been installed, it may be installed from the "Add-ins." Use the Microsoft Excel Help under the heading "Load or unload add-in programs" to find and install this program.

In cell [B7] calculate the flow that leaves through the surface of the tank.

$$= \frac{[B6]}{2} = 0.150 \text{ m}^3/\text{s}$$

Note that because only one-half of the flow leaves through the surface of the tank (the remainder leaves through the bottom as Q_r), only one-half of the hydraulic load is used to compute the overflow rate.

In cell [B8] enter a first guess for the overflow rate (v_o). It should be in the range given for the average flow in a conventional process, i.e., between 16 and 30 m/d.

In cell [B9] enter the solids loading rate, i.e., 2,860, mg/L.

In cell [B13] write an equation to calculate the required surface area for the specified overflow rate

$$= \frac{[B7]*86400}{[B8]}$$

The value 86,400 s/d is to convert the numerator to compatible units with the overflow rate.

In cell [B15] write an equation to calculate the diameter of the tank

$$= \left(\frac{(4*[B13])}{3.14159} \right)^{\wedge}0.5$$

In cell [B17] write an equation to calculate the solids loading rate

$$= \left(\frac{([B6]*[B9])}{((3.14.159*[B15]^\wedge 2)/4)} \right)*0.001*86400$$

The value 0.001 is to convert g to kg. The value 86,400 s/d is to convert m/s to m/d.

Activate the dialog box for Solver and designate the target cell as [B17], i.e., the one for the solids loading rate.

Set *Equal to* to "Min."

Set *By changing* to the cell containing the overflow rate (v_o) , i.e., [B8].

Add the following *constraints* in the dialog box:

[B17] < = 150
[B17] > = 100
[B8] < = 30
[B8] > = 16

Execute the *solve* function to find the solution shown in Figure 8-19. The tank diameter is 30.7 m.

b. Using the diameter of 30.7 m, select a side water depth of 4.3 m from Table 8-11.

c. Calculate the weir length.

Assuming the weir is located on the perimeter of the tank, the weir loading rate for a single weir located on the periphery of the tank is

$$\text{WL} = \frac{0.150 \text{ m}^3/\text{s} \times 86,400 \text{ s/d}}{\pi(30.7 \text{ m})} = 134.37 \text{ or } 134 \text{ m}^3/\text{d} \cdot \text{m}$$

This is less than the prescribed loading of 250 m^3/d · m given by GLUMRB so the weir length is 30.7 m.

Comment:

1. The final selection of the diameter must accommodate manufacturer's standard sludge scraping equipment sizes.

Sludge problems.   A *bulking sludge* is one that has poor settling characteristics and poor compactability. There are two principal types of sludge bulking. The first is caused by the growth of filamentous organisms, and the second is caused by water trapped in the bacterial floc, thus reducing the density of the agglomerate and resulting in poor settling.

Filamentous bacteria have been blamed for much of the bulking problem in activated sludge. Although filamentous organisms are effective in removing organic matter, they have poor floc-forming and settling characteristics. Bulking may also be caused by a number of other factors, including long, slow-moving collection-system transport; low available ammonia nitrogen when the organic load is high; low pH, which could favor acid-favoring fungi; and the lack of macronutrients, which stimulates predomination of the filamentous actinomycetes over the normal floc-forming bacteria. The lack of nitrogen also favors slime-producing bacteria, which have a low specific gravity, even though they are not filamentous. The multicellular fungi cannot compete with the bacteria normally, but can compete under specific environmental conditions, such as low pH, low nitrogen, low oxygen, and high carbohydrates. As the pH decreases below 6.0, the fungi are less affected than the bacteria and tend to predominate. As the nitrogen concentrations drop below a BOD$_5$: N ratio of 20:1, the fungi, which have a lower protein level than the bacteria, are able to produce normal protoplasm, while the bacteria produce nitrogen-deficient protoplasm.

A sludge that floats to the surface after apparently good settling is called a *rising sludge*. Rising sludge results from denitrification, that is, reduction of nitrates and nitrites to nitrogen gas in the sludge blanket (layer). Much of this gas remains trapped in the sludge blanket, causing globs of sludge to rise to the surface and float over the weirs into the receiving stream.

Rising-sludge problems can be overcome by increasing the rate of return sludge flow (Q_r), by increasing the speed of the sludge-collecting mechanism, by decreasing the mean cell residence time, and, if possible, by decreasing the flow from the aeration tank to the offending tank.

Trickling Filters

A typical process flow diagram for a trickling filter plant is shown in Figure 8-20. The trickling filter itself consists of a bed of coarse material, such as stones, slats, or plastic materials (*media*), over which wastewater is applied. Because the microorganisms that biodegrade the waste form a film on the media, this process is known as an *attached growth* process. Trickling filters have been a popular biologic treatment process. The most widely used design for many years was simply a bed of stones from 1 to 3 m deep through which the wastewater passed. The wastewater is typically

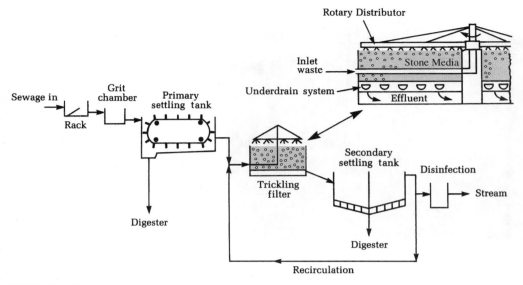

FIGURE 8-20
Trickling-filter plant with enlargement of trickling filter.

distributed over the surface of the rocks by a rotating arm (Figure 8-21). Rock filter diameters may range up to 60 m.

As the wastewater trickles through the bed, a microbial growth establishes itself on the surface of the stone or packing in a fixed film. The wastewater passes over the stationary microbial population, providing contact between the microorganisms and the organic contaminants.

Trickling filters are not primarily a filtering or straining process as the name implies. The rocks in a rock filter are 25 to 100 mm in diameter and hence have openings too large to strain out solids. They are a means of providing large amounts of surface area where the microorganisms cling and grow in a slime on the rocks as they feed on the organic matter.

Excess growths of microorganisms wash from the rock media and would cause undesirably high levels of suspended solids in the plant effluent if not removed. Thus, the flow from the filter is passed through a sedimentation basin to allow these solids to settle out. As in the case of the activated sludge process this sedimentation basin is referred to as a *secondary clarifier,* or *final clarifier.*

Although rock trickling filters have performed well for years, they have certain limitations. Under high organic loadings, the slime growths can be so prolific that they plug the void spaces between the rocks, causing flooding and failure of the system. Also, the volume of void spaces is limited in a rock filter, which restricts the circulation of air and the amount of oxygen available for the microbes. This limitation, in turn, restricts the amount of wastewater that can be processed. Other problems include odors and filter flies.

(a)

(b)

FIGURE 8-21
Trickling filters, rock media (a) and synthetic media (b). (*Sources:* (a) Wyoming, MI WWTP, photo by M. L. Davis and (b) Brentwood Industries.)

To overcome these limitations, other materials have become popular for filling the trickling filter. These materials include modules of corrugated plastic sheets and plastic rings. These media offer larger surface areas for slime growths (typically 90 square meters of surface area per cubic meter of bulk volume, as compared to 40 to 60 square meters per cubic meter for 75 mm rocks) and greatly increase void ratios for increased air flow. The materials are also much lighter than rock (by a factor of about 30), so that the trickling filters can be much taller without facing structural problems. While rock in filters is usually not more than 3 m deep, synthetic media depths may reach 12 m, thus reducing the overall space requirements for the trickling-filter portion of the treatment plant.

Trickling filters are classified according to the applied hydraulic and organic load. The hydraulic load may be expressed as cubic meters of wastewater applied per day per square meter of bulk filter surface area ($m^3/d \cdot m^3$) or, preferably, as the depth of water applied per unit of time (mm/s or m/d). Organic loading is expressed as kilograms of BOD_5 per day per cubic meter of bulk filter volume ($kg/d \cdot m^3$). Common hydraulic and organic loadings for the the various filter classifications are summarized in Table 8-12.

An important element in trickling filter design is the provision for return of a portion of the effluent to flow through the filter. This practice is called *recirculation*. The ratio of the returned flow to the incoming flow is called the *recirculation ratio (r)*. Recirculation is practiced in stone filters for the following reasons:

1. To increase contact efficiency by bringing the waste into contact more than once with active biological material.

TABLE 8-12
Comparison of different types of trickling filters[a]

Design characteristics	Trickling filter classification				
	Low or standard rate	Intermediate rate	High rate (stone media)	Super rate (plastic media)	Roughing
Hydraulic loading, m/d	1 to 4	4 to 10	10 to 40	15 to 90[b]	60 to 180[b]
Organic loading, kg BOD_5/d · m^3	0.08 to 0.32	0.24 to 0.48	0.32 to 1.0	0.32 to 1.0	Above 1.0
Recirculation ratio	0	0 to 1	1 to 3	0 to 1	1 to 4
Filter flies	Many	Varies	Few	Few	Few
Sloughing	Intermittent	Varies	Continuous	Continuous	Continuous
Depth, m	1.5 to 3	1.5 to 2.5	1 to 2	Up to 12	1 to 6
BOD_5 removal, %	80 to 85	50 to 70	40 to 80	65 to 85	40 to 85
Effluent quality	Well nitrified	Some nitrification	Nitrites	Limited nitrification	No nitrification

[a]*Source:* Adapted from WEF, 1992.
[b]Not including recirculation.

2. To dampen variations in loadings over a 24-hour period. The strength of the recirculated flow lags behind that of the incoming wastewater. Thus, recirculation dilutes strong influent and supplements weak influent.

3. To raise the DO of the influent.

4. To improve distribution over the surface, thus reducing the tendency to clog and also reduce filter flies.

5. To prevent the biological slimes from drying out and dying during nighttime periods when flows may be too low to keep the filter wet continuously.

Recirculation may or may not improve treatment efficiency. The more dilute the incoming wastewater, the less likely it is that recirculation will improve efficiency.

Recirculation is practiced for plastic media to provide the desired wetting rate to keep the microorganisms alive. Generally, increasing the hydraulic loading above the minimum wetting rate does not increase BOD_5 removal. The minimum wetting rate normally falls in the range of 25 to 60 m/d.

Two-stage trickling filters (Figure 8-22) provide a means of improving the performance of filters. The second stage acts as a polishing step for the effluent from the primary stage by providing additional contact time between the waste and the microorganisms. Both stages may use the same media or each stage may have different media as shown in Figure 8-22. The designer will select the types of media and their arrangement based on the desired treatment efficiencies and an economic analysis of the alternatives.

Design Formulas. Numerous investigators have attempted to correlate operating data with the bulk design parameters of trickling filters. Rather than attempt a comprehensive review of these formulations, we have selected the National Research Council (NRC, 1946) equations and Schulze's equation (Schulze, 1960) as illustrations. A thorough review of several of the more important equations is given in the Water Environment Federation's publication on wastewater treatment plant design (WEF, 1992).

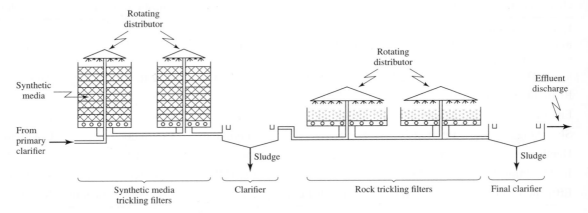

FIGURE 8-22
Two-stage trickling-filter plant.

During World War II, the NRC made an extensive study of the operation of trickling filters serving military installations. From this study, empirical equations were developed to predict the efficiency of the filters based on the BOD load, the volume of the filter media, and the recirculation. For a single-stage filter or the first stage of a two-stage filter, the efficiency is

$$E_1 = \frac{1}{1 + 4.12\left(\dfrac{QC_{\text{in}}}{\Psi F}\right)^{0.5}} \tag{8-40}$$

where E_1 = fraction of BOD$_5$ removal for first stage at 20°C, including recirculation and sedimentation
Q = wastewater flow rate, m^3/s
C_{in} = influent BOD$_5$, mg/L
Ψ = volume of filter media, m^3
F = recirculation factor

The recirculation factor is

$$F = \frac{1 + R}{(1 + 0.1\,R)^2} \tag{8-41}$$

where R = recirculation ratio = Q_r/Q
Q_r = recirculation flow rate, m^3/s
Q = wastewater flow rate, m^3/s

The recirculation factor represents the average number of passes of the raw wastewater BOD through the filter. The factor $0.1\,R$ is to account for the empirical observation that the biodegradability of the organic matter decreases as the number of passes increases. For the second stage filter, the efficiency is

$$E_2 = \frac{1}{1 + \dfrac{4.12}{1 - E_1}\left(\dfrac{QC_e}{\Psi F}\right)^{0.5}} \tag{8-42}$$

where E_2 = fraction of BOD$_5$ removal for second stage filter at 20°C, including recirculation and sedimentation
E_1 = fraction of BOD$_5$ removed in first stage
C_e = effluent BOD$_5$ from first stage, mg/L

The effect of temperature on the efficiency may be estimated from the following equation:

$$E_T = E_{20}\theta^{(T-20)} \tag{8-43}$$

where a value of 1.035 is used for θ.

Some care should be used in applying the NRC equations. Military wastewater during this period (World War II) had a higher strength than domestic wastewater today. The filter media was rock. Clarifiers associated with the trickling filters were shallower and carried higher hydraulic loads than current practice would permit. The second stage filter is assumed to be preceded by an intermediate settling tank (see Figure 8-22).

Example 8-12. Using the NRC equations, determine the BOD_5 of the effluent from a single-stage, low-rate trickling filter that has a filter volume of 1,443 m³, a hydraulic flow rate of 1,900 m³/d, and a recirculation factor of 2.78. The influent BOD_5 is 150 mg/L.

Solution. To use the NRC equation, the flow rate must first be converted to the correct units.

$$Q = (1,900 \text{ m}^3/\text{d})\left(\frac{1}{86,400 \text{ s/d}}\right) = 0.022 \text{ m}^3/\text{s}$$

The efficiency of a single-stage filter is

$$E_1 = \frac{1}{1 + 4.12\left(\dfrac{(0.022)(150)}{(1,443)(2.78)}\right)^{0.5}} = 0.8943$$

The concentration of BOD_5 in the effluent is then

$$C_e = (1 - 0.8943)(150) = 15.8 \text{ mg/L}$$

Schulze (1960) proposed that the time of wastewater contact with the biological mass in the filter is directly proportional to the depth of the filter and inversely proportional to the hydraulic loading rate:

$$t = \frac{CD}{(Q/A)^n} \tag{8-44}$$

where t = contact time, d
$\quad\quad C$ = mean active film per unit volume
$\quad\quad D$ = filter depth, m
$\quad\quad Q$ = hydraulic loading, m³/d
$\quad\quad A$ = filter area over which wastewater is applied, m²
$\quad\quad n$ = empirical constant based on filter media

The mean active film per unit volume may be approximated by

$$C \simeq \frac{1}{D^m} \tag{8-45}$$

where m is an empirical constant that is an indicator of biological slime distribution. It is normally assumed that the distribution is uniform and that $m = 0$. Thus, C is 1.0.

Schulze combined his relationship with Velz's (1948) first-order equation for BOD removal

$$\frac{S_t}{S_o} = \exp\left[-\frac{KD}{(Q/A)^n}\right] \tag{8-46}$$

where K is an empirical rate constant with the units of

$$\frac{(m/d)^n}{m}$$

The values of K and n determined by Shulze were 0.69 $(m/d)^n/m$ and 0.67 at 20°C. The temperature correction for K may be computed with Equation 8-43 if K_T is substituted for E_T and K_{20} is substituted for E_{20}.

Example 8-13. Determine the BOD_5 of the effluent from a low-rate trickling filter that has a diameter of 35.0 m and a depth of 1.5 m if the flow rate is 1,900 m^3/d and the influent BOD_5 is 150.0 mg/L. Assume the rate constant is 2.3 $(m/d)^n/m$ and $n = 0.67$.

Solution. We begin by computing the area of the filter.

$$A = \frac{\pi(35.0)^2}{4}$$

$$= 962.11\,m^2$$

This area is then used to compute the loading rate.

$$\frac{Q}{A} = \frac{1{,}900\ m^3/d}{962.11\ m^2}$$

$$= 1.97\ m^3/d \cdot m^2$$

Now we can compute the effluent BOD using Equation 8-46

$$S_t = (150)\exp\left[\frac{-(2.3)(1.5)}{(1.97)^{0.67}}\right]$$

$$= 16.8\,mg/L$$

Oxidation Ponds

Treatment ponds have been used to treat wastewater for many years, particularly as wastewater treatment systems for small communities (Benefield and Randall, 1980). Many terms have been used to describe the different types of systems employed in wastewater treatment. For example, in recent years, *oxidation pond* has been widely used as a collective term for all types of ponds. Originally, an oxidation pond was a pond that received partially treated wastewater, whereas a pond that received raw wastewater was known as a *sewage lagoon. Waste stabilization pond* has been used as an all-inclusive term that refers to a pond or lagoon used to treat organic waste by biological and physical processes. These processes would commonly be referred to as self-purification if they took place in a stream. To avoid confusion, the classification to be employed in this discussion will be as follows (Caldwell et al., 1973):

1. *Aerobic ponds:* Shallow ponds, less than 1 m in depth, where dissolved oxygen is maintained throughout the entire depth, mainly by the action of photosynthesis.

2. *Facultative ponds:* Ponds 1 to 2.5 m deep, which have an anaerobic lower zone, a facultative middle zone, and an aerobic upper zone maintained by photosynthesis and surface reaeration.

3. *Anaerobic ponds:* Deep ponds that receive high organic loadings such that anaerobic conditions prevail throughout the entire pond depth.

4. *Maturation or tertiary ponds:* Ponds used for polishing effluents from other biological processes. Dissolved oxygen is furnished through photosynthesis and surface reaeration. This type of pond is also known as a *polishing pond.*

5. *Aerated lagoons:* Ponds oxygenated through the action of surface or diffused air aeration.

Aerobic Ponds.   The aerobic pond is a shallow pond in which light penetrates to the bottom, thereby maintaining active algal photosynthesis throughout the entire system. During the daylight hours, large amounts of oxygen are supplied by the photosynthesis process; during the hours of darkness, wind mixing of the shallow water mass generally provides a high degree of surface reaeration. Stabilization of the organic material entering an aerobic pond is accomplished mainly through the action of aerobic bacteria.

Anaerobic Ponds.   The magnitude of the organic loading and the availability of dissolved oxygen determine whether the biological activity in a treatment pond will occur under aerobic or anaerobic conditions. A pond may be maintained in an anaerobic condition by applying a BOD_5 load that exceeds oxygen production from photosynthesis. Photosynthesis can be reduced by decreasing the surface area and increasing the depth. Anaerobic ponds become turbid from the presence of reduced metal sulfides. This restricts light penetration to the point that algal growth becomes negligible. Anaerobic treatment of complex wastes involves three stages. In the first stage organic matter is hydrolyzed. In the second stage (known as acid fermentation), complex organic materials are broken down mainly to short-chain acids and alcohols. In the third stage (known as methane fermentation), these materials are converted to gases, primarily methane and carbon dioxide. The proper design of anaerobic ponds must result in environmental conditions favorable to methane fermentation.

Anaerobic ponds are used primarily as a pretreatment process and are particularly suited for the treatment of high-temperature, high-strength wastewaters. However, they have been used successfully to treat municipal wastewaters as well.

Facultative Ponds.   Of the five general classes of lagoons and ponds, facultative ponds are by far the most common type selected as wastewater treatment systems for small communities. Approximately 25 percent of the municipal wastewater treatment plants in the United States are ponds and about 90 percent of these ponds are located in communities of 5,000 people or fewer. Facultative ponds are popular for such treatment situations because long retention times facilitate the management of large fluctuations in wastewater flow and strength with no significant effect on effluent quality. Also capital, operating, and maintenance costs are less than those of other biological systems that provide equivalent treatment.

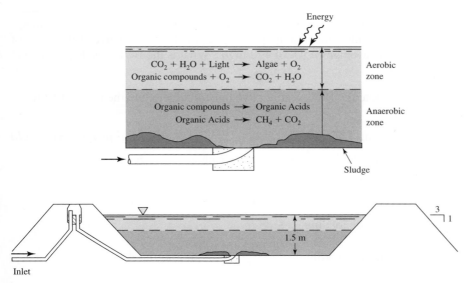

FIGURE 8-23
Schematic diagram of facultative pond relationships.

A schematic representation of a facultative pond operation is given in Figure 8-23. Raw wastewater enters at the center of the pond. Suspended solids contained in the wastewater settle to the pond bottom, where an anaerobic layer develops. Microorganisms occupying this region do not require molecular oxygen as an electron acceptor in energy metabolism, but rather use some other chemical species. Both acid fermentation and methane fermentation occur in the bottom sludge deposits.

The facultative zone exists just above the anaerobic zone. This means that molecular oxygen will not be available in the region at all times. Generally, the zone is aerobic during the daylight hours and anaerobic during the hours of darkness.

Above the facultative zone, there exists an aerobic zone that has molecular oxygen present at all times. The oxygen is supplied from two sources. A limited amount is supplied from diffusion across the pond surface. However, the majority is supplied through the action of algal photosynthesis.

Two rules of thumb commonly used in Michigan in evaluating the design of facultative lagoons are as follows:

1. The BOD_5 loading rate should not exceed 22 kg/ha · d on the smallest lagoon cell.

2. The detention time in the lagoon (considering the total volume of all cells but excluding the bottom 0.6 m in the volume calculation) should be six months.

The first criterion is to prevent the pond from becoming anaerobic. The second criterion is to provide enough storage to hold the wastewater during winter months when the receiving stream may be frozen or during the summer when the flow in the stream might be too low to absorb even a small amount of BOD.

Example 8-14.   A lagoon having three cells, each 115,000 m^2 in area, a minimum operating depth of 0.6 m, and a maximum operating depth of 1.5 m, receives 1,900 m^3/d of wastewater having an average BOD$_5$ of 122 mg/L. What is the BOD$_5$ loading and what is the detention time?

Solution.   To compute the BOD loading, we must first compute the mass of BOD$_5$ entering each day.

$$BOD_5 \text{ mass} = (122 \text{ mg/L})(1,900 \text{ m}^3)(1,000 \text{ L/m}^3)(1 \times 10^{-6} \text{ kg/mg}) = 231.8 \text{ kg/d}$$

Then, we must convert the area into hectares. Using only one cell,

$$\text{Area} = (115,000 \text{ m}^2)(1 \times 10^{-4} \text{ ha/m}^2)$$
$$= 11.5 \text{ ha each}$$

Now we can compute the loading.

$$BOD_5 \text{ loading} = \frac{231.8 \text{ kg/d}}{11.5 \text{ ha}} = 20.2 \text{ kg/ha} \cdot \text{d}$$

This loading rate is acceptable.

The detention time is simply the working volume between the minimum and maximum operating levels divided by the average daily flow.

$$\text{Detention time} = \frac{(115,000 \text{ m}^2)(3 \text{ lagoons})(1.5 - 0.6 \text{ m})}{1,900 \text{ m}^3/\text{d}}$$
$$= 163.4 \text{ days}$$

This is less than the desired 180 days.

Comment: We have ignored the slope of the lagoon walls in this calculation. For large lagoons, this is probably acceptable. In small lagoons, the slope should be considered.

Rotating Biological Contactors (RBCs)

The RBC process consists of a series of closely spaced discs (3 to 3.5 m in diameter) mounted on a horizontal shaft and rotated, while about one-half of their surface area is immersed in wastewater (Figure 8-24). The discs are typically constructed of light-weight plastic. The speed of rotation of the discs is adjustable.

When the process is placed in operation, the microbes in the wastewater begin to adhere to the rotating surfaces and grow there until the entire surface area of the discs is covered with a 1- to 3-mm layer of biological slime. As the discs rotate, they carry a film of wastewater into the air; this wastewater trickles down the surface of the discs, absorbing oxygen. As the discs complete their rotation, the film of water mixes with the reservoir of wastewater, adding to the oxygen in the reservoir and mixing the treated and partially treated wastewater. As the attached microbes pass through the reservoir, they absorb other organics for breakdown. The excess growth of microbes is sheared from the discs as they move through the reservoir. These

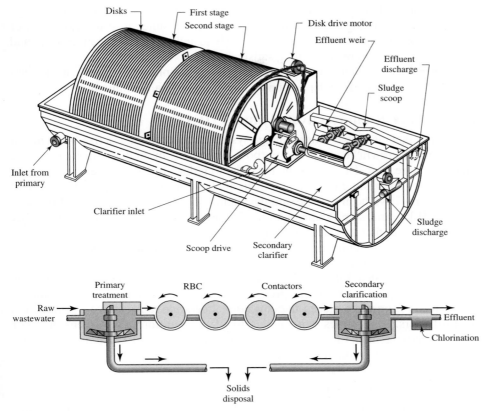

FIGURE 8-24
Rotating Biological Contactor (RBC) and process arrangement. (*Source:* U.S. EPA, 1977.)

dislodged organisms are kept in suspension by the moving discs. Thus, the discs serve several purposes:

1. They provide media for the buildup of attached microbial growth.

2. They bring the growth into contact with the wastewater.

3. They aerate the wastewater and the suspended microbial growth in the reservoir.

The attached growths are similar in concept to a trickling filter, except the microbes are passed through the wastewater rather than the wastewater passing over the microbes. Some of the advantages of both the trickling filter and activated sludge processes are realized.

As the treated wastewater flows from the reservoir below the discs, it carries the suspended growths out to a downstream settling basin for removal. The process can achieve secondary effluent quality or better. By placing several sets of discs in series, it is possible to achieve even higher degrees of treatment, including biological conversion of ammonia to nitrates.

The history of RBC installations has not been exemplary. Poor mechanical design and lack of understanding of the biological process has resulted in structural failure of shafts and disks. Many of the problems with early installations have been resolved and numerous installations are performing satisfactorily.

The principal advantages of the RBC process are simplicity of operation and relatively low energy costs. They have found application in small communities.

Integrated Fixed-Film Activated Sludge (IFAS)

This category includes any activated sludge system that incorporates a fixed-film media in a suspended growth reactor. The purpose of fixed-film media is to increase the biomass in the reactor. This offers the potential to reduce the basin size or to increase the capacity of an existing basin in a retrofit application. Various types of suspended growth systems have been used. Examples include conventional, modified Ludzack-Ettinger (MLE), and step denitrification. These processes differ from the MBBR described below in that they use a return sludge flow.

A number of proprietary media types have been used including rope (no longer in use), sponge, plastic carriers, and a honeycomb polyester fabric called BioWeb™. The media that are fixed in a frame are preferred because they require fewer appurtenances and are less susceptible to hydraulic problems that result from free-floating media.

The media frames are placed in conventional aeration tanks above a grid of fine bubble diffusers. The frames vary in size to fit the aeration tank dimensions. For example, an arrangement of two units with dimensions of 3.8 m $\times$ 3.8 m $\times$ 4 m high is set side by side across the flow path of the reactor.

Moving Bed Biofilm Reactor (MBBR)

This process uses small, plastic elements (on the order of 7 to 22 mm effective diameter) to support the growth of biofilm in the rector. The suspended growth portion of the hybrid is designed as a complete mix reactor. It is commonly mixed with aeration but may also be mixed with a mechanical mixer. The process does not use a return sludge flow.

The media (typically polyethylene) is formed in a geometry to provide a high surface area ($250–515$ m^2/m^3). It has a density near that of water (≈ 0.96 g/cm^3). The reactors are normally filled from one-third to two-thirds of their empty volume with media. Because of their shape, less than 15 percent of the water is displaced. A screen across the outlet is used to prevent the media from leaving the aeration tank. Aeration is typically by coarse bubble diffusers.

8-8   DISINFECTION

The last treatment step in a secondary plant is the addition of a disinfectant to the treated wastewater. The addition of chlorine gas or some other form of chlorine is the process most commonly used for wastewater disinfection in the United States. Chlorine is injected into the wastewater by automated feeding systems. Wastewater then flows into a basin, where it is held for about 15 minutes to allow the chlorine to react with the pathogens.

There is concern that wastewater disinfection may do more harm than good. Early U.S. Environmental Protection Agency rules calling for disinfection to achieve 200 fecal coliforms per 100 mL of wastewater have been modified to a requirement for disinfection only during the summer season when people may come into contact with contaminated water. There were three reasons for this change. The first was that the use of chlorine and, perhaps, ozone causes the formation of organic compounds that are carcinogenic. The second was the finding that the disinfection process was more effective in killing the predators to cysts and viruses than it was in killing the pathogens themselves. The net result was that the pathogens survived longer in the natural environment because there were fewer predators. The third reason was that chlorine is toxic to fish.

8-9 TERTIARY WASTEWATER TREATMENT

The need for treatment of wastewater beyond that which can normally be accomplished in secondary treatment is based on the recognition of one or more of the following:

1. Increasing population pressures result in increasing loads of organic matter and suspended solids to rivers, streams, and lakes.

2. The need to increase the removal of suspended solids to provide more efficient disinfection.

3. The need to remove nutrients to limit eutrophication of sensitive water bodies.

4. The need to remove constituents that preclude or inhibit water reclamation.

Initially, in the 1970s, these processes were called "advanced wastewater treatment" because they employed techniques that were more advanced than secondary treatment methods. In the last three decades many of these technologies have either been directly incorporated into the secondary processes, for example nutrient removal, or they are so inherent in meeting stringent discharge standards that they have become conventional. These processes include chemical precipitation, granular filtration, membrane filtration, and carbon adsorption. As conventional processes, they are better termed *tertiary treatment* processes rather than an advanced treatment process. In current practice, the employment of air stripping, ion exchange, NF or RO treatment, and other similar processes to meet water quality requirements is correctly termed *advanced wastewater treatment*. Advanced wastewater treatment technologies are, fundamentally, those employed to treat water for reuse.

This discussion focuses on the following tertiary treatment processes: granular filtration, membrane filtration, carbon adsorption, chemical phosphorus removal, biological phosphorus removal, and nitrogen control.

In addition to solving difficult pollution problems, these processes improve the effluent quality to the point that it is adequate for many reuse purposes, and may convert what was originally a wastewater into a valuable sustainable resource too good to throw away.

Filtration

Secondary treatment processes, such as the activated-sludge process, are highly efficient for removal of biodegradable colloidal and soluble organics. However, the typical effluent contains a much higher BOD_5 than one would expect from theory. The typical BOD is approximately 20 to 50 mg/L. This is principally because the secondary clarifiers are not perfectly efficient at settling out the microorganisms from the biological treatment processes. These organisms contribute both to the suspended solids and to the BOD_5 because the process of biological decay of dead cells exerts an oxygen demand.

Granular Filtration. By using a filtration process similar to that used in water treatment plants, it is possible to remove the residual suspended solids, including the unsettled microorganisms. Removing the microorganisms also reduces the residual BOD_5. Conventional sand filters identical to those used in water treatment can be used, but they often clog quickly, thus requiring frequent backwashing. To lengthen filter runs and reduce backwashing, it is desirable to have the larger filter grain sizes at the top of the filter. This arrangement allows some of the larger particles of biological floc to be trapped at the surface without plugging the filter. Multimedia filters accomplish this by using low-density coal for the large grain sizes, medium-density sand for intermediate sizes, and high-density garnet for the smallest size filter grains. Thus, during backwashing, the greater density offsets the smaller diameter so that the coal remains on top, the sand remains in the middle, and the garnet remains on the bottom.

Typically, plain filtration can reduce activated sludge effluent suspended solids from 25 to 10 mg/L. Plain filtration is not as effective on trickling filter effluents because trickling filter effluents contain more dispersed growth. However, the use of coagulation and sedimentation followed by filtration can yield suspended solids concentrations that are virtually zero. Typically, filtration can achieve 80 percent suspended solids reduction for activated sludge effluent and 70 percent reduction for trickling filter effluent.

Membrane Filtration. The alternative membrane processes have been discussed in Chapter 6. Of the five processes, the one most commonly used in tertiary treatment is microfiltration (MF). It is used as a replacement for granular filtration. MF processes have achieved BOD removals of 75–90 percent and total suspended solids removals of 95–98 percent. Performance is highly site-specific. Membrane fouling is of particular concern and on-site pilot testing is highly recommended (Metcalf & Eddy, 2003).

Carbon Adsorption

Even after secondary treatment, coagulation, sedimentation, and filtration, soluble organic materials that are resistant to biological breakdown will persist in the effluent. The persistent materials are often referred to as *refractory organics*. Refractory organics can be detected in the effluent as soluble COD. Secondary effluent COD values are often 30 to 60 mg/L.

The most practical available method for removing refractory organics is by *adsorbing* them on *activated carbon* (U.S. EPA, 1979a). *Adsorption* is the accumulation of materials at an *interface*. The interface, in the case of wastewater and activated carbon, is the liquid/solid boundary layer. Organic materials accumulate at the interface because of physical binding of the molecules to the solid surface. Carbon is activated by heating in the absence of oxygen. The activation process results in the formation of many pores within each carbon particle. Since adsorption is a surface phenomenon, the greater the surface area of the carbon, the greater its capacity to hold organic material. The vast areas of the walls within these pores account for most of the total surface area of the carbon, which makes it so effective in removing organics.

After the adsorption capacity of the carbon has been exhausted, it can be restored by heating it in a furnace at a temperature sufficiently high to drive off the adsorbed organics. Keeping oxygen at very low levels in the furnace prevents carbon from burning. The organic matter is passed through an afterburner to prevent air pollution. In small plants where the cost of an on-site regeneration furnace cannot be justified, the spent carbon is shipped to a central regeneration facility for processing.

Chemical Phosphorus Removal

All the polyphosphates (molecularly dehydrated phosphates) gradually hydrolyze in aqueous solution and revert to the ortho form (PO_4^{3-}) from which they were derived. Phosphorus is typically found as mono-hydrogen phosphate (HPO_4^{2-}) in wastewater.

The removal of phosphorus by chemical precipitation is typically accomplished by using one of three compounds. The precipitation reactions for each are shown below.

Using ferric chloride:

$$FeCl_3 + HPO_4^{2-} \rightleftharpoons FePO_4\downarrow + H^+ + 3Cl^- \qquad (8\text{-}47)$$

Using alum:

$$Al_2(SO_4)_3 + 2HPO_4^{2-} \rightleftharpoons 2AlPO_4\downarrow + 2H^+ + 3SO_4^{2-} \qquad (8\text{-}48)$$

Using lime:

$$5Ca(OH)_2 + 3HPO_4^{2-} \rightleftharpoons Ca_5(PO_4)_3OH\downarrow + 3H_2O + 6OH^- \qquad (8\text{-}49)$$

You should note that ferric chloride and alum reduce the pH while lime increases it. The effective range of pH for alum and ferric chloride is between 5.5 and 7.0. If there is not enough naturally occurring alkalinity to buffer the system to this range, then lime must be added to counteract the formation of H^+.

The precipitation of phosphorus requires a reaction basin and a settling tank to remove the precipitate. When ferric chloride and alum are used, the chemicals may be added directly to the aeration tank in the activated sludge system. The aeration tank serves as a reaction basin. The precipitate is then removed in the secondary clarifier. This is not possible with lime because the high pH required to form the precipitate is detrimental to the activated sludge organisms. In some wastewater treatment plants, the $FeCl_3$ (or alum) is added before the wastewater enters the primary sedimentation tank. This improves the efficiency of the primary tank, but may deprive the biological processes of needed nutrients.

Example 8-15. If a wastewater has a soluble orthophosphate concentration of 4.00 mg/L as P, what *theoretical* amount of ferric chloride will be required to remove it completely?

Solution. From Equation 8-47, we see that one mole of ferric chloride is required for each mole of phosphorus to be removed. The pertinent gram molecular weights are as follows:

$$FeCl_3 = 162.2\,g$$
$$P = 30.97\,g$$

With a PO_4-P of 4.00 mg/L, the theoretical amount of ferric chloride would be

$$4.00 \times \frac{162.2}{30.97} = 20.95,\ or\ 21.0\ mg/L$$

Comment: Because of side reactions, solubility product limitations, and day-to-day variations, the actual amount of chemical to be added must be determined by jar tests on the wastewater. You can expect that the actual ferric chloride dose will be 1.5 to 3 times the theoretically calculated amount. Likewise, the actual alum dose will be 1.25 to 2.5 times the theoretical amount.

Biological Phosphorus Removal

In biological phosphorus removal (BPR or Bio-P), or enhanced biological phosphorous removal (EBPR) as it is sometimes called, the phosphorus in the wastewater is incorporated into cell mass in excess of levels needed for cell synthesis and maintenance. This is accomplished by moving the biomass from an anaerobic to an aerobic environment. The phosphorus contained in the biomass is removed from the process as biological sludge.

Microbiology. The original work on enhanced Bio-P identified *Acinetobacter* as the responsible genus. Subsequent work has identified Bio-P bacteria in other genera such as *Arthrobacter*, *Aeromonas*, *Nocardia*, and *Pseudomonas*. The Bio-P organisms in these genera are referred to as *phosphorus accumulating organisms* (PAOs).

Based on the work of Comeau et al. (1986), Wentzel et al. (1986) developed a mechanistic model used to explain BPR. This model proposes that (Stephens and Stensel, 1998):

> Chemical oxygen demand (COD) is fermented to acetate by facultative bacteria under anaerobic conditions. The bacteria assimilate acetate in the anaerobic zone and convert it to polyhydroxybutyrate (PHB). Stored polyphosphate is degraded to provide adenosine triphosphate (ATP) necessary for PHB formation, and the polyphosphate degradation is accomplished by the release of orthophosphorus and magnesium, potassium, and calcium. Under aerobic conditions, the PHB is oxidized to synthesize new cells and to produce reducing equivalents needed for ATP formation. Phosphate and inorganic cations are taken up, reforming polyphosphate granules. The amount of phosphate taken up under aerobic conditions exceeds the phosphorus released during anaerobic conditions, resulting in excess phosphorus removal.

Stoichiometry. Common heterotrophic bacteria in activated sludge have a phosphorus composition of 0.01 to 0.02 g P/g biomass. PAOs are capable of storing phosphorus in the form of phosphates. In the PAOs, the phosphorus content may be as high as 0.2 to 0.3 g P/g biomass.

Acetate (i.e., acetic acid – CH_3COOH) uptake is critical in determining the amount of PAOs and, thus, the amount of phosphorus that can be removed by this pathway. If significant amounts of DO or nitrate enter the anaerobic zone, the acetate will be depleted before it is taken up by the PAOs. Bio-P removal is not used in systems that are designed for nitrification without providing a means of denitrification.

The amount of phosphorus removal can be estimated from the amount of soluble COD in the wastewater influent. The following assumptions are used to evaluate the stoichiometry of biological phosphorus removal: (1) 1.06 g of acetate/g of COD will be produced as the COD is fermented to volatile fatty acids (VFAs), (2) a cell yield of 0.3 g VSS/g of acetate, and (3) a cell phosphorus content of 0.3 g of P/g VSS. With these assumptions, it is estimated that 10 g of COD is required to remove 1 g of phosphorus (Metcalf & Eddy, 2003).

Processes. Process flow diagrams for several Bio-P processes may be found at the text website: www.mhhe.com/davis. Design principles and practice are presented for Bio-P processes in *Water and Wastewater Engineering* (Davis, 2010).

Nitrogen Control

Nitrogen in any soluble form (NH_3, NH_4^+, NO_2^-, and NO_3^-, but not N_2 gas) is a nutrient and may need to be removed from wastewater to help control algal growth in the receiving body. In addition, nitrogen in the form of ammonia exerts an oxygen demand and can be toxic to fish. Removal of nitrogen can be accomplished either biologically or chemically. The biological processes are called *nitrification/denitrification* (NDN or BNDN) or sometimes biological nitrogen removal (BNR). The chemical process is called *ammonia stripping.*

Nitrification/Denitrification. The natural nitrification process can be forced to occur in the activated-sludge system by maintaining a cell detention time (θ_c) of 15 days in moderate climates and over 20 days in cold climates. The nitrification step is expressed in chemical terms as follows:

$$NH_4^+ + 2O_2 \xrightleftharpoons{\text{bacteria}} NO_3^- + H_2O + 2H^+ \qquad (8\text{-}50)$$

Of course, bacteria must be present to cause the reaction to occur. This step satisfies the oxygen demand of the ammonium ion. If the nitrogen level is not of concern for the receiving body, the wastewater can be discharged after settling. If nitrogen is of concern, the nitrification step must be followed by anoxic denitrification by bacteria:

$$2NO_3^- + \text{Organic matter} \xrightarrow{\text{bacteria}} N_2 + CO_2 + H_2O \qquad (8\text{-}51)$$

As indicated by the chemical reaction, organic matter is required for denitrification. Organic matter serves as an energy source for the bacteria. The organic matter may be obtained from within or outside the cell. In multistage nitrogen-removal systems, because the concentration of BOD_5 in the flow to the denitrification process is usually quite low, a supplemental organic carbon source is required for rapid denitrification. (BOD_5 concentration is low because the wastewater previously has undergone carbonaceous BOD removal and the nitrification process.) The organic matter may be either raw, settled sewage or a synthetic material such as methanol (CH_3OH). Raw, settled sewage may adversely affect the effluent quality by increasing the BOD_5 and ammonia content.

Process flow diagrams for several biological nitrification/denitrification processes may be found at the text website: www.mhhe.com/davis. Design principles and practices for BNR processes are presented in *Water and Wastewater Engineering* (Davis, 2010).

Ammonia Stripping. Nitrogen in the form of ammonia can be removed chemically from water by raising the pH to convert the ammonium ion into ammonia, which can then be stripped from the water by passing large quantities of air through the water. The process has no effect on nitrate, so the activated sludge process must be operated at a short cell-detention time to prevent nitrification. The ammonia stripping reaction is

$$NH_4^+ + OH^- \rightleftharpoons NH_3 + H_2O \tag{8-52}$$

The hydroxide is usually supplied by adding lime. The lime also reacts with CO_2 in the air and water to form a calcium carbonate scale, which must be removed periodically. Low temperatures cause problems with icing and reduced stripping ability. The reduced stripping ability is caused by the increased solubility of ammonia in cold water.

8-10   LAND TREATMENT FOR SUSTAINABILITY

This discussion on land treatment follows two EPA publications: *Environmental Control Alternatives: Municipal Wastewater* and *Land Treatment of Municipal Wastewater Effluents, Design Factors I* (Pound et al., 1976).

An alternative to the previously discussed tertiary treatment processes for producing an extremely high-quality effluent is offered by an approach called *land treatment*. Land treatment is the application of effluents, usually following secondary treatment, on the land by one of the several available conventional irrigation methods. This approach uses wastewater, and often the nutrients it contains, as a resource rather than considering it as a disposal problem. Treatment is provided by natural processes as the effluent moves through the natural filter provided by soil and plants. Part of the wastewater is lost by evapotranspiration, while the remainder returns to the hydrologic cycle through overland flow or the groundwater system. Most of the groundwater eventually returns, directly or indirectly, to the surface water system.

Land treatment of wastewaters can provide moisture and nutrients necessary for crop growth. In semiarid areas, insufficient moisture for peak crop growth and limited water supplies make this water especially valuable. The primary nutrients (nitrogen, phosphorus, and potassium) are reduced only slightly in conventional secondary treatment processes, so that most of these elements are still present in secondary effluent. Soil nutrients that are consumed each year by crop removal and by losses through soil erosion may be replaced by the application of wastewater.

Land application is the oldest method used for treatment and disposal of wastes. Cities have used this method for more than 400 years. Historically, several major cities, including Berlin, Melbourne, and Paris, have used "sewage farms" for waste treatment and disposal. Approximately 1,500 communities in the United States reuse municipal wastewater treatment plant effluent in surface irrigation.

Land treatment systems use one of the three basic approaches:

1. Slow rate

2. Overland flow

3. Rapid infiltration

Each method, shown schematically in Figure 8-25, can produce renovated water of different quality, can be adapted to different site conditions, and can satisfy different overall objectives.

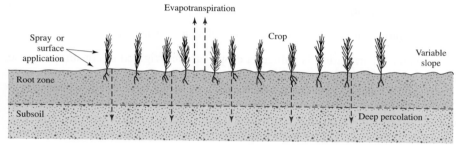

Slow Rate

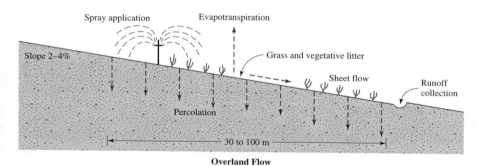

Overland Flow

Rapid Infiltration

FIGURE 8-25
Methods of land application. (*Source:* Pound et al., 1976.)

Slow Rate

Irrigation, the predominant land application method in use today, involves the application of effluent to the land for treatment and for meeting the growth needs of plants. The applied effluent is treated by physical, chemical, and biological means as it seeps into the soil. Effluent can be applied to crops or vegetation (including forestland) either by sprinkling or by surface techniques, for purposes such as:

1. Avoidance of surface discharge of nutrients

2. Economic return from use of water and nutrients to produce marketable crops

3. Water conservation by exchange when lawns, parks, or golf courses are irrigated

4. Preservation and enlargement of greenbelts and open space

Where water for irrigation is valuable, crops can be irrigated at consumptive use rates (3.5 to 10 mm/d, depending on the crop), and the economic return from the sale of the crop can be balanced against the increased cost of the land and distribution system. On the other hand, where water for irrigation is of little value, hydraulic loadings can be maximized (provided that renovated water quality criteria are met), thereby minimizing system costs. Under high-rate irrigation (10 to 15 mm/d), water-tolerant grasses with high nutrient uptake become the crop of choice.

Overland Flow

Overland flow is essentially a biological treatment process in which wastewater is applied over the upper reaches of sloped terraces and allowed to flow across the vegetated surface to runoff collection ditches. Renovation is accomplished by physical, chemical, and biological means as the wastewater flows in a thin sheet down the relatively impervious slope.

Overland flow can be used as a secondary treatment process where discharge of a nitrified effluent low in BOD is acceptable or as a tertiary wastewater treatment process. The latter objective will allow higher rates of application (18 mm/d or more), depending on the degree of wastewater treatment required. Where a surface discharge is prohibited, runoff can be recycled or applied to the land in irrigation or infiltration-percolation systems.

Rapid Infiltration

In infiltration-percolation systems, effluent is applied to the soil at higher rates by spreading in basins or by sprinkling. Treatment occurs as the water passes through the soil matrix. System objectives can include:

1. Groundwater recharge

2. Natural treatment followed by pumped withdrawal or the use of underdrains for recovery

3. Natural treatment where renovated water moves vertically and laterally in the soil and recharges a surface watercourse

Where groundwater quality is being degraded by salinity intrusion (see Chapter 7), groundwater recharge can reverse the hydraulic gradient and protect the existing groundwater. Where existing groundwater quality is not compatible with expected renovated quality, or where existing water rights control the discharge location, a return of renovated water to surface water can be designed, using pumped withdrawal, underdrains, or natural drainage. In Phoenix, Arizona, for example, the native groundwater quality is poor, and the renovated water is to be withdrawn by pumping, with discharge into an irrigation canal.

8-11 SLUDGE TREATMENT

In the process of purifying the wastewater, another problem is created: sludge. The higher the degree of wastewater treatment, the larger the residue of sludge that must be handled. The exceptions to this rule are where land applications or polishing lagoons are used. Satisfactory treatment and disposal of the sludge can be the single most complex and costly operation in a municipal wastewater treatment system (U.S. EPA, 1979b). The sludge is made of materials settled from the raw wastewater and of solids generated in the wastewater treatment processes.

The quantities of sludge involved are significant. For primary treatment, they may be 0.25 to 0.35 percent by volume of wastewater treated. The activated sludge process increases the volume to 1.5 to 2.0 percent. Use of chemicals for phosphorus removal can add another 1.0 percent. The sludges withdrawn from the treatment processes are still largely water, as much as 97 percent. Sludge treatment processes, then, are concerned with separating the large amounts of water from the solid residues. The separated water is returned to the wastewater plant for processing.

The basic processes for sludge treatment are as follows:

1. *Thickening:* Separating as much water as possible by gravity or flotation.

2. *Stabilization:* Converting the organic solids to more refractory (inert) forms so that they can be handled or used as soil conditioners without causing a nuisance or health hazard. These biochemical oxidation processes are called *digestion.*

3. *Conditioning:* Treating the sludge with chemicals or heat so that the water can be readily separated.

4. *Dewatering:* Separating water by subjecting the sludge to vacuum, pressure, or drying.

5. *Reduction:* Converting the solids to a stable form by wet oxidation or incineration. (These are chemical oxidation processes; they decrease the volume of sludge, hence the term reduction.)

Although a large number of alternative combinations of equipment and processes are used for treating sludges, the basic alternatives are fairly limited. The ultimate depository of the materials contained in the sludge must either be land, air, or water. Current policies discourage practices such as ocean dumping of sludge. Air pollution considerations necessitate air pollution control facilities as part of the sludge incineration process.

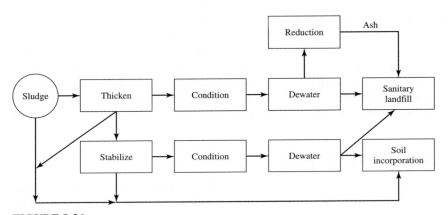

FIGURE 8-26
Basic sludge handling alternatives.

The following sections discuss the processes commonly used. The basic alternative routes by which these processes may be employed are shown in Figure 8-26.

Sources and Characteristics of Various Sludges

Before we begin the discussion of the various treatment processes, it is worthwhile to recapitulate the sources and nature of the sludges that must be treated.

Grit.   The sand, broken glass, nuts, bolts, and other dense material that is collected in the grit chamber is not true sludge in the sense that it is not fluid. However, it still requires disposal. Because grit can be drained of water easily and is relatively stable in terms of biological activity (it is not biodegradable), it is generally trucked directly to a landfill without further treatment.

Primary or Raw Sludge.   Sludge from the bottom of the primary clarifiers contains from 3 to 8 percent solids (1 percent solids $\simeq$ 1 g solids/100 mL sludge volume). The solid fraction is approximately 70 percent organic matter. This sludge rapidly becomes anaerobic and is highly odiferous.

Secondary Sludge.   This sludge consists of microorganisms and inert materials that have been wasted from the secondary treatment processes. Thus, the solids are about 90 percent organic matter. When the supply of air is removed, this sludge also becomes anaerobic, creating noxious conditions if not treated before disposal. The solids content depends on the source. Wasted activated sludge is typically 0.5 to 2 percent solids, while trickling filter sludge contains 2 to 5 percent solids. In some cases, secondary sludges contain large quantities of chemical precipitates because the aeration tank is used as the reaction basin for the addition of chemicals to remove phosphorus.

Tertiary Sludges.   The characteristics of sludges from the tertiary treatment processes depend on the nature of the process. For example, phosphorus removal results in a chemical sludge that is difficult to handle and treat. When phosphorus removal occurs in the activated sludge process, the chemical sludge is combined with the biological sludge, making the latter more difficult to treat. Nitrogen removal by denitrification results in a biological sludge with properties very similar to those of waste activated sludge.

Solids Computations

Volume–Mass Relationships. Because most WWTP sludges are primarily water, the volume of the sludge is primarily a function of the water content. Thus, if we know the percent solids and the specific gravity of the solids we can estimate the volume of the sludge. The solid matter in wastewater sludge is composed of fixed (mineral) solids and volatile (organic) solids. The volume of the total mass of solids may be expressed as

$$\mathcal{V}_{\text{solids}} = \frac{M_s}{S_s \rho} \tag{8-53}$$

where M_s = mass of solids, kg
 S_s = specific gravity of solids
 ρ = density of water = 1,000 kg/m³

Because the total mass is composed of fixed and volatile fractions, Equation 8-53 may be rewritten as:

$$\frac{M_s}{S_s \rho} = \frac{M_f}{S_f \rho} + \frac{M_v}{S_v \rho} \tag{8-54}$$

where M_f = mass of fixed solids, kg
 M_v = mass of volatile solids, kg
 S_f = specific gravity of fixed solids
 S_v = specific gravity of volatile solids

The specific gravity of the solids may be expressed in terms of the specific gravities of the fixed and solid fractions by solving Equation 8-54 for S_s:

$$S_s = M_s \left[\frac{S_f S_v}{M_f S_v + M_v S_f} \right] \tag{8-55}$$

The specific gravity of sludge (S_{sl}) may be estimated by recognizing that, in a similar fashion to the fractions of solids, the sludge is composed of solids and water so that

$$\frac{M_{sl}}{S_{sl} \rho} = \frac{M_s}{S_s \rho} + \frac{M_w}{S_w \rho} \tag{8-56}$$

where M_{sl} = mass of sludge, kg
 M_w = mass of water, kg
 S_{sl} = specific gravity of sludge
 S_w = specific gravity of water

It is customary to report solids concentrations as percent solids, where the fraction of solids (P_s) is computed as

$$P_s = \frac{M_s}{M_s + M_w} \tag{8-57}$$

and the fraction of water (P_w) is computed as

$$P_w = \frac{M_w}{M_s + M_w} \qquad (8\text{-}58)$$

Thus, it is more convenient to solve Equation 8-53 in terms of percent solids. If we divide each term in Equation 8-56 by $(M_s + M_w)$ and recognize that $M_{sl} = M_s + M_w$, then Equation 8-56 may be expressed as

$$\frac{1}{S_{sl}\rho} = \frac{P_s}{S_s\rho} + \frac{P_w}{S_w\rho} \qquad (8\text{-}59)$$

If the specific gravity of water is taken as 1.0000, as it can be without appreciable error, then solving for S_{sl} yields

$$S_{sl} = \frac{S_s}{P_s + (S_s)(P_w)} \qquad (8\text{-}60)$$

With these expressions in hand, or at least where you can find them, you can calculate the volume of sludge (V_{sl}) with the following equation:

$$V_{sl} = \frac{M_s}{(\rho)(S_{sl})(P_s)} \qquad (8\text{-}61)$$

Example 8-16. Using the following primary settling-tank data, determine the daily sludge production.

Operating Data:
Flow = 0.150 m³/s
Influent SS = 280.0 mg/L = 280.0 g/m³
Removal efficiency = 59.0%
Sludge concentration = 5.00%
Volatile solids = 60.0%
Specific gravity of volatile solids = 0.990
Fixed solids = 40.0%
Specific gravity of fixed solids = 2.65

Solution. We begin by calculating S_s. We can do this without calculating M_s, M_f, and M_v directly by recognizing that they are proportional to the percent composition. With

$$M_s = M_f + M_v$$
$$= 0.400 + 0.600 = 1.00$$

Then Equation 8-55 gives the following:

$$S_s = \frac{(2.65)(0.990)}{[(0.400)(0.990)] + [(0.600)(2.65)]}$$
$$= 1.321, \text{ or } 1.32$$

The specific gravity of the sludge is calculated with Equation 8-60:

$$S_{sl} = \frac{1.321}{0.05 + (1.321 \times 0.950)}$$
$$= 1.012 \, \text{or} \, 1.01$$

The mass of the sludge is estimated from the incoming suspended solids concentration and the removal efficiency of the primary tank.

$$M_s = 0.59 \times 280.0 \, \text{g/m}^3 \times 0.15 \, \text{m}^3/\text{s} \times 86,400 \, \text{s/d} \times 10^{-3} \, \text{kg/g}$$
$$= 2.14 \times 10^3 \, \text{kg/d}$$

The sludge volume is then calculated with Equation 8-61:

$$V_{sl} = \frac{2.14 \times 10^3 \, \text{kg/d}}{1,000 \, \text{kg/m}^3 \times 1.012 \times 0.05}$$
$$= 42.29, \, \text{or} \, 42.3 \, \text{m}^3/\text{d}$$

Comment: Note that 95 percent, or 40.2 m³/d of the sludge is water. To reduce the ultimate sludge disposal costs, many of the sludge handling steps are focused on removing water (See Figure 8-26).

Mass Balance. Quantitative estimates of sludge production may be made using mass balance techniques. The fundamental equation is:

$$\frac{dS}{dt} = M_{in} - M_{out} \qquad (8\text{-}62)$$

where M_{in} and M_{out} refer to the mass of dissolved chemicals, solids, or gas entering and leaving a process or group of processes. If we assume steady-state conditions, then $dS/dt = 0$ and Equation 8-62 reduce to the following:

$$M_{in} = M_{out} \qquad (8\text{-}63)$$

Several interrelated processes are examined together in the flowsheet shown in Figure 8-27. When labeled with mass flows, the flowsheet may be called a *quantitative flow diagram* (QFD). The solids mass balance can be an important aid to a designer in predicting long-term average solids loadings on sludge treatment components. This allows the designer to establish such factors as operating costs and quantities of sludge for ultimate disposal. However, it does not establish the solids loading that each equipment item must be capable of processing. A particular component should be sized to handle the most rigorous loading conditions it is expected to encounter. This loading is usually not determined by applying steady-state models because of storage and plant scheduling considerations. Thus, the rate of solids reaching any particular piece of equipment does not usually rise and fall in direct proportion to the rate of solids arriving at the plant headworks.

The mass balance calculation is carried out in a step-by-step procedure:

1. Draw the flowsheet (as in Figure 8-27).
2. Identify all streams. For example, Stream A contains raw sewage solids plus chemical solids generated by dosing the sewage with chemicals. Let the *mass flow rate* of solids in Stream A be equal to A kg per day.

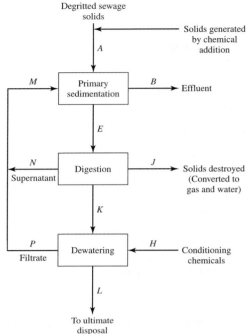

FIGURE 8-27
Primary WWTP flowsheet. (*Source:* U.S. EPA, 1979b.)

3. For each processing unit, identify the relationship of entering and leaving streams to one another in terms of mass. For example, for the primary sedimentation tank, let the ratio of solids in the tank underflow (E) to entering solids ($A + M$) be equal to η_E. η_E is actually an indicator of solids separation efficiency. The general form in which such relationships are expressed is:

$$\eta_i = \frac{\text{Mass of solids in stream } i}{\text{Mass of solids entering the unit}} \tag{8-64}$$

For example,

$$\eta_p = \frac{P}{K + H}; \eta_j = \frac{J}{E}$$

The processing unit's performance is specified when a value is assigned to η_i.

4. Combine the mass balance relationships so as to reduce them to one equation describing a specific stream in terms of given or known quantities, or ones that can be calculated from a knowledge of the process behavior.

Example 8-17. Using Figure 8-27 and assuming that A, η_E, η_j, η_N, η_P, and η_H are known or can be determined from a knowledge of water chemistry and an understanding of the general solids separation/destruction efficiencies of the processing involved, derive an expression for E, the mass flow out of the primary sedimentation tank.

Solution. The derivation is carried out as follows.

a. Define M by solids balances on streams around the primary sedimentation tank:

$$\eta_E = \frac{E}{A + M} \tag{i}$$

Therefore,

$$M = \frac{E}{\eta_E} - A \tag{ii}$$

b. Define M by balances on recycle streams:

$$M = N + P \tag{iii}$$

$$N = \eta_N E \tag{iv}$$

$$P = \eta_P(H + K) \tag{v}$$

$$H = \eta_H K \tag{vi}$$

Therefore,

$$P = \eta_P(1 + \eta_H)K \tag{vii}$$

$$K + J + N = E \tag{viii}$$

Therefore,

$$K = E - J - N = E - \eta_j E - \eta_N E = E(1 - \eta_j - \eta_N) \tag{ix}$$

and

$$P = \eta_P E (1 - \eta_j - \eta_N)(1 + \eta_H) \tag{x}$$

Therefore,

$$M = E[\eta_N + \eta_P (1 - \eta_j - \eta_N)(1 + \eta_H)] \tag{xi}$$

c. Equate equations (ii) and (xi) to eliminate M:

$$\frac{E}{\eta_E} - A = E[\eta_N + \eta_P(1 - \eta_j - \eta_N)(1 + \eta_H)]$$

$$E = \frac{A}{\dfrac{1}{\eta_E} - \eta_N - \eta_P (1 - \eta_j - \eta_N)(1 + \eta_H)}$$

E is expressed in terms of assumed or known influent solids loadings and solids separation/destruction efficiencies.

Comment: Once the equation for E is derived, equations for other streams follow rapidly; in fact, most have already been derived. These are summarized in Table 8-13.

TABLE 8-13
Mass balance equations for Figure 8-27

$$E = \frac{A}{\dfrac{1}{\eta_E} - \eta_N - \eta_P(1 - \eta_j - \eta_N)(1 + \eta_H)}$$

$$M = \frac{E}{\eta_E} - A$$

$$B = (1 - \eta_E)(A + M)$$

$$J = \eta_J E$$

$$N = \eta_N E$$

$$K = E(1 - \eta_J - \eta_N)$$

$$H = \eta_H K$$

$$P = \eta_P(1 + \eta_H)K$$

$$L = K(1 + \eta_H)(1 - \eta_P)$$

Source: U.S. EPA, 1979b.

The example just worked was relatively simple. A more complex system is illustrated in Figure 8-28. Mass balance equations for this system are summarized in Table 8-14 on page 540. For this flowsheet the following information must be specified:

A = influent solids

X = effluent solids, that is, overall suspended solids removal must be specified

$\eta_E, \eta_G, \eta_J, \eta_N, \eta_R,$ and η_T = assumptions about the degree of solids removal, addition, or destruction

η_D = describes the net solids destruction reduction or the net solids synthesis in the biological system, and must be estimated from yield data. A positive η_D signifies net solids destruction. A negative η_D signifies net solids growth. In this example, 8 percent of the solids entering the biological process are assumed destroyed, that is, converted to gas or liquified.

Note that alternative processing schemes can be evaluated simply by manipulating appropriate variables. For example:

Filtration can be eliminated by setting η_R to zero.

Thickening can be eliminated by setting η_G to zero.

Digestion can be eliminated by setting η_J to zero.

Dewatering can be eliminated by setting η_P to zero.

A system without primary sedimentation can be simulated by setting η_E equal to approximately zero, for example, 1×10^{-8}. η_E cannot be set equal to exactly zero, since division by η_E produces indeterminate solutions when computing E.

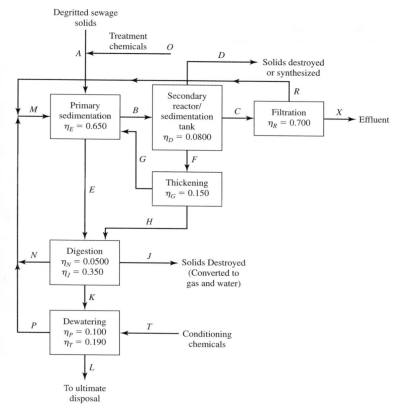

FIGURE 8-28
Flowsheet for a complex WWTP. (*Source:* U.S. EPA, 1979b.)

A set of different mass balance equations must be derived if flow paths between processing units are altered. For example, the equations of Table 8-14 do not describe operations in which the dilute stream from the thickener (Stream G) is returned to the secondary reactor instead of the primary sedimentation tank.

Thickening

Thickening is usually accomplished in one of two ways: the solids are floated to the top of the liquid (*flotation*) or are allowed to settle to the bottom (*gravity thickening*).

The goal is to remove as much water as possible before final dewatering or digestion of the sludge. The processes involved offer a low-cost means of reducing sludge volumes by a factor of two or more. The costs of thickening are usually more than offset by the resulting savings in the size and cost of downstream sludge processing equipment.

Flotation. In the flotation thickening process (Figure 8-29) air is injected into the sludge under pressure (275 to 550 kPa). Under this pressure, a large amount of air can be dissolved in the sludge. The sludge then flows into an open tank where, at atmospheric

TABLE 8-14
Mass balance equations for Figure 8-28

$$E = \frac{A - \left(\dfrac{X}{1 - \eta_R}\right)(\gamma - \eta_R)}{\dfrac{1}{\eta_E} - \alpha - \beta(\gamma)}$$

Where $\alpha = \eta_P(1 - \eta_J - \eta_N)(1 + \eta_T) + \eta_N$

$$\beta = \frac{(1 - \eta_E)(1 - \eta_D)}{\eta_E}$$

$$\gamma = \eta_G + \alpha(1 - \eta_G)$$

$$B = \frac{(1 - \eta_E)E}{\eta_E}$$

$$C = \frac{X}{1 - \eta_R}$$

$$D = \eta_D B$$

$$F = \beta E - \frac{X}{1 - \eta_R}$$

$$G = \eta_G F$$

$$H = (1 - \eta_G)F$$

$$J = \eta_J(E + H)$$

$$K = (1 - \eta_J - \eta_N)(E + H)$$

$$L = K(1 + \eta_T)(1 - \eta_P)$$

$$M = \frac{E}{\eta_E} - G - A$$

$$N = \eta_N(E + H)$$

$$P = \eta_P(1 + \eta_T)K$$

$$R = \frac{\eta_R}{1 - \eta_R}X$$

$$T = \eta_T K$$

Source: U.S. EPA, 1979b.

pressure, much of the air comes out of solution as minute bubbles. The bubbles attach themselves to sludge solids particles and float them to the surface. The sludge forms a layer at the top of the tank; this layer is removed by a skimming mechanism for further processing. The process typically increases the solids content of activated sludge from 0.5–1 percent to 3–6 percent. Flotation is especially effective on activated sludge, which is difficult to thicken by gravity.

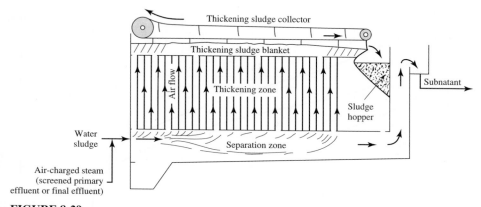

FIGURE 8-29
Air flotation thickener.

Gravity Thickening. Gravity thickening is a simple and inexpensive process that has been used widely on primary sludges for many years. It is essentially a sedimentation process similar to that which occurs in all settling tanks. Sludge flows into a tank that is very similar in appearance to the circular clarifiers used in primary and secondary sedimentation (Figure 8-30); the solids are allowed to settle to the bottom where a heavy-duty mechanism scrapes them to a hopper from which they are withdrawn for further processing. The type of sludge being thickened has a major effect on performance. The best results are obtained with purely primary sludges. As the proportion of activated sludge increases, the thickness of settled sludge solids decreases. Purely primary sludges can be thickened from 1–3 percent to 10 percent solids.

Dick has described a graphical procedure for sizing gravity thickeners using a *batch flux curve** (Yoshioka et al., 1957, and Dick, 1970). *Flux* is the term used to describe the

---

* The original development of this method was by N. Yoshioka and others.

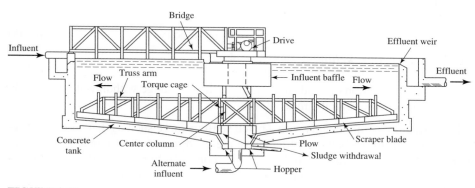

FIGURE 8-30
Gravity thickener.

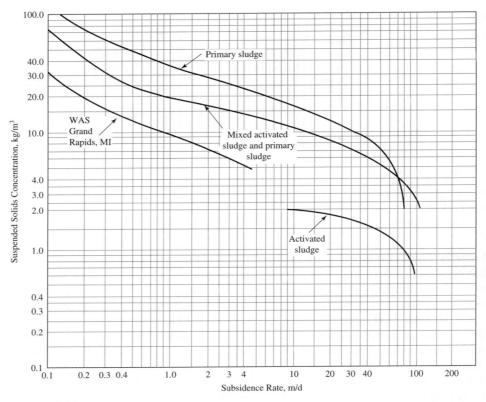

FIGURE 8-31
Batch settling curve.

rate of settling of solids. It is defined as the mass of solids that pass through a horizontal unit area per unit of time (kg/d · m²). This may be expressed mathematically as follows:

$$F_s = (C_u)(v)$$
$$= (C_s)(\text{zone settling velocity}) \qquad (8\text{-}65)$$

where F_s = solids flux, kg/m² · d
C_s = suspended solids concentration, kg/m³
C_u = concentration of solids in underflow, that is, sludge withdrawal pipe, kg/m³
v = underflow velocity, m/d

The sizing procedure begins with a batch settling curve such as that shown in Figure 8-31. Data from the batch settling curve are used to construct a batch flux curve (Figure 8-32). Knowing the desired underflow concentration, a line through the desired concentration and tangent to the batch flux curve is constructed. The extension of this line to the axis of ordinates yields the design flux. From this flux and the inflow solids concentration, the surface area may be determined.

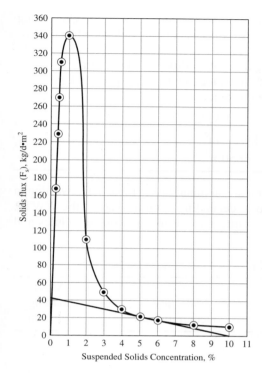

FIGURE 8-32
Batch flux curve.

Example 8-18. A gravity thickener is to be designed to thicken the sludge from the primary tank described in Example 8-16. The thickened sludge should have an underflow solids concentration of 10.0 percent. Assume that the sludge yields a batch settling curve such as that shown in Figure 8-31.

Solution. First we must compute the solids flux for several arbitrarily selected suspended solids concentrations.

SS, kg/m^3	v, m/d	F_s, kg/d · m^2	SS, kg/m^3	v, m/d	F_s, kg/d · m^2
100	0.125	12.5	20	5.30	106.
80	0.175	14.0	10	34.0	340.
60	0.30	18.	5	62.0	310.
50	0.44	22.	4	68.0	272.
40	0.78	31.	3	76.0	228.
30	1.70	51.	2	83.0	166.

The data in the first column were selected arbitrarily. The data in the second column were read from Figure 8-31 at the abscissa points noted in the first column. The data in the third column are the products of the first and second column, that is, $100.0 \times 0.125 = 12.5$, $80.0 \times 0.175 = 14.0$, and so on.

The percent solids concentration is simply 0.10 times the SS in kg/m^3. Converting the first column to percent and plotting it versus the last column yields the batch flux curve (Figure 8-31).

The tangent line from 10.0 percent yields a solids flux of 43 $kg/d \cdot m^2$.

From Example 8-16, we find the solids mass loading to be 2.14×10^3 kg/d. Therefore, the required surface area of the thickener is

$$A_s = \frac{2.14 \times 10^3 \, kg/d}{43 \, kg/d \cdot m^2}$$
$$= 49.77 \text{ or } 50 \, m^2$$

Typical gravity-thickener design criteria are summarized in Table 8-15. Wasting to the thickener may or may not be continuous, depending upon the size of the WWTP. Frequently, smaller plants will waste intermittently because of work schedules and

TABLE 8-15
Typical gravity-thickener design criteria

Sludge Source	Influent SS, %	Expected underflow concentration, %	Mass loading kg/h · m²
Individual sludges			
PS	2–7	5–10	4–6
TF	1–4	3–6	1.5–2.0
RBC	1–3.5	2–5	1.5–2.0
WAS	0.5–1.5	2–3	0.5–1.5
Tertiary sludges			
High CaO	3–4.5	12–15	5–12
Low CaO	3–4.5	10–12	2–6
Fe	0.5–1.5	3–4	0.5–2.0
Combined sludges			
PS + WAS	0.5–4	4–7	1–3.5
PS + TF	2–6	5–9	2–4
PS + RBC	2–6	5–8	2–3
PS + Fe	2	4	1
PS + Low CaO	5	7	4
PS + High CaO	7.5	12	5
PS + (WAS + Fe)	1.5	3	1
PS + (WAS + Al)	0.2–0.4	4.5–6.5	2–3.5
(PS + Fe) + TF	0.4–0.6	6.5–8.5	3–4
(PS + Fe) + WAS	1.8	3.6	1
WAS + TF	.5–2.5	2–4	0.5–1.5

(*Source:* U.S. EPA, 1979b.)

Legend: PS = primary sedimentation; TF = trickling filter; RBC = rotating biological contactor; WAS = waste activated sludge; High CaO = high lime; Low CaO = low lime; Fe = iron; Al = alum; + = mixture of sludges from processes indicated; ( ) = chemical added to process is within parentheses.

TABLE 8-16
Reported operation results for gravity thickeners

Location	Sludge source	Influent SS, %	Mass loading, kg/h · m²	Underflow concentration, %	Overflow SS, mg/L
Port Huron, MI	PS + WAS	0.6	1.7	4.7	2,500
Sheboygan, WI	PS + TF	0.3	2.2	8.6	400
	PS + (TF + Al)	0.5	3.6	7.8	2,400
Grand Rapids, MI	WAS	1.2	2.1	5.6	140
Lakewood, OH	PS + (WAS + Al)	0.3	2.9	5.6	1,400

(*Source:* U.S. EPA, 1979b.)
(*Note:* Values shown are average values only.)

lower volumes of sludge. Some examples of thickener performance are listed in Table 8-16. You should note that the supernatant suspended solids levels are quite high. Thus, the supernatant must be returned to the head end of the WWTP.

Example 8-19. One hundred cubic meters per day (100.0 m³/d) of mixed sludge at 4.0 percent solids is to be thickened to 8.0 percent solids. What is the approximate volume of the sludge after thickening?

Solution. A "4.0 percent sludge" contains 4.0 percent by mass of solids and 96.0 percent by mass of water. Assuming that the specific gravity is not appreciably different from that of water, we can approximate the relationship between volume and percent solids as follows:

$$\frac{V_1}{V_2} = \frac{P_2}{P_1}$$

In this case then, the volume of sludge after thickening would be

$$\frac{100.0 \text{ m}^3/\text{d}}{V_2} = \frac{0.080}{0.040}$$

$$V_2 = 50.0 \text{ m}^3/\text{d}$$

Comment: There is a substantial reduction in the volume that must be handled by thickening the sludge from 4 to 8 percent solids.

Stabilization

The principal purposes of sludge stabilization are to break down the organic solids biochemically so that they are more stable (less odorous and less putrescible) and more dewaterable, and to reduce the mass of sludge. If the sludge is to be dewatered and burned, stabilization is not used. There are two basic stabilization processes in use. One is carried out in closed tanks devoid of oxygen and is called *anaerobic digestion*. The other approach injects air into the sludge to accomplish *aerobic digestion*.

Aerobic Digestion.   The aerobic digestion of biological sludges is nothing more than a continuation of the activated sludge process. When a culture of aerobic heterotrophs is placed in an environment containing a source of organic material, the microorganisms remove and utilize most of this material. A fraction of the organic material removed will be used for the synthesis of new biomass. The remaining material will be channeled into energy metabolism and oxidized to carbon dioxide, water, and soluble inert material to provide energy for both synthesis and maintenance (life-support) functions. Once the external source of organic material is exhausted, however, the microorganisms enter into endogenous respiration, where cellular material is oxidized to satisfy the energy of maintenance (that is, energy for life-support requirements). If this condition is continued over an extended period of time, the total quantity of biomass will be considerably reduced. Furthermore, that portion remaining will exist at such a low energy state that it can be considered biologically stable and suitable for disposal in the environment. This forms the basic principle of aerobic digestion.

Aerobic digestion is accomplished by aerating the organic sludges in an open tank resembling an activated sludge aeration tank. Like the activated sludge aeration tank, the aerobic digestor must be followed by a settling tank unless the sludge is to be disposed of on land in liquid form. Unlike the activated sludge process, the effluent (supernatant) from the clarifier is recycled back to the head end of the plant. This is because the supernatant is high in suspended solids (100 to 300 mg/L), BOD_5 (to 500 mg/L), TKN (to 200 mg/L), and total P (to 100 mg/L).

Approximately 20 to 35 percent of waste-activated sludge from plants with primary treatment is not biodegradable. The goal of aerobic digestion is 38 percent reduction in volatile solids. Both the liquid temperature and SRT control the degree of solids reduction. A plot of degree-days (temperature times SRT) versus volatile solids reduction (Figure 8-33) reveals that 38 percent reduction can be achieved above approximately 400 degree-days. To produce well-stabilized biosolids, at least 550 degree-days are recommended.

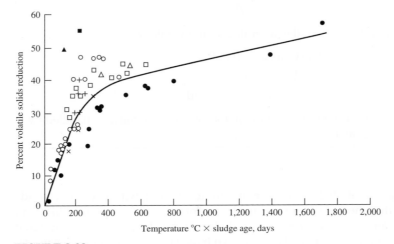

FIGURE 8-33
Volatile solids reduction as a function of digester liquid temperature and digester sludge age. (*Source:* U.S. EPA, 1979b.)

The volume of the digester tank may be estimated with the following equation (WEF, 1998):

$$V = \frac{Q(X_i + FS)}{X(k_d P_v + 1/\text{SRT})}$$ (8-66)

where V = volume of aerobic digester, m^3
Q = average flowrate to digester, m^3/d
X_i = influent suspended solids, mg/L
F = fraction of influent BOD that is raw primary solids
S = digester influent BOD, mg/L
X = digester suspended solids, mg/L
k_d = reaction rate constant, d^{-1}
P_v = volatile fraction of digester suspended solids
SRT = solids retention time, d

The term FS can be ignored if primary sludge is not included in the sludge load to the digester. Representative values for k_d range linearly from 0.02 d^{-1} at 10°C to 0.14 d^{-1} at 25°C for waste activated sludge (U.S. EPA, 1979b). Bench-scale or pilot-scale studies are recommended to obtain site-specific decay coefficients.

Because the fraction of volatile matter is reduced, the specific gravity of the digested sludge solids will be higher than it was before digestion. Thus, the sludge settles to a more compact mass, and the clarifier underflow concentration can be expected to reach 3 percent.

The literature addressing dewatering of aerobically digested sludge is contradictory. Good results can be obtained with sand drying beds. Mixed results with mechanical devices leads to the recommendation to conduct a thorough on-site investigation with pilot-scale devices.

Anaerobic Digestion. The anaerobic treatment of complex wastes involves three distinct stages. In the first stage, complex waste components, including fats, proteins, and polysaccharides, are hydrolyzed to their component subunits. This is accomplished by a heterogeneous group of facultative and anaerobic bacteria. These bacteria then subject the products of hydrolysis (triglycerides, fatty acids, amino acids, and sugars) to fermentation and other metabolic processes leading to the formation of simple organic compounds and hydrogen in a process called *acidogenesis* or *acetogenesis*. The organic compounds are mainly short-chain (volatile) acids and alcohols. The second stage is commonly referred to as *acid fermentation*. In this stage, organic material is simply converted to organic acids, alcohols, and new bacterial cells, so that little stabilization of BOD or COD is realized. In the third stage, the end products of the first stage are converted to gases (mainly methane and carbon dioxide) by several different species of strictly anaerobic bacteria. Thus, it is here that true stabilization of the organic material occurs. This stage is generally referred to as *methane fermentation*. The stages of anaerobic waste treatment are illustrated in Figures 8-34 and 8-35. Even though the anaerobic process is presented as being sequential in nature, all stages take place simultaneously and synergistically. The primary acid produced during acid fermentation is acetic acid. The significance of this acid as a precursor for methane formation is illustrated in Figure 8-35.

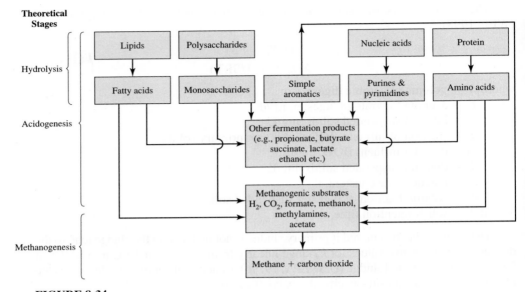

FIGURE 8-34

Schematic diagram of the patterns of carbon flow in anaerobic digestion.

The bacteria responsible for acid fermentation are relatively tolerant to changes in pH and temperature and have a much higher rate of growth than the bacteria responsible for methane fermentation. As a result, methane fermentation is generally assumed to be the rate-controlling step in anaerobic waste treatment processes.

Considering 35°C as the optimum temperature for anaerobic waste treatment, Lawrence proposes that, in the range of 20 to 35°C, the kinetics of methane fermentation of long- and short-chain fatty acids will adequately describe the overall kinetics of

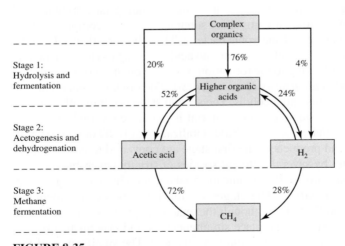

FIGURE 8-35

Steps in anaerobic digestion process with energy flow.

anaerobic treatment (Lawrence and Milnes, 1971). Thus, the kinetic equations we presented to describe the completely mixed activated sludge process are equally applicable to the anaerobic process.

The mesophilic anaerobic digesters are described as standard-rate, two-stage, separate digesters, and high-rate digesters.

The standard-rate process does not employ sludge mixing, but rather allows the digester contents to stratify into zones. The major disadvantage of the standard-rate process is the large tank volume required because of long retention times, low loading rates, and thick scum-layer formation. Only about one-third of the tank volume is utilized in the digestion process. The remaining two-thirds of the tank volume contains the scum layer, stabilized solids, and the supernatant. It is seldom used for digester design today.

The two-stage system evolved as a result of efforts to improve the standard-rate unit. Although many units are now in operation, it is seldom used in modern digester design. In this process, two digesters operating in series separate the functions of fermentation and solids/liquid separation. The contents of the first-stage, high-rate unit are thoroughly mixed and the sludge is heated to increase the rate of fermentation. Because the contents are thoroughly mixed, temperature distribution is more uniform throughout the tank volume. Sludge feeding and withdrawal are continuous or nearly so.

The primary functions of the second-stage digester are solids/liquid separation and residual gas extraction. First-stage digesters may be equipped with fixed or floating covers. Second-stage digester covers are often of the floating type. Second-stage units are generally not heated.

Separate sludge digestion employs a separate digester for primary sludge and for activated sludge. This arrangement is to improve the separation of the sludge solids from the liquid after digestion. Design criteria for this process are very limited.

The most common design today is a single-stage, high-rate digester (Figure 8-36). It is characterized by heating, auxiliary mixing, uniform feeding, and thickening of the feed stream. Uniform feeding of the sludge is very important to the operation of the digester. For economical anaerobic digestion, a feed concentration of at least 4 percent total solids

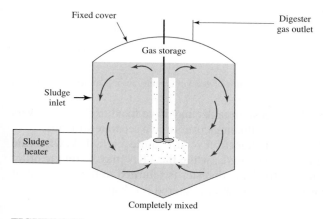

FIGURE 8-36
Schematic of a high-rate anaerobic digester.

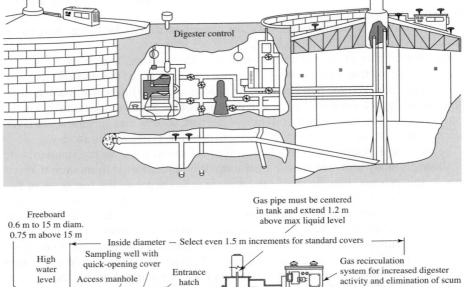

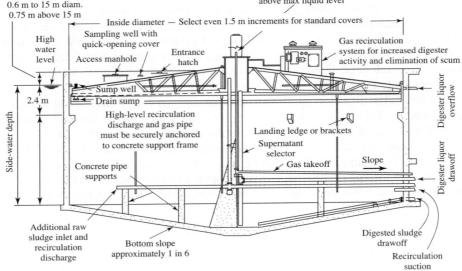

FIGURE 8-37
Phantom view of high-rate anaerobic digester and cross section of detail of floating cover. (Courtesy of US Filter.)

is desirable (Shimp, et al., 1995). The digestion tanks may have fixed or floating covers (Figure 8-37). Gas may be stored under the floating cover or in a separate structure.

The BOD remaining at the end of digestion is still quite high. Likewise, the suspended solids may be as high as 12,000 mg/L, while the TKN may be on the order of 1,000 mg/L. Thus, the supernatant from the digester is returned to the head end of the WWTP. The settled sludge is conditioned and dewatered for disposal.

Sludge Conditioning

Chemical Conditioning. Several methods of conditioning sludge to facilitate the separation of the liquid and solids are available. One of the most commonly used is the addition of coagulants such as ferric chloride, lime, or organic polymers. Ash from incinerated sludge has also found use as a conditioning agent. As happens when coagulants are added to turbid water, chemical coagulants act to clump the solids together so that they are more easily separated from the water. In recent years, organic polymers have become increasingly popular for sludge conditioning. Polymers are easy to handle, require little storage space, and are very effective. The conditioning chemicals are injected into the sludge just before the dewatering process and are mixed with the sludge.

Heat Treatment.   Another conditioning approach is to heat the sludge at high temperatures (175 to 230°C) and pressures (1,000 to 2,000 kPa). Under these conditions, much like those of a pressure cooker, water that is bound up in the solids is released, improving the dewatering characteristics of the sludge. Heat treatment has the advantage of producing a sludge that dewaters better than chemically conditioned sludge. The process has the disadvantages of relatively complex operation and maintenance and the creation of highly polluted cooking liquors that when recycled to the treatment plant impose a significant added treatment burden.

Sludge Dewatering

Sludge Drying Beds.   The most popular method of sludge dewatering in the past has been the use of sludge drying beds. These beds are especially popular in small plants because of their simplicity of operation and maintenance. Seventy-seven percent of all United States wastewater treatment plants utilized drying beds; one-half of all the municipal sludge produced in the United States was dewatered by this method (U.S. EPA, 1981). Most of these plants are located in small- and medium-sized communities, with an average flow rate of less than 0.10 m^3/s. Some larger cities, such as Albequerque, Fort Worth, Phoenix, and Salt Lake City, use sand drying beds. Although the use of drying beds might be expected in the warmer, sunny regions, they are also used in several large facilities in northern climates.

Operational procedures common to all types of drying beds involve the following steps:

1. Pump 0.20 to 0.30 m of stabilized liquid sludge onto the drying bed surface.

2. Add chemical conditioners continuously, if conditioners are used, by injection into the sludge as it is pumped onto the bed.

3. When the bed is filled to the desired level, allow the sludge to dry to the desired final solids concentration. (This concentration can vary from 18 to 60 percent, depending on several factors, including type of sludge, processing rate needed, and degree of dryness required for lifting. Nominal drying times vary from 10 to 15 d under favorable conditions, to 30 to 60 d under barely acceptable conditions.)

4. Remove the dewatered sludge either mechanically or manually.

5. Repeat the cycle.

Sand drying beds are the oldest, most commonly used type of drying bed. Many design variations are possible, including the layout of drainage piping, thickness and type of gravel and sand layers, and construction materials. Sand drying beds for wastewater

FIGURE 8-38
Continuous belt filter press. (Courtesy of Komline-Sanderson Engineering Corporation.)

sludge are constructed in the same manner as water treatment plant sludge-drying beds. Current U.S. practice was discussed and illustrated in Chapter 6.

Sand drying beds can be built with or without provision for mechanical sludge removal, and with or without a roof. When the cost of labor is high, newly constructed beds are designed for mechanical sludge removal.

Continuous Belt Filter Presses (CBFP).   The CBFP equipment used in treating wastewater sludges is the same as that used for water treatment plant sludges (Figure 8-38).

The CBFP is successful with many normal mixed sludges. Typical dewatering results for digested mixed sludges with initial feed solids of 5 percent give a dewatered cake of 19 percent solids at a rate of 32.8 kg/h · m^2. In general, most of the results with these units closely parallel those achieved with rotary vacuum filters. An advantage of CBFPs is that they do not have the sludge pickup problem that sometimes occurs with rotary vacuum filters. Additionally, they have a lower energy consumption.

Reduction

Incineration.   If sludge use as a soil conditioner is not practical, or if a site is not available for landfill using dewatered sludge, cities may turn to the alternative of sludge reduction. Incineration completely evaporates the moisture in the sludge and combusts the organic solids to a sterile ash. To minimize the amount of fuel used, the sludge must be dewatered as completely as possible before incineration. The exhaust gas from an incinerator must be treated carefully to avoid air pollution.

8-12   ALTERNATIVE SLUDGE DISPOSAL TECHNIQUES

"Ultimate disposition" of biosolids or residue (i.e., ash from incineration) falls into four general categories: land application, landfilling, dedicated land disposal, and utilization. Land application is discussed in the next section of this chapter.

Landfilling

When there is an acceptable, convenient site, the landfill is typically selected for "ultimate disposal" of biosolids, grit, screenings, and other solids. Landfilling of biosolids and/or ash in a sanitary landfill with municipal solid waste is regulated by the U.S. EPA under 40 CFR 258.

Dewatering is typically required and stabilization may be required before the landfill can be used. If methane recovery is practiced at the landfill site, the addition of biosolids may be welcome as it will increase gas production.

Dedicated Land Disposal

Dedicated land disposal means the application of heavy sludge loadings to some finite land area that has limited public access and has been set aside or dedicated "for all time" to the disposal of wastewater sludge. Dedicated land disposal does not mean in-place utilization. No crops may be grown. Dedicated sites typically receive liquid sludges. While application of dewatered sludges is possible, it is not common. In addition, disposal of dewatered sludge in landfills is generally more cost-effective.

One of the common sites for dedicated land disposal is a location where surface mining has taken place. The biosolids improve the recovery of the land by providing organic matter and nutrients for plant growth.

Utilization

Wastewater solids may sometimes be used beneficially in ways other than as a soil nutrient. Of the several methods worthy of note, composting and co-firing with municipal solid waste are two that have received increasing amounts of interest in the last few years. The recovery of lime and the use of the sludge to form activated carbon have also been in practice to a lesser extent.

Land Application of Biosolids

One of the methods for disposal of biosolids/wastewater sludge is by land application. Land application is defined as the spreading of biosolids on or just below the soil surface. The application to land for agricultural purposes is beneficial because the organic matter improves soil structure, soil aggregation, water holding capacity, water infiltration, and soil aeration. In addition, macronutrients (such as nitrogen, phosphorus, and potassium) and micronutrinents (such as iron, manganese, copper, and zinc) aid plant growth. These contributions also serve as a partial replacement for chemical fertilizers.

To qualify for application to agricultural and nonagricultural land, the biosolids must, at a minimum, meet the pollutant ceiling concentrations, class B requirements for pathogens, and vector attraction requirements. For biosolids processed for application to lawns and gardens, class A criteria and one of the vector-attraction reduction requirements must be met. These are discussed later in this section.

Site Selection. A critical step in land application of biosolids is the identification of a suitable site. Among the factors that must be considered are topography (for erosion potential), soil characteristics, depth to the groundwater, accessibility, proximity to critical areas (such as domestic water supply, property boundaries, public access), and haul distance. The employment of a soil scientist to assist in the assessment is critical. The plan that determines the selection process should involve all the stakeholders.

Design Loading Rates. Nitrogen and heavy metals concentrations in the sludge are two of the major concerns in determining the sludge loading rate. The nitrogen limit is typically determined on an annual basis. Heavy-metal loadings are based on long-term averages.

The nitrogen loading rate is typically set to match the available nitrogen provided by commercial fertilizers. It is dependent on the crop and can vary from 120 to 245 kg/ha · y for field crops (corn, wheat, and soybeans) and from 175 to 670 kg/ha · y for forage crops (alfalfa and grasses). Extensive soil testing and analysis by a soil scientist is essential in determining an appropriate loading rate.

Application Methods. The application methods are broadly classified as liquid application and dewatered biosolids application.

Liquid biosolids application is attractive because of its simplicity. Dewatering processes are not required. The solids concentrations range from 1 to 10 percent. The application method may be by vehicular application or by irrigation.

Vehicular application may be either surface distribution or subsurface distribution. Special vehicles are used. They have wheels designed to minimize compaction and to improve mobility. For surface distribution, rear-mounted spray manifolds, nozzles, or guns are used. For subsurface injection, two alternatives are commonly used. Injection shanks force the liquid into the ground directly. Alternatively, plows or discs with manifolds apply the biosolids, which are then incorporated immediately after injection by covering spoons.

Injection below the soil surface is preferred as it minimizes odors, reduces vector attraction, minimizes ammonia loss, eliminates surface runoff, and minimizes visibility, which leads to better public acceptance. However, this method is not suitable for all crops.

Irrigation may be by sprinkling or furrow irrigation. These methods find application in locations isolated from public view and access. They have the following disadvantages: high power costs for the pumps, contact of all parts of the crop with the biosolids, potential odors, vector attraction, and high visibility to the public.

Application of dewatered solids is similar to the application of semisolid animal manure. Typical solids concentrations are in the range of 15 to 20 percent. It must be followed by incorporation. This method has the potential to generate dust and odors as well as being an attraction to vectors. Public acceptance of this application method may be difficult to achieve.

Sludge Disposal Regulations

On February 19, 1993, the EPA promulgated risk-based regulations that govern the use or disposal of sewage sludge. These regulations are codified as 40 CFR Part 503 and have become known as the "503 Regulations." The regulations apply to sewage sludge generated from the treatment of domestic sewage that is land-applied, placed on a surface disposal site, or incinerated in an incinerator that accepts only sewage sludge. The regulations do not apply to sludge generated from treatment of industrial process wastes at an industrial facility, hazardous sewage sludge, sewage sludge with polychlorinated biphenyl (PCB) concentrations of 50 mg/L or greater, or drinking water sludge.

Figure 8-39 summarizes the sludge quality requirements for use or disposal. The regulation establishes two levels of sewage sludge quality with respect to

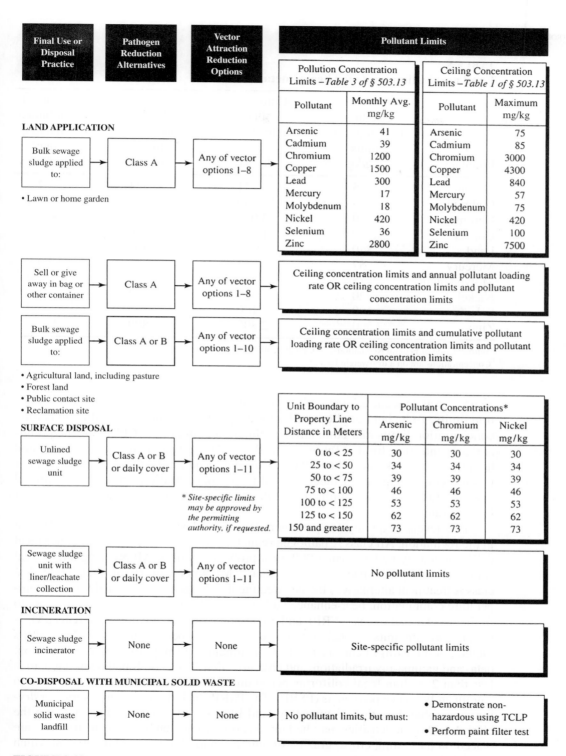

FIGURE 8-39
Sludge quality requirements for use or disposal practices.

TABLE 8-17
Land application limits for heavy metals[a,b]

Pollutant	Ceiling concentration limits, mg/kg	Cumulative pollutant loading rates, kg/ha	Pollutant concentration limits, mg/kg	Annual pollutant loading rates, kg/ha · y
Arsenic	75	41	41	2.0
Cadmium	85	39	39	1.9
Chromium	3,000	3,000	1,200	150
Copper	4,300	1,500	1,500	75
Lead	840	300	300	15
Mercury	57	17	17	0.85
Molybdenum	75	18	18	0.90
Nickel	420	420	420	21
Selenium	100	100	36	5.0
Zinc	7,500	2,800	2,800	140

[a]*Source:* Code of Federal Regulations, 40 CFR Part 503.13
[b]Concentrations are on a dry-weight basis

heavy-metal concentrations: ceiling concentration limits and pollution concentration limits. To be land-applied, bulk sewage sludge must meet the pollutant ceiling concentration limits *and* cumulative pollutant loading rates (CPLR) *or* the pollutant concentration limits (Table 8-17). Bulk sewage sludge applied to lawns and home gardens must meet the pollutant concentration limits. Sewage sludge sold or given away in bags must meet the pollutant concentration limits *or* the annual sewage sludge product application rates that are based on the annual pollutant loading rates.

Two levels of quality for pathogen densities (class A and class B) are defined in the regulation. All class A pathogen reduction alternatives require that either fecal coliform density be less than 1,000 most probable number (MPN) per gram of total solids, or *Salmonella* bacteria be less than 3 MPN per 4 grams of total solids. The class A treatment alternatives include treating the sludge for a specified time and temperature combination, heat-enhanced alkaline stabilization, treatment in a process to further reduce pathogens (PFRP), and use of processes that are proven to reduce virus plaque-forming units and helminth ova to less than 1 per 4 grams of sludge. PFRPs include composting, heat drying, heat treatment, thermophilic aerobic digestion, beta- and gamma-ray irradiation, and pasteurization. The class B pathogen standard is less than 2 million fecal coliforms per gram of sludge or treatment in a process to significantly reduce pathogens (PSRP). The PSRPs include aerobic digestion, air drying, anaerobic digestion, composting, and lime stabilization. Sludges meeting the class A pathogen densities may be land-disposed immediately. Time restrictions are placed on harvesting crops, grazing of animals, and public access to sites on which class B sludge is applied.

Vectors are insects (or other animals) that transmit disease. The organic nature of sludge often attracts vectors after the sludge is land-applied. The 503 regulations provide 11 alternatives to reduce vector attraction. Some of the alternatives are: volatile solids reduction of 38 percent of more, achieving a standard oxygen uptake rate of less than 1.5 mg O_2 per hour per gram of dry solids at 20°C, aerobic treatment at greater than 40°C with an average temperature greater than 45°C for 14 days, alkaline stabilization, sludge drying, surface incorporation, and soil cover.

The 503 regulations are "self-implementing" in that permits are not required to require conformance.

8-13 CHAPTER REVIEW

When you have completed studying this chapter, you should be able to do the following without the aid of your textbook or notes:

1. List BOD_5 values for strong, medium, and weak domestic wastewater.

2. Explain the difference between pretreatment, primary treatment, secondary treatment, and tertiary treatment, and show how they are related.

3. Sketch a graph showing the average variation of daily flow at a municipal wastewater treatment plant (WWTP).

4. Define and explain the purpose of equalization.

5. For each type of decomposition (aerobic, anoxic, and anaerobic), list the electron acceptor, important end products, and relative advantages and disadvantages as a waste treatment process.

6. List the growth requirements of bacteria and explain why a bacterium needs them.

7. Sketch and label the bacterial growth curve for a pure culture. Define or explain each phase labeled on the curve.

8. Define θ_c, SRT, and sludge age, and explain their use in regulating the activated sludge process.

9. Explain the purpose of the F/M ratio and define F and M in terms of BOD_5 and mixed liquor volatile suspended solids.

10. Explain the relationship between F/M and θ_c.

11. Explain how cell production is regulated using F/M and/or θ_c.

12. Compare two systems operating at two different F/M ratios.

13. Define SVI.

14. Explain the difference between bulking sludge and rising sludge and what circumstances cause each to occur.

15. Sketch, label, and explain the function of the parts of an activated sludge plant and a trickling filter plant.

16. List and explain the relationship of the five types of oxidation ponds to oxygen.

17. Explain what an RBC is and how it works.

18. Compare the positive and negative effects of disinfection of wastewater effluents.

19. List the four common tertiary wastewater treatment processes and the pollutants they remove.

20. Explain why removal of residual suspended solids effectively removes residual BOD_5.

21. Describe refractory organic compounds and the method used to remove them.

22. List three chemicals used to remove phosphorus from wastewaters.

23. Explain biological nitrification and denitrification either in words or with an equation.

24. Explain ammonia stripping either in words or with an equation.

25. Describe the three basic approaches to land treatment of wastewater.

26. State the two major purposes of sludge stabilization.

27. Explain the purpose of each of the sludge treatment steps and describe the major processes used.

28. Describe the locations for ultimate disposal of sludges and the treatment steps needed prior to ultimate disposal.

With the aid of this text, you should be able to do the following:

29. Determine the required volume of an equalization basin to dampen a given periodic flow.

30. Determine the effect of equalization on mass loading of a pollutant.

31. Evaluate or size primary and secondary sedimentation tanks with respect to detention time, overflow rate, solids loading, and weir loading.

32. Calculate the bacterial population at a time t, given the initial population and the number of generations.

33. Estimate the soluble BOD_5 in the effluent from a completely mixed or plug-flow activated sludge plant; determine the mean cell residence time or the hydraulic detention time to achieve a desired degree of treatment; determine the "wasting" flow rate to achieve a desired mean cell residence time or F/M ratio.

34. Calculate the F/M ratio given an influent BOD_5, flow, and detention time, or calculate the volume of the aeration basin given F/M, BOD_5, and flow.

35. Calculate SVI and utilize it to determine return sludge concentration and/or flow rate.

36. Calculate the required mass of sludge to be wasted from an activated sludge process given the appropriate data.

37. Calculate the theoretical mass of oxygen required and the amount of air required to supply it given the appropriate data.

38. Use the appropriate trickling filter equation to determine one or more of the following, given the appropriate data: treatment efficiency, filter volume, filter depth, hydraulic loading rate.

39. Perform a sludge mass balance, given the separation efficiencies and appropriate mass flow rate.

8-14   PROBLEMS

8-1. A treatment plant being designed for Cynusoidal City requires an equalization basin to even out flow and BOD variations. The average daily flow is 0.400 m^3/s. The following flows and BOD_5 have been found to be typical of the average variation over a day. What size equalization basin, in cubic meters, is required to provide for a uniform outflow equal to the average daily flow? Assume the flows are hourly averages.

Time	Flow, m^3/s	BOD_5, mg/L	Time	Flow, m^3/s	BOD_5, mg/L
0000	0.340	123	1200	0.508	268
0100	0.254	118	1300	0.526	282
0200	0.160	95	1400	0.530	280
0300	0.132	80	1500	0.552	268
0400	0.132	85	1600	0.570	250
0500	0.140	95	1700	0.596	205
0600	0.160	100	1800	0.604	168
0700	0.254	118	1900	0.570	140
0800	0.360	136	2000	0.552	130
0900	0.446	170	2100	0.474	146
1000	0.474	220	2200	0.412	158
1100	0.482	250	2300	0.372	154

Answer: ∀ = 6,105.6 plus 25% excess = 7,630 m^3

8-2. A treatment plant being designed for Metuchen requires an equalization basin to even out flow and BOD variations. The following flows and BOD_5 have been found to be typical of the average variation over a day.

What size equalization basin, in cubic meters, is required to provide for a uniform outflow equal to the average daily flow? Assume the flows are hourly averages.

Time	Flow, m³/s	BOD₅, mg/L	Time	Flow, m³/s	BOD₅, mg/L
0000	0.0875	110	1200	0.135	160
0100	0.0700	81	1300	0.129	150
0200	0.0525	53	1400	0.123	140
0300	0.0414	35	1500	0.111	135
0400	0.0334	32	1600	0.103	130
0500	0.0318	42	1700	0.104	120
0600	0.0382	66	1800	0.105	125
0700	0.0653	92	1900	0.116	150
0800	0.113	125	2000	0.127	200
0900	0.131	140	2100	0.128	215
1000	0.135	150	2200	0.121	170
1100	0.137	155	2300	0.110	130

8-3. A treatment plant being designed for the village of Excel requires an equalization basin to even out flow and BOD variations. The following flows and BOD₅ have been found to be typical of the average variation over a day. What size equalization basin, in cubic meters, is required to provide for a uniform outflow equal to the average daily flow?

Time	Flow, m³/s	BOD₅, mg/L	Time	Flow, m³/s	BOD₅, mg/L
0000	0.0012	50	1200	0.0041	290
0100	0.0011	34	1300	0.0041	290
0200	0.0009	30	1400	0.0042	275
0300	0.0009	30	1500	0.0038	225
0400	0.0009	33	1600	0.0033	170
0500	0.0013	55	1700	0.0039	180
0600	0.0018	73	1800	0.0046	190
0700	0.0026	110	1900	0.0046	190
0800	0.0033	150	2000	0.0044	190
0900	0.0039	195	2100	0.0034	160
1000	0.0047	235	2200	0.0031	125
1100	0.0044	265	2300	0.0020	80

8-4. Compute and plot the unequalized and the equalized hourly BOD mass loadings to the Cynusoidal City WWTP (Problem 8-1). Using the plot and computations, determine the following ratios for BOD mass loading: peak to average; minimum to average; peak to minimum.

Answers:

	Unequalized	Equalized
P/A	1.97	1.47
M/A	0.14	0.63
P/M	14.05	2.34

8-5. Repeat Problem 8-4 using the data from Problem 8-2.

8-6. Repeat Problem 8-4 using the data from Problem 8-3.

8-7. Using an overflow rate of 26.0 m/d and a detention time of 2.0 h, size a primary sedimentation tank for the average flow at Cynusoidal City (Problem 8-1). What would the overflow rate be for the unequalized maximum flow? Assume 15 sedimentation tanks with length-to-width ratio of 4.7.

> *Answers:* Tank dimensions = 15 tanks at 2.17 m deep by 4.34 m by 20.4 m. Maximum overflow rate = 39.3 m/d

8-8. Determine the surface area of a primary settling tank sized to handle a maximum hourly flow of 0.570 m^3/s at an overflow rate of 60.0 m/d. If the effective tank depth is 3.0 m, what is the effective theoretical detention time?

> *Answers:* Surface area = 820.80 or 821 m^2; t_o = 1.2 h

8-9. If an equalization basin is installed ahead of the primary tank in Problem 8-8, the average flow to the tank is reduced to 0.400 m^3/s. What is the new overflow rate and detention time?

8-10. The influent BOD_5 to a primary settling tank is 345 mg/L. The average flow rate is 0.050 m^3/s. If the BOD_5 removal efficiency is 30 percent, how many kilograms of BOD_5 are removed in the primary settling tank each day?

8-11. The influent suspended solids concentration to a primary settling tank is 435 mg/L. The average flow rate is 0.050 m^3/s. If the suspended solids removal efficiency is 60 percent, how many kilograms of suspended solids are removed in the primary settling tank each day?

8-12. If the population of microorganisms is 3.0×10^5 at time t_o and 36 hours later it is 9.0×10^8, how many generations have occurred?

> *Answer:* n = 11.55 or 12 generations

8-13. The following data were gathered in a bacterial growth experiment. Plot a semilogarithm graph of the data, using a computer spreadsheet you have written. At approximately what time did log growth start and terminate? How many generations occurred during log growth?

Time, h	Bacterial count
0	1×10^3
5	1×10^3
10	1.5×10^3
15	5.4×10^3
20	2.0×10^4
25	7.5×10^4
30	2.85×10^5
35	1.05×10^5
40	1.15×10^5
45	1.15×10^5

8-14. The following data were gathered by Kajima (1923) in an *E. coli* growth experiment. Plot a semilogarithm graph of the data using a computer spreadsheet you have written. Label the following phases on the graph: log growth, stationary, and death. Note that there is no lag phase or acclimation phase. Also note the change in pH with time as a result of the accumulation of by-products of metabolism.

Time, h	Bacterial count	pH
0	50×10^3	7.2
6	175×10^6	6.9
12	320×10^6	6.8
18	538×10^6	7.2
24	609×10^6	7.6
36	559×10^6	7.9
48	493×10^6	8.2
96	330×10^6	8.3
192	53×10^6	8.5
240	7.5×10^6	8.7

8-15. Using the assumptions given in Example 8-4, the rule-of-thumb values for growth constants, and the further assumption that the influent BOD_5 was reduced by 32.0 percent in the primary tank, estimate the liquid volume of a completely mixed activated sludge aeration tank required to treat the wastewater in Problem 8-1. Assume an MLVSS of 2,000 mg/L.

Answer: Volume = 4,032, or 4,000 m^3

8-16. Repeat Problem 8-15 using the wastewater in Problem 8-2.

8-17. Repeat Problem 8-15 using the wastewater in Problem 8-3.

8-18. The town of Camp Verde has been directed to upgrade its primary WWTP to a secondary plant that can meet an effluent standard of 25.0 mg/L BOD_5 and 30 mg/L suspended solids. They have selected a completely mixed activated sludge system for the upgrade. The existing primary treatment plant has a flow rate of 0.029 m^3/s. The effluent from the primary tank has a BOD_5 of 240 mg/L. Using the following assumptions, estimate the required volume of the aeration tank:

1. BOD_5 of the effluent suspended solids is 70 percent of the allowable suspended solids concentration.

2. Growth constants values are estimated to be: K_s = 100 mg/L BOD_5; k_d = 0.025 d^{-1}; μ_m = 10 d^{-1}; Y = 0.8 mg VSS/mg BOD_5 removed.

3. The design MLVSS is 3,000 mg/L.

8-19. Using a spreadsheet program you have written, rework Example 8-4 using the following MLVSS concentrations instead of the 2,000 mg/L used in the example: 1,000 mg/L; 1,500 mg/L; 2,500 mg/L; and 3,000 mg/L.

8-20. Using a spreadsheet program you have written, determine the effect of MLVSS concentration on the effluent soluble BOD_5 (S) using the data in Example 8-4. Assume the volume of the aeration tank remains constant at 970 m^3. Use the same MLVSS values used in Problem 8-19.

8-21. If the F/M of a 0.4380 m^3/s activated sludge plant is 0.200 mg/mg · d, the influent BOD_5 after primary settling is 150 mg/L and the MLVSS is 2,200 mg/L, what is the volume of the aeration tank?

Answer: Volume = 1.29 × 10^4 m^3

8-22. Two activated sludge aeration tanks at Turkey Run, Indiana, are operated in series. Each tank has the following dimensions: 7.0 m wide by 30.0 m long by 4.3 m effective liquid depth. The plant operating parameters are as follows:

Flow = 0.0796 m^3/s

Soluble BOD_5 after primary settling = 130 mg/L

MLVSS = 1,500 mg/L

MLSS = 1.40 (MLVSS)

Settled sludge volume after 30 min = 230.0 mL/L

Aeration tank liquid temperature = 15°C

Determine the following: aeration period, F/M ratio, SVI, solids concentration in the return sludge.

> *Answers:* Aeration period = 6.3 h; F/M = 0.33; SVI = 110 m/g;
> X = 9,130 mg/L

8-23. The 500-bed Lotta Hart Hospital has a small activated sludge plant to treat its wastewater. The average daily hospital discharge is 1,500 L per day per bed, and the average soluble BOD_5 after primary settling is 500 mg/L. The aeration tank has effective liquid dimensions of 10.0 m wide by 10.0 m long by 4.5 m deep. The plant operating parameters are as follows:

MLVSS = 2,500 mg/L

MLSS = 1.20 (MLVSS)

Settled sludge volume after 30 min = 200 mL/L

Determine the following: aeration period, F/M ratio, SVI, solids concentration in return sludge.

8-24. The Jambalaya shrimp processing plant generates 0.012 m^3/s of wastewater each day. The wastewater is treated in an activated sludge plant. The average BOD_5 of the raw wastewater before primary settling is 1,400 mg/L. The aeration tank has effective liquid dimensions of 8.0 m wide by 8.0 m long by 5.0 m deep. The plant operating parameters are as follows:

Soluble BOD_5 after primary settling = 966 mg/L

MLVSS = 2,000 mg/L

MLSS = 1.25 (MLVSS)

Settled sludge volume after 30 min = 225.0 mL/L

Aeration tank liquid temperature = 15°C

Determine the following: aeration period, F/M ratio, SVI, solids concentration in the return sludge.

8-25. Using the following assumptions, determine the sludge age, cell wastage flow rate, and the return sludge flow rate for the Turkey Run WWTP (Problem 8-22). Assume:

Suspended solids in the effluent are negligible

Wastage is from the aeration tank

Yield coefficient = 0.40 mg VSS/mg BOD_5 removed

Decay rate of microorganisms = 0.040 d^{-1}

Effluent BOD_5 = 5.0 mg/L (soluble)

> *Answers:* θ_c = 11.50 d; Q_w = 0.00182 m^3/s; Q_r = 0.0214 m^3/s

8-26. Using the following assumptions, determine the solids retention time, the cell wastage flow rate, and the return sludge flow rate for the Lotta Hart Hospital WWTP (Problem 8-23). Assume:

> Allowable BOD_5 in effluent = 25.0 mg/L
>
> Suspended solids in effluent = 25.0 mg/L
>
> Wastage is from the return sludge line
>
> Yield coefficient = 0.60 mg VSS/mg BOD_5 removed
>
> Decay rate of microorganisms = 0.060 d^{-1}
>
> Inert fraction of suspended solids = 66.67%

8-27. Using the following assumptions, determine the solids retention time, the cell wastage flow rate, and the return sludge flow rate for the Jambalaya shrimp processing plant WWTP (Problem 8-24). Assume:

> Allowable BOD_5 in effluent = 25.0 mg/L
>
> Suspended solids in effluent = 30.0 mg/L
>
> Wastage is from the return sludge line
>
> Yield coefficient = 0.50 mg VSS/mg BOD_5 removed
>
> Decay rate of microorganisms = 0.075 d^{-1}
>
> Inert fraction of suspended solids = 30.0%

8-28. The two secondary settling tanks at Turkey Run (Problem 8-22) are 16.0 m in diameter and 4.0 m deep at the side wall. The effluent weir is a single launder set on the tank wall. Evaluate the overflow rate, depth, solids loading, and weir length of this tank for conformance to standard practice.

> *Answers:* v_o = 17.1 m/d < 33 m/d; OK
>
> SWD > 3.7 m recommended depth; OK
>
> SL = 45.57 kg/m^2 · d ≪ 250 kg/m^2 · d; OK
>
> WL = 68.4 m^3/d · m, which is acceptable

8-29. The single secondary settling tank at the Lotta Hart Hospital WWTP (Problem 8-23) is 10.0 m in diameter and 3.4 m deep at the side wall. The effluent weir is a single launder set on the tank wall. Evaluate the overflow rate, depth, solids loading, and weir length for conformance to standard practice.

8-30. The single secondary settling tank at the Jambalaya shrimp processing WWTP (Problem 8-24) is 5.0 m in diameter and 2.5 m deep at the side wall. The effluent weir is a single launder set on the tank wall. Evaluate the overflow rate, depth, solids loading, and weir length for conformance to standard practice.

8-31. Envirotech Systems markets synthetic media for use in the construction of trickling filters. Envirotech uses the following formula to determine BOD removal efficiency:

$$\frac{L_e}{L_i} = \exp\left[-\frac{k\theta D}{Q^n} \right]$$

where L_e = BOD$_5$ of effluent, mg/L
L_i = BOD$_5$ of influent, mg/L
k = treatability factor, (m/d)$^{0.5}$/m
θ = temperature correction factor
= $(1.035)^{T-20}$
T = wastewater temperature, °C
D = media depth, m
Q = hydraulic loading rate, m/d
n = 0.5

Using the following data for domestic wastewater, determine the treatability factor k.

Wastewater temperature = 13°C

Hydraulic loading rate = 41.1 m/d

% BOD remaining	Media depth, m
100.0	0.00
80.3	1.00
64.5	2.00
41.6	4.00
17.3	8.00

Answer: k = 1.79 (m/d)$^{0.5}$/m at 20°C

8-32. Using the Envirotech systems equation and the treatability factor from Problem 8-31, estimate the depth of filter required to achieve 82.7 percent BOD$_5$ removal if the wastewater temperature is 20°C and the hydraulic loading rate is 41.1 m/d.

8-33. Koon, et al. (1976), suggest that recirculation for a synthetic media filter may be considered by the following formula:

$$\frac{L_e}{L_i} = \frac{\exp\left[-\dfrac{k\theta D}{Q^n}\right]}{(1 + r) - r\exp\left[-\dfrac{k\theta D}{Q^n}\right]}$$

where r = recirculation ratio and all other terms are as described in Problem 8-31.

Use this equation to determine the efficiency of a 1.8-m-deep synthetic media filter loaded at a hydraulic loading rate of 5.00 m/d with a recirculation ratio of 2.00. The wastewater temperature is 16°C and the treatability factor is 1.79 (m/d)$^{0.5}$/m at 20°C.

8-34. Determine the concentration of the effluent BOD_5 for the two-stage trickling filter described below. The wastewater temperature is 17°C. Assume the NRC equations apply.

>   Design flow = 0.0509 m³/s
>
>   Influent BOD_5 (after primary treatment) = 260 mg/L
>
>   Diameter of each filter = 24.0 m
>
>   Depth of each filter = 1.83 m
>
>   Recirculation flow rate for each filter = 0.0594 m³/s

8-35. Using a computer spreadsheet program you have written, plot a graph of the final effluent BOD_5 of the two-stage filter described in Problem 8-34 as a function of the influent flow rate. Assume that the recirculation flow remains constant. Use flow rates of 0.02, 0.03, 0.04, 0.05, 0.06, 0.07, 0.08, 0.09, and 0.10 m³/s for the influent.

8-36. Determine the diameter of a single-stage rock media filter to reduce an applied BOD_5 of 125 mg/L to 25 mg/L. Use a flow rate of 0.14 m³/s, a recirculation ratio of 12.0, and a filter depth of 1.83 m. Assume the NRC equations apply and that the wastewater temperature is 20°C.

8-37. Using a spreadsheet program you have written, plot a graph of the final effluent BOD_5 of the single-stage filter described in Problem 8-36 as a function of the influent flow rate. Assume that the ratio of recirculation flow to influent flow remains constant and that the filter diameter is 35.0 m. Use hydraulic loading rates of 10, 12, 14, 16, 18, and 20 m³/m² · d.

8-38. An oxidation pond having a surface area of 90,000 m² is loaded with a waste flow of 500 m³/d containing 180 kg of BOD_5. The operating depth is from 0.8 to 1.6 m. Using the Michigan rules of thumb, determine whether this design is acceptable.

>   *Answers:* Loading rate = 20.0 kg/ha · d
>
>   Detention time = 180 d
>
>   This design is acceptable.

8-39. Determine the required surface area and the loading rate for a facultative oxidation pond to treat a waste flow of 3,800 m³/d with a BOD_5 of 100.0 mg/L.

8-40. Rework Example 8-15 using alum $[Al_2(SO_4)_3 \cdot 18\ H_2O]$ to remove the phosphorus.

>   *Answer:* 86.1 mg/L of alum

8-41. Rework Example 8-15 using lime (CaO) to remove the phosphorus.

8-42. Prepare a monthly water balance and estimate of the storage volume required (in m³) for a spray irrigation system being designed for Wheatville, Iowa.

The design population is 1,000 and the design wastewater generation rate is 280.0 Lpcd. Based on a nitrogen balance, the allowable application rate is 27.74 mm/mo. The area available is 40.0 ha. The percolation rate during the spray season is 150 mm/mo. Assume that the runoff is contained and reapplied. Assume "spray season" is when temperature is above 0°C. In fact, spraying can continue to about −4°C but once spraying has stopped, it may not recommence until temperatures exceed +4°C. The following climatological data (from Kansas City) may be used. The water balance is a direct application of the hydrologic balance equation. The mass balance equation may be rewritten as:

$$\frac{dS}{dt} = P + WW - ET - G - R$$

where dS/dt = change in storage, mm/mo
  P = precipitation, mm/mo
  WW = wastewater application rate, mm/mo
  ET = evapotranspiration, mm/mo
  G = groundwater infiltration, mm/mo
  R = runoff, mm/mo

The storage volume required may be estimated from a mass balance of the form used for determining storage volume for reservoirs in Chapter 4.

Climatological data from Kansas City, Missouri

Month	Average temperature, °C	Evapotranspiration,[a] mm	Precipitation, mm
Jan	−0.2	23	36
Feb	2.1	28	32
Mar	6.3	43	63
Apr	13.2	79	90
May	18.7	112	112
Jun	24.4	155	116
Jul	27.5	203	81
Aug	26.6	198	96
Sep	21.8	152	83
Oct	15.7	114	73
Nov	7.0	64	46
Dec	2.1	25	39

[a]Estimated.

8-43. Prepare a monthly water balance and estimate the storage volume required for Flushing Meadows. The area available for spraying is 125 ha. The design population for Flushing Meadows is 8,880. The average wastewater generation rate is 485.0 Lpcd. The percolation rate is 200 mm/mo during the spray season. Assume that runoff is to be contained and reapplied. Assume also that the following climatological data apply. Assume "spray season" is when temperature is above 0°C. In fact, spraying can continue to about −4°C but once spraying has stopped, it may not recommence until temperatures exceed +4°C. See Problem 8-42 for hints.

Climatological data from Columbus, Ohio

Month	Average temperature, °C	Evapotranspiration,[a] mm	Precipitation, mm
Jan	−1.2	15	80
Feb	−0.5	20	59
Mar	3.8	28	80
Apr	10.4	58	59
May	16.4	89	102
Jun	21.9	117	106
Jul	23.8	142	100
Aug	22.9	130	73
Sep	18.8	104	67
Oct	12.3	76	54
Nov	5.2	41	63
Dec	−0.3	15	59

[a]Estimated.

8-44. Prepare a monthly water balance for Weeping Water. The design population for Weeping Water is 10,080. The average wastewater generation rate is 385.0 Lpcd. The area available is 200.0 ha. The available lagoon volume is 300,000 m³. The percolation rate is 150 mm/mo during the spray season. Assume that runoff is to be contained and reapplied. Assume also that the following climatological data apply. Assume "spray season" is when temperature is above 4°C. In fact, spraying can continue to about −4°C but once spraying has stopped, it may not recommence until temperatures exceed +4°C. Is the lagoon volume sufficient? If not, how much additional volume is needed? See Problem 8-42 for hints.

Climatological data from Helena, Montana[a]

Month	Average temperature, °C	Evapotranspiration,[b] mm	Precipitation, mm
Jan	−6.6	3	13
Feb	−3.1	5	10
Mar	1.7	5	16
Apr	6.7	35	23
May	11.6	102	45
Jun	16.2	178	46
Jul	19.9	180	34
Aug	19.3	163	32
Sep	13.4	76	27
Oct	8.4	5	16
Nov	4.3	5	12
Dec	3.1	3	12

[a] http//cdo.ncdc.noaa.gov/ancsum/ACS.
[b] Estimated.

8-45. Determine the daily and annual primary sludge production for a WWTP having the following operating characteristics:

> Flow = 0.0500 m³/s
> Influent suspended solids = 155.0 mg/L
> Removal efficiency = 53.0%
> Volatile solids = 70.0%
> Specific gravity of volatile solids = 0.970
> Fixed solids = 30.0%
> Specific gravity of fixed solids = 2.50
> Sludge concentration = 4.50%
>
> *Answer:* V_{sl} = 7.83 m³/d

8-46. Repeat Problem 8-45 using the following operating data:

> Flow = 2.00 m³/s
> Influent suspended solids = 179.0 mg/L
> Removal efficiency = 47.0%
> Specific gravity of fixed solids = 2.50
> Specific gravity of volatile solids = 0.999
> Fixed solids = 32.0%

Volatile solids = 68.0%

Sludge concentration = 5.20%

8-47. Using a spreadsheet you have written, and the data in Problem 8-46, determine the daily and annual sludge production at the following removal efficiencies: 40, 45, 50, 55, 60, and 65 percent. Plot annual sludge production as a function of efficiency.

8-48. Using Figure 8-26, Table 8-14, and the following data, determine B, E, J, K, and L in megagrams per day (Mg/d).

$A = 185.686$ Mg/d

$\eta_E = 0.900$; $\eta_j = 0.250$; $\eta_N = 0.00$; $\eta_P = 0.150$; $\eta_H = 0.190$

Answers: $B = 21.112$ or 21.1 Mg/d

$E = 190.011$ or 190 Mg/d

$J = 47.503$ or 47.5 Mg/d

$K = 142.509$ or 143 Mg/d

$L = 144.147$ or 144 Mg/d

8-49. Rework Problem 8-48 assuming that the digestion solids are not dewatered prior to ultimate disposal; that is, $K = L$.

8-50. The value for η_E in Problem 8-48 is quite high. Rework the problem with a more realistic value of $\eta_E = 0.50$.

8-51. The flowsheet for the Doubtful WWTP is shown in Figure P-8-51. Assuming that the appropriate values of η given in Figure 8-28 may be used when needed and that $A = 7.250$ Mg/d, $X = 1.288$ Mg/d, and $N = 0.000$ Mg/d, what is the mass flow (in kg/d) of sludge to be sent to ultimate disposal?

Answer: $L = K = 3.743$ Mg/d, or 3,743 kg/d

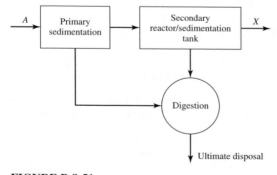

FIGURE P-8-51
Flowsheet for Doubtful WWTP.

8-52. Using the following mass flow data from the Doubtful WWTP (Problem 8-51) determine η_E, η_D, η_N, η_J, η_X. Mass flows for Doubtful WWTP in

Mg/d: $A = 7.280$, $B = 7.798$, $D = 0.390$, $E = 8.910$, $F = 6.940$, $J = 4.755$, $K = 6.422$, $N = 9.428$, $X = 0.468$

8-53. The city of Doubtful (Problem 8-51) is considering the installation of thickening and dewatering facilities. The revised flow diagram for Doubtful to include thickening and dewatering with appropriate return lines is shown in Figure P-8-53. Calculate a value for L in Mg/d. Assume that the appropriate values of η given in Figure 8-28 maybe used when needed and that $A = 7.250$ Mg/d, $X = 1.288$ Mg/d.

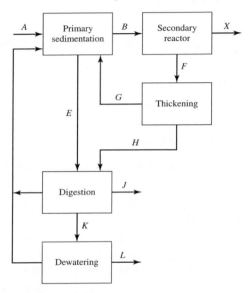

FIGURE P-8-63
Revised flowsheet for Doubtful WWTP.

8-54. Determine the surface area required for the gravity thickeners (assume that no thickener is greater than 30.0 m in diameter) to thicken the waste activated sludge (WAS) at Grand Rapids, Michigan, from 10,600 mg/L to 2.50 percent solids. The waste activated sludge flow is 3,255 m³/d. Assume that the batch settling curves of Figure 8-31 apply. Use a spreadsheet program you have written to plot the data and fit the tangent line.

Answer: $A_s = 2,851.4$ or 2,850 m² depending on graph reading.
Thus, choose four thickeners at 30 m diameter.

8-55. Determine the surface area required for the gravity thickeners of Problem 8-54 if 710 m³/d of primary sludge is mixed with the WAS to form a sludge having 2.00 percent solids. The final sludge is to have a solids concentration of 5.00 percent. The batch settling curve for mixed WAS and PS in Figure 8-31 is assumed to apply. Because of the additional sludge, assume five thickeners will be used. Use a spreadsheet program you have written to plot the data and fit the tangent line.

8-56. Settling test data from the Little Falls WWTP are shown below. Determine the surface area for a gravity thickener for 733 m^3/d of waste activated sludge. The final sludge concentration is to be 3.6 percent. Use a spreadsheet program you have written to plot the data and fit the tangent line.

Suspended solids concentration, g/L	Initial settling velocity, m/d
4.0	58.5
6.0	36.6
8.0	24.1
14.0	8.1
29.0	2.2
41.0	0.73

8-57. The Pomdeterra wastewater treatment plant produces thickened sludge that has a suspended solids concentration of 3.8 percent. They are investigating a filter press that will yield a solids concentration of 24 percent. If they now produce 33 m^3/d of sludge, what annual volume savings will they achieve if they install the press?

8-58. Ottawa's anaerobic digester produces 13 m^3/d of sludge with a suspended solids concentration of 7.8 percent. What volume of sludge must they dispose of each year if their sand drying beds yield a solids concentration of 35 percent?

8-59. Weed Patch's digester produces 30 m^3/mo of sludge with a suspended solids concentration of 2.5 percent. What solids concentration must their drying facility achieve to reduce the volume to 3 m^3/mo?

8-15 DISCUSSION QUESTIONS

8-1. You are touring the research labs of the environmental engineers at your university. Two biological reactors are in a controlled temperature room that has a temperature of 35°C. Reactor A has a strong odor. Reactor B has virtually no odor. What electron acceptors are being used in each reactor?

8-2. If the state regulatory agency requires tertiary treatment of a municipal wastewater, what, if any, processes would you expect to find preceding the tertiary process?

8-3. What is the purpose of recirculation in a trickling filter plant and how does it differ from return sludge in an activated sludge plant?

8-4. In which of the following cases is the cost of sludge disposal higher?

a. $\theta_c = 3$ days

b. $\theta_c = 10$ days

8-5. Would an industrial wastewater containing only NH_4 at a pH of 7.00 be denitrified if pure oxygen was bubbled into it? Explain your reasoning.

8-16   FE EXAM FORMATTED PROBLEMS

8-1. A 2-liter graduated cylinder was used to determine the SVI of an activated sludge sample. The settled volume was 850 mL and the MLSS was 3,000 mg/L. What was the SVI?

a. 283 mL/g   b. 850 mL/g

c. 425 mL/g   d. 142 mL/g

8-2. Estimate the biomass concentration in a CSTR aeration tank with the following operating conditions: hydraulic residence time = 3 h; mean cell residence time = 6 d; yield coefficient = 0.6 mg VSS/mg BOD; decay rate = $0.10\ d^{-1}$; influent soluble BOD = 200 mg/L; effluent BOD = 2.0 mg/L.

a. 3,600 mg/L   b. 150 mg/L

c. 210 mg/L   d. 1,200 mg/L

8-3. When the soluble BOD in the influent to the treatment plant rises from 133 mg/L to 222 mg/L, the operator called to ask your advice on a new MLVSS for the plant. It has been operating with an F/M of 0.31 mg/mg · d and the operator would like to continue using the same F/M. The plant data are as follows: flow rate = 7,630 mg/L; aeration tank volume = $1,270\ m^3$; mixed liquor volatile suspended solids = 2,600 mg/L.

a. 2,600 mg/L   b. 4,300 mg/L

c. 2,400 mg/L   c. 5,800 mg/L

8-4. Determine the SRT of an aerobic digester to treat 9,500 ft^3 of sludge. The sludge temperature is 10°C and the "503" rules require a 38 percent reduction in volatile solids. Assume a 25 percent safety factor in estimating the SRT.

a. 32 d   b. 50 d

c. 40 d   d. 63 d

8-17   REFERENCES

APHA (2005) *Standard Methods for the Examination of Waste and Wastewater,* 21st ed., American Public Health Association, Washington, DC.

Bailey, J. E., and D. F. Ollis (1977) *Biochemical Engineering Fundamentals,* McGraw-Hill, New York, p. 222.

Benefield, L. D., and C. W. Randall (1980) *Biological Process Design for Wastewater Treatment,* Prentice Hall, Upper Saddle River, NJ, pp. 322–324, 338–340, 353–354.

Caldwell, D. H., D. S. Parker, and W. R. Uhte (1973) *Upgrading Lagoons,* U.S. Environmental Protection Agency Technology Transfer Publication, Washington, DC.

CFR (2005a) *Code of Federal Regulations,* 40 CFR §413.02, 464.02, 467.02, and 469.12.

CFR (2005b) *Code of Federal Regulations*, 40 CFR §133.102.

Comeau, Y., K. J. Hall, R. E. W. Hancock, and W. K. Oldham (1986) "Biochemical Model for Enhanced Biological Phosphorus Removal," *Water Research*, vol. 20, no. 12, pp. 1511–1522.

Curds, C. R. (1973) "A Theoretical Study of Factors Influencing the Microbial Population Dynamics of the Activated Sludge Process—I," *Water Research,* vol. 7, pp. 1269–1284.

Davis, M.L. (2010) *Water and Wastewater Engineering*, McGraw-Hill, New York, pp. 18-2–18-9.

Dick, R. I. (1970) "Thickening," in E. F. Gloyna and W. W. Eckenfelder, (eds.), *Advances in Water Quality Improvement—Physical and Chemical Processes,* University of Texas Press, Austin, p. 380.

Dick, R.I. (1976) "Folklore in the Design of Final Settling Tanks," *Journal of Water Pollution Control Federation*, vol. 48, no. 4, pp. 633–644.

GLUMRB (2004) *Recommended Standards for Sewage Works,* Great Lakes–Upper Mississippi River Board of State and Provincial Public Health and Environmental Managers, Albany, NY, p. 70–74.

Grady, C. P. L., and H. C. Lim (1980) *Biological Wastewater Treatment, Theory and Applications,* Marcel Dekker, New York.

Heukelekian, H., H. Orford, and R. Maganelli (1951) "Factors Affecting the Quantity of Sludge Production in the Activated Sludge Process," *Sewage & Industrial Wastes,* vol. 23, p. 8.

Kajima (1923) *Scientific Reports, Government Institute of Infectious Diseases,* Tokyo, vol. 2, p. 305.

Koon, J. H., R. F. Curran, C. E. Adams, and W. W. Eckenfelder (1976) *Evaluation and Upgrading of a Multistage Trickling Filter Facility,* U.S. Environmental Protection Agency, Publication No. EPA 600/2-76-193), Washington, DC.

Lawrence, A. W., and P. L. McCarty (1970) "A Unified Basis for Biological Treatment Design and Operation," *Journal of the Sanitary Engineering Division*, American Society of Civil Engineers, vol. 96, no. SA3.

Lawrence, A. W., and T. R. Milnes (1971) "Discussion Paper," *Journal of the Sanitary Engineering Division,* American Society of Civil Engineers, vol. 97, p. 121.

McKinney, R. E. (1962) *Microbiology for Sanitary Engineers*, McGraw-Hill, New York, p. 40.

Metcalf & Eddy, Inc. (1979) *Wastewater Engineering: Treatment, Disposal, Reuse,* revised by G. Tchobanoglous, McGraw-Hill, New York, pp. 328, 395.

Metcalf & Eddy, Inc. (1991) *Wastewater Engineering: Treatment, Disposal, Reuse,* revised by G. Tchobanoglous and F. L. Burton, McGraw-Hill, New York, pp. 394, 540–542.

Metcalf & Eddy, Inc. (2003) *Wastewater Engineering: Treatment and Reuse*, revised by G. Tchobanoglous, F. L. Burton, and H. D. Stensel, McGraw-Hill, New York, p. 408.

Metcalf & Eddy (2007) *Water Reuse: Issues, Technologies, and Applications,* McGraw-Hill, New York, pp. 927–947.

Monod, J. (1949) "The Growth of Bacterial Cultures," *Annual Review of Microbiology,* vol. 3, pp. 371–394.

NRC (1946) "Sewage Treatment at Military Installations," *Sewage Works Journal,* vol. 18, p. 787.

Parker, D. S., D. J. Kinnea, and E. J. Wahlberg (2001) "Review of Folklore in Design and Operation of Secondary Clarifiers," *Journal of Environmental Engineering,* American Society of Civil Engineers, vol. 127, pp. 476-484.

Pelczar, M. J., and R. D. Reid (1958) *Microbiology,* McGraw-Hill, New York, p. 424.

Pound, C. E., R. W. Crites, and D. A. Griffes (1976) *Land Treatment of Municipal Wastewater Effluents, Design Factors I,* U.S. Environmental Protection Agency Technology Transfer Seminar Publication, Washington, DC.

Sawyer, C. N. (1956) "Bacterial Nutrition and Synthesis," *Biological Treatment of Sewage and Industrial Wastes,* vol. I, p. 3.

Schulze, K. L. (1960) "Load and Efficiency of Trickling Filters," *Journal of Water Pollution Control Federation,* vol. 32, p. 245.

Shahriari, H., C. Eskicioglu, and R. L. Droste (2006) "Simulating Activated Sludge System by Simple-to-Advanced Models," *Journal of Environmental Engineering,* American Society of Civil Engineers, vol. 132, pp. 42–50.

Shimp, G. F., D. M. Bond, J. Sandino, and D. W. Oerke (1995) "Biosolids Budgets," *Water Environment & Technology,* November, pp. 44–49.

Stephens, H. L., and H. D. Stensel (1998) "Effect of Operating Conditions on Biological Phosphorus Removal," *Water Environment Research,* vol. 70, no. 3, pp. 362–369.

U.S. EPA (1977) *Process Design Manual: Wastewater Treatment Facilities for Sewered Small Communities,* U.S. Environmental Protection Agency, EPA Pub. No. 625/1-77-009, Washington. DC, p. 8–12.

U.S. EPA (1979a) *Environmental Pollution Control Alternatives: Municipal Wastewater,* U.S. Environmental Protection Agency, EPA Pub. No. 625/5-79-012, Washington, DC, pp. 33–35, 52–55.

U.S. EPA (1979b) *Process Design Manual, Sludge Treatment and Disposal,* U.S. Environmental Protection Agency, Publication No. 625/1-79-011, Washington DC, pp. 3–19, 3–21, 3–25, 3–26, 5–7, 5–8.

U.S. EPA (1981) *The 1980 Needs Survey. Conveyance, Treatment, and Control of Municipal Wastewater, Combined Sewer Outflows, and Stormwater Runoff, Summaries of Technical Data,* U.S. Environmental Protection Agency, EPA-430/9-81-008, Washington, DC.

U.S. EPA (1996) *NPDES Permit Writers' Manual,* U.S. Environmental Protection Agency, EPA Pub. No. 833-B-96-003, Washington, DC, pp. 77–78.

U.S. EPA (2005) http://www.epa.gov. Search: Region 10 => Homepage => NPDES Permits => Current NPDES Permits in Pacific Northwest and Alaska => Current Individual NPDES Permits in Idaho.

Velz, C. J. (1948) "A Basic Law for the Performance of Biological Filters," *Sewage Works,* vol. 20, p. 607.

WEF (1992) *Design of Municipal Wastewater Treatment Plants,* Vol. I, Manual of Practice No. 8, Joint Task Force of the Water Environment Federation and American Society of Civil Engineers, Alexandria, VA, pp. 528, 529, 530, 586, 595, 705–714.

WEF (1998) *Design of Municipal Wastewater Treatment Plants,* Vol. 2, Manual of Practice No. 8, Joint Task Force of the Water Environment Federation and American Society of Civil Engineers, Alexandria, VA, pp. 11–100, 11–106.

Wentzel, M. C., L. H. Lotter, R. E. Loewenthal, and G. v. R. Marais (1986) "Metabolic Behavior of *Acinetobacter* sp. In Enhanced Biological Phosphorous Removal—A Biochemical Model," *Water*, South Africa, vol. 12, p. 209.

Yoshioka, N., et al. (1957) "Continuous Thickening of Homogenous Flocculated Slurries," *Chemical Engineering,* Tokyo (in Japanese).

CHAPTER

9

AIR POLLUTION

9-1   AIR POLLUTION PERSPECTIVE

Air pollution is of public health concern on several scales: micro, meso, and macro. Indoor air pollution results from products used in construction materials, inadequacy of general ventilation, and geophysical factors that may result in exposure to naturally occurring radioactive materials. Industrial and mobile sources contribute to meso-scale air pollution that contaminates the ambient air that surrounds us outdoors. Macro-scale effects include transport of ambient air pollutants over large distances and global impact. Examples of macro-scale impacts include acid rain and ozone pollution. Global impacts of air pollution result from sources that may potentially change the upper atmosphere. Examples include depletion of the ozone layer and global warming. While micro- and macro-scale effects are of concern, our focus will predominately be on meso-scale air pollution.

Applications

After a brief review of the ideal gas law, this chapter provides an overview of air pollution standards for ambient air, the origin and fate of a selected list of air pollutants, and air pollution at various scales ranging from indoor pollutants to global warming. In addition, the fundamental design equations for several air pollution control devices are presented. This chapter will provide you with tools to do the following:

- Estimate the amount of sulfur dioxide that will be released from burning coal or fuel oil with a given sulfur content.

- Determine the concentration of a pollutant at a downwind point that results from a stack emission or automobiles on a highway.

- Estimate the maximum amount of pollutant that may be emitted and still meet air quality standards.

- Size a device for control of a specific pollutant.

9-2   PHYSICAL AND CHEMICAL FUNDAMENTALS

Ideal Gas Law

Although polluted air may not be "ideal" from the biological point of view, we may treat its behavior with respect to temperature and pressure as if it were ideal. Thus, we assume that at the same temperature and pressure, different kinds of gases have densities proportional to their molecular masses. This may be written as

$$\rho = \frac{1}{R} \frac{PM}{T} \tag{9-1}$$

where ρ = density of gas, kg/m^3
 P = absolute pressure, kPa*
 M = molecular mass, grams/mole

---

*1 Pascal (P) = 1 Newton/m^2; 1 Newton = 1kg $\cdot$ m/s^2

T = absolute temperature, K
R = universal gas constant = 8.3143 J/K · mole = 8.3143 Pa · m³/K · mole

Because density is mass per unit volume, or the number of moles per unit volume, n/V, the expression may be rewritten in the general form as

$$PV = nRT \qquad\qquad \text{(9-2)}$$

where V is the volume occupied by n moles of gas. At 273.15 K and 101.325 kPa, one mole of an ideal gas occupies 22.414 L.

Dalton's Law of Partial Pressures

Stack and exhaust sampling measurements are made with instruments calibrated with air. Because combustion products have an entirely different composition than air, the readings must be adjusted ("corrected" in sampling parlance) to reflect this difference. Dalton's law forms the basis for the calculation of the correction factor. Dalton found that the total pressure exerted by a mixture of gases is equal to the sum of the pressures that each type of gas would exert if it alone occupied the container. In mathematical terms,

$$P_t = P_1 + P_2 + P_3 + \cdots \qquad\qquad \text{(9-3)}$$

where P_t = total pressure of mixture
P_1, P_2, P_3 = pressure of each gas if it were in container alone, that is, *partial pressure*

Dalton's law also may be written in terms of the ideal gas law:

$$P_t = n_1\frac{RT}{V} + n_2\frac{RT}{V} + n_3\frac{RT}{V} + \cdots$$

$$= (n_1 + n_2 + n_3 + \cdots)\frac{RT}{V}$$

Adiabatic Expansion and Compression

Air pollution meteorology is, in part, a consequence of the thermodynamic processes of the atmosphere. One such process is adiabatic expansion and contraction. An *adiabatic* process is one that takes place with no addition or removal of heat and with sufficient slowness so that the gas can be considered to be in equilibrium at all times.

As an example, let us consider the piston and cylinder in Figure 9-1. The cylinder and piston face are assumed to be perfectly insulated. The gas is at pressure P. A force, F, equal to PA must be applied to the piston to maintain equilibrium. If the force is

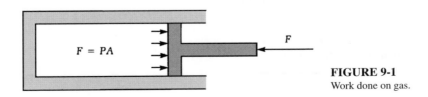

FIGURE 9-1
Work done on gas.

increased and the volume is compressed, the pressure will increase and work will be done on the gas by the piston. Because no heat enters or leaves the gas, the work will go into increasing the thermal energy of the gas in accordance with the first principle of thermodynamics, that is,

(Heat added to gas) = (Increase in thermal energy)
+ (External work done by or on the gas)

Because the left side of the equation is zero (because it is an adiabatic process), the increase in thermal energy is equal to the work done. The increase in thermal energy is reflected by an increase in the temperature of the gas. If the gas is expanded adiabatically, its temperature will decrease.

Units of Measure

The three basic units of measure used in reporting air pollution data are *micrograms per cubic meter* ($\mu g/m^3$), *parts per million* (ppm), and the *micron* (μ) or, preferably, its equivalent, the *micrometer* (μm). Micrograms per cubic meter and parts per million are measures of concentration. Both $\mu g/m^3$ and ppm are used to indicate the concentration of a gaseous pollutant. However, the concentration of particulate matter may be reported only as $\mu g/m^3$. The μm is used to report particle size.

There is an advantage to the unit ppm that frequently makes it the unit of choice. The advantage results from the fact that ppm is a volume-to-volume ratio. (Note that this is different than ppm in water and wastewater, which is a mass-to-mass ratio.) Changes in temperature and pressure do not change the ratio of the volume of pollutant gas to the volume of air that contains it. Thus, it is possible to compare ppm readings from Denver and Washington, DC, without further conversion.

Converting $\mu g/m^3$ to ppm. The conversion between $\mu g/m^3$ and ppm is based on the fact that at standard conditions (0°C and 101.325 kPa), one mole of an ideal gas occupies 22.414 L. Thus, we may write an equation that converts the mass of the pollutant M_p in grams to its equivalent volume V_p in liters at standard temperature and pressure (STP):

$$V_p = \frac{M_p}{GMW} \times 22.414 \text{ L/GM}$$

(9-4)

where GMW is the gram molecular weight of the pollutant. For readings made at temperatures and pressures other than standard conditions, the standard volume, 22.414 L/GM, must be corrected. We use the ideal gas law to make the correction:

$$22.414 \text{ L/gram mole} \times \frac{T_2}{273 \text{ K}} \times \frac{101.325 \text{ kPa}}{P_2}$$

(9-5)

where T_2 and P_2 are the absolute temperature and absolute pressure at which the readings were made. Since ppm is a volume ratio, we may write

$$\text{ppm} = \frac{V_p}{V_a}$$

(9-6)

where V_a is the volume of air in cubic meters at the temperature and pressure of the reading. We then combine Equations 9-4, 9-5, and 9-6 to form Equation 9-7.

$$ppm = \frac{\frac{M_P}{GMW} \times 22.414 \times \frac{T_2}{273\,K} \times \frac{101.325\,kPa}{P_2}}{V_a \times 1,000\,L/m^3} \tag{9-7}$$

where M_p is in μg. The factors converting μg to g and L to millions of L cancel one another. Unless otherwise stated, it is assumed that $V_a = 1.00\,m^3$

Example 9-1.   A one-cubic-meter sample of air was found to contain 80 $\mu g/m^3$ of SO_2. The temperature and pressure were 25°C and 103.193 kPa when the air sample was taken. What was the SO_2 concentration in ppm?

Solution.   First we must determine the GMW of SO_2. From the chart inside the front cover, we find

$$GMW\ of\ SO_2 = 32.07 + 2(16.00) = 64.07$$

Next we must convert the temperature to absolute temperature. Thus,

$$25°C + 273\,K = 298\,K$$

Now we may make use of Equation 9-7.

$$ppm = \frac{\frac{80\mu g}{64.07} \times 22.414 \times \frac{298}{273} \times \frac{101.325}{103.193}}{1.00\,m^3 \times 1,000\,L/m^3} = 0.0300\ ppm\ of\ SO_2$$

Relativity.   Let's take a moment to look at the relationship of a ppm and a μm to something relevant to daily life. Four crystals of common table salt in one cup of granulated sugar is approximately equal to 1 ppm on a volume-to-volume basis. Figure 9-2 should help you visualize the size of a μm. Note that a hair has an average diameter of approximately 80 μm.

9-3   AIR POLLUTION STANDARDS

The 1970 Clean Air Act (CAA) required the U.S. Environmental Protection Agency (EPA) to investigate and describe the environmental effects of any air pollutant emitted by stationary or mobile sources that may adversely affect human health or the environment. The EPA used these studies to establish the National Ambient Air Quality Standards (NAAQS). These standards are for the ambient air, that is, the outdoor air that normally surrounds us. EPA calls the pollutants listed in Table 9-1 on page 584 *criteria pollutants* because they were developed on health-based criteria. The *primary standard* was established to protect human health with an "adequate margin of safety." The *secondary standards* are intended to prevent environmental and property damage. In 1987, the EPA revised the NAAQS. The standard for hydrocarbons was dropped and the standard for Total Suspended Particulates (TSP) was replaced with a particulate standard based on the mass of particulate matter with an aerodynamic diameter less than or equal to 10 μm. This standard is referred to as the PM_{10} standard. It was

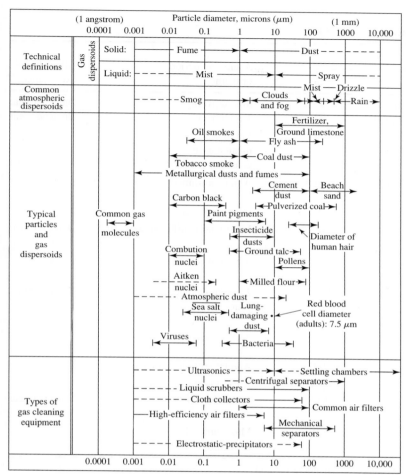

FIGURE 9-2

Characteristics of particles and particle dispersoids. (*Source:* Lapple, 1951.)

replaced by a standard for particulate matter with an aerodynamic diameter less than or equal to 2.5 μm.

States are divided into *Air Quality Control Regions (AQRs)*. An AQR that has air quality equal to or better than the primary standard is called an *attainment area*. Those areas that do not meet the primary standard are called *nonattainment areas*.

Under the 1970 CAA, the EPA was directed to establish regulations for *hazardous air pollutants (HAPs)* using a risk-based approach. These were called NESHAPs—national emission standards for hazardous air pollutants. Because EPA had difficulty defining "an ample margin of safety" as required by the law, only seven HAPs were regulated between 1970 and 1990: asbestos, arsenic, benzene, beryllium, mercury, vinyl chloride, and radionuclides. The Clean Air Act Amendments of 1990 directed EPA to establish a HAP emissions control program based on technology for 189 chemicals*

---

*Subsequently modified. As of July 2005 there were 188 HAPs.

TABLE 9-1
National Ambient Air Quality Standards (NAAQS)

Criteria pollutant	Standard type	Concentration $\mu g/m^3$	Concentration ppm	Averaging period or method	Allowable exceedances[a]
CO	Primary	10,000	9	8-hour average	Once per year
	Primary	40,000	35	1-hour average	Once per year
Lead	Primary and secondary	0.15	N/A	Maximum arithmetic mean measured over a calendar quarter	[f]
NO_2	Primary and secondary	100	0.053	Annual arithmetic mean	
NO_2	Primary	189	0.100	1-hour average	[e]
Ozone	Primary and secondary	235	0.12	Maximum hourly average[b]	Once per year
Ozone	Primary and secondary	147	0.075	8-hour average	[c]
Particulate matter $(PM_{10})^d$	Primary and secondary	150	N/A	24-hour average year	One day per
$(PM_{2.5})$	Primary and secondary	35	N/A	24-hour average	[e]
		15	N/A	Annual arithmetic mean	[f]
SO_2	Primary	80	0.03	Annual arithmetic mean	
	Primary	365	0.14	Maximum 24-hour concentration	Once per year
SO_2	Primary	1,950	0.75	Maximum 1-hour concentration	[g]

[a]Allowable exceedances may actually be an average value over a multi-year period.
[b]The 1-hour NAAQS will no longer apply to an area one year after the effective date of the designation of that area for the 8-hour ozone NAAQS. For most areas, the date of designation was June 15, 2004.
[c]Average fourth highest concentration over 3-year period.
[d]Particulate matter standard applies to particles with an aerodynamic diameter $\leq 10\ \mu$m.
[e]Three-year average of 98th percentile 24-hour concentration.
[f]Three-year average of weighted annual mean.
[g]Three-year average of 99th percentile of daily maximum 1-hour average.
(*Source: Code of Federal Regulations* 40 CFR 50.4–50.12 and 69 FR 23996.)

(see Table 9-2 for the list). EPA will establish emission allowances based on *Maximum Achievable Control Technology (MACT)* for 174 categories of industrial sources that potentially emit 9.08 megagrams (Mg) per year of a single HAP or 22.7 Mg per year of a combination of HAPs. A MACT can include process changes, material substitutions, or air pollution control equipment.

Emission standards place a limit on the amount or concentration of one or more specified contaminants that may be emitted from a source. In 1971, the U.S. Environmental Protection Agency published final standards for the first of many stationary

TABLE 9-2
Hazardous air pollutants (HAPs)

Acetaldehyde
Acetamide
Acetonitrile
Acetophenone
2-Acetylaminofluorene
Acrolein
Acrylamide
Acrylic acid
Acrylonitrile
Allyl chloride
4-Aminobiphenyl
Aniline
o-Anisidine
Asbestos
Benzene (including benzene from gasoline)
Benzidine
Benzotrichloride
Benzyl chloride
Biphenyl
Bis(2-ethylhexyl)phthalate (DEHP)
Bis(chloromethyl)ether
Bromoform
1,3-Butadiene
Calcium cyanamide
Caprolactam (deleted 1996)
Captan
Carbaryl
Carbon disulfide
Carbon tetrachloride
Carbonyl sulfide
Catechol
Chloramben
Chlordane
Chlorine
Chloroacetic acid
2-Chloroacetophenone
Chlorobenzene
Chlorobenzilate
Chloroform
Chloromethyl methyl ether
Chloroprene
Cresols/Cresylic acid (isomers and mixture)
o-Cresol
m-Cresol
p-Cresol
Cumene
2,4-D, salts and esters
DDE
Diazomethane
Dibenzofurans
1,2-Dibromo-3-chloropropane
Dibutylphthalate

1,4-Dichlorobenzene(p)
3,3-Dichlorobenzidene
Dichloroethyl ether [Bis(2-chloroethyl)ether]
1,3-Dichloropropene
Dichlorvos
Diethanolamine
N,N-Diethyl aniline (N,N-Dimethylaniline)
Diethyl sulfate
3,3-Dimethoxybenzidine
Dimethyl aminoazobenzene
3,3'-Dimethyl benzidine
Dimethyl carbamoyl chloride
Dimethyl formamide
1,1-Dimethyl hydrazine
Dimethyl phthalate
Dimethyl sulfate
4,6-Dinitro-o-cresol, and salts
2,4-Dinitrophenol
2,4-Dinitrotoluene
1,4-Dioxane (1,4-Diethyleneoxide)
1,2-Diphenylhydrazine
Epichlorohydrin (l-chloro-2,3-epoxypropane)
1,2-Epoxybutane
Ethyl acrylate
Ethyl benzene
Ethyl carbamate (Urethane)
Ethyl chloride (Chloroethane)
Ethylene dibromide (Dibromoethane)
Ethylene dichloride (1,2-Dichloroethane)
Ethylene glycol
Ethylene imine (Aziridine)
Ethylene oxide
Ethylene thiourea
Ethylidene dichloride (1,1-Dichloroethane)
Formaldehyde
Heptachlor
Hexachlorobenzene
Hexachlorobutadiene
Hexachlorocyclopentadiene
Hexachloroethane
Hexamethylene-1,6-diisocyanate
Hexamethylphosphoramide
Hexane
Hydrazine
Hydrochloric acid
Hydrogen fluoride (Hydrofluoric acid)
Hydrogen sulfide (clerical error; deleted 1991)
Hydroquinone
Isophorone
Lindane (all isomers)
Maleic anhydride
Methanol

Methoxychlor
Methyl bromide (Bromomethane)
Methyl chloride (Chloromethane)
Methyl chloroform (1,1,1-Trichloroethane)
Methyl ethyl ketone (2-Butanone)
Methyl hydrazine
Methyl iodide (Iodomethane)
Methyl isobutyl ketone (Hexone)
Methyl isocyanate
Methyl methacrylate
Methyl tert butyl ether
4,4-Methylene bis(2-chloroaniline)
Methylene chloride (Dichloromethane)
Methylene diphenyl diisocyanate (MDI)
4,4'-Methylenedianiline
Naphthalene
Nitrobenzene
4-Nitrobiphenyl
4-Nitrophenol
2-Nitropropane
N-Nitroso-N-methylurea
N-Nitrosodimethylamine
N-Nitrosomorpholine
Parathion
Pentachloronitrobenzene (Quintobenzene)
Pentachlorophenol
Phenol
p-Phenylenediamine
Phosgene
Phosphine
Phosphorus
Phthalic anhydride
Polychlorinated biphenyls (Aroclors)
1,3-Propane sultone
beta-Propiolactone
Propionaldehyde
Propoxur (Baygon)
Propylene dichloride (1,2-Dichloropropane)
Propylene oxide
1,2-Propylenimine (2-Methyl aziridine)
Quinoline
Quinone
Styrene
Styrene oxide
2,3,7,8-Tetrachlorodibenzo-p-dioxin
1,1,2,2-Tetrachloroethane
Tetrachloroethylene (Perchloroethylene)
Titanium tetrachloride
Toluene
2,4-Toluene diamine
2,4-Toluene diisocyanate
o-Toluidine

(continued)

TABLE 9-2

Hazardous air pollutants (HAPs) (*continued*)

Toxaphene (chlorinated camphene)	Vinylidene chloride (1,1-Dichloroethylene)	Coke oven emissions
1,2,4-Trichlorobenzene	Xylenes (isomers and mixture)	Cyanide compounds[1]
1,1,2-Trichloroethane	o-Xylenes	Glycol ethers[2]
Trichloroethylene	m-Xylenes	Lead compounds
2,4,5-Trichlorophenol	p-Xylenes	Manganese compounds
2,4,6-Trichlorophenol	Antimony compounds	Mercury compounds
Triethylamine	Arsenic compounds (inorganic, including	Fine mineral fibers[3]
Trifluralin	arsine)	Nickel compounds
2,2,4-Trimethylpentane	Beryllium compounds	Polycyclic organic matter[4]
Vinyl acetate	Cadmium compounds	Radionuclides (including radon)[5]
Vinyl bromide	Chromium compounds	Selenium compounds
Vinyl chloride	Cobalt compounds	

NOTE: For all listings above which contain the word "compounds" and for glycol ethers, the following applies: Unless otherwise specified, these listings are defined as including any unique chemical substance that contains the named chemical (i.e., antimony, arsenic, etc.) as part of that chemical's infrastructure.

[1]X'CN where X = H' or any other group where a formal dissociation may occur. For example KCN or Ca(CN)$_2$

[2]Includes mono- and di- ethers of ethylene glycol, diethylene glycol, and triethylene glycol R-(OCH2CH2)$_n$-OR' where

n = 1, 2, or 3

R = alkyl or aryl groups

R' = R, H, or groups which, when removed, yield glycol ethers with the structure: R-(OCH2CH)$_n$-OH. Polymers are excluded from the glycol category. Ethylene glycol monobutyl ether and surfactant alcohol ethoxylates and derivatives delisted November 29, 2004, 69 FR 692988.

[3]Includes mineral fiber emissions from facilities manufacturing or processing glass, rock, or slag fibers (or other mineral derived fibers) of average diameter 1 micrometer or less.

[4]Includes organic compounds with more than one benzene ring, and which have a boiling point greater than or equal to 100°C.

[5]A type of atom which spontaneously undergoes radioactive decay.

Source: Public Law 101-549, Nov. 15, 1990, 40 CFR 63.60

sources. The initial five industries that were regulated under the New Source Performance Standards (NSPS) included electric steam generating units, Portland cement plants, incinerators, nitric acid plants, and sulfuric acid plants. As an example of the NSPS, those for large electric utility steam generating units are summarized in Table 9-3.

TABLE 9-3

New source performance standards for coal-fired electric utility steam generating units of more than 73 MW—Summary[a]

SO$_2$ Standard
Emission limit: 90 percent reduction of potential SO$_2$ emissions and a limit of SO$_2$ emissions to 1.2 lb$_m$/million Btu of heat input (516 g/million kJ)

Particulate Standard
Emission limit: 0.03 lb$_m$/million Btu heat input (13 g/million kJ)

NO$_x$ Standard
Emission limit for sub-bituminous coal: 0.50 lb$_m$/million Btu heat input (210 g/million kJ)
Emission limit for anthracite coal: 0.60 lb$_m$/million Btu heat input (260 g/million kJ)

[a]*Federal Register,* vol. 45, February 1980, pp. 8210–8215

TABLE 9-4
Federal motor vehicle emission standards in g/mile[a]

Bin	NO_x	NMOG[b]	CO	HCHO[c]	PM[d]
8	0.20	0.125	4.2	0.018	0.02
7	0.15	0.090	4.2	0.018	0.02
6	0.10	0.090	4.2	0.018	0.01
5	0.07	0.090	4.2	0.018	0.01
4	0.04	0.070	2.1	0.011	0.01
3	0.03	0.055	2.1	0.011	0.01
2	0.02	0.010	2.1	0.004	0.01
1	0.00	0.000	0.0	0.000	0.00

[a]For 2004–2009 model years; full useful life
[b]NMOG = non-methane organic matter
[c]HCHO = formaldehyde
[d]PM = particulate matter
Source: Code of Federal Regulations, 40 CFR 86.1811-04, 2 NOV 2010.

Federal motor vehicle standards are expressed in terms of grams of pollutant per mile of driving. These standards were divided into two tiers. Tier I for 1994–1997 model years and Tier II for the 2004–2009 model years. The emission standards are applicable to light duty vehicles (LDVs), light duty trucks (LDTs) and medium duty passenger vehicles (MDPVs). Tier II is subdivided into bins to allow manufacturer's to classify their production (called a *fleet*) for the purpose of calculations to meet the standard. By 2008, the original 11 bins had been reduced to the 8 shown in Table 9-4. Tier II, Bin 5 roughly defines what the fleet average should be. Because emissions vary as driving conditions change, a standard driving cycle is defined for testing vehicles for compliance. The original driving cycle, called the Federal Test Procedure (FTP), was modified in 1996 by the Supplemental Federal Test Procedure (SFTP) to account for more aggressive driving behavior, the impact of air conditioning, and emissions after the engine is turned off.

Greenhouse gas air pollutant emission standards for passenger automobiles are defined as follows (40 CFR 86.1818-12):

- nitrous oxide $\leq$ 0.010 g/mile

- methane $\leq$ 0.030 g/mile

For carbon dioxide there are target values based on the *footprint* of the vehicle and the model year. For example, a passenger car with a footprint less than or equal to 41 square feet has the following target values by model year:

- 2012 $\rightarrow$ 244.0 g/mile

- 2013 $\rightarrow$ 237.0 g/mile

- 2014 $\rightarrow$ 228.0 g/mile

- 2014 → 217.0 g/mile

- 2016 and later → 206.0 g/mile

where footprint = wheel base × track width

9-4   EFFECTS OF AIR POLLUTANTS

Effects on Materials

Mechanisms of Deterioration.   Five mechanisms of deterioration have been attributed to air pollution: abrasion, deposition and removal, direct chemical attack, indirect chemical attack, and electrochemical corrosion (Yocom and McCaldin, 1968).

Solid particles of large enough size and traveling at high enough speed can cause deterioration by abrasion. With the exception of soil particles in dust storms and lead particles from automatic weapons fire, most air pollutant particles either are too small or travel at too slow a speed to be abrasive.

Small liquid and solid particles that settle on exposed surfaces do not cause more than aesthetic deterioration. For certain monuments and buildings, such as the White House, this form of deterioration is in itself quite unacceptable. For most surfaces, it is the cleaning process that causes the damage. Sandblasting of buildings is an obvious case in point. Frequent washing of clothes weakens their fiber, while frequent washing of painted surfaces dulls their finish.

Solubilization and oxidation/reduction reactions typify direct chemical attack. Frequently, water must be present as a medium for these reactions to take place. Sulfur dioxide and SO_3 in the presence of water react with limestone ($CaCO_3$) to form calcium sulfate ($CaSO_4$) and gypsum ($CaSO_4 \cdot 2H_2O$). Both $CaSO_4$ and $CaSO_4 \cdot 2H_2O$ are more soluble in water than $CaCO_3$, and both are leached away when it rains. The tarnishing of silver by H_2S is a classic example of an oxidation/reduction reaction.

Indirect chemical attack occurs when pollutants are absorbed and then react with some component of the absorbent to form a destructive compound. The compound may be destructive because it forms an oxidant, reductant, or solvent. Further, a compound can be destructive by removing an active bond in some lattice structure. Leather becomes brittle after it absorbs SO_2, which reacts to form sulfuric acid because of the presence of minute quantities of iron. The iron acts as a catalyst for the formation of the acid. A similar result has been noted for paper.

Oxidation/reduction reactions cause local chemical and physical differences on metal surfaces. These differences, in turn, result in the formation of microscopic anodes and cathodes. Electrochemical corrosion results from the potential that develops in these microscopic batteries.

Factors that Influence Deterioration.   Moisture, temperature, sunlight, and position of the exposed material are among the more important factors that influence the rate of deterioration.

Moisture, in the form of humidity, is essential for most of the mechanisms of deterioration to occur. Metal corrosion does not appear to occur even at relatively high SO_2

pollution levels until the relative humidity exceeds 60 percent. On the other hand, humidities above 70 to 90 percent will promote corrosion without air pollutants. Rain reduces the effects of pollutant-induced corrosion by dilution and washing away of the pollutant.

Higher air temperatures generally result in higher reaction rates. However, when low air temperatures are accompanied by cooling of surfaces to the point where moisture condenses, then the rates may be accelerated.

In addition to the oxidation effect of its ultraviolet wave lengths, sunlight stimulates air pollution damage by providing the energy for pollutant formation and cyclic reformation. The cracking of rubber and the fading of dyes have been attributed to ozone produced by these photochemical reactions.

The position of the exposed surface influences the rate of deterioration in two ways. First, whether the surface is vertical or horizontal or at some angle affects deposition and wash-off rates. Second, whether the surface is an upper or lower one may alter the rate of damage. When the humidity is sufficiently high, the lower side usually deteriorates faster because rain does not remove the pollutants as efficiently.

Effects on Vegetation

Cell and Leaf Anatomy. Because the leaf is the primary indicator of the effects of air pollution on plants, we shall define some terms and explain how the leaf functions. A typical plant cell (Figure 9-3) has three main components: the cell wall, the protoplast, and the inclusions. Much like human skin, the cell wall is thin in young plants and gradually thickens with age. Protoplast is the term used to describe the protoplasm of one cell. It consists primarily of water, but it also includes protein, fat,

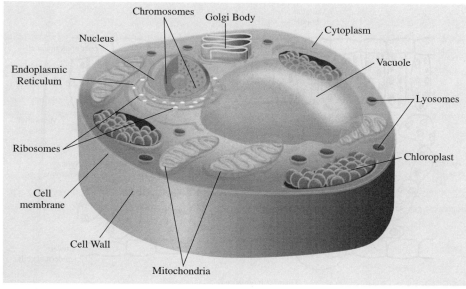

FIGURE 9-3
Typical plant cell.

and carbohydrates. The nucleus contains the hereditary material (DNA), which controls the operation of the cell. The protoplasm located outside the nucleus is called cytoplasm. Within the cytoplasm are tiny bodies or plastids. Examples include chloroplasts, leucoplasts, chromoplasts, and mitochondria. Chloroplasts contain the chlorophyll that manufactures the plant's food through photosynthesis. Leucoplasts convert starch into starch grains. Chromoplasts are responsible for the red, yellow, and orange colors of fruit and flowers.

A cross section through a typical mature leaf (Figure 9-4) reveals three primary tissue systems: the epidermis, the mesophyll, and the vascular bundle (veins). Chloroplasts are usually not present in epidermal cells. The opening in the underside of the leaf is called a stoma. (The plural of stoma is stomata.) The mesophyll, which includes both the palisade parenchyma and the spongy parenchyma, contains chloroplasts. It is the food production center. The vascular bundles carry water, minerals, and food throughout the leaf and to and from the main stem of the plant.

The guard cells regulate the passage of gases and water vapor in and out of the leaf. When it is hot, sunny, and windy, the processes of photosynthesis and respiration are increased. The guard cells open, which allows increased removal of water vapor that otherwise would accumulate because of the increased transport of water and minerals from the roots.

Pollutant Damage.   Ozone injures the palisade cells (Hindawi, 1970). The chloroplasts condense and ultimately the cell walls collapse. This results in the formation of red-brown spots that turn white after a few days. The white spots are called *fleck*. Ozone injury appears to be the greatest during midday on sunny days. The guard cells are more likely to be open under these conditions and thus allow pollutants to enter the leaf.

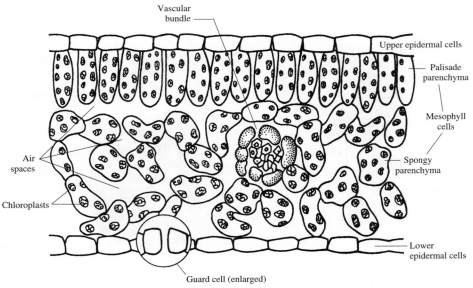

FIGURE 9-4
Cross section of intact leaf. (*Source:* Hindawi, 1970.)

Plant growth may be inhibited by continuous exposure to 0.5 ppm of NO_2. Levels of NO_2 in excess of 2.5 ppm for periods of four hours or more are required to produce *necrosis* (surface spotting due to plasmolysis or loss of protoplasm).

Sulfur dioxide injury is also typified by necrosis, but at much lower levels. A concentration of 0.3 ppm for eight hours is sufficient (O'Gara, 1922). Lower levels for longer periods of exposure will produce a diffuse *chlorosis* (bleaching).

The net result of air pollutant damage goes beyond the apparent superficial damage to the leaves. A reduction in surface area results in less growth and small fruit. For commercial crops this results in a direct reduction in income for the farmer. For other plants the net result is likely to be an early death.

Fluoride deposition on plants not only causes them damage but may result in a second untoward effect. Grazing animals may accumulate an excess of fluoride that mottles their teeth and ultimately causes them to fall out.

Problems of Diagnosis.   Various factors make it difficult to diagnose actual air pollution damage. Droughts, insects, diseases, herbicide overdoses, and nutrient deficiencies all can cause injury that resembles air pollution damage. Also, combinations of pollutants that alone cause no damage are known to produce acute effects when combined (Hindawi, 1970). This effect is known as *synergism.*

Effects on Health

Susceptible Population.   It is difficult at best to assess the effects of air pollution on human health. Personal pollution from smoking results in exposure to air pollutant concentrations far higher than the low levels found in the ambient atmosphere. Occupational exposure may also result in pollution doses far above those found outdoors. Tests on rodents and other mammals are difficult to interpret and apply to human anatomy. Tests on human subjects are usually restricted to those who would be expected to survive. This leads us to a question of environmental ethics. If the allowable concentration levels (standards) are based on results from tests on rodents, they would be rather high. If the allowable concentration levels must also protect those with existing cardiorespiratory ailments, they should be lower than those resulting from the observed effects on rodents.

We noted earlier that the air quality standards were established to protect public health with an "adequate margin of safety." In the opinion of the Administrator of the EPA, the standards must protect the most sensitive responders. Thus, as you will note in the following paragraphs, the standards have been set at the lowest level of observed effect. This decision has been attacked by some theorists. They say it would make better economic sense to build more hospitals (Connolly, 1972). However, one also might apply this kind of logic in establishing speed limits for highways, that is, raise the speed limit and build more hospitals, junk yards, and cemeteries!

Anatomy of the Respiratory System.   The respiratory system is the primary indicator of air pollution effects in humans. The major organs of the respiratory system are the nose, pharynx, larynx, trachea, bronchi, and lungs (Figure 9-5). The nose, pharynx, larynx, and trachea together are called the *upper respiratory tract* (URT). The primary effects of air pollution on the URT are aggravation of the sense of smell and inactivation of the sweeping motion of cilia, which remove mucus and entrapped particles.

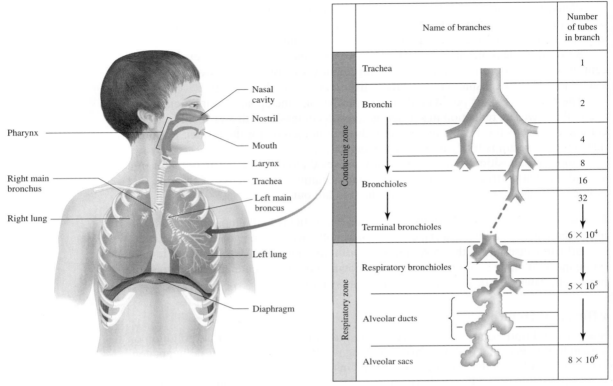

	Name of branches	Number of tubes in branch
	Trachea	1
	Bronchi	2
		4
		8
	Bronchioles	16
		32
	Terminal bronchioles	6×10^4
	Respiratory bronchioles	5×10^5
	Alveolar ducts	
	Alveolar sacs	8×10^6

(Left column spanning labels: Conducting zone — Trachea through Terminal bronchioles; Respiratory zone — Respiratory bronchioles through Alveolar sacs.)

Diagram labels: Nasal cavity, Nostril, Mouth, Larynx, Trachea, Left main broncus, Left lung, Diaphragm, Pharynx, Right main bronchus, Right lung.

FIGURE 9-5
The respiratory system. (*Source:* Davis and Masten, 2009.)

The *lower respiratory tract* (LRT) consists of the branching structures known as bronchi and the lung itself, which is composed of grape-like clusters of sacs called *alveoli*. The alveoli are approximately 300 μm in diameter. The walls of alveoli are lined with capillaries. Carbon dioxide diffuses through the capillary wall into the alveolus, while oxygen diffuses out of the alveolus into the blood cell. The difference in partial pressure of each of the gases causes it to move from the higher to lower partial pressure.

Inhalation and Retention of Particles. The degree of penetration of particles into the LRT is primarily a function of the size of the particles and the rate of breathing. Particles greater than 5 to 10 μm are screened out by the hairs in the nose. Sneezing also helps the screening process. Particles in the 1 to 2 μm size range penetrate to the alveoli. These particles are small enough to bypass screening and deposition in the URT, however they are big enough that their terminal settling velocity allows them to deposit where they can do the most damage. Particles that are 0.5 μm in diameter do not have a large enough terminal settling velocity to be removed efficiently. Smaller particles diffuse to the alveolar walls. Refer to Figure 9-2 and note that the size of "Lung Damaging Dust" falls in the critical particle size range.

Chronic Respiratory Disease.   Several long-term diseases of the respiratory system are seriously aggravated by and perhaps may be caused by air pollution. *Airway resistance* is the narrowing of air passages because of the presence of irritating substances. The result is that breathing becomes difficult. *Bronchial asthma* is a form of airway resistance that results from an allergy. An asthma "attack" is the result of the narrowing of the bronchioles because of a swelling of the mucous membrane and a thickening of the secretions. The bronchioles return to normal after the attack. *Chronic bronchitis* is currently defined to be present in a person when excess mucus in the bronchioles results in a cough for three months a year for two consecutive years and lung infections, tumors, and heart disease must be absent. *Pulmonary emphysema* is characterized by a breakdown of the alveoli. The small grape-like clusters become a large nonresilient balloon-like structure. The amount of surface area for gas exchange is reduced drastically. *Cancer of the bronchus* (lung cancer) is characterized by abnormal, disorderly new cell growth originating in the bronchial mucous membrane. The growth closes off the bronchioles. It is usually fatal.

Carbon Monoxide (CO).   This colorless, odorless gas is lethal to humans within a few minutes at concentrations exceeding 5,000 ppm. CO reacts with hemoglobin in the blood to form carboxyhemoglobin (COHb). Hemoglobin has a greater affinity for CO than it does for oxygen. Thus, the formation of COHb effectively deprives the body of oxygen. At COHb levels of 5 to 10 percent, visual perception, manual dexterity, and ability to learn are impaired. A concentration of 50 ppm of CO for eight hours will result in a COHb level of about 7.5 percent. At COHb levels of 2.5 to 3 percent, people with heart disease are not able to perform certain exercises as well as they might in the absence of COHb. A concentration of 20 ppm of CO for eight hours will result in a COHb level of about 2.8 percent (Ferris, 1978). (We should note here that the average concentration of CO inhaled in cigarette smoke is 200 to 400 ppm!) The sensitive populations are those with heart and circulatory ailments, chronic pulmonary disease, developing fetuses, and those with conditions that cause increased oxygen demand, such as fever from an infections disease.

Hazardous Air Pollutants (HAPs).   Most of the information on the direct human health effects of hazardous air pollutants (also known as air toxics) comes from studies of industrial workers. Exposure to air toxics in the workplace is generally much higher than in the ambient air. We know relatively little about the specific effects of the HAPs at the low levels normally found in ambient air.

The HAPs regulated under the NESHAP program were identified as causal agents for a variety of diseases. For example, asbestos, arsenic, benzene, coke oven emissions, and radionuclides may cause cancer. Beryllium primarily causes lung disease but also affects the liver, spleen, kidneys, and lymph glands.

Mercury has been especially targeted for regulation because it is released during the combustion of coal. Thus, it is one of the few HAPs that is widespread in the environment. Of particular concern are children who are exposed to methyl mercury prenatally. They are at increased risk of poor performance on neurobehavioral tasks such as those measuring attention, fine motor function, language skills, visual-spatial abilities and verbal memory (U.S. EPA, 1997, and U.S. EPA, 2004).

Lead (Pb). In contrast to the other criteria air pollutants, lead is a cumulative poison. A further difference is that it is ingested in food and water, as well as being inhaled. Of that portion taken by ingestion, approximately 5 to 10 percent is absorbed in the body. Between 20 and 50 percent of the inspired portion is absorbed. Those portions that are not absorbed are excreted in the feces and urine. Lead is measured in the urine and blood for diagnostic evidence of lead poisoning.

An early manifestation of acute lead poisoning is a mild anemia (deficiency of red blood cells). Fatigue, irritability, mild headache, and pallor indistinguishable from other causes of anemia occur when the blood level of lead increases to 60 to 120 μg/100 g of whole blood. Blood levels in excess of 80 μg/100 g result in constipation and abdominal cramps. When an acute exposure results in blood levels of lead greater than 120 μg/100 g, acute brain damage (encephalopathy) may result (Goyer and Chilsolm, 1972). Such acute exposure results in convulsions, coma, cardiorespiratory arrest, and death. Acute exposures may occur over a period of one to three weeks.

Chronic exposure to lead may result in brain damage characterized by seizures, mental incompetence, and highly active aggressive behavior. Weakness of extensor muscles of the hands and feet and eventual paralysis may also result. Canfield et al. (2003) found a decline in intelligence quotient (IQ) of 7.4 points for a lifetime blood lead concentration of up to 10 μg per deciliter. For a lifetime average blood lead concentration ranging from more than 10 μg per deciliter to 30 μg per deciliter, a more gradual decrease of 2.5 IQ points was observed.

Atmospheric lead occurs as a particulate. The particle size range is between 0.16 and 0.43 μm. Nonsmoking residents of suburban Philadelphia exposed to approximately 1 μg/m^3 of lead in air have blood levels averaging 11 μg/100 g. Nonsmoking residents of downtown Philadelphia exposed to approximately 2.5 μg/m^3 of lead have blood levels averaging 20 μg/100 g (U.S. PHS, 1965). In the early 1990s, 4.4 percent of U.S. children ages 1 to 5 had elevated lead levels. The percentage dropped to 1.6 percent by 2002. The U.S. Centers for Disease Control and Prevention attributed this drop to the removal of lead from gasoline as well as other efforts to screen and treat children for lead exposure (U.S. CDC, 2005).

Nitrogen Dioxide (NO$_2$). Exposure to NO$_2$ concentrations above 5 ppm for 15 minutes results in cough and irritation of the respiratory tract. Continued exposure may produce an abnormal accumulation of fluid in the lung (pulmonary edema). The gas is reddish brown in concentrated form and gives a brownish yellow tint at lower concentrations. At 5 ppm it has a pungent sweetish odor. The average NO$_2$ concentration in tobacco smoke is approximately 5 ppm. Slight increases in respiratory illness and decrease in pulmonary function have been associated with concentrations of about 0.10 ppm (Ferris, 1978). You should note that these concentrations are very high with respect to the NAAQS in Table 9-1.

Photochemical Oxidants. Although the photochemical oxidants include peroxyacetyl nitrate (PAN), acrolein, peroxybenzoyl nitrates (PBzN), aldehydes, and nitrogen oxides, the major oxidant is ozone (O$_3$). Ozone is commonly used as an indicator of the total amount of oxidant present. Oxidant concentrations above 0.1 ppm result in eye irritation. At a concentration of 0.3 ppm, cough and chest discomfort are increased. Those people who suffer from chronic respiratory disease are particularly susceptible.

PM_{10}. As noted earlier, large particles are not inhaled deeply into the lungs. This is why EPA switched from an air quality standard based on total suspended matter to one based on particles with an aerodynamic diameter less than 10 μm (PM_{10}). Studies in the United States, Brazil, and Germany have related higher levels of particulates to increased risk of respiratory, cardiovascular, and cancer-related deaths, as well as pneumonia, lung function loss, hospital admissions, and asthma (Reichhardt, 1995).

Particles 2.5 μm in aerodynamic diameter have been identified as a major contributor to elevated death rates in polluted cities (Pope et al., 1995). One hypothesized biological mechanism is pollution-induced lung damage resulting in declines in lung function, in respiratory distress, and in cardiovascular disease potentially related to hypoxemia (Pope et al., 1999).

Sulfur Oxides (SO_x) and Total Suspended Particulates (TSP). The sulfur oxides include sulfur dioxide (SO_2), sulfur trioxide (SO_3), their acids, and the salts of their acids. Rather than try to separate the effects of SO_2 and SO_3, they are usually treated together. There is speculation that a definite synergism exists whereby fine particulates carry absorbed SO_2 to the LRT. The SO_2 in the absence of particulates would be absorbed in the mucous membranes of the URT.

Patients suffering from chronic bronchitis have shown an increase in respiratory symptoms when the TSP levels exceeded 350 $\mu g/m^3$ and the SO_2 level was above 0.095 ppm. Studies made in Holland at an interval of three years showed that pulmonary function improved as SO_2 and TSP levels dropped from 0.10 ppm and 230 $\mu g/m^3$ to 0.03 ppm and 80 $\mu g/m^3$, respectively.

Air Pollution Episodes. The word *episode* is used as a refined form of the word *disaster.** Indeed, it was the shock of these disasters that stimulated the first modern legislative action to require control of air pollutants. The characteristics of the three major episodes are summarized in Table 9-5. Careful study of the table will reveal that all of the episodes had some things in common. Comparison of these situations and others where no episode occurred (that is, where the number of dead and ill was considerably less) has revealed that four ingredients are essential for an episode. If one ingredient is omitted, fewer people will get sick and only a few people can be expected to die. The crucial ingredients are: (1) a large number of pollution sources, (2) a restricted air volume, (3) failure of officials to recognize that anything is wrong, and (4) the presence of water droplets of the "right" size (Goldsmith, 1968).

Although a sufficient quantity of any pollutant is lethal by itself, it is generally agreed that some mix is required to achieve the results seen in these episodes. Atmospheric levels of individual pollutants seldom rise to lethal levels without an explosion or transportation accident. However, the proper combination of two or more pollutants will yield untoward symptoms at much lower levels. The sulfur oxides and particulates were the most suspect in the three major episodes.

---

*In the nuclear power business, they would call it an "incident."

TABLE 9-5
Three major air pollution episodes

	Meuse Valley, 1930 (Dec. 1)	Donora, 1948 (Oct. 26–31)	London, 1952 (Dec. 5–9)
Population	No data	12,300	8,000,000
Weather	Anticyclone, inversion, and fog	Anticyclone, inversion, and fog	Anticyclone, inversion, and fog
Topography	River valley	River valley	River plain
Most probable source of pollutants	Industry (including steel and zinc plants)	Industry (including steel and zinc plants)	Household coal-burning
Nature of the illnesses	Chemical irritation of exposed membranous surfaces	Chemical irritation of exposed membranous surfaces	Chemical irritation of exposed membranous surfaces
No. of deaths	63	17	4,000
Time of deaths	Began after second day of episode	Began after second day of episode	Began on first day of episode
Suspected proximate cause of irritation	Sulfur oxides with particulates	Sulfur oxides with particulates	Sulfur oxides with particulates

(*Source:* WHO, 1961.)

The meteorology must be such that there is little air movement. Thus, the pollutants cannot be diluted. Although a valley is most conducive to a stagnation effect, the London episode proved that it isn't necessary. The stagnant conditions must persist for several days. Three days appears to be the minimum.

Tragically, each of these hazardous air pollution conditions became lethal because of the failure of city officials to notice anything strange. If they have no measurements of pollution levels or reports from hospitals and morgues, city authorities have no reason to alert the public, shut down factories, or restrict traffic.

The last and, perhaps, most crucial element is fog.* The fog droplets must be of the "right" size, namely, in the 1 to 2 μm diameter range or, perhaps, in the range below 0.5 μm. As mentioned earlier, these particle sizes are most likely to penetrate into

---

*The word "smog" is a term coined by Londoners before World War I to describe the combination of smoke and fog that accounted for much of their weather. Los Angeles smog is a misnomer because little smoke and no fog is present. In fact, as we shall see later, Los Angeles smog cannot occur without a lot of sunshine. "Photochemical smog" is the correct term to describe the Los Angeles haze.

the LRT. Pollutants that dissolve into the fog droplet are thus carried deep into the lungs and deposited there.

9-5   ORIGIN AND FATE OF AIR POLLUTANTS

Carbon Monoxide

Incomplete oxidation of carbon results in the production of carbon monoxide. The natural anaerobic decomposition of carbonaceous material by microorganisms releases approximately 160 teragrams* (T_g) of methane (CH_4) to the atmosphere each year worldwide (IPCC, 1995). The natural formation of CO results from an intermediate step in the oxidation of the methane. The hydroxyl radical $(OH\cdot)$ serves as the initial oxidizing agent. It combines with CH_4 to form an alkyl radical (Wofsy et al., 1972).

$$CH_4 + OH\cdot \rightarrow CH_3\cdot + H_2O \qquad (9\text{-}8)$$

This reaction is followed by a complex series of 39 reactions, which we have over-simplified to the following:

$$CH_3\cdot + O_2 + 2(h\upsilon) \rightarrow CO + H_2 + OH\cdot \qquad (9\text{-}9)$$

This says that $CH_3\cdot$ and O_2 are each zapped by a photon of light energy $(h\upsilon)$. The symbol υ stands for the frequency of the light. The h is Planck's constant $= 6.626 \times 10^{-34}$ J/Hz.

Anthropogenic sources (those associated with the activities of human beings) include motor vehicles, fossil fuel burning for electricity and heat, industrial processes, solid waste disposal, and miscellaneous burning of such things as leaves and brush. Approximately 600–1250 Tg of CO are released by these sources (IPCC, 1995). Motor vehicles account for more than 60 percent of the emission.

No significant change in the global atmospheric CO level has been observed over the past 20 years. Yet the worldwide anthropogenic contribution of combustion sources has doubled over the same time period. Because there is no apparent change in the atmospheric concentration, a number of mechanisms (*sinks*) have been proposed to account for the missing CO. The two most probable are

1. Reaction with hydroxyl radicals to form carbon dioxide

2. Removal by soil microorganisms

It has been estimated that these two sinks annually consume an amount of CO that just equals the production (Seinfeld, 1975).

Hazardous Air Pollutants (HAPs)

The EPA has identified 166 categories of major sources and 8 categories of area sources for the HAPs listed in Table 9-2 (57 FR 31576). The source categories represent a wide range of industrial groups: fuel combustion, metal processing, petroleum

---

*One teragram $= 1 \times 10^{12}$ grams.

and natural gas production and refining, surface coating processes, waste treatment and disposal processes, agricultural chemicals production, and polymers and resins production. There are also a number of miscellaneous source categories, such as dry cleaning and electroplating.

In addition to these direct emissions, air toxics can result from chemical formation reactions in the atmosphere. These reactions involve chemicals emitted to the atmosphere that are not listed HAPs and may not be toxic themselves, but can undergo atmospheric transformations to generate HAPs. For organic compounds present in the gas phase, the most important transformation processes involve photolysis and chemical reactions with ozone, hydroxyl radicals (OH·), and nitrate radicals (Kao, 1994). *Photolysis* is the chemical fragmentation or rearrangement of a chemical upon the adsorption of radiation of the appropriate wavelength. Photolysis is only important during the daytime for those chemicals that absorb strongly within the solar radiation spectrum. Otherwise, reaction with OH· or O_3 is likely to predominate. The HAPs most often formed are formaldehyde and acetaldehyde.

The major removal mechanisms appear to be OH abstraction or addition. The reaction products lead to the formation of CO and CO_2. Eighty-nine of the 188 HAPs have atmospheric lifetimes of less than one day.

Lead

Volcanic activity and airborne soil are the primary natural sources of atmospheric lead. Smelters and refining processes, as well as incineration of lead-containing wastes, are major point sources of lead. Approximately 70 to 80 percent of the lead that used to be added to gasoline was discharged to the atmosphere.

Submicron lead particles, which are formed by volatilization and subsequent condensation, attach to larger particles or they form nuclei before they are removed from the atmosphere. Once they have attained a size of several microns, they either settle out or are washed out by rain.

Nitrogen Dioxide

Bacterial action in the soil releases nitrous oxide (N_2O) to the atmosphere. In the upper troposphere and stratosphere, atomic oxygen reacts with the nitrous oxide to form nitric oxide (NO).

$$N_2O + O \rightarrow 2NO \qquad (9\text{-}10)$$

The atomic oxygen results from the dissociation of ozone. The nitric oxide further reacts with ozone to form nitrogen dioxide (NO_2).

$$NO + O_3 \rightarrow NO_2 + O_2 \qquad (9\text{-}11)$$

The global formation of NO_2 by this process is estimated to be 0.45 petagrams* (Pg) annually (Seinfeld, 1975).

Combustion processes account for 96 percent of the anthropogenic sources of nitrogen oxides. Although nitrogen and oxygen coexist in our atmosphere without

---

*One petagram $= 1 \times 10^{15}$ grams.

reaction, their relationship is much less indifferent at high temperatures and pressures. At temperatures in excess of 1,600 K, they react.

$$N_2 + O_2 \overset{\Delta}{\rightleftharpoons} 2NO \tag{9-12}$$

If the combustion gas is rapidly cooled after the reaction by exhausting it to the atmosphere, the reaction is quenched and NO is the byproduct. The NO in turn reacts with ozone or oxygen to form NO_2. The anthropogenic contribution to global emission of NO_x amounted to 32 Tg/y (as N) in 1995 (IPCC, 1995). Between 40 and 45 percent of the NO_x emissions in the United States come from transportation, 30 to 35 percent from power plants, and 20 percent from industrial sources (Seinfeld and Pandis, 1998).

The U.S. EPA emission factors provide an example of a method for estimating emissions from coal-fired electric utility boilers. For pulverized coal, dry bottom, wall-fired boilers using bituminous and sub-bituminous coal:

- Pre-NSPS standards—22 lb_m of NO_x/ton of bituminous coal

- Pre-NSPS standards—12 lb_m of NO_x/ton of sub-bituminous coal

- After NSPS standards—12 lb_m of NO_x/ton of bituminous coal

- After NSPS standards—7.4 lb_m of NO_x/ton of sub-bituminous coal

where a "ton" is defined as 2,000 lb_m.

Ultimately, the NO_2 is converted to either NO_2^- or NO_3^- in particulate form. The particulates are then washed out by precipitation. The dissolution of nitrate in a water droplet allows for the formation of nitric acid (HNO_3). This, in part, accounts for "acid" rain found downwind of industrialized areas.

Photochemical Oxidants

Unlike the other pollutants, the photochemical oxidants result entirely from atmospheric reactions and are not directly attributable to either people or nature. Thus, they are called *secondary pollutants*. They are formed through a series of reactions that are initiated by the absorption of a photon by an atom, molecule, free radical, or ion. Ozone is the principal photochemical oxidant. Its formation is usually attributed to the nitrogen dioxide photolytic cycle. Hydrocarbons modify this cycle by reacting with atomic oxygen to form free radicals (highly reactive organic species). The hydrocarbons, nitrogen oxides, and ozone react and interact to produce more nitrogen dioxide and ozone. This cycle is represented in summary form in Figure 9-6. The whole reaction sequence depends on an abundance of sunshine. A result of these reactions is the photochemical "smog" for which Los Angeles is famous.

Sulfur Oxides

Sulfur oxides may be both primary and secondary pollutants. Power plants, industry, volcanoes, and the oceans emit SO_2, SO_3, and SO_4^{2-} directly as primary pollutants. In addition, biological decay processes and some industrial sources emit H_2S, which is oxidized to form the secondary pollutant SO_2. In terms of sulfur, approximately 30 Tg

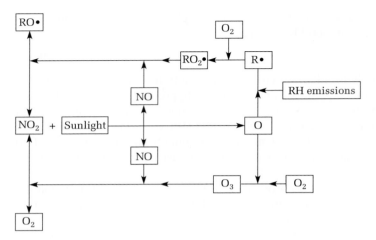

FIGURE 9-6
Interaction of hydrocarbons with atmospheric nitrogen oxide photolytic cycle.
(*Source:* NAPCA, 1970.)

are emitted annually by natural sources. Approximately 75 Tg of sulfur may be attributed to anthropogenic sources each year (Seinfeld and Pandis, 1998).

The most important oxidizing reaction for H_2S appears to be one involving ozone:

$$H_2S + O_3 \rightarrow H_2O + SO_2 \qquad (9\text{-}13)$$

The combustion of fossil fuels containing sulfur yields sulfur dioxide in direct proportion to the sulfur content of the fuel:

$$S + O_2 \rightarrow SO_2 \qquad (9\text{-}14)$$

This reaction implies that for every gram of sulfur in the fuel, two grams of SO_2 are emitted to the atmosphere. Because the combustion process is not 100 percent efficient, we generally assume that 5 percent of the sulfur in the fuel ends up in the ash, that is, 1.90 g SO_2 per gram of sulfur in the fuel is emitted.

EPA uses emission factors for estimating emissions from coal-fired electric utility boilers. For pulverized coal, dry bottom, wall-fired boilers using bituminous and sub-bituminous coal:

- Pre-NSPS standards emission factor for SO_2 from bituminous coal = 38S

- Pre-NSPS standards emission factor for SO_2 from sub-bituminous coal = 35S

- After NSPS standards emission factor for SO_2 from bituminous coal = 38S

- After NSPS standards emission factor for SO_2 from sub-bituminous coal = 35S

The quantity "S" is the weight percent sulfur content of the coal. For example, if the fuel is 1.2 percent sulfur, then S = 1.2 and the emission factor for bituminous coal is $(38)(1.2) = 45.6$ lb_m/ton where a "ton" is defined as 2,000 lb_m.

Example 9-2. An Illinois coal is burned at a rate of 1.00 kg per second. If the analysis of the coal reveals a sulfur content of 3.00 percent, what is the annual rate of emission of SO_2?

Solution. Using the mass balance approach, we begin by drawing a mass balance diagram:

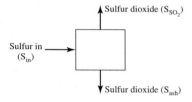

The mass balance equation may be written as

$$S_{in} = S_{ash} + S_{SO_2}$$

From the problem data, the mass of "sulfur in" is

$$S_{in} = 1.00 \text{ kg/s} \times 0.030 = 0.030 \text{ kg/s}$$

In one year,

$$S_{in} = 0.030 \text{ kg/s} \times 86,400 \text{ s/d} \times 365 \text{ d/y} = 9.46 \times 10^5 \text{ kg/y}$$

The sulfur in the ash is 5 percent of the input sulfur:

$$S_{ash} = (0.05)(9.46 \times 10^5 \text{ kg/y}) = 4.73 \times 10^4 \text{ kg/y}$$

The amount of sulfur available for conversion to SO_2:

$$S_{SO_2} = S_{in} - S_{ash} = 9.46 \times 10^5 - 4.73 \times 10^4 = 8.99 \times 10^5 \text{ kg/y}$$

The amount of sulfur dioxide formed is determined from the proportional weights of the oxidation reaction (Equation 9-14):

$$S + O_2 \rightarrow SO_2$$
$$GMW = 32 + 32 = 64$$

The amount of sulfur dioxide formed is then 64/32 of the sulfur available for conversion:

$$S_{SO_2} = \frac{64}{32} (8.99 \times 10^5 \text{ kg/y}) = 1.80 \times 10^6 \text{ kg/y}$$

The ultimate fate of most of the SO_2 in the atmosphere is conversion to sulfate salts, which are removed by sedimentation or by washout with precipitation. The conversion to sulfate is by either of two routes: catalytic oxidation or photochemical oxidation. The first process is most effective if water droplets containing Fe^{3+}, Mn^{2+}, or NH_3 are present:

$$2SO_2 + 2H_2O + O_2 \xrightarrow{\text{catalyst}} 2H_2SO_4 \tag{9-15}$$

At low relative humidities, the primary conversion process is photochemical oxidation. The first step is photoexcitation* of the SO_2.

$$SO_2 + h\nu \rightarrow \overset{*}{S}O_2 \tag{9-16}$$

The excited molecule then readily reacts with O_2 to form SO_3:

$$\overset{*}{S}O_2 + O_2 \rightarrow SO_3 + O \tag{9-17}$$

The trioxide is very hygroscopic and consequently is rapidly converted to sulfuric acid:

$$SO_3 + H_2O \rightarrow H_2SO_4 \tag{9-18}$$

This reaction in large part accounts for acid rain (that is, precipitation with a pH value less than 5.6) found in industrialized areas. Normal precipitation has a pH of 5.6, due to the carbonate buffer system.

Particulates

Sea salt, soil dust, volcanic particles, and smoke from forest fires account for 2.9 Pg of particulate emissions each year. Anthropogenic emissions from fossil fuel burning and industrial processes account for emissions of 110 Tg per year (Kiehl and Rodhe, 1995). For example, the emission factor for a pulverized coal-fired, dry bottom, wall-fired boiler is given as 10A where "A" is the percent ash in the coal. For example, if the fuel is 8 percent ash, then A = 8 and the emission factor for bituminous or sub-bituminous coal is $(10)(8) = 80 \text{ lb}_m/\text{ton}$ where a "ton" is defined as $2,000 \text{ lb}_m$.

Secondary sources of particulates include the conversion of H_2S, SO_2, NO_x, NH_3, and hydrocarbons. H_2S and SO_2 are converted to sulfates. NO_x and NH_3 are converted to nitrates. The hydrocarbons react to form products that condense to form particles at atmospheric temperatures. Natural sources of secondary pollutants yield about 240 Tg annually. Anthropogenic sources yield about 340 Tg annually (Kiehl and Rodhe, 1995).

Dust particles that are *entrained* (picked up) by the wind and carried over long distances tend to sort themselves out to the sizes between 0.5 and 50 μm in diameter. Sea salt nuclei have sizes between 0.05 and 50 μm. Particles formed as a result of photochemical reactions tend to have very small diameters ($< 0.4 \mu$m). Smoke and fly ash particles cover a wide range of sizes from 0.05 to 200 μm or more. Particle mass distributions in urban atmospheres generally exhibit two maxima. One is between 0.1 and 1 μm in diameter. The other is between 1 and 30 μm. The smaller fraction is the result of condensation. The coarse fraction consists of fly ash and dust generated by mechanical abrasion.

Small particles are removed from the atmosphere by accretion to water droplets, which grow in size until they are large enough to precipitate. Larger particles are removed by direct washout by falling raindrops.

---

*Photoexcitation is the displacement of an electron from one shell to another, thereby storing energy in the molecule. Photoexcitation is represented in reactions by an asterisk.

Example 9-3. Determine whether or not a pulverized coal, dry bottom, wall-fired boiler using bituminous coal at a power plant rate at 61 MW meets the NSPS for SO_2, particulate matter, and NO_x. The power plant burns bituminous coal with a sulfur content of 1.8 percent and ash content of 6.2 percent. The coal has a heating value of 14,000 Btu/lb. The boiler efficiency is 35 percent. Use the EPA emission factors to estimate the emissions. Assume the efficiency of SO_2 control is 85 percent and the efficiency of particulate control equipment is 99 percent.

Solution. Begin by calculating the coal firing rate for 61 MW at a boiler efficiency of 35 percent.

$$\frac{61 \text{ MW}}{0.35} = 174.3 \text{ MW, or } 174.3 \times 10^6 \text{ W}$$

Using a one-hour time increment, convert W-h to Btu with the conversion factor from Appendix C.

$$(174.3 \times 10^6 \text{ W})(1 \text{ h})(3.4144 \text{ Btu/W-h}) = 5.95 \times 10^8 \text{ Btu}$$

The mass of coal burned in an hour is

$$\left(\frac{5.95 \times 10^8 \text{ Btu}}{14,000 \text{ Btu/lb}_m}\right)\left(\frac{1 \text{ ton}}{2,000 \text{ lb}_m}\right) = 21.25 \text{ tons}$$

a. Check the SO_2 emission rate.

Using the EPA emission factor of 38S for bituminous coal:

$$\text{Uncontrolled } SO_2 \text{ emission rate} = (38)(1.8) = 68.4 \text{ lb}_m/\text{ton of coal}$$

The estimated SO_2 emission rate with 85 percent control is

$$(68.4 \text{ lb}_m/\text{ton of coal})(21.25 \text{ tons/h})(1 - 0.85) = 218.03 \text{ lb}_m$$

The SO_2 emission rate per million Btu is

$$\frac{218.03 \text{ lb}_m}{5.95 \times 10^8 \text{ Btu}} = 3.66 \times 10^{-7} \text{ lb}_m/\text{Btu}$$

Or on a million Btu basis,

$$(3.66 \times 10^{-7} \text{ lb}_m/\text{Btu})(10^6) = 0.37 \text{ lb}_m/\text{million Btu}$$

This meets the standard of 1.2 lb_m/million Btu but does not meet the 90 percent reduction requirement given in Table 9-3.

b. Check the particulate emission rate.

Using the EPA emission factor of "10A" for a pulverized coal, dry bottom, wall-fired boiler:

$$\text{Uncontrolled particulate emission} = (10)(6.2) = 62.0 \text{ lb}_m/\text{ton of coal burned}$$

The estimated particulate emission with 99 percent control is

$$(62.0 \text{ lb}_m/\text{ton of coal burned})(21.25 \text{ tons})(1 - 0.99) = 13.2 \text{ lb}_m$$

The particulate emission rate per million Btu is

$$\frac{13.2 \text{ lb}_m}{5.95 \times 10^8 \text{ Btu}} = 2.23 \times 10^{-8} \text{ lb}_m/\text{Btu}$$

On a million Btu basis,

$$(2.23 \times 10^{-8} \text{ lb}_m/\text{Btu})(10^6) = 0.022 \text{ lb}_m/10^6 \text{ Btu}$$

This meets the standard of 0.03 lbm/10^6 Btu (see Table 9-3).

c. Check the NO_x emission rate.

Using the EPA emission factor of 22 lb$_m$/ton, the estimated emission is

$$(22 \text{ lb}_m/\text{ton})(21.25 \text{ tons}) = 467.5 \text{ lb}_m$$

The NO_x emission rate per million Btu is

$$\frac{467.5 \text{ lb}_m}{5.95 \times 10^8 \text{ Btu}} = 7.86 \times 10^{-7} \text{ lb}_m/\text{Btu}$$

On a million Btu basis,

$$(7.86 \times 10^{-7} \text{ lb}_m/\text{Btu})(10^6) = 0.79 \text{ lb}_m/10^6 \text{ Btu}$$

The standard for bituminous coal is 0.60 lb$_m$/10^6 Btu. The power plant does not meet the NO_x standard noted in Table 9-3.

Comments:

1. The substitution of sub-bituminous or lignite coal for the bituminous coal is one alternative method to achieve the standard. In general, they have a lower sulfur content and a similar or lower ash content. A coal analysis is required to verify this general assumption.

2. Modification of the burner will be required to meet the NO_x standard.

9-6   MICRO AND MACRO AIR POLLUTION

Air pollution problems may occur on three scales: micro, meso, and macro. Micro-scale problems range from those covering less than a centimeter to those the size of a house or slightly larger. Meso-scale air pollution problems are those of a few hectares up to the size of a city or county. Macro-scale problems extend from counties to states, nations, and in the broadest sense, the globe. Much of the remaining discussion in this chapter is focused on the meso-scale problem. In this section we will address the general micro-scale and macro-scale problems recognized today.

Indoor Air Pollution

People who live in urban, cold climates may spend more than 90 percent of their time indoors (Lewis, 2001). In the last three decades, researchers have identified sources, concentrations, and impacts of air pollutants that arise in conventional domestic residences. The startling results indicate that, in certain instances, indoor air may be substantially more polluted than outdoor air.

TABLE 9-6
Tested combustion sources and their emission rates

Source	NO	NO$_2$	NO$_x$ (as NO$_2$)	CO	SO$_2$
			Range of emission rates,[a] mg/MJ		
Range-top burner[b]	15–17	9–12	32–37	40–244	—[c]
Range oven[d]	14–29	7–13	34–53	12–19	—
Pilot light[e]	4–17	8–12	[f]	40–67	—
Gas space heaters[g]	0–15	1–15	1–37	14–64	—
Gas dryer[h]	8	8	20	69	—
Kerosene space heaters[i]	1–13	3–10	5–31	35–64	11–12
Cigarette smoke[j]	2.78	0.73	[f]	88.43	—

[a]The lowest and highest mean values of emission rates for combustion sources tested in milligrams per mega-Joule (mg/MJ). Note: It takes 4.186 Joules to raise the temperature of 1.0 g of water from 14.5°C to 15.5°C at 100 percent efficiency.
[b]Three ranges were evaluated. Reported results are for blue flame condition.
[c]Dash (–) means combustion source is not emitting the pollutant.
[d]Three ranges were evaluated. Ovens were operated for several different settings (bake, broil, self-clean cycle, etc.).
[e]One range was evaluated with all three pilot lights, two top pilots, and a bottom pilot.
[f]Emission rates not reported.
[g]Three space heaters including one convective, radiant, and catalytic were tested.
[h]One gas dryer was evaluated.
[i]Two kerosene heaters including a convective and radiant type were tested.
[j]One type of cigarette. Reported emission rates are in mg/cigarette (800 mg tobacco/cigarette).
(*Source:* D. J. Moschandreas et al., 1985.)

Carbon monoxide from improperly operating furnaces has long been a serious concern. In numerous instances, people have died from furnace malfunction. More recently, chronic low levels of CO pollution have been recognized. Gas ranges, ovens, pilot lights, gas and kerosene space heaters, and cigarette smoke are sources of CO (Table 9-6).

Nitrogen oxide sources are also shown in Table 9-6. NO$_2$ levels have been found to range from 70 $\mu g/m^3$ in air-conditioned houses with electric ranges to 182 $\mu g/m^3$ in non-air-conditioned houses with gas stoves (Hosein et al., 1985). The latter value is quite high in comparison to the national ambient air quality limits. SO$_2$ levels were found to be very low in all houses investigated.

Over 800 volatile organic compounds (VOCs) have been identified in indoor air (Hines et al., 1993). Aldehydes, alkanes, alkenes, ethers, ketones, and polynuclear aromatic hydrocarbons (PAHs) are among them. Although they are not all present all the time, frequently there are several present at the same time. Typical sources of these compounds are listed in Table 9-7.

Between 1979 and 1987, the EPA investigated personal exposures of the general public to VOCs. These studies, titled the Total Exposure Assessment Methodology (TEAM), revealed that personal exposures exceeded median outdoor air concentrations

TABLE 9-7
Common volatile organic compounds and their sources[a]

Volatile organic compounds	Major indoor sources of exposure
Acetaldehyde	Paint (water-based), sidestream smoke
Alcohols (ethanol, isopropanol)	Spirits, cleansers
Aromatic hydrocarbons (ethylbenzene toluene, xylenes, trimethylbenzenes)	Paints, adhesives, gasoline, combustion sources
Aliphatic hydrocarbons (octane, decane, undecane)	Paints, adhesives, gasoline, combustion sources
Benzene	Sidestream smoke
Butylated hydroxytoluene (BHT)	Urethane-based carpet cushions
Chloroform	Showering, washing clothes, washing dishes
p-Dichlorobenzene	Room deodorizers, moth cakes
Ethylene glycol	Paints
Formaldehyde	Sidestream smoke, pressed wood products, photocopier
Methylene chloride	Paint stripping, solvent use
Phenol	Vinyl flooring
Styrene	Smoking, photocopier
Terpenes (limonene, α-pinene)	Scented deodorizers, polishes, fabric softeners
Tetrachloroethylene	Wearing/storing dry-cleaned clothes
Tetrahydrofuran	Sealer for vinyl flooring
Toluene	Photocopier, sidestream smoke, synthetic carper fiber
1,1,1-Trichlorotheane	Aerosol sprays, solvents

[a]Compiled from Tucker, 2001, and Wallace, 2001

by a factor of 2 to 5 for nearly all of the 19 VOCs investigated. Traditional sources (automobiles, industry, petrochemical plants) contributed only 20 to 25 percent of the total exposure to most of the target VOCs (Wallace, 2001).

Formaldehyde (CH_2O) has been singled out as one of the more prevalent, as well as one of the more toxic, compounds (Hines et al., 1993). Formaldehyde may not be generated directly by the activity of the homeowner. It is emitted by a variety of consumer products and construction materials including pressed wood products, insulation materials [urea-formaldehyde foam insulation (UFFI) in trailers has been particularly suspect], textiles, and combustion sources. In a composite of several studies, CH_2O concentrations ranged from 0.01 to 5.52 ppm, with median concentration of approximately 0.18 ppm (Godish, 1989). The highest values were for manufactured homes and conventional houses in cold climates (examples included Minnesota and Indiana). For comparison, the American Society of Heating, Refrigeration and Air Conditioning Engineers (ASHRAE, 1981) set a guideline concentration of 0.1 ppm.

Unlike the other air pollution sources that continue to emit as long as there is anthropogenic activity (or in the case of radon, for geologic time), CH_2O is not regenerated unless new materials are brought into the residence. If the house is ventilated over a period of time, the concentration will drop.

The primary source of heavy metals indoors is from infiltration of outdoor air and soil and dust that is tracked into the building. Arsenic, cadmium, chromium, mercury, lead, and nickel have been measured in indoor air. Lead and mercury may be generated from indoor sources such as paint. Old lead paint is a source of particulate lead as it is abraded or during removal. Mercury vapor is emitted from latex-based paints that contain diphenyl mercury dodecenyl succinate to prevent fungus growth.

Although little or no effort has been exerted to reduce or eliminate the danger from ranges, ovens, and so on, the public has come to expect that the recreational habits of smokers should not interfere with the quality of the air others breathe. The results of a general ban on cigarette smoking in one office are shown in Table 9-8. Smokers were allowed to smoke only in the designated lounge area. Period 1 was prior to the implementation of the new policy. It is obvious that the new policy had a positive effect outside of the lounge. On the other hand, respirable particulate matter (RSP) was found to increase with one smoker and to rise dramatically with two.

Indoor tobacco smoking is of particular concern because of the carcinogenic properties of the smoke. While *mainstream smoking* (taking a puff) exposes the smoker to large quantities of toxic compounds, the smoldering cigarette in the ashtray (*sidestream smoke*) adds a considerable burden to the room environment. Table 9-9 illustrates the emission rates of mainstream and sidestream smoke.

In the early 1990s and in 2002, the U.S. Centers for Disease Control and Prevention (CDC) tested nonsmokers for levels of cotinine, a product of nicotine metabolism. The 2002 serum levels of cotinine were 75 percent less in adults and 68 percent less in children than 10 years previously. CDC attributed this dramatic decrease to restrictions to reduce secondhand smoke. Yet, more needs to be done. The levels in children were more than twice those of nonsmoking adults (U.S. CDC, 2005).

TABLE 9-8

Mean respirable particulates (RSP), and carbon monoxide measured on a test floor

	RSP ($\mu g/m^3$)	Carbon monoxide ($\mu g/m^3$)
Period 1	26	1,908
Period 2	18	1,245

(*Data Source:* Lee et al., April 1985.)

TABLE 9-9
Emission of chemicals from mainstream and sidestream smoke

Chemicals	Mainstream (μg/cigarette)	Sidestream (μg/cigarette)
Particulates		
Aniline	0.36	16.8
Benzo (a) pyrene	20–40	68–136
Methyl naphthalene	2.2	60
Naphthalene	2.8	4.0
Nicotine	100–2,500	2,700–6,750
Nitrosonornicotine	0.1–0.55	0.5–2.5
Pyrene	50–200	180–420
Total phenols	228	603
Total suspended particles	36,200	25,800
Gas and Vapor Phase		
Acetaldehyde	18–1,400	40–3,100
Acetone	100–600	250–1,500
Acrolein	25–140	55–130
Ammonia	10–150	980–150,000
Carbon dioxide	20,000–60,000	160,000–480,000
Carbon monoxide	1,000–20,000	25,000–50,000
Dimethylnitrosamine	10–65	520–3,300
Formaldehyde	20–90	1,300
Hydrogen cyanide	430	110
Methyl chloride	650	1,300
Nitric oxide	10–570	2,300
Nitrogen dioxide	0.5–30	625
Nitrosopyrolidine	10–35	270–945
Pyridine	9–93	90–930

(*Data Sources:* HEW, 1979; Hoegg, 1972; Wakeham, 1972.)

Bacteria, viruses, fungi, mites, and pollen are collectively referred to as *bioaerosols*. They require a reservoir (for storage), an amplifier (for reproduction), and a means of dispersal. Most bacteria and viruses in indoor air come from humans and pets. Other microorganisms and pollen are introduced from the ambient air through either natural ventilation or through the intakes of building air handling systems. Humidifiers, air-conditioning systems, and other places where water accumulates are potential reservoirs for bioaerosols.

Radon is not regulated as an ambient air pollutant but has been found in dwellings at alarmingly high concentrations. We will address the radon issue in depth in Chapter 14, which can be found at the text's website: www.mhhe.com/davis. Suffice it to say at this juncture that radon is a radioactive gas that emanates from natural geologic formations and, in some cases, from construction materials. It is not generated from the activities of the householder, unlike the pollutants discussed above.

It is doubtful that there will be any regulatory effort to reduce the emissions of indoor air pollutants in the near future. Thus the house or apartment dweller has little recourse other than to replace gas appliances, remove or cover formaldehyde sources, and put out the smokers.

Acid Rain

Unpolluted rain is naturally acidic because CO_2 from the atmosphere dissolves to a sufficient extent to form carbonic acid (see Chapter 5). The equilibrium pH for pure rainwater is about 5.6. Measurements taken over North America and Europe have revealed lower pH values. In some cases individual readings as low as 3.0 have been recorded. The average pH in rain weighted by the amount of precipitation over the United States and lower Canada in 1997 is shown in Figure 7-15.

Chemical reactions in the atmosphere convert SO_2, NO_x, and volatile organic compounds (VOCs) to acidic compounds and associated oxidants (Figure 9-7). The primary conversion of SO_2 in the eastern United States is through the aqueous phase reaction with hydrogen peroxide (H_2O_2) in clouds. Nitric acid is formed by the reaction of NO_2 with OH radicals formed photochemically. Ozone is formed and then protected by a series of reactions involving both NO_x and VOCs.

As discussed in Chapter 7, the concern about acid rain relates to potential effects of acidity on aquatic life, damage to crops and forests, and damage to building materials. Lower pH values may affect fish directly by interfering with reproductive cycles or by releasing otherwise insoluble aluminum, which is toxic. Dramatic dieback of trees in Central Europe has stimulated concern that similar results could occur in North America. It is hypothesized that the acid rain leaches calcium and magnesium from the soil (see Figure 6-10). This lowers the molar ratio of calcium to aluminum which, in turn, favors the uptake of aluminum by the fine roots, that ultimately leads to their deterioration.

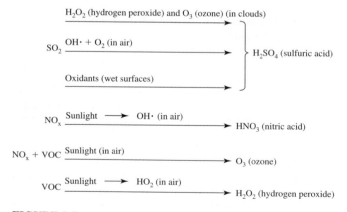

FIGURE 9-7
Acid rain precursors and products.

In 1980, Congress authorized a 10-year study to assess the causes and effects of acidic deposition. This study was titled the National Acid Precipitation Assessment Program (NAPAP). In September 1987, the NAPAP released an interim report that indicated that acidic precipitation appeared to have no measurable and consistent effect on crops, tree seedlings, or human health, and that a small percentage of lakes across the United States were experiencing pH values lower than 5.0 (Lefohn and Krupa, 1988). On the other hand, oxidant damage was measurable.

Approximately 70 percent of the SO_2 emissions in the United States are attributable to electric utilities. In order to decrease the SO_2 emissions, the Congress developed a two-phase control program under Title IV of the 1990 Clean Air Act Amendments. Phase I sets emission allowances for 110 of the largest emitters in the Eastern half of the U.S. Phase II will include smaller utilities. The utilities may buy or sell allowances. Each allowance is equal to about 1 Mg of SO_2 emissions. If a company does not expend its maximum allowance, it may sell it to another company. This program is called a *market-based system*. As a result of this program utility emissions have decreased by 9 Tg.

In 2003, the EPA reported on the long-term response of surface water chemistry to the 1990 Clean Air Act Amendments (U.S. EPA, 2003a). Eighty-one selected sites in the Northeast and Upper Midwest have been monitored for acidity since the early 1980s. The EPA's estimate of changes in the number and proportion of acidic surface waters is summarized in Table 9-10. In two areas, the New England lakes and the Blue Ridge Province streams, there is little evidence of reduction in acidity over the last decade. Sulfate levels have decreased significantly while nitrate levels have not changed appreciably. The widespread decrease in sulfate concentration parallels the general decrease in national emissions of sulfur dioxide since 1980. The EPA concluded from its analysis that surface waters have responded relatively rapidly to the decline in sulfate deposition and that additional reductions in deposition will result in additional declines in sulfate concentration.

TABLE 9-10

Estimates of change in number and proportion of acidic surface waters in acid-sensitive regions of the North and East[a]

Region	Population or size	Number acidic in past surveys[b]	Estimated number currently acidic	Percent Change
New England	6,834 lakes	386	374	−3
Adirondacks	1,830 lakes	238	149	−38
Northern Appalachians	42,426 km^2	5,014	3,600	−28
Blue Ridge	32,687 km^2	1,634	1,634	0
Upper Midwest	8,574 lakes	800	251	−68

[a]Adapted from U.S. EPA, 2003a.
[b]Survey dates range from 1984 in Upper Midwest to 1993–1994 in the Northern Appalachians.

In its report EPA also noted that, in many cases, sites that are not chronically acidic do undergo short-term episodic acidification during spring snow melt or during intense rain events.

Ozone Depletion

Without ozone, every living thing on the earth's surface would be incinerated. (On the other hand, as we have already noted, ozone can be lethal.) The presence of ozone in the upper atmosphere (20 to 40 km and up) provides a barrier to ultraviolet (UV) radiation. The small amounts that do seep through provide you with your summer tan. Too much UV will cause skin cancer. Although oxygen also serves as a barrier to UV radiation, it absorbs only over a narrow band centered at a wavelength of 0.2 μm. The photochemistry of these reactions is shown in Figure 9-8. The M refers to any third body (usually N_2).

In 1974, Molina and Rowland (1974) revealed a potential air pollution threat to this protective ozone shield. It is noteworthy that they, along with Paul Crutzen, jointly received the Nobel Prize in chemistry for their research. They hypothesized that chlorofluorocarbons (CF_2Cl_2 and $CFCl_3$—often abbreviated as CFC), which are used as aerosol propellants and refrigerants, react with ozone (Figure 9-9). The frightening aspects of this series of reactions are that the chlorine atom removes ozone from the system, and that the chlorine atom is continually recycled to convert more ozone to oxygen. It has been estimated that a 5 percent reduction in ozone could result in nearly a 10 percent increase in skin cancer (ICAS, 1975). Thus, CFCs that are rather inert compounds in the lower atmosphere become a serious air pollution problem at higher elevations.

By 1987, the evidence that CFCs destroy ozone in the stratosphere above Antarctica every spring had become irrefutable. In 1987, the ozone hole was larger than ever. More than half of the total ozone column was wiped out and essentially all ozone disappeared from some regions of the stratosphere.

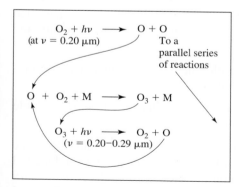

FIGURE 9-8
Photoreactions of ozone.

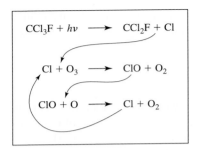

FIGURE 9-9
Ozone destruction by chlorofluoromethane.

Research confirmed that the ozone layer, on a worldwide basis, shrunk approximately 2.5 percent in the preceding decade (Zurer, 1988). Initially, it was believed that this phenomenon was peculiar to the geography and climatology of Antarctica and that the warmer northern hemisphere was strongly protected from the processes that lead to massive ozone losses. Studies of the North Pole stratosphere in the winter of 1989 revealed that this is not the case (Zurer, 1989).

In September 1987, the Montreal Protocol on Substances That Deplete the Ozone Layer was developed. The Protocol, which was initially ratified by 36 countries and became effective in January 1989, proposed that CFC production first be frozen and then reduced 50 percent by 1998. Yet, under the terms of the Protocol, the chlorine content of the atmosphere would continue to grow because the fully halogenated CFCs have such long lifetimes in the atmosphere. CF_2Cl_2, for example, has a lifetime of 110 years (Reisch and Zurer, 1988). Eighty countries met at Helsinki, Finland, in the spring of 1989 to assess new information. The delegates gave their unanimous assent to a five-point "Helsinki Declaration":

1. All join the 1985 Vienna Convention for the Protection of the Ozone Layer and the follow-up Montreal Protocol.

2. Phase out production and consumption of ozone-depleting CFCs no later than 2000.

3. Phase out production and consumption as soon as feasible of halons and such chemicals as carbon tetrachloride and methyl chloroform that also contribute to ozone depletion.

4. Commit themselves to accelerated development of environmentally acceptable alternative chemicals and technologies.

5. Make relevant scientific information, research results, and training available to developing countries (Sullivan, 1989).

The Montreal Protocol was strengthened in 1990, 1992, 1997, and 1999. The current terms of the treaty ban production of CFCs, carbon tetrachloride, and methyl chloroform as of January 1996. A ban on halon production took effect in January 1995 (Zurer, 1994). As of September 2002, 183 countries have ratified the Protocol (UNDP, 2005).

A number of alternatives to the fully chlorinated and, hence, more destructive CFCs have been developed. The two groups of compounds that emerged as significant replacements for the CFCs are hydrofluorocarbons (HFCs) and hydrochlorofluorocarbons (HCFCs). In contrast to the CFCs, HFCs and HCFCs contain one or more C-H bonds. This makes them susceptible to attack by OH radicals in the lower atmosphere. Because HFCs do not contain chlorine, they do not have the ozone depletion potential associated with the chlorine cycle shown in Figure 9-9. Although HCFCs contain chlorine, this chlorine is not transported to the stratosphere because OH scavenging in the troposphere is relatively efficient.

The implementation of the Montreal Protocol appears to be working. The use of CFCs has been reduced to one-tenth of the 1990 levels (UN, 2005). Total tropospheric chlorine from the long- and short-lived chlorocarbons was about 5 percent lower in 2000 than that observed at its peak in 1992–1994. The rate of change in 2000 was about -22 parts per trillion per year. Total chlorine from CFCs is no longer increasing, in contrast to the slight increase noted in 1998. Total tropospheric bromine from halons continues to increase at about 3 percent per year, which is about two-thirds of the 1996 rate (UNEP/WHO, 2002).

The issues of ozone depletion and climate change are interconnected. As the atmospheric abundance of CFCs declines, their contribution to global warming will decline. On the other hand, the use of HFCs and HCFCs as substitutes for CFCs will contribute to increases in global warming. Because ozone depletion tends to cool the earth's climate system, recovery of the ozone layer will tend to warm the climate system (UNEP/WHO, 2002).

Global Warming

Scientific Basis.   The case for global warming has grown very strong over the last two decades. As shown in Figure 9-10, the 5-year running average temperature in 2000 was almost 0.6°C above the 1951–1980 average (Hansen and Sato, 2004). Mann and Jones (2003) have compiled proxy temperature data from sediments, ice cores, and tree-ring temperature reconstruction over the past two millennia. Their research shows (Figure 9-11) the average global surface temperature has been increasing for the last 100 years and was higher in 2000 than in any time in the past 2,000 years.

To understand the physics of global warming we will use the simplest model of energy balance. It does not take into account location on the planet, time, precipitation, wind, ocean currents, soil moisture, or any of a number of other variables. The model is a simple radiation balance based on the principles described in Chapter 2. It equates the solar energy absorbed by the earth from the sun with the energy radiated back into space from the earth.

We begin with a definition of the *solar constant*. The solar constant is the average annual intensity of the radiation that is intercepted by the cross section of a sphere equivalent to the earth's diameter, normal to the incident radiation just outside of the earth's atmosphere (Figure 9-12). The solar constant has been evaluated over many years and, as a working estimate, it may be taken to be 1,400 W/m^2 (various sources report values from 1,379 to 1,396 W/m^2). The rate at which solar energy is radiated on the earth is the product of the flux of energy (W/m^2) and the area of the intercepting cross section.

$$E = S\pi r^2 \qquad (9\text{-}19)$$

where E = energy intercepted by earth, W
S = solar constant, W/m^2
r = radius of earth, m)

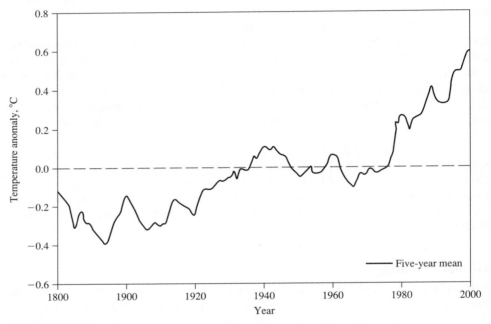

FIGURE 9-10
Global average surface temperature. Temperature anomaly is departure above and below the 1951–1980 average temperature, shown by dashed line. (*Source:* Hansen and Sato, 2004.)

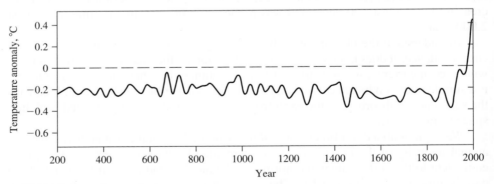

FIGURE 9-11
Global average surface temperature reconstruction. Temperature anomaly is departure from 1961–1990 instrumental reference period, for which the average is shown by dashed line. (*Source:* Mann and Jones, 2003.)

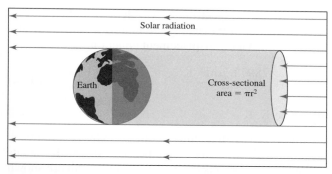

FIGURE 9-12
Cross-sectional area of sphere that intercepts incident radiation from the sun outside of the earth's atmosphere. The earth's radius is r. (*Source:* Davis & Masten, 2009.)

A fraction of the radiation that reaches the earth is reflected back into space. The ratio of the radiation reflected by an object to that absorbed by it is called the *albedo*. The earth's albedo is taken to be 0.3. The energy absorbed by earth is then

$$E_{abs} = (1 - \alpha)S\pi r^2 \qquad (9\text{-}20)$$

where α = albedo

We assume that the earth radiates as a blackbody (see Chapter 2) at a temperature T. The energy emitted by a unit area in a unit time is given by Equation 2-57. The total energy emitted by the surface of the earth is

$$E_{emit} = \sigma T_e^4 4\pi r^2 \qquad (9\text{-}21)$$

where σ = Stefan-Boltzman constant = 5.67×10^{-8} W/m$^2 \cdot$ K^4
T_e = earth's blackbody temperature, K
$4\pi r^2$ = surface area of sphere

If we assume steady-state conditions, that is, that over the millennia the earth's temperature has not changed appreciably with time, we can assume:

$$E_{abs} = E_{emit} \qquad (9\text{-}22)$$

and solve for the earth's blackbody temperature:

$$T_e \approx \left[\frac{(1 - \alpha)S}{4\sigma} \right]^{1/4} \qquad (9\text{-}23)$$

$$\approx \left[\frac{(1 - 0.3)(1{,}400 \text{ W/m}^2)}{4(5.67 \times 10^{-8} \text{ W/m}^2 \cdot \text{K}^4)} \right]^{1/4}$$

$$\approx 256.3 \text{ or } 256 \text{ K or } -16.6 \text{ or } -17°\text{C}$$

This result is at great variance from the actual value of the earth's average surface temperature of 288 K (+15°C). The actual temperature differs from the blackbody temperature because of the *greenhouse effect*. To understand the greenhouse effect we must first review the relationship between the spectrum of wavelengths radiated by an object and its temperature. The wavelength of the energy emitted by a blackbody at a given temperature can be estimated with Wien's displacement equation:

$$\lambda_{max} = \frac{2897.8 \ \mu m \cdot K}{T} \tag{9-24}$$

where T = absolute temperature of body, K

The sun is assumed to be a blackbody with a temperature of 6,000 K and a peak intensity at about 0.5 μm. The earth, with a blackbody temperature of 288 K, has its peak at about 10 μm. The blackbody emission spectra for the sun and earth are shown in Figure 9-13a. The spectra for the sun shows incoming "short-wave" radiation. The spectra for the earth shows outgoing "long-wave" radiation. Note that the abscissa is a logarithmic scale.

As radiant energy enters our atmosphere, it is affected by aerosols and atmospheric gases. Some of the constituents scatter the radiation by reflection, some stop it by adsorption, and some let is pass unchanged. The key phenomenon of interest in causing the greenhouse effect is the ability of gases to absorb radiant energy. As the atoms in gas vibrate and rotate, they absorb and radiate energy in specific wavelengths. If the frequency of the molecular oscillations is close to the frequency of the passing radiant energy, the molecule can absorb that energy. The absorption occurs over a limited range of frequencies. It is different for each molecule. A plot of the percent absorption of solar radiation versus wavelength is called the *absorption spectrum* (Figure 9-13b). The sum of the absorption of the gases at ground level is shown in the bottom frame of Figure 9-13b. The shaded areas show radiation that is absorbed. The unshaded areas show radiation that is transmitted. These areas are often referred to as radiation "windows".

From Figure 9-13b, it is evident that essentially all of the incoming solar radiation at wavelengths in the ultraviolet range ($< 0.3 \ \mu$m) is absorbed by oxygen and ozone. This absorption occurs in the stratosphere. It shields us from harmful ultraviolet radiation.

At the other end of the spectrum, radiatively active gases that absorb at wavelengths greater than 4 μm are called *greenhouse gases* (GHGs). This absorption heats the atmosphere, which radiates energy back to earth as well as into space. The GHGs act much like the glass on a greenhouse (thus, the name *greenhouse gases*): They let in short-wave (ultraviolet) radiation from the sun that heats the ground surface, but restrict the loss of heat by radiation from the ground surface. The more GHGs in the atmosphere, the more effective it is in restricting the outflow of long-wave (infrared) radiation. These greenhouse gases act as a blanket that raises the earth's temperature above the 256 K calculated from the radiation balance.

To elaborate on the radiation balance, we must take into account reflection by clouds and aerosols, evapotranspiration, latent heat release, and convective heat transfer. The simple global mean energy balance shown in Figure 9-14 summarizes the major energy flows.

The hypothesis is that increasing levels of certain gases (the so-called greenhouse gases, GHGs) leads to global warming. Unlike ozone, the greenhouse gases are relatively transparent to short-wave ultraviolet light from the sun. They do, however,

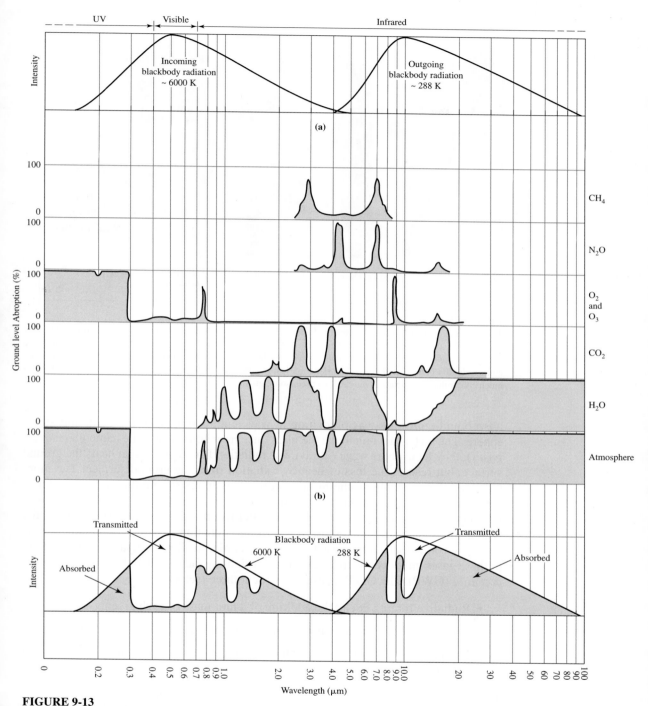

FIGURE 9-13

(a) Blackbody radiation curves for the sun (6000 K) and earth (288 K) (b) Absorption curves for various gases. The bottom frames show a total atmospheric absorption and the overlay of absorption on the blackbody radiation. The shaded areas depict absorption. The unshaded areas depict transmission.

617

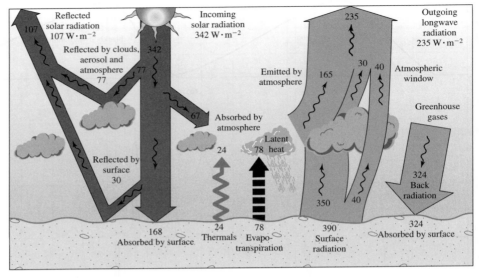

FIGURE 9-14

The earth's annual and global mean energy balance. Units are W/m². Of the incoming solar radiation, 49 percent (l68 W/m²) is absorbed by the surface. That heat is returned to the atmosphere as sensible heat, as evapotranspiration (latent heat), and as thermal infrared radiation. Most of this radiation is absorbed by the atmosphere, which in turn emits radiation both up and down. The radiation lost to space comes from cloud tops and atmospheric regions much colder than the surface. This causes a greenhouse effect.

absorb and emit long-wave radiation at wavelengths typical of the earth and atmosphere. The GHGs act much like the glass on a greenhouse (thus, the name *greenhouse gases*): they let in short-wave (ultraviolet) radiation from the sun that heats the ground surface, but restrict the loss of heat by radiation from the ground surface. The more GHGs in the atmosphere, the more effective it is in restricting the outflow of long-wave (infrared) radiation. This is called *radiative forcing*. CO_2 has been identified as the major GHG because of its abundance and its strong absorption spectrum in the region where the earth emits most of its infrared radiation.

Gases other than CO_2 also act as GHGs. In order to allow comparison of the effect of these gases on global warming, we use a weighting factor called the *global warming potential* (GWP). The GWP takes into account three factors:

- Radiative forcing due to the addition of a unit mass of each greenhouse gas.

- Estimates of the rate at which the injected unit mass decays over time.

- Estimates of the cumulative radiative forcing that the unit mass addition will have over time.

The GWPs of a few selected chemical species are listed in Table 9-11.

Since the first systematic measurements were made at Mauna Loa in Hawaii in 1958, CO_2 levels have risen from 316 ppm to 389 ppm (Keeling and Whorf, 2005; Pittman, 2011). From analysis of air trapped in ice cores in Greenland and Antarctica,

TABLE 9-11
Global warming potentials (GWPs) relative to carbon dioxide over 20-year time period

Chemical species	Lifetime, y	Global warming potential, kg of CO_2/kg of gas
Carbon dioxide (CO_2)	30–200	1
Methane (CH_4)	12	62
Nitrous oxide (N_2O)	114	275
CFC-12 (CF_2Cl_2)	100	10,200
HCFC-22 (CHF_3Cl)	12	4,800
Tetrafluoromethane (CF_4)	50,000	3,900
Sulfur hexafluoride (SF_6)	3,200	15,100

Source: IPCC, 2000

we know that preindustrial levels of CO_2 were about 280 ppm. The ice core records indicate that, over the last 160,000 years, no fluctuations of CO_2 have been larger than 70 ppm (Hileman, 1989) and that the current concentrations are higher than any level attained in the past 650,000 years (Hileman, 2005). It is estimated that the atmospheric CO_2 concentration has increased 30 percent since 1750 and that the present concentration has not been exceeded during the past 420,000 years and likely not during the past 20 million years (IPCC, 2001a). Several gases have been recognized as contributing to the greenhouse effect. Methane (CH_4), nitrous oxide (N_2O), and CFCs are similar to CO_2 in their radiative behavior. Even though their concentrations are much lower than CO_2, these gases are now estimated to trap about 60 percent as much long-wave radiation as CO_2.

In 2007, the United Nations Intergovernmental Panel on Climate Change* (IPCC) declared (IPCC, 2007a): "Most of the observed increase in globally averaged temperatures since the mid-20th century is *very likely*[†] due to the observed increase in greenhouse gas concentrations. . . . discernable human influences now extend to other aspects of climate, including ocean warming, continental-average temperatures, temperature extremes, and wind patterns."

About three-quarters of the anthropogenic emissions of CO_2 that have been added to the atmosphere over the past 20 years is attributed to the combustion of fossil fuel (IPCC, 2001a). In the 1980s, massive deforestation was identified as a possible contribution. Both the burning of timber and the release of carbon from bacterial degradation contribute. Perhaps more important, deforestation removes a mechanism for removing CO_2 from the atmosphere (commonly referred to as a *sink*). In normal respiration, green plants utilize CO_2 much as a carbon source. This CO_2 is fixed in the biomass by photosynthetic processes. A rapidly growing rain forest can fix between 1 and 2 kg per year of carbon per square meter of ground surface. Cultivated fields, in contrast, fix only about 0.2 to 0.4 kg/m^2—and this amount is recycled by bioconsumption and conversion to CO_2.

---

*The IPCC is composed of over 673 scientists and 420 expert reviewers from around the world.
[†]*Very likely* is defined by the IPCC as a "9 out of 10 chance of being correct."

Impacts. Attempts to understand the consequences of global warming are based on mathematical models of the global circulation of the atmosphere and oceans. Using best estimates, the IPCC estimates that the globally averaged surface temperature will rise 1.8 to 4.0°C by 2100 (IPCC, 2007a). To date these models have a "good news-bad news" conclusion. Based on the 1.4 to 5.8°C rise in global temperature, the following is predicted for North America (IPCC, 2007b):

1. A decrease in heating costs (partly offset by increased air-conditioning cost).

2. Potential increased food production in areas of Canada and an increase in warm-temperature mixed forest production with modest warming; with severe warming, crop production could possibly become a net loss.

3. Much easier navigation in the Arctic seas.

4. Drier crop conditions in the Midwest and Great Plains, requiring more irrigation.

5. Warming in western mountains is projected to cause decreased snowpack, more winter flooding, and reduced summer flows exacerbating competition for overall water resources.

6. Increasing impacts on forests from pests, diseases, and fire.

7. Widespread melting of permanently frozen ground with adverse effects on animal and plant life as well as building technology in Alaska and northern Canada.

8. A rise in sea level between 0.18 and 0.57 m that would result in an increase in the severity of flooding, damage to coastal structures, destruction of wetlands, and saltwater intrusion into drinking water supplies in coastal areas, particularly in Florida and much of the Atlantic coast.*

As shown in Table 9-12, on a global scale the impacts will range from severe to catastrophic.

Kyoto Protocol. The framework convention for the Protocol was signed in 1992. In 1997 the Protocol that set targets for industrialized countries to reduce their GHG emissions was finalized. To become legally binding two conditions had to be fulfilled:

- Ratification by 55 countries

- Ratification by nations accounting for at least 55 percent of emissions from 38 industrialized countries plus Belarus, Turkey, and Kazakhstan

---

*Contraction of the Greenland ice sheet is projected to contribute to sea level rise after 2100 even if radiative forcing is stabilized at 1.8°C above the current value. The Greenland ice sheet would be virtually eliminated and the resulting contribution to sea level rise would be about 7 m!

TABLE 9-12

Key impacts as a function of increasing global average temperature change

Global Mean Annual Temperature Change Relative to 1980–1999 (°C)

	0	1	2	3	4	5°C

WATER
- Increased water availability in moist tropics and high latitudes – – – – – – – – – – – – – – – – →
- Decreasing water availability and increasing drought in mid-latitudes and semi-arid low latitudes – – – →
- Hundreds of millions of people exposed to increased water stress – – – – – – – – – – – – – – – →

ECOSYSTEMS
- —————— Up to 30% of species at ——————— Significant[†] extinctions – – – →
  increasing risk of extinction   around the globe
- Increased coral bleaching ———— Most corals bleached ———— Widespread coral mortality – – – →
- Terrestrial biosphere tends toward a net carbon source as:
  ~15% ——————— ~40% of ecosystems affected – – – – →
- Increasing species range shifts and wildfire risk
- Ecosystem changes due to weakening of the meridional – – →
  overturning circulation

FOOD
- Complex, localized negative impacts on small holders, subsistence farmers and fishers – – – – – – – →
- Tendencies for cereal productivity ———— Productivity of all cereals – – – – →
  to decrease in low latitudes   decreases in low latitudes
- Tendencies for some cereal productivity ——— Cereal productivity to
  to increase at mid- to high latitudes   decrease in some regions

COASTS
- Increased damage from floods and storms – – – – – – – – – – – – – – – – – – – →
- About 30% of
  global coastal – – – – – – – – – – – →
  wetlands lost[‡]
- Millions more people could experience – – – – – – – – – →
  coastal flooding each year

HEALTH
- Increasing burden from malnutrition, diarrheal, cardio-respiratory, and infectious diseases – – →
- Increased morbidity and mortality from heat waves, floods, and droughts – – – – – – – – – →
- Changed distribution of some disease vectors – – – – – – – – – – – – – – – →
- Substantial burden on health services – →

[†]Significant is defined here as more than 40%.

[‡]Based on average rate of sea level rise of 4.2 mm · year^{-1} from 2000 to 2080.

Note: The black lines link impacts, dotted arrows indicate impacts continuing with increasing temperature. Entries are placed so that the left hand side of text indicates approximate onset of a given impact. Adaptation to climate change is not included in these estimations. All entries are from published studies recorded in the chapters of the Assessment. Confidence levels for all statements are high.

Source: IPCC, 2007 b.

The first condition was met in 2002. Following the decision of the United States and Australia not to ratify, Russia's position became crucial to fulfill the second condition. On 18 November 2004, Russia ratified the Kyoto Protocol. It came into force 90 days later on 16 February 2005. At that time the targets for reducing emissions became binding on the countries that ratified the Protocol. The agreement set levels to reduce emissions by 5 percent from the 1990 baseline level. As of December 2005, 157 nations had ratified the accord and the United States remained unwilling to make any commitments to reduce greenhouse emissions (AP, 2005a).

Russia's ratification of the treaty sounded a bell for a new international financial market in which companies buy and sell what amounts to global-warming pollution permits. The Protocol mandates emission reductions only from industrialized countries. However, it allows the industrialized countries to finance projects that reduce their emissions in developing countries and, thus, to generate credits toward their quotas. The theory is that, because global warming is global, the atmosphere doesn't care where the emissions or the emission reductions occur. Because financing an emissions-reduction project in a developing country is cheaper than in an industrialized country, there is a great incentive to put together investment funds.

Although the United States did not a ratify the Protocol, 136 U.S. mayors representing more than 30 million people have signed an agreement to meet the goals spelled out in the treaty (AP, 2005b). On December 20, 2005, seven northeastern states (Connecticut, Delaware, Maine, New Hampshire, New Jersey, New York, and Vermont) signed an agreement to establish carbon dioxide emissions caps for electric utilities in their states (C&EN, 2006).

In 2005, Massachusetts petitioned the U.S. EPA to regulate CO_2 emissions from automobiles. EPA declined, saying that the Clean Air Act did not authorize it to issue mandatory regulations to address global climate change, that even if it was authorized, it would be unwise because the link between GHG and global warming is not unequivocally established, and that regulation of automobile emissions would be a piecemeal approach and would conflict with the president's comprehensive approach using additional support for technological innovation and nonregulatory programs. On April 2, 2007, the Supreme Court of the United States ruled that the harms associated with climate change are serious and well recognized, that while automobile emissions may not by itself reverse global warming, it does not follow that the court lacks jurisdiction to decide whether EPA has a duty to take steps to slow or reduce it. The court decided that because greenhouse gases fit well within the act's capacious definition of "air pollutant," EPA has statutory authority to regulate emission of such gases from new motor vehicles. The court further noted that under the act's clear terms, EPA can avoid promulgating regulations only if it determines that greenhouse gases do not contribute to climate change (Supreme Court, 2007).

A Rationale for Action.   While there is still considerable disagreement about the potential for global warming, the consequences of ignoring these trends are sufficiently dramatic that intensive research must continue in the decades to come. Even without the risks of climate change, improvements in energy efficiency to reduce greenhouse gas emissions are amply justified from two points of view: economics and sustainability. Higher energy efficiency will yield economic benefit in reducing the cost of electricity and transportation. Higher efficiency will contribute to sustainability by extending the availability of finite energy resources. The expectation of damages from climate change provides extra incentive for pursuing these programs vigorously.

9-7 AIR POLLUTION METEOROLOGY

The Atmospheric Engine

The atmosphere is somewhat like an engine. It is continually expanding and compressing gases, exchanging heat, and generally raising chaos. The driving energy for this unwieldy machine comes from the sun. The difference in heat input between the equator and the poles provides the initial overall circulation of the earth's atmosphere. The rotation of the earth coupled with the different heat conductivities of the oceans and land produce weather.

Highs and Lows. Because air has mass, it also exerts pressure on things under it. Like water, which we intuitively understand to exert greater pressures at greater depths, the atmosphere exerts more pressure at the surface than it does at higher elevations. The highs and lows depicted on weather maps are simply areas of greater and lesser pressure. The elliptical lines shown on more detailed weather maps are lines of constant pressure, or *isobars*. A two-dimensional plot of pressure and distance through a high- or low-pressure system would appear as shown in Figure 9-15.

The wind flows from the higher pressure areas to the lower pressure areas. On a nonrotating planet, the wind direction would be perpendicular to the isobars (Figure 9-16a). However, since the earth rotates, an angular thrust called the Coriolis effect is added to this motion. The resultant wind direction in the northern hemisphere

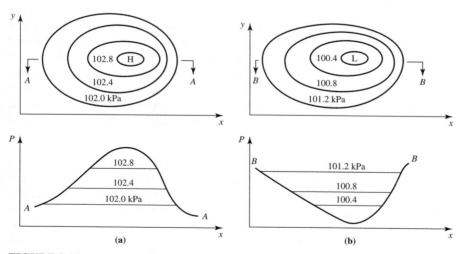

FIGURE 9-15
High (a) and low (b) pressure systems.

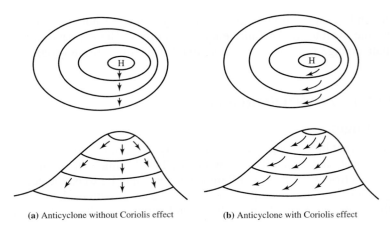

(**a**) Anticyclone without Coriolis effect   (**b**) Anticyclone with Coriolis effect

FIGURE 9-16
Wind flow due to pressure gradient.

is as shown in Figure 9-16*b*. The technical names given to these systems are *anticyclones* for highs and *cyclones* for lows. Anticyclones are associated with good weather. Cyclones are associated with foul weather. Tornadoes and hurricanes are the foulest of the cyclones.

Wind speed is in part a function of the steepness of the pressure surface. When the isobars are close together, the pressure gradient (slope) is said to be steep and the wind speed relatively high. If the isobars are well spread out, the winds are light or nonexistent.

Turbulence

Mechanical Turbulence.   In its simplest terms, we may consider turbulence to be the addition of random fluctuations of wind velocity (that is, speed and direction) to the overall average wind velocity. These fluctuations are caused, in part, by the fact that the atmosphere is being sheared. The shearing results from the fact that the wind speed is zero at the ground surface and rises with elevation to near the speed imposed by the pressure gradient. The shearing results in a tumbling, tearing motion as the mass just above the surface falls over the slower moving air at the surface. The swirls thus formed are called *eddies*. These small eddies feed larger ones. As you might expect, the greater the mean wind speed, the greater the mechanical turbulence. The more mechanical turbulence, the easier it is to disperse and spread atmospheric pollutants.

Thermal Turbulence.   Like all other things in nature, the rather complex interaction that produces mechanical turbulence is confounded and further complicated by a third party. Heating of the ground surface causes turbulence in the same fashion that heating the bottom of a beaker full of water causes turbulence. At some point below boiling, you can see density currents rising off the bottom. Likewise, if the earth's surface is heated strongly and in turn heats the air above it, thermal turbulence will be generated. Indeed, the "thermals" sought by glider pilots and hot air balloonists are these thermal currents rising on what otherwise would be a calm day.

The converse situation can arise during clear nights when the ground radiates its heat away to the cold night sky. The cold ground, in turn, cools the air above it, causing a sinking density current.

Stability

The tendency of the atmosphere to resist or enhance vertical motion is termed *stability*. It is related to both wind speed and the change of air temperature with height (*lapse rate*). For our purpose, we may use the lapse rate alone as an indicator of the stability condition of the atmosphere.

There are three stability categories. When the atmosphere is classified as *unstable,* mechanical turbulence is enhanced by the thermal structure. A *neutral* atmosphere is one in which the thermal structure neither enhances nor resists mechanical turbulence. When the thermal structure inhibits mechanical turbulence, the atmosphere is said to be *stable.* Cyclones are associated with unstable air. Anticyclones are associated with stable air.

Neutral Stability.   The lapse rate for a neutral atmosphere is defined by the rate of temperature increase (or decrease) experienced by a parcel of dry air that expands (or contracts) *adiabatically* (without the addition or loss of heat) as it is raised through the atmosphere. This rate of temperature decrease (dT/dz) is called the *dry adiabatic lapse rate.* It is designated by the Greek letter gamma (Γ). It has a value of approximately $-1.00°C/100$ m*. (Note that this is not a slope in the normal sense, that is, it is not dy/dx.) In Figure 9-17a, the dry adiabatic lapse rate of a parcel of air is shown as a dashed line and the temperature of the atmosphere (ambient lapse rate) is shown as a solid line. Since the ambient lapse rate is the same as Γ, the atmosphere is said to have a *neutral stability.*

Unstable Atmosphere.   If the temperature of the atmosphere falls at a rate greater than Γ, the lapse rate is said to be *superadiabatic,* and the atmosphere is unstable. Using Figure 9-17b, we can see that this is so. The actual lapse rate is shown by the solid line. If we capture a balloon full of polluted air at elevation A and adiabatically displace it 100 m vertically to elevation B, the temperature of the air inside the balloon will decrease from 21.15° to 20.15°C. At a lapse rate of $-1.25°C/100$ m, the temperature of the air outside the balloon will decrease from 21.15° to 19.90°C. The air inside the balloon will be warmer than the air outside; this temperature difference gives the balloon buoyancy. It will behave as a hot gas and continue to rise without any further mechanical effort. Thus, mechanical turbulence is enhanced and the atmosphere is unstable. If we adiabatically displace the balloon downward to elevation C, the temperature inside the balloon would rise at the rate of the dry adiabat. Thus, in moving 100 m, the temperature will increase from 21.15° to 22.15°C. The temperature outside the

---

*The value for Γ for dry air is 9.76°C/km. In practice, this usually is rounded to 10°C/km. The dry adiabatic lapse rate is a theoretical construct that is used to examine atmospheric behavior. The U.S Standard Atmosphere describes a more realistic average mean temperature profile. A discussion of the Standard Atmosphere may be found at the text website: www.mhhe.com/davis.

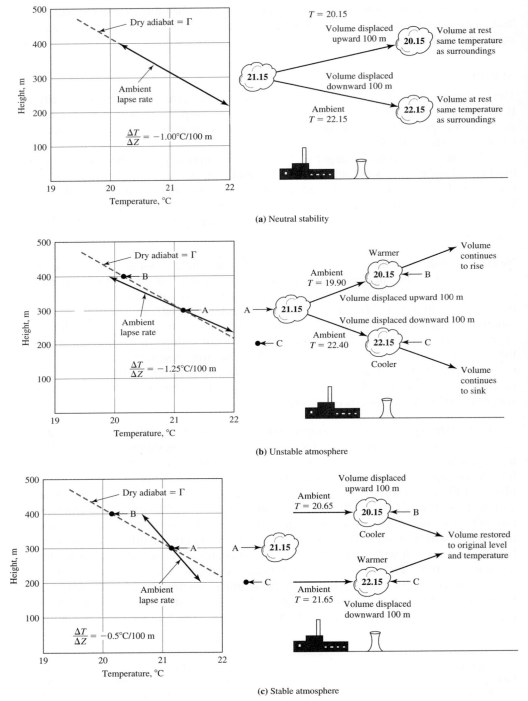

FIGURE 9-17
Lapse rate and displaced air volume. (*Source:* AEC, 1968.)

balloon will increase at the superadiabatic lapse rate to 22.40°C. The air in the balloon will be cooler than the ambient air and the balloon will have a tendency to sink. Again, mechanical turbulence (displacement) is enhanced.

Stable Atmosphere.   If the temperature of the atmosphere falls at a rate less than Γ, it is called *subadiabatic,* and the atmosphere is stable. If we again capture a balloon of polluted air at elevation A (Figure 9-17c) and adiabatically displace it vertically to elevation B, the temperature of the polluted air will decrease at a rate equal to the dry adiabatic rate. Thus, in moving 100 m, the temperature will decrease from 21.15° to 20.15°C as before. However, because the ambient lapse rate is $-0.5°C/100$ m, the temperature of the air outside the balloon will have dropped to only 20.65°C. Because the air inside the balloon is cooler than the air outside the balloon, the balloon will have a tendency to sink. Thus, the mechanical displacement (turbulence) is inhibited.

In contrast, if we displace the balloon adiabatically to elevation C, the temperature inside the balloon would increase to 22.15°C, while the ambient temperature would increase to 21.65°C. In this case, the air inside the balloon would be warmer than the ambient air and the balloon would tend to rise. Again, the mechanical displacement would be inhibited.

There are two special cases of subadiabatic lapse rate. When there is no change of temperature with elevation, the lapse rate is called *isothermal.* When the temperature increases with elevation, the lapse rate is called an *inversion.* The inversion is the most severe form of a stable temperature profile. It is often associated with restricted air volumes that cause air pollution episodes.

Example 9-4.   Given the following temperature and elevation data, determine the stability of the atmosphere.

Elevation, m	Temperature, °C
2.00	14.35
324.00	11.13

Solution.   Begin by determining the existing lapse rate:

$$\frac{\Delta T}{\Delta Z} = \frac{T_2 - T_1}{Z_2 - Z_1}$$

$$= \frac{11.13 - 14.35}{324.00 - 2.00} = \frac{-3.22}{322.00}$$

$$= -0.0100°C/m = -1.00°C/100 \text{ m}$$

Now we compare this with Γ and find that they are equal. Thus, the atmospheric stability is neutral.

Comment: Temperature measurements are typically not measured to two decimal places.

Plume Types. The smoke trail or plume from a tall stack located on flat terrain has been found to exhibit a characteristic shape that is dependent on the stability of the atmosphere. The six classical plumes are shown in Figure 9-18, along with the corresponding temperature profiles. In each case, Γ is given as a broken line to allow comparison with the actual lapse rate, which is given as a solid line. In the bottom three

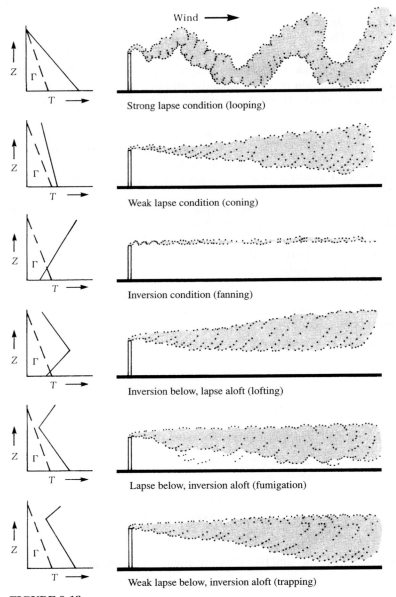

FIGURE 9-18
Six types of plume behavior. (*Source:* USAEC, 1955.)

cases, particular attention should be given to the location of the inflection point with respect to the top of the stack.

Terrain Effects

Heat Islands. A heat island results from a mass of material, either natural or anthropogenic, that absorbs and reradiates heat at a greater rate than the surrounding area. This causes moderate to strong vertical convection currents above the heat island. The effect is superimposed on the prevailing meteorological conditions. It is nullified by strong winds. Large industrial complexes and small to large cities are examples of places that may have a heat island.

Because of the heat island effect, atmospheric stability will be less over a city than it is over the surrounding countryside. Depending upon the location of the pollutant sources, this can be either good news or bad news. First, the good news: For ground level sources such as automobiles, the bowl of unstable air that forms will allow a greater air volume for dilution of the pollutants. Now the bad news: Under stable conditions, plumes from tall stacks would be carried out over the countryside without increasing ground level pollutant concentrations. Unfortunately, the instability caused by the heat island mixes these plumes to the ground level.

Land/Sea Breezes. Under a stagnating anticyclone, a strong local circulation pattern may develop across the shoreline of large water bodies. During the night, the land cools more rapidly than the water. The relatively cooler air over the land flows toward the water (a land breeze, Figure 9-19). During the morning the land heats faster than water. The air over the land becomes relatively warm and begins to rise. The rising air is replaced by air from over the water body (a sea or lake breeze, Figure 9-20).

The effect of the lake breeze on stability is to impose a surface-based inversion on the temperature profile. As the air moves from the water over the warm ground, it is

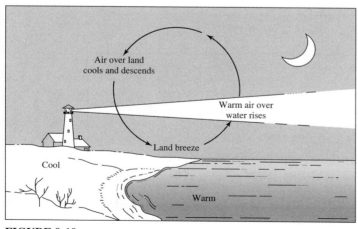

FIGURE 9-19
Land breeze during the night.

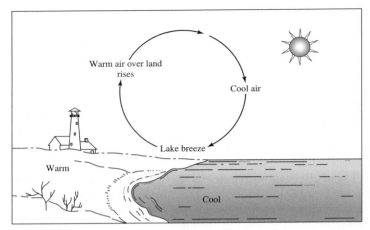

FIGURE 9-20
Lake breeze during the day.

heated from below. Thus, for stack plumes originating near the shoreline, the stable lapse rate causes a fanning plume close to the stack (Figure 9-21). The lapse condition grows to the height of the stack as the air moves inland. At some point inland, a fumigation plume results.

Valleys. When the general circulation imposes moderate to strong winds, valleys that are oriented at an acute angle to the wind direction channel the wind. The valley effectively peels off part of the wind and forces it to follow the direction of the valley floor (Figure 9-22).

Under a stagnating anticyclone, the valley will set up its own circulation. Warming of the valley walls will cause the valley air to be warmed. It will become more buoyant and flow up the valley. At night the cooling process will cause the wind to flow down the valley.

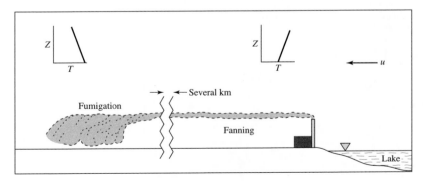

FIGURE 9-21
Effect of lake breeze on plume dispersion.

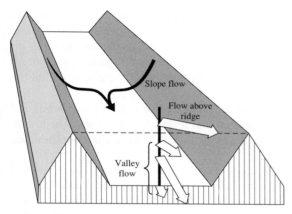

FIGURE 9-22
Idealized representation of the circulation that might be expected
in a typical valley on a clear night. (*Source:* AEC, 1968.)

Valleys oriented in the north-south direction are more susceptible to inversions
than level terrain. The valley walls protect the floor from radiative heating by the sun.
Yet the walls and floor are free to radiate heat away to the cold night sky. Thus, under
weak winds, the ground cannot heat the air rapidly enough during the day to dissipate
the inversion that formed during the night.

9-8   ATMOSPHERIC DISPERSION

Factors Affecting Dispersion of Air Pollutants

The factors that affect the transport, dilution, and dispersion of air pollutants can generally be categorized in terms of the emission point characteristics, the nature of the
pollutant material, meteorological conditions, and effects of terrain and anthropogenic
structures. We have discussed all of these except the source conditions. Now we wish
to integrate the first and third factors to describe the qualitative aspects of calculating
pollutant concentrations. We shall follow this with simple quantitative models for a
point source and a line source. More complex models for point sources (in rough terrain,
in industrial settings, or for long time periods), area sources, and mobile sources are
left for more advanced texts.

Source Characteristics.   Most industrial effluents are discharged vertically into the
open air through a stack or duct. As the contaminated gas stream leaves the discharge
point, the plume tends to expand and mix with the ambient air. Horizontal air movement will tend to bend the discharge plume toward the downwind direction. At some
point between 300 and 3,000 m downwind, the effluent plume will level off. While the
effluent plume is rising, bending, and beginning to move in a horizontal direction,
the gaseous effluents are being diluted by the ambient air surrounding the plume. As
the contaminated gases are diluted by larger and larger volumes of ambient air, they
are eventually dispersed toward the ground.

The plume rise is affected by both the upward inertia of the discharge gas stream and by its buoyancy. The vertical inertia is related to the exit gas velocity and mass. The plume's buoyancy is related to the exit gas mass relative to the surrounding air mass. Increasing the exit velocity or the exit gas temperature will generally increase the plume rise. The plume rise, together with the physical stack height, is called the *effective stack height.*

The additional rise of the plume above the discharge point as the plume bends and levels off is a factor in the resultant downwind ground level concentrations. The higher the plume rises initially, the greater distance there is for diluting the contaminated gases as they expand and mix downward.

For a specific discharge height and a specific set of plume dilution conditions, the ground level concentration is proportional to the amount of contaminant materials discharged from the stack outlet for a specific period of time. Thus, when all other conditions are constant, an increase in the pollutant discharge rate will cause a proportional increase in the downwind ground level concentrations.

Downwind Distance.   The greater the distance between the point of discharge and a ground level receptor downwind, the greater will be the volume of air available for diluting the contaminant discharge before it reaches the receptor.

Wind Speed and Direction.   The wind direction determines the direction in which the contaminated gas stream will move across local terrain. Wind speed affects the plume rise and the rate of mixing or dilution of the contaminated gases as they leave the discharge point. An increase in wind speed will decrease the plume rise by bending the plume over more rapidly. The decrease in plume rise tends to increase the pollutant's ground level concentration. On the other hand, an increase in wind speed will increase the rate of dilution of the effluent plume, tending to lower the downwind concentrations. Under different conditions, one or the other of the two wind speed effects becomes the predominant effect. These effects, in turn, affect the distance downwind of the source at which the maximum ground level concentration will occur.

Stability.   The turbulence of the atmosphere follows no other factor in power of dilution. The more unstable the atmosphere, the greater the diluting power. Inversions that are not ground based, but begin at some height above the stack exit, act as a lid to restrict vertical dilution.

Dispersion Modeling

General Considerations and Use of Models.   A dispersion model is a mathematical description of the meteorological transport and dispersion process that is quantified in terms of source and meteorologic parameters during a particular time. The resultant numerical calculations yield estimates of concentrations of the particular pollutant for specific locations and times.

To verify the numerical results of such a model, actual measured concentrations of the particular atmospheric pollutant must be obtained and compared with the calculated values by means of statistical techniques. The meteorological parameters required for use of the models include wind direction, wind speed, and atmospheric stability. In some models, provisions may be made for including lapse rate and vertical mixing height. Most models will require data about the physical stack height, the diameter of the stack at the emission discharge point, the exit gas temperature and velocity, and the mass rate of emission of pollutants.

Models are usually classified as either short-term or climatological models. Short-term models are generally used under the following circumstances: (1) to estimate ambient concentrations where it is impractical to sample, such as over rivers or lakes, or at great distances above the ground; (2) to estimate the required emergency source reductions associated with periods of air stagnations under air pollution episode alert conditions; and (3) to estimate the most probable locations of high, short-term, ground-level concentrations as part of a site selection evaluation for the location of air monitoring equipment.

Climatological models are used to estimate mean concentrations over a long period of time or to estimate mean concentrations that exist at particular times of the day for each season over a long period of time. Long-term models are used as an aid for developing emissions standards. We will be concerned only with short-term models in their most simple application.

Basic Point Source Gaussian Dispersion Model. The basic Gaussian diffusion equation assumes that atmospheric stability is uniform throughout the layer into which the contaminated gas stream is discharged. The model assumes that turbulent diffusion is a random activity and hence the dilution of the contaminated gas stream in both the horizontal and vertical direction can be described by the Gaussian or normal equation. The model further assumes that the contaminated gas stream is released into the atmosphere at a distance above ground level that is equal to the physical stack height plus the plume rise (ΔH). The model assumes that the degree of dilution of the effluent plume is inversely proportional to the wind speed (u). The model also assumes that pollutant material that reaches ground level is totally reflected back into the atmosphere like a beam of light striking a mirror at an angle. Mathematically, this ground reflection is accounted for by assuming a virtual or imaginary source located at a distance of $-H$ with respect to ground level, and emitting an imaginary plume with the same source strength as the real source being modeled. The same general idea can be used to establish other boundary layer conditions for the equations, such as limiting horizontal or vertical mixing.

The Model. We have selected the model* equation in the form presented by D. B. Turner (1967). It gives the ground level concentration (χ) of pollutant at a point (coordinates x and y) downwind from a stack with an effective height (H) (Figure 9-23). The standard deviation of the plume in the horizontal and vertical directions is designated by s_y and s_z, respectively. The standard deviations are functions of the

---

Note: Turner provides guidelines on the accuracy of this model. It is an estimating tool and not a definitive model to be used indiscriminately.

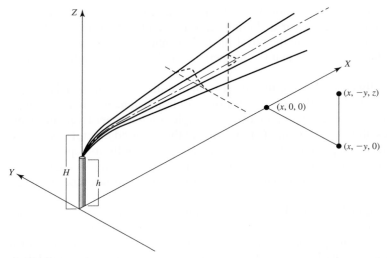

FIGURE 9-23
Plume dispersion coordinate system. (*Source:* Turner, 1967.)

downward distance from the source and the stability of the atmosphere. The equation is as follows:

$$\chi_{(x,y,0,H)} = \left[\frac{E}{\pi s_y s_z u}\right]\left\{\exp\left[-\frac{1}{2}\left(\frac{y}{s_y}\right)^2\right]\right\}\left\{\exp\left[-\frac{1}{2}\left(\frac{H}{s_z}\right)^2\right]\right\} \qquad (9\text{-}25)$$

where $\chi_{(x,y,0,H)}$ = downwind concentration at ground level, g/m^3
E = emission rate of pollutant, g/s
s_y, s_z = plume standard deviations, m
u = wind speed, m/s
$x, y, z,$ and H = distances, m
$\exp$ = exponential e such that terms in brackets immediately following are powers of e, that is, $e^{[\ ]}$ where e = 2.7182

The value for the effective stack height is the sum of the physical stack height (h) and the plume rise ΔH:

$$H = h + \Delta H \qquad (9\text{-}26)$$

ΔH may be computed from Holland's formula as follows (Holland, 1953):

$$\Delta H = \frac{v_s d}{u}\left\{1.5 + \left[2.68 \times 10^{-2}(P)\left(\frac{T_s - T_a}{T_s}\right)d\right]\right\} \qquad (9\text{-}27)$$

where v_s = stack velocity, m/s
d = stack diameter, m
u = wind speed, m/s
P = pressure, kPa
T_s = stack temperature, K
T_a = air temperature, K

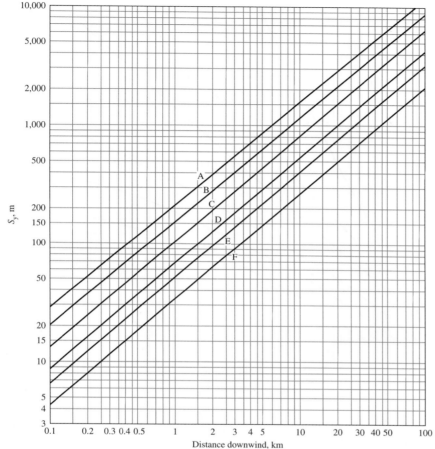

FIGURE 9-24
Horizontal dispersion coefficient. (*Source:* Turner, 1967).

The values of s_y and s_z depend upon the turbulent structure or stability of the atmosphere. Figures 9-24 and 9-25 provide graphical relationships between the downwind distance x in kilometers and values of s_y and s_z in meters. The curves on the two figures are labeled A through F. The label A refers to very unstable atmospheric conditions, B to unstable atmospheric conditions, C to slightly unstable conditions, D to neutral conditions, E to stable atmospheric conditions, and F to very stable atmospheric conditions. Each of these stability parameters represents an averaging time of approximately 3 to 15 min.

Other averaging times may be approximated by multiplying by empirical constants, for example, 0.36 for 24 hours. Turner presented a table and discussion that allows an estimate of stability based on wind speed and the conditions of solar radiation. This is given in Table 9-13.

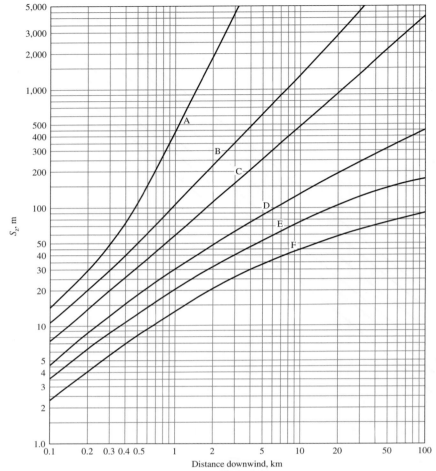

FIGURE 9-25
Vertical dispersion coefficient. (*Source:* Turner, 1967.)

For computer solutions of the dispersion model, it is convenient to have an algorithm to express the stability class lines in Figures 9-24 and 9-25. D. O. Martin (1976) has developed the following equations that provide an approximate fit.

$$s_y = ax^{0.894} \tag{9-28}$$

$$s_z = cx^d + f \tag{9-29}$$

where the constants a, c, d, and f are defined in Table 9-14. These equations were developed to yield s_y and s_z in meters for downwind distance x in kilometers.

As noted above, the wind speed varies with height. Unless the wind speed at the effective height of the plume (H) is known, the wind speed must be corrected to account for the change in speed with elevation. For elevations up to a few hundred meters, a

TABLE 9-13
Key to stability categories 🏳️

Surface wind speed (at 10 m) (m/s)	Day[a] Incoming solar radiation Strong	Moderate	Slight	Night[a] Thinly overcast or ≥ 1/2 Low cloud	≤ 3/8 Cloud
<2	A	A–B	B	—	—
2–3	A–B	B	C	E	F
3–5	B	B–C	C	D	E
5–6	C	C–D	D	D	D
>6	C	D	D	D	D

[a]The neutral class, D, should be assumed for overcast conditions during day or night. Note that "thinly overcast" is not equivalent to "overcast."
Notes: Class A is the most unstable and class F is the most stable class considered here. Night refers to the period from one hour before sunset to one hour after sunrise. Note that the neutral class, D, can be assumed for overcast conditions during day or night, regardless of wind speed.

"Strong" incoming solar radiation corresponds to a solar altitude greater than 60° with clear skies; "slight" insolation corresponds to a solar altitude from 15° to 35° with clear skies. Table 170, Solar Altitude and Azimuth, in the Smithsonian Meteorological Tables, can be used in determining solar radiation. Incoming radiation that would be strong with clear skies can be expected to be reduced to moderate with broken (5/8 to 7/8 cloud cover) middle clouds and to slight with broken low clouds.
(*Source:* Turner, 1967.)

TABLE 9-14
Values of a, c, d, and f for calculating s_y and s_z

Stability class	a	$x \leq 1$ km c	d	f	$x > 1$ km c	d	f
A	213	440.8	1.941	9.27	459.7	2.094	−9.6
B	156	100.6	1.149	3.3	108.2	1.098	2
C	104	61	0.911	0	61	0.911	0
D	68	33.2	0.725	−1.7	44.5	0.516	−13.0
E	50.5	22.8	0.678	−1.3	55.4	0.305	−34.0
F	34	14.35	0.74.0	−0.35	62.6	0.18	−48.6

(*Source:* Martin, 1976.)

power law expression of the following form may be used to estimate the wind speed at heights other than that of the measurement:

$$u_2 = u_1 \left(\frac{z_2}{z_1} \right)^p \qquad (9\text{-}30)$$

where u_2 is the windspeed at elevation z_2 and u_1 is the windspeed at elevation z_1. The exponent p is a function of the terrain roughness and the stability. EPA's recommended values for p are shown in Table 9-15.

TABLE 9-15
Exponent p values for rural and urban regimes

Stability class	Rural	Urban
A	0.07	0.15
B	0.07	0.15
C	0.10	0.20
D	0.15	0.25
E	0.35	0.30
F	0.55	0.30

(*Source:* U.S. EPA, 1995.)

Example 9-5.   It has been estimated that the emission of SO_2 from a coal-fired power plant is 1,656.2 g/s. At 3 km downwind on an overcast summer afternoon, what is the centerline concentration of SO_2 if the wind speed is 4.50 m/s? (Note: "centerline" implies $y = 0$.)

> *Stack parameters:*
> Height = 120.0 m
> Diameter = 1.20 m
> Exit velocity = 10.0 m/s
> Temperature = 315°C
> *Atmospheric conditions:*
> Pressure = 95.0 kPa
> Temperature = 25.0°C

Solution.   We begin by determining the effective stack height (H).

$$\Delta H = \frac{(10.0)(1.20)}{4.50}\left[1.5 + \left(2.68 \times 10^{-2}(95.0)\frac{588 - 298}{588}1.20\right)\right]$$

$$\Delta H = 8.0 \text{ m}$$
$$H = 120.0 + 8.0 = 128.0 \text{ m}$$

Next, we must determine the atmospheric stability class. The footnote to Table 9-13 indicates that the D class should be used for overcast conditions.

From Equations 9-28 and 9-29 we can determine that, at 3 km downwind with a D stability, the plume standard deviations are as follows:

$$s_y = 68(3)^{0.894} = 181.6 \text{ m}$$
$$s_z = 44.5(3)^{0.516} + (-13) = 65.4 \text{ m}$$

Thus,

$$\chi = \left[\frac{1,656.2}{\pi(181.6)(65.4)(4.50)}\right]\left\{\exp\left[-\frac{1}{2}\left(\frac{0}{181.5}\right)^2\right]\right\}\left\{\exp\left[-\frac{1}{2}\left(\frac{128.0}{65.4}\right)^2\right]\right\}$$

$$= 1.45 \times 10^{-3} \text{ g/m}^3, \text{ or } 1.5 \times 10^{-3} \text{ g/m}^3, \text{ of } SO_2$$

Comments:

1. The plume standard deviations shown in Figures 9-24 and 9-25 are based on field measurements using a 10-minute sampling time. The field measurements were conducted in relatively open country. Application of the equations to other settings is not recommended.

2. This concentration is about 0.56 ppm. Several authors have proposed rules of thumb to correct for the averaging time difference between the standards and the estimates provided by Equation 9-25 and the standard deviation figures. These corrections range from 0.33 to 0.63 for an hourly average. Using these rules, the maximum one-hour concentration would be estimated to be between 0.19 ppm and 0.35 ppm. These estimates would imply that the NAAQS one-hour standard would not be exceeded. Extrapolation of the rules of thumb to longer time periods is hard to justify because the wind seldom maintains "a direction" for more than a few minutes.

3. The stack temperature used for this example is a bit high for coal-fired power plants. For brick and lined stacks the temperature drop along the height of the stack is about 0.9°C/m (0.5°F/ft). The temperature of the gas entering the stack is on the order of 320°C (Fryling, 1967). Thus, a more realistic temperature would have been about 210°C.

Inversion Aloft. When an inversion is present, the basic diffusion equation must be modified to take into account the fact that the plume cannot disperse vertically once it reaches the inversion layer. The plume will begin to mix downward when it reaches the base of the inversion layer (Figure 9-26). The downward mixing will begin at a distance x_L downwind from the stack. The x_L distance is a function of the stability in the layer below the inversion. It has been determined empirically that the vertical standard deviation of the plume can be calculated with the following formula at the distance x_L:

$$s_z = 0.47(L - H) \tag{9-31}$$

where L = height to bottom of inversion layer, m
H = effective stack height, m

When the plume reaches twice the distance to initial contact with the inversion base, the plume is said to be completely mixed throughout the layer below the inversion.

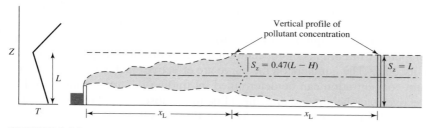

FIGURE 9-26
Effect of elevated inversion on dispersion.

Beyond a distance equal to $2x_L$ the centerline concentration of pollutants may be estimated by using the following equation:

$$\chi = \frac{E}{(2\pi)^{1/2}s_y(u)(L)} \tag{9-32}$$

Note that s_y is determined by the stability of the layer below the inversion and the distance to the receptor. We call this the "inversion" or "short form" of the dispersion equation.

Example 9-6.   Determine the distance downwind from a stack at which we must switch to the "inversion form" of the dispersion model given the following meteorologic situation:

> Effective stack height: 50 m
>
> Inversion base: 350 m
>
> Wind speed: 7.3 m/s
>
> Cloud cover: none
>
> Time: 1130 h
>
> Season: summer

Solution.   Determine the stability class using Table 9-13. At > 6 m/s with strong radiation, the stability class is C.
   Calculate the value of s_z.

$$s_z = 0.47(350 \text{ m} - 50 \text{ m}) = 141 \text{ m}$$

Using Figure 9-26, find x_L. With $s_z = 141$, draw a horizontal line to stability class C. Drop a vertical line to the "distance downwind." Find $x_L = 2.5$ km.
   Therefore, at any distance equal to or greater than 5 km downwind ($2x_L$), use the "inversion form" of the equation (Equation 9-32).
   For distances less than 5 km, we use Equation 9-25 with s_z determined from the distance to the point of interest and the stability. Thus, in no case do we use s_z computed from Equation 9-30 to calculate χ.

Peak Concentration.   One method of identifying the location and concentration of the peak downwind concentration is by the use of a spreadsheet program. Equation 9-25 is solved for a set of downwind distances for a given set of meteorological conditions. The values of s_y and s_z are calculated using Equations 9-28, 9-29, and Table 9-13.
   A short-cut method for a quick estimate is to use Figure 9-27 (Turner, 1967). This figure was prepared from graphs of concentration versus distance. Each curve represents a stability class. The numbers on the curve are effective stack heights in meters. The maximum concentration can be determined by finding the appropriate stability class and effective stack height and drawing a line to the abscissa and reading a value for the product $\chi u/E$. The terms are as defined for Equation 9-25. The units are m^{-2}. The maximum concentration (χ_{max}) is found by multiplying this value by the emission rate (E) and dividing by the wind speed (u). The distance from the stack to the point of

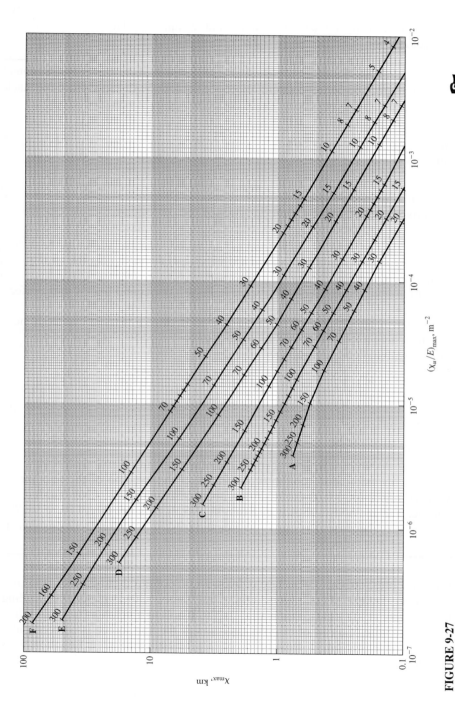

FIGURE 9-27
Distance of maximum concentration and maximum $\chi u/E$ as a function of stability (curves) and effective height of emission in meters (numbers). (*Source:* Turner, 1967.)

maximum concentration is found by finding the appropriate stability class and effective stack height and drawing a line to the ordinate and reading x_{max}. The procedure is illustrated in Example 9-7.

Example 9-7. Given the meteorological and power plant data shown below, determine the maximum downwind concentration of SO_2 and the distance from the stack that it occurs.

Meteorological conditions:
  Wind speed at 10 m = 4 m/s
  Sunny summer afternoon

Power plant conditions:
  SO_2 emission rate = 151 g/s
  Effective stack height = 150 m

Solution. On a sunny summer afternoon the incoming solar radiation may be assumed to be strong. From Table 9-13 at a wind speed of 4 m/s under the strong solar radiation column find the stability class is B.

  Using Figure 9-27 and the curve labeled B, draw a horizontal line from the effective stack height of 150 m to the ordinate and read $x_{max} \approx 1$ km.

  Again, using Figure 9-27 and the curve labeled B, draw a vertical line from the effective stack height to the abscissa and read $\chi u / E = 7.5 \times 10^{-6}$. Calculate the maximum downwind concentration as

$$\chi_{max} = \left(\frac{\chi u}{E}\right)\left(\frac{E}{u}\right) = (7.5 \times 10^{-6} \text{ m}^{-2})\left(\frac{151 \text{ g/s}}{4 \text{ m/s}}\right) = 2.8 \times 10^{-4} \text{ g/m}^3 \text{ of } SO_2$$

Comments:

1. This concentration is about 0.1 ppm.
2. Figure 9-27 does not take into account the effects of an inversion aloft. If there is an inversion aloft, one must determine if the location of the maximum downwind concentration is less than the distance at which the inversion has an influence. The method is shown in Example 9-6. If it is not then Figure 9-27 cannot be used because it is based on Equation 9-25. Equation 9-32 governs if the inversion has influence.

Continuous Line Source. Motor vehicles traveling along a straight section of a highway are an example of continuous infinite line source. Others include the leading edge of a burning field, a row of industrial stacks, a spill of a volatile chemical into a river, and the rupture of a row of railroad tank cars. If the line source is perpendicular to the wind, concentrations downwind of it may be estimated with Equation 9-33 (Turner, 1967):

$$\chi = \frac{2E_{\text{line}}}{(2\pi)^{0.5}(s_z)(u)} \exp\left[-(0.5)\left(\frac{H}{(s_z)}\right)\right] \tag{9-33}$$

Where E_{line} = emission rate per unit distance, g/m · s and other terms are as defined for Equation 9-25. If the source is at ground level, $H = 0$ and the exponential term equals 1. If the line source is not perpendicular to the wind, the result of Equation 9-33 is divided by the sine of the angle between the wind direction and the line source. If the angle is less than 45° this equation should *not* be used. The case of automobiles is illustrated in Example 9-8.

Example 9-8. Estimate the concentration of nonmethane hydrocarbons (NMHC) 300 m downwind of an interstate highway at 5:30 PM on an overcast day with a wind speed of 4 m/s. The interstate runs east-west and the wind is from due south. The measured traffic flow is 8,000 vehicles per hour. The average speed is 40 miles per hour (mph) and the emission rate is 0.25 g/mile.

Solution. Estimate the number of vehicles per second.

$$(8{,}000 \text{ vehicles/h})\left(\frac{1 \text{ h}}{3{,}600 \text{ s}}\right) = 2.22 \text{ vehicles/s}$$

Estimate the emission rate.

$$(0.25 \text{ g/mile})(2.22 \text{ veh/s})\left(\frac{1 \text{ mile}}{1{,}609 \text{ m}}\right) = 3.45 \times 10^{-4} \text{ g/s} \cdot \text{m}$$

Estimate the plume standard deviations using Tables 9-13 and 9-14. The footnote to Table 9-13 suggests that on an overcast day the stability class should be D. With D stability, and $c = 3.2$, $d = 0.725$, and $f = -1.7$ (from Table 9-14), the plume standard deviation is estimated to be

$$s_z = cx^d + f = 33.2(0.3 \text{ km})^{0.725} - 1.7 = 13.87 - 1.7 = 12.17 \text{ m}$$

Using Equation 9-33 with $H = 0$

$$\chi = \frac{2(3.45 \times 10^{-4} \text{ g/s} \cdot \text{m})}{(2\pi)^{0.5}(12.17 \text{ m})(4 \text{ m/s})} = \frac{6.9 \times 10^{-4}}{122.01} = 5.65 \times 10^{-6} \text{ g/m}^3 \text{ of NMHC}$$

Comment: Because the east-west highway is perpendicular to the wind direction there is no need to correct for the wind's angle with respect to the highway.

9-9   INDOOR AIR QUALITY MODEL

If we envision a house or room in a house or other enclosed space as a simple box (Figure 9-28), then we can construct a simple mass balance model to explore the behavior of the indoor air quality as a function of infiltration of outdoor, indoor sources and sinks, and leakage to the outdoor air. If we assume the contents of the box are well mixed, then

$$
\begin{array}{llllll}
\text{Rate of} & & \text{Rate of} & \text{Rate of} & \text{Rate of} & \text{Rate of} \\
\text{pollutant} & & \text{pollutant} & \text{pollutant} & \text{pollutant} & \text{pollutant} \\
\text{increase} & = & \text{entering} & + & \text{entering box} & - & \text{leaving box} & - & \text{leaving box} \\
\text{in box} & & \text{box from} & & \text{from indoor} & & \text{by leakage} & & \text{by decay} \\
& & \text{outdoors} & & \text{emissions} & & \text{to outdoors} & &
\end{array}
\quad (9\text{-}34)
$$

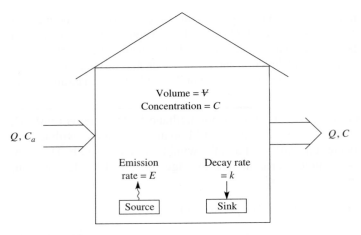

FIGURE 9-28
Mass balance model for indoor air pollution.

or

$$V \frac{dC}{dt} = QC_a + E - QC - kCV \qquad (9\text{-}35)$$

where V = volume of box, m^3
C = concentration of pollutant, g/m^3
Q = rate of infiltration of air into and out of box, m^3/s
C_a = concentration of pollutant in outdoor air, g/m^3
E = emission rate of pollutant into box from indoor source, g/s
k = pollutant decay rate or reaction rate coefficient, s^{-1}

Reaction rate coefficients for a selected list of pollutants are given in Table 9-16. Emission factors for selected indoor air pollution sources are listed in Table 9-17.

TABLE 9-16
Reaction rate coefficients for selected pollutants

Pollutant	k, s^{-1}
CO	0.0
CH$_2$O	1.11×10^{-4}
NO	0.0
NO$_x$ (as N)	4.17×10^{-5}
Particulates ($< 0.5\ \mu$m)	1.33×10^{-4}
Radon	2.11×10^{-6}
SO$_2$	6.39×10^{-5}

(*Data Source:* Traynor et al., 1982.)

TABLE 9-17

Emission factors for selected indoor air pollution sources and pollutants[a]

Pollutant	Emission factor, $\mu g/h \cdot m^2$, at various times after being put into use				
	1 h	1 day	1 week	1 month	1 year
Floor materials: carpet, synthetic fiber					
Formaldehyde	15	10	5	2	1
Styrene	50	20	6	3	2
Toluene	300	40	20	10	1
TVOC[b]	600	80	20	10	5
Paints and coatings: solvent-based paint					
Decane	200,000	2,000	0	0	0
Nonane	100,000	100	0	0	0
Pentylcyclohexane	10,000	3,000	0	0	0
Undecane	100,000	10,000	0	0	0
m,p Xylenes	50,000	5	0	0	0
TVOC	3×10^6	200,000	0	0	0
Paints and coatings: water-based paints					
Acetaldehyde	100	10	2	1	0
Ethylene glycol	20,000	20,000	15,000	4,000	0
Formaldehyde	40	100	2	1	0
TVOC	50,000	40,000	20,000	200	20

	Photocopiers: dry-process, $\mu g/h$ per machine	
	Machine in standby mode	Machine making copies
Ethylebenzene	10	30,000
Styrene	500	7,000
m,p Xylenes	200	20,000

Formaldehyde emission factors, $\mu g/d \cdot m^2$	
Medium density fiberboard	17,600–55,000
Particleboard	2,000–25,000
Paper products	260–280
Fiberglass products	400–470
Clothing	35–570

[a]Compiled from Godish, 2001, and Tucker, 2001.
[b]TVOC = total volatile organic compounds.

The general solution for Equation 9-35 is

$$C_t = \frac{\frac{E}{V} + C_a\frac{Q}{V}}{\frac{Q}{V} + k}\left\{1 - \exp\left[-\left(\frac{Q}{V} + k\right)t\right]\right\} + C_o\exp\left[-\left(\frac{Q}{V} + k\right)t\right] \qquad (9\text{-}36)$$

The steady-state solution for Equation 9-35 may be found by setting $dC/dt = 0$ and solving for C:

$$C = \frac{QC_a + E}{Q + kV} \qquad (9\text{-}37)$$

When the pollutant is conservative and does not decay with time or have a significant reactivity, $k = 0$. In the special case when the pollutant is conservative and the ambient concentration is negligible and the initial indoor concentration is zero, Equation 9-35 reduces to:

$$C_t = \frac{E}{Q}\left\{1 - \exp\left[-\left(\frac{Q}{V}\right)t\right]\right\} \qquad (9\text{-}38)$$

Example 9-9.   An unvented kerosene heater is operated for one hour in an apartment having a volume of 200 m³. The heater emits SO_2 at a rate of 50 μg/s. The ambient air concentration (C_a) and the initial indoor air concentration (C_o) of SO_2 are 100 μg/m³. If the rate of ventilation is 50 L/s, and the apartment is assumed to be well mixed, what is the indoor air concentration of SO_2 at the end of one hour?

Solution.   The concentration may be determined using the general solution form of the indoor air quality model (Equation 9-36). The decay rate for SO_2 from Table 9-16 is 6.39×10^{-5} s^{-1} and 50 L/s is equivalent to 0.050 m³/s.

$$C_t = \frac{\dfrac{50\ \mu g/s}{(200\ m^3)} + 100\ \mu g/m^3\ \dfrac{0.050\ m^3/s}{200\ m^3}}{\dfrac{0.050\ m^3/s}{200\ m^3} + 6.39 \times 10^{-5}\ s^{-1}}$$

$$\times \left\{1 - \exp\left[-\left(\frac{0.050\ m^3/s}{200\ m^3} + 6.39 \times 10^{-5}\ s^{-1}\right)(3600\ s)\right]\right\}$$

$$+ (100\ \mu g/m^3)\ \exp\left[-\left(\frac{0.050\ m^3/s}{200\ m^3} + 6.39 \times 10^{-5}\ s^{-1}\right)(3600\ s)\right]$$

$$= 876.08(1 - \exp(-1.13)) + 100\ \exp(-1.13) = 876.08(1 - 0.323) + 100(0.323)$$

$$= 593.09 + 32.3 = 625.39,\ \text{or } 630\ \mu g/m^3$$

In addition to the mass balance model, statistical and *computational fluid dynamics* (CFD) models have been developed. Sparks (2001) provides an overview of the different types, their advantages and disadvantages, and a list of complex computer models that are available.

9-10 AIR POLLUTION CONTROL OF STATIONARY SOURCES

Gaseous Pollutants

Absorption. Control devices based on the principle of absorption attempt to transfer the pollutant from a gas phase to a liquid phase. This is a *mass transfer* process in which the gas dissolves in the liquid. The dissolution may or may not be accompanied by a reaction with an ingredient of the liquid. Mass transfer is a diffusion process wherein the pollutant gas moves from points of higher concentration to points of lower concentration. The removal of the pollutant gas takes place in three steps:

1. Diffusion of the pollutant gas to the surface of the liquid

2. Transfer across the gas/liquid interface (dissolution)

3. Diffusion of the dissolved gas away from the interface into the liquid

Structures such as *spray chambers* (Figure 9-29) and *towers* or *columns* (Figure 9-30) are two classes of devices employed to absorb pollutant gases. In scrubbers, which are a type of spray chamber, liquid droplets are used to absorb the gas. In towers, a thin film of liquid is used as the absorption medium. Regardless of the type of device, the

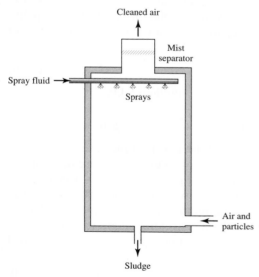

FIGURE 9-29
Spray chamber.

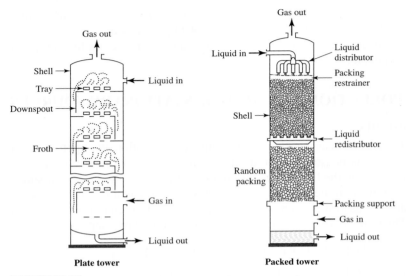

FIGURE 9-30
Absorption systems.

solubility of the pollutant in the liquid must be relatively high. If water is the solute, this generally limits the application to a few inorganic gases such as NH_3, Cl_2, and SO_2. Scrubbers are relatively inefficient absorbers but have the advantage of being able to simultaneously remove particulates. Towers are much more efficient absorbers but they become plugged by particulate matter.

The amount of absorption that can take place for a nonreactive solution is governed by the partial pressure of the pollutant. For dilute solutions, as we have in pollution control systems, the relationship between partial pressure and the concentration of the gas in solution is given by *Henry's law:*

$$P_g = K_H C_{equil} \tag{9-39}$$

where P_g = partial pressure of gas in equilibrium with liquid, kPa
   K_H = Henry's law constant, kPa · m^3/g
   C_{equil} = concentration of pollutant gas in the liquid phase, g/m^3

Equation 9-39 implies that the partial pressure of the gas must increase as the liquid accumulates more pollutant or else it will come out of solution. Because the liquid is removing pollutant from the gas phase, this means the partial pressure is decreasing as the gas is cleaned. This is just the reverse of what we want to happen. The easiest way to get around this problem is to run the gas and liquid in opposite directions. This is called *countercurrent flow*. In this manner, the high concentration gas is absorbed into a liquid with a high pollutant concentration. The lower concentration gas is absorbed by liquid with no pollutants in it.

A mass balance diagram of a countercurrent flow absorption column is shown in Figure 9-31. The mass balance equation is

$$(G_{m1})(y_1) - (G_{m2})(y_2) = (L_{m1})(x_1) - (L_{m2})(x_2) \tag{9-40}$$

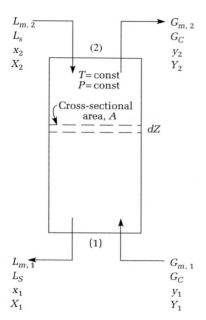

$L_{m,2}$
L_S
x_2
X_2

(2)

$T = \text{const}$
$P = \text{const}$

Cross-sectional area, A

dZ

(1)

$L_{m,1}$
L_S
x_1
X_1

$G_{m,2}$
G_C
y_2
Y_2

$G_{m,1}$
G_C
y_1
Y_1

FIGURE 9-31
Notation for a countercurrent flow packed absorption tower.

where G_{m1}, G_{m2} = total gas flow (air plus pollutant) into and out of the column respectively, kg · mole/h

y_1, y_2 = mole fraction of pollutant in the gas phase at inlet and outlet of column, respectively*

L_{m1}, L_{m2} = total liquid flow (solvent plus absorbed pollutant) out of and into the column respectively, kg · mole/h

x_1, x_2 = mole fraction of pollutant in the liquid phase out of and into the column, respectively

Three variables of interest in the design of a packed tower are the gas flow rate, the liquid flow rate, and the height of the tower. As you might expect, the three are related. If we consider a differential height of the absorber, dZ, as shown in Figure 9-31, the total interfacial area open to mass transfer is defined as

$$\text{area for mass transfer} = (a)(A)(dZ) \qquad (9\text{-}41)$$

where a = area per unit volume of packing
A = cross-sectional area of column

---

*A mole fraction is defined as follows:

$$y = \frac{P}{P_t}$$

P = partial pressure of gas
P_t = total pressure of gas
$y^* = (P^*/P_t)$

We may describe the rate of mass transfer of a gas, i, into solution (N_i) by the following differential equation:

$$N_i = \frac{dC}{dt} = K_y(y - y^*) \tag{9-42}$$

where K_y = overall mass transfer coefficient for gas
  y, y^* = mole fraction of gaseous pollutant and equilibrium mole fraction, respectively

The rate of transfer of species i then is

$$\text{rate of mass transfer} = (N_i)(A)(a)(dZ) \tag{9-43}$$

This mass is equal to the mass loss from the gas phase as it passes through the differential height dZ:

$$\text{mass loss} = d(G_m y) \tag{9-44}$$

We may expand this expression by defining two new terms: the mass flow rate per unit area G'_m and the mole ratio:

$$Y = \frac{y}{1 - y} \tag{9-45}$$

and noting that

$$G_c Y = G_m y \tag{9-46}$$

where G_c is the mass flow of the carrier gas without the pollutant. Equating the mass transfer (Equation 9-42) with the mass loss (Equation 9-44) and making substitutions from Equations 9-43, 9-45, and 9-46 yields

$$K_y a(y - y^*) \, dZ = \frac{G'_m dy}{1 - y} \tag{9-47}$$

or

$$dZ = \frac{G'_m \, dy}{K_y a(y - y^*)(1 - y)} \tag{9-48}$$

The overall driving force $(y - y^*)$ at any location in the tower may be written in the form

$$y - y^* = (1 - y^*) - (1 - y) \tag{9-49}$$

It is convenient then to define the log-mean value of $(1 - y^*)$ and $(1 - y)$:

$$(1 - y)_{LM} = \frac{(1 - y^*) - (1 - y)}{\ln[(1 - y^*)/(1 - y)]} \tag{9-50}$$

Multiplying the numerator and denominator of Equation 9-48 by $(1 - y)_{LM}$, we obtain

$$dZ = \left(\frac{G'_m}{K_y a(1 - y)_{LM}}\right)\left(\frac{(1 - y)_{LM} dy}{(y - y^*)(1 - y)}\right) \tag{9-51}$$

Although G'_m, K_ya, and $(1 - y)_{LM}$ vary along the absorption column, the first term of this equation is reasonably constant. This quantity is called the *overall height of a transfer unit* (H_{og}). As a first approximation to the height of the column, Equation 9-51 may be rewritten as

$$Z = (H_{og}) \int_{y_2}^{y_1} \frac{(1 - y)_{LM} dy}{(y - y^*)(1 - y)} \tag{9-52}$$

The integral is called the *number of transfer units* (N_{og}). The height of the tower is computed from the following equation:

$$Z_t = (H_{og})(N_{og}) \tag{9-53}$$

For dilute solutions that obey Henry's law, the number of overall gas transfer units may be calculated as follows (Treybal, 1968):

$$N_{og} = \frac{\ln\left[\left(\frac{y_1 - mx_2}{y_2 - mx_2}\right)(1 - A) + A\right]}{1 - A} \tag{9-54}$$

where y_1, y_2 = mole fraction of pollutant in the gas phase at inlet and outlet of tower, respectively

m = slope of equilibrium curve defined by Henry's law = y^*/x^* in mole fraction units (m has no units)

x_2 = mole fraction of pollutant in the liquid phase entering the tower

$A = mQ_g/Q_l$

Q_l = liquid flow rate, kg $\cdot$ mole/h $\cdot$ m^2

Q_g = gas flow rate, kg $\cdot$ mole/h $\cdot$ m^2

The height of a single overall mass transfer unit (HTU) may also be expressed as the sum of the gas and liquid HTUs.

$$H_{og} = H_g + AH_l \tag{9-55}$$

where H_g and H_l are complex functions of the flow rate, surface area of the packing, viscosity of the liquid and air, and the diffusivity of the pollutant gas.

Example 9-10. Determine the height of a packed tower that is to reduce NH_3 in air from a concentration of 0.10 kg/m^3 to a concentration of 0.0005 kg/m^3 given the following data:

Column diameter = 3.00 m

Operating temperature = 20.0°C

Operating pressure = 101.325 kPa

H_g = 0.438 m

H_l = 0.250 m

$Q_g = Q_l$ = 10.0 kg/s

Incoming liquid is water free of NH_3

Solution.   We begin by converting to mole fractions. NH_3 has a GMW of 17.03. For air we assume a GMW of 28.970 and a density of 1.185 kg/m^3 at 25°C. Because the operating temperature is 20°C, we correct the density of the air:

$$1.185 \times \frac{298}{293} = 1.205 \text{ kg/m}^3$$

Now we compute the mole fractions at the inlet (y_1) and outlet (y_2):

$$y_1 = \frac{\dfrac{0.10 \text{ kg/m}^3}{17.03 \text{ GMW NH}_3}}{\dfrac{1.205 \text{ kg/m}^3}{28.970 \text{ GMW air}}} = \frac{0.005872}{0.04159} = 0.14118$$

In a like manner, $y_2 = 0.000706$. Since the incoming liquid has no NH_3, the mole fraction is zero, that is, $x_2 = 0.0$.

The Henry's law constant in mole fraction units must be determined from experimental data. From the *Chemical Engineers' Handbook* we find the following data (Perry and Chilton, 1973):

P_{NH_3}, kPa	kg NH_3 per 100 kg H_2
15.199	15
9.319	10
4.266	5
1.600	1

If we convert each value to mole fractions and plot x^* versus y^* (the asterisk refers to the steady-state condition), the slope of the line will be m. An example calculation is shown for the first value of x^* and y^*. The total pressure is taken to be 101.325 kPa. The GMW of H_2O is 18.015. For 15 kg NH_3 per 100 kg H_2O:

$$x^* = \frac{\dfrac{15 \text{ kg}}{17.030 \text{ GMW NH}_3}}{\dfrac{15 \text{ kg}}{17.03 \text{ GMW NH}_3} + \dfrac{100 \text{ kg}}{18.015 \text{ GMW H}_2\text{O}}}$$

$$x^* = 0.1369$$

$$y^* = \frac{15.199 \text{ kPa}}{101.325 \text{ kPa}}$$

$$y^* = 0.1500$$

The value of m is then found by a least squares linear regression fit of a line through the four pairs of x^* and y^* values. The slope of the line is m.

$$m = 1.068$$

The value of A is computed in mole units as follows:

$$A = \frac{1.068\left[\dfrac{10.0 \text{ kg/s of air}}{28.97 \text{ GMW of air}}\right]}{\dfrac{10.0 \text{ kg/s of H}_2\text{O}}{18.015 \text{ GMW of H}_2\text{O}}}$$

$$= 0.6641$$

The number of gas transfer units is then

$$N_{og} = \frac{\ln\left[\dfrac{0.14118 - 1.068(0)}{0.000706 - 1.068(0)}(1 - 0.6641) + 0.6641\right]}{1 - 0.6641}$$

$$= 12.5545$$

The height of an individual gas transfer unit is

$$H_{og} = 0.438 + 0.6641(0.250) = 0.6040$$

The height of the tower is then

$$Z_t = (0.6040)(12.5545) = 7.5832$$

Since the limiting concentration data were given to only one significant figure, the answer would be

$$Z_t = 8 \text{ m}$$

Comment: Before we leave this example, we should look back and see what we have wrought. Because the absorption tower neither creates nor destroys matter, the mass of NH_3 entering and leaving the column must be the same. If we assume isothermal, steady-state conditions (that is, gas and liquid rates in and out are equal), we can solve the mass balance equation (Equation 9-40) for x_1. After some calculations we find $x_1 = 0.08734$. This is 90,300 mg/L of NH_3. This is a classic example of a multimedia problem. In solving an air pollution problem, we have created a serious water pollution problem.

Adsorption. This is a *mass-transfer* process in which the gas is bonded to a solid. It is a surface phenomenon. The gas (the *adsorbate*) penetrates into the pores of the solid (the *adsorbent*) but not into the lattice itself. The bond may be physical or chemical. Electrostatic forces hold the pollutant gas when physical bonding is significant. Chemical bonding is by reaction with the surface. Pressure vessels having a fixed bed are used to hold the adsorbent (Figure 9-32). Active carbon (activated charcoal), molecular sieves, silica gel, and activated alumina are the most common adsorbents. Active carbon is manufactured from nut shells (coconuts are great) or coal subjected to heat treatment in a reducing atmosphere. Molecular sieves are dehydrated zeolites (alkali-metal silicates). Sodium silicate is reacted with sulfuric acid to make silica gel. Activated alumina is a porous hydrated aluminum oxide. The common property of these adsorbents is a large "active" surface area per unit volume after treatment. They are

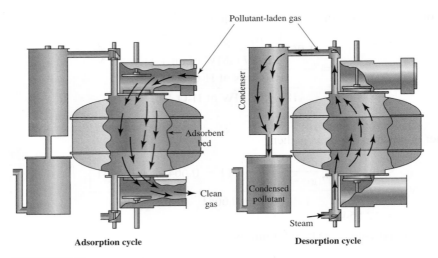

FIGURE 9-32
Adsorption system.

very effective for hydrocarbon pollutants. In addition, they can capture H_2S and SO_2. One special form of molecular sieve can also capture NO_2. With the exception of the active carbons, adsorbents have the drawback that they preferentially select water before any of the pollutants. Thus, water must be removed from the gas before it is treated. All of the adsorbents are subject to destruction at moderately high temperatures (150°C for active carbon, 600°C for molecular sieves, 400°C for silica gel, and 500°C for activated alumina). They are very inefficient at these high temperatures. In fact, their activity is regenerated at these temperatures!

The relation between the amount of pollutant adsorbed and the equilibrium pressure at constant temperature is called an *adsorption isotherm*. The equation that best describes this relation for gases is the one derived by Langmuir (Buonicore and Theodore, 1975).

$$W = \frac{aC_R^*}{1 + bC_g^*} \tag{9-56}$$

where W = amount of gas per unit mass of adsorbent, kg/kg
   a, b = constants determined by experiment
   C_g^* = equilibrium concentration of gaseous pollutant, g/m^3

In the analysis of experimental data, Equation 9-56 is rewritten as follows:

$$\frac{C_g^*}{W} = \frac{1}{a} + \frac{b}{a}\,C_g^* \tag{9-57}$$

In this arrangement, a plot of (C_g^*/W) versus C_g^* should yield a straight line with a slope of (b/a) and an intercept equal to $(1/a)$.

In contrast to absorption towers where the collected pollutant is continuously removed by flowing liquid, the collected pollutant remains in the adsorption bed. Thus, while the bed has sufficient capacity, no pollutants are emitted. At some point in time, the bed will become saturated with pollutant. As saturation is approached, pollutant will begin to leak out of the bed. This is called *breakthrough.* When the bed capacity is exhausted, the influent and effluent concentration will be equal. A typical break-through curve is shown in Figure 9-33. In order to allow for continuous operation, two beds are provided (Figure 9-32). While one is collecting pollutant, the other is being regenerated. The concentrated gas released during regeneration is usually returned to the process as recovered product. The critical factor in the operation of the bed is the length of time it can operate before breakthrough occurs. The time to breakthrough may be calculated from the following (Crawford, 1976):

$$t_B = \frac{Z_t - \delta}{v_f} \tag{9-58}$$

where Z_t = height of bed, m
 δ = width of adsorption zone, m
 v_f = velocity of adsorption zone as defined by Equation 9-60, m/s

The height of the adsorption bed (Z_t) can be determined in the same manner as it was for absorption towers, with a few exceptions. The value of N_{og} must be determined by

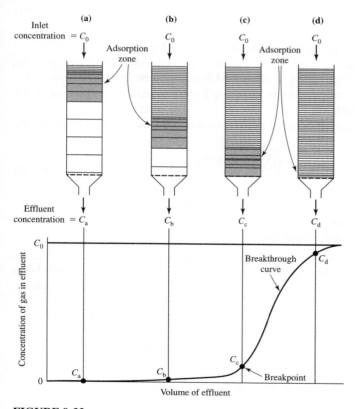

FIGURE 9-33
Adsorption wave and breakthrough curve. (*Source:* Treybal, 1968.)

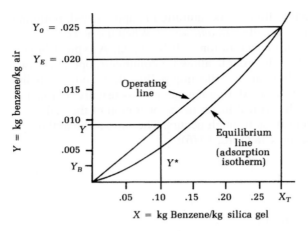

FIGURE 9-34

Equilibrium and operating lines for adsorption of benzene on silica gel. (*Source:* Seinfeld, 1975.)

integration of the following expression (Treybal, 1968):

$$N_{og} = \int_{c_2}^{c} \frac{dC}{C - C_g^*} \tag{9-59}$$

where C_g^* = the equilibrium partial pressure described by Equation 9-56 and C is described by the operating line (Figure 9-34). The H_{og} equation is modified by replacing H_l with H_s. The value of slope m in Equation 9-54 is determined from Equation 9-57.

The width of the adsorption zone is shown in Figure 9-33. It is a function of the shape of the adsorption isotherm.

The velocity of the adsorption zone may be calculated from the properties of the system:

$$v_f = \frac{(Q_g)(1 + bC_g^*)}{a\rho_s\rho_g A_c} \tag{9-60}$$

where ρ_s, ρ_g = density of solid and gas, kg/m^3 (Note that ρ_s is the density of the absorbent "as packed.")

A_c = cross-sectional area of bed, m^2

Example 9-11. Determine the breakthrough time for an adsorption bed that is 0.50 m thick and 10 m^2 in cross section. The operating parameters for the bed are as follows:

Gas flow rate = 1.3 kg/s of air

Gas temperature = 25°C

Gas pressure = 101.325 kPa

Bed density as packed = 420 kg/m^3

Inlet pollutant concentration = 0.0020 kg/m^3

Langmuir parameters: $a = 18$; $b = 124$

Width of adsorption zone = 0.03 m

Solution. Using Table A-5 in Appendix A, we find $\rho_g = 1.185 \text{ kg/m}^3$. Then the face velocity of the adsorption wave is

$$v_f = \frac{(1.3 \text{ kg/s})[1 + 124(0.0020 \text{ kg/m}^3)]}{(18)(420 \text{ kg/m}^3)(1.185 \text{ kg/m}^3)(10 \text{ m}^2)}$$
$$= 1.8 \times 10^{-5} \text{ m/s}$$

The breakthrough time is calculated directly from Equation 9-58:

$$t_B = \frac{0.50 \text{ m} - 0.03 \text{ m}}{1.8 \times 10^{-5} \text{ m/s}}$$
$$= 2.6 \times 10^4 \text{ s or } 7.2 \text{ h}$$

Comments:

1. The time to breakthrough can be adjusted by changing either the depth of the bed or the area.

2. If the bed is too shallow, breakthrough may be much sooner than predicted by these equations.

Combustion. When the contaminant in the gas stream is oxidizable to an inert gas, combustion is a possible alternative method of control. Typically, CO and hydrocarbons fall into this category. Both direct flame combustion by afterburners (Figure 9-35) and catalytic combustion (Figure 9-36) have been used in commercial applications.

Direct flame incineration is the method of choice if two criteria are satisfied. First, the gas stream must have a *net heating value* (NHV) greater than 3.7 MJ/m³. At this NHV, the gas flame will be *autogenous* (self-supporting after ignition). Below this point, supplementary fuel is required. The second requirement is that none of the by-products of combustion be toxic. In some cases the combustion by-product may be more toxic than the original pollutant gas. For example, the combustion of trichloroethylene

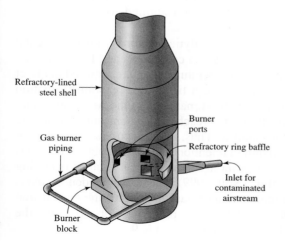

Refractory-lined steel shell

Gas burner piping

Burner ports

Refractory ring baffle

Inlet for contaminated airstream

Burner block

FIGURE 9-35
Direct flame incineration.

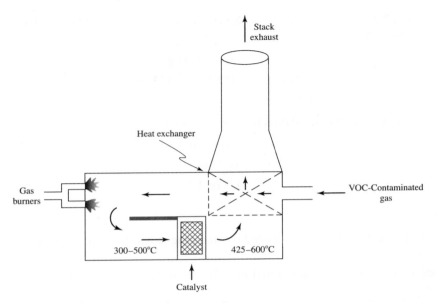

FIGURE 9-36
Catalytic incinerator.

produces phosgene, which was used as a poison gas in World War I. Direct flame incineration has been successfully applied to varnish-cooking, meat-smokehouse, and paint bake-oven emissions.

Some catalytic materials enable oxidation to be carried out in gases that have an NHV of less than 3.7 MJ/m^3. Conventionally, the catalyst is placed in beds similar to adsorption beds. Frequently, the active catalyst is a platinum or palladium compound. The supporting lattice is usually a ceramic. Aside from expense, a major drawback of the catalysts is their susceptibility to poisoning by sulfur and lead compounds in trace amounts. Catalytic combustion has successfully been applied to printing-press, varnish-cooking, and asphalt-oxidation emissions.

The fundamental problem in the design of a catalytic reactor is to determine the volume and dimensions of the catalyst bed for a given conversion and flow rate. The catalyst increases the rate of reaction at lower temperatures than are required in direct flame incineration. The reaction is assumed to be a first-order reaction (Equation 2-14). While the reaction rate constant k may be estimated from the Arrhenius equation for flame incineration (Beard et al., 1980), the reaction rate constant for catalytic incineration is highly dependent on the catalyst. Thus, manufacturers' data or pilot plant data must be used to estimate the required retention time in the catalyst bed. Gas velocities in the range of 6 to 12 m/s are commonly used. Typical catalyst operating temperatures are in the range of 250–550°C (Noll, 1999). The actual residence time is estimated from the total gas flow rate (contaminated gas stream plus the combustion gases) at the operating temperature of the catalyst. The procedure for estimating the dimension and volume of the catalyst bed is shown in Example 9-12.

Example 9-12. Determine the cross-sectional area and depth of catalyst to meet a 95 percent destruction efficiency for methyl ethyl ketone (MEK). The exhaust gas flow rate entering the control equipment is 5.66 m³/s. Combustion air is supplied at a rate of 0.60 m³/s. Both the exhaust gas and the combustion air flow rates are at 20°C. The manufacturers' specifications require that the catalyst be operated at 480°C and that the bed gas velocity be limited to 6.0 m/s. Assume that the MEK combustion reaction follows first order kinetics and that the rate constant at 480°C is 100 s^{-1}.

Solution.
Calculate the volumetric gas flow rate at the catalyst operating temperature.

$$(5.66 \text{ m}^3/\text{s} + 0.60 \text{ m}^3/\text{s}) \left(\frac{480 + 273}{20 + 273} \right) = 16.09 \text{ m}^3/\text{s}$$

The cross-sectional area is computed from the bed gas velocity and the gas flow rate:

$$\text{Area} = \frac{16.09 \text{ m}^3/\text{s}}{6.0 \text{ m/s}} = 2.68 \text{ m}^2$$

Based on first order kinetics and 95 percent destruction efficiency, the desired retention time in the bed is calculated by using Equation 2-14:

$$C_t = C_0 \exp(-kt)$$

with $C_t/C_0 = 0.05$ for 95 percent destruction

$$0.05 = \exp(-100t)$$

Taking the natural logarithm of both sides of this equation and solving for t gives

$$-2.9957 = -100t$$
$$t = 0.030 \text{ s}$$

The depth of the catalyst is then

$$\text{Depth} = (6.0 \text{ m/s})(0.030 \text{ s}) = 0.18 \text{ m}$$

Comment: Because the gas composition is often not pure, pilot plant data should be gathered to determine the rate constant.

Flue Gas Desulfurization (FGD)

Flue gas desulfurization systems fall into two broad categories: nonregenerative and regenerative. Nonregenerative means that the reagent used to remove the sulfur oxides from the gas stream is used and discarded. Regenerative means that the reagent is recovered and reused. In terms of the number and size of systems installed, nonregenerative systems dominate.

Nonregenerative Systems. There are nine commercial nonregenerative systems (Hance and Kelly, 1991). All have reaction chemistries based on lime (CaO), caustic soda (NaOH), soda ash (Na_2CO_3), or ammonia (NH_3).

The SO_2 removed in a lime/limestone-based FGD system is converted to sulfite. The overall reactions are generally represented by (Karlsson and Rosenberg, 1980):

$$SO_2 + CaCO_3 \rightarrow CaSO_3 + CO_2 \tag{9-61}$$

$$SO_2 + Ca(OH)_2 \rightarrow CaSO_3 + H_2O \tag{9-62}$$

when using limestone and lime, respectively. Part of the sulfite is oxidized with the oxygen content in the flue gas to form sulfate:

$$CaSO_3 + \frac{1}{2}O_2 \rightarrow CaSO_4 \tag{9-63}$$

Although the overall reactions are simple, the chemistry is quite complex and not well defined. The choice between lime and limestone, the type of limestone, and method of calcining and slaking can influence the gas-liquid-solid reactions taking place in the absorber.

The principal types of absorbers used in the wet scrubbing systems include venturi scrubber/absorbers, static packed scrubbers, moving-bed absorbers, tray towers, and spray towers (Black & Veatch, 1983).

Spray dryer-based FGD systems consist of one or more spray dryers and a particulate collector.* The reagent material is typically a slaked lime slurry or a slurry of lime and recycled material. Although lime is the most common reagent, soda ash has also been used. The reagent is injected in droplet form into the flue gas in the spray dryer. The reagent droplets absorb SO_2 while simultaneously being dried. Ideally, the slurry or solution droplets are completely dried before they impact the wall of the dryer vessel. The flue gas stream becomes more humidified in the process of evaporation of the reagent droplets, but it does not become saturated with water vapor. This is the single most significant difference between spray dryer FGD and wet scrubber FGD. The humidified gas stream and a significant portion of the particulate matter (fly ash, FGD reaction products, and unreacted reagent) are carried by the flue gas to the particulate collector located downstream of the spray dryer vessel (Cannell and Meadows, 1985). Generally, larger units firing high-sulfur coals use wet FGD. Smaller units use spray dryers.

Control Technologies for Nitrogen Oxides

Almost all nitrogen oxide (NO_x) air pollution results from combustion processes. They are produced from the oxidation of nitrogen bound in the fuel, from the reaction of molecular oxygen and nitrogen in the combustion air at temperatures above 1,600 K (see Equation 9-12), and from the reaction of nitrogen in the combustion air with hydrocarbon radicals. Control technologies for NO_x are grouped into two categories: those that prevent the formation of NO_x during the combustion process and

---

*Historically, from a mass transfer point of view, spray drying refers to the evaporation of a solvent from an atomized spray. Simultaneous diffusion of a gaseous species into the evaporating droplet is not true spray drying. Nonetheless, many authors have adopted the term "spray drying" as synonymous with dry scrubbing.

those that convert the NO_x formed during combustion into nitrogen and oxygen (Prasad, 1995).

Prevention. The processes in this category employ the fact that reduction of the peak flame temperature in the combustion zone reduces NO_x formation. Nine alternatives have been developed to reduce flame temperature: (1) minimizing operating temperatures, (2) fuel switching, (3) low excess air, (4) flue gas recirculation, (5) lean combustion, (6) staged combustion, (7) low NO_x burners, (8) secondary combustion, and (9) water/steam injection.

Routine burner tune-ups and operation with combustion zone temperatures at minimum values reduce the fuel consumption and NO_x formation. Converting to a fuel with a lower nitrogen content or one that burns at a lower temperature will reduce NO_x formation. For example, petroleum coke has a lower nitrogen content and burns with a lower flame temperature than coal. On the other hand, natural gas has no nitrogen content but burns at a relatively high flame temperature and, thus, produces more NO_x than coal.

Low excess air and flue gas recirculation work on the principle that reduced oxygen concentrations lower the peak flame temperatures. In contrast, in lean combustion, additional air is introduced to cool the flame.

In staged combustion and low NO_x burners, initial combustion takes place in a fuel-rich zone that is followed by the injection of air downstream of the primary combustion zone. The downstream combustion is completed under fuel-lean conditions at a lower temperature.

Staged combustion consists of injecting part of the fuel and all of the combustion air into the primary combustion zone. Thermal NO_x production is limited by the low flame temperatures that result from high excess air levels.

Water/steam injection reduces thermal NO_x emissions by lowering the flame temperature.

Post-Combustion. Three processes may be used to convert NO_x to nitrogen gas: selective catalytic reduction (SCR), selective noncatalytic reduction (SNCR), and nonselective catalytic reduction (NSCR).

The SCR process uses a catalyst bed (usually vanadium-titanium, or platinum-based and zeolite) and anhydrous ammonia (NH_3). After the combustion process, ammonia is injected upstream of the catalyst bed. The NO_x reacts with the ammonia in the catalyst bed to form N_2 and water.

In the SNCR process ammonia or urea is injected into the flue gas at an appropriate temperature (870 to 1,090°C). The urea is converted to ammonia, which reacts to reduce the NO_x to N_2 and water.

NSCR uses a three-way catalyst similar to that used in automotive applications. In addition to NO_x control, hydrocarbons and carbon monoxide are converted to CO_2 and water. These systems require a reducing agent similar to CO and hydrocarbons upstream of the catalyst. Larger boilers that have post combustion NO_x controls are generally equipped with SCR.

Typical reduction capabilities of the NO_x techniques range from 30 to 60 percent for the "prevention" methods, 30 to 50 percent for SNCR, and 70 to 90 percent for the SCR systems (Srivastava et al., 2005).

Cleaned gas

Dirty gas

Dust

FIGURE 9-37
Reverse flow cyclone. (*Source:* Crawford, 1976.)

Particulate Pollutants

Cyclones. For particle sizes greater than about 10μ m in diameter, the collector of choice is the cyclone (Figure 9-37). This is an inertial collector with no moving parts. The particulate-laden gas is accelerated through a spiral motion, which imparts a centrifugal force to the particles. The particles are hurled out of the spinning gas and impact on the cylinder wall of the cyclone. They then slide to the bottom of the cone. Here they are removed through an airtight valving system. The standard single-barrel cyclone will have dimensions proportioned as shown in Figure 9-38.

The efficiency of collection of various particle sizes (η) can be determined from an empirical expression and graph (Figure 9-39) developed by Lapple (1951):

$$d_{0.5} = \left[\frac{9\mu B^2}{\rho_p Q_g} \frac{H}{\theta} \right]^{1/2} \qquad \text{(9-64)}$$

where $d_{0.5}$ = cut diameter, the particle size for which the collection efficiency is 50 percent

μ = dynamic viscosity of gas, Pa · s

B = width of entrance, m

H = height of entrance, m

ρ_p = particle density, kg/m³s

Q_g = gas flow rate, m³/s

θ = effective number of turns made in traversing the cyclone as defined in Equation 9-65.

The value of θ may be determined approximately by the following:

$$\theta = \frac{\pi}{H} (2L_1 + L_2) \qquad \text{(9-65)}$$

where L_1 and L_2 are the length of the cylinder and cone, respectively.

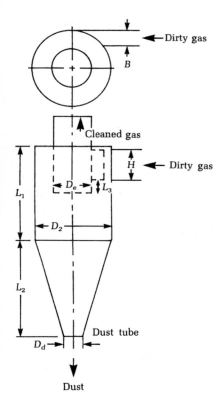

FIGURE 9-38
Standard reverse flow cyclone proportions.
Note: Standard cyclone proportions are as follows:

Length of cylinder, $L_1 = 2D_2$

Length of cone, $L_2 = 2D_2$

Diameter of exit, $D_e = 0.5D_2$

Height of entrance, $H = 0.5D_2$

Width of entrance, $B = 0.25D_2$

Diameter of dust exit, $D_d = 0.25D_2$

Length of exit duct, $L_3 = 0.125D_2$

(*Source:* Crawford, 1976.)

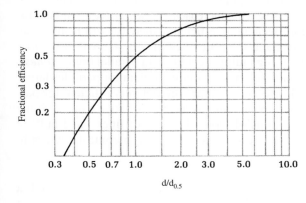

FIGURE 9-39
Empirical cyclone collection efficiency.
(*Source:* Lapple, 1951.)

Example 9-13. Determine the efficiency of a "standard" cyclone having the following characteristics for particles 10 μm in diameter with a density of 800 kg/m³:

Cyclone barrel diameter = 0.50 m

Gas flow rate = 4.0 m³/s

Gas temperature = 25°C

Solution.   From the standard cyclone dimensions we can calculate the following:

$$B = (0.25)(0.50 \text{ m}) = 0.13 \text{ m}$$
$$H = (0.50)(0.50 \text{ m}) = 0.25 \text{ m}$$
$$L_1 = L_2 = (2.00)(0.50 \text{ m}) = 1.0 \text{ m}$$

The number of turns θ is then

$$\theta = \frac{\pi}{0.25}[2(1.0) + 1.0]$$
$$= 37.7$$

From the gas temperature and Table A-4 of Appendix A, we find the dynamic viscosity is 18.5 μPa · s. The cut diameter is then

$$d_{0.5} = \left[\frac{9(18.5 \times 10^{-6} \text{ Pa} \cdot \text{s})(0.13 \text{ m})^2(0.25 \text{ m})}{(800 \text{ kg/m}^3)(4.0 \text{ m}^3/\text{s})(37.7)} \right]^{1/2}$$
$$= 2.41 \times 10^{-6} \text{ m} = 2.41 \ \mu\text{m}$$

The ratio of particle sizes is

$$\frac{d}{d_{0.5}} = \frac{10 \ \mu\text{m}}{2.41 \ \mu\text{m}} = 4.15$$

From Figure 9-39 we find that the collection efficiency is about 95 percent.

As the diameter of the cyclone is reduced, the efficiency of collection is increased. However, the pressure drop also increases. This increases the power requirements for moving the gas through the collector. Because an efficiency increase will result, even if the tangential velocity remains constant, the efficiency may be increased without increasing the power consumption by using multiple cyclones in parallel (*multiclones*).

From the example, you can see that cyclones are quite efficient for particles larger than 10 μm. Conversely, you should note that cyclones are not very efficient for particles 1 μm or less in diameter. Thus, they are employed only for coarse dusts. Some applications include controlling emissions of wood dust, paper fibers, and buffing fibers. Multiclones are frequently used as precleaners for fly-ash control devices in power plants.

Filters.   When high efficiency control of particles smaller than 5 μm is desired, a filter may be selected as the control method. Two types are in use: (1) the deep bed filter, and (2) the baghouse (Figure 9-40). The deep bed filter resembles a furnace filter. A packing of fibers is used to intercept particles in the gas stream. For relatively clean gases and low volumes, such as air conditioning systems, these are quite effective. For dirty industrial gas with high volumes, the baghouse is preferable.

The fundamental mechanisms of collection include screening or sieving (where the particles are larger than the openings between the fibers), interception by the fibers themselves, and electrostatic attraction (because of the difference in static charge on

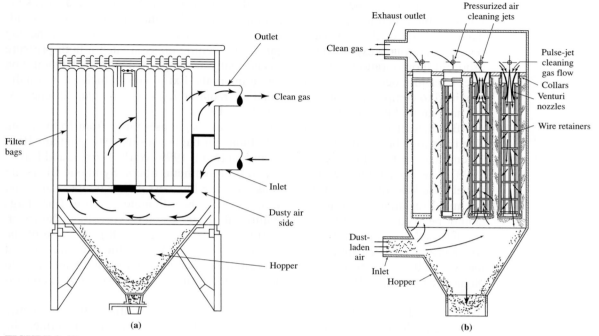

FIGURE 9-40
Mechanically cleaned (shaker) baghouse (*a*) and pulse-jet-cleaned baghouse (*b*). Pulse-jet baghouse shows normal operation for three left-hand-side bags and pulse-jet cleaning for the bag on the right-hand side. (*Source*: Walsh, 1967.)

the particle and fiber). Once a dust cake begins to form on the fabric, sieving is probably the dominant mechanism. As particulate matter collects on the bag, the collection efficiency increases. The buildup of the dust cake also increases the resistance to gas flow.

At some point the pressure drop across the filter bags reduces the gas flow rate to an unacceptable level and the filter bags must be cleaned. The three methods used to clean the bags are mechanical shaking, reverse air flow, and pulse-jet cleaning.

Mechanically cleaned baghouses operate by directing the dirty gas into the inside of the bag. The particulate matter is collected on the inside of the bag much in the same manner as a vacuum cleaner bag. The bags are hung on a frame that oscillates. They are shaken at periodic intervals, ranging from 30 minutes to more than 2 hours. The bags are arranged in groups in separate compartments that are taken off line during cleaning.

In reverse air flow cleaning, a compartment is isolated and a large volume of gas flow is forced countercurrent to normal operation. The dust cake is removed by collapsing or flexing the bag. The reverse flow combined with the inward collapse of the bag causes the collected dust cake to fall into the hopper below.

Pulse-jet baghouses are designed with frame structures, called cages, that support the bags. In contrast to the other two cleaning methods, the particulate matter is collected on the outside of the bag instead of the inside of the bag. The dust cake is removed by directing a pulsed jet of compressed air into the bag. This causes a sudden expansion of the bag. Dust is removed primarily by inertial forces as the bag reaches maximum expansion. The pulse of cleaning air is at such a high pressure drop and short duration that cleaning is normally accomplished with the baghouse on line. Cleaning occurs at 2- to 15-minute intervals. Extra bags, which are normally provided to compensate for the bags that are required in the other cleaning schemes, are not required in pulse-jet baghouses (Noll, 1999).

Bag diameters for shaker and reverse air flow baghouses range from 15 to 45 cm. The bags may be up to 12 m in length. Pulse-jet baghouses use bags that are 10 to 15 cm in diameter with lengths less than 5 m (Noll, 1999; Wark, Warner, and Davis, 1998). The bags are made of either natural or synthetic fibers. Synthetic fibers are widely used as filtration fabrics because of their low cost, better temperature- and chemical-resistance characteristics, and small fiber diameter. Cotton fiber bags cannot be used for sustained temperatures above 90°C. Glass fiber bags, however, can be used at temperatures up to 260°C (McKenna et al., 2000). Because of the stress produced in cleaning, only woven fibers are used when the bags are cleaned mechanically or by reverse air flow. Felted fabrics are used in pulse-jet-cleaned baghouse (Noll, 1999). Bag life varies between 1 and 5 years. Two years is considered normal.

The fundamental design parameter for baghouses is the ratio of the volumetric flow rate of the gas to be cleaned to the area of filter fabric. This ratio is termed the *air-to-cloth ratio*.* It has units of $m^3/s \cdot m^2$ or m/s. Typical air-to-cloth ratios are shown in Tables 9-18 and 9-19.

Baghouses have found a wide variety of applications. Examples include the carbon black and gypsum industries, cement crushing, feed and grain handling, limestone crushing, sanding machines, and coal-fired utility boilers. Of all of the particulate control devices, filtration is the only technology that has the potential to include the addition of adsorption media to facilitate concurrent removal of gas phase contaminants.

TABLE 9-18
Typical air-to-cloth ratios[a]

Baghouse cleaning method	Air-to-cloth ratio, m/s
Shaking	0.010 to 0.017
Reverse air	0.010 to 0.020
Pulse jet	0.033 to 0.083

[a]Compiled from Davis, 2000; Noll, 1999; Wark, Warner, and Davis, 1998.

---

*It may also be called the *gas-to-cloth ratio, filtration velocity,* or the *face velocity.*

TABLE 9-19
Air-to-cloth ratios for bag filters 🏴

Dust	Shaker/woven, reverse, $m^3/min \cdot m^2$	Pulse jet, $m^3/min \cdot m^2$
Coal	0.8	2.4
Cement	0.6	2.4
Fly ash	0.8	1.5
Grain	1.1	4.3
Iron ore	0.9	3.4
Lead oxide	0.8	2.1
Lime	0.8	3.0
Paper	1.1	3.0
Sand	0.8	3.0

Source: U.S. EPA, 1990.

Example 9-14. An aggregate plant at Lime Ridge has been found to be in violation of particulate discharge standards. A mechanical shaker baghouse has been selected for particulate control. Estimate the number of bags required for a gas flow rate of 20 m^3/s if each bag is 15 cm in diameter and 12 m in length. One-eighth of the bags are taken off line for cleaning. The manufacturer's recommended air-to-cloth ratio for aggregate plants is 0.010 m/s.

Solution.
Noting that the air-to-cloth ratio units of m/s are equivalent to $m^3/s \cdot m^2$, calculate the net cloth area required with one compartment off line for cleaning:

$$A = \frac{Q}{v} = \frac{20 \text{ m}^3/\text{s}}{0.010 \text{ m}^3/\text{s} \cdot \text{m}^2} = 2,000 \text{ m}^2$$

The net number of bags is the total area divided by the area of one bag:

$$\frac{2,000 \text{ m}^2}{(\pi)(0.15 \text{ m})(12 \text{ m}) \text{ per bag}} = 353.67 \text{ or } 354 \text{ bags}$$

With one-eighth of the bags off line, an additional one-eighth of the net number is required:

$$\frac{354 \text{ bags}}{8} = 44.25 \text{ or } 44 \text{ bags}$$

The total number of bags is $354 + 44 = 398$.

Comment: In order to have an equal number of bags in each compartment, the total number of bags will have to be slightly larger (398 bags/8 compartments = 49.75 bags per compartment). With 50 bags per compartment, the total will be 400 bags.

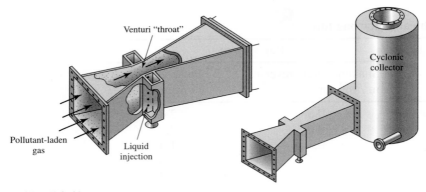

FIGURE 9-41
Venturi scrubber.

Liquid Scrubbing.   When the particulate matter to be collected is wet, corrosive, or very hot, the fabric filter may not work. Liquid scrubbing might. Typical scrubbing applications include control of emission of talc dust, phosphoric acid mist, foundry cupola dust, and open hearth steel furnace fumes.

Liquid scrubbers vary in complexity. Simple spray chambers are used for relatively coarse particle sizes. For high efficiency removal of fine particles, the combination of a venturi scrubber followed by a cyclone would be selected (Figure 9-41). The underlying principle of operation of the liquid scrubbers is that a differential velocity between the droplets of collecting liquid and the particulate pollutant allows the particle to impinge onto the droplet. Because the droplet-particle combination is still suspended in the gas stream, an inertial collection device is placed downstream to remove it. Because the droplet enhances the size of the particle, the collection efficiency of the inertial device is higher than it would be for the original particle without the liquid drop.

The most popular collection efficiency equation is that proposed by Johnstone, Field, and Tassler (1954):

$$\eta = 1 - \exp(-\kappa R \sqrt{\psi}) \tag{9-66}$$

where η = efficiency
   exp = exponential to base e
   κ = correlation coefficient, m^3 of gas/m^3 of liquid
   R = liquid flow rate, m^3/m^3 of gas
   ψ = inertial impaction parameter defined by Equation 9-67

The inertial impaction parameter (ψ) relates the particle and droplet sizes and relative velocities:

$$\psi = \frac{C\rho_p v_g (d_p)^2}{18 d_d \mu} \tag{9-67}$$

where C = Cunningham correction factor defined by Equation 9-68, unitless
   ρ_p = particle density, kg/m^3
   v_g = speed of gas at throat, m/s

d_p = diameter of particle, m
d_a = diameter of droplet, m
μ = dynamic viscosity of gas, Pa · s

The Cunningham correction factor accounts for the fact that very small particles do not obey Stokes' settling equation. They tend to "slip" between the gas molecules. Thus, the drag coefficient (C_D) is reduced and the particles fall faster than otherwise would be expected. This is particularly true for particles less than 1 μm in diameter. The Cunningham factor may be approximated with the following equation (Hesketh, 1977):

$$C = 1 + \frac{6.21 \times 10^{-4}(T)}{d_p} \tag{9-68}$$

where T = absolute temperature, K
d_p = diameter of particle, μm

Example 9-15.   Given the scrubber described below, write an expression for collection efficiency that is a function of particle size. Assume the particles are fly ash with a density of 700 kg/m^3 and a minimum size of 10 μm diameter.

Venturi characteristics:
Throat area = 1.00 m^2
Gas flow rate = 94.40 m^3/s
Gas temperature = 150°C
Liquid flow rate = 0.13 m^3/s
Coefficient κ = 200
Droplet diameter = 100 μm

Solution.   We begin by determining the value of the Cunningham correction factor for the smallest particle to see if the d_p term in the denominator must be retained.

$$C = 1 + \frac{6.21 \times 10^{-4}(423 \text{ K})}{10 \text{ } \mu\text{m}}$$

$$= 1 + 0.0263$$

For this we can see that the term containing d_p will be small for all particles greater than 10 μm and we can use the approximation:

$$C = 1$$

Before we can proceed to calculate a value for ψ, we must determine the gas velocity at the throat:

$$v_g = \frac{Q_g}{A_t}$$

where A_t = cross-sectional area of throat

$$v_g = \frac{94.40 \text{ m}^3/\text{s}}{1.00 \text{ m}^2} = 94.40 \text{ m/s}$$

The dynamic viscosity of the gas is determined from Table A-4 of Appendix A and from the temperature of the gas (150°C). It is 25.2 μPa $\cdot$ s.

Now we can calculate in terms of d_p in μm. Note that $C = 1$ and that 18 is a constant.

$$\psi = \frac{(1)(700 \text{ kg/m}^3)(94.40 \text{ m}^3/\text{s})(1 \times 10^{-12} \, \mu\text{m}^2/\text{m}^2)(d_p)^2}{(18)(100 \times 10^{-6} \text{ m})(25.2 \times 10^{-6} \text{ Pa} \cdot \text{s})}$$

$$= (1.46)(d_p)^2$$

Taking the square root of ψ and computing R as 0.13/94.40, the expression for efficiency as a function of diameter is then

$$\eta = 1 - \exp[-(200)(1.38 \times 10^{-3})(1.21)d_p]$$
$$= 1 - \exp[-0.33(d_p)]$$

Electrostatic Precipitation (ESP).   High efficiency, dry collection of particles from hot gas streams can be obtained by electrostatic precipitation of the particles. The ESP is usually constructed of alternating plates and wires (Figure 9-42). A large direct current potential (30–75 kV) is established between the plates and wires. This results in the creation of an ion field between the wire and plate (Figure 9-43a). As the particle-laden gas stream passes between the wire and the plate, ions attach to the particles, giving them a net negative charge (Figure 9-43b). The particles then migrate toward the positively charged plate where they stick (Figure 9-43c). The plates are rapped at frequent intervals and the agglomerated sheet of particles falls to a hopper.

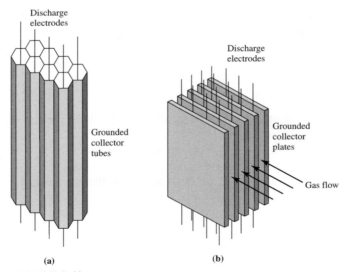

(a)   **(b)**

FIGURE 9-42
Electrostatic precipitator with (a) wire in tube, (b) wire and plate.
(*Source:* NCAPCA, 1967.)

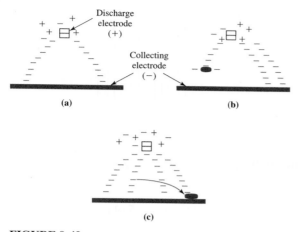

FIGURE 9-43
Particle charging and collection in ESP. (*Source:* NCAPCA, 1967.)

Unlike the baghouse, the gas flow between the plates is not stopped during cleaning. The gas velocity through the ESP is kept low (less than 1.5 m/s) to allow particle migration. Thus, the terminal settling velocity of the sheet is sufficient to carry it to the hopper before it exits the precipitator.

The classic ESP efficiency equation is the one proposed by Deutsch (1922).

$$\eta = 1 - \exp\left(-\frac{Aw}{Q_g}\right) \qquad \text{(9-69)}$$

where A = collection area of plates, m^2
  w = migration velocity of particles, m/s
  Q_g = gas flow rate, m^3/s

The migration velocity of the particles is a function of the electrostatic force. The migration velocity is described by the following equation:

$$w = \frac{qE_pC}{6\pi r\mu} \qquad \text{(9-70)}$$

where q = charge, coulombs (C)
  E_p = collection field intensity, volts/m
  r = particle radius, m
  μ = dynamic viscosity of gas, Pa · s
  C = Cunningham correction factor

Example 9-16. Determine the collection efficiency of the electrostatic precipitator described below for a particle 154 μm in diameter having a drift velocity of 0.184 m/s. What is the effect of reducing the plate spacing to one-half of its current value and doubling the number of plates?

ESP specifications:

Height = 7.32 m

Length = 6.10 m

Number of plates = 5

Plate spacing = 0.28 m

Gas flow rate = 19.73 m³/s

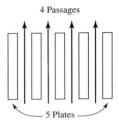

4 Passages

5 Plates

Solution.   First we calculate the area of the plates. For a single plate,

$$A = 7.32 \times 6.1 = 44.65 \text{ m}^2$$

Since there are eight collecting surfaces (two for each passage, 5 plates form 4 passages):

$$A = 44.65 \text{ m}^2 \times 8 = 357.2 \text{ m}^2$$

The efficiency is then calculated in a straightforward manner using Equation 9-69.

$$\eta = 1 - \exp\left[-\frac{(357.2)(0.184)}{19.73}\right]$$

$$= 0.964$$

Therefore the efficiency is 96.4 percent. Now what is the effect of reducing the plate spacing? The spacing enters into the efficiency equation through the calculation of the collection field intensity (E_p). Treating everything else in Equation 9-70 as a constant, we can write the following equation:

$$w = KE_p$$

where E_p is measured in volts per meter. If the distance between the plates is reduced, the collection field intensity is proportionately increased:

$$w = KE_p \frac{0.28}{0.14} = 2KE_p$$

Thus, w increases by a factor of two. In order to maintain the same gas velocity, the number of plates and, hence, the surface area (A) must double. The new efficiency would then be:

$$\eta = 1 - \exp\left[-\frac{(714.4)(0.368)}{19.73}\right]$$

$$= 1 - 0.0000016$$

$$= 0.999998, \text{ or } 1.00$$

Comment: The efficiency would be increased to 100 percent. This plate spacing may not be feasible because of sparkover problems.

One operational problem of ESPs is of particular note. *Fly ash* is a generic term used to describe the particulate matter carried in the effluent gases from furnaces burning fossil fuels. ESPs often are used to collect fly ash. The strongest force holding fly ash to the collection plate is electrostatic and is caused by the flow of current through the fly ash. The fly ash acts like a resistor and, hence, resists the flow of current. This resistance to current flow is called the resistivity of the fly ash. It is measured in units of ohm · cm. If the resistivity is too low (less than 10^4 ohm · cm), not enough charge will be retained to produce a strong force and the particles will not "stick" to the plate. Conversely, and often more importantly, if the resistivity is too high (greater than 10^{10} ohm · cm), there is an insulating effect. The layer of fly ash breaks down locally and a local discharge of current (*back corona*) from the normally passive collection electrode occurs. This discharge lowers the sparkover voltage and produces positive ions that decrease particle charging and, hence, collection efficiency.

The presence of SO_2 in the gas stream reduces the resistivity of the fly ash. This makes particle collection relatively easy. However, the mandate to reduce SO_2 emissions has frequently been satisfied by switching to low sulfur coal. The result has been increased particulate emissions. This problem can be resolved by adding conditioners such as SO_3 or NH_3 to reduce the resistivity or by building larger precipitators.

Electrostatic precipitators have been used to control air pollution from electric power plants, Portland cement kilns, blast furnace gas, kilns and roasters for metallurgical processes, and mist from acid production facilities.

Control Technologies for Mercury

During combustion, the mercury in coal is volatilized and converted to Hg^0 vapor. As the flue gas cools, a series of complex reactions convert Hg^0 to Hg^{2+} and particulate Hg compounds (Hg_p). The presence of chlorine favors the formation of mercuric chloride. In general, the majority of gaseous mercury in bituminous coal-fired boilers is Hg^{2+}. The majority of gaseous mercury in sub-bituminous- and lignite-fired boilers is Hg^0.

Existing boiler control equipment achieves some ancillary removal mercury compounds. Hg_p is collected in particulate control equipment. Soluble Hg^{2+} compounds are collected in FGD systems. Particulate control equipment has achieved a range of emission reductions from 0 to 90 percent. Of these units, fabric filters obtained the highest levels of control. Dry scrubbers achieve average total mercury (particulate plus compounds) ranging from 0 to 98 percent. Wet FGD scrubber efficiencies were similar. Higher efficiencies were achieved at boilers using bituminous coal than at those using sub-bituminous and lignite coal. EPA estimates that existing controls remove about 36 percent of the 75 Mg of mercury input with coal in U.S. coal-fired boilers (Srivastava et al., 2005).

There are two broad approaches being developed to control mercury emissions: powdered activated carbon (PAC) injection and enhancement of existing control devices. The leading candidates for top efficiency (90 percent) are PAC with pulse-jet fabric filters and FGD (wet or dry) with fabric filters (U.S. EPA, 2003b).

9-11 AIR POLLUTION CONTROL OF MOBILE SOURCES

Engine Fundamentals

Before we examine some cures for the pollution from the common gasoline auto engine, it may be useful to compare the three familiar types of engines: the gasoline engine, the diesel engine, and the jet engine.

The Gasoline Engine. Each of the four strokes of the engine is diagrammed in Figure 9-44. In the typical automobile engine with no air pollution controls, a mixture of fuel and air is fed into a cylinder and is compressed and ignited by a spark from the spark plug. The explosive energy of the burning mixture moves the pistons. The pistons' motion is transmitted to the crankshaft that drives the car. The burnt, spent mixture passes out of the engine and out through the tail pipe.

As noted in Figure 9-44, a mixture of air and fuel is drawn into the cylinder. The ratio of air-to-fuel is the single most important factor in determining emissions from a four-stroke internal combustion engine. An estimate of the theoretical mass of air required to burn the fuel may be made using C_7H_{13} to represent the blend of hydrocarbons that we call gasoline. The complete combustion of C_7H_{13} in pure oxygen may be expressed by the following stoichiometric equation:

$$C_7H_{13} + 10.25O_2 \rightarrow 7CO_2 + 6.5H_2O \qquad (9\text{-}71)$$

Because air rather than pure oxygen is used, we use the molar ratio of nitrogen to air, that is, 3.76 moles of N_2 for each mole of O_2. Thus, Equation 9-71 is rewritten as

$$C_7H_{13} + 10.25O_2 + 38.54N_2 \rightarrow 7CO_2 + 6.5H_2O + 38.54N_2 \qquad (9\text{-}72)$$

The oxidation of nitrogen to nitrogen oxides has been ignored in this reaction. The calculation of the stoichiometric air-to-fuel ratio for this reaction is illustrated in Example 9-17.

Example 9-17. Determine the stoichiometric air-to-fuel ratio for C_7H_{13}. Ignore constituents other than oxygen and nitrogen in air and ignore the oxidation of nitrogen to nitrogen oxides.

Solution. From Equation 9-72, 10.25 moles of oxygen and 38.54 moles of nitrogen react with each mole of C_7H_{13}. Calculate the molar masses of each constituent as follows:

1 mole of C_7H_{13} = (7 mole $\times$ 12 g/mole) + (13 moles $\times$ 1 g/mole) = 97 g

10.25 moles of O_2 = 10.25 $\times$ 2 moles $\times$ 16 g/mole) = 328 g

38.54 moles of N_2 = 38.54 $\times$ 2 moles $\times$ 14 g/mole = 1,079 g

The stoichiometric air-to-fuel ratio is

$$\frac{\text{Air}}{\text{Fuel}} = \frac{(328 \text{ g} + 1,079 \text{ g})}{97 \text{ g}} = 14.5$$

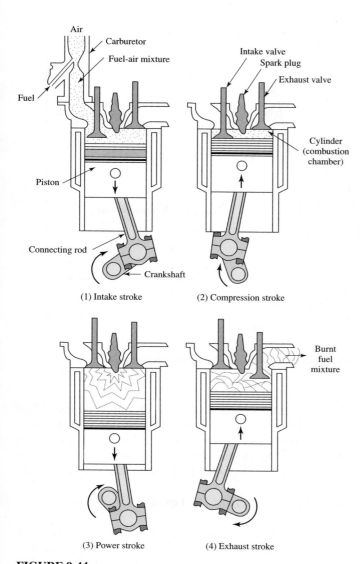

Air
Carburetor
Fuel-air mixture
Intake valve
Spark plug
Exhaust valve
Fuel
Cylinder
(combustion
chamber)
Piston
Connecting rod
Crankshaft

(1) Intake stroke   (2) Compression stroke

Burnt
fuel
mixture

(3) Power stroke   (4) Exhaust stroke

FIGURE 9-44

Combustion in an automobile engine. On the intake stroke (*1*), the piston moves down and a mixture of fuel and air is drawn into the cylinder past the open intake valve. With the compression stroke (*2*), the intake valve closes and the piston moves and compresses the air-fuel mixture. On the power stroke (*3*), a spark from the spark plug ignites the heated, compressed mixture, which begins to burn, expands, and pushes the piston down. For the exhaust stroke (*4*), the exhaust valve opens, the spent, burned mixture exits with its pollutants, and the piston returns to the top of the cylinder. (*Source:* NTRDA, 1969.)

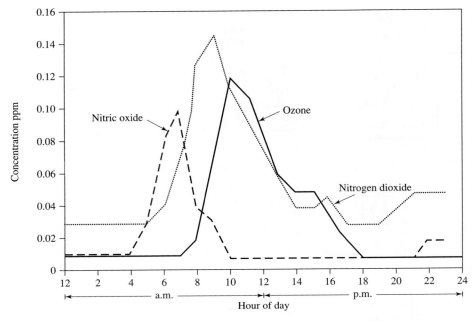

FIGURE 9-45
Diurnal variation of NO, NO_2, and O_3 concentrations in Los Angeles on July 19, 1965.
(*Source:* NAPCA, 1970.)

For maximum power, however, the proportion of air to fuel must be less. Most driving takes place at less than the 15-to-1 *air-to-fuel* ratio. Combustion is incomplete, and substantial amounts of material other than carbon dioxide and water are discharged through the tail pipe. One result of having an inadequate supply of air is the emission of carbon monoxide instead of carbon dioxide. Other by-products are unburned gasoline and hydrocarbons.

Because of the high temperatures and pressures that exist in the cylinder, copious amounts of NO_x are formed (see Equation 9-12).

The role of automobile emissions in the formation of ozone is illustrated in Figure 9-45. As the workday begins and automobiles take to the road NO concentrations increase. NO is oxidized to NO_2, which increases in concentration while the NO concentration declines. Photolysis by sunlight decomposes NO_2 to NO plus O. The atomic oxygen combines with diatomic oxygen in the atmosphere to form ozone. Ozone can then convert NO back to NO_2. These reactions are summarized in the following equations:

$$\text{During combustion: } N_2 + O_2 \rightarrow 2NO \tag{9-73}$$

$$\text{Oxidation of NO in atmosphere: } 2NO + O_2 \rightarrow 2NO_2 \tag{9-74}$$

$$\text{Photolysis: } NO_2 + h\nu \rightarrow NO + O \tag{9-75}$$

$$\text{Formation of ozone: } O + O_2 + M \rightarrow O_3 + M \tag{9-76}$$

$$\text{Conversion of NO back to } NO_2\text{: } O_3 + NO \rightarrow NO_2 + O_2 \tag{9-77}$$

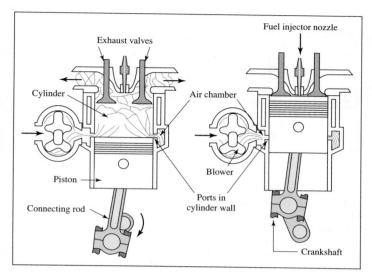

FIGURE 9-46

Combustion in a two-stroke diesel engine. With the piston at the bottom of the cylinder (shown left), and exhaust valves and ports open, fresh air is forced into the cylinder by the blower, and the used air-fuel mixture—along with any polluting byproducts—left from the previous stroke is forced out. On the second stroke (shown right), the exhaust valves close, the piston rises—shutting off the ports—and compresses the air. When the piston reaches a position near the top of the cylinder, fuel is injected into the now highly compressed, heated air. This heated air ignites the fuel without a spark, and the resulting combustion forces the piston down to its first position. (*Source:* NTRDA, 1969.)

Where hv represents a photon and M represents another molecule such as nitrogen. These reaction products in combination with reaction products from hydrocarbons and ozone are the constituents of photochemical smog.

The Diesel Engine.   As shown in Figure 9-46, the diesel engine differs from the four-stroke engine in two respects.

First, the air supply is unthrottled; that is, its flow into the engine is unrestricted. Thus, a diesel normally operates at a higher air-to-fuel ratio than does a gasoline engine.

Second, there is no spark ignition system. The air is heated by compression. That is, the air in the engine cylinder is squeezed until it exerts a pressure high enough to raise the air temperature to about 540°C, which is enough to ignite the fuel oil as it is injected into the cylinder.

A well-designed, well-maintained, and properly adjusted diesel engine will emit less CO and hydrocarbons than the four-stroke engine because of the diesel's high air-to-fuel ratio. However, the higher operating temperatures lead to substantially higher NO_x emissions. In addition, when the engine is overloaded during acceleration from a stop, CO, VOCs, odors and particulate matter (smoke) may be emitted in large quantities (Cooper and Alley, 2002).

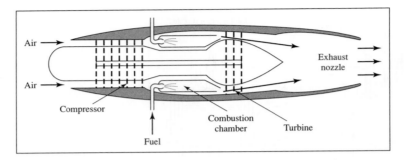

FIGURE 9-47

Combustion in a jet engine. Air enters through the front and goes to a compressor, where it is increasingly compressed and forced into combustion chambers that are arranged in circles around the engine. Fuel is sprayed into the front end of the combustion chamber in a steady stream so that it ignites and burns continuously. The burning air-fuel mixture expands and pushes toward the rear. (On the way, it hits turbine wheel blades and forces them to rotate. This rotation drives the compressor.) As the expanded mixture moves toward the tailpipe, the areaway narrows and the stream of burning air-fuel mixture is compressed into the exceedingly strong jet stream that shoots out of the rear of the plane. (*Source:* NTRDA, 1969.)

The Jet Engine. Large commercial aircraft that utilize the thrust of compressed gases for propulsion may contribute significant amounts of particulates and NO_x to urban atmospheres. Their largest emission rate is on takeoff and climb-out. However, on an annual basis, emissions from jet engines are small relative to those from highway vehicles.

As shown in Figure 9-47, air drawn into the front of the engine is compressed and then heated by burning fuel. The expanding gas passes through turbine blades, which drive the compressor. The gas then exits the engine through an exhaust nozzle.

Effect of Design and Operating Variables on Emissions. The list of variables that affect internal combustion (automobile) emissions includes the following (Patterson and Henein, 1972):

1. Air-to-fuel ratio

2. Load or power level

3. Speed

4. Spark timing

5. Exhaust back pressure

6. Valve overlap

7. Intake manifold pressure

8. Combustion chamber deposit buildup

9. Surface temperature

10. Surface-to-volume ratio

11. Combustion chamber design

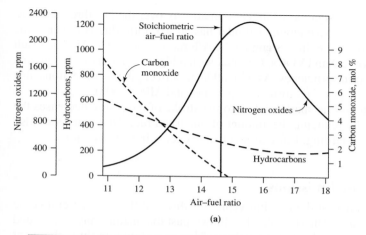

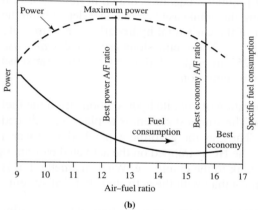

FIGURE 9-48
Effect of air-to-fuel ratio (*a*) on emissions, and (*b*) on power and economy. (*Source:* Seinfeld, 1975.)

12. Stroke-to-bore ratio

13. Displacement per cylinder

14. Compression ratio

A discussion of all of these items is beyond the scope of an introductory text such as this. Therefore, we shall restrict ourselves to a few of the variables that serve to illustrate the kinds of problems encountered in trying to design pollution out of an internal combustion engine.

The *air-to-fuel* ratio (A/F) is fairly easy to regulate. As we noted previously, it has a direct effect on all three emissions. As shown in Figure 9-48*a*, the A/F of 14.6 is the *stoichiometric* mixture for complete combustion.* At lower ratios, both CO and HC emissions increase. At higher ratios, to about 15.5, NO_x emissions increase. At very lean mixtures (high A/F ratios), the NO_x emission begins to decrease.

---

*Note that stoichiometric (stoi-chio-met-ric) means "combined in exactly the proper proportions according to their molecular weight."

Combustion chamber design has changed dramatically over the last three decades. For example, the *extra-lean-burn engine* cylinder has an A/F ratio that is richer at the spark plug and leaner elsewhere. In this engine the A/F ratio is as high as 25. The result is a very low emission of CO and VOCs. Coupled with a higher drive ratio, this engine also gives improved gasoline efficiency. As with all lean-burn engine modifications, this one has the drawback of higher NO formation (Cooper and Alley, 2002).

Retarding the timing of the spark relative to the stroke of the piston decreases the hydrocarbon emissions by reducing the amount of unburned fuel. NO_x emissions also decrease with increased retarding. Little or no change occurs in CO emissions.

Control of Automobile Emissions

Blowby.   The flow of air past the moving vehicle is directed through the crankcase in order to rid it of any gas-air mixture that has blown past the pistons, any evaporated lubricating oil, and any escaped exhaust products. The air is drawn in through a vent and emitted through a tube extending from the crankcase at a rate that depends on the speed of the car. About 20 to 40 percent of the car's total hydrocarbon emissions are sent into the atmosphere from the crankcase. These emissions are called *crankcase blowby*. All vehicles manufactured after 1963 are required to have a positive crankcase ventilation (PCV) valve to eliminate blowby emissions.

Fuel Tank Evaporation Losses.   Evaporation of volatile hydrocarbons from the fuel tank is controlled by one of two systems. The simplest system is to place an activated charcoal adsorber in the tank vent line. Thus, as the gasoline expands during warm weather and forces vapor out of the vent, the HC is trapped on the activated carbon.

An alternative system is to vent the tank to the crankcase. With this method, it is more difficult to achieve 100 percent control than with the activated charcoal system.

Engine Exhaust.   Because engine modifications alone are not sufficient to meet stringent emission standards, an external catalytic reactor (commonly referred to as the *catalytic converter*) is placed on the exhaust system. The function of the catalytic converter is to promote reactions that convert NO_x to N_2, CO to CO_2, and hydrocarbons to CO_2 and H_2O. A *three-way catalyst* (TWC) that simultaneously oxidizes the hydrocarbons and CO and reduces the NO_x is employed. The catalyst is a precious metal (for example platinum/rhodium) on an alumina support structure. The catalyst must operate in a narrow band of A/F ratios that is centered about the stoichiometric point (Figure 9-48). In addition, the gases entering the catalyst bed must have a specific composition and the catalyst temperature must be carefully controlled. A sophisticated computerized electronic control system maintains the correct A/F and temperature (Figure 9-49).

The major problems with the catalysts are their susceptibility to "poisoning" by lead, phosphorus, and sulfur, and their poor wear characteristics under thermal cycling. The poisoning problem is solved by removing the lead, phosphorus, and sulfur from the fuel.

Another approach being implemented is fuel modification. The use of lead in fuels was completely phased out by January 1996. In addition, diesel fuel refining is being changed so that it will contain less sulfur and emit 20 percent less VOC's. Lowering

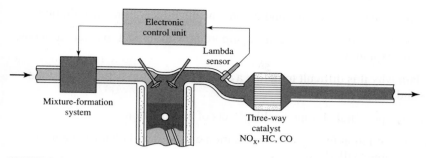

FIGURE 9-49
Single-bed, three-way catalyst with electronic control systems. (*Source:* Bosch, 1988.)

the gasoline vapor pressure (called the *Reid vapor pressure*) reduces hydrocarbon emissions. *Oxyfuel* is yet another alternative. Oxyfuel is one with more oxygen to allow the fuel to burn more efficiently. Other alternatives include alcohols, liquified petroleum gas, and natural gas.

Inspection/Maintenance (I/M) Programs. The devices installed by automobile manufacturers are extremely successful in minimizing the pollution from the exhaust and from evaporating fuel. However, as with other aspects of running an automobile, these devices wear out and fail. Because their failure does not inhibit the operation of the automobile, they are not likely to be repaired by the owner. In those areas that have exceeded the NAAQS (nonattainment areas), inspection/maintenance programs have been implemented to ensure that the control devices are in good working order. These programs require periodic checks of the exhaust and, in some instances, the evaporative controls. If the vehicle fails the inspection, the owner is required to provide the required maintenance and have the vehicle reinspected. Failure to pass the inspection may be cause to deny the issuance of license plates or tags.

9-12   CHAPTER REVIEW

When you have completed studying this chapter, you should be able to do the following without the aid of your textbook or notes:

1. List the six criteria air pollutants for which the U.S. Environmental Protection Agency has designated National Ambient Air Quality Standards (NAAQS).

2. List and define three units of measure used to report air pollution data (that is, ppm, $\mu g/m^2$, and μm).

3. Explain the difference between ppm in air pollution and ppm in water pollution.

4. Explain the effect of temperature and pressure on readings made in ppm.

5. Explain the influence of moisture, temperature, and sunlight on the severity of air pollution effects on materials.

6. Differentiate between acute and chronic health effects from air pollution.

7. State which particle sizes are more important with respect to alveolar deposition and explain why.

8. Explain why it is difficult to define a causal relationship between air pollution and health effects.

9. List three potential chronic health effects of air pollution.

10. List four common features of air pollution episodes and identify the locations of three "killer" episodes.

11. Discuss the natural and anthropogenic origin of the six criteria air pollutants and identify the likely mechanisms for their removal from the atmosphere.

12. Identify one indoor air pollution source for each of the following pollutants: CH_2O, CO, NO_x, Rn, respirable particulates, and SO_x.

13. Define the term "acid rain" and explain how it occurs.

14. Discuss the photochemistry of ozone in the upper atmosphere using the pertinent chemical reactions. Discuss the effect of chlorofluorocarbons on these reactions.

15. Explain the term "greenhouse effect," its hypothesized cause, and why it is being debated, pro and con.

16. Determine the stability (ability to dissipate pollutants) of the atmosphere from vertical temperature readings.

17. Explain why valleys are more susceptible to inversions than is flat terrain.

18. Explain why lake breezes and land breezes occur.

19. Explain how a lake breeze adversely affects the dispersion of pollutants.

20. State the theoretical principle on which each of the following air pollution control devices operates: (*a*) absorption column (either a packed tower or plate tower), (*b*) adsorption column, (*c*) either afterburner or catalytic combustor, (*d*) cyclone, (*e*) baghouse, (*f*) venturi scrubber, and (*g*) electrostatic precipitator.

21. Select the correct air pollution control device for a given pollutant and source.

22. Discuss the pros and cons of FGD and the problem of fly ash resistivity.

23. Explain the difference between prevention and post-combustion techniques for reduction of nitrogen oxide emissions and give one example of each.

24. Graph the relationship between air-to-fuel ratio and emission of CO, HC, and NO_x from automobiles.

25. Explain how evaporative emissions are commonly controlled.

26. Explain how exhaust emissions are commonly controlled and the role of computerized control systems in making them work.

With the aid of this text, you should be able to do the following:

27. Solve gas law problems.

28. Convert parts per million (ppm) to micrograms per cubic meter ($\mu g/m^3$) and vice versa.

29. Calculate the amount of SO_2 that will be released from burning coal or fuel oil with a given sulfur content in percent.

30. Calculate the ground level concentration of air pollutants released from a stationary elevated source or the emission rate (QE) for a given ground level concentration.

31. Use the air pollution control equations to analyze the performance and modify the design of absorption and adsorption control devices, cyclones, scrubbers, and ESPs.

9-13   PROBLEMS

9-1. What is the density of oxygen at a temperature of 273.0 K and a pressure of 98.0 kPa?

> *Answer:* 1.382 kg/m^3

9-2. Determine the density of carbon monoxide gas at a pressure of 102.0 kPa and a temperature of 298.0 K.

9-3. Calculate the density of methane at a temperature of 273.0 K and at a pressure of 101.325 kPa?

9-4. Show that 1 mole of any ideal gas will occupy 22.414 L at standard temperature and pressure (STP). (STP is 273.16 K and 101.325 kPa.)

9-5. What volume would 1 mole of an ideal gas occupy at 20°C and 101.325 kPa?

9-6. A sample of air contains 8.583 moles/m^3 of oxygen and 15.93 moles/m^3 of nitrogen at STP. Determine the partial pressures of oxygen and nitrogen in 1.0 m^3 of the air.

> *Answer:* 19.45 kPa; 36.18 kPa

9-7. A 1.000 m^3 volume tank contains a gas mixture of 8.32 moles of oxygen, 16.40 moles of nitrogen, and 16.15 moles of carbon dioxide. What is the partial pressure of each component in the gas mixture at 25.0°C?

9-8. A 1.000 m^3 volume tank contains a gas mixture of oxygen, nitrogen, and carbon dioxide. How many moles are there of each of these components of the gas mixture at 25.0°C if the partial pressures of each gas are as shown below?

$$P_{O_2} = 45.39 \text{ kPa}$$
$$P_{N_2} = 40.63 \text{ kPA}$$
$$P_{CO_2} = 15.24 \text{ kPa}$$

9-9. Calculate the volume occupied by 5.2 kg of carbon dioxide at 152.0 kPa and 315.0 K.

Answer: 2,036 L

9-10. Determine the mass of oxygen contained in a 5.0 m^3 volume under a pressure of 568.0 Pa and at a temperature of 263.0 K.

9-11. Calculate the volume occupied by 235 μg of O$_3$ at STP. If this volume is contained in 1.00 m^3 of air, what is the volumetric ratio (that is volume of O$_3$ per volume of air)?

9-12. A gas mixture at 0°C and 108.26 kPa contains 250 mg/L of H$_2$S gas. What is the partial pressure exerted by this gas?

Answer: 16.7 kPa/L

9-13. A 28-L volume of gas at 300.0 K contains 11 g of methane, 1.5 g of nitrogen, and 16 g of carbon dioxide. Determine the partial pressure exerted by each gas.

9-14. Given the gas mixture of Problem 9-13, how many moles of each gas are present in the 28-L volume?

Answer: 0.688 moles of CH$_4$; 0.054 moles of N$_2$; 0.364 moles of CO$_2$

9-15. The partial pressures of the gases in a 22.414 L volume of air at STP are: oxygen, 21.224 kPa; nitrogen, 79.119 kPa; argon 0.946 kPa; and carbon dioxide, 0.036 kPa. Determine the gram-molecular weight of air.

9-16. Using the data in Problem 9–15, determine the gram–molecular weight of air at 500°C and 101.325 kPa.

9-17. Convert 1,950 μg/m^3 of SO$_2$ to ppm at 25°C and 101.325 kPa pressure.

Answer: 0.75 ppm

9-18. Convert 0.55 ppm of NO$_2$ to μg/m^3 at -17.7°C and 100.0 kPa pressure.

9-19. Convert 370 ppm of CO$_2$ to μg/m^3 at 20°C and 101.325 kPa.

9-20. Given the following temperature profiles, determine whether the atmosphere is unstable, neutral, or stable. Show all work and explain choices.

a. Z, m	T, °C
2	−3.05
318	−6.21

b. Z, m	T, °C
10	6.00
202	3.09

c. Z, m T, °C

18	14.03
286	16.71

Answers: (a) neutral; (b) unstable; (c) stable (inversion)

9-21. Determine the atmospheric stability for each of the following temperature profiles. Show all work and explain choices.

a. Z, m T, °C

1.5	−4.49
339	0.10

b. Z, m T, °C

12	28.05
279	19.67

c. Z, m T, °C

8	19.55
339	18.93

9-22. Determine the atmospheric stability for each of the following temperature profiles. Show all work and explain choices.

a. Z, m T, °C

2.00	5.00
50.00	4.52

b. Z, m T, °C

2.00	5.00
50.00	5.00

c. Z, m T, °C

2.00	−21.01
50.00	−25.17

9-23. Given the following observations, use the key to stability categories (Table 9-13) to determine the stability.
a. Clear winter morning at 9:00 A.M.; wind speed of 5.5 m/s
b. Overcast summer afternoon at 1:30 P.M.; wind speed of 2.8 m/s
c. Clear winter night at 2:00 A.M.; wind speed of 2.8 m/s
d. Summer morning at 11:30 A.M.; wind speed of 4.1 m/s

Answers: (a) D; (b) D; (c) F; (d) B

9-24. Determine the atmospheric stability category of the following observations.
a. Clear summer afternoon at 1:00 P.M.; wind speed of 1.6 m/s
b. Overcast summer night at 1:30 A.M.; wind speed of 2.1 m/s
c. Clear winter morning at 9:30 A.M.; wind speed of 6.6 m/s
d. Thinly overcast winter night at 8:00 P.M.; wind speed of 2.4 m/s

9-25. Determine the atmospheric stability category of the following observations.
a. Clear summer afternoon at 1:00 P.M.; wind speed of 5.6 m/s
b. Clear summer night at 1:30 A.M.; wind speed of 2.1 m/s
c. Overcast winter afternoon at 2:30 P.M.; wind speed of 6.6 m/s
d. Summer afternoon at 1:00 P.M. with broken low clouds; wind speed of 5.2 m/s

9-26. A power plant in a college town is burning coal on a cold, clear winter morning at 8:00 A.M. with a wind speed of 2.6 m/s and an inversion layer with its base at a height of 697 m. The effective stack height is 30 m. Calculate the distance downwind x_L at which the plume released will reach the inversion layer and begin to mix downward.

   Answer: 5.8 km

9-27. A factory releases a plume into the atmosphere on an overcast summer afternoon. At what distance downwind will the plume begin mixing downward if an inversion layer exists at a base height of 414 m and the wind speed is 1.8 m/s? The effective stack height is 45 m.

9-28. At what distance downwind will the plume from a stack begin mixing downward if an inversion layer exists at a base height of 265 m and the wind speed is 4.0 m/s on an overcast summer afternoon? The effective stack height is 85 m.

9-29. Given the same power plant and conditions that were found in Example 9-5, determine the concentration of SO_2 at a point 4 km downwind and 0.2 km perpendicular to the plume centerline ($y = 0.2$ km) if there is an inversion with a base height of 328 m.

   Answer: 1.16×10^{-3} g/m^3

9-30. On a clear summer afternoon with a wind speed of 3.20 m/s, the particulate concentration was found to be 1,520 μg/m^3 at a point 2 km downwind and 0.5 km perpendicular to the plume centerline from a coal-fired power plant. Given the following parameters and conditions, determine the particulate emission rate of the power plant:

   Stack parameters:
   Height = 75.0 m
   Diameter = 1.50 m
   Exit velocity = 12.0 m/s
   Temperature = 322°C

   Atmospheric conditions:
   Pressure = 100.0 kPa
   Temperature = 28.0°C

9-31. Calculate the downwind concentration at 30 km ($y = 0$) in g/m^3 resulting from an emission of 1,976 g/s of SO_2 into a 2.5 m/s wind at 1:00 A.M. on a clear winter night. Assume an effective stack height of 85 m and an inversion layer at 185 m. Identify the stability class and show all work.

9-32. Using a computer spreadsheet program you have written, determine the maximum concentration of SO_2 (in ppm) and the distance downwind of a power plant stack that the maximum concentration of SO_2 occurs for the following conditions:

Power plant data:

Coal specifications
Bituminous (Saginaw No. 1, Belmont, OH)
Sulfur: 2.80 percent
Ash: 9.8 percent

Burning rate: 28.82 megagrams per hour
Stack conditions:
Height: 40.0 m
Inside diameter: 1.8 m
Exit velocity: 10.5 m/s
Exit temperature: 297°C

Meteorological conditions to be investigated:

a. Wind speed: 3.8 m/s
b. Inversion base: 170.0 m above ground surface
c. Ambient temperature: −11°C
d. Ambient pressure: 103.285 kPa
e. Thinly overcast winter night (midnight to 4:00 A.M.)

Your solution should include the microcomputer spreadsheet and a graph of the ground level concentration of SO_2 in g/m^3 versus distance from the stack. Provide a table of values for distances from 0.1 km to 100 km in the following steps:

From 0.1 to 1.0 km in 0.1-km steps
From 1.0 to 10.0 km in 1.0-km steps
From 10.0 to 100.0 km in 10.0-km steps

HINTS: Some of the initial and final values may be very small and may be ignored in the plot. Plot enough values to identify the maximum point and to show the effect of the elevated inversion. A log–log graph will be required to include enough points on a reasonable scale. You will have to determine the distance downwind where the short form of the dispersion equation is to be used (i.e., $2 x_L$) and switch equations at that point in your spreadsheet calculations.

9-33. Anna Lytical purchased a mobile home last year. She has been suffering severe allergic symptoms and has detected a strong odor of formaldehyde. An analysis of the mobile home air has revealed that the formaldehyde concentration is 0.28 ppm. Anna's friend, Sybil Injuneer, has measured the air flow in the ventilation system and found that the ventilation rate is 0.56 air changes per hour (ach). She has recommended increasing the ventilation rate to reduce the formaldehyde concentration below the threshold odor level of 0.05 ppm (Lee et al., 2001). Assuming the mobile home volume is 148 m^3 and that the outdoor air concentration is 0.0 ppm, estimate the ach required to achieve the threshold odor level.

9-34. A manufacturer of carbon monoxide detector/alarms has asked you to perform an analysis of the time to achieve various levels of CO in a standard house so it can set the detection level of the monitor. The standard house has a volume of 540 m^3 and a ventilation rate of 100 m^3/h. The manufacturer uses an assumption of a furnace flue malfunction that results in an emission of 3.0 mg/s of CO into the house. They use the following World Health Organization guidelines to prevent excess levels of COHb (WHO, 1987): 10 mg/m^3 for 8 h, 30 mg/m^3 for 1 h, 60 mg/m^3 for 30 min, and 100 mg/m^3 for 15 min. Using a spreadsheet program you have written, estimate the time it will take to achieve each of these concentrations. What safety factors (time to achieve a given level divided by the allowable time) will be achieved if the alarm is set at each of these levels. Assume that the outdoor and indoor air concentration equals the 8-hour NAAQS for CO and the starting concentration is 1.0 mg/m^3.

9-35. Determine the slope of the equilibrium curve defined by Henry's law for HCl gas at 20°C from the following data:

P_{HCl}, kPa	kg HCl per 100 kg H_2O
0.6533	38.9
0.0871	31.6
0.02733	25.0

Answer: Henry's law is not followed well. By linear regression, $m = 0.120$.

9-36. Find the slope of the equilibrium curve defined by Henry's law for SO_2 gas at 30°C from the following data:

P_{SO_2}, kPa	Kg SO_2 per 100 kg H_2O
10.532	1.0
6.933	0.7
4.800	0.5
2.626	0.3

9-37. Determine the height of a packed tower that is to reduce the concentration in air of H_2S from 0.100 kg/m^3 to 0.005 kg/m^3 given the following data:

> Incoming liquid is water free of H_2S
> Operating temperature = 25.0°C
> Operating pressure = 101.325 kPa
> Henry's law constant, m = 5.522 mole fraction units
> H_g = 0.444 m
> H_l = 0.325 m
> Liquid flow rate = 20.0 kg/s
> Gas flow rate = 5.0 kg/s
>
> *Answer:* 7 m

9-38. Neighbors are complaining about odor from the facility that produces the H_2S in Problem 9-37. Using the data in Problem 9-37, determine the height of a packed tower that is to reduce the concentration in air of H_2S from 0.100 kg/m^3 to the threshold odor concentration of 0.0002 mg/L of air. Is the tower height practical?

9-39. The concentration of chlorine gas in air must be reduced from 10.0 mg/m^3 to 2.95 mg/m^3. Determine the height of the packed tower that should be used if the following parameters apply:

> Incoming liquid is water free of Cl_2
> Operating temperature = 20.0°C
> Operating pressure 101.325 kPa
> H_g = 0.662 m
> H_l = 0.285 m
> Henry's law constant, m = 6.820 mole fraction units
> Liquid flow rate = 15.0 kg/s
> Gas flow rate = 3.00 kg/s

9-40. Determine the Langmuir constants a and b for the following isotherm data for adsorption of H_2S on a molecular sieve. Use a computer spreadsheet program to plot the data.

P_{H_2S}, kPa	W, g H_2S/g sieve
0.840	0.082
1.667	0.1065
2.666	0.118
3.333	0.122

Answer: a = 20; b = 135

9-41. Determine the Langmuir constants a and b for the adsorption of benzene on activated carbon, given the following isotherm data.

$P_{C_6H_6}$ kPa	W, kg C_6H_6/kg carbon
0.027	0.129
0.067	0.170
0.133	0.204
0.266	0.240

9-42. When the isotherm data are nonlinear, the Freundlich model may be used to fit the data:

$$q_e = KP^n$$

where q_e = mg of pollutant/g of adsorbent
   K, n = curve fitting constants
      P = pollutant partial pressure

The following adsoption data were obtained using beaded activated carbon to remove tetrachloroethylene (Noll et al., 1992).

q_e, mg/g of carbon	C_e, ppm
520	70
550	170
640	700
690	1,750
740	4,000
780	7,000

Use a spreadsheet to plot the data and fit a curve. Use the spreadsheet "trendline" to determine the Freundlich curve-fitting constants.

9-43. Determine the breakthrough time for toluene on an adsorption bed of activated carbon that is 0.75 m thick and 5.0 m² in cross section. The operating parameters for the bed are as follows:

   Gas flow rate = 1.185 kg/s

   Gas temperature = 25°C

   Bed density = 450 kg/m³

   Inlet pollutant concentration = 0.00350 kg/m³

   Langmuir parameters: $a = 465$; $b = 3,000$

   Width of adsorption zone = 0.045 m

Answer: 17.8 h

9-44. What thickness of molecular sieve adsorption bed is required for the following system to ensure an SO_2 breakthrough time (t_B) of not less than 8.00 h?

> Gas flow rate = 2.36 m^3/s of air
>
> Gas temperature = 25.0°C
>
> Gas pressure = 105.0 kPa
>
> Bed density as packed = 390 kg/m^3
>
> Inlet pollutant concentration = 3,000 ppm
>
> Langmuir parameters: $a = 400$; $b = 900$
>
> Width of adsorption zone = 0.028 m
>
> Bed diameter = 3.00 m

9-45. Determine the cross-sectional area and depth of catalyst to reduce an inlet concentration of toluene from 1.87 g/m^3 to 0.00187 g/m^3. The exhaust gas flow rate entering the control equipment is 16.33 m^3/s. Combustion air is supplied at a rate of 1.80 m^3/s. Both the exhaust gas and the combustion air flow rates are at 20°C. The manufacturer's specifications require that the catalyst be operated at 510°C and that the bed gas velocity be limited to 7.5 m/s. Assume that the toluene combustion reaction follows first-order kinetics and that the rate constant at 510°C is 120 s^{-1}.

> *Answers:* area = 6.5 m^2, depth = 0.43 m

9-46. Hexane (C_6H_{14}) is emitted from a baking oven at a rate of 454 g/min. The exhaust gas flow rate is 7.1 m^3/s at a temperature of 315°C. Determine the cross-sectional area and depth of catalyst to produce an exhaust concentration of 100 ppm at STP. Combustion air is supplied at 0.70 m^3/s at 20°C. The manufacturer's specifications require that the catalyst be operated at 550°C and that the bed gas velocity be limited to 9.5 m/s. Assume that the hexane combustion reaction follows first order kinetics and that the rate constant at 550°C is 55 s^{-1}.

9-47. Calculate the efficiency of removal of a 2.50-μm-diameter particle having a density of 1,250 kg/m^3 for a cyclone barrel diameter of 1.0 m. The gas flow rate is 2.80 m^3/s and the gas temperature is 25°C.

> *Answer:* $\eta \approx 14\%$

9-48. Because the efficiency of the large-barrel-diameter cyclone in Problem 9-47 is low for fine particles, a multiclone consisting of 10 barrels has been proposed as an alternative. Each barrel is to be 0.10 m in diameter. Calculate the efficiency of the multiclone using the particle and gas data given Problem 9-47.

9-49. Determine the efficiency of the cyclone in Example 9-13 for particles having a density of 1,000 kg/m^3 and radii of 1.00, 5.00, 10.00, and 25.00 μm. Using a spreadsheet, plot the efficiency as a function of particle diameter for the specified cyclone and gas conditions.

9-50. A consultant has proposed that a pulse-jet baghouse with bags that are 15 cm in diameter and 5 m in length be used instead of the mechanical shaker bag system proposed in Example 9-14. Estimate the net number of bags required if the manufacturer's recommended air-to-cloth ratio for aggregate plants is 0.050 m/s.

9-51. A green coffee bean screening and handling operation emits 0.75 g/m^3 of fine particulate matter. A reverse-air baghouse is being proposed for controlling the particulate emissions. The gas handling system has an exhaust flow rate of 3.3 m^3/s. A manufacturer has supplied the following data:

> Bag diameter = 20 cm
>
> Bag length = 12 m
>
> Air-to-cloth ratio = 0.010
>
> Bag cleaning = 0.5

Estimate the number of bags required and the mass of particulate matter collected each day if the efficiency is 99 percent. Assume 24-hour operation.

9-52. Calculate the overall mass efficiency (η) of the venturi described in Example 9-15 for the following particle size distribution

Average diameter, μm	% of total mass
2.5	25
7.5	20
15.0	15
25.0	15
35.0	10
50.0	15

Answer: overall η = 87.73 or 88 percent

9-53. Calculate the venturi throat area required to achieve 99.0 percent removal of a 1.25-μm-radius particle having a density of 1,400 kg/m^3 for the following gas stream and venturi characteristics.

> Gas flow rate = 10.0 m^3/s
>
> Gas temperature = 180°C
>
> Droplet diameter = 100 μm
>
> Liquid flow rate = 0.100 m^3/s
>
> Coefficient κ = 200

9-54. Using a spreadsheet progrm you have written, calculate the overall mass efficiency (η) of the venturi described in Problem 9-53 for a throat velocity of 26.3 m/s and the following fly ash particle size distribution (after Noll, 1999).

Average diameter, μm	% of total mass
0.05	0.01
0.3	0.21
0.8	0.78
3.0	13.0
8.0	16.0
13.0	12.0
18.0	8.0
80.0	50.0

9-55. Determine the collection efficiency of an electrostatic precipitator (ESP) tube that is 0.300 m in diameter and 2.00 m in length for particles that are 1.00 μm in diameter. The flow rate is 0.150 m³/s, the collection field intensity is 100,000 V/m, the particle charge is 0.300 femtocoulombs (fC), and the gas temperature is 25°C.

> *Answer:* 92.4 percent

9-56. Rework Problem 9-55 with the gas flow rate reduced to 0.075 m³/s.

9-14   DISCUSSION QUESTIONS

9-1. A gas sample is collected in a special gas sampling bag that does not react with the pollutants collected but is free to expand and contract. When the sample was collected, the atmospheric pressure was 103.0 kPa. At the time the sample was analyzed the atmospheric pressure was 100.0 kPa. The bag was found to contain 0.020 ppm of SO_2. Would the original concentration of SO_2 be more, less, or the same? Explain.

9-2. Under which of the following conditions would you expect the strongest inversion (largest positive lapse rate) to form?
a. Foggy day in the fall after the leaves have fallen
b. Clear winter night with fresh snow on the ground
c. Clear summer morning just before sunrise
Explain why.

9-3. Cement dust is characterized by very fine particulates. The exhaust gas temperatures from a cement kiln are very hot. Which of the following air pollution control devices would appear to be appropriate? Explain the reasoning for your selection.
a. Venturi scrubber
b. Baghouse
c. Electrostatic precipitator

9-4. Photochemical oxidants are not directly attributable to either people or natural sources. Why, then, are automobiles singled out as the major cause of the formation of ozone?

9-5. Explain why the $PM_{2.5}$ standard is more appropriate than a "Total Suspended Particulate" for protection of human health.

9-15   FE EXAM FORMATTED PROBLEMS

9-1. A power plant emits 600 g of SO_2/s from an effective stack height of 300 m. The wind speed at this height is 5.0 m/s. The stability class is C. What is the maximum downwind concentration?

a. 2.4×10^{-4} g/m^3
b. 1.5×10^{-6} g/m^3
c. 1.8×10^{-4} g/m^3
d. 1.9×10^{-5} g/m^3

9-2. Select an appropriate air-to-cloth ratio for fly ash and estimate the number of bags required for a gas flow rate of 15 m^3/s. Assume each bag is 15 cm in diameter and 5 m in length and that bag cleaning is by pulse jet.

a. 255 bags
b. 800 bags
c. 1,700 bags
d. 50 bags

9-3. Estimate the stoichiometric air-to-fuel ratio for a 100 percent ethanol (CH_3CH_2OH) fuel. One mole of $CH_3CH_2OH = 46$ g; one mole of $O_2 = 32$ g; one mole of $N_2 = 28$ g. Assume the following reaction:

$$6CH_3CH_2OH + 36O_2 + 135.36N_2 \rightarrow 12CO_2 + 18H_2O + 135.36N_2$$

a. 25
b. 107
c. 15
d. 18

9-4. Calculate the rate of emission of SO_2 in g/s that results in a centerline ($y = 0$) concentration at ground level of 1.412×10^{-3} g/m^3 one kilometer from the stack. The time of the measurement was 1 PM on a clear summer afternoon. The wind speed was 1.8 m/s measured at a height of 10 m. The effective stack height is 94 m. No inversion is present.

a. 790 g/s
b. 860 g/s
c. 280 g/s
d. 440 g/s

9-16   REFERENCES

AEC (1968) *Meteorology and Atomic Energy,* 1968, U.S. Atomic Energy Commission, USAEC Division of Technical Information Extension, Oak Ridge, TN.

AP (2005a) Associated Press, *Lansing State Journal,* December 18, p. 14A.

AP (2005b) Associated Press, www.myrtlebeachonline.com.

ASHRAE (1981) *Ventilation for Acceptable Air Quality,* American Society of Heating, Refrigerating, and Air Conditioning Engineers, Standard 62-1981, Atlanta.

Beard, J., F. Lachetta, and L. Lilleleht (1980) *Combustion Evaluation,* U.S. Environmental Protection Agency Report No. 4500/2-80-063.

Black & Veatch Consulting Engineers (1983) *Lime FGD Systems Data Book,* 2nd ed., EPRI Publication No. CS-2781.

Bosch, R. (1988) *Automotive Electric/Electronic Systems,* Society of Automotive Engineers, Warrendale, PA.

Buonicore, A. J., and L. Theodore (1975) *Industrial Control Equipment for Gaseous Pollutants,* Vol. I, CRC Press, Cleveland, pp. 149–1500.

Canfield, R. L., C. R. Henderson, D. A. Cory-Slechta, et al. (2003) "Intellectual Impairment in Children with Blood Lead Concentrations Below 10 μg per Deciliter," *The New England Journal of Medicine,* vol. 348, pp. 1517–1526.

Cannell, A. L., and M. L. Meadows (1985) "Effects of Recent Operating Experience on the Design of Spray Dryer FGD Systems," *Journal of the Air Pollution Control Association*, vol. 35 (7), pp. 782–789.

C&EN (2006) "Government Concentrates: Seven States Agree to Cut CO_2 Emissions," January 2, p. 16.

Connolly, C. H. (1972) *Air Pollution and Public Health,* Holt, Rinehart & Winston, New York, p. 7.

Cooper, C. D., and F. C. Alley (2002) *Air Pollution Control: A Design Approach,* Waveland Press, Long Grove, IL, p. 547.

Crawford, M. (1976) *Air Pollution Control Theory,* McGraw-Hill, New York, p. 516.

Davis, M. L., and S. J. Masten (2004) Principles of Environmental Engineering and Science, McGraw-Hill Higher Education, Dubuque, IA, p. 532.

Davis, W. T. (2000) *Air Pollution Engineering Manual,* 2nd edition, John Wiley & sons, Inc., New York, p. 112.

Deutsch, W. (1922) "Motion and Charge of a Charged Particle in the Cylindrical Condenser," *Annals of Physics,* vol. 68, pp. 335–344s.

Ferris, B. G. (1978) "Health Effects of Exposure to Low Levels of Regulated Air Pollutants," *Journal of the Air Pollution Control Association,* vol. 28, pp. 482–497.

Fryling, G. R. (1967) *Combustion Engineering,* Combustion Engineering, Inc., New York, pp. 21–28.

Godish, T. (1989) *Indoor Air Pollution Control,* Lewis Publishers, Chelsea, MI.

Godish, T. (2001) "Aldehydes," in J. D. Spengler, J. M. Samet, and J. F. McCarthy (eds.), *Indoor Air Quality Handbook,* McGraw-Hill, New York, pp. 32.1–32.22.

Goldsmith, J. R. (1968) "Effects of Air Pollution on Human Health," in A. C. Stern (ed.), *Air Pollution,* Academic Press, New York, pp. 554–557.

Goyer, R. A., and J. J. Chilsolm (1972) "Lead," in D. K. K. Lee (ed.), *Metallic Contaminants and Human Health,* Academic Press, New York, pp. 57–95.

Hance, S. B., and J. L. Kelly (1991) "Status of Flue Gas Desulfurization Systems," Paper No. 91-157.3, 84th Annual Meeting of the Air and Waste Management Association.

Hansen, J., and M. Sato (2004) "Temperature Trends: 2004 Summation," http://www. giss.nasa.gov/data/update/gistemp/2004/.

Hesketh, H. E. (1977) *Fine Particles in Gaseous Media,* Ann Arbor Science, Ann Arbor, MI, p. 19.

HEW (1979) *Smoking and Health: A Report of the Surgeon General of the Public Health Service*, U.S. Department of Health Education and Welfare, Pub. No. 79-50066, Washington, D.C.

Hileman, B. (1989) "Global Warming," *C&E News,* March 13, pp. 25–44.

Hileman, B. (2005) "Ice Core Record Extended," *C&E News,* November 28, p. 7.

Hindawi, I. (1970) *Air Pollution Injury to Plants,* U.S. Department of Health, Education, and Welfare, National Air Pollution Control Administration Publication No. AP-71, Washington, DC, p. 13.

Hines, A. L., T. K. Ghosh, S. K. Loyalka, and R. C. Warder (1993) *Indoor Air Quality & Control,* PTR Prentice Hall, Englewood Cliffs, NJ, pp. 21, 22, 34.

Hoegg, V. R. (1972) "Cigarette Smoke in Closed Spaces," *Environmental Health Perspectives*, vol. 2, p. 117.

Holland, J. Z. (1953) *A Meteorological Survey of the Oak Ridge Area,* U.S Atomic Energy Commission Report No. ORO-99, Washington, DC, p. 540.

Hosein, R., F. Silverman, P. Coreg, et al. (1985) "The Relationship Between Pollutant Levels in Homes and Potential Sources," *Transactions, Indoor Air Quality in Cold Climates, Hazards and Abatement Measures,* Air Pollution Control Association, Pittsburgh, pp. 250–260.

ICAS (1975) *The Possible Impact of Fluorocarbons and Hydrocarbons on Ozone,* Interdepartmental Committee for Atmospheric Sciences, Federal Council for Science and Technology, National Science Foundation Publication No. ICAS 18a-FY 75, Washington, DC, p. 3.

IPCC (1995) *Climate Change1994: Radiative Forcing of Climate Change and an Evaluation of the IPCC 1992 Emission Scenarios,* Intergovernmental Panel on Climate Change, Cambridge University Press, Cambridge, U.K.

IPCC (2000) *IPCC Special Report: Emissions Scenarios,* Intergovernmental Panel on Climate Change, Cambridge University Press, Cambridge, U.K.

IPCC (2001a) *Climate Change 2001: The Scientific Basis, Summary for Policymakers,* Intergovernmental Panel on Climate Change, Cambridge University Press, Cambridge, U.K., pp. 1–18.

IPCC (2007a) *Climate Change 2007: The Physical Science Basis, Summary for Policymakers,* Intergovernmental Panel on Climate Change, Cambridge University Press, Cambridge, U.K., pp. 1–18.

IPCC (2007b) *Climate Change 2007: Impacts, Adaptation and Vulnerability The Physical Science Basis, Summary for Policymakers,* Intergovernmental Panel on Climate Change, Cambridge University Press, Cambridge, U.K., pp. 1–18.

Johnstone, H. F., R. B. Field, and M. C. Tassler (1954) "Gas Absorption and Aerosol Collection in a Venturi Atomizer," *Industrial Engineering Chemistry,* vol. 46, pp. 1601–1608.

Karlsson, H. T., and H. S. Rosenberg, (1980) "Technical Aspects of Lime/Limestone Scrubbers for Coal fired Power Plants, Part I, Process Chemistry and Scrubber Systems," *Journal of the Air Pollution Control Association,* vol. 30(6), pp. 710–714.

Kao, A. S. (1994) "Formation and Removal Reactions of Hazardous Air Pollutants," *Journal of the Air Pollution Control Association,* vol. 44, pp. 683–696.

Keeling, C. M., and T. P. Whorf (2005) "Atmospheric Carbon Dioxide Record from Mauna Loa," http://www.mlo.noaa.gov.

Kiehl, J. T., and H. Rodhe (1995) "Modeling Geographic and Seasonal Forcing Due to Aerosols," in R. J. Charlson and J. Heintzenberg (eds.), *Aerosol Forcing of Climate,* John Wiley & Sons, New York, pp. 281–296.

Lapple, C. E. (1951) "Processes Use Many Collection Types," *Chemical Engineer,* vol. 58, pp. 144–151.

Lee, H. K., T. K. McKenna, L. N. Renton, et al. (1985) "Impact of a New Smoking Policy on Office Air Quality," *Indoor Air Quality in Cold Climates, Transactions of the Air Pollution Control Association,* Pittsburgh, pp. 307–322.

Lee, W. G., H. Chen, and C. Wu (2001) "Emission of VOCs from Wooden Building Materials in Indoor Environment," *Proceedings of the Air & Waste Management Association 94th Annual Conference & Exhibition,* Orlando, June 24–28, 2001.

Lefohn, A. S., and S. V. Krupa (1988) "Acidic Precipitation, A Technical Amplification of NAPAP's Findings," *Proceedings of APCA International Conference,* Pittsburgh, p. 1.

Lewis, R. G. (2001) "Pesticides," in J. D. Spengler, J. M. Samet, and J. F. McCarthy (eds.) *Indoor Air Quality Handbook,* McGraw-Hill, New York, p. 35.14.

Mann, M. E., and P. D. Jones (2003) "Global Surface Temperatures Over the Past Two Millennia," *Geophysical Research Letters,* vol. 30, no. 15, pp. CLM 5-1– CLM 5-4.

Martin, D. O. (1976) "Comment on the Change of Concentration Standard Deviations with Distance," *Journal of the Air Pollution Control Association,* vol. 26, pp. 145–146.

McKenna, J. D., A. B. Nunn, and D. A. Furlong (2000) "Fabric Filters," in W. T. Davis (ed.), *Air Pollution Engineering Manual,* 2nd ed., Air Pollution Control Association and John Wiley & Sons, New York, p. 104.

Molina, M. J., and F. S. Rowland (1974) "Stratospheric Sink for Chlorofloromethanes; Chlorine Atom Catalysed Destruction of Ozone," *Nature,* vol. 248, pp. 810–812.

Moschandreas, D. J., J. D. Zabpansky, and S. D. Pelta, (1985) *Characteristics of Emissions from Indoor Combustion Sources,* Gas Research Institute Report No. 85/0075, Chicago, IL.

NAPCA (1970) *Air Quality Criteria for Photochemical Oxidants,* U.S. Department of Health, Education, and Welfare, National Air Pollution Control Administration Publication No. AP-63, Washington, DC.

NCAPCA (1967) *Control of Particulate Emissions,* National Center for Air Pollution Control Training Manual, U.S. Department of Health, Education, and Welfare, Cincinnati, OH.

Noll, K. E., V. Gounaris, and Wain-Sun Hou (1992) *Adsorption for Air and Water Pollution Control,* Lewis Publishers, Chelsea, MI, pp. 74–79.

Noll, K. (1999) *Fundamentals of Air Quality Systems: Design of Air Pollution Control Devices,* American Academy of Environmental Engineers, Annapolis, MD, pp. 228, 402–403.

NTRDA (1969) *Air Pollution Primer,* National Tuberculosis and Respiratory disease Association,

O'Gara, P. J. (1922) "Sulfur Dioxide and Fume Problems and Their Solutions," *Industrial Engineering Chemistry,* vol. 14, p. 744.

Patterson, D. J., and N. A. Henein (1972) *Emissions from Combustion Engines and Their Control,* Ann Arbor Science, Ann Arbor, MI, p. 143.

Perry, R. H., and C. H. Chilton (eds.) (1973) *Chemical Engineers Handbook,* 5th ed., McGraw-Hill, New York, pp. 3–96.

Pittman, D. (2011) "Monitoring a Troubling Trend," *C&E News,* January 10, p. 26.

Pope, C. A., M. J. Thun, M. M. Namboodri, et al. (1995) "Particulate Air Pollution as a Predictor of Mortality in a Prospective Study of U.S. Adults," *American Journal of Respiratory and Critical Care Medicine,* vol. 151, pp. 669–674.

Pope, C. A., D. W. Dockery, R. E. Kanner, G. M. Villegas and J. Schwartz (1999) "Oxygen Saturation, Pulse Rate, and Particulate Air Pollution: A Daily Time-series Panel Study," *American Journal of Respiratory and Critical Care Medicine,* vol. 159, pp. 365–372.

Prasad, A. (1995) "Air Pollution Control Technologies for Nitrogen Oxides," *The National Environmental Journal,* May/June, pp. 46–50.

Reichhardt, T. (1995) "Weighing the Health Risks of Airborne Particulates," *Environmental Science and Technology,* vol. 29, pp. 360A–364A.

Reisch, M., and P. S. Zurer (1988) "CFC Production: DuPont Seeks Total Phaseout," *C&E News,* April 4, p. 4.

Seinfeld, J. H. (1975) *Air Pollution, Physical and Chemical Fundamentals,* McGraw-Hill, New York, p. 71.

Seinfeld, J. H., and S. N. Pandis (1998) *Atmospheric Chemistry and Physics,* John Wiley & Sons, New York, pp. 59, 71.

Sparks, L. E. (2001) "Indoor Air Quality Modeling," in J. D. Spengler, J. M. Samet, and J. F. McCarthy (eds.), *Indoor Air Quality Handbook,* McGraw-Hill, New York, pp. 58.1–58.28.

Srivastava, R. K., J. E. Staudt, and W. Josewicz (2005) "Preliminary Estimates of Performance and Cost of Mercury Emission Control Technology Applications on Electric Utility Boilers: An Update," *Environmental Progress,* vol. 24, no. 2, pp. 198–213.

Sullivan, D. A. (1989) "International Gathering Plans Ways to Safeguard Atmospheric Ozone," *C&E News,* June 26, pp. 33–36.

Supreme Court (2007) *Massachusetts et al. v. Environmental Protection Agency et al.,* Supreme Court of the United States, Syllabus No. 05-1120, Argued November 29, 2006. Decided April 2, 2007.

Traynor, G. W., J. R. Allen, and M. G. Apte (1982) *Indoor Air Pollution from Portable Kerosene-Fired Space Heaters, Woodburning Stoves and Woodburning Furnaces,* Lawrence Berkeley Laboratory Report No. LBL-14027.

Treybal, R. E. (1968) *Mass Transfer Operations,* McGraw-Hill, New York, pp. 253, 535.

Tucker, W. G. (2001) "Volatile Organic Compounds," in J. D. Spengler, J. M. Samet, and J. F. McCarthy (eds.), *Indoor Air Quality Handbook,* McGraw-Hill, New York, pp. 31.1–31.20.

Turner, D. B. (1967) *Workbook of Atmospheric Dispersion Estimates,* U.S. Department of Health, Education, and Welfare, U.S. Public Health Service Publication No. 999-AP-28, Washington, DC.

UN (2005) *The Millennium Development Goals Report: 2005,* United Nations, New York, p. 32.

UNDP (2005) *The Montreal Protocol,* http://www.undp.org/seed/eap/montreal/montreal.htm.

UNEP/WHO (2002) "Executive Summary," *Scientific Assessment of Ozone Depletion: 2002,* United Nations Environmental Programme/World Health Organization, New York.

U.S. AEC (1955) "Meteorology and Atomic Energy," U.S. Atomic Energy Commission, Washington, D.C., 1955, p. 59.

U.S. CDC (2005) *Third National Report on Human Exposure to Environmental Chemicals,* U.S. Centers for Disease Control and Prevention, National Center for Environmental Health, NCEH Pub. No. 05-0570, pp. 41, 74–75.

U.S. EPA (1990) *OAQPS Control Cost Manual,* U.S. Environmental Protection Agency, Publication No. 450/3-90-006.

U.S. EPA (1995) *User's Guide for ISC3 Dispersion Models,* Vol. II, EPA-454/B-95-003b, U.S. Environmental Protection Agency, Research Triangle Park, NC.

U.S. EPA (1997) *1997 Mercury Study Report to Congress,* U.S. Environmental Protection Agency, http://www.epa.gov.

U.S. EPA (2003a) *Response of Surface Water Chemistry to the Clean Air Act Amendments of 1990,* U.S. Environmental Protection Agency, Report No. 620/R-03/001, Research Triangle Park, NC, pp. 59–62.

U.S. EPA (2003b) *Performance and Cost of Mercury Emission Control Technology Applications on Electric Utility Boilers,* Report No. 600/R-03/1100, Research Triangle Park, NC.

U.S. EPA (2004) *EPA Fact Sheet,* U.S. Environmental Protection Agency, http://www.epa.gov.

U.S. PHS (1965) *Survey of Lead in the Atmosphere of Three Urban Communities,* U.S. Department of Health, Education, and Welfare, U.S. Public Health Service Publication No. 999-AP-12, Washington, DC.

Wakeham, H. (1972) "Recent Trends in Tobacco Smoke Research," in I. Schmeltz (ed.) The Chemistry of Tobacco Smoke, Plenum Press, New York.

Wallace, L. A. (2001) "Assessing Human Exposure to Volatile Organic Compounds," in J. D. Spengler, J. M. Samet, and J. F. McCarthy (eds.), *Indoor Air Quality Handbook,* McGraw-Hill, New York, p. 33.1–33.35.

Walsh, G. W. (1967) "Fabric Filtration," in *Control of Particulate Emissions Training Course Manual in Air Pollution,* National Center for Air Pollution Control, U.S. Public Health Service, Cincinnati, p. 9.

Wark, K., C. F. Warner, and W. T. Davis (1998) *Air Pollution. Its Origin and Control,* 3rd ed., Addison-Wesley, Menlo Park, CA, pp. 250–251.

Wofsy, S. C., J. C. McConnell, and M. B. McElroy (1972) "Atmospheric CH_4, CO, and CO_2," *Journal of Geophysical Research,* vol. 67, pp. 4477–4493.

WHO (1961) *Air Pollution,* World Health Organization, Geneva, Switzerland, p. 180.

WHO (1987) "Carbon Monoxide," in *Air Quality Guidelines for Europe,* World Health Organization Regional Office for Europe, European Series 23, Copenhagen, pp. 210–220.

Yocom, J. E., and R. O. McCaldin (1968) "Effects on Materials and the Economy," in A. C. Stern (ed.), *Air Pollution,* vol. I, 2nd ed., Academic Press, New York, pp. 617–654.

Zurer, P. S. (1988) "Studies on Ozone Destruction Expand Beyond Antarctic," *C&E News,* May 30, pp. 18–25.

Zurer, P. S. (1989) "Scientists Find Arctic May Face Ozone Hole," *C&E News,* February 27, p. 5.

Zurer, P. S. (1994) "Scientists Expect Ozone Loss to Peak About 1998," *C&E, News,* September 12, p. 5.

10-1 INTRODUCTION

Noise, commonly defined as unwanted sound, is an environmental phenomenon to which we are exposed before birth and throughout life. Noise is an environmental pollutant, a waste product generated in conjunction with various anthropogenic activities. Under the latter definition, noise is any sound—independent of loudness—that can produce an undesired physiological or psychological effect in an individual, and that may interfere with the social ends of an individual or group. These social ends include all of our activities—communication, work, rest, recreation, and sleep.

As waste products of our way of life, we produce two general types of pollutants. The general public has become well aware of the first type—the mass residuals associated with air and water pollution—that remain in the environment for extended periods of time. However, only recently has attention been focused on the second general type of pollution, the energy residuals such as the waste heat from manufacturing processes that creates thermal pollution of our streams. Energy in the form of sound waves constitutes yet another kind of energy residual, but, fortunately, one that does not remain in the environment for extended periods of time. The total amount of energy dissipated as sound throughout the earth is not large when compared with other forms of energy; it is only the extraordinary sensitivity of the ear that permits such a relatively small amount of energy to adversely affect us and other biological species.

It has long been known that noise of sufficient intensity and duration can induce temporary or permanent hearing loss, ranging from slight impairment to nearly total deafness. In general, a pattern of exposure to any source of sound that produces high enough levels can result in temporary hearing loss. If the exposure persists over a period of time, this can lead to permanent hearing impairment. It has been estimated that 1.7 million workers in the United States between 50 and 59 years of age have enough hearing loss to be awarded compensation. The potential cost to U.S. industry could be in excess of $1 billion* (Olishifshi and Harford, 1975). Short-term, but frequently serious, effects include interference with speech communication and the perception of other auditory signals, disturbance of sleep and relaxation, annoyance, interference with an individual's ability to perform complicated tasks, and general diminution of the quality of life.

Beginning with the technological expansion of the Industrial Revolution and continuing through a post-World War II acceleration, environmental noise in the United States and other industrialized nations has been gradually and steadily increasing, with more geographic areas becoming exposed to significant levels of noise. Where once noise levels sufficient to induce some degree of hearing loss were confined to factories and occupational situations, noise levels approaching such intensity and duration are today being recorded on city streets and, in some cases, in and around the home.

There are valid reasons why widespread recognition of noise as a significant environmental pollutant and potential hazard or, as a minimum, a detractor from the quality of life, has been slow in coming. In the first place, noise, if defined as unwanted sound, is a subjective experience. What is considered noise by one listener may be considered desirable by another.

---

*In 2005 dollars.

Secondly, noise has a short decay time and thus does not remain in the environment for extended periods, as do air and water pollution. By the time the average individual is spurred to action to abate, control, or, at least, complain about sporadic environmental noise, the noise may no longer exist.

Thirdly, the physiological and psychological effects of noise on us are often subtle and insidious, appearing so gradually that it becomes difficult to associate cause with effect. Indeed, to those persons whose hearing may already have been affected by noise, it may not be considered a problem at all.

Further, the typical citizen is proud of this nation's technological progress and is generally happy with the things that technology delivers, such as rapid transportation, labor-saving devices, and new recreational devices. Unfortunately, many technological advances have been associated with increased environmental noise, and large segments of the population have tended to accept the additional noise as part of the price of progress.

In the last three decades, the public has begun to demand that the price of progress not fall to them. They have demanded that the environmental impact of noise be mitigated. The cost of mitigation is not trivial. The average cost of soundproofing each of 600 suburban houses around the Chicago O'Hare airport was about $27,500 in 1997 (Sylvan, 2000). Through 2001, the Boston Logan airport had spent about $99 million and the Los Angeles International airport had allocated about $119 million for soundproofing and land acquisition. At the end of 2001, the total amount spent in the United States for noise mitigation exceeded $5.2 billion (de Neufville and Odoni, 2003). The cost to retrofit and replace airplanes to reduce noise probably exceeds $3.6 billion* (Achitoff, 1973). Traffic noise reduction programs have been in place since the first noise barrier was built in 1963. As of 2001, departments of transportation in 44 states and the Commonwealth of Puerto Rico had constructed more than 2,900 linear kilometers of noise barriers at a cost of more than $2.8 billion* (FHWA, 2005).

The engineering and scientific community has already accumulated considerable knowledge concerning noise, its effects, and its abatement and control. In that regard, noise differs from most other environmental pollutants. Generally, the technology exists to control most indoor and outdoor noise. As a matter of fact, this is one instance in which knowledge of control techniques exceeds the knowledge of biological and physical effects of the pollutant.

This chapter will provide you with the tools to:

- Calculate the cumulative noise level from several sources

- Estimate the potential for violation of environmental noise standards

- Estimate the noise level at a specified distance from a noise source

- Evaluate strategies for noise impact reduction for

    Workers

    Work space

    Communities near highways

---

*In 2005 dollars.

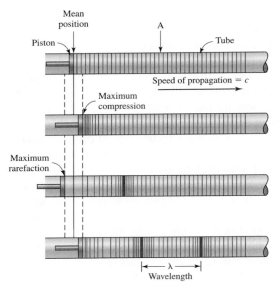

FIGURE 10-1
Alternating compression and rarefaction of air molecules resulting from a vibrating piston.

Properties of Sound Waves

Sound waves result from the vibration of solid objects or the separation of fluids as they pass over, around, or through holes in solid objects. The vibration and/or separation causes the surrounding air to undergo alternating compression and rarefaction, much in the same manner as a piston vibrating in a tube (Figure 10-1). The compression of the air molecules causes a local increase in air density and pressure. Conversely, the rarefaction causes a local decrease in density and pressure. These alternating pressure changes are the sound detected by the human ear.

Let us assume that you could stand at Point A in Figure 10-1. Also let us assume that you have an instrument that will measure the air pressure every 0.000010 seconds and plot the value on a graph. If the piston vibrates at a constant rate, the condensations and rarefactions will move down the tube at a constant speed. That speed is the *speed of sound* (*c*). The rise and fall of pressure at point A will follow a cyclic or wave pattern over a "period" of time (Figure 10-1). The wave pattern is called *sinusoidal*. The time between successive peaks or between successive troughs of the oscillation is called the *period* (*P*). The inverse of this, that is, the number of times a peak arrives in one second of oscillations, is called the *frequency* (*f*). Period and frequency are then related as follows:

$$P = \frac{1}{f} \tag{10-1}$$

Since the pressure wave moves down the tube at a constant speed, you would find that the distance between equal pressure readings would remain constant. The distance

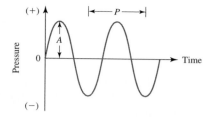

FIGURE 10-2
Sinusoidal wave that results from alternating compression and
rarefaction of air molecules. The amplitude is shown as A and
the period is P.

between adjacent crests or troughs of pressure is called the *wavelength* (λ). Wavelength and frequency are then related as follows:

$$\lambda = \frac{c}{f} \tag{10-2}$$

The *amplitude* (A) of the wave is the height of the peak or depth of the trough measured from the zero pressure line (Figure 10-2). From Figure 10-2, we can also note that
the average pressure could be zero if an averaging time was selected that corresponded to
the period of the wave. This would result regardless of the amplitude! This, of course, is
not an acceptable state of affairs. The root mean square (rms) sound pressure (p_{rms}) is used
to overcome this difficulty.* The rms sound pressure is obtained by squaring the value of
the amplitude at each instant in time; summing the squared values; dividing the total by
the averaging time; and taking the square root of the total. The equation for rms is

$$p_{rms} = \left(\overline{p^2}\right)^{1/2} = \left[\frac{1}{T}\int_0^T P^2(t)\,dt\right]^{1/2} \tag{10-3}$$

where the overbar refers to the time-weighted average and T is the time period of the
measurement.

Sound Power and Intensity

Work is defined as the product of the magnitude of the displacement of a body and the
component of force in the direction of the displacement. Thus, traveling waves of
sound pressure transmit energy in the direction of propagation of the wave. The rate at
which this work is done is defined as the *sound power* (W).

Sound intensity (I) is defined as the time-weighted average sound power per unit
area normal to the direction of propagation of the sound wave. Intensity and power are
related as follows:

$$I = \frac{W}{A} \tag{10-4}$$

where A is a unit area perpendicular to the direction of wave motion. Intensity, and
hence, sound power, is related to sound pressure in the following manner:

$$I = \frac{(p_{rms})^2}{\rho c} \tag{10-5}$$

---

*Sound pressure = (total atmospheric pressure) − (barometric pressure).

where I = intensity, W/m^2
p_{rms} = root mean square sound pressure, Pa
ρ = density of medium, kg/m^3
c = speed of sound in medium, m/s

Both the density of air and speed of sound are a function of temperature. Given the temperature and pressure, the density of air (1.185 kg/m^3 at 101.325 kPa and 298 K) may be determined using the gas laws. The speed of sound in air at 101.325 kPa may be determined from the following equation:

$$c = 20.05\sqrt{T} \tag{10-6}$$

where T is the absolute temperature in kelvins (K) and c is in m/s.

Levels and the Decibel

The sound pressure of the faintest sound that a normal healthy individual can hear is about 0.00002 pascal. The sound pressure produced by a Saturn rocket at liftoff is greater than 200 pascal. Even in scientific notation this is an "astronomical" range of numbers.

In order to cope with this problem, a scale based on the logarithm of the ratios of the measured quantities is used. Measurements on this scale are called *levels*. The unit for these types of measurement scales is the *bel*, which was named after Alexander Graham Bell:

$$L' = \log \frac{Q}{Q_o} \tag{10-7}$$

where L' = level, bels
Q = measured quantity
Q_o = reference quantity
$\log$ = logarithm in base 10

A bel turns out to be a rather large unit, so for convenience it is divided into 10 subunits called *decibels* (dB). Levels in dB are computed as follows:

$$L = 10 \log \frac{Q}{Q_o} \tag{10-8}$$

The dB does not represent any physical unit. It merely indicates that a logarithmic transformation has been performed.

Sound Power Level.   If the reference quantity (Q_o) is specified, then the dB takes on physical significance. For noise measurements, the reference power level has been established as 10^{-12} watts. Thus, sound power level may be expressed as

$$L_w = 10 \log \frac{W}{10^{-12}} \tag{10-9}$$

Sound power levels computed with Equation 10-9 are reported as dB re: 10^{-12} W.

Sound Intensity Level.   For noise measurements, the reference sound intensity (Equation 10-4) is 10^{-12} W/m². Thus, the sound intensity level is given as

$$L_I = 10 \log \frac{I}{10^{-12}} \tag{10-10}$$

Sound Pressure Level.   Because sound-measuring instruments measure the root mean square pressure, the sound pressure level is computed as follows:

$$L_P = 10 \log \frac{(p_{\text{rms}})^2}{(p_{\text{rms}})_o^2} \qquad \text{🚩 (10-11)}$$

which, after extraction of the squaring term, is given as

$$L_P = 20 \log \frac{p_{\text{rms}}}{(p_{\text{rms}})_o} \tag{10-12}$$

The reference pressure has been established as 20 micropascals (μPa). A scale showing some common sound pressure levels is shown in Figure 10-3.

Combining Sound Pressure Levels.   Because of their logarithmic heritage, decibels don't add and subtract the way apples and oranges do. Remember: adding the logarithms of numbers is the same as multiplying them. If you take a 60-decibel noise (re: 20 μPa) and add another 60-decibel noise (re: 20 μPa) to it, you get a 63-decibel

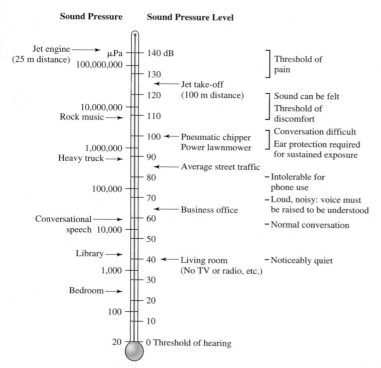

FIGURE 10-3
Relative scale of sound pressure levels.

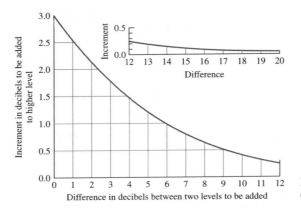

FIGURE 10-4
Graph for solving decibel addition problems.

noise (re: 20 μPa). If you're strictly an apple-and-orange mathematician, you may take this on faith. For skeptics, this can be demonstrated by converting the dB to sound power level, adding them, and converting back to dB. Figure 10-4 provides a graphical solution for this type of problem. For noise pollution work, results should be reported to the nearest whole number. When there are several levels to be combined, they should be combined two at a time, starting with lower-valued levels and continuing two at a time with each successive pair until one number remains. Henceforth, in this chapter we will assume levels are all "re: 20 μPa" unless stated otherwise.

Example 10-1. What sound power level results from combining the following three levels: 68 dB, 79 dB, and 75 dB?

Solution. This problem can be worked by converting the readings to sound power level, adding them, and converting back to dB.

$$L_w = 10 \log \Sigma 10^{(68/10)} + 10^{(75/10)} + 10^{(79/10)}$$
$$= 10 \log(117{,}365{,}173)$$
$$= 80.7 \text{ dB}$$

Rounding off to the nearest whole number yields an answer of 81 dB re: 20 μPa.

An alternative solution technique using Figure 10-4 begins by selecting the two lowest levels: 68 dB and 75 dB. The difference between the values is $75 - 68 = 7.00$. Using Figure 10-4, draw a vertical line from 7.00 on the abscissa to intersect the curve. A horizontal line from the intersection to the ordinate yields about 0.8 dB. Adding this value to the highest value, the combination of 68 dB and 75 dB results in a level of 75.8 dB. This, and the remainder of the computation, is shown diagrammatically below.

Characterization of Noise

Weighting Networks.   Because our reasons for measuring noise usually involve people, we are ultimately more interested in the human reaction to sound than in sound as
a physical phenomenon. Sound pressure level, for instance, can't be taken at face value
as an indication of loudness because the frequency (or pitch) of a sound has quite a bit
to do with how loud it sounds. For this and other reasons, it often helps to know something about the frequency of the noise you're measuring. Weighting networks are used
to account for the frequency of a sound. They are electronic filtering circuits built into
the meter to attenuate certain frequencies. They permit the sound level meter to respond
more to some frequencies than to others with a prejudice something like that of the human ear. Writers of the acoustical standards have established three weighting characteristics: A, B, and C. The chief difference among them is that very low frequencies are
filtered quite severely by the A network, moderately by the B network, and hardly at all
by the C network. Therefore, if the measured sound level of a noise is much higher on
C weighting than on A weighting, much of the noise is probably of low frequency.
If you really want to know the frequency distribution of a noise (and most serious noise
measurers do), it is necessary to use a *sound analyzer.* But if you are unable to justify
the expense of an analyzer, you can still find out something about the frequency of a
noise by shrewd use of the weighting networks of a sound level meter.

Figure 10-5 shows the response characteristics of the three basic networks as prescribed by the American National Standards Institute (ANSI) specification number
S1.4–1971. When a weighting network is used, the sound level meter electronically
subtracts or adds the number of dB shown at each frequency shown in Table 10-1
from or to the actual sound pressure level at that frequency. It then sums all the resultant numbers by logarithmic addition to give a single reading. Readings taken
when a network is in use are said to be "sound levels" rather than "sound pressure
levels." The readings taken are designated in decibels in one of the following forms:
dB(A); dBa; dBA; dB(B); dBb; dBB; and so on. Tabular notations may refer to
L_A, L_B, L_C.

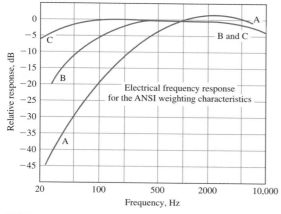

FIGURE 10-5
Response characteristics of the three basic weighting networks.

TABLE 10-1
Sound level meter network weighting values

Frequency (Hz)	Curve A (dB)	Curve B (dB)	Curve C (dB)
10	−70.4	−38.2	−14.3
12.5	−63.4	−33.2	−11.2
16	−56.7	−28.5	−8.5
20	−50.5	−24.2	−6.2
25	−44.7	−20.4	−4.4
31.5	−39.4	−17.1	−3.0
40	−34.6	−14.2	−2.0
50	−30.2	−11.6	−1.3
63	−26.2	−9.3	−0.8
80	−22.5	−7.4	−0.5
100	−19.1	−5.6	−0.3
125	−16.1	−4.2	−0.2
160	−13.4	−3.0	−0.1
200	−10.9	−2.0	0
250	−8.6	−1.3	0
315	−6.6	−0.8	0
400	−4.8	−0.5	0
500	−3.2	−0.3	0
630	−1.9	−0.1	0
800	−0.8	0	0
1,000	0	0	0
1,250	0.6	0	0
1,600	1.0	0	−0.1
2,000	1.2	−0.1	−0.2
2,500	1.3	−0.2	−0.3
3,150	1.2	−0.4	−0.5
4,000	1.0	−0.7	−0.8
5,000	0.5	−1.2	−1.3
6,300	−0.1	−1.9	−2.0
8,000	−1.1	−2.9	−3.0
10,000	−2.5	−4.3	−4.4
12,500	−4.3	−6.1	−6.2
16,000	−6.6	−8.4	−8.5
20,000	−9.3	−11.1	−11.2

Example 10-2. A new Type 2 sound level meter is to be tested with two pure tone sources that emit 90 dB. The pure tones are at 1,000 Hz and 100 Hz. Estimate the expected readings on the A, B, and C weighting networks.

Solution. From Table 10-1 at 1,000 Hz, we note that the relative response (correction factor) for each of the weighting networks is zero. Thus for the pure tone at 1,000 Hz we would expect the readings on the A, B, and C networks to be 90 dB.

From Table 10-1 at 100 Hz, the relative response for each weighting network differs. For the A network, the meter will subtract 19.1 dB from the actual reading, for the B network, the meter will subtract 5.6 dB from the actual reading, and for the C network, the meter will subtract 0.3 dB. Thus, the anticipated readings would be:

A network: 90 − 19.1 = 70.9 or 71 dB(A)

B network: 90 − 5.6 = 84.4 or 84 dB(B)

C network: 90 − 0.3 = 89.7 or 90 dB(C)

Example 10-3. The following sound levels were measured on the A, B, and C weighting networks:

Source 1: 94 dB(A), 95 dB(B), and 96 dB(C)

Source 2: 74 dB(A), 83 dB(B), and 90 dB(C)

Characterize the sources as "low frequency" or "mid/high frequency."

Solution. From Figure 10-5, we can see that readings on the A, B, and C networks will be close together if the source emits noise in the frequency range above about 500 Hz. This range may be classified "mid/high frequency" because we cannot distinguish between "mid" and "high" frequency using a Type 2 sound level meter. Likewise, we can see that below 200 Hz (low frequency), readings on the A, B, and C scale will be substantially different. The readings from the A network will be lower than the readings from the B network, and readings from both the A and B networks will be lower than those from the C network.

Source 1: Note that the sound levels on each of the weighting networks differ by 1 dB. From Figure 10-5, it appears that the sound level will be in the mid/high frequency range.

Source 2: Note that the sound levels on each of the weighting networks differ by several dB and that the reading from the A network is lower than that from the B network and both are below that from the C network. From Figure 10-5, it appears that the sound level will be in the low frequency range.

Octave Bands. To completely characterize a noise, it is necessary to break it down into its frequency components or spectra. Normal practice is to consider 8 to 11 octave bands.* The standard octave bands and their geometric mean frequencies (center band frequencies) are given in Table 10-2. Octave analysis is performed with a combination precision sound level meter and an octave filter set.

While octave band analysis is frequently satisfactory for community noise control (that is, identifying violators), more refined analysis is required for corrective action and design. One-third octave band analysis provides a slightly more refined picture of the noise source than the full octave band analysis (Figure 10-6a). This improved

---

*An octave is the frequency interval between a given frequency and twice that frequency. For example, given the frequency 22 Hz, the octave band is from 22 to 44 Hz. A second octave band would then be from 44 to 88 Hz.

TABLE 10-2
Octave bands

Octave frequency range (Hz)	Geometric mean frequency (Hz)
22–44	31.5
44–88	63
88–175	125
175–350	250
350–700	500
700–1,400	1,000
1,400–2,800	2,000
2,800–5,600	4,000
5,600–11,200	8,000
11,200–22,400	16,000
22,400–44,800	31,500

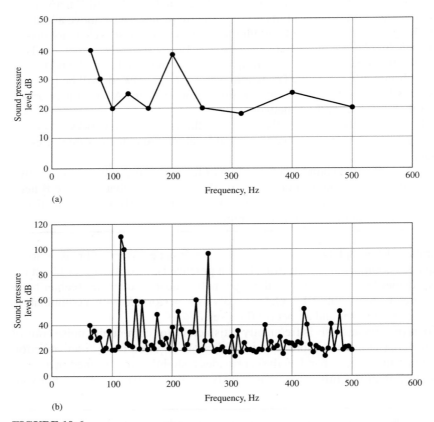

(a)

(b)

FIGURE 10-6
(*a*) One-third octave band analysis of a small electric motor. (*b*) Narrowband analysis of a small electric motor.

resolution is usually sufficient for determining corrective action for community noise problems. Narrow band analysis is highly refined and may imply band widths down to 2 Hz (Figure 10-6b). This degree of refinement is only justified in product design and testing or in troubleshooting industrial machine noise and vibration.

Averaging Sound Pressure Levels.   Because of the logarithmic nature of the dB, the average value of a collection of sound pressure level measurements cannot be computed in the normal fashion. Instead, the following equation must be used:

$$\overline{L}_p = 20 \log \frac{1}{N} \sum_{j=1}^{N} 10^{(L_j/20)} \qquad \text{(10-13)}$$

where L_p = average sound pressure level, dB re: 20 μPa
   N = number of measurements
   L_j = the jth sound pressure level, dB re: 20 μPa
   $j = 1, 2, 3 \ldots, N$

This equation is equally applicable to sound levels in dBA. It may also be used to compute average sound power levels if the factors of 20 are replaced with 10s.

Example 10-4.   Compute the mean sound level from the following four readings (all dBA): 38, 51, 68, and 78.

Solution.   First we compute the sum:

$$\sum_{j=1}^{4} = 10^{(38/20)} + 10^{(51/20)} + 10^{(68/20)} + 10^{(78/20)}$$

$$= 1.09 \times 10^4$$

Now we complete the computation:

$$\overline{L}_p = 20 \log \frac{1.09 \times 10^4}{4}$$

$$= 68.7 \text{ or } 69 \text{ dBA}$$

Straight arithmetic averaging would yield 58.7 or 59 dB.

Types of Sounds.   Patterns of noise may be qualitatively described by one of the following terms: *steady-state* or *continuous; intermittent;* and *impulse* or *impact.* Continuous noise is an uninterrupted sound level that varies less than 5 dB during the period of observation. An example is the noise from a household fan. Intermittent noise is a continuous noise that persists for more than one second that is interrupted for more than one second. A dentist's drilling would be an example of an intermittent noise. Impulse noise is characterized by a change of sound pressure of 40 dB or more within 0.5 second with a duration of less than one second.* The noise from firing a weapon would be an example of an impulsive noise.

---

*The Occupational Safety and Health Administration (OSHA) classifies repetitive events, including impulses, as steady noise if the interval between events is less than 0.5 seconds.

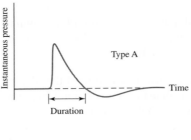

FIGURE 10-7
Type A impulse noise.

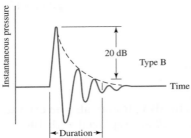

FIGURE 10-8
Type B impulse noise.

Two types of impulse noise generally are recognized. The type A impulse is characterized by a rapid rise to a peak sound pressure level followed by a small negative pressure wave or by decay to the background level (Figure 10-7). The type B impulse is characterized by a damped (oscillatory) decay (Figure 10-8). Where the duration of the type A impulse is simply the duration of the initial peak, the duration of the type B impulse is the time required for the envelope to decay to 20 dB below the peak. Because of the short duration of the impulse, a special sound-level meter must be employed to measure impulse noise. You should note that the peak sound pressure level is different than the impulse sound level because of the time-averaging used in the latter.

10-2   EFFECTS OF NOISE ON PEOPLE

For the purpose of our discussion, we have classified the effects of noise on people into the following two categories: auditory effects and psychological/sociological effects. Auditory effects include both hearing loss and speech interference. Psychological/sociological effects include annoyance, sleep interference, effects on performance, and acoustical privacy.

The Hearing Mechanism

Before we can discuss hearing loss, it is important to outline the general structure of the ear and how it works.

Anatomically, the ear is separated into three sections: the outer ear, the middle ear, and the inner ear (Figure 10-9). The outer and middle ear serve to convert sound pressure to vibrations. In addition, they perform the protective role of keeping debris and objects from reaching the inner ear. The Eustachian tube extends from the middle ear space to the upper part of the throat behind the soft palate. The tube is normally closed.

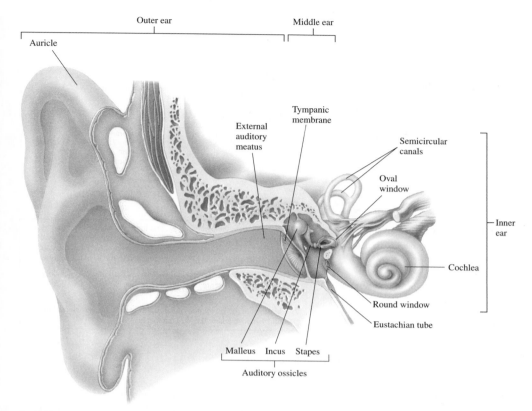

FIGURE 10-9
Anatomical divisions of the ear. (*Source:* Seeley et al., 2003.)

Contraction of the palate muscles during yawning, chewing, or swallowing opens the tubes. This allows the middle ear to ventilate and equalize pressure. If external air pressure changes rapidly, for example, by a sudden change in elevation, the tube is opened by involuntary swallowing or yawning to equalize the pressure.

The sound transducer mechanism is housed in the middle ear.* It consists of the *tympanic membrane* (eardrum) and three *ossicles* (bones) (Figure 10-10). The ossicles are supported by ligaments and may be moved by two muscles or by deflection of the tympanic membrane. The muscle movement is involuntary. Loud sounds cause these muscles to contract. This stiffens and diminishes the movement of the ossicular chain (Borg and Counter, 1989). The discussion on the middle ear that follows is excerpted from Clemis (1975).

> The primary function of the middle ear in the hearing process is to transfer sound energy from the outer to the inner ear. As the eardrum vibrates, it transfers its motion to the malleus. Since

---

*A transducer is a device that transmits power from one system to another. In this case, sound power is converted to mechanical displacement, which is later measured and interpreted by the brain.

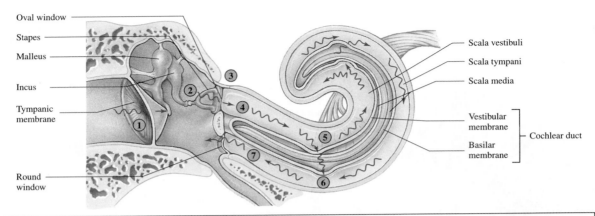

1. Sound waves strike the tympanic membrane and cause it to vibrate.

2. Vibration of the tympanic membrane causes the three bones of the middle ear to vibrate.

3. The foot plate of the stapes vibrates in the oval window.

4. Vibration of the foot plate causes the perilymph in the scala vestibuli to vibrate.

5. Vibration of the perilymph causes displacement of the basilar membrane. Short waves (high pitch) cause displacement of the basilar membrane near the oval window, and longer waves (low pitch) cause displacement of the basilar membrane some distance from the oval window. Movement of the basilar membrane is detected in the hair cells of the spiral organ, which are attached to the basilar membrane.

6. Vibrations of the perilymph in the scala vestibuli and of the endolymph in the cochlear duct are transferred to the perilymph of the scala tympani.

7. Vibrations in the perilymph of the scala tympani are transferred to the round window, where they are dampened.

FIGURE 10-10
The sound transducer mechanism housed in the middle ear. (*Source:* Seeley et al., 2003.)

the bones of the ossicular chain are connected to one another, the movements of the malleus are passed on to the incus, and finally to the stapes, which is imbedded in the oval window.

As the stapes moves back and forth in a rocking motion, it passes the vibrations into the inner ear through the oval window. Thus, the mechanical motion of the eardrum is effectively transmitted through the middle ear and into the fluid of the inner ear.

The sound-conducting transducer amplifies sound by two main mechanisms. First, the large surface area of the drum as compared to the small surface area of the base of the stapes (footplate) results in a hydraulic effect. The eardrum has about 25 times as much surface area as the oval window. All of the sound pressure collected on the eardrum is transmitted through the ossicular chain and is concentrated on the much smaller area of the oval window. This produces a significant increase in pressure.

The bones of the ossicular chain are arranged in such a way that they act as a series of levers. The long arms are nearest the eardrum, and the shorter arms are toward the oval window. The fulcrums are located where the individual bones meet. A small pressure on the long arm of the lever produces a much stronger pressure on the shorter arm. Since the longer arm is attached to the eardrum and the shorter arm is attached to the oval window, the ossicular chain acts as an amplifier of sound pressure. The magnification effect of the entire sound-conducting mechanism is about 22-to-1.

The inner ear houses both the balance receptors and the auditory receptors. The auditory receptors are in the *cochlea*. It is a bone shaped like a snail coiled two and one-half times around its own axis (Figure 10-9). A cross section through the cochlea

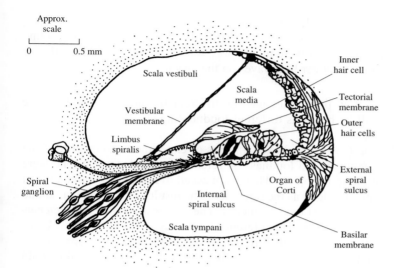

FIGURE 10-11
Cross section through the cochlea.

(Figure 10-11) reveals three compartments: the *scala vestibuli;* the *scala media;* and the *scala tympani.* The scala vestibuli and the scala tympani are connected at the apex of the cochlea. They are filled with a fluid called *perilymph,* in which the scala media floats. The hearing organ, the *organ of Corti,* is housed in the scala media. The scala media contains a different fluid, *endolymph,* which bathes the organ of Corti.

The scala media is triangular in shape and is about 34 mm in length. As shown in Figure 10-11, there are cells growing up from the *basilar membrane.* They have a tuft of hair at one end and are attached to the hearing nerve at the other end. A gelatinous membrane (*tectorial membrane*) extends over the hair cells and is attached to the *limbus spiralis.* The hair cells are embedded in the tectorial membrane.

Vibration of the oval window by the stapes causes the fluids of the three scalae to develop a wave-like motion. The movement of the basilar membrane and the tectorial membrane in opposite directions causes a shearing motion on the hair cells. The dragging of the hair cells sets up electrical impulses in the auditory nerves, which are transmitted to the brain.

The nerve endings near the oval and round windows are sensitive to high frequencies. Those near the apex of the cochlea are sensitive to low frequencies.

Normal Hearing

Frequency Range and Sensitivity. The ear of the young, audiometrically healthy, adult male responds to sound waves in the frequency range of 20 to 16,000 Hz. Young children and women often have the capacity to respond to frequencies up to 20,000 Hz. The speech zone lies in the frequency range of 500 to 2,000 Hz. The ear is most sensitive in the frequency range from 2,000 to 5,000 Hz. The smallest perceptible sound pressure in this frequency range is 20 μPa.

A sound pressure of 20 μPa at 1,000 Hz in air corresponds to a 1.0 nm displacement of the air molecules. The thermal motion of the air molecules corresponds to a sound pressure of about 1 μPa. If the ear were much more sensitive, you would hear the air molecules crashing against your ear like waves on the beach!

Loudness.   In general, two pure tones having different frequencies but the same sound pressure level will be heard as different loudness levels. Loudness level is a psychoacoustic quantity.

Fletcher and Munson (1935) conducted a series of experiments to determine the relationship between frequency and loudness. A reference tone and a test tone were presented alternately to the test subjects. They were asked to adjust the sound level of the test tone until it sounded as loud as the reference. The results were plotted as sound pressure level in dB versus the test tone frequency. The curves are called the Fletcher-Munson or *equal loudness contours*. The reference frequency is 1,000 Hz. The curves are labeled in *phons,* which are the sound pressure levels of the 1,000 Hz pure tone in dB. The lowest contour represents the "threshold of hearing." The actual threshold may vary by as much as ± 10 dB between individuals with normal hearing.

Audiometry.   Hearing tests are conducted with a device known as an *audiometer.* Basically, it consists of a source of pure tones with variable sound pressure level output into a pair of earphones. If the instrument also automatically prepares a graph of the test results (an *audiogram*), then it will include a weighting network called the *hearing threshold level* (HTL) scale.

The HTL scale is one in which the loudness of each pure tone is adjusted by frequency such that "0" dB is the level just audible for the average normal young ear. Two reference standards are in use: ASA–1951 and ANSI–1969. The ANSI reference values are shown in Figure 10-12. Note the similarity to the Fletcher-Munson contours. The initial audiogram prepared for an individual may be referred to as the baseline HTL or simply as the HTL.

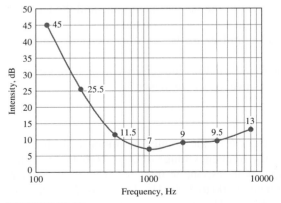

FIGURE 10-12
The ANSI reference values for hearing threshold level.

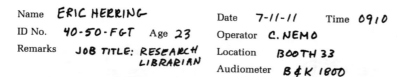

Name ERIC HERRING Date 7-11-11 Time 0910
ID No. 40-50-FGT Age 23 Operator C. NEMO
Remarks JOB TITLE: RESEARCH Location BOOTH 33
LIBRARIAN Audiometer B&K 1800

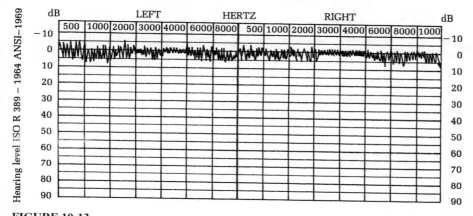

FIGURE 10-13

An audiogram illustrating excellent hearing response.

The audiogram shown in Figure 10-13 reflects excellent hearing response. The average normal response may vary ±10 dB from the "0" dB value. As noted on the audiogram, this test was conducted with the ANSI–1969 weighting network.

You may have noted that we keep stressing young in our references to normal hearing. This is because there is hearing loss due to the aging process. This type of loss is called *presbycusis*. The average amount of loss as a function of age is shown in Figure 10-14.

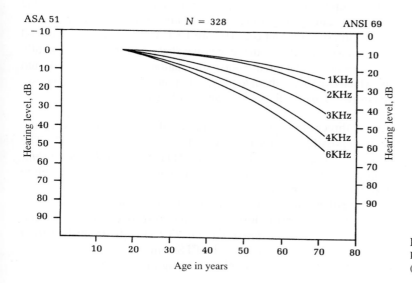

FIGURE 10-14

Hearing loss as a result of presbycusis.
(*Source:* Olishifski and Harford, 1975.)

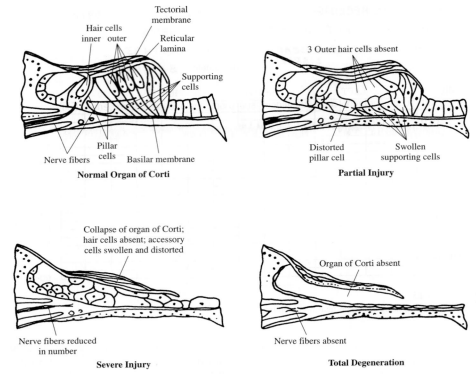

FIGURE 10-15
Various degrees of injury to the hair cells.

Hearing Impairment

Mechanism. With the exception of eardrum rupture from intense explosive noise, the outer and middle ear rarely are damaged by noise. More commonly, hearing loss is a result of neural damage involving injury to the hair cells (Figure 10-15). Two theories are offered to explain noise-induced injury. The first is that excessive shearing forces mechanically damage the hair cells. The second is that intense noise stimulation forces the hair cells into high metabolic activity, which overdrives them to the point of metabolic failure and consequent cell death. Once destroyed, hair cells are not capable of regeneration.

Measurement. Because direct observation of the organ of Corti in persons having potential hearing loss is impossible, injury is inferred from losses in their HTL. The increased sound pressure level required to achieve a new HTL is called *threshold shift*. Obviously, any measurement of threshold shift is dependent upon having a baseline audiogram taken before the noise exposure.

Hearing losses may be either temporary or permanent. Noise-induced losses must be separated from other causes of hearing loss such as age (presbycusis), drugs, disease, and blows on the head. *Temporary threshold shift* (TTS) is distinguished *from permanent threshold shift* (PTS) by the fact that in TTS removal of the noise overstimulation will result in a gradual return to baseline hearing thresholds.

Factors Affecting Threshold Shift. Important variables in the development of temporary and permanent hearing threshold changes include the following (NIOSH, 1972).

1. Sound level: Sound levels must exceed 60 to 80 dBA before the typical person will experience TTS.

2. Frequency distribution of sound: Sounds having most of their energy in the speech frequencies are more potent in causing a threshold shift than are sounds having most of their energy below the speech frequencies.

3. Duration of sound: The longer the sound lasts, the greater the amount of threshold shift.

4. Temporal distribution of sound exposure: The number and length of quiet periods between periods of sound influences the potentiality of threshold shift.

5. Individual differences in tolerance of sound may vary greatly among individuals.

6. Type of sound—steady-state, intermittent, impulse, or impact: The tolerance to peak sound pressure is greatly reduced by increasing the duration of the sound.

Temporary Threshold Shift (TTS). TTS is often accompanied by a ringing in the ear, muffling of sound, or discomfort of the ears. Most of the TTS occurs during the first two hours of exposure. Recovery to the baseline HTL after TTS begins within the first hour or two after exposure. Most of the recovery that is going to be attained occurs within 16 to 24 hours after exposure.

Permanent Threshold Shift (PTS). There appears to be a direct relationship between TTS and PTS. Noise levels that do not produce TTS after two to eight hours of exposure will not produce PTS if continued beyond this time. The shape of the TTS audiogram will resemble the shape of the PTS audiogram.

Noise-induced hearing loss generally is first characterized by a sharply localized dip in the HTL curve at the frequencies between 3,000 and 6,000 Hz. This dip commonly occurs at 4,000 Hz (Figure 10-16). This is the *high frequency notch*. The progress from TTS to PTS with continued noise exposure follows a fairly regular pattern. First, the high frequency notch broadens and spreads in both directions. While substantial losses may occur above 3,000 Hz, the individual will not notice any change in hearing. In fact, the individual will not notice any hearing loss until the speech frequencies between 500 and 2,000 Hz average more than a 25 dB increase in HTL on the ANSI–1969 scale. The onset and progress of noise-induced permanent hearing loss is slow and insidious. The exposed individual is unlikely to notice it. Total hearing loss from noise exposure has not been observed.

Acoustic Trauma. The outer and middle ear rarely are damaged by intense noise. However, explosive sounds can rupture the tympanic membrane or dislocate the ossicular chain. The permanent hearing loss that results from very brief exposure to a very loud noise is termed *acoustic trauma* (Davis, 1958). Damage to the outer and middle ear may or may not accompany acoustic trauma.

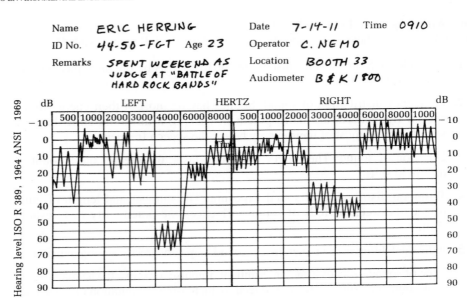

FIGURE 10-16
An audiogram illustrating hearing loss at the high frequency notch.

Protective Mechanisms. Although the extent and mechanisms are not clear, it appears that the structures of the middle ear offer some protection to the delicate sensory organs of the inner ear (Borg and Counter, 1989). One mechanism of protection is a change in the mode of vibration of the stapes. As noted earlier, there is evidence that the muscles of the middle ear contract reflexively in response to loud noise. This contraction results in a reduction in the amplification that this series of levers normally produces. Changes in transmission may be on the order of 20 dB. However, the reaction time of the muscle/bone structure is on the order of 100–200 milliseconds. Thus, this protection is not effective against steep acoustic wave fronts that are characteristic of impact or impulsive noise.

Damage-Risk Criteria

A damage-risk criterion specifies the maximum allowable exposure to which a person may be exposed if risk of hearing impairment is to be avoided. The American Academy of Ophthalmology and Otolaryngology has defined hearing impairment as an average HTL in excess of 25 dB (ANSI–1969) at 500, 1,000, and 2,000 Hz. This is called the *low fence*. Total impairment is said to occur when the average HTL exceeds 92 dB. Presbycusis is included in setting the 25 dB ANSI low fence. Two criteria have been set to provide conditions under which nearly all workers may be repeatedly exposed without adverse effect on their ability to hear and understand normal speech.

Continuous or Intermittent Exposure. The National Institute for Occupational Safety and Health (NIOSH) has recommended that occupational noise exposure be

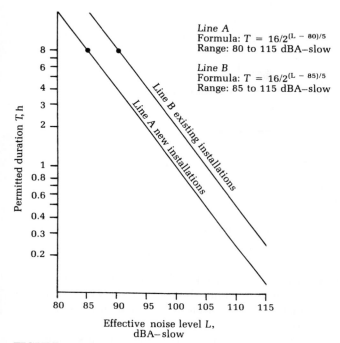

Line A
Formula: $T = 16/2^{(L - 80)/5}$
Range: 80 to 115 dBA–slow

Line B
Formula: $T = 16/2^{(L - 85)/5}$
Range: 85 to 115 dBA–slow

FIGURE 10-17
NIOSH occupational noise exposure limits for continuous or intermittent noise exposure.

controlled so that no worker is exposed in excess of the limits defined by line B in Figure 10-17. In addition, NIOSH recommends that new installations be designed to hold noise exposure below the limits defined by line A in Figure 10-17. The Walsh-Healey Act, which was enacted by Congress in 1969 to protect workers, used a damage-risk criterion equivalent to the line A criterion.

Speech Interference

As we all know, noise can interfere with our ability to communicate. Many noises that are not intense enough to cause hearing impairment can interfere with speech communication. The interference, or *masking,* effect is a complicated function of the distance between the speaker and listener and the frequency components of the spoken words. The Speech Interference Level (SIL) was developed as a measure of the difficulty in communication that could be expected with different background noise levels (Beranek, 1954). It is now more convenient to talk in terms of A-weighted background noise levels and the quality of speech communication (Figure 10-18).

Example 10-5. Consider the problem of a speaker in a quiet zone who wishes to speak to a listener operating a 4.5 Mg (megagram) truck 6.0 m away. The sound level in the truck cab is about 73 dBA.

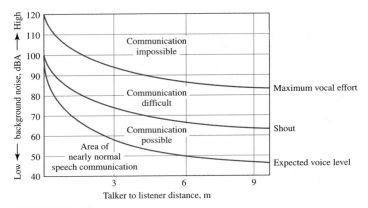

FIGURE 10-18
Quality of speech communication as a function of sound level and distance.
(*Source:* Miller, 1971.)

Solution. Using Figure 10-18, we can see that she is going to have to shout very loudly to be heard. However, if she moved to within about 1.0 m, she would be able to use her "expected" voice level, that is, the unconscious slight rise in voice level that one would normally use in a noisy situation.

It can be seen that at distances not uncommon in living rooms or classrooms (4.5 to 6.0 m), the A-weighted background level must be below about 50 dB for normal conversation.

Annoyance

Annoyance by noise is a response to auditory experience. Annoyance has its base in the unpleasant nature of some sounds, in the activities that are disturbed or disrupted by noise, in the physiological reactions to noise, and in the responses to the meaning of "messages" carried by the noise (Miller, 1971). For example, a sound heard at night may be more annoying than one heard by day, just as one that fluctuates may be more annoying than one that does not. A sound that resembles another sound that we already dislike and that perhaps threatens us may be especially annoying. A sound that we know is mindlessly inflicted and will not be removed soon may be more annoying than one that is temporarily and regretfully inflicted. A sound, the source of which is visible, may be more annoying than one with an invisible source. A sound that is new may be less annoying. A sound that is locally a political issue may have a particularly high or low annoyance (May, 1978).

The degree of annoyance and whether that annoyance leads to complaints, product rejection, or action against an existing or anticipated noise source depend upon many factors. Some of these factors have been identified, and their relative importance has been assessed. Responses to aircraft noise have received the greatest attention. There is less information available concerning responses to other noises, such as those of surface transportation and industry, and those from recreational activities (Miller, 1971). Many of the noise rating or forecasting systems that are now in existence were developed in an effort to predict annoyance reactions.

Sonic Booms. One noise of special interest with respect to annoyance is called *sonic boom* or, more correctly as we shall see, sonic booms.

The flow of air around an aircraft or other object whose speed exceeds the speed of sound (supersonic) is characterized by the existence of discontinuities in the air known as *shock wave*. These discontinuities result from the sudden encounter of an impenetrable body with air. At subsonic speeds, the air seems to be forewarned; thus, it begins its outward flow before the arrival of the leading edge. At supersonic speeds, however, the air in front of the aircraft is undisturbed, and the sudden impulse at the leading edge creates a region of overpressure (Figure 10-19) where the pressure is higher than atmospheric pressure. This overpressure region travels outward with the speed of sound, creating a conically shaped shock wave called the *bow wave* that changes the direction of airflow. A second shock wave, the *tail wave,* is produced by the tail of the aircraft and is associated with a region where the pressure is lower than normal. This underpressure discontinuity causes the air behind the aircraft to move sideways.

Major pressure changes are experienced at the ear as the bow and tail shock waves reach an observer. Each of these pressure deviations produces the sensation of an explosive sound (Minnix, 1978).

You should note that the pressure wave and, hence, the sonic boom exist whenever the aircraft is at supersonic speed and not "just when it breaks the sound barrier."

Both the loudness of the noise and the startling effect of the impulse (it makes us "jump") are found to be very annoying. Apparently we can never get used to this kind of noise. Supersonic flight by commercial aircraft is forbidden in the airspace above the United States. Supersonic flight by military aircraft is restricted to sparsely inhabited areas.

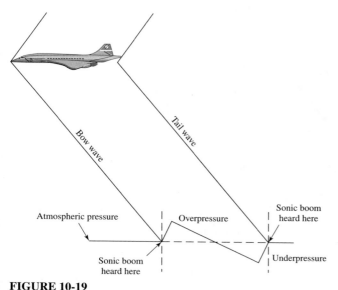

FIGURE 10-19
Sonic booms resulting from bow wave and tail wave set in motion by supersonic flight.

Sleep Interference

Sleep interference is a special category of annoyance that has received a great deal of attention and study. Almost all of us have been wakened or kept from falling asleep by loud, strange, frightening, or annoying sounds. It is commonplace to be wakened by an alarm clock or clock radio. But it also appears that one can get used to sounds and sleep through them. Possibly, environmental sounds only disturb sleep when they are unfamiliar. If so, disturbance of sleep would depend only on the frequency of unusual or novel sounds. Everyday experience also suggests that sound can help to induce sleep and, perhaps, to maintain it. The soothing lullaby, the steady hum of a fan, or the rhythmic sound of the surf can serve to induce relaxation. Certain steady sounds can serve as an acoustical shade and mask disturbing transient sounds.

Common anecdotes about sleep disturbance suggest an even greater complexity. A rural person may have difficulty sleeping in a noisy urban area. An urban person may be disturbed by the quiet when sleeping in a rural area. And how is it that a parent may wake to a slight stirring of his or her child, yet sleep through a thunderstorm? These observations all suggest that the relations between exposure to sound and the quality of a night's sleep are complicated.

The effects of relatively brief noises (about three minutes or less) on a person sleeping in a quiet environment have been studied the most thoroughly. Typically, presentations of the sounds are widely spaced throughout a sleep period of 5 to 7 hours. A summary of some of these observations is presented in Figure 10-20. The dashed lines are hypothetical curves that represent the percent of awakenings under conditions in which the subject is a normally rested young adult male who has been adapted for several nights to the procedures of a quiet sleep laboratory. He has been instructed to press an easily reached button to indicate that he has awakened, and had been moderately motivated to awake and respond to the noise.

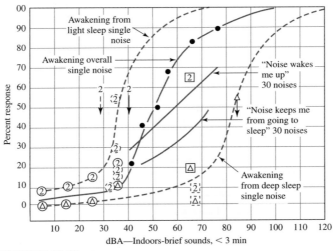

FIGURE 10-20
Effects of brief noise on sleep (*Source:* Miller, 1971.)

While in light sleep, subjects can awake to sounds that are about 30–40 decibels above the level at which they can be detected when subjects are conscious, alert, and attentive. While in deep sleep, the stimulus may have to be 50–80 decibels above the level at which they can be detected by conscious, alert, attentive subjects before they will awaken the sleeping subject.

The solid lines in Figure 10-20 are data from questionnaire studies of persons who live near airports. The percentage of respondents who claim that flyovers wake them or keep them from falling asleep is plotted against the A-weighted sound level of a single flyover. These curves are for the case of approximately 30 flyovers spaced over the normal sleep period of six to eight hours. The filled circles represent the percentage of sleepers that awake to a three-minute sound at each A-weighted sound level (dBA) or lower. This curve is based on data from 350 persons, each tested in his or her own bedroom. These measures were made between 2:00 and 7:00 AM. It is reasonable to assume that most of the subjects were roused from a light sleep.

Effects on Performance

When a task requires the use of auditory signals, speech or nonspeech, then noise at any intensity level sufficient to mask or interfere with the perception of these signals will interfere with the performance of the task.

Where mental or motor tasks do not involve auditory signals, the effects of noise on their performance have been difficult to assess. Human behavior is complicated, and it has been difficult to discover exactly how different kinds of noises might influence different kinds of people doing different kinds of tasks. Nonetheless, the following general conclusions have emerged. Steady noises without special meaning do not seem to interfere with human performance unless the A-weighted noise level exceeds about 90 decibels. Irregular bursts of noise (intrusive noise) are more disruptive than steady noises. Even when the A-weighted sound levels of irregular bursts are below 90 decibels, they may sometimes interfere with performance of a task. High-frequency components of noise, above about 1,000–2,000 hertz, may produce more interference with performance than low-frequency components of noise. Noise does not seem to influence the overall rate of work, but high levels of noise may increase the variability of the rate of work. There may be "noise pauses" followed by compensating increases in work rate. Noise is more likely to reduce the accuracy of work than to reduce the total quantity of work. Complex tasks are more likely to be adversely influenced by noise than are simple tasks.

Acoustic Privacy

Without opportunity for privacy, either everyone must conform strictly to an elaborate social code or everyone must adopt highly permissive attitudes. Opportunity for privacy avoids the necessity for either extreme. In particular, without opportunity for acoustical privacy, one may experience all of the effects of noise previously described and, in addition, one is constrained because one's own activities may disturb others. Without acoustic privacy, sound, like a faulty telephone exchange, reaches the "wrong number." The result disturbs both the sender and the receiver.

10-3   RATING SYSTEMS

Goals of a Noise-Rating System

An ideal noise-rating system is one that allows measurements by sound level meters or analyzers to be summarized succinctly and yet represent noise exposure in a meaningful way. In our previous discussions on loudness and annoyance, we noted that our response to sound is strongly dependent on the frequency of the sound. Furthermore, we noted that the type of noise (continuous, intermittent, or impulsive) and the time of day that it occurred (night being worse than day) were significant factors in annoyance.

Thus, the ideal system must take frequency into account. It should differentiate between daytime and nighttime noise. And, finally, it must be capable of describing the cumulative noise exposure. A statistical system can satisfy these requirements.

The practical difficulty with a statistical rating system is that it would yield a large set of parameters for each measuring location. A much larger array of numbers would be required to characterize a neighborhood. It is literally impossible for such an array of numbers to be used effectively in enforcement. Thus, there has been a considerable effort to define a single number measure of noise exposure. The following paragraphs describe two of the systems now being used.

The L_N Concept

The parameter L_N is a statistical measure that indicates how frequently a particular sound level is exceeded. If, for example, we write $L_{30} = 67$ dBA, then we know that 67 dB(A) was exceeded for 30 percent of the measuring time. A plot of L_N against N where $N = 1$ percent, 2 percent, 3 percent, and so forth, would look like the cumulative distribution curve shown in Figure 10-21.

Allied to the cumulative distribution curve is the probability distribution curve. A plot of this will show how often the noise levels fall into certain class intervals. In Figure 10-22 we can see that 35 percent of the time the measured noise levels ranged between 65 and 67 dBA; for 15 percent of the time they ranged between 67 and 69 dBA; and so on. The relationship between this picture and the one for L_N is really

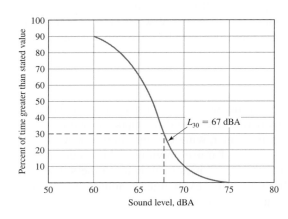

FIGURE 10-21
Cumulative distribution curve.

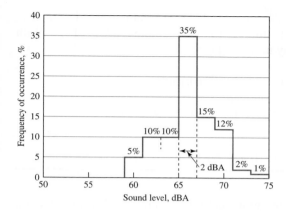

FIGURE 10-22
Probability distribution plot.

quite simple. By adding the percentages given in successive class intervals from right to left, we can arrive at a corresponding L_N where N is the sum of the percentages and L is the lower limit of the left-most class interval added, thus, L_{30}

$$L (1 + 2 + 12 + 15) = 67 \text{ dBA}$$

The L_{eq} Concept

The equivalent continuous equal energy level (L_{eq}) can be applied to any fluctuating noise level. It is that constant noise level that, over a given time, expends the same amount of energy as the fluctuating level over the same time period. It is expressed as follows:

$$L_{eq} = 10 \log \frac{1}{t} \int_0^t 10^{L(t)/10} \, dt \qquad (10\text{-}14)$$

where t = the time over which L_{eq} is determined
  $L(t)$ = the time varying noise level in dBA

Generally speaking, there is no well-defined relationship between $L(t)$ and time, so a series of discrete samples of $L(t)$ have to be taken. This modifies the expression to:

$$L_{eq} = 10 \log \sum_{i=t}^{i=n} (10^{L_i/10})(t_i) \qquad (10\text{-}15)$$

where n = the total number of samples taken
  L_i = the noise level in dBA of the ith sample
  t_i = fraction of total sample time

Example 10-6.   Consider the case where a noise level of 90 dBA exists for 10 minutes and is followed by a reduced noise level of 70 dBA for 30 minutes. What is the equivalent continuous equal energy level for the 40-minute period? Assume a five-minute sampling interval.

Solution. If the sampling interval is five minutes, then the total number of samples (n) is 8, and the fraction of total sample time (t_i) for each sample is $1/8 = 0.125$. With these preliminary calculations, we may now compute the sum:

$$\sum_{t=1}^{2} = (10^{90/10})(0.250) + (10^{70/10})(0.750)$$

$$= (2.50 \times 10^8) + (7.50 \times 10^6) = 2.58 \times 10^8$$

And finally, we take the log to find

$$L_{eq} = 10 \log{(2.58 \times 10^8)} = 84.11, \text{ or } 84, \text{ dBA}$$

The example calculation is depicted graphically in Figure 10-23. From this you may note that great emphasis is put on occasional high noise levels.

The equivalent noise level was introduced in 1965 in Germany as a rating specifically to evaluate the impact of aircraft noise upon the neighbors of airports (Burck et al., 1965). It was almost immediately recognized in Austria as appropriate for evaluating the impact of street traffic noise in dwellings and schoolrooms. It has been embodied in the National Test Standards of Germany for rating the subjective effects of fluctuating noises of all kinds, such as from street and road traffic, rail traffic, canal and river ship traffic, aircraft, industrial operations (including the noise from individual machines), sports stadiums, playgrounds, and the like.

The L_{dn} Concept

The L_{dn} is the L_{eq} computed over a 24-hour period with a "penalty" of 10 dBA for a designated nighttime period. Thus, it is a day-night average and the subscript "dn" is assigned instead of "eq." In applications to airport noise, the L_{dn} may be referred to as DNL. The nighttime period is from 10 PM. to 7 AM. The L_{dn} equation is derived from the L_{eq} equation with the time increment specified as 1 second. Because the

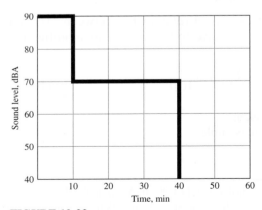

FIGURE 10-23
Graphical illustration of L_{eq} computation given in Example 10-6.

time over which the L_{dn} is computed is a day, the total time period is 86,400 seconds. Equation 10–15 is then written as

$$L_{dn} = 10 \log \left[\frac{1}{86,400} \sum 10^{Li/10} t_i + \sum 10^{(L_j + 10)/10} t_i \right] \quad (10\text{-}16)$$

Because $10(\log 86,400) \approx 49.4$, the day-night average sound level may be written as

$$L_{dn} = 10 \log \left[\sum 10^{Li/10} t_i + \sum 10^{(L_j + 10)/10} t_i \right] - 49.4 \quad (10\text{-}17)$$

10-4 COMMUNITY NOISE SOURCES AND CRITERIA

It is not our intent to provide a detailed discussion of the noise characteristics of all community noise sources. Likewise, we have not attempted to provide a comprehensive list of noise criteria. Rather, we have selected a few examples to provide you with a feeling for the magnitude and range of the numbers.

Transportation Noise

Aircraft Noise. The noise spectra of a wide body fan jet (for example, the Boeing 747) reveal that sound pressure levels are higher on takeoff than during the approach to land. This is typical of all aircraft. With the notable exception of the turbojets, smaller aircraft have lower sound pressure levels.

The annoyance criteria for aircraft operations are based on extensive field measurements and opinion surveys. The results of annoyance surveys at nine airports in the United States and Great Britain are summarized in Figure 10-24.

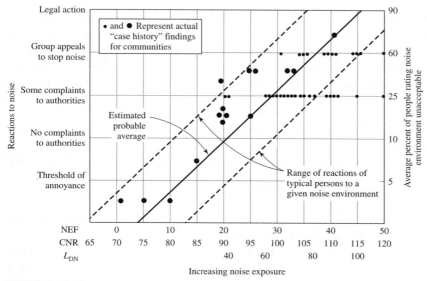

FIGURE 10-24
Relationship between exposure to aircraft noise and annoyance. (*Source:* Kryter et al., 1971.)

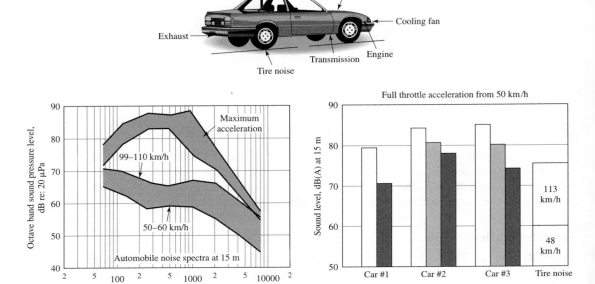

FIGURE 10-25
Typical noise spectra of automobiles. (*Source:* U.S. EPA, 1971.)

Highway Vehicle Noise. For most automobiles, exhaust noise constitutes the predominant source for normal operation below about 55 km/h (Figure 10-25). Although tire noise is much less of a problem in automobiles than in trucks, it is the dominant noise source at speeds above 80 km/h. While not as noisy as trucks, the total contribution of automobiles to the noise environment is significant because of the very large number in operation.

Diesel trucks are 8 to 10 dB noisier than gasoline-powered ones. At speeds above 80 km/h, tire noise often becomes the dominant noise source on the truck. The "cross-bar" tread is the noisiest.

Motorcycle noise is highly dependent on the speed of the vehicle. The primary source of noise is the exhaust. The noise spectra of two-cycle and four-cycle engines are of somewhat different character. The two-cycle engines exhibit more high frequency spectra energy content.

In 1968, Griffiths and Langdon (1968) reported on the results of an extensive attitude survey on traffic noise. They correlated their results with the Traffic Noise Index rating system (Figure 10-26). The U.S. Federal Highway Administration has developed the standards shown in Table 10-3. The levels are above those that would be expected to yield no problems but are below those of many existing highways.

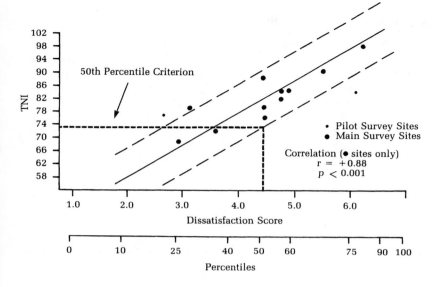

FIGURE 10-26
Annoyance as a function of the Traffic Noise Index (TNI). (*Source:* Alexandre et al., 1975.)

TABLE 10-3
FHA noise standards for new construction[a]

Land use category	Exterior design noise level dBA[b]		Description of land use category
	L_{eq}	L_{10}	
A	57	60	Tracts of lands in which serenity and quiet are of extraordinary significance and serve an important public need, and where the preservation of those qualities is essential if the area is to continue to serve its intended purpose. For example, such areas could include amphitheaters, particular parks or portions of parks, or open spaces, which are dedicated or recognized by appropriate local officials for activities requiring special qualities of serenity and quiet.
B	67	70	Residences, motels, hotels, public meeting rooms, schools, churches, libraries, hospitals, picnic areas, recreation areas, playgrounds, active sports areas, and parks.
C	72	75	Developed lands, properties, or activities not included in categories A and B above.
D	Unlimited	Unlimited	Undeveloped lands.
E	52 (Interior)	55 (Interior)	Public meeting rooms, schools, churches, libraries, hospitals, and other such public buildings.

[a]FHWA, 1973.
[b]Either L_{eq} or L_{10} may be used, but not both. The levels are to be based on a 1-hour sample.

TABLE 10-4
Summary of noise characteristics of internal combustion engines

Source	A-weighted noise energy (kw · h/d)[a]	Typical A-weighted noise level at 15.2 m [dB(A)]	8-hr exposure level [db(A)][b] Average	8-hr exposure level [db(A)][b] Maximum	Typical exposure time (h)
Lawn mowers	63	74	74	82	1.5
Garden tractors	63	78	N/A	N/A	N/A
Chain saws	40	82	85	95	1
Snow blowers	40	84	61	75	1
Lawn edgers	16	78	67	75	0.5
Model aircraft	12	78	70[c]	79[c]	0.25
Leaf blowers	3.2	76	67	75	0.25
Generators	0.8	71	—	—	—
Tillers	0.4	70	72	80	1

[a]Based on estimates of the total number of units in operation per day.
[b]Equivalent level for evaluation of relative hearing damage risk.   [c]During engine trimming operation.
(*Source:* U.S. EPA, 1971.)

Other Internal Combustion Engines

Because of their ubiquitous nature and the general interest they stimulate, the combustion engines listed in Table 10-4 are included at this point. "In general, these devices are not significant contributors to average residential noise levels in urban areas. However, the relative annoyance of most of the equipment tends to be high" (U.S. EPA, 1971). The eight-hour exposure level is in reference to the equipment operator.

Construction Noise

The range of sound levels found for 19 common types of construction equipment is shown in Figure 10-27. Although the sample was limited, the data appear to be reasonably accurate. The noise produced by the interaction of the machine and the material on which it acts often contributes greatly to the sound level.

It is difficult, at best, to quantify the annoyance that results from construction noise. The following generalizations appear to hold:

1. Single house construction in suburban communities will generate sporadic complaints if the boundary line eight-hour L_{eq} exceeds 70 dBA.

2. Major excavation and construction in a normal suburban community will generate threats of legal action if the boundary line eight-hour L_{eq} exceeds 85 dBA.

Zoning and Siting Considerations

The U.S. Department of Housing and Urban Development (HUD) set out guideline criteria for noise exposure at residential sites for new construction (Table 10-5).

The Federal Aviation Administration (FAA) specifies the L_{dn} for various types of land use compatibility (Table 10-6). These guidelines, and those given on page 733 for traffic noise (Table 10-3), if followed in zoning and siting, will minimize annoyance and complaints.

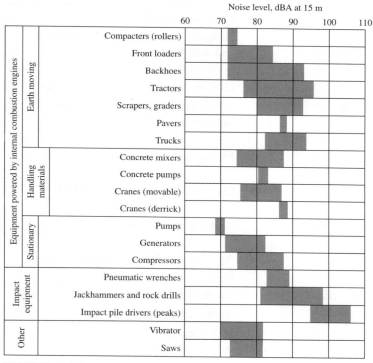

FIGURE 10-27

Range of sound levels from various types of construction equipment (based on limited available data samples). (*Source:* U.S. EPA, 1972.)

TABLE 10-5

HUD noise assessment criteria for new residential construction

General external exposures	Assessment
Exceeds 89 dBA 60 minutes per 24 hours Exceeds 75 dBA 8 hours per 24 hours	Unacceptable
Exceeds 65 dBA 8 hours per 24 hours Loud repetitive sounds on site	Discretionary: normally unacceptable
Does not exceed 65 dBA more than 8 hours per 24 hours	Discretionary: normally acceptablez
Does not exceed 45 dBA more than 30 minutes per 24 hours	Acceptable

TABLE 10-6
FAA land use compatability[a]

Land use category	Exterior L_{dn}, dBA yearly average	Description of land use category
Residential	<65	Dwellings schools
Public	<65	Hospitals, nursing homes, churches and auditoriums[b]
Public	65–70	Government services[c]
Commercial	65–70	Offices, retail trade, communication[c]
Commercial	80–85	Wholesale and retail equipment, utilities
Manufacturing and Production	60–75	Photographic and optical[c]
Manufacturing and Production	70–75	Livestock farming and breeding
Manufacturing and Production	80–85	General manufacturing
Manufacturing and Production	>85	Agriculture, forestry, mining, and fishing
Recreational	<65	Outdoor amphitheaters
Recreational	65–70	Nature exhibits and zoos
Recreational	65–70	Golf course, riding stables[c]
Recreational	70–75	Outdoor sports arena
Recreational	>85	Amusement parks and camps

[a]Adapted from FAA Advisory Circular AC150/5020-1.
[b]L_{dn} of 65–70 if 25 dB reduction indoors is provided; 70–75 if 30 dB reduction is provided.
[c]L_{dn} of 70–75 if 25 dB reduction indoors is provided; 75–80 if 30 dB reduction is provided.

Levels to Protect Health and Welfare

In accordance with the directive from Congress, the U.S. Environmental Protection Agency published noise criteria levels that it deemed necessary to protect the health and welfare of U.S. citizens (Table 10-7) (U.S. EPA, 1974). The EPA maintained that a quiet residential environment is necessary in both urban and rural areas to prevent activity interference and annoyance and to permit the hearing mechanism an opportunity to recuperate if it is exposed to high levels during the day. The L_{dn} of 45 provides a fair margin of safety.

TABLE 10-7

Yearly energy average L_{eq} identified as requisite to protect the public health and welfare with an adequate margin of safety

	Measure	Indoor			Outdoor		
		Activity interference	Hearing loss consideration	To protect against both effects (b)	Activity interference	Hearing loss consideration	To protect against both effects (b)
Residential with outside space and farm residences	L_{dn} $L_{eq(24)}$	45	70	45	55	70	55
Residential with no outside space	L_{dn} $L_{eq(24)}$	45	70	45			
Commercial	$L_{eq(24)}$	(a)	70	70(c)	(a)	70	70(c)
Inside transportation	$L_{eq(24)}$	(a)	70	(a)			
Industrial	$L_{eq(24)(d)}$	(a)	70	70(c)	(a)	70	70(c)
Hospitals	L_{dn} $L_{eq(24)}$	45	70	45	55	70	55
Educational	$L_{eq(24)}$ $L_{eq(24)(d)}$	45	70	45	55	70	55
Recreational areas	$L_{eq(24)}$	(a)	70	70(c)	(a)	70	70(c)
Farm land and general unpopulated land	$L_{eq(24)}$	(a)			(a)	70	70(c)

Code:

(a) Since different types of activities appear to be associated with different levels, identification of a maximum level for activity interference may be difficult except in those circumstances where speech communication is a critical activity.

(b) Based on lowest level.

(c) Based only on hearing loss.

(d) An $L_{eq(8)}$ of 75 dB may be identified in these situations so long as the exposure over the remaining 16 hours per day is low enough to result in negligible contribution to the 24-hour average, that is, no greater than an L_{eq} of 60 dB.

Note: Explanation of identified level for hearing loss: The exposure period that results in hearing loss at the identified level is a period of 40 years.

(*Source:* U.S. EPA, 1974).

10-5 TRANSMISSION OF SOUND OUTDOORS

Inverse Square Law

If a sphere of radius δ vibrates with a uniform radial expansion and contraction, sound waves radiate uniformly from its surface. If the sphere is placed such that no sound waves are reflected back in the direction of the source, and if the product $\kappa\delta$, where κ is the wave number, is much less than 1, then the sound intensity at any radial distance r from the sphere is inversely proportional to the square of distance, that is*:

$$I = \frac{W}{4\pi r^2}$$
(10-18)

where I = sound intensity, watts/m^2
W = sound power of source, watts

This is the *inverse square law*. It explains that portion of the reduction of sound intensity with distance that is due to wave divergence (Figure 10-28). For a line source such as a roadway or a railroad, the reduction of sound intensity is inversely proportional to r rather than r^2. If we measure sound power level (L_w, re: 10^{-12} W) rather than sound power (W), we can rewrite Equation 10-18 in terms of sound pressure level[†]:

$$L_P \cong L_w - 20 \log r - 11 \tag{10-19}$$

where $\quad L_P$ = sound pressure level, dB re: 20 μPa
L_w = sound power level, dB re: 10^{-12} W
r = distance between source and receiver, m
$20 \log r$ = decibel transform = $10 \log r^2$
11 = decibel transform $\cong [10 \log (4\pi) = 10.99]$

The tilde ($\sim$), indicating "approximately," results from the assumptions used above. L_w should be computed for all frequency bands of interest.

From a practical point of view it is difficult, if not impossible, to measure the sound power of the source. In such instances we measure the sound pressure level at some known distance from the source and then use the inverse square law or radial dependence relationships to estimate the sound pressure level at some other distance. For example, using the inverse square law, the sound pressure level L_{p2} at a distance r_2 from the source may be determined if the sound pressure level L_{p1} at some closer point r_1 is known:

$$L_{p2} = L_{p1} - 10\log\left(\frac{r_2}{r_1}\right)^2$$
(10-20)

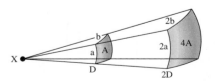

FIGURE 10-28
Illustration of inverse square law.

---

*$\kappa = 2\pi/\lambda$, where λ = wavelength, κ has units of reciprocal length, m^{-1}.
[†]This can be proved by using Equations 10-4, 10-5, 10-9, 10-10, and 10-11, and the assumption that $\rho c = 400$ kg/m$^2 \cdot$ s.

For a line source, the sound pressure level L_{p2} at a distance r_2 from the source may be determined at some closer point r_1 by a similar equation:

$$L_{p2} = L_{p1} - 10\log\left(\frac{r_2}{r_1}\right)$$

🏴 (10-21)

See Problems 10-29 and 10-30 for an alternate way to express Equations 10-20 and 10-21.

Radiation Fields of a Sound Source

The character of the wave radiation from a noise source will vary with distance from the source (Figure 10-29). At locations close to the source, the *near field*, the particle velocity is not in phase with the sound pressure. In this area, L_p fluctuates with distance and does not follow the inverse square law. When the particle velocity and sound pressure are in phase, the location of the sound measurement is said to be in the *far field*. If the sound source is in free space, that is, there are no reflecting surfaces, then measurements in the far field are also *free field measurements*. If the sound source is in a highly reflective space, for example, a room with steel walls, ceiling, and floor, then measurements in the far field are also *reverberant field measurements*. The shaded area in the far field of Figure 10-29 shows that L_p does not follow the inverse square law in the reverberant field.

Directivity

Most real sources do not radiate sound uniformly in all directions. If you were to measure the sound pressure level in a given frequency band at a fixed distance from a real source, you would find different levels for different directions. If you plotted these data in polar coordinates, you would obtain the directivity pattern of the source.

The *directivity factor* is the numerical measure of the directivity of a sound source. In logarithmic form the directivity factor is called the *directivity index*. For a spherical source it is defined as follows:

$$DI_\theta = L_{p\theta} - L_{ps}$$

(10-22)

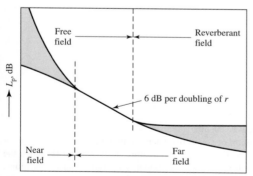

FIGURE 10-29

Variation of sound-pressure level in an enclosure along radius r from a noise source. (*Source:* Beranek, Noise and Vibration Control, Institute of Noise Control Engineering, 1988.)

where $L_{p\theta}$ = sound pressure level measured at distance r' and angle $0°$ from a directive source radiating power W into an echo-free (*anechoic*) space, dB

L_{ps} = sound pressure level measured at distance r' from a nondirective source radiating power W into anechoic space,* dB

For a source located on or near a hard, flat surface, the directivity index takes the following form:

$$DI_\theta = L_{p\theta} - L_{ps} + 3 \qquad (10\text{-}23)$$

The 3 dB addition is made because the measurement is made over a hemisphere instead of a sphere. That is, the intensity at a radius, r, is twice as large if a source radiates into a hemisphere rather than the ideal sphere we have used up to this point. Each directivity index is applicable only to the angle at which $L_{p\theta}$ was measured and only for the frequency at which it was measured.

We assume that the directivity pattern does not change its shape regardless of the distance from the source. This allows us to apply the inverse square law to directive sources simply by adding the directivity index:

$$L_{p\theta} \cong L_w + DI_\theta - 20\log r - 11 \qquad (10\text{-}24)$$

You should note that it is not possible to reduce the equation by using the equality given in Equation 10-22. The values of $L_{p\theta}$ are at a distance r, which is different than the r' in Equation 10-22.

Favorable propagation conditions are those shown in Figure 10-30a. They are environmentally relevant. These conditions are also stable in the sense that they are suitable for reproducible measurements. It has become standard practice to restrict prediction of airborne transmission of noise to conditions favorable for propagation. These are specified as (ISO, 1989, 1990):

1. The wind direction is within an angle of $45°$ of the direction connecting the center of the sound source and the center of the specified area, with the wind blowing from the source to the receiver.

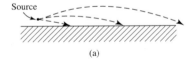

(a)

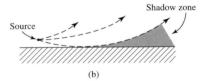

(b)

FIGURE 10-30
Refraction of sound (*a*) when the propagation is downwind or under conditions of a temperature inversion and (*b*) when the propagation is upwind or under temperature lapse conditions. (*Source:* Piercy and Daigle, 1991.)

---

*This is the same source as the directive source, but acting in the ideal fashion that we assumed in developing the inverse square law.

2. The wind speed is between approximately 1 and 5 m/s measured at a height of 3–11 m.

3. Propagation in any near horizontal direction is under a well developed ground-based inversion.

Airborne Transmission

Effects of Atmospheric Conditions. Sound energy is absorbed in quiet isotropic air by molecular excitation and relaxation of oxygen molecules and, at very low temperatures, by heat conduction and viscosity in the air. Molecular excitation is a complex function of the frequency of noise, humidity, and temperature. In general, we may say that as the humidity decreases, sound absorption increases. As the temperature increases to about 10 to 20°C (depending upon the noise frequency), absorption increases. Above 25°C, absorption decreases. Sound absorption is higher at higher frequencies.

The vertical temperature profile greatly alters the propagation paths of sound. If a superadiabatic lapse rate exists, sound rays bend upward and noise shadow zones are formed (Figure 10-30b). If an inversion exists, sound rays are bent back toward the ground (Figure 10-30a). This results in an increase in the sound level. These effects are negligible for short distances but may exceed 20 dB at distances over 800 m.

In a similar fashion, wind speed gradients alter the way noise propagates. Sound traveling with the wind is bent down, while sound traveling against the wind is bent upward. When sound waves are bent down, there is little or no increase in sound levels. But when sound waves are bent upward, there can be a noticeable reduction in sound levels.

Basic Point Source Model. A point source is one for which $\kappa\delta \ll 1$ and for which Equation 10-18 holds. According to Magrab (1975),

> In practice most noise sources cannot be classified as simple point sources. However, the sound field of a complicated sound source will look as if it were a point source if the following two conditions are met: (1) $r/\delta \gg 1$, that is, the distance from the source is large compared to its characteristic dimension, and (2) $\delta/\lambda \ll r/\delta$, that is, the ratio of the size of the source to the wavelength of sound in the medium is small compared to the ratio of the distance from the source to its characteristic dimension. Recall that $r/\delta \gg 1$ from the first condition. A value of $r/\delta > 3$ is a sufficient approximation; therefore, $\delta\lambda \ll 3$.

The basic point source equation is

$$L_p \cong L_w - 20 \log r - 11 - A_e \qquad (10\text{-}25)$$

where L_p = the desired SPL (re: (20 μPa) at angle θ and distance r from source, dB
L_w = the measured sound power level (re: 10^{-12} W) at angle θ, dB
A_e = attenuation for the distance r, dB

With the exception of the last term (A_e), it is the inverse square law (Equations 10-18 and 10-19). The A_e term is the excess attenuation beyond wave divergence. It is caused by environmental conditions and has units of dB.

The A_e term may be further divided into five terms as follows:

A_{e1} = attenuation by absorption in the air, dB

A_{e2} = attenuation by the ground, dB

A_{e3} = attenuation by barriers, dB

A_{e4} = attenuation by foliage, dB

A_{e5} = attenuation by houses, dB

Because of the introductory nature of this text, we have chosen to limit the following discussion to the first two terms, A_{e1} and A_{e2}. In addition, we will consider only the case of ground attenuation ranges greater than 100 m. For detailed examination of the other cases we recommend that you consult Piercy and Daigle (1991).

The attenuation of sound by air absorption is given as (Piercy and Daigle, 1991):

$$A_{e1} = \frac{\alpha d}{1000 \text{ m/km}} \tag{10-26}$$

where α = air attenuation coefficient, dB/km

d = distance, m

The air attenuation coefficient α as a function of temperature and humidity is listed in Table 10-8.

The sound above a reflecting ground surface arrives at a receiver R from a source S by two paths (Figure 10-31). The path r_d is a direct ray. The path r_r is the reflection of the sound from the ground. The attenuation A_{e2} is a result of interference between the reflected ray and the direct ray. It strongly depends on the type of ground surface, the grazing angle ψ, the path length difference ($r_r - r_d$), and the frequency of the sound. The ground surface may be classified as follows:

- Hard: asphalt or concrete pavement, water, and all surfaces with low porosity.

- Soft: ground covered by grass or other vegetation and all surfaces suitable for growth of vegetation.

- Mixed: surface that includes both hard and soft conditions.

Very soft ground such as snow cover is not considered in the following analysis.

As noted above, for long range propagation (>100 m) the ground attenuation (A_{e2}) is calculated for atmospheric conditions favorable to propagation. At distances less than 100 m the results obtained by the following method will differ slightly from the short range technique (Piercy and Daigle, 1991).

Figure 10-32 is used to define the zones and terms used in calculating A_{e2}. Each zone is assigned a *ground factor* according to the following rules:

1. The *source zone* (A_s) extends from the source S toward the receiver R a distance of $30h_s$, with a maximum of r.

2. The *receiver zone* (A_r) extends from the receiver R toward the source S a distance of $30h_r$ with a maximum of r.

3. The *middle zone* (A_m) lies between the source and receiver zones. If $r < 30(h_s + h_r)$, then the source and receiver zones overlap and there is no middle zone.

TABLE 10-8
Air attenuation coefficient, dB/km, for an ambient pressure of 101.3 kPa (one standard sea-level atmosphere) for sound propagation in open air

Temperature	Relative humidity, %	Frequency, Hz					
		125	250	500	1,000	2,000	4,000
30°C	10	0.96	1.8	3.4	8.7	29	96
	20	0.73	1.9	3.4	6.0	15	47
	30	0.54	1.7	3.7	6.2	12	33
	50	0.35	1.3	3.6	7.0	12	25
	70	0.26	0.96	3.1	7.4	13	23
	90	0.20	0.78	2.7	7.3	14	24
20°C	10	0.78	1.6	4.3	14	45	109
	20	0.71	1.4	2.6	6.5	22	74
	30	0.62	1.4	2.5	5.0	14	49
	50	0.45	1.3	2.7	4.7	9.9	29
	70	0.34	1.1	2.8	5.0	9.0	23
	90	0.27	0.97	2.7	5.3	9.1	20
10°C	10	0.79	2.3	7.5	22	42	57
	20	0.58	1.2	3.3	11	36	92
	30	0.55	1.1	2.3	6.8	24	77
	50	0.49	1.1	1.9	4.3	13	47
	70	0.41	1.0	1.9	3.7	9.7	33
	90	0.35	1.0	2.0	3.5	8.1	26
0°C	10	1.3	4.0	9.3	14	17	19
	20	0.61	1.9	6.2	18	35	47
	30	0.47	1.2	3.7	13	36	69
	50	0.41	0.82	2.1	6.8	24	71
	70	0.39	0.76	1.6	4.6	16	56
	90	0.38	0.76	1.5	3.7	12	43

(*Source:* ISO, 1990.)

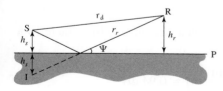

FIGURE 10-31
Paths for propagation from source S to receiver R. The direct ray is r_d and the ray reflected from plane P (which effectively comes from image source I) is r_r. (*Source:* Piercy and Daigle, 1991.)

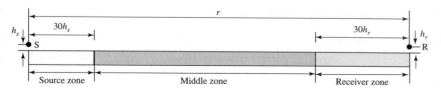

FIGURE 10-32
Three zones between a source S and receiver R separated by distance r, used in determining the ground attenuation A_{ground} at long ranges. (*Source:* Piercy and Daigle, 1991.)

The ground factor G for each zone is based on the surface characteristics:

- Hard ground: $G = 0$
- Soft ground: $G = 1$
- Mixed ground: G equals the fraction of the ground that is soft [for example if 25 percent of the ground is soft, then $G = (0.25)(1) = 0.25$]

The ground attenuation for an octave band is calculated for each zone by using the equations and data in Table 10-9. The value for e for the middle zone calculation in the table is determined from the following equation:

$$e = 1 - \left[\frac{30(h_s + h_r)}{r}\right] \qquad (10\text{-}27)$$

The total ground attenuation is then $\qquad A_{e2} = A_s + A_r + A_m \qquad (10\text{-}28)$

For a complete analysis, the attenuation must computed for each relevant octave band.

Example 10-7. The sound power level (re: 10^{-12} W) of a compressor is 124.5 dB at 1,000 Hz. Determine the SPL 200 m downwind on a clear summer afternoon if the wind speed is 5 m/s, the temperature is 20°C, the relative humidity is 50 percent, and the barometric pressure is 101.325 kPa. The heights of the compressor and the receiver are 1.2 m. The ground surface characteristics are shown in the sketch below.

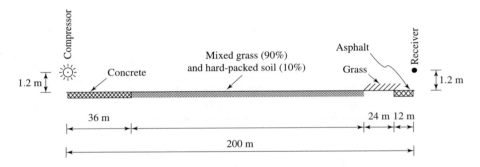

Solution. The attenuation by air absorption (A_{e1}) is calculated directly from Table 10-8 with the distance being 200 m.

$$A_{e1} = (4.7 \text{ dB/km})\left(\frac{200 \text{ m}}{1,000 \text{ m/km}}\right) = 0.94 \text{ dB}$$

Calculate the ground attenuation in four steps.

1. The source zone attenuation extends a distance

$$30h_s = (30)(1.2 \text{ m}) = 36 \text{ m}$$

TABLE 10-9

Expressions to be used in calculating the octave-band ground attenuation (A_{ground}) in decibels at long range

Octave-band frequency, Hz	A_s or A_r, dB	A_m, dB
63	-1.5	$-3e$
125	$(a \cdot G) - 1.5$	$-3e(1 - G)$
250	$(b \cdot G) - 1.5$	$-3e(1 - G)$
500	$(c \cdot G) - 1.5$	$-3e(1 - G)$
1,000	$(d \cdot G) - 1.5$	$-3e(1 - G)$
2,000	$(1 - G) - 1.5$	$-3e(1 - G)$
4,000	$(1 - G) - 1.5$	$-3e(1 - G)$
8,000	$(1 - G) - 1.5$	$-3e(1 - G)$

	Source or receiver height, m				
Distance, m	0.5	1.5	3.0	6.0	>10.0
Factor a					
50	1.7	2.0	2.7	3.2	1.6
100	1.9	2.2	3.2	3.8	1.6
200	2.3	2.7	3.6	4.1	1.6
500	4.6	4.5	4.6	4.3	1.6
>1,000	7.0	6.6	5.7	4.4	1.7
Factor b					
50	6.8	5.9	3.9	1.7	1.5
100	8.8	7.6	4.8	1.8	1.5
>200	9.8	8.4	5.3	1.8	1.5
Factor c					
50	9.4	4.6	1.6	1.5	1.5
100	12.3	5.8	1.7	1.5	1.5
>200	13.8	6.5	1.7	1.5	1.5
Factor d					
50	4.0	1.9	1.5	1.5	1.5
>100	5.0	2.1	1.5	1.5	1.5

(*Source:* ISO, 1989.)

From the sketch we note that the source zone is 100 percent hard and that $G = 0$. From Table 10-9, the equation for the source zone at 1,000 Hz is

$$A_s = [(d)(G)] - 1.5$$

From Table 10-9, for distances >100 m and a source height of 1.5 m, select $d = 2.1$. The ground attenuation is then

$$A_s = [(2.1)(0)] - 1.5 = -1.5 \text{ dB}$$

(Note that we did not interpolate for the height because the product would obviously be zero.)

2. The receiver zone extends a distance

$$30h_r = (30)(1.2 \text{ m}) = 36 \text{ m}$$

From the sketch we note that 12 m is hard and $36 - 12 = 24$ m is soft. The fraction that is soft is then $12/36 = 0.33$. The G value for "soft" is 1.0. From Table 10-9, the equation for receiver attenuation at 1,000 Hz is

$$A_r = [(d)(G)] - 1.5$$

The d value is the same as that for the source zone. The G value must be multiplied by the fraction of receiver zone that is soft. The receiver zone attenuation is

$$A_r = [(2.97)(0.33)(1.0)] - 1.5 = 0.98 - 1.5 = -0.52 \text{ dB}$$

(Note that in this case we did interpolate the receiver height to obtain 2.97 for the factor d.)

3. The middle zone is 90 percent covered in grass, so $G = (0.90)(1.0) = 0.90$. From Table 10-9, the equation for the middle zone attenuation is

$$A_m = -3e(1 - G)$$

The value for e is calculated by using Equation 10-27:

$$e = 1 - \left[\frac{30(1.2 + 1.2)}{200} \right] = 1 - 0.36 = 0.64$$

The attenuation in the middle zone is:

$$A_m = -3(0.64)(1 - 0.90) = -0.19 \text{ dB}$$

4. The total ground attenuation is

$$A_{e2} = -1.5 - 0.52 - 0.19 = -2.21 \text{ dB}$$

Note that the attenuation is negative. Thus, the ground surface reflection actually increases the SPL.

Using the basic point source model (Equation 10-25) gives the SPL at the receiver as

$$L_p = 124.5 - 20 \log (200) - 11 - 0.94 - (-2.21)$$
$$= 124.5 - 46 - 11 - 0.94 + 2.21 = 68.77 \text{ or } 69 \text{ dB at } 1,000 \text{ Hz}$$

10-6 TRAFFIC NOISE PREDICTION

National Cooperative Highway Research Program 174

The National Cooperative Highway Research Program has developed a series of documents (NCHRP 117, NCHRP 144, and NCHRP 174) that provide design guidance for the prediction and control of highway noise (Kugler et al., 1976). These documents have been used widely because of their simplicity and relatively high success in making accurate noise predictions. The NCHRP 174 procedure is the last revision in the series. It contains a four-step procedure for the prediction and control of highway noise. We have limited ourselves to the first prediction step, that is, the "short method." The Federal Highway Administration (FHWA) has developed sophisticated computer models to replace the NCHRP 174 manual technique used in this text for illustration purposes. The FHWA released Version 2.5 of the Traffic Noise Model—TNM in 2004.*

The objective of the "short method" is to obtain a quick and gross (always overpredicting) prediction of the expected noise levels. This is necessary because the prediction of true highway noise levels is a rather complicated subject. In many instances it is desirable to first obtain a rough idea of the potential problem areas before full knowledge of the horizontal and vertical roadway design parameters has been gained. Such is the case, for example, of a location study where a number of alignments must be considered. Also, this first step helps to eliminate areas that do not represent a problem in terms of noise levels, thus simplifying further evaluation.

The "short method" prediction can be performed quickly through use of two *nomographs* and knowledge of a few traffic and roadway parameters.[†] By its design, the "short method" requires many assumptions and approximations and should not be used as a final tool.

The second step (the "complete method") utilizes a microcomputer program to refine the predictions made in the first step. The third step is the selection of a noise control design. The fourth step is to redo the second step and check the design solution. In the following paragraphs we have reproduced the short method as it appears in NCHRP 174, with the addition of clarifying comments and modification to SI units.

Methodology. The flow diagram that illustrates the methodology of the short method is shown in Figure 10-33. The method assumes that the roadway can be approximated by one infinite element with constant traffic parameters and roadway characteristics.

The initial step in using the short method consists of defining an infinite straight-line approximation to the real highway configuration. On-ramps, off-ramps, and interchange ramps are omitted from the short method analysis.

---

*TNM 2.5.

[†]A nomograph is a graph that provides the solution to an equation or series of equations containing three or more variables (see Figure 10-37).

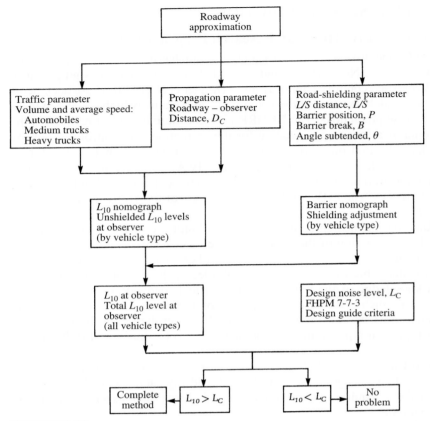

FIGURE 10-33

Flow diagram of methodology for applying NCHRP 174 method for estimating L_{10} from traffic. (*Source:* Kugler et al., 1976.)

Once the approximate roadway has been chosen, the following parameters must be computed or estimated: (*a*) the traffic parameters, which include the speed and volume of each class of vehicles; (*b*) the propagation characteristics, which describe the location of the receiver relative to the roadway; and (*c*) the roadway-shielding parameters, which describe the shielding provided by the roadway, if any. [Only barriers located within the right-of-way (ROW) may be considered.]

These parameters are used in two operations. First, the traffic and propagation parameters are combined in the L_{10} nomograph to determine, for each type of source, the unshielded L_{10} level at the observer.

The final result is then compared to the criteria level, L_c, at the observer (see, for example, Table 10-3) to define a "no problem" or "potential problem" condition. If a potential problem is identified, the observer location in question should be evaluated using the complete method.

Procedure.   The step-by-step procedures necessary to calculate noise levels by the short method are presented in the following numbered paragraphs. In addition to the

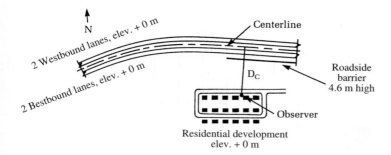

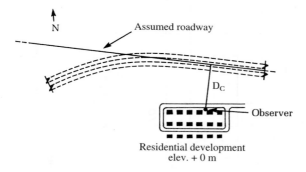

(a) Route map showing observer location and observer -
roadway centerline distance, D_C

(b) Model of assumed roadway alignment

FIGURE 10-34

(a) Roadway and (b) roadway approximation. (*Source:* Kugler et al., 1976.)

nomographs, the method uses a noise prediction worksheet to aid the user in the sequential steps. A blank worksheet and larger scale nomographs are included in Appendix B.

1. Observer identification: On a route map of convenient scale, identify all observer locations at which analysis is desired.

2. Roadway approximation: Approximate the roadway alignment by a straight, infinite line. The procedure is as follows: Determine and measure the nearest perpendicular distance, D_c, between the roadway centerline and observer, as shown in Figure 10-34a. Enter on line 4 on the noise prediction worksheet (Figure 10-35). Note that the infinite roadway approximation automatically assumes a line perpendicular to the centerline distance, D_c. There is no need to draw this line on the route map for computation reasons. An illustration of this assumption is shown in Figure 10-34b. Note that for each observer location, a

Project __BRISTOL HWY.__ Date __13 AUG. 2011__ Engineer __I. THOMPSON__

Step			Ex. 9-10a A	T_M	T_H	Ex. 9-10b A	T_M	T_H	A	T_M	T_H	A	T_M	T_H	A	T_M	T_H
1	Traffic	Vehicle Volume, V(Vph)	2000	100	100	2000	100	100									
2		Vehicle Av. Speed, S(km/h)	80.5	80.5	80.5	80.5	80.5	80.5									
3		Combined Veh. Vol.*, V_C(Vph)	3000		■	3000		■			■			■			■
4	Prop.	Observer-Roadway Dist., D_C(m)	60			60											
5	Shielding	Line-of-Sight Dist., L/S(m)	—			60											
6		Barrier Position Dist., P(m)	—			15.2											
7		Break in Barrier., B(m)	—			4.6		2.7									
8		Angle Subtended, θ (deg)	170			170											
9	Prediction**	Unshield L_{10} Level (dBA)	66	—	68	66	—	68									
10		Shielding Adjust. (dBA)	0		0	13		10									
11		L_{10} at Observer (By Veh. Class)	66	—	68	53	—	58									
12		L_{10} at Observer – Total	70 dBA			59.25 or 59 dBA											

Code:

A = Automobiles, T_M = Medium Trucks, T_H = Heavy Trucks

* Applies only when automobile and medium truck average speeds are equal. $V_C = V_A + (10)V_{T_M}$

** If automobile-medium truck volume V_C is combined, use L_{10} nomograph prediction only once for these two vehicle classes

FIGURE 10-35

Noise prediction worksheet. (*Source:* Kugler et al., 1976.)

different roadway approximation might result. Note also that the noise prediction worksheet (NPWS) allows for computations for six different observer locations by entering an observer location identification at the head of each column. Similarly, the NPWS can be used to calculate six different traffic conditions for the same observer location.

3. Traffic parameters: Determine the vehicle operating conditions by using the traffic parameters at the roadway point nearest the observer (if these parameters vary along the roadway). The procedure is as follows:

   a. Determine the automobile volume (vph) and average speed (km/h) and enter them on lines 1 and 2 under automobiles (A).

   b. Determine the medium truck volume (vph) and average speed (km/h) and enter them on lines 1 and 2 under medium trucks (T_M).

   c. Determine the heavy truck volume (vph) and average speed (km/h) and enter them on lines 1 and 2 under heavy trucks (T_H).

   d. If the automobile and medium truck speeds are the same, multiply the medium truck volume by 10 and add to the automobile volume. Enter combined volume V_c on line 3 of the NPWS. If the automobile and medium truck volumes are combined, in subsequent operations consider the two vehicle classes as one source. If the speed of the medium trucks differs from that of the automobiles, the volumes are *not* combined *but* the medium truck volume is still multiplied by 10 for the determination of the L_{10} value for the reason noted in the footnote below.*

4. Roadway-shielding parameters: If the roadway cross section at the nearest point is not at grade (either elevated or depressed), or if a roadside barrier (on the roadway right-of-way) is present, determine the roadway-shielding parameters. If the elevation, depression, or roadside barrier is less than 1.5 m high (compared to the surrounding terrain), disregard it. The procedure is as follows: Determine the barrier parameters and enter on lines 5 through 8 of the NPWS. Use Figure 10-36 for definitions of parameters. The parameters that must be measured are: (*a*) line-of-sight distance, L/S, (*b*) break in line of sight, (*c*) barrier position distance, and (*d*) angle subtended, θ (degree). The angle subtended is measured from the ends of the barrier with respect to the position of the observer. For an observer placed equidistant from the ends of the barrier, the angle may be determined from simple trigonometric principles. Graphical methods may be more appropriate for other configurations. Note that as the barrier length increases, the angle approaches 180 degrees.

5. Unshielded L_{10} level at observer location: Determine the unshielded L_{10} level at the observer location for all three traffic sources (automobiles, medium trucks, and heavy trucks) using the L_{10} nomograph (Figure 10-37 on page 753).

---

*The medium truck volume is multiplied by 10 because this traffic noise source behaves similarly to automobile noise, but the overall level is 10 dBA higher than for automobiles.

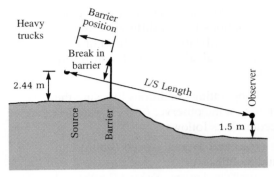

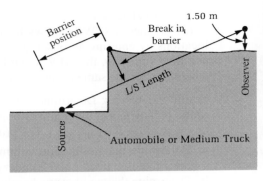

(a) Barrier parameters for simple barrier, section view

(b) Barrier parameters for depressed raodway, section view

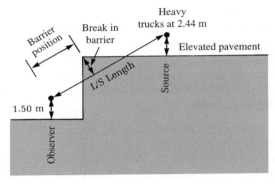

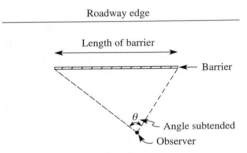

(c) Barrier parameters for elevated roadway, section view

(d) Barrier parameters, plan view

FIGURE 10-36

Definitions of barrier parameters. (*Source:* Kugler et al., 1976.)

Note that if the automobile and medium truck speeds are equal, these two sources may be evaluated together using the combined volume, V_c, and average speed, S_A or S_M on lines 3 and 2 of NPWS. The procedure is as follows:

a. Automobiles (and medium trucks): Using the vehicle volume, V_A (this corresponds to V_c, the combined auto and medium truck volumes, when the speeds of these two populations are equal), and the average speed, S_A (or S_M), enter the L_{10} nomograph and determine the unshielded L_{10} noise level at the observer. Enter on line 9 of the NPWS.

b. Medium trucks: Using the vehicle volume, V_M, *multiplied by ten,* and the average speed, S_M, enter the L_{10} nomograph and determine the unshielded L_{10} noise level at the observer. Enter on line 9 of the NPWS. If automobiles and medium trucks were combined in step a, this step should be omitted.

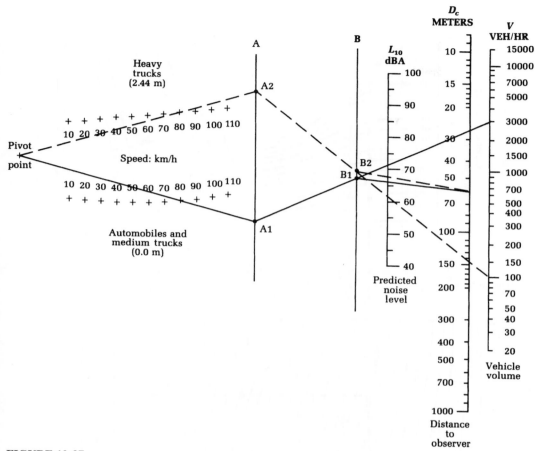

FIGURE 10-37
L_{10} nomograph.

    c. Heavy trucks: Using the vehicle volume, V_T, and the average speed, S_T, enter the L_{10} nomograph and determine the unshielded L_{10} noise level at the observer. Enter on line 9 of the NPWS.

6. Shielding adjustment: Determine the noise reduction afforded by the roadway geometry using the barrier nomograph (Figure 10-38) and the roadway parameters listed in the NPWS. This procedure must be performed twice: once for the 0 m source elevation (automobiles and medium trucks) and once for the 2.44 m elevation (heavy trucks).

    a. Low sources (0 m): Using the line-of-sight distance, L/S, barrier position distance, P, break in L/S distance, B (for sources at ground level), and the angle subtended, θ, enter the barrier nomograph and calculate the shielding adjustment. Enter on line 10 of the NPWS under automobile and medium trucks.

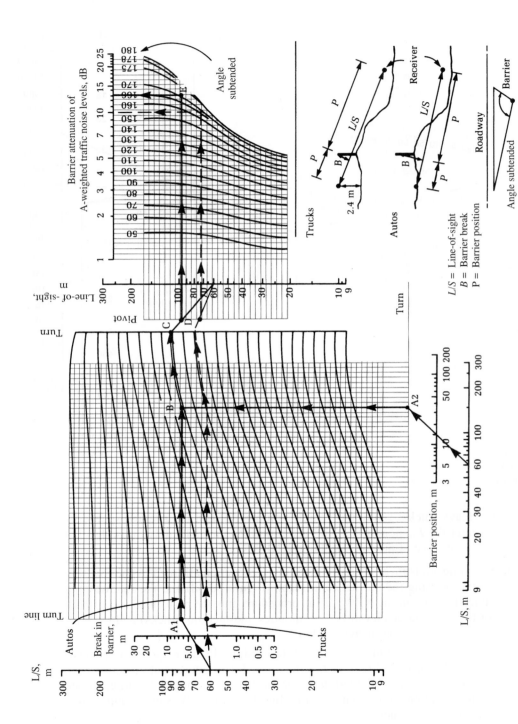

FIGURE 10-38

Barrier nomograph. (*Source:* Kugler et al., 1976.)

b. High sources (2.44 m): Using the line-of-sight distance, L/S, barrier position distance, P, break in L/S distance, B (for 2.44 m sources), and the angle subtended, θ, enter the barrier nomograph and calculate the shielding adjustment. Enter on line 10 of the NPWS under heavy trucks.

7. L_{10} at observer, by vehicle type: Calculate the L_{10} noise level at the observer for each individual source by subtracting the shielding adjustment (line 10) from the unshielded L_{10} level at the observer (line 9), and enter the result in line 11. Note that the shielding adjustment is always negative and can be subtracted algebraically from line 9.

8. Total L_{10} level at observer: Determine the total L_{10} noise level at the observer and enter on line 12. This is done by logarithmically adding (decibel addition) the contributions from automobiles, medium trucks, and heavy trucks computed in line 11. Taking two L_{10} levels at a time, find the difference between them and enter the addition scale provided in Figure 10-4 to find the "adjustment." The "adjustment" should be added to the higher of the two L_{10} levels. The operation is then repeated, two levels at a time, until only one level remains. The lowest levels should be added first for maximum accuracy.

Example 10-8. The county road commissioner has requested a noise evaluation of a proposed highway near Bristol. The proposed highway is to be routed such that the centerline of the roadway will be 60 m from a school. Using the following data, determine whether or not the FHA criterion will be met:

Average vehicle speed = 80.5 km/h for all vehicles
Automobiles = 2,000/h
Medium trucks = 100/h
Heavy trucks = 100/h

The terrain is level and no shielding is present.

Solution. According to step 3d, when the speed of the medium trucks and the cars is the same, the volumes can be combined by multiplying the medium truck volume by ten, and then combining it with the car volume. The combined volume, V_c, for cars and medium trucks is 2,000 + 10 (100) = 3,000 vph. Using the L_{10} nomograph shown in Figure 10-37, proceed as follows:

1. Draw a straight line from the left pivot point through the 80.5 km/h point on the automobile speed scale. Extend the straight line to turn line A. The intersection is marked A1.

2. Draw a second straight line from the intersection point A1 to the 3,000 vph point on the volume scale on the far right of the figure. The intersection of this straight line with turn line B is marked B1.

3. Draw a third straight line from point B1 to the point on the D_c scale (60 m). The intersection of this third line with the L_{10} scale gives the predicted A-weighted L_{10} level at the observer. For this example, the predicted L_{10} level is 66 dBA. This value is entered on line 9 of the NPWS (Figure 10-35).

Now repeat the procedure for the heavy trucks. It is shown by the dashed line in Figure 10-37. The predicted L_{10} for heavy trucks is 68 dBA.

The combined level of the automobiles and trucks is found by "decibel addition" (Example 10-1) to be 70 dBA. This just meets the FHA design level for land use category B (Table 10-3). Let us see what effect a barrier will have. We have arbitrarily selected the following characteristics for the barrier:

Height = 5.0 m, which yields a "Break in Barrier" of 4.6 m for automobiles and 2.6 m for heavy trucks

Position = 15.2 m from centerline of roadway

Subtended angle = 170°

Note: The "Break in Barrier" is determined by constructing a scale drawing of the roadway and receiver as shown in Figure 10-36. Because the automobile noise is assumed to originate at an elevation of 0.0 m, the "Break in Barrier" is greater than for the heavy trucks, where the noise is asssumed to originate at an elevation of 2.44 m.

Using the barrier nomograph shown in Figure 10-38, proceed as follows:

1. Starting from the vertical *L/S* scale on the left: From the 60-m point on the *L/S* scale, draw a straight line going through the 4.6-m point on the "Break in barrier" to the "Turn line." Note that the scale is logarithmic. From the intersection, called A_1, draw a straight horizontal line.

2. Starting from the horizontal *L/S* scale on the bottom: Draw a straight line through the 60-m point on the *L/S* scale, and the 15.2-m point on the "Barrier position" scale to the "Turn line." From the intersection, called A_2, draw a straight vertical line until it intersects (at B) with the horizontal drawn from A_1.

3. From B, move to the right upward following the nearest curve to turn line C. (If B is on one of the curves, simply follow it; if B is between curves, follow parallel to the nearest curve upward to the right until it intersects with the turn line.)

4. From C, draw a straight line to the line-of-sight (*L/S*) distance (60 m) point on the vertical *L/S* scale. It will intersect with the pivot line at D.

5. From D, draw a horizontal line to the right until it intersects the curve corresponding to the subtended angle (170°) at E.

6. Finally, draw a vertical line from E upward until it intersects the barrier attenuation scale. The attenuation can now be read, that is, 13 dBA for automobiles and medium trucks.

The same procedure is followed for heavy trucks using 2.6 m for the "Break in Barrier" because of the higher noise emission elevation from the heavy trucks. Note from Figure 10-36 that this distance is perpendicular to *L/S* and not perpendicular to the horizontal. The attenuation is about 10 dBA.

The revised overall combined L_{10} would then be 60 dBA. This is very acceptable for Class B land use category. All of the tabulations are summarized in Figure 10-35.

L_{eq} Prediction

At about the same time that the NCHRP 174 report was being finalized, the Ontario Ministry of Transportation and Communications completed development of a predictive equation based on the L_{eq} concept (Hajek, 1977). The empirical equation they developed is as follows:

$$L_{eq} = 42.3 + 10.2 \log{(V_c + 6V_t)} - 13.9 \log D + 0.13S \qquad (10\text{-}29)$$

where L_{eq} = energy equivalent sound level during one hour, dBA
V_c = volume of automobiles (four tires only), veh/h
V_t = volume of trucks (six or more tires), veh/h
D = distance from edge of pavement to receiver, m
S = average speed of traffic flow during one hour, km/h

The simplicity of the equation is an obvious advantage over the NCHRP method. It does have the restriction that it does not account for barriers. A nomograph technique similar to the NCHRP method is available to take barriers into account.

L_{dn} Prediction

As a direct extension of the L_{eq} methodology, the Ontario method was extended to enable the calculation of L_{dn}. The modified model has the following form:

$$L_{dn} = 31.0 + 10.2 \log{[\text{AADT} + (\text{T\% AADT}/20)]} - 13.9 \log D + 0.13\,S \quad (10\text{-}30)$$

where L_{dn} = equivalent A-weighted sound level during 24-hour time period with 10 dBA weighting applied to 2200–0700 h, dBA
AADT = annual average daily traffic, veh/d
$\%T$ = average percentage of trucks during a typical day, %

Equation 10-30 has the same advantages and disadvantages as Equation 10-29.

10-7 NOISE CONTROL

Source-Path-Receiver Concept

If you have a noise problem and want to solve it, you have to find out something about what the noise is doing, where it comes from, how it travels, and what can be done about it. A straightforward approach is to examine the problem in terms of its three basic elements: that is, sound arises from a source, travels over a path, and affects a receiver or listener.*

The source may be one or any number of mechanical devices that radiate noise or vibratory energy. Such a situation occurs when several appliances or machines are in operation at a given time in a home or office.

The most obvious transmission path by which noise travels is simply a direct line-of-sight air path between the source and the listener. For example, aircraft flyover

---

*This discussion in large part was taken from Berendt, Corliss, and Ojalvo, 1976.

noise reaches an observer on the ground by the direct line-of-sight air path. Noise also travels along structural paths. Noise can travel from one point to another via any one path or a combination of several paths. Noise from a washing machine operating in one apartment may be transmitted to another apartment along air passages such as open windows, doorways, corridors, or duct work. Direct physical contact of the washing machine with the floor or walls sets these building components into vibration. This vibration is transmitted structurally throughout the building, causing walls in other areas to vibrate and to radiate noise.

The receiver may be, for example, a single person, a classroom of students, or a suburban community.

Solution of a given noise problem might require alteration or modification of any or all of these three basic elements:

1. Modifying the source to reduce its noise output

2. Altering or controlling the transmission path and the environment to reduce the noise level reaching the listener

3. Providing the receiver with personal protective equipment

Control of Noise Source by Design

Reduce Impact Forces. Many machines and items of equipment are designed with parts that strike forcefully against other parts, producing noise. Often, this striking action or impact is essential to the machine's function. A familiar example is the typewriter—its keys must strike the ribbon and paper in order to leave an inked impression. But the force of the key also produces noise as the impact falls on the ribbon, paper, and platen.

Several steps can be taken to reduce noise from impact forces. The particular remedy to be applied will be determined by the nature of the machine in question. Not all of the steps listed below are practical for every machine and for every impact-produced noise. But application of even one suggested measure can often reduce the noise appreciably.

Some of the more obvious design modifications are as follows:

1. Reduce the weight, size, or height of fall of the impacting mass.

2. Cushion the impact by inserting a layer of shock-absorbing material between the impacting surfaces. (For example, insert several sheets of paper in the typewriter behind the top sheet to absorb some of the noise-producing impact of the keys.) In some situations, you could insert a layer of shock-absorbing material behind each of the impacting heads or objects to reduce the transmission of impact energy to other parts of the machine.

3. Whenever practical, one of the impact heads or surfaces should be made of nonmetallic material to reduce resonance (ringing) of the heads.

4. Substitute the application of a small impact force over a long time period for a large force over a short period to achieve the same result.

5. Smooth out acceleration of moving parts by applying accelerating forces gradually. Avoid high, jerky acceleration or jerky motion.

6. Minimize overshoot, backlash, and loose play in cams, followers, gears, linkages, and other parts. This can be achieved by reducing the operational speed of the machine, better adjustment, or by using spring-loaded restraints or guides. Machines that are well made, with parts machined to close tolerances, generally produce a minimum of such impact noise.

Reduce Speeds and Pressures.   Reducing the speed of rotating and moving parts in machines and mechanical systems results in smoother operation and lower noise output. Likewise, reducing pressure and flow velocities in air, gas, and liquid circulation systems lessens turbulence, resulting in decreased noise radiation. Some specific suggestions that may be incorporated in design are the following:

1. Fans, impellers, rotors, turbines, and blowers should be operated at the lowest bladetip speeds that will still meet job needs. Use large-diameter, low-speed fans rather than small-diameter, high-speed units for quiet operation. In short, maximize diameter and minimize tip speed.

2. All other factors being equal, centrifugal squirrel-cage type fans are less noisy than vane axial or propeller type fans.

3. In air ventilation systems, a 50 percent reduction in the speed of the air flow may lower the noise output by 10 to 20 dB, or roughly one-quarter to one-half of the original loudness. Air speeds less than 3 m/s measured at a supply or return grille produce a level of noise that usually is unnoticeable in residential or office areas. In a given system, reduction of air speed can be achieved by operating at lower motor or blower speeds, installing a greater number of ventilating grilles, or increasing the cross-sectional area of the existing grilles.

Reduce Frictional Resistance.   Reducing friction between rotating, sliding, or moving parts in mechanical systems frequently results in smoother operation and lower noise output. Similarly, reducing flow resistance in fluid distribution systems results in less noise radiation.

Four of the more important factors that should be checked to reduce frictional resistance in moving parts are the following:

1. Alignment: Proper alignment of all rotating, moving, or contacting parts results in less noise output. Good axial and directional alignment in pulley systems, gear trains, shaft couplings, power transmission systems, and bearing and axle alignment are fundamental requirements for low noise output.

2. Polish: Highly polished and smooth surfaces between sliding, meshing, or contacting parts are required for quiet operation, particularly where bearings, gears, cams, rails, and guides are concerned.

3. Balance: Static and dynamic balancing of rotating parts reduces frictional resistance and vibration, resulting in lower noise output.

4. Eccentricity (out-of-roundness): Off-centering of rotating parts such as pulleys, gears, rotors, and shaft/bearing alignment causes vibration and noise.

Likewise, out-of-roundness of wheels, rollers, and gears causes uneven wear, resulting in flat spots that generate vibration and noise.

The key to effective noise control in fluid systems is *streamline flow.* This holds true regardless of whether one is concerned with air flow in ducts or vacuum cleaners, or with water flow in plumbing systems. Streamline flow is simply smooth, nonturbulent, low-friction flow.

The two most important factors that determine whether flow will be streamline or turbulent are the speed of the fluid and the cross-sectional area of the flow path, that is, the pipe or duct diameter. The rule of thumb for quiet operation is to use a low-speed, large-diameter system to meet a specified flow capacity requirement. However, even such a system can inadvertently generate noise if certain aerodynamic design features are overlooked or ignored. A system designed for quiet operation will employ the following features:

1. Low fluid speed: Low fluid speeds avoid turbulence, which is one of the main causes of noise.

2. Smooth boundary surfaces: Duct or pipe systems with smooth interior walls, edges, and joints generate less turbulence and noise than systems with rough or jagged walls or joints.

3. Simple layout: A well-designed duct or pipe system with a minimum of branches, turns, fittings, and connectors is substantially less noisy than a complicated layout.

4. Long-radius turns: Changes in flow direction should be made gradually and smoothly. It has been suggested that turns should be made with a curve radius equal to about five times the pipe diameter or major cross-sectional dimension of the duct.

5. Flared sections: Flaring of intake and exhaust openings, particularly in a duct system, tends to reduce flow speeds at these locations, often with substantial reductions in noise output.

6. Streamline transition in flow path: Changes in flow path dimensions or cross-sectional areas should be made gradually and smoothly with tapered or flared transition sections to avoid turbulence. A good rule of thumb is to keep the cross-sectional area of the flow path as large and as uniform as possible throughout the system.

7. Remove unnecessary obstacles: The greater the number of obstacles in the flow path, the more tortuous, turbulent, and hence noisier, the flow. All other required and functional devices in the path, such as structural supports, deflectors, and control dampers, should be made as small and as streamlined as possible to smooth out the flow patterns.

Reduce Radiating Area.   Generally speaking, the larger the vibrating part or surface, the greater the noise output. The rule of thumb for quiet machine design is to minimize the effective radiating surface areas of the parts without impairing their operation or structural strength. This can be done by making parts smaller, removing excess material, or by cutting openings, slots, or perforations in the parts. For example, replacing a large, vibrating sheet-metal safety guard on a machine with a guard made of wire mesh or metal webbing might result in a substantial reduction in noise because of the drastic reduction in surface area of the part.

Reduce Noise Leakage.   In many cases, machine cabinets can be made into rather effective soundproof enclosures through simple design changes and the application of some sound-absorbing treatment. Substantial reductions in noise output may be achieved by adopting some of the following recommendations:

1. All unnecessary holes or cracks, particularly at joints, should be caulked.

2. All electrical or plumbing penetrations of the housing or cabinet should be sealed with rubber gaskets or a suitable nonsetting caulk.

3. If practical, all other functional or required openings or ports that radiate noise should be covered with lids or shields edged with soft rubber gaskets to effect an airtight seal.

4. Other openings required for exhaust, cooling, or ventilation purposes should be equipped with mufflers or acoustically lined ducts.

5. Openings should be directed away from the operator and other people.

Isolate and Dampen Vibrating Elements.   In all but the simplest machines, the vibrational energy from a specific moving part is transmitted through the machine structure, forcing other component parts and surfaces to vibrate and radiate sound—often with greater intensity than that generated by the originating source itself.

Generally, vibration problems can be considered in two parts. First, we must prevent energy transmission between the source and surfaces that radiate the energy. Second, we must dissipate or attenuate the energy somewhere in the structure. The first part of the problem is solved by *isolation*. The second part is solved by *damping*.

The most effective method of vibration isolation involves the resilient mounting of the vibrating component on the most massive and structurally rigid part of the machine. All attachments or connections to the vibrating part, in the form of pipes, conduits, and shaft couplers, must be made with flexible or resilient connectors or couplers. For example, pipe connections to a pump that is resiliently mounted on the structural frame of a machine should be made of resilient tubing and be mounted as close to the pump as possible. Resilient pipe supports or hangers may also be required to avoid bypassing the isolated system (Figure 10-39).

Damping material or structures are those that have some viscous properties. They tend to bend or distort slightly, thus consuming part of the noise energy in molecular motion. The use of spring mounts on motors and laminated galvanized steel and plastic in air-conditioning ducts are two examples.

When the vibrating noise source is not amenable to isolation, as, for example, in ventilation ducts, cabinet panels, and covers, then damping materials can be used to reduce the noise.

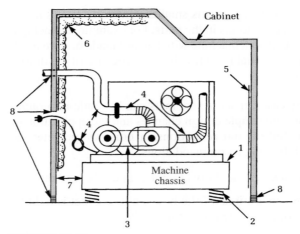

FIGURE 10-39
Examples of vibration isolation.

Code:
1. Motors, pumps, and fans installed on most massive part of the machine
2. Resilient mounts or vibration isolators used for the installation
3. Belt-drive or roller-drive systems used in place of gear trains
4. Flexible hoses and wiring used instead of rigid piping and stiff wiring
5. Vibration-damping materials applied to surfaces undergoing most vibration
6. Acoustical lining installed to reduce noise buildup inside machine
7. Mechanical contact minimized between the cabinet and the machine chassis
8. Openings at the base and other parts of the cabinet scaled to prevent noise leakage

(*Source:* Berendt et al., 1976.)

The type of material best suited for a particular vibration problem depends on factors such as size, mass, vibrational frequency, and operational function of the vibrating structure. Generally speaking, the following guidelines should be observed in the selection and use of such materials to maximize vibration damping efficiency:

1. Damping materials should be applied to those sections of a vibrating surface where the most flexing, bending, or motion occurs. These usually are the thinnest sections.

2. For a single layer of damping material, the stiffness and mass of the material should be comparable to that of the vibrating surface to which it is applied. This means that single-layer damping materials should be about two or three times as thick as the vibrating surface to which they are applied.

3. Sandwich materials (*laminates*) made up of metal sheets bonded to mastic (sheet metal viscoelastic composites) are much more effective vibration dampers than single-layer materials; the thickness of the sheet-metal constraining layer and the viscoelastic layer should each be about one-third the thickness of the vibrating surface to which they are applied. Ducts and panels can be purchased already fabricated as laminates.

Provide Mufflers/silencers.   There is no real distinction between mufflers and silencers. They are often used interchangeably. They are, in effect, acoustical filters and are used when fluid flow noise is to be reduced. The devices can be classified into two fundamental groups: *absorptive mufflers* and *reactive mufflers.* An absorptive muffler is one whose noise reduction is determined mainly by the presence of fibrous or porous materials, which absorb the sound. A reactive muffler is one whose noise reduction is determined mainly by geometry. It is shaped to reflect or expand the sound waves with resultant self-destruction.

Although there are several terms used to describe the performance of mufflers, the most frequently used appears to be *insertion loss* (IL). Insertion loss is the difference between two sound pressure levels that are measured at the same point in space before and after a muffler has been inserted. Because each muffler's IL is highly dependent on the manufacturer's selection of materials and configuration, we will not present general IL prediction equations.

Noise Control in the Transmission Path

After you have tried all possible ways of controlling the noise at the source, your next line of defense is to set up devices in the transmission path to block or reduce the flow of sound energy before it reaches your ears. This can be done in several ways: (*a*) absorb the sound along the path, (*b*) deflect the sound in some other direction by placing a reflecting barrier in its path, or (*c*) contain the sound by placing the source inside a sound-insulating box or enclosure.

Selection of the most effective technique will depend upon various factors, such as the size and type of source, intensity and frequency range of the noise, and the nature and type of environment.

Separation.   We can make use of the absorptive capacity of the atmosphere, as well as divergence, as a simple, economical method of reducing the noise level. Air absorbs high-frequency sounds more effectively than it absorbs low-frequency sounds. However, if enough distance is available, even low-frequency sounds will be absorbed appreciably.

If you can double your distance from a point source, you will have succeeded in lowering the sound pressure level by 6 dB. It takes about a 10 dB drop to halve the loudness. If you have to contend with a line source such as a railroad train, the noise level drops by only 3 dB for each doubling of distance from the source. The main reason for this lower rate of attenuation is that line sources radiate sound waves that are cylindrical in shape. The surface area of such waves only increases two-fold for each doubling of distance from the source. However, when the distance from the train becomes comparable to its length, the noise level will begin to drop at a rate of 6 dB for each subsequent doubling of distance.

Indoors, the noise level generally drops only from 3 to 5 dB for each doubling of distance in the near vicinity of the source. However, further from the source, reductions of only 1 or 2 dB occur for each doubling of distance due to the reflections of sound off hard walls and ceiling surfaces.

Absorbing Materials.   Noise, like light, will bounce from one hard surface to another. In noise control work, this is called *reverberation*. If a soft, spongy material is placed on the walls, floors, and ceiling, the reflected sound will be diffused and soaked up (absorbed). Sound-absorbing materials are rated either by their *Sabin absorption coefficients* (α_{SAB}) at 125, 500, 1,000, 2,000, and 4,000 Hz or by a single number rating called the *noise reduction coefficient* (NRC). If a unit area of open window is assumed to transmit all and reflect none of the acoustical energy that reaches it, it is assumed to be 100 percent absorbent. This unit area of totally absorbent surface is called a "sabin" (Sabin, 1942). The absorptive properties of acoustical materials are then compared with this standard. The performance is expressed as a fraction or percentage of the sabin (α_{SAB}). The NRC is the average of the α_{SAB}s at 250, 500, 1,000, and 2,000 Hz rounded to the nearest multiple of 0.05. The NRC has no physical meaning. It is a useful means of comparing similar materials.

Sound-absorbing materials such as acoustical tile, carpets, and drapes placed on ceiling, floor, or wall surfaces can reduce the noise level in most rooms by about 5 to 10 dB for high-frequency sounds, but only by 2 or 3 dB for low-frequency sounds. Unfortunately, such treatment provides no protection to an operator of a noisy machine who is in the midst of the direct noise field. For greatest effectiveness, sound-absorbing materials should be installed as close to the noise source as possible.

If you have a small or limited amount of sound-absorbing material and wish to make the most effective use of it in a noisy room, the best place to put it is in the upper trihedral corners of the room, formed by the ceiling and two walls. Due to the process of reflection, the concentration of sound is greatest in the trihedral corners of a room. Additionally, the upper corner locations also protect the lightweight fragile materials from damage.

Because of their light weight and porous nature, acoustical materials are ineffectual in preventing the transmission of either airborne or structure-borne sound from one room to another. In other words, if you can hear people walking or talking in the room or apartment above, installing acoustical tile on your ceiling will not reduce the noise transmission.

Acoustical Lining.   Noise transmitted through ducts, pipe chases, or electrical channels can be reduced effectively by lining the inside surfaces of such passageways with sound-absorbing materials. In typical duct installations, noise reductions on the order of 10 dB/m for an acoustical lining 2.5 cm thick are well within reason for high-frequency noise. A comparable degree of noise reduction for the lower frequency sounds is considerably more difficult to achieve because it usually requires at least a doubling of the thickness and/or length of acoustical treatment.

Barriers and Panels.   Placing barriers, screens, or deflectors in the noise path can be an effective way of reducing noise transmission, provided that the barriers are large enough in size, and depending upon whether the noise is high frequency or low frequency. High-frequency noise is reduced more effectively than low-frequency noise.

The effectiveness of a barrier depends on its location, its height, and its length. Referring to Figure 10-40, we can see that the noise can follow five different paths.

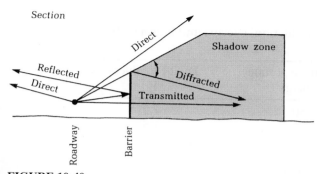

Section

Direct

Reflected

Direct

Shadow zone

Diffracted

Transmitted

Roadway

Barrier

FIGURE 10-40
Noise paths from a source to a receiver. (*Source:* Kugler et al., 1976.)

First, the noise follows a direct path to receivers who can see the source well over the top of the barrier. The barrier does not block their line of sight (*L/S*) and therefore provides no attenuation. No matter how absorptive the barrier is, it cannot pull the sound downward and absorb it.

Second, the noise follows a diffracted path to receivers in the shadow zone of the barrier. The noise that passes just over the top edge of the barrier is diffracted (bent) down into the apparent shadow shown in the figure. The larger the angle of diffraction, the more the barrier attenuates the noise in this shadow zone. In other words, less energy is diffracted through large angles than through smaller angles.

Third, in the shadow zone, the noise transmitted directly through the barrier may be significant in some cases. For example, with extremely large angles of diffraction, the diffracted noise may be less than the transmitted noise. In this case, the transmitted noise compromises the performance of the barrier. It can be reduced by constructing a heavier barrier. The allowable amount of transmitted noise depends on the total barrier attenuation desired. More is said about this transmitted noise later.

The fourth path shown in Figure 10-40 is the reflected path. After reflection, the noise is of concern only to a receiver on the opposite side of the source. For this reason, acoustical absorption on the face of the barrier may sometimes be considered to reduce this reflected noise; however, this treatment will not benefit any receivers in the shadow zone. It should be noted that in most practical cases the reflected noise does not play an important role in barrier design. If the source of noise is represented by a line of noise, another short-circuit path is possible. Part of the source may be unshielded by the barrier. For example, the receiver might see the source beyond the ends of the barrier if the barrier is not long enough. This noise from around the ends may compromise, or short-circuit, barrier attenuation. The required barrier length depends on the total net attenuation desired. When 10 to 15 dB attenuation is desired, barriers must, in general, be very long. Therefore, to be effective, barriers must not only break the line of sight to the nearest section of the source, but also to the source far up and down the line.

Of these four paths, the noise diffracted over the barrier into the shadow zone represents the most important parameter from the barrier design point of view. Generally, the determination of barrier attenuation or barrier noise reduction involves only calculation

TABLE 10-10
Relation between sound level reduction, energy, and loudness for line sources

To reduce A-level by dB	Remove portion of energy (%)	Divide loudness by
3	50	1.2
6	75	1.5
10	90	2
20	90	4
30	99.9	8
40	99.99	16

(*Source:* Kugler et al., 1976.)

of the amount of energy diffracted into the shadow zone. The procedures presented in the barrier nomograph used to predict highway noise are based on this concept.

Another general principle of barrier noise reduction that is worth reviewing at this point is the relation between noise attenuation expressed in (1) decibels, (2) energy terms, and (3) subjective loudness. Table 10-10 gives these relationships for line sources. As indicated in the loudness column, a barrier attenuation of 3 dB will be barely discerned by the receiver. However, to attain this reduction, 50 percent of the acoustical energy must be removed. To cut the loudness of the source in half, a reduction of 10 dB is necessary. That is equivalent to eliminating 90 percent of the energy initially directed toward the receiver. As indicated previously, this drastic reduction in energy requires very long and high barriers. In summary, when designing barriers, you can expect the complexity of the design to be about as follows:

Attenuation (dB)	Complexity
5	Simple
10	Attainable
15	Very difficult
20	Nearly impossible

Roadside barriers can be designed using the barrier nomograph in reverse order. A set of typical solutions is summarized in Table 10-11. The noise reduction at 152 m is less than that at 30 m because the barrier does not cast as large a shadow at a distance. The effectiveness of the barrier is reduced for trucks because of the elevated nature of the source.

Transmission Loss.   When the position of the noise source is very close to the barrier, the diffracted noise is less important than the transmitted noise. If the barrier is in fact a wall panel that is sealed at the edges, the transmitted noise is the only one of concern.

TABLE 10-11
Noise reductions for various highway configurations

Highway configuration[a]		Height or depth (m)	Truck mix (%)	Noise reduction[b] at distance from ROW (dBA)	
Sketch	Description			30 m	152 m
	Roadside barriers 7.6 m from edge of shoulders; ROW = 78 m wide	6.1	0	13.9	13.3
			5	13.0	12.1
			10	12.6	11.7
			20	12.3	11.3
	Depressed roadway w/2:1 slopes; ROW = 102 m	6.1	0	9.9	11.4
			5	8.8	10.3
			10	8.4	9.8
			20	8.1	9.4
	Fill elevated roadway w/2:1 slopes; ROW = 102 m	6.1	0	9.0	6.3
			5	7.6	2.7
			10	7.1	1.8
			20	6.7	1.1
	Elevated structure; ROW = 78 m	7.3	0	9.8	6.0
			5	9.6	2.4
			10	9.3	1.5
			20	8.8	0.8

[a]Assumes divided 8 lanes with 9.1 m median.
[b]Based on observed 1.5 m above grade.
(*Source:* Kugler et al., 1976.)

The ratio of the sound energy incident on one surface of a panel to the energy radiated from the opposite surface is called the *sound transmission loss* (TL). The actual energy loss is partially reflected and partially absorbed. Since TL is frequency-dependent, only a complete octave or one-third octave band curve provides a full description of the performance of the barrier.

Enclosures. Sometimes it is much more practical and economical to enclose a noisy machine in a separate room or box than to quiet it by altering its design, operation, or component parts. The walls of the enclosure should be massive and airtight to contain the sound. Absorbent lining on the interior surfaces of the enclosure will reduce the reverberant buildup of noise within it. Structural contact between the noise source and the enclosure must be avoided, so that the source vibration is not transmitted to the enclosure walls, thus short-circuiting the isolation. For maximum effective noise control, all of the techniques illustrated in Figure 10-41 must be employed.

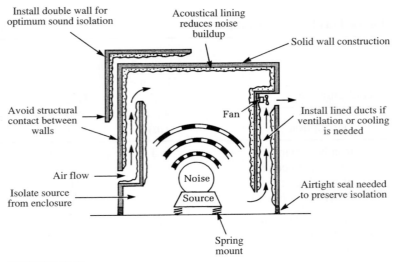

FIGURE 10-41
Enclosures for controlling noise. (*Source:* Berendt et al., 1976.)

Control of Noise Source by Redress

The best way to solve noise problems is to design them out of the source. However, we are frequently faced with an existing source that, either because of age, abuse, or poor design, is a noise problem. The result is that we must redress, or correct, the problem as it currently exists. The following sections identify some measures that might apply if you are allowed to tinker with the source.

Balance Rotating Parts.   One of the main sources of machinery noise is structural vibration caused by the rotation of poorly balanced parts, such as fans, fly wheels, pulleys, cams, shafts, and so on. Measures used to correct this condition involve the addition of counterweights to the rotating unit or the removal of some weight from the unit. You are probably familiar with noise caused by imbalance in the high-speed spin cycle of washing machines. The imbalance results from clothes not being distributed evenly in the tub. By redistributing the clothes, balance is achieved and the noise ceases. This same principle of balance can be applied to furnace fans and other common sources of such noise.

Reduce Frictional Resistance.   A well-designed machine that has been poorly maintained can become a serious source of noise. General cleaning and lubrication of all rotating, sliding, or meshing parts at contact points should go a long way toward fixing the problem.

Apply Damping Materials.   Because a vibrating body or surface radiates noise, the application of any material that reduces or restrains the vibrational motion of that body will decrease its noise output. Three basic types of redress vibration damping materials are available:

   1. Liquid mastics, which are applied with a spray gun and harden into relatively solid materials, the most common being automobile "undercoating"

2. Pads of rubber, felt, plastic foam, leaded vinyls, adhesive tapes, or fibrous blankets, which are glued to the vibrating surface

3. Sheet metal viscoelastic laminates or composites, which are bonded to the vibrating surface

Seal Noise Leaks.   Small holes in an otherwise noise-tight structure can reduce the effectiveness of the noise control measures. As you can see in Figure 10-42, if the designed transmission loss of an acoustical enclosure is 40 dB, an opening that comprises only 0.1 percent of the surface area will reduce the effectiveness of the enclosure by 10 dB.

Perform Routine Maintenance.   We all recognize the noise of a worn muffler. Likewise, studies of automobile tire noise in relation to pavement roughness show that maintenance of the pavement surface is essential to keep noise at minimum levels. Normal road wear can yield noise increases on the order of 6 dBA.

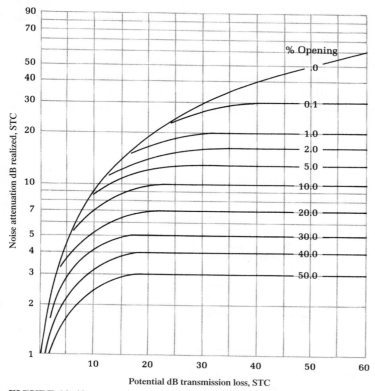

FIGURE 10-42

Transmission loss potential versus transmission loss realized for various opening sizes as a percent of total wall area. STC = sound transmission coefficient (*Source:* Adapted from Warnock and Quirt, 1991.)

Protect the Receiver

When All Else Fails.   When exposure to intense noise fields is required and none of the measures discussed so far is practical, as, for example, for the operator of a chain saw or pavement breaker, then measures must be taken to protect the receiver. The following two techniques are commonly employed.

Alter Work Schedule.   Limit the amount of continuous exposure to high noise levels. In terms of hearing protection, it is preferable to schedule an intensely noisy operation for a short interval of time each day over a period of several days rather than a continuous eight-hour run for a day or two.

In industrial or construction operations, an intermittent work schedule would benefit not only the operator of the noisy equipment, but also other workers in the vicinity. If an intermittent schedule is not possible, then workers should be given relief time during the day. They should take their relief time at a low-noise-level location, and should be discouraged from trading relief time for dollars, paid vacation, or an "early out" at the end of the day!

Inherently noisy operations, such as street repair, municipal trash collection, factory operation, and aircraft traffic, should be curtailed at night and early morning to avoid disturbing the sleep of the community. Remember: operations between 10 P.M. and 7 A.M. are effectively 10 dBA higher than the measured value.

Ear Protection.   Molded and pliable earplugs, cup-type protectors, and helmets are commercially available as hearing protectors. Such devices may provide noise reductions ranging from 15 to 35 dB (Figure 10-43). Earplugs are effective only if they are properly fitted by medical personnel. As shown in Figure 10-43, maximum protection

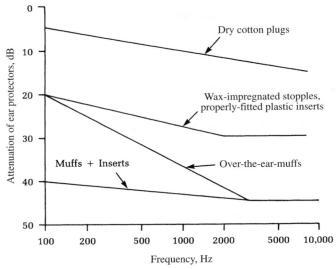

FIGURE 10-43
Attenuation of ear protectors at various frequencies. (*Source:* Berendt et al., 1976.)

can be obtained when both plugs and muffs are employed. Only muffs that have a certification stipulating the attenuation should be used.

These devices should be used only as a last resort, after all other methods have failed to lower the noise level to acceptable limits. Ear protection devices should be used while operating lawn mowers, mulchers, and chippers, and while firing weapons at target ranges. It should be noted that protective ear devices do interfere with speech communication and can be a hazard in some situations where warning calls may be a routine part of the operation (for example, TIMBERRRR!). A modern ear-destructive device is a portable digital music player that uses earphones. In this "reverse" muff, high noise levels are directed at the ear without attenuation. If you can hear someone else's music player, that person is subjecting him or herself to noise levels in excess of 90–95 dBA!

10-8 CHAPTER REVIEW

When you have completed studying this chapter, you should be able to do the following without the aid of your textbook or notes:

1. Define frequency, based on a sketch of a harmonic wave you have drawn, and state its units of measure (namely, hertz, Hz).

2. State the basic unit of measure used in measuring sound energy (namely, the decibel) and explain why it is used.

3. Define sound pressure level in mathematical terms, that is,

$$SPL = 20 \log \frac{P_{rms}}{(P_{rms})_0}$$

4. Explain why a weighting network is used in a sound level meter.

5. List the three common weighting networks and sketch their relative frequency response curves. (Label frequencies, that is, 20, 1,000, and 10,000 Hz; and relative response, that is, 0, -5, -20, and -45 dB, as in Figure 10-5.)

6. Differentiate between a mid/high-frequency noise source and a low-frequency noise source on the basis of A, B, and C scale readings.

7. Explain the purpose of octave band analysis.

8. Differentiate between continuous, intermittent, and impulsive noise.

9. Sketch the curves and label the axes of the two typical types of impulsive noise.

10. Sketch a Fletcher-Munson curve, label the axes, and explain what the curve depicts.

11. Define "phon."

12. Explain the mechanism by which hearing damage occurs.

13. Explain what hearing threshold level (HTL) is.

14. Define presbycusis and explain why it occurs.

15. Distinguish between temporary threshold shift (TTS), permanent threshold shift (PTS), and acoustic trauma with respect to cause of hearing loss, duration of exposure, and potential for recovery.

16. Explain why impulsive noise is more dangerous than steady-state noise.

17. Explain the relationship between the allowable duration of noise exposure and the allowable level for hearing protection, that is, damage-risk criteria.

18. List five effects of noise other than hearing damage.

19. List the three basic elements that might require alteration or modification to solve a noise problem.

20. Describe two techniques to protect the receiver when design and/or redress are not practical, that is, when all else fails.

With the aid of this text, you should be able to do the following:

21. Calculate the resultant sound pressure level from a combination of two or more sound pressure levels.

22. Determine the A-, B-, and C-weighted sound levels from octave band readings.

23. Compute the mean sound level from a series of sound level readings.

24. Compute the following noise statistics if you are provided the appropriate data: L_N and/or L_{eq}; L_{dn}.

25. Determine whether or not a noise level will be acceptable given a series of measurements and the criteria listed in Tables 10-3, 10-5, 10-6, 10-7, and/or Figures 10-24 and 10-26.

26. Calculate the sound level at a receptor site after transmission through the atmosphere.

27. Estimate the noise level L_{10} that might be expected for a given roadway configuration and traffic pattern.

10-9 PROBLEMS

10-1. A building located near a road is 6.92 m high. How high is the building in terms of wavelengths of a 50.0-Hz sound? Assume that the speed of sound is 346.12 m/s.

Answer: One wavelength

10-2. Repeat Problem 10-1 for a 500-Hz sound if the temperature is 25.0°C.

10-3. Determine the sum of the following sound levels (all in dB): 68, 82, 76, 68, 74, and 81.

Answer: 85.5, or 86, dB

10-4. A motorcyclist is warming up his racing cycle at a racetrack approximately 200 m from a sound level meter. The meter reading is 56 dBA. What meter reading would you expect if 15 of the motorcyclist's friends join him with motorcycles having exactly the same sound emission characteristics? You may assume that the sources may be treated as ideal point sources located at the same point.

10-5. A sound power level reading of 127 dB was taken near a construction site where chippers were being used. When all but one of the chippers stopped working, the sound power level reading was 120 dB. Estimate the number of chippers in operation when the reading of 127 was obtained. You may assume that the sources may be treated as ideal point sources located at the same point.

10-6. A law enforcement officer has taken the following readings with her sound level meter. Is the noise source a predominantly low- or middle-frequency emitter? Readings: 80 dBA, 84 dBB, and 90 dBC.

>   *Answer:* Predominantly low frequency

10-7. The following readings have been made outside the open stage door of the opera house: 109 dBA, 110 dBB, and 111 dBC. Is the singer a bass or a soprano? Explain how you arrived at your answer.

10-8. Convert the following octave band measurements to an equivalent A-weighted sound level.

Band center frequency (Hz)	Band level (dB)
31.5	78
63	76
125	78
250	82
500	81
1,000	80
2,000	80
4,000	73
8,000	65

>   *Answer:* 85.5, or 86, dBA

10-9. The following noise spectrum was obtained from a jet aircraft flying overhead at an altitude of 250 m. Compute the equivalent A-weighted sound level using sound power level addition in a spreadsheet program you have written.

Band center frequency (Hz)	Band level (dB)
125	85
250	88
500	96
1,000	100
2,000	104
4,000	101

10-10. Using the typical noise spectrum for automobiles traveling at 50 to 60 km/h, determine the equivalent A-weighted level using sound power level addition in a spreadsheet program you have written. The following band levels were estimated from Figure 10-25.

Band center frequency (Hz)	Band level (dB)
63	67
125	64
250	58
500	59
1,000	59
2,000	55
4,000	51
8,000	45

10-11. You have been asked to evaluate the A-weighted sound level of a new model lawn mower and make a recommendation on an acceptable noise spectrum to achieve 74 dBA. Three approaches are being considered by the manufacturer: (1) an improved muffler that will reduce the sound level 3 dB in each frequency band, (2) a reduction in the speed of the mower that will reduce the sound level 5 dB in each frequency band, and (3) an engine redesign that will reduce the sound level 15 dB in the five highest frequency bands. Using a spreadsheet program you have written, compute the A-weighted sound level for the sound spectrum shown on the following page and develop a recommended noise spectrum based on the manufacturer's alternatives that results in a sound level of less than 74 dBA. Assume that each of the alternative reductions may be added together (by decibel addition) in each frequency band in which it is applicable.

Band center frequency (Hz)	Band level (dB)
63	78
125	76
250	76
500	77
1,000	79
2,000	80
4,000	78
8,000	70

10-12. Compute the average sound pressure level of the following readings by simple arithmetic averaging and by logarithmic averaging (Equation 10-13) (all readings in dB): 42, 50, 65, 71, and 47. Does arithmetic averaging underestimate or overestimate the sound pressure level?

Answers: $\bar{x} = 55.00$ or 55 dB $L_p = 61.57$, or 62, dB

10-13. Repeat Problem 10-12 for the following data (all in dB): 76, 59, 35, 69, and 72.

10-14. The following noise record was obtained in the front yard of a home. Is this a relatively quiet or a relatively noisy neighborhood? Determine the equivalent continuous equal energy level.

Time (h)	Sound level (dBA)
0000–0600	42
0600–0800	45
0800–0900	50
0900–1500	47
1500–1700	50
1700–1800	47
1800–0000	45

Answers: It is a quiet neighborhood. $L_{eq} = 46.2$, or 46, dBA

10-15. A developer has proposed putting a small shopping mall next to a very quiet residential area in Nontroppo, Michigan. Based on the measurements given on the following page, which were taken at a similar size mall in a similar setting, should the developer expect complaints or legal action? Calculate L_{eq}.

Time (h)	Sound level (dBA)
0000–0600	42
0600–0800	55
0800–1000	65
1000–2000	70
2000–2200	68
2200–0000	57
1800–0000	45

10-16. The U.S. EPA (1974) estimated that the following was a typical noise exposure pattern for a factory worker living in an urban area. Estimate the L_{dn} for the exposure shown.

Time (h)	Sound level (dBA)
0000–0500	52
0500–0700	78
0700–1130	90
1130–1200	70
1200–1530	90
1530–1800	52
1800–2200	60
2200–0000	52

10-17. The U.S. EPA (1974) estimated that the following was a typical noise exposure pattern for a middle school student living in an urban area. Estimate the L_{dn} for the exposure shown.

Time (h)	Sound level (dBA)
0000–0700	52
0700–0900	82
0900–1200	60
1200–1300	65
1300–1500	60
1500–1700	75
1700–1800	90
1800–2100	60
2100–0000	52

10-18. Two oil-fired boilers for a 600 megawatt (MW) power plant produce a sound power level of 139 dB (re: 10^{-12} W) at 4,000 Hz, from the induced draft fans. Determine the sound pressure level 408.0 m downwind on a clear winter night when the wind speed is 4.50 m/s, the temperature is 0.0°C, the relative humidity is 30.0 percent, and the barometric pressure is 101.3 kPa. The height of the boiler is 12 m. The height of the receiver is 1.5 m. The ground surface characteristics are shown in Figure P-10-18.

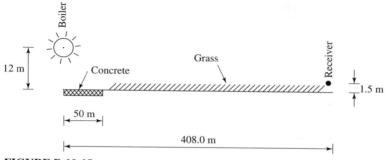

FIGURE P-10-18
Ground surface characteristics for Problem 10-18.

Answer: SPL at 408.0 m = 50.50 or 50 dB at 4,000 Hz

10-19. The 125-Hz sound power level (re: 10^{-12} W) from a jet engine test cell is 149 dB. What is the 125-Hz sound pressure level 1,200 m downwind on a clear summer morning during an inversion when the wind speed is 1.50 m/s, the temperature is 25.0°C, the relative humidity is 70.0 percent, and the barometric pressure is 101.3 kPa? The height of the engine and the receiver are 1.5 m. The ground surface characteristics are shown in Figure P-10-19.

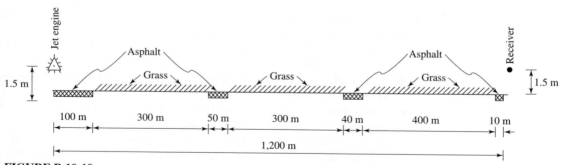

FIGURE P-10-19
Ground surface characteristics for Problem 10-19.

10-20. Using a spreadsheet you have written, calculate the A-weighted sound level produced by a jet airplane that is 2,000 m from the receiver. The sound power level in each octave band for the jet (at the source) is tabulated below. The meteorological conditions are as follows: clear summer morning during an inversion when the wind speed is 2.50 m/s, the temperature is 20.0°C, the relative humidity is 70.0 percent, and the barometric pressure is 101.3 kPa. The source and the receiver heights are 1.5 m. The ground surface characteristics are shown in Figure P-10-20.

Band center frequency (Hz)	Band level, re: 10^{-12} W
125	144
250	148
500	155
1,000	160
2,000	165
4,000	168

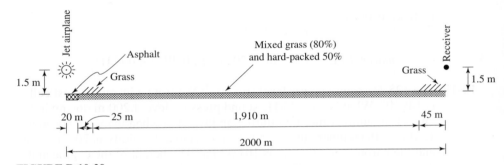

FIGURE P-10-20
Ground surface characteristics for Problem 10-20.

10-21. Consider an ideal single lane of road that carries 1,200 vehicles per hour uniformly spaced along the road and determine the following:

   a. The average center-to-center spacing of the vehicles for an average traffic speed of 40.0 km/h.

   b. The number of vehicles in a 1 km length of the lane when the average speed is 40.0 km/h.

   c. The sound level (dBA) 60.0 m from a 1 km length of this roadway with automobiles emitting 71 dBA at the edge of an 8.0 m-wide roadway. Assume that the autos travel at a speed of 40.0 km/h, that the sound radiates ideally from a hemisphere, and that contributions of less than 0.3 dBA may be ignored.

Answers:
   a. Average center-to-center spacing = 33.3 m
   b. Number of vehicles in a 1 km length = 30 vehicles/km
   c. L_p = 47.47 or 48 dBA

10-22. Repeat Problem 10-21 if the vehicle speed is increased to 80.0 km/h and the spacing is decreased or increased appropriately to maintain 1,200 vehicles per hour.

10-23. Determine the L_{10} noise level for the following highway traffic situation:

Autos = 1,500 per hour   Speed = 60 km/h

Medium trucks = 150 per hour Speed = 60 km/h

Heavy trucks = 0

Observer-roadway distance = 40.0 m

Line-of-sight distance = 40.0 m

Barrier position = 3.0 m

Break in line-of-sight distance = 6.0 m

Angle subtended = 170°

Answer: L_{10} = 51 dBA

10-24. Determine the L_{10} noise level for the following highway traffic situation:

Autos = 2,000 per hour   Speed = 70 km/h

Medium trucks = 200 per hour Speed = 70 km/h

Heavy trucks = 0

Observer-roadway distance = 60.0 m

Line-of-sight distance = 60.0 m

Barrier position = 5.0 m

Break in line-of-sight distance = 1.0 m

Angle subtended = 160°

10-25. In preparation for a public hearing on a proposed interstate bypass at Nontroppo, Michigan (Figure P-10-25a), the County Road Commission has requested that you prepare an estimate of the potential for violation of FHA noise standards 75 m from the interstate. The city engineer has supplied sketch maps (see Figure P-10-25b) and data summary for your use.

Data for I-481 at Pianissimo Avenue

Estimated traffic:
 Automobiles: 7,800 per hour at 88.5 km/h
 Medium trucks: 520 per hour at 80.5 km/h
 Heavy trucks: 650 per hour at 80.5 km/h

Roadway configuration: Depressed

Section length: 857.25 m east and 857.25 m west of center line of Pianissimo Avenue

Assume that the receiver is located on the center line of Pianissimo Avenue 75.00 m from the center line of I-481.

Answer: L_{10} at observer = 68.6, or 69, dBA

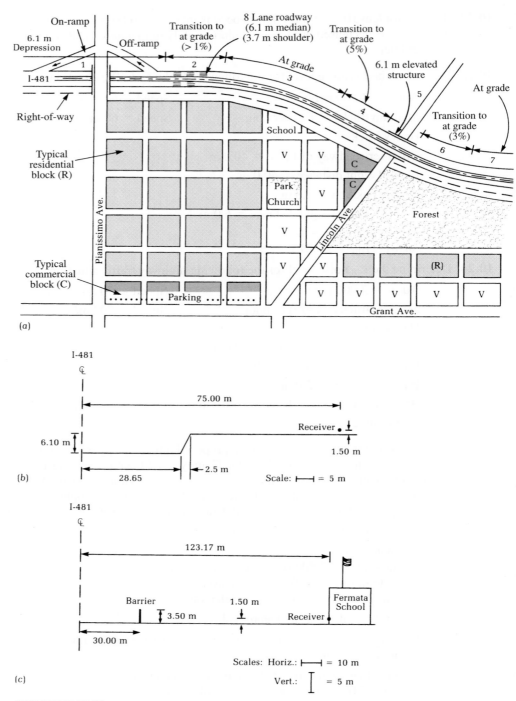

FIGURE P-10-25

Sketch maps for proposed bypass. (*a*) plan view, (*b*) cross section along Pianissimo Avenue, (*c*) cross section at Fermata School. (See Problems 10-25 and 10-26.)

10-26. Determine the potential for violation of FHA noise standards at the north side of Fermata School. The city engineer has supplied sketch maps (see Figure P-10-25c) and data for your use.

> Data for I-481 at Fermata School
>
> Estimated traffic: Same as at Pianissimo Avenue
>
> Roadway configuration: At grade
>
> Barrier length: 199.80 m east and 199.80 m west of Fermata School
>
> Assume that the receiver is located just outside of the north side of the school, 123.17 m from the center line of I-481, and is 1.5 m above the ground.

10-27. Using the data from Problem 10-25, compute the *unattenuated* L_{eq}, at the receiver for autos only. Assume the edge of the roadway is at the "toe" of the road cut.

> *Answer*: $L_{eq} = 70$ dBA

10-28. Rework Problem 10-27 using the data from Problem 10-26. Assume the edge of the roadway is at the barrier.

10-29. The following equation is found in the *Fundamentals of Engineering Supplied-Reference Handbook.* Demonstrate that it is equivalent to Equation 10-20.

$$\Delta \text{SPL (dB)} = 10 \log \left(\frac{R_1}{R_2} \right)^2 \qquad \text{🏳}$$

10-30. The following equation is found in the *Fundamentals of Engineering Supplied-Reference Handbook.* Demonstrate that it is equivalent to Equation 10-21.

$$\Delta \text{SPL (dB)} = 10 \log \left(\frac{R_1}{R_2} \right) \qquad \text{🏳}$$

10-10 DISCUSSION QUESTIONS

10-1. Classify each of the following noise sources by "type," that is, continuous, intermittent or impulse. (Not all sources fit these three classifications.)
(a) electric saw
(b) air conditioner
(c) alarm clock (bell type)
(d) punch press

10-2. Is the following statement true or false? If it is false, correct it in a nontrivial manner.

> "A sonic boom occurs when an aircraft breaks the sound barrier."

10-3. Is the following statement true or false? If it is false, correct it in a nontrivial manner.

> "Excessive continuous noise causes hearing damage by breaking the stapes."

10-4. As the safety officer of your company, you have been asked to determine the feasibility of reducing exposure time as a method of reducing hearing damage for the following situation:

> The worker is operating a high-speed grinder on steel girders for a high-rise building. The effective noise level at the operator's ear is 100 dBA. She cannot wear protective ear devices because she must communicate with others.

> What amount of exposure time would you set as the limit?

10-5. In Figure 10-41, identify where the following noise-control techniques are applied: isolation and/or damping, reduction in noise leakage, use of absorbing materials, use of acoustical lining, enclosure.

10-11 FE EXAM FORMATTED PROBLEMS

10-1. Assuming the basic point source model applies (i.e., ignoring attenuation factors), at what distance is a 100 dBA sound pressure level measured 2 m from a source attenuated to 70 dBA?

a. 15 m
b. 63 m
c. 80 m
d. 5,000 m

10-2. What sound power level results from combining the following three levels: 72 dB, 88 dB, 90 dB?

a. 96 dB
b. 250 dB
c. 110 dB
d. 92 dB

10-3. Compute the mean sound pressure level of the following three readings: 36 dBA, 76 dBA, 83 dBA

a. 77 dBA
b. 65 dBA
c. 86 dBA
d. 80 dBA

10-4. The noise level outside of a school should not exceed 70dBA. If the sound pressure level 5 m from the edge of a roadway is 80 dBA, what is the sound pressure level outside of a school that is 30 m from the edge of the roadway? Assume that the roadway is a continuous line source.

a. 64 zdBA
b. 72 dBA
c. 88 dBA
d. 62 dBA

10-12 REFERENCES

Achitoff, L. (1973) "Aircraft Noise—A Threat to Aviation," *Journal of Water, Air and Soil Pollution*, vol. 2, no. 3, pp. 357–363.

Alexandre, Barde, Lamure, and Langdon (1975) *Road Traffic Noise*, Applied Science Publishers.

Beranek, L. L. (1954) *Acoustics,* McGraw-Hill, New York.

Beranek, L. L. (1988) "Noise and Vibration Control," Institute of Noise Control Engineering, Washington, DC.

Berendt, R. D., E. L. R. Corliss, and M. S. Ojalro, (1976) A Practical Guide to Noise Control, *National Bureau of Standards Handbook* 119, U.S. Department of Commerce, pp. 16-41.

Borg, E., and S. A. Counter (1989) "The Middle-Ear Muscles," *Scientific American,* August, pp. 74–79.

Burck, W., et al. (1965.) "Gutachten erstatet im Auftrag des Bundesministers für Gesundheits wesen,*" Flugärm.*

Clemis, J. D. (1975) "Anatomy, Physiology, and Pathology of the Ear," in J. B. Olishifski and E. R. Harford (eds.), *Industrial Noise and Hearing Conservation,* National Safety Council, Chicago, p. 213.

Davis, H. (1958) "Effects of High Intensity Noise on Navy Personnel," *U.S. Armed Forces Medical Journal,* vol. 9, pp. 1027–1047.

de Neufville, R., and A. R. Odoni (2003) *Airport Systems: Planning, Design, and Management*, McGraw-Hill, New York, p. 198.

FHWA (1973) *Policy and Procedure Memorandum 90-2, Noise Standards and Procedures,* U.S. Department of Transportation, Washington, DC, http://www.fhwa.dot.gove/environment.

FHWA (2005) "Priority, Market-Ready Technologies and Innovations, FHWA Traffic Noise Model®, Version 2.1," Federal Highway Administration, U.S. Department of Transportation.

Fletcher, H., and W. A. Munson (1935) "Loudness, Its Definition, Measurement and Calculation," *Journal of Acoustic Society of America,* vol. 5, October, pp. 82–105.

Griffiths, I. D., and F. J. Langdon (1968) "Subjective Response to Road Noise," *Journal of Sound & Vibration,* vol. 8, pp. 16–32.

Hajek, J. (1977) "L_{eq} Traffic Noise Prediction Method," *Environmental and Conservation Concerns in Transportation: Energy, Noise and Air Quality,* (Transportation Research Record No. 648), Transportation Research Board, National Academy of Sciences, pp. 48–53.

ISO (1989) *Acoustics—Attenuation of Sound During Propagation Outdoors, Part 2, A General Calculation*, International Organization for Standardization, ISO/DIS 9613-2, Geneva.

ISO (1990) *Acoustics—Attenuation of Sound During Propagation Outdoors, Part 1, Calculation of Absorption of Sound by the Atmosphere*, International Organization for Standardization, ISO/DIS 9613-1, Geneva.

Kryter, K. D. et al. (1971) *Non-Auditory Effects of Noise,* Report WG-63, National Academy of Science, Washington, DC.

Kugler, B. A., D. E. Commins, and W. J. Galloway (1976) *Highway Noise: A Design Guide for Prediction and Control,* National Cooperative Highway Research Program Report.

Magrab, E. B. (1975) *Environmental Noise Control,* John Wiley & Sons, New York.

May, D. (1978) *Handbook of Noise Assessment,* Van Nostrand Reinhold, New York, p. 5.

Miller, J. D. (1971) *Effects of Noise on People*, U.S. Environmental Protection Agency Publication No. NTID 300.7, Washington, DC, p. 93.

Minnix, R. B. (1978) "The Nature of Sound," in D. M. Lipscomb and A. C. Taylor (eds.), *Noise Control: Handbook of Principles and Practices*, Van Nostrand Reinhold, New York, pp. 29–30.

NIOSH (1972) *Criteria for a Recommended Standard: Occupational Exposure to Noise*, National Institute for Occupational Safety and Health, U.S. Department of Health, Education, and Welfare, Washington, DC.

Olishifski, J. B., and E. R. Harford (eds.), (1975) *Industrial Noise and Hearing Conservation*, National Safety Council, Chicago, pp. 7, 340.

Piercy, J. E., and G. A. Daigle (1991) "Sound Propagation in the Open Air," in C. M. Harris (ed.), *Handbook of Acoustical Measurements and Noise Control*, McGraw-Hill, New York, pp. 3.1–3.26.

Sabin, H. J. (1942) "Notes on Acoustic Impedance Measurement," *Journal of the Acoustical Society of America*, vol. 14, p. 143.

Seeley, R., T. Stephens, and P. Tate (2003) *Anatomy and Physiology*, 6th ed., McGraw-Hill, New York.

Sylvan, S. (2000) *Best Environmental Practices in Europe and North America*, County Administration of Vastra Gotaland, Sweden.

U.S. EPA (1971) *Transportation Noise and Noise from Equipment Powered by Internal Combustion Engines*, U.S. Environmental Protection Agency Publication No. NTID 300.13, Washington, DC, p. 230.

U.S. EPA (1972) *Report to the President and Congress on Noise*, U.S. Environmental Protection Agency, Washington, DC.

U.S. EPA (1974) *Information on Levels of Environmental Noise Requisite to Protect Public Health and Welfare With an Adequate Margin of Safety*, U.S. Environmental Protection Agency, Publication No. 550/9-74-004, Washington, DC, pp. 29, B9, B10.

Ward, W. D., and A. Glorig (1961) "A Case of Firecracker-Induced Hearing Loss," *Laryngoscope*, vol. 71, pp. 1590–1596.

Warnock, A. C. C., and J. D. Quirt (1991) "Noise Control in Buildings," in C. M. Harris (ed.), *Handbook of Acoustical Measurements and Noise Control*, McGraw-Hill, New York, p. 33.13.

CHAPTER
11

SOLID WASTE MANAGEMENT

11-1 PERSPECTIVE

Solid waste is a generic term used to describe the things we throw away. It includes things we commonly describe as garbage, refuse, and trash. The U.S. Environmental Protection Agency's (EPA) regulatory definition is broader in scope. It includes any discarded item; things destined for reuse, recycle, or reclamation; sludges; and hazardous wastes. The regulatory definition specifically excludes radioactive wastes and *in situ* mining wastes. The Resource Conservation and Recovery Act (RCRA) is the federal act that regulates the disposal of solid waste. The act has two subtitles that have become a shorthand means of identifying the type of solid waste. "Subtitle C" wastes are hazardous wastes. "Subtitle D" wastes are all other solid wastes that are not hazardous or radioactive.

We have limited the discussion in this chapter to solid wastes generated from residential and commercial sources. Sludges were discussed in Chapters 6 and 8. Hazardous waste will be discussed in Chapter 12, and radioactive waste will be discussed in Chapter 14, which can be found at the text's website: www.mhhe.com/davis.

Magnitude of the Problem

Solid waste disposal creates a problem primarily in highly populated areas. The more concentrated the population, the greater the problem becomes. Various estimates have been made of the quantity of solid waste generated and collected per person per day. In 2009, the EPA estimated that the national average rate of solid waste generated was 2.0 kg/capita · day (U.S. EPA, 2010). On this basis, in 2009, the U.S. produced 221 teragrams (Tg) of solid waste.* This is a 60 percent increase over the 1980 estimate of 137.8 Tg and a nearly 175 percent increase over the 1960 estimate of 80.1 Tg. The EPA estimates that between 55 and 65 percent of the waste stream comes from residential sources, and the remainder is from commercial sources. Individual cities may vary greatly from these estimates. For example, Los Angeles, California, generates about 3.18 kg/capita · day while the rural community of Wilson, Wisconsin, generates about 1.0 kg/capita · day.

Figure 11-1 shows solid waste production rates. Averages are subject to adjustment depending on many local factors. Studies show there are wide differences in amounts collected by municipalities because of differences in climate, living standards, time of year, education, location, and collection and disposal practices.

Recycling Trends in Waste Management

While the amount of solid waste produced in the United States has increased, the recycling rate has also increased. In 1980, less than 10 percent of the solid waste produced was recycled. In 2009, the percentage of waste recycled increased to 34 percent. An additional 12 percent of the solid waste produced in 2009 was combusted with energy recovery, resulting in 54 percent of the solid waste generated in the United States in 2009 discarded in landfills. Therefore, although the amount of solid waste

---

*In keeping with correct SI notation, we use teragrams (1×10^{12} grams). One Tg is equivalent to 1×10^9 kilograms (kg) or 1×10^6 megagrams (Mg). The megagram is often referred to as the "metric ton."

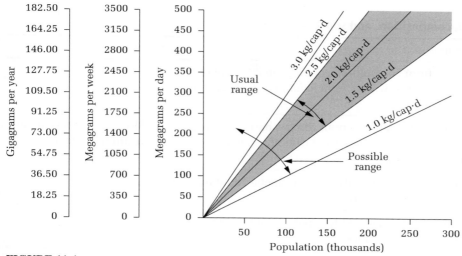

FIGURE 11-1
Solid waste produced: varying per capita figures.

produced each year from 1980 to 2009 increased by 60 percent, the amount of waste discarded in landfills remained about the same because of combustion and recycle of solid waste.

EPA also estimates the percentage of certain types of solid waste recycled. In 2009, for example, 74 percent of office paper was recycled, as were 96 percent of auto batteries, 88 percent of newspapers, 63 percent of junk mail, and 54 percent of magazines. Large percentages of aluminum and steel cans, yard trimmings, tires, and glass containers were also recycled.

Characteristics of Solid Waste

The terms *refuse* and *solid waste* are used more or less synonymously, although the latter term is preferred. The common materials of solid waste can be classified in several different ways. The point of origin is important in some cases, so classification as domestic, institutional, commercial, industrial, street, demolition, or construction may be useful. The nature of the material may be important, so classification can be made on the basis of organic, inorganic, combustible, noncombustible, putrescible, and nonputrescible fractions. One of the most useful classifications is based on the kinds of materials as shown in Table 11-1. Another classification system that is similar to this is the one used by the Incinerator Institute of America (Table 11-2). This is based primarily on the heat content of the waste.

Garbage is the animal and vegetable waste resulting from the handling, preparation, cooking, and serving of food. It is composed largely of putrescible organic matter and moisture; it includes a minimum of free liquids. The term does not include food processing wastes from canneries, slaughterhouses, packing plants, and similar facilities, or large quantities of condemned food products. Garbage originates primarily in home kitchens, stores, markets, restaurants, and other places where food is stored,

TABLE 11-1
Refuse materials by kind, composition, and sources

Kind	Composition	Sources
Garbage	Wastes from preparation, cooking, and serving of food; market wastes; wastes from handling, storage, and sale of produce	Households, restaurants, institutions, stores, markets
Rubbish	Combustible: paper, cartons, boxes, barrels, wood, excelsior, tree branches, yard trimmings, wood furniture, bedding, dunnage	
	Noncombustible: metals, tin cans, metal furniture, dirt, glass, crockery, minerals	
Ashes	Residue from fires used for cooking and heating and from on-site incineration	
Street refuse	Sweepings, dirt, leaves, catch basin dirt, contents of litter receptacles	Streets, sidewalks, alleys, vacant lots
Dead animals	Cats, dogs, squirrels, deer	
Abandoned vehicles	Unwanted cars and trucks left on public property	
Industrial wastes	Food-processing wastes, boiler house cinders, lumber scraps, metal scraps, shavings	Factories, power plants
Demolition wastes	Lumber, pipes, brick, masonry, and other construction materials from razed buildings and other structures	Demolition sites to be used for new buildings, renewal projects, expressways
Construction wastes	Scrap lumber, pipe, other construction materials	New construction, remodeling
Special wastes	Hazardous solids and liquids; explosives, pathological wastes, radioactive materials	Households, hotels, hospitals, institutions, stores, industry
Sewage treatment residue	Solids from coarse screening and from grit chambers; septic tank sludge	Sewage treatment plants, septic tanks

(*Source:* ISW, 1970.)

TABLE 11-2
Incinerator Institute of America waste classification

Classification of wastes to be incinerated						
Classification of Wastes		Principal components	Approximate composition % by weight	Moisture content %	Incombustible solids %	MJ heat value/kg of refuse as fired
Type	Description					
[a]0	Trash	Highly combustible waste, paper, wood, cardboard cartons, including up to 10% treated papers, plastic or rubber scraps; commercial and industrial sources	Trash 100%	10%	5%	19.8
[a]1	Rubbish	Combustible waste, paper, cartons, rags, wood scraps, combustible floor sweepings; domestic, commercial, and industrial sources	Rubbish 80% Garbage 20%	25%	10%	15.1
[a]2	Refuse	Rubbish and garbage; residential sources	Rubbish 50% Garbage 50%	50%	7%	10.0
[a]3	Garbage	Animal and vegetable wastes; restaurants, hotels, markets; institutional, commercial, and club sources	Garbage 65% Rubbish 35%	70%	5%	5.8
4	Animal solids and organic wastes	Carcasses, organs, solid organic wastes; hospital, laboratory, abattoirs, animal pounds, and similar sources	100% Animal and human tissue	85%	5%	2.3
5	Gaseous, liquid, or semi-liquid wastes	Industrial process wastes	Variable	Dependent on predominant components	Variable according to wastes survey	Variable according to wastes survey
6	Semi-solid and solid wastes	Combustibles requiring hearth, retort, or grate burning equipment	Variable	Dependent on predominant components	Variable according to wastes survey	Variable according to wastes survey

[a]The above figures on moisture content, ash, and MJ as fired have been determined by analysis of many samples. They are recommended for use in computing heat release, burning rate, velocity, and other details of incinerator designs. Any design based on these calculations can accomodate minor variations. (*Data Source:* IIA, 1968.)

prepared, or served. Garbage decomposes rapidly, particularly in warm weather, and may quickly produce disagreeable odors. There is some commercial value in garbage as animal food and as a base for commercial feeds. However, this use may be precluded by health considerations.

Rubbish consists of a variety of both combustible and noncombustible solid wastes from homes, stores, and institutions, but does not include garbage. Trash is synonymous with rubbish in some parts of the country, but trash is technically a subcomponent of rubbish. Combustible rubbish (the "trash" component of rubbish) consists of paper, rags, cartons, boxes, wood, furniture, tree branches, yard trimmings,

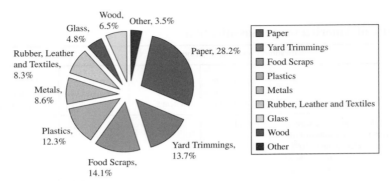

FIGURE 11-2
Materials generated in municipal solid waste (percent by mass), 2009. (*Source:* U.S. EPA, 2010.)

and so on. Some cities have separate designations for yard wastes. Combustible rubbish is not putrescible and may be stored for long periods of time. Noncombustible rubbish is material that cannot be burned at ordinary incinerator temperatures of 700 to 1,100°C. It is the inorganic portion of refuse, such as tin cans, heavy metals, glass, ashes, and so on.

The average municipal solid waste composition in the United States in 2003 is shown in Figure 11-2.

The density of loose combustible refuse is approximately 115 kg/m^3, while the density of collected solid waste is 235 to 300 kg/m^3.

Solid Waste Management Overview

The first objective of solid waste management is to remove discarded materials from inhabited places in a timely manner to prevent the spread of disease, to minimize the likelihood of fires, and to reduce aesthetic insults arising from putrifying organic matter. The second objective, which is equally important, is to dispose of the discarded materials in a manner that is environmentally responsible.

Policy Making.   Solid waste system policy making is primarily a function of the public sector rather than the private sector. The goal of a private firm is to minimize a well-defined cost function or to maximize profits. These are generally not the only, or even the primary, constraints of the public sector. The public objective function is more vague and difficult to express formally.

Constraints on the public sector, especially those of a political or a social nature, are difficult to measure, and criteria of effectiveness may not exist in units that can be quantified. Criteria of effectiveness against which public efficiency might be measured include such things as the frequency of collection, types of waste collected, location from which waste is collected, method of disposal, location of disposal site, environmental acceptability of disposal system, and the level of satisfaction of the customers. Public receptivity of a solid waste management system also depends on even less quantifiable parameters, which we group under the term *institutional factors*. Institutional factors include such things as political feasibility of the system, legislative constraints, and administrative simplicity.

Additional constraints on decision making in the public sector are environmental factors and resource conservation. Environmental factors are most important in the areas of waste storage and disposal because these functions represent prolonged exposure of wastes to the environment. Resource conservation is considered seriously by local governments as we become increasingly conscious of the limits of our natural resources.

Decisions in solid waste management policy formulation must be made in four basic areas: collection, transport, processing, and disposal. The flowchart in Figure 11-3 illustrates the decisions that must be made from the point of generation to the ultimate disposal of residential solid waste.

In designing a solid waste collection system, one of the first decisions to be made is where the waste will be picked up: the curb or the backyard. This is an important decision because it affects many other collection variables, including choice of storage containers, crew size, and the selection of collection trucks. Backyard service, once the predominant method of pickup, is still used by some communities. It is generally more costly, but it eliminates the need for scheduled pickups.

Another key decision is frequency of collection. Both point of collection and frequency of collection should be evaluated in terms of their impact on collection costs. Because collection costs generally account for 70 to 85 percent of total solid waste management costs, and labor represents 60 to 75 percent of collection costs, increases in the productivity of collection personnel can dramatically reduce overall costs. Most communities offer collection once or twice a week, with once per week being the most common schedule (U.S. EPA, 1995).

Systems with once-a-week curbside collection help maximize labor productivity and result in significantly lower costs than systems with more frequent collection and/or backyard pickup. The main reason many communities retain twice-a-week backyard service is that the citizens demand this convenience and are willing to pay for it. In warmer regions of the country, twice-a-week service may be deemed essential to prevent gross odors and to break the fly-breeding cycle. The egg-larvae-adult cycle is about 4–5 days.

The choice of solid waste storage containers must be evaluated in terms of both environmental effects and costs. From the environmental standpoint, some storage containers can present health and safety problems to the collectors, as well as to the general public. Therefore, the decision facing a community is which storage system is both environmentally sound and most economical, given the collection system characteristics. For example, paper and plastic bags are superior to many other containers from a health and esthetic standpoint and can increase productivity when used in conjunction with curbside collection. However, with backyard collection systems, bags have little effect on productivity.

The type of container used may also be dictated by the type of collection. If solid waste is collected manually, then plastic bags or cans can be used. Some communities have recently begun to sell special plastic bags or stickers to put on plastic bags that include the cost of the bag as well as the disposal fee. If the system is

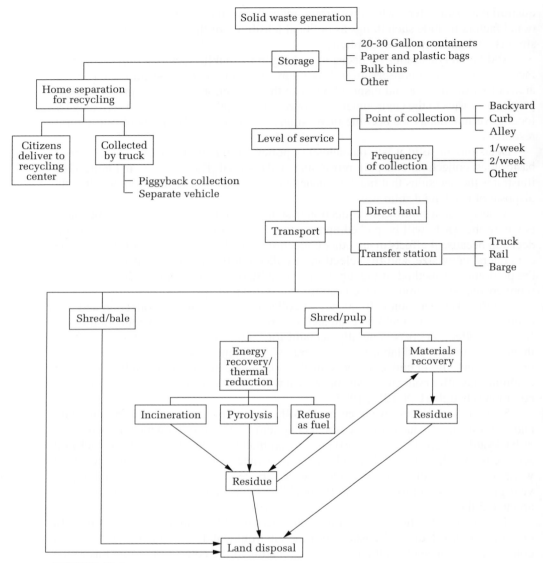

FIGURE 11-3
Solid waste management decision alternatives. (*Source:* U.S. EPA, 1974.)

automated or semiautomated, then the container must be specifically designed to fit the truck-mounted loading system. The containers typically hold from 1 to 20 cubic meters of waste.

Another factor to be considered in examining storage alternatives is home separation of various materials for recovery. The collection of materials for recovery/recycle is a growing practice that many cities are implementing. The technique of greatest interest to municipal decision makers is home separation and collection by either the regular collection truck equipped with special bins or by separate trucks.

One of the primary factors to consider in implementing a separate collection system is whether the benefits of recovery outweigh the costs involved. The economic viability of separate collection depends primarily on the local market price for the material and the degree of participation by the citizens. If these factors are positive, it may be possible to implement a recovery system with no increase, and possibly a savings, in collection operating costs; often no additional capital expenditure is required. Another factor to be considered is the expectation of the community that the municipality be actively involved in recycling. Most people perceive recycling as an environmentally friendly practice and so expect municipalities to provide the opportunity.

The distance between the disposal site and the center of the city will determine the advisability of including a *transfer station* in the transport system.* In addition to distance traveled to the disposal site, the time required for the transport is a key factor, especially in traffic-congested large cities.

The tradeoffs involved in transfer station operations are the capital and operating costs of the transfer station as compared to the cost (mostly labor) of having route collection vehicles travel excessive distances to the disposal site. These tradeoffs can be computed to find the point at which transfer becomes economically advantageous.

The sheer quantities of solid waste to be disposed of daily makes the problem of what to do with the waste, once it has been collected, among the most difficult problems confronting community officials. A crisis situation can develop very quickly, for example, in the case of an incinerator or land disposal site forced to shut down because of failure to meet newly passed environmental regulations. Alternatively, a crisis can build gradually over a period of time if needed new facilities are not properly planned for and put into service.

There are three basic alternatives for disposal. Some have subalternatives. The major alternatives are: (1) direct disposal of unprocessed waste in a municipal solid waste landfill, (2) processing of waste followed by land disposal, and (3) processing of waste to recover resources (materials and/or energy) with subsequent disposal of the residues. Most municipal solid waste is landfilled, but the amount landfilled declined from 73 percent in 1988 to 56 percent in 2003. Fourteen percent of the waste was incinerated in 2003 and 30 percent was recycled or composted. EPA projects the increase in waste incineration and recycling to continue.

Direct haul to a sanitary landfill (with or without transfer and long haul) is usually the cheapest disposal alternative in terms of both operating and capital costs. In 1988, it was estimated that about 8,000 landfills were in operation, but by 2002 the number had decreased to about 1,800. Many were closed as a result of regulatory restrictions. Municipalities own 75 percent of the sites (Wolpin, 1994). With rising *tipping fees* (the cost to dump solid waste at a disposal facility), a surplus of disposal capacity has replaced the late 1980s predictions of lack of landfill space.

---

*A transfer station is a place where trucks dump their loads into a larger vehicle where it is compacted. By combining loads, the cost per Mg $\cdot$ km for transport to the landfill is reduced.

With the second alternative, processing prior to land disposal, the primary objective is to reduce the volume of wastes. Such volume reduction has definite advantages because it reduces hauling costs and ultimate disposal cost, both of which are, to some extent, a function of waste volume. However, the capital and operating cost to achieve this volume reduction are significant and must be balanced against the savings achieved.

An additional consideration is the environmental benefit that might be derived from the volume reduction process. In some cases, shredding and baling may reduce the chances for water pollution from leachate. This alternative is more conserving of land than sanitary landfilling of unprocessed wastes, but by itself provides no opportunity for material or energy recovery.

The third category of disposal alternatives includes those processes that recover energy or materials from solid waste and leave only a residue for ultimate land disposal. There are significant capital and operating costs associated with all these energy and/or materials recovery systems. However, if markets are available, both energy and materials can be sold to reduce the net costs of recovery.

While resource recovery techniques may be more costly than other disposal alternatives, they do achieve the goal of resource conservation while enhancing sustainability and the residuals of the processes require much less space for land disposal than unprocessed wastes.

Affecting all four major functions are basic decisions regarding how the solid waste system will be managed and operated. This includes how the system will be financed, which level of government will administer it, and whether a public agency or private firm will operate the collection, transport, processing, and disposal functions. The criteria most relevant for making these decisions are the institutional factors of political feasibility and legislative constraints.

Integrated Solid Waste Management (ISWM).   The selection of a combination of techniques, technologies, and management programs to achieve waste management objectives is called *integrated solid waste management* (ISWM). This approach has made major strides in recent years. The EPA proposed a hierarchy of actions to implement ISWM: source reduction (including reuse and waste reduction), recycling and composting, and disposal in combustion facilities and landfills (U.S. EPA, 1995). The most obvious effect of the integrated approach is to reduce the size of the incineration facility. This reduces the capital cost of the incineration facility. Although the energy output is also reduced, the waste that remains has a higher energy content so that the reduction in energy output is less than the reduction in plant size. Recycling also reduces waste elements that can damage the boilers and removes those components that slag in the furnace and foul it (Shortsleeve and Roche, 1990).

11-2   COLLECTION

The solid waste collection policies of a city begin with decisions made by elected representatives about whether collection is to be made by: (1) city employees (municipal collection), (2) private firms that contract with city government (contract collection), or (3) private firms that contract with private residents (private collection). Many communities have moved away from exclusive municipal collection and toward a combined system.

More and more communities are moving toward mandatory recycling of materials such as paper, plastic, and glass. In these situations, separation of waste is required.

Elected officials may also determine what type of solid wastes are to be collected and from whom. In some municipalities broad classes of solid wastes (such as rubbish) are not accepted for collection. In others, certain materials (such as tires, grass trimmings, furniture, or dead animals) may be excluded. Hazardous wastes are generally excluded from regular collections because of disposal and collection dangers. The nature of the service may be governed by limitations of disposal facilities or by the opinion of the legislative body as to what service should be performed. A city may collect garbage only or it may collect everything but garbage. Almost all municipal systems collect residential waste, but only about one-third collect industrial waste.

The final decision concerning collection, which is made by the elected officials, is the frequency of collection. The proper frequency for the most satisfactory and economical service is governed by the amount of solid waste that must be collected and by climate, cost, and public requests. For the collection of solid waste that contains garbage, the maximum period should not be greater than

1. The normal time for the accumulation of the amount that can be placed in containers of reasonable size.

2. The time it takes for fresh garbage to putrefy and emit foul odors under average storage conditions.

3. The length of the fly-breeding cycle, which, during the hot summer months, is less than seven days.

In the last three decades the prevailing frequency of collection has changed from twice a week pickup to once a week. The increased use of once per week service is due to two factors. First, unit costs are reduced when frequency is cut from twice to once per week. Second, the increased percentage of paper and decreased percentage of garbage in the solid waste permit longer periods of acceptable storage.

Once policy has been set, the actual method of collection is determined by engineers or managers. Major considerations include how the solid waste will be collected, how the crews will be managed, how the trucks will be routed, and the type of equipment to be used.

Collection Methods

The first decision to be made is how the solid waste container will get from the residence to the collection vehicle. The three basic methods are: (1) *curbside* or alley pickup, (2) *set-out, set-back collection,* and (3) *backyard pickup,* or the tote barrel method. Most urban and suburban areas utilize curbside pickup, but a few communities still use backyard pickup. In some less populated areas, municipal waste collection is sometimes accomplished by requiring residents to transport waste to a specified point. This point may be a transfer station or the disposal site. This is the least expensive method for a municipality, but it is the least convenient method for the homeowner.

The quickest and most economical point of collection is from curbs or alleys using standard containers. It is the most common type of collection used. It costs only about one-half as much as backyard collection. Usually the city designates what type of containers are to be used. The crews simply empty the containers into the collection vehicles. Whenever possible the crews collect from both sides of the street at the same time. Municipal ordinances or administrative regulations usually specify when the containers must be placed at the curb or in the alley for pickup and also how long they may remain after pickup. Common limits are out by 7 A.M. and back by 7 P.M. When solid wastes are loaded from curbs or alleys, work progresses rapidly. A typical crew consists of a driver and two collectors. Some crews still have three or even four collectors, but the trend is toward fewer collectors. Recent studies indicate that small crews are more efficient than larger ones, because labor costs are a major element of the total cost. Aside from the cost advantage of this method, it also eliminates the need for the collectors to enter private property, and the amount of service given each homeowner is relatively uniform. However, many citizens dislike having to set their solid wastes out at certain times and object to the unsightly appearance on the streets. Some surveys have shown that many homeowners would prefer to pay more in order to receive backyard service.

When curbside removal is chosen, automatic and semiautomatic collection vehicles can be utilized. In an automated system, residents are provided with large specialized containers (approximately 90 gallons), which they roll to the curb. These containers are then lifted by powerful hydraulic arms that empty the contents of the container into the truck's hopper. The crew, or often just the driver, performs the operation from inside the cab of the collection vehicle. A typical side-loading vehicle with a hydraulic arm is shown in Figure 11-4. A fully automated system can be the most economical for a

FIGURE 11-4
Side-loading refuse collection vehicle with hydraulic lift arm. In this model, the tractor-trailer configuration allows for additional maneuverability. (*Source:* Heil Environmental, 2006.)

community, particularly if the community also uses this single truck to collect recyclables. The city of Los Angeles converted to such a system and in 2000 collected 712 Gg of refuse, recyclables, and yard waste with automated sideloading trucks. The waste is then transported to a waste processing facility where the materials are separated.

However, many communities cannot accommodate these large vehicles in their existing residential neighborhoods. They therefore use some combination of automatic and semiautomatic vehicles. In a semiautomatic system, the crew wheels the cart to the collection vehicle, lines the cart up with the lifting device and activates the lifter. A hydraulic device lifts and tips the cart, allowing the contents to fall into the hopper of the truck.

The existence of cul de sacs, alleys, and narrow streets as well as low-hanging utility lines may dictate the type of vehicle selected. For example, the city of Houston uses three different types of vehicles in its fleet of 200 vehicles. The city uses automated sideloaders to pick up curbside trash as well as recyclables, semiautomatic rear loaders to pick up yard waste, and a combination of rear loaders and a one-operator heavy-duty vehicle equipped with a grapple to pick up heavy trash to deposit in the rear loader (Bader, 2001, and Luken and Bush, 2002).

The set-out, set-back method eliminates most of the disadvantages of the curb method, but it does require the collector to enter private property. This method consists of the following operations: (1) the set-out crew carries the full containers from the residential storage location to the curb or alley before the collection vehicle arrives, (2) the collection crew loads the refuse in the same manner as the curb method, and (3) the set-back crew returns the empty cans. Any of the crew may be required to do more than one step or the homeowner may be required to do one of the steps. This method has not been shown to be more economical or advantageous than the backyard method, and it is more costly and time-consuming than curbside pickup.

Backyard pickup is usually accomplished by the use of tote barrels. In this method, the collector enters the resident's property, dumps the container into a tote barrel, carries it to the truck, and dumps it. The collector may collect refuse from more than one house before returning to the truck to dump. The primary advantage of this system is in the convenience to the homeowner. The major disadvantage is the high cost. Many homeowners object to having the collectors enter their private property. With this collection method, a rear-loading vehicle, such as the one shown in Figure 11-5, is used.

Cost analyses have revealed that 70 to 85 percent of the cost of solid waste collection and disposal can be attributed to the collection phase. For this reason, it would seem that a great deal of municipal effort should be directed to studying collection alternatives to determine the most efficient system. However, many analyses begin their studies assuming that waste loads are already collected and waiting for disposal. There are two major reasons why the collection system is not studied more often. First, the collection system is a complex and expensive system to analyze. The primary reasons for this are that it involves people, equipment, and levels of service, plus the possibility of numerous variations in secondary factors such as collection methodology; quantity, nature, and the method of storage of refuse; location of pickup point; equipment type and characteristics of operation; road factors; service density; route topography; climatic factors; and human factors. Human factors would include morale, incentive, fatigue, and other variables that influence the time required to complete a

FIGURE 11-5
Typical rear-loading refuse collection vehicle. (*Source:* Heil Environmental, 2006.)

given task. Secondly, most cities are already collecting refuse in some manner, and the cliche "leave well enough alone" often prevails. It is generally on the disposal system that the public is placing pressure for improvement, rather than the collection system.

Most changes in collection systems will require a great deal of investigation and testing. Even if the change is an obvious one, often "proof" of some sort is needed to convince the elected officials. The most important thing to realize about the solid waste collection system is that it is too big, complex, and vital to allow actual experimentation except on a very small scale. Coupled with this are all the other problems peculiar to studying large-scale public systems. A relevant database is probably nonexistent. The political implications of control of the system and cost distribution may override an otherwise practical solution. A large investment will have already been made in the existing system and the designer is not allowed the luxury of starting at the beginning, but must start with a system that may be founded on a pyramid of errors.

EPA suggests a method that can be used to estimate the time requirements of a waste collection system in order to evaluate and subsequently optimize the system (U. S. EPA, 1995). The steps included in a time study are shown in Table 11-3.

TABLE 11-3
Steps for conducting a time study

1. Select crew(s) representative of average level and skill level.

2. Determine the best method (series of movements) for conducting the work.

3. Set up a data sheet that can be used to record the following information: date, name of crew members and time recorder, type of collection method and equipment (including loading mechanism), specific area of municipality, and distance between collection points.

4. Divide loading activity into elements that are appropriate for the type of collection service. For example, the following elements might be appropriate for a study of residential collection loading times:

   • Time to travel from last loading point to next one
   • Time to get out of vehicle and carry container to the loading area
   • Time to load vehicle
   • Time to return container to the collection point and return to the vehicle.

5. Using a stop watch, record the time required to complete each element for a representative number of repetitions. Time may be measured using one of the following two methods:

   • *Snapback method:* The time recorder records the time after each element and then resets watch to zero for measurement of the next element.
   • *Continuous method:* The time recorder records the time after each element but does not reset the watch so that it moves continuously until the last elements is completed.

   Because the continuous method requires the time recorder to perform fewer movements and no time is lost for watch resetting, the continuous method is usually recommended.

   The number of repetitions that will be representative depends on the time required to complete the overall activity (cycle). The following numbers of repetitions have been suggested as sufficient:

Number of Repetitions	Minutes Per Cycle	Number of Repetitions	Minutes Per Cycle
60	0.50	20	2.0
40	0.75	15	5.0
30	1.00	10	10.5

6. Determine the average time recorded (T_o) and adjust it for "normal" conditions.

   In the case of waste collection, adjustments should be made for delays and for crew fatigue. These adjustments are typically in terms of the percent of time spent in a workday. The delay allowance (D) should include time for traffic conditions, equipment failures, and other uncontrollable delays. Crew fatigue allowance (F) should include adequate rest time for recovery from heavy lifting, extreme hot and cold weather conditions, and other circumstances encountered in waste collection. The allowance factors (D and F) along with the average observed time (T_o), can be used to estimate the "normal" time (T_n):

$$T_n = (T_o) \times [1 + (F + D)/100]$$

   This "normal" time is the loading time required for the particular area and collection system.

   For other activities, adjustments are also made for personal time (bathroom breaks). In this case, adjustment for personal time is made when calculating the number of loads/crew/day.

(*Source:* U.S. EPA, 1995.)

Waste Collection System Design Calculations

Often, it is desirable to calculate "quick and dirty" estimates of such things as crew size, desired truck capacity, and labor and capital costs. Simple formulas have been developed that enable such calculations. The formulas are based on crude averages regarding collection times, and they make broad assumptions. An example of a not-always-justifiable assumption is that if one collector can collect a house in one minute, then two can do it in one-half minute. Several such equations follow.

Estimating Truck Capacity.   Given that you are able to estimate a large number of factors, the following equation will allow you to estimate the volume of solid waste a truck must be able to carry.

$$V_T = \frac{V_p}{rt_p}\left[\frac{H}{N_d} - \frac{2x}{s} - 2t_d - t_u - \frac{B}{N_d}\right] \tag{11-1}$$

where V_T = volume of solid waste carried per trip by truck at a mean density, D_T, m^3
   V_p = volume of solid waste per pickup location or stop, m^3/stop
   r = compaction ratio
   t_p = mean time per collection stop plus the mean time to reach the next stop, h
   H = length of working day,* h
   N_d = number of trips to the disposal site per day
   x = one-way distance to disposal site, km
   s = average haul speed to and from disposal site, km/h
   t_d = one-way delay time, h/trip
   t_u = unloading time at disposal site, h/trip
   B = off route time per day, h

The factor of two in Equation 11-1 accounts for travel both to and from the disposal site. The average haul speed is a function of the total round-trip distance to the disposal site (Figure 11-6). As noted in the definitions, the volume carried presumes a mean density, D_T. This is the density that results after the waste has been compacted in the truck. The compaction ratio (r) is the ratio of the density after compaction to that before compaction. Typical densities "as discarded" are given for several solid waste components in Table 11-4. If, for example, paper waste was compacted to a density of 163.4 kg/m^3, the compaction ratio would be two to one. Compactor trucks can achieve densities ranging from 300 to 600 kg/m^3.

A value for t_p can be estimated from empirical data (U. Calif., 1952; Stone, 1969). The data may be approximated by linear equations of the following form:

$$t_p' = t_{b_p} + a(C_n) + b(PRH) \tag{11-2}$$

---

*We should note that it is standard practice to allow two fifteen-minute breaks during the day. Because the crew is paid for this, the number of hours in the workday (H) are unchanged. However, some allowance must be made for it. Hence the off route time (B) is included in the equation.

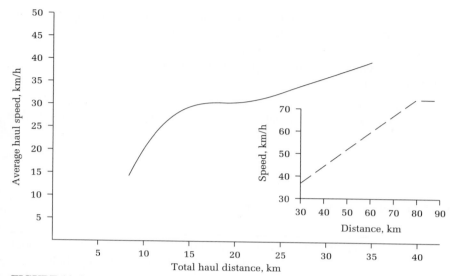

FIGURE 11-6
Effect of haul distance on average haul speed. (Adapted from U. Calif., 1952.)

TABLE 11-4
Typical properties of uncompacted solid waste as discarded in Davis, California

Component	Mass (kg)	Density (kg/m^3)	Volume (m^3)
Food wastes	4.3	288	0.0149
Paper	19.6	81.7	0.240
Cardboard[a]	2.95	99.3	0.0297
Plastics	0.82	64	0.013
Textiles	0.091	64	0.0014
Rubber	—	128	—
Leather	0.68	160	0.0043
Garden trimmings	6.5	104	0.063
Wood	1.59	240	0.00663
Glass	3.4	194	0.018
Tin cans	2.36	88.1	0.0268
Nonferrous metals	0.68	160	0.0043
Ferrous metals	1.95	320	0.00609
Dirt, ashes, brick	0.50	480	0.0010
Total	45.4		0.429

[a]Cardboard partially compressed by hand before being placed in container.
(*Source:* Tchobanoglous et al., 1977.)

where t'_p = mean time per collection stop plus mean time to reach next stop, min/stop
t_{b_p} = mean time between collection stops, min/stop
a, b = coefficients of regression fit to data points
C_n = mean number of containers at each pickup location
PRH = rear of house pickup locations, %

To convert t'_p to t_p, we must divide by 60 min/h.

The number of pickup locations that can be handled by a given crew is simply the available time after haul divided by the mean pickup time:

$$N_p = \frac{\frac{H}{N_d} - \frac{2x}{s} - 2t_d - t_u - \frac{B}{N_d}}{t_p} \tag{11-3}$$

where N_p = number of pickup locations per load

Example 11-1. The solid waste collection vehicle of Watapitae, Michigan, is about to expire, and city officials are in need of advice on the size of truck they should purchase. The compactor trucks available from a local supplier are rated to achieve a density (D_T) of 400 kg/m^3 and a dump time of 6.0 minutes. In order to ensure once-a-week pickup the truck must service 250 locations per day. The disposal site is 6.4 km away from the collection route. From past experience, a delay time of 13 minutes can be expected. The data given in Table 11-4 have been found to be typical for the entire city. Each stop typically has three cans containing 4 kg each. About 10 percent of the stops are backyard pickups. Assume that two trips per day will be made to the disposal site. Also assume that the crew size will be two and that the empirical equation of Tchobanoglous, Theisen, and Eliassen for a two-person crew applies (1977). That equation is given as follows:

$$t'_p = 0.72 + 0.18(C_n) + 0.014(PRH)$$
$$t'_p = 0.72 + 0.54 + 0.14 = 1.40 \text{ min/stop}$$
$$t_p = \frac{1.40 \text{ min}}{60 \text{ min/h}} = 0.0233 \text{ h}$$

Solution. Using Table 11-4 we determine the mean density of the uncompacted solid waste to be

$$D_u = \frac{\text{Total Mass}}{\text{Total Volume}} = \frac{45.4 \text{ kg}}{0.429 \text{ m}^3} = 105.83 \text{ or } 106 \text{ kg/m}^3$$

The volume per pickup is then

$$V_p = \frac{(3 \text{ cans})(4 \text{ kg/can})}{106 \text{ kg/m}^3} = 0.11 \text{ m}^3$$

The compaction ratio is determined from the densities:

$$r = \frac{D_T}{D_u} = \frac{400 \text{ kg/m}^3}{106 \text{ kg/m}^3} = 3.77$$

The average haul speed is determined from Figure 11-6. Because the graph is for total haul distance, we enter with $(2)(6.4) = 12.8$ km and determine that $s = 27$ km/h. All of the other required data were given; thus, we can now use Equation 11-1. The factor of 60 is to convert minutes to hours. For two 15-minute breaks, $B = 0.50$.

$$V_t = \frac{0.11}{(3.77)(0.0233)}\left[\frac{8}{2} - \frac{(2)(6.4)}{27} - 2\frac{13 \text{ min}}{60 \text{ min/h}} - \frac{6 \text{ min}}{60 \text{ min/h}} - \frac{0.50}{2}\right]$$

$$= (1.25)(2.74) = 3.43 \text{ m}^3$$

The number of stops that can be handled is given by Equation 11-3:

$$N_p = \frac{2.74}{0.0233} = 117.60, \text{ or } 118, \text{ pickups per load}$$

The smallest compactor truck available is one that will hold 4.0 m³. Obviously, this will be satisfactory. However, the crew will not be able to reach the required 250 stops per day. Thus, some other alternative must be considered. One would be to extend the workday by 30 minutes.

Estimating Costs. Most of the decisions involved in the collection of solid waste are based on economic considerations rather than technical ones. The costs are considered on the basis of a unit mass of solid waste to facilitate comparison between different size vehicles, crews, and the like. Furthermore, truck costs are considered separately from labor costs.

Truck costs include depreciation of the initial capital investment plus the *operating and maintenance (O & M)* costs.*

The following equation may be used to estimate the annual cost per Mg (U. Calif., 1952):

$$A_T = \frac{1{,}000(F)}{V_T D_T N_T Y}\left[1 + \frac{i(Y + 1)}{2}\right] + \frac{1{,}000(X_t)(OM)}{V_T D_T} \tag{11-4}$$

where A_T = annual truck cost, $/Mg
$\quad\ F$ = initial (first) cost of truck, $
$\quad V_T$ = volume of solid waste carried per trip by the truck
$\quad D_T$ = mean density of solid waste in truck, kg/m³
$\quad N_T$ = number of trips per year
$\quad\ Y$ = useful life of truck, y
$\quad\ \ i$ = interest rate on capital
$\quad X_t$ = distance per trip, pickup plus haul, km
$\quad OM$ = operating and maintenance cost, $/km

The factor of 1,000 is to convert kg to Mg.

---

*Government-operated collection systems, by the nature of their operation, do not actually depreciate purchases. First of all, they get no tax credit for doing so and, secondly, they do not save or put aside money in a bank and therefore cannot draw interest. In spite of all this, good engineering economics demands that capital costs be depreciated in order to allow valid comparisons between alternatives.

Labor costs consist of direct wages plus some overhead costs for such things as supervision, secretarial support, phone, utilities, insurance, and fringe benefits. Equation 11-5 can be used to estimate the annual labor cost per Mg:

$$A_L = \frac{1,000(CS)(W)(H)}{V_T D_T N_d}[1 + (OH)]$$
(11-5)

where A_L = annual labor cost, \$/Mg
CS = average crew size
W = average hourly wage rate, \$/h
OH = overhead as a fraction of wages*

Again, the factor of 1,000 is to convert kg to Mg.

Example 11-2.   Estimate the customer service charge for the situation of Example 11-1. The initial truck cost of a 4.0 m^3 compactor truck is \$104,000, and the average O & M cost over the five-year life of the truck is expected to be \$5.50/km. The interest rate is 8.25 percent. The average route length is 6.3 km. The average hourly wage rate is \$13.50 per hour with time and a half for overtime. The overhead rate is 125 percent of the hourly wage rate.

Solution.   Assuming a five-day workweek and ignoring holidays, the number of trips per year would be

$$N_t = N_d(5)(52) = 2(5)(52) = 520$$

Because the average route length is 6.3 km and the average haul distance from Example 11-1 is 2(6.4) = 12.8 km, then

$$X_t = 6.3 + 12.8 = 19.1 \text{ km}$$

For the extended workday proposed at the end of Example 11-1, the volume of solid waste per trip would be

$$V_T = (1.25)(2.74 + 1/2(0.5)) = 3.74 \text{ m}^3$$

The factor of one-half times the extra half hour was selected because we assumed the time to be equally divided between each of the two trips. Note that we do not use the actual volume of the truck, which is somewhat larger than V_T. (The truck size is the nearest standard size.) Now we may compute the annualized truck cost.

$$A_T = \frac{1,000(104,000)}{(3.74)(400)(520)(5)}\left[1 + \frac{0.0825(5+1)}{2}\right] + \frac{1,000(19.1)(5.50)}{(3.74)(400)}$$
$$= (26.74)(1.25) + 70.22 = \$103.65/\text{Mg}$$

---

*OH is not a product. It is shorthand for "overhead".

Because we have planned for an extra half hour of work each workday, we must adjust the hourly wage rate accordingly before we can use Equation 11-5. The adjustment is simply a determination of the weighted average rate.

$$W = \frac{(\text{reg. shift hours})(\text{wage}) + (\text{overtime hours})(\text{OT rate})(\text{wage})}{\text{total hours}}$$

$$= \frac{8(13.50) + 0.5(1.5)(13.50)}{8.5} = \$13.90/\text{h}$$

Now we may apply Equation 11-5 directly.

$$A_L = \frac{(1{,}000)(2)(13.90)(8.5)}{(3.74)(400)(2)}[1 + 1.25] = \$177.70/\text{Mg}$$

The total annual cost is then

$$A_{tot} = \$103.65 + \$177.70 = \$281.35/\text{Mg}$$

From Example 11-1, we know that each service stop averages three cans per week at 4 kg per can. Thus, each service stop contributes $3(4)(52) = 624$ kg or 0.624 Mg per year. The annual cost per service stop should be $(\$281.35/\text{Mg})(0.624\ \text{Mg}) = \$175.56$. For 52 pickups per year, this is an average cost of about \$3.38 per week (that is, \$175.56/52).

Truck Routing

The routing of trucks may follow one of four methods. The first possibility is the daily route method. In this method the crew has a definite route that must be finished before going home. When the route is finished the crew can leave, but if necessary, they must work overtime to finish the route. This is the simplest method and the most common. The advantages of this method are as follows:

1. The homeowner knows when the refuse will be picked up.

2. The route sizes can be adjusted for the load to maximize crew and truck utilization.

3. The crew likes the method because it provides an incentive to get done early.

The disadvantages include:

1. If the route is not finished, the crew will work overtime, which will increase the expense.

2. The crew may have a tendency to become careless as they try to finish the job sooner.

3. Frequently the result is underutilization of the crew and equipment due to the increased incentive of the crew.

4. A breakdown seriously affects operations.

5. It is hard to plan routes if the load is variable, because of the disposal of yard wastes and the like.

The next method is the large route method. In this scheme the crew has enough work to last the entire week. The route must be completed in one week. The crew is left on its own to decide when to pick up the route. Usually some time off at the end of the week is the goal of the crew. This method is only good for backyard pickup because the residents don't know when pickup will be. The same advantages and disadvantages apply to this method as to the daily route method.

In the single load method, the routes are planned to get a full truck load. Each crew is assigned as many loads as it can collect per day. The biggest advantage of this method is that it can minimize travel time. The method must consider size of crew, capacity of truck, length of travel, refuse generated, and similar variables. Other advantages include:

1. A full day's work can be provided for maximum utilization of the crew and equipment.

2. It can be used for any type of pickup.

The major disadvantage is that it is hard to predict the number of homes that can be serviced before the truck is filled.

The last method is the definite working day method. As its name implies, the crew works for its assigned number of hours and quits. This method predominates in areas where unions are strong. With this method, the crew and the equipment get maximum utilization. Regularity is sacrificed with this method, and residents have little idea when pickup will occur.

Having determined the method by which the trucks will be managed, it is still necessary to find the actual route the truck will follow through the city. The purpose of routing and districting is to subdivide the community into units that will permit collection crews to work efficiently. No matter what the size of the community, it can be divided into districts, with each district constituting one day's work for the crew. The route is the detailed path of travel for the collection vehicle. The size of each route depends upon various factors as discussed earlier. The Office of Solid Waste Management Programs of the U.S. Environmental Protection Agency has developed a simple, noncomputerized "heuristic" (rule-of-thumb) approach to routing based on logical principles. The goal is to minimize deadheading, delay, and left turns. This method relies on developing, recognizing, and using certain patterns that repeat themselves in every municipality. Routing skills can be quickly acquired by applying the rules and developing experience. The following rules are taken from an EPA publication (Shuster and Schur, 1974).

1. Routes should not be fragmented or overlapped. Each route should be compact, consisting of street segments clustered in the same geographical area.

2. Total collection plus haul times should be reasonably constant for each route in the community (equalized workloads).

3. The collection route should be started as close to the garage or motor pool as possible, taking into account heavily traveled and one-way streets. (See rules 4 and 5.)

4. Heavily traveled streets should not be collected during rush hours.

5. In the case of one-way streets, it is best to start the route near the upper end of the street, working down it through the looping process.

6. Services on dead-end streets can be considered as services on the street segment that they intersect, because they can only be collected by passing down that street segment. To keep left turns at a minimum, collect the dead-end streets when they are to the right of the truck. They must be collected by walking down, backing down, or making a U-turn.

7. When practical, service stops on steep hills should be collected on both sides of the street while the vehicle is moving downhill for safety, ease, speed of collection, wear on vehicle, and conservation of gas and oil.

8. Higher elevations should be at the start of the route.

9. For collection from one side of the street at a time, it is generally best to route with many clockwise turns around blocks. (Authors' note: Heuristic rules 8 and 9 emphasize the development of a series of clockwise loops in order to minimize left turns, which generally are more difficult and time-consuming than right turns. Especially for right-hand-drive vehicles, right turns are safer.)

10. For collection from both sides of the street at the same time, it is generally best to route with long, straight paths across the grid before looping clockwise.

11. For certain block configurations within the route, specific routing patterns should be applied.

See Figure 11-7 for an example of the heuristic routing procedure.

Crew Integration

Another area of consideration is the integration of several crews. There are four ways of managing crews; usually some combination of the four is employed by any given city.

The swing crew method utilizes an extra crew as standby for heavy pickups, breakdown, or illness. Many times this crew will not report until noon to begin its day.

Crew sizes may be varied because of heavy loads, rain, different route sizes, and other factors. This is referred to as the variable crew method.

With the interroute relay method, when a crew member finishes one job, he or she is put on another route that needs additional help. This method requires more administration to operate, but results in better utilization of personnel and helps ensure that all routes will be completed during the day. Some form of this method has found wide acceptance with good results. Management must be sure that the workload is being balanced fairly and that a faster worker doesn't have to carry the load for others.

The last possibility is the reservoir route method. In this method, the crews work around a central core. When they have finished the route, the crews go to the core and begin picking up there. The core is usually an every day pickup, such as a park or a downtown area.

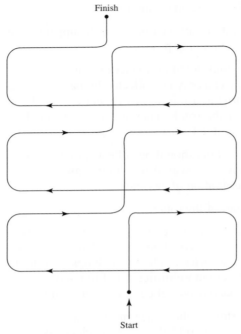

Finish

Start

FIGURE 11-7
Arrows show heuristic routing pattern developed for a
north-south, one-way street combined with east-west,
two-way streets. If both sides of the one-way street
cannot be collected in one pass, it is necessary to loop
back to the upper end and make a straight pass down
the other side.

11-3 INTERROUTE TRANSFER

It is not always economical, or even possible, to haul the solid waste directly to the
disposal site in the collection vehicle. In these cases, the solid waste is transferred
from several collection vehicles to a larger vehicle, which then carries it to the disposal
site. The larger vehicle (*transfer vehicle*) may be a tractor-trailer, railroad car, or barge.
A special facility, called a *transfer station,* must be constructed to permit this exchange
in a rapid and sanitary fashion.

Among the more important considerations in planning and designing a transfer
station are location, type of station, sanitation, access, and accessories such as weighing
scales and fences. The use of a transfer station may also provide for present or future
resource recovery facilities.

Maximum Haul Time

As in estimating collection times, it is possible to use average values to evaluate tradeoffs in transfer station effectiveness. One such method is to compute the travel time

available to the crew to travel to the disposal site and still collect the appointed route. This can be done by rearranging Equation 11-3:

$$T_H = \frac{H}{N_d} - t_p N_p - 2t_d - t_u - \frac{B}{N_d}$$ (11-6)

where T_H = maximum available haul time, h.

If the maximum available haul time is less than the round trip distance divided by the average route speed ($2x/s$), then you have a problem. Up to a point, changes in t_d, t_u, B, and/or H may alleviate the situation.

Economical Haul Time

The travel time in and of itself is not usually the prime consideration. Cost is usually the prime consideration. Costs are saved when a transfer operation is used because

1. The nonproductive time of collectors is reduced, because they no longer ride to and from the disposal site. It may be possible to reduce the number of collection crews needed because of increased productive collection time.

2. Any reduction in mileage traveled by the collection trucks results in a savings in operating costs.

3. The maintenance requirements for collection trucks can be reduced when these vehicles are no longer required to drive into the landfill site. Much of the damage to suspensions, drive trains, and tires occurs at landfills.

4. The capital cost of collection equipment may be reduced; because the trucks will be traveling only on improved roads, lighter duty, less expensive models can be used (U.S. EPA, 1995).

In order to compare "direct haul" with "transfer" costs, the costs are computed on the basis of $/Mg · km or, preferably, $/Mg · min. The time-based comparison is preferred because the average haul speed of the collection vehicle will often be greater than that of the transfer vehicle. Because it is time, not distance, that costs money, this gives a fairer comparison. In addition to the travel cost of operating the transfer vehicle, there are fixed costs for the construction and operation of the transfer station and for maneuvering and unloading the transfer vehicle. Figure 11-8 may be used to estimate the cost of the transfer station.

Example 11-3. The disposal site for Watapitae will be closed in two years because of the lack of capacity. An alternative disposal site will be available when the present site is closed. It will be a countywide regional system that will be 32.5 km from the collection route. Using the data from Examples 11-1 and 11-2 and the following assumptions, determine the maximum haul time for the collection vehicle and the cost for collection vehicle and transfer vehicle haul: $N_d = 1$, $B = 0.50$ h, and the amortized capital cost and operating cost for the transfer station is approximately $37/Mg.

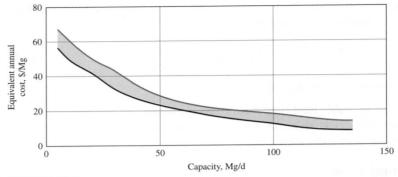

FIGURE 11-8

Transfer station equivalent annual cost as a function of capacity. Costs adjusted to 2006. (*Data Source:* Zuena, 1987.)

Solution.   First we must determine whether or not the collection vehicle has the time to get to the disposal site while still making all of its pickups.

$$T_H = \frac{8.5}{1} - (0.0233)(250) - 2\frac{13}{60} - \frac{6}{60} - \frac{0.5}{1}$$

$$= 1.64 \text{ h or } 98.5 \text{ min}$$

We now note that the round trip distance is two times the distance from the collection route. The average haul speed can be determined from Figure 11-6. The average haul speed is 64 km/h. Thus, we find the round trip travel time to the regional facility to be

$$\frac{2(32.5 \text{ km})}{64 \text{ km/h}} = 1.02 \text{ h or } 61 \text{ min}$$

The collection vehicle can make it to the disposal site. However, because we have reduced the number of trips to the disposal site, we must either provide an additional vehicle of the same size or replace the existing one with one that is twice as large. Because the existing crew size can handle the 250 pickups per day, the more logical choice would seem to be to choose the larger vehicle. (This is especially true because the existing one is about to expire.) Let us assume the new vehicle will have a capacity of 10.0 m^3.

Now let us examine the comparative haul costs. First we will look at the collection vehicle. We will take the annual cost for a new vehicle exclusive of O & M to be $29,851. Assuming eight hours of operation per day for five days a week for 52 weeks per year, the annual cost per minute of operation is

$$\frac{\$29,851}{(8 \text{ h/d})(60 \text{ min/h})(5 \text{ d/w})(52 \text{ wk/y})} = \$0.2392/\text{min}$$

With the effective wage rate of $13.90 per hour from Example 11-2, the cost of wages and 125 percent overhead is

$$\frac{(\$13.90 \times 2.25)}{60 \text{ min/h}} = \$0.5213/\text{min}$$

per worker or $1.0425/min for the crew. The operating cost will be about $5.50 per kilometer. For travel to the disposal site, the cost per minute would be

$$\frac{(\$5.50/km)(32.5\ km)(2)}{61\ min} = \$5.8607/min$$

The factor of two is for the round trip to the disposal site. The total haul cost per trip would be

$$61[(\$0.2392) + (\$1.0425) + (\$5.8607)] = \$435.69$$

The mass of solid waste hauled per trip is

$$(V_T)(D_T) = mass$$
$$(7.48\ m^3)(400\ kg/m^3) = 2{,}992\ kg,\ or\ 3.0,\ Mg$$

Note that the volume is twice that of a single trip (Example 11-2), but is considerably less than the capacity of the new vehicle. The unit cost of the haul would then be

$$\frac{\$435.69}{3.0\ Mg} = 145.23,\ or\ \$145/Mg$$

Now let us look at the transfer vehicle. Assume that a tractor-trailer rig having a capacity of 46 m³ has an annual cost exclusive of O & M of $37,601. The cost per minute is then

$$\frac{\$37{,}601}{(8\ h/d)(60\ min/h)(5\ d/wk)(52\ wk/y)} = \$0.3013/min$$

Because the tractor-trailer rig requires an operator with higher skill, the wage rate will be higher. Using a rate of $19.85 per hour and an overhead rate of 125 percent of wages, the cost per minute is

$$\frac{(\$19.85 \times 2.25)}{60\ min/h} = \$0.7444/min$$

In contrast to the collection vehicle, the crew is comprised of only the operator. Thus, the crew cost is $0.7444/min.

The operating cost will be about $6.50 per kilometer. The time for the rig to travel to the disposal site will be about 25 percent more than the collection vehicle. The travel cost would then be

$$\frac{(\$6.50)(32.5)(2)}{61 \times 1.25} = \$5.541/min$$

The total haul cost per trip would be

$$(1.25)(61)[(\$0.3013) + (\$0.7444) + (\$5.541)] = \$502.23$$

Because the capacity of the rig is four times that of the collection vehicle, the mass hauled per trip is

$$4(3.0) = 12\ Mg$$

The unit cost of the haul, including the cost of building and operating the transfer station (approximately \$37/Mg), would be

$$\frac{\$502.23}{12} + \$37 = 78.83, \text{ or } \$79/\text{Mg}$$

Obviously, consideration should be given to the construction and operation of a transfer station as an alternative to direct haul.

11-4   DISPOSAL BY MUNICIPAL SOLID WASTE LANDFILL

A municipal solid waste (MSW) landfill is defined as a land disposal site employing an engineered method of disposing of solid wastes on land in a manner that minimizes environmental hazards by spreading the solid wastes to the smallest practical volume, and applying and compacting cover material at the end of each day.

Site Selection

Site location is perhaps the most difficult obstacle to overcome in the development of a MSW landfill. Opposition by local citizens eliminates many potential sites. In choosing a location for a landfill, consideration should be given to the following variables:

1. Public opposition

2. Proximity of major roadways

3. Speed limits

4. Load limits on roadways

5. Bridge capacities

6. Underpass limitations

7. Traffic patterns and congestion

8. Haul distance (in time)

9. Detours

10. Hydrology

11. Availability of cover material

12. Climate (for example, floods, mud slides, snow)

13. Zoning requirements

14. Buffer areas around the site (for example, high trees on the site perimeter)

15. Historic buildings, endangered species, wetlands, and similar environmental factors.

In October of 1991, under Subtitle D of the Resource Conservation and Recovery Act (RCRA), the EPA promulgated new federal regulations for landfills. These regulations are known as the Criteria for Municipal Solid Waste Landfills (MSWLF Criteria). EPA also published a companion document to assist owners and municipalities comply with these criteria (U.S. EPA, 1998). These included siting criteria that specify restrictions on distances from airports, flood plains, and fault areas, as well as limitations on construction in wetlands, seismic impact areas, and other areas of unstable geology such as landslide areas and those susceptible to sink holes. Other restrictions may apply. For example, a landfill should be more than:

30 m from streams,

160 m from drinking water wells,

65 m from houses, schools, and parks, and

3,000 m from airport runways.

Site Preparation

The plans and specifications for a MSW landfill should require that certain steps be carried out before operations begin. These steps include grading the site area, constructing access roads and fences, and installing signs, utilities, and operating facilities.

On-site access roads should be of all-weather construction and wide enough to permit two-way truck travel (7.3 m). Grades should not exceed equipment limitations. For loaded vehicles, most uphill grades should be less than 7 percent, and downhill grades should be less than 10 percent.

All MSW landfill sites should have electric, water, and sanitary services. Remote sites may have to use acceptable substitutes, for example, portable chemical toilets, trucked-in drinking water, and electric generators. Water should be available for drinking, fire-fighting, dust control, and sanitation. Telephone or radio communications are desirable.

A small MSW landfill operation will usually require only a small building for storing hand tools and equipment parts and a shelter with sanitary facilities. A single building may serve both purposes. Buildings may be temporary and preferably movable.

Equipment

The size, type, and amount of equipment required at an MSW landfill depends on the size and method of operation, quantities and time of solid waste deliveries, and, to a degree, the experience and preference of the designer and equipment operators. Another factor to be considered is the availability and dependability of service from the equipment.

The most common equipment used on MSW landfills is the crawler or rubber-tired tractor (Figure 11-9). The tractor can be used with a dozer blade, trash blade, or a front-end loader. A tractor is versatile and can perform a variety of operations: spreading,

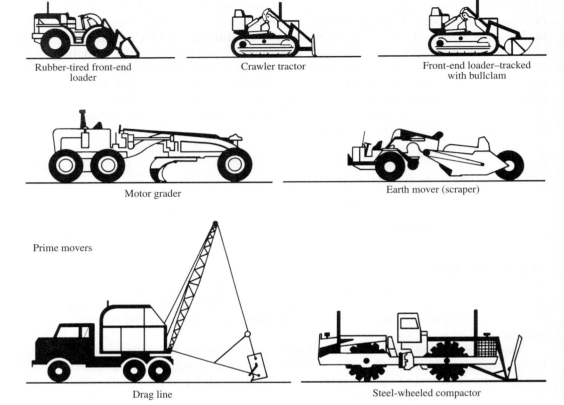

Rubber-tired front-end loader

Crawler tractor

Front-end loader–tracked with bullclam

Motor grader

Earth mover (scraper)

Prime movers

Drag line

Steel-wheeled compactor

FIGURE 11-9
Municipal solid waste landfill equipment.

compacting, covering, trenching, and even hauling the cover material. The decision on whether to select a rubber-tired or a crawler-type tractor, and a dozer blade, trash blade, or front-end loader must be based on the conditions at each individual site (see Table 11-5).

The crawler dozer is excellent for grading and can be economically used for dozing solid waste or soil over distances up to 100 m. The larger trash or landfill blade can be used in lieu of a straight dozer blade, thereby increasing the volume of solid waste that can be dozed. The crawler loader has the capability to lift materials off the ground for carrying. It is an excellent excavator, well suited for trench operations.

Rubber-tired machines are generally faster than crawler machines. Because their loads are concentrated more, rubber-tired machines have less flotation and traction than crawler machines. Rubber-tired machines can be economically operated at distances of up to 200 m.

Steel-wheeled compactors are finding increased application at MSW landfills. In basic design, compactors are similar to rubber-tired tractors. The unique feature of

TABLE 11-5
Performance characteristics of landfill equipment[a]

Equipment	Spreading	Compacting	Excavating	Spreading	Compacting	Hauling	Density of compacted solid waste (kg/m³)
Crawler dozer	E	G	E	E	G	NA	750
Crawler loader	G	G	E	G	G	NA	—
Rubber-tired dozer	E	G	F	G	G	G	733
Rubber-tired loader	G	G	F	G	G	G	—
Steel-wheeled compactor	E	E	P	E	E	NA	809
Scraper	NA	NA	G	E	NA	E	NA
Dragline	NA	NA	E	F	NA	NA	NA

[a]*Basis of evaluation*: Easily workable soil and cover material haul distance greater than 300 m.
Rating key: E, excellent; G, good; F, fair; P, poor; NA, not applicable.
Note: Density of "well-compacted" solid waste resulting from four passes over each square meter. Density measured after daily soil cover emplaced but not including soil in volume and weight measurements.
(*Source*: Data from Stone and Conrad, 1969, and O'Leary and Walsh, 2002.)

compactors is the design of their wheels, which are steel and equipped with teeth or lugs of varying shape and configuration. This design is employed to impart greater crushing and demolition forces to the solid waste. Use of compactors should be restricted to solid waste, because their design does not lend them to application of a smooth layer of compacted cover material. Thus, compactors are best used in conjunction with tracked or rubber-tired machines that can be used for cover material application.

Other equipment used at MSW landfills are scrapers, water wagons, drag-lines, dump trucks, and graders. This type of equipment is normally found only at large solid waste landfills where specialized equipment increases the overall efficiency.

Equipment size depends on the size of the operation. Small landfills for communities of 15,000 or less, or landfills handling 50 Mg of solid wastes per day or less, can operate successfully with one tractor in the 20 to 30 Mg range. Heavier equipment in the 30 to 45 Mg range, or larger, can handle more waste and achieve better compaction. Heavy equipment is recommended for MSW landfill sites serving more than 15,000 people or handling more than 50 Mg per day. MSW landfills serving 50,000 people or less or handling no more than about 150 Mg of solid waste per day normally can manage well with one piece of heavy equipment (30 to 45 Mg range).

Operation

Although various titles are used to describe the operating methods employed at MSW landfills, only two basic techniques are involved. They are termed the *area method* (Figure 11-10) and the *trench method* (Figure 11-11). At many sites, both methods are used, either simultaneously or sequentially.

In the area method, the solid waste is deposited on the surface, compacted, then covered with a layer of compacted soil at the end of the working day. Use of the area method is seldom restricted by topography; flat or rolling terrain, canyons, and other types of depressions are all acceptable. The cover material may come from on- or off-site.

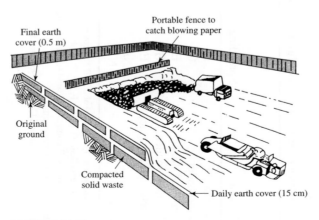

FIGURE 11-10
The area method.

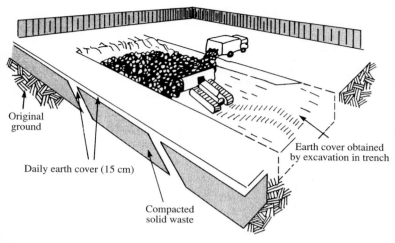

Original
ground

Daily earth cover (15 cm)

Earth cover obtained
by excavation in trench

Compacted
solid waste

FIGURE 11-11
The trench method.

The trench method is used on level or gently sloping land where the water table is low. In this method a trench is excavated; the solid waste is placed in it and compacted; and the soil that was taken from the trench is then laid on the waste and compacted. The advantage of the trench method is that cover material is readily available as a result of trench excavation. Stockpiles can be created by excavating long trenches, or the material can be dug up daily. The depth depends on the location of the groundwater and/or the character of the soil. Trenches should be at least twice as wide as the compacting equipment so that the treads or wheels can compact all the material on the working area.

A MSW landfill does not need to be operated by using only the area or trench method. Combinations of the two are possible. The methods used can be varied according to the constraints of the particular site.

A profile view of a typical landfill is shown in Figure 11-12. The waste and the daily cover placed in a landfill during one operational period form a *cell*. The operational period is usually one day. The waste is dumped by the collection and transfer vehicles onto the working *face*. It is spread in 0.4 to 0.6 m layers and compacted by driving a crawler tractor or other compaction equipment over it. At the end of each day *cover* material is placed over the cell. The cover material may be native soil or other approved materials. Its purpose is to prevent fires, odors, blowing litter, and scavenging. The federal regulations also permit the state regulatory authority to allow the use of alternative daily covers (ADC) if the owner of the landfill can demonstrate that the alternative material functions as well as the earthen cover without presenting a threat to human health or the environment. Some landfills have successfully demonstrated that diverted wastes such as chipped tires, yard waste, shredded wood waste, and petroleum-contaminated soils can be used effectively as ADCs. Using these waste products as ADCs presents a cost savings for the landfill and also increases the landfill's available space. The use of manufactured ADCs such as colored tarps is also

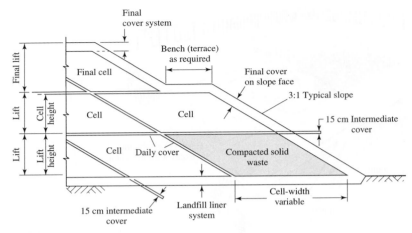

FIGURE 11-12
Sectional view through a MSW landfill. (*Source:* Tchobanoglous et al., 1993.)

being accepted in some localities. Recommended depths of cover for various exposure periods are given in Table 11-6. The dimensions of a cell are determined by the amount of waste and the operational period.

A *lift* may refer to the placement of a layer of waste or the completion of the horizontal active area of the landfill. In Figure 11-12 a lift is shown as the completion of the active area of the landfill. An extra layer of intermediate cover may be provided if the lift is exposed for long periods. The active area may be up to 300 m in length and width. The side slopes typically range from 1.5:1 to 2:1. Trenches vary in length from 30 to 300 m with widths of 5 to 15 m. The trench depth may be 3 to 9 m (Tchobanoglous et al., 1993).

Benches are used where the height of the landfill exceeds 15 to 20 m. They are used to maintain the slope stability of the landfill, for the placement of surface water drainage channels, and for the location of landfill gas collection piping.

Final cover is applied to the entire landfill site after all landfilling operations are complete. A modern final cover will contain several different layers of material to perform different functions. These are discussed more fully in the landfill design section of this chapter.

TABLE 11-6
Recommended depths of cover

Type of cover	Minimum depth (m)	Exposure time (d)
Daily	0.15	< 7
Intermediate	0.30	7 to 365
Final	0.60	> 365

Additional considerations in the operation of the landfill are those required by the 1991 Subtitle D regulations promulgated by EPA. These require exclusion of hazardous waste, use of cover materials, disease vector control, explosive gas control, air quality measurements, access control, runoff and run-on controls, surface water and liquids restrictions, and groundwater monitoring, as well as record keeping (40 CFR 257 and 258; FR 9 OCT 1991).

Environmental Considerations

Vectors (carriers of disease) and water and air pollution should not be a problem in a properly operated and maintained landfill. Good compaction of the waste, daily covering of the solid waste with good compaction of the cover, and good housekeeping are musts for control of flies, rodents, and fires.

Burning, which may cause air pollution, is never permitted at a MSW landfill. If accidental fires should occur, they should be extinguished immediately using soil, water, or chemicals. Odors can be controlled by covering the wastes quickly and carefully, and by sealing any cracks that may develop in the cover.

Landfill Gases.   The principal gaseous products emitted from a landfill (methane and carbon dioxide) are the result of microbial decomposition. Typical concentrations of landfill gases and their characteristics are summarized in Table 11-7. During the early life of the landfill, the predominant gas is carbon dioxide. As the landfill matures, the gas is composed almost equally of carbon dioxide and methane. Because the methane is explosive, its movement must be controlled. The heat content of this landfill gas mixture

TABLE 11-7
Typical constituents found in MSW landfill gas

Component	Percent (dry volume basis)
Methane	45–60
Carbon dioxide	40–60
Nitrogen	2–5
Oxygen	0.1–1.0
Sulfides, disulfides, mercaptans, etc.	0–1.0
Ammonia	0.1–1.0
Hydrogen	0–0.2
Carbon monoxide	0–0.2
Trace constituents	0.01–0.06

Characteristic	Value
Temperature, °C	35–50
Specific gravity	1.02–1.05
Moisture content	Saturated
High heating value, kJ/m^3	16,000–20,000

(*Source:* G. Tchobanoglous et al., 1993.)

(16,000 to 20,000 kJ/m^3), although not as substantial as methane alone (37,000 kJ/m^3), has sufficient economic value that many landfills have been tapped with wells to collect it. At the end of 2004, there were 378 landfill gas (LFG) recovery projects in the United States. This is a four-fold increase over the 86 LFG projects operating in 1990.

Because of their toxicity, trace gas emissions from landfills are of concern. More than 150 compounds have been measured at various landfills. Many of these may be classified as volatile organic compounds (VOCs). The occurrence of significant VOC concentrations is often associated with older landfills that previously accepted industrial and commercial wastes containing these compounds. The concentrations of 10 compounds measured in landfill gases from several California sites are shown in Table 11-8.

Leachate

Liquid that passes through the landfill and that has extracted dissolved and suspended matter from it is called *leachate*. The liquid enters the landfill from external sources such as rainfall, surface drainage, groundwater, and the liquid in and produced from the decomposition of the waste.

Leachate Quantity. The amount of leachate generated from a landfill site may be estimated using a hydrologic mass balance for the landfill. Those portions of the global hydrologic cycle (see Chapter 4) that typically apply to a landfill site include precipitation, surface runoff, evaporation, transpiration (when the landfill cover is completed), infiltration, and storage. Precipitation may be estimated in the conventional fashion from climatological records. Surface runoff or run-on may be estimated using the rational formula (Equation 4-19 or 4-20). Evaporation and transpiration are often lumped together as *evapotranspiration*. It may be estimated from regional data such as that provided by the U.S. Geologic Service *Water Atlas*. Infiltration (and exfiltration) may be estimated using Darcy's law (Equation 4-27). Until the landfill becomes saturated, some of the water infiltration will be stored in both the cover material and the waste. The quantity of water that can be held against the pull of gravity is referred to as *field capacity* (Figure 11-13 on page 822). Theoretically, when the landfill reaches its field capacity, leachate will begin to be produced. Then, the potential quantity of leachate is the amount of moisture within the landfill in excess of the field capacity. In reality, leachate will begin to be produced almost immediately because of channeling in the waste. The following equation may be used to estimate the field capacity of the waste (Tchobanoglous et al., 1993):

$$FC = 0.6 - 0.55\left(\frac{2.205W}{10,000 + 2.205W}\right) \tag{11-7}$$

where FC = field capacity (fraction of water in the waste based on dry weight of the waste)

W = overburden mass of waste calculated at midheight of the lift in question, kg

The EPA and the Waterways Experiment Station of the U.S. Army Corps of Engineers developed a microcomputer model of the hydrologic balance called the Hydrologic Evaluation of Landfill Performance (HELP) (Schroeder et al., 1984). The

TABLE 11-8
Concentrations of specified air contaminants measured in landfill gases (in parts per billion)

				Landfill Site			
Compound	Yolo Co.	City of Sacramento	Yuba Co.	El Dorado Co.	L.A.-Pacific (Ukiah)	City of Clovis	City of Willits
Vinyl chloride	6,900	1,850	4,690	2,200	<2	66,000	7.5
Benzene	1,860	289	963	328	<2	895	<18
Ethylene dibromide	1,270	<10	<50	<1	<1	<1	<0.5
Ethylene dichloride	nr	nr	nr	<20	0.2	<20	4
Methylene chloride	1,400	54	4,500	12,900	<1	41,000	<1
Perchloroethylene	5,150	92	140	233	<0.2	2,850	8.1
Carbon tetrachloride	13	<5	<7	<5	<0.2	<5	<0.2
1,1,1-TCA[1]	1,180	6.8	<60	3,270	0.52	113	0.8
TCE[2]	1,200	470	65	900	<0.6	895	8
Chloroform	350	<10	<5	120	<0.8	1,200	<0.8
Methane	nr	nr	nr	nr	0.11%	17%	0.14%
Carbon dioxide	nr	nr	nr	nr	0.12%	24%	<0.1%
Oxygen	nr	nr	nr	nr	nr	10%	21%

nr: Not reported by operator

[1]1,1,1-TCA: 1,1,1-trichloroethane, methyl chloroform

[2]TCE: Trichloroethene, trichloroethylene

(*Data Source:* CARB, 1988.)

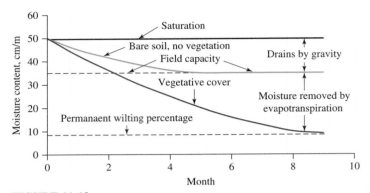

FIGURE 11-13
Soil moisture relationships.

program contains extensive data on the characteristics of various soil types, precipitation patterns, and evapotranspiration-temperature relationships as well as the algorithms to perform a routing of the moisture flow through the landfill.

Leachate Composition.   Solid wastes placed in a sanitary landfill may undergo a number of biological, chemical, and physical changes. Aerobic and anaerobic decomposition of the organic matter results in both gaseous and liquid end products. Some materials are chemically oxidized. Some solids are dissolved in water percolating through the fill. A range of leachate compositions is listed in Table 11-9. The VOCs in the landfill gas often contribute to contamination of groundwater because they dissolve in the leachate as it passes through the landfill. Henry's law (see Chapter 5) may be used to estimate the VOC concentrations that might occur in the leachate. Because of the differential heads (slope of the piezometric surface), the water containing dissolved substances moves into the groundwater system. The result is gross pollution of the groundwater.

Bioreactor Landfills

The implementation of RCRA Subtitle D resulted in more stringent protection of the environment, particularly the groundwater resources. The future trend in landfill design appears to be the development of engineered systems that optimize waste degradation and so minimize the amount of land needed for waste disposal. One technology that shows a lot of promise is bioreactor landfills. EPA has initiated a number of studies and partnerships with waste management companies to fully investigate the potential of this technology.

In traditional municipal solid waste landfills, organic waste eventually decomposes and stabilizes. These processes are controlled by microorganisms. In bioreactor landfills, biological decomposition is accelerated by enhancing the conditions necessary for these microorganisms to flourish. This is accomplished by the controlled addition of supplemental air and water. The degradation and stabilization of organic waste is then accelerated.

TABLE 11-9

Typical data of the composition of leachate from new and mature landfills

	Value, mg/L		
	New landfill (less than 2 years)		Mature landfill (greater than 10 years)
Constituent	Range	Typical	
BOD$_5$ (5-day biochemical oxygen demand)	2,000–30,000	10,000	100–200
TOC (total organic carbon)	1,500–20,000	6,000	80–160
COD (chemical oxygen demand)	3,000–60,000	18,000	100–500
Total suspended solids	200–2,000	500	100–400
Organic nitrogen	10–800	200	80–120
Ammonia nitrogen	10–800	200	20–40
Nitrate	5–40	25	5–10
Total phosphorus	5–100	30	5–10
Ortho phosphorus	4–80	20	4–8
Alkalinity as CaCO$_3$	1,000–10,000	3,000	200–1,000
pH (no units)	4.5–7.5	6	6.6–7.5
Total hardness as CaCO$_3$	300–10,000	3,500	200–500
Calcium	200–3,000	1,000	100–400
Magnesium	50–1,500	250	50–200
Potassium	200–1,000	300	50–400
Sodium	200–2,500	500	100–200
Chloride	200–3,000	500	100–400
Sulfate	50–1,000	300	20–50
Total iron	50–1,200	60	20–200

(*Source:* Tchobanoglous et al., 1993.)

EPA defines a bioreactor landfill as "any permitted Subtitle D landfill (under RCRA) or landfill cell where liquid or air is injected in a controlled fashion into the waste mass in order to accelerate or enhance biostabilization of the waste" (40 CFR 257 and 258, FR, 9 OCT 1991.) In these landfills additional moisture is introduced to the waste, typically by recirculating the leachate and adding additional moisture such as stormwater, wastewater, and wastewater treatment plant sludge. The goal is to provide enough moisture to the waste to maintain the optimal moisture content for microbial decomposition, typically 35 to 65 percent moisture.

One of the benefits of this system is that the decomposition rate is increased, so complete decomposition can occur in years instead of decades. The waste density is increased, so over the life of the landfill, 15 to 30 percent additional space is available. Also, the cost of the leachate disposal is reduced, because it is recirculated. And there is a significant increase in the landfill gas that is generated. If this is captured on-site, then it can be used to produce energy.

These systems have a higher initial cost to build and operate, because extensive recirculation and monitoring is required. Bioreactor landfills can be designed to use aerobic, anaerobic, or facultative microorganisms.

Phases of Bioreaction.   Five more or less sequential phases of bioreaction are thought to occur in a landfill. In the *initial adjustment phase,* the organic biodegradable components in the MSW undergo aerobic biodegradation because some air is trapped when the waste is placed in the landfill. In a conventional landfill, the principle source of microorganisms is the soil material that is used as daily and final cover. Digested wastewater treatment plant sludge as well as recycled leachate are also sources of microorganisms. In a bioreactor landfill, the latter sources provide a means of accelerating the decomposition process.

The second phase is called the *transitional phase.* Oxygen is depleted and anoxic and anaerobic conditions begin to develop. As the landfill becomes anaerobic, nitrate and sulfate serve as electron acceptors. Nitrogen, hydrogen, and hydrogen sulfide are products of the decomposition process. As the conversion process proceeds, the microbial community responsible for conversion of organic material to methane and carbon dioxide begin the three-step process described in Chapter 8 (Figure 8-35).

In the *acid phase,* the anaerobic microbial activity initiated in the second phase accelerates. Significant amounts of organic acids are produced and the production of hydrogen decreases. Carbon dioxide is the principle gas produced in this phase. The pH of the leachate will often drop to 5 or lower (Tchobanoglous et al., 1993).

The fourth phase is called the *methane fermentation phase. Methanogens* convert the acetic acid and hydrogen gas produced by the acid formers into methane (CH_4) and CO_2. The pH of the leachate rises to more neutral values in the range 6 to 8.

The *maturation phase* begins after the readily available biodegradable organic matter has been converted to CH_4 and CO_2. The rate of landfill gas generation decreases dramatically.

Volume of Gas Produced.   Cossu et al. (1996) present the following reaction representing the overall methane fermentation process:

$$C_aH_bO_cN_d + nH_2O \rightarrow x\,CH_4 + y\,CO_2 + w\,NH_3 = z\,C_5H_7O_2N + \text{energy} \quad (11\text{-}8)$$

where $C_aH_bO_cN_d$ is the empirical formula for the biodegradable organic matter and $C_5H_7O_2N$ is the empirical chemical formula of bacterial cells.

The maximum theoretical landfill gas yield (neglecting bacterial cell conversion) may be estimated as (Tchobanoglous et al., 1993):

$$C_aH_bO_cN_d + \left(\frac{4a - b - 2c + 3d}{4}\right)H_2O \rightarrow \left(\frac{4a + b - 2c - 3d}{8}\right)CH_4$$

$$+ \left(\frac{4a - b + 2c + 3d}{8}\right)CO_2 + dNH_3 \quad (11\text{-}9)$$

For the purpose of analysis, the MSW may be divided into two classes: rapidly biodegradable and slowly biodegradable. Food waste, newspaper, office paper, cardboard, leaves, and leafy yard trimmings fall into the first category. Textiles, rubber, leather, tree branches, and wood fall into the second category.

Tchobanoglous et al. (1993) developed empirical chemical formulas for typical U.S. MSW as collected in 1990 for each of these categories:

- Rapidly decomposable = $C_{68}H_{111}O_{50}N$

- Slowly decomposable = $C_{20}H_{29}O_9N$

These formulas may be used to estimate the maximum theoretical gas production. Actual quantities of gas will be lower because (1) all of the biodegradable organic matter is not available for decomposition, (2) the biodegradability is less for organic wastes with high lignin content, and (3) moisture may be limiting. The construction and operation of a bioreactor landfill is designed to minimize these limitations. Actual gas production rates from typical MSW landfills ranges from 40 to 400 m^3/Mg of MSW.

The rate of decomposition that is reflected in gas production of MSW is highly variable. Most models use first-order equations in two stages to describe the gas production as it rises to some peak value and then falls (Cossu et al., 1996).

Gas Flux.   The rate of evolution of gas from the landfill cover is called the *gas flux*. It may be estimated with the following equation:

$$F_A = \frac{D_A \, \eta^{4/3} \, (C_{A\text{-}atmos} - C_{A\text{-}fill})}{T}$$

(11-10)

Where F_A = gas flux of compound A, g/$m^2 \cdot$ min
η = landfill cover porosity
D_A = diffusion coefficient of compound A, m^2/min
$C_{A\text{-}atmos}$ = concentration of compound A at surface of landfill cover, g/m^3
$C_{A\text{-}fill}$ = concentration of compound A at bottom of landfill cover, g/m^3
T = depth of landfill cover, m

Landfill Design

The design of the landfill has many components including site preparation, buildings, monitoring wells, size, liners, leachate collection system, final cover, and gas collection system. Figure 11-14 shows a schematic of a typical municipal solid waste landfill with all of these components shown. In the following discussion we will limit ourselves to introductory consideration of the design of the size of the landfill, the selection of a liner system, the design of a leachate collection system, and a discussion of the final cover system.

Volume Required.   To estimate the volume required for a landfill, it is necessary to know the amount of refuse being produced and the density of the in-place, compacted refuse. The volume of refuse differs markedly from one city to another because of local conditions.

Salvato recommends a formula of the following form for estimating the annual volume required (Salvato, 1972).

$$V_{LF} = \frac{PEC}{D_c}$$

(11-11)

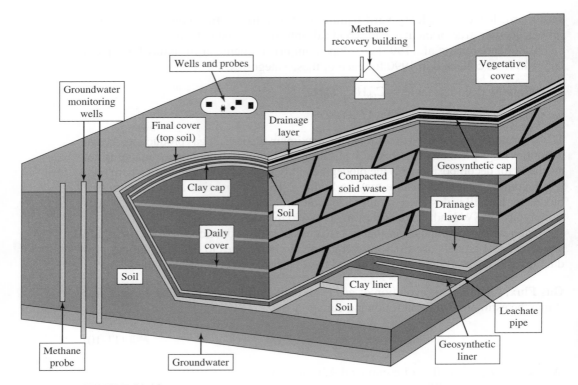

FIGURE 11-14
Schematic of a typical municipal solid waste landfill.
(*Source:* U.S. EPA, 1995.)

where V_{LF} = volume of landfill, m^3
$\quad P$ = population
$\quad E$ = ratio of cover (soil) to compacted fill

$$= \frac{V_{sw} + V_c}{V_{sw}}$$

$\quad V_{sw}$ = volume of solid waste, m^3
$\quad V_c$ = volume of cover, m^3
$\quad C$ = average mass of solid waste collected per capita per year, kg/person
$\quad D_c$ = density of compacted fill, kg/m^3

The density of the compacted fill is somewhat dependent on the equipment used at the landfill site and the moisture content of the waste. Compacted solid waste densities vary from 300 to 700 kg/m^3. Nominal values are generally in the range of 475 to 600 kg/m^3. The compaction ratios given in Table 11-10 may be used for estimating the density of the compacted fill.

TABLE 11-10
Typical compaction ratios[a]

Component	Poorly compacted	Normal compaction	Well-compacted
Food wastes	2.0	2.8	3.0
Paper	2.5	5.0	6.7
Cardboard	2.5	4.0	5.8
Plastics	5.0	6.7	10.0
Textiles	2.5	5.8	6.7
Rubber, leather, wood	2.5	3.3	3.3
Garden trimmings	2.0	4.0	5.0
Glass	1.1	1.7	2.5
Nonferrous metal	3.3	5.6	6.7
Ferrous metal	1.7	2.9	3.3
Ashes, masonry	1.0	1.2	1.3

[a] The ratio of the density after compaction to that as discarded, that is, before pickup by collection vehicle.
(*Source:* Tchobanoglous et al., 1977)

Example 11-4. How much landfill space does Watapitae require for 20 years of operation? Assume that the village will use a cell height of 2.4 m and that it will follow normal practice and use 0.15 m of soil for daily cover; 0.3 m to complete the cell; and a final cover of 0.6 m for every stack of three cells. Assume that compaction will be "normal."

Solution. Although we do not know the population or per capita waste generation rate, we can estimate the mass generated per year from other data. From Example 11-1 we know that 1,250 service stops must be collected each week. From Example 11-2 we know that each service stop contributes an average of 0.624 Mg per year. Then the annual mass generation rate is

$$\text{Mass} = (1,250 \text{ stops}) \times (0.624 \text{ Mg/y stop}) = 780 \text{ Mg/y}$$

This is equivalent to the product $(P)(C)$ in Equation 11-11.

In Example 11-1 we determined that the mean density of the uncompacted solid waste was 106 kg/m^3. Using the fractional mass composition of the waste as given in Table 11-4 and the "normal" compaction ratios in Table 11-10, we can determine the weighted compaction ratio by multiplying the fractional mass by the compaction ratio (Table 11-11).

With a compaction ratio of 4.18, the density of the compacted fill is estimated to be

$$D_c = (106 \text{ kg/m}^3) \times (4.18) = 443 \text{ kg/m}^3 \text{ or } 0.443 \text{ Mg/m}^3$$

Note that this implies that waste dumped at the face of the fill in a 1.25-m layer would have to be compressed to a depth of 0.3 m, that is,

$$\left(\frac{1}{4.18}\right)(1.25 \text{ m})$$

TABLE 11-11
Weighted compaction ratios for Example 11-4

Component	Mass fraction	Weighted compaction ratio
Food wastes	0.0947	0.27
Paper	0.4317	2.16
Cardboard	0.0650	0.26
Plastics	0.0181	0.12
Textiles	0.0020	0.01
Rubber	—	—
Leather	0.0150	0.05
Garden trimmings	0.1432	0.57
Wood	0.0350	0.12
Glass	0.0749	0.12
Tin cans	0.0520	0.29
Nonferrous metals	0.0150	0.08
Ferrous metals	0.0430	0.12
Dirt, ashes, brick	0.0110	0.01
Total	1.0006	4.18

Before we can estimate E, we must determine the daily volume of solid waste and the area over which it will be spread. For a five-day week, the daily volume is determined as follows:

$$V = \frac{780 \text{ Mg/y}}{0.443 \text{ Mg/m}^3} \times \frac{1}{52 \text{ wk/y}} \times \frac{1}{5 \text{ d/wk}} = 6.77 \text{ m}^3/\text{d}$$

If this is spread in a 0.3-m layer, then the area would be

$$\frac{6.77 \text{ m}^3}{0.3 \text{ m}} = 22.57 \text{ m}^2/\text{d}$$

This is equivalent to a square 4.75 m on each side. This seems reasonable for a small community.

If 0.15 m of soil is used as cover each day, then 0.45 m will be placed each day and it will take

$$\frac{2.4 \text{ m} - 0.15 \text{ m}}{0.45 \text{ m/day}} = 5.00 \text{ days}$$

to complete the cell. (The 0.15 m is the addition to daily cover to complete the cell with 0.3 m of cover.) At this rate we will complete a stack of three cells every three weeks (15 working days).

The soil volume separating a stack of three cells will be about

$$0.3 \text{ m thick} \times 2.4 \text{ m high} \times 4.75 \text{ m long} \times 3 \text{ cells} = 10.26 \text{ m}^3$$

To account for two sides of the cell, this number needs to be multiplied by two.

$$10.26 \text{ m}^3 \times 2 = 20.52 \text{ m}^3$$

If we ignore this volume, E can be calculated as

$$E = \frac{0.3 + (0.15 + 0.03 + 0.02)}{0.3} = 1.67$$

The terms in the brackets account for the daily cover of 0.15 m; the cell cover of an additional 0.15 m each five days or 0.03 m per day; and the final stack cover of an additional 0.3 m to the three-cell cover each 15 days or 0.02 m per day.

If we do not ignore the soil separating the cells, then the soil volume per stack of three cells as shown in Figure 11-15 is calculated as follows:

$$(3 \text{ cells/stack})(5 \text{ lifts/cell})(22.57 \text{ m}^2)(0.15 \text{ m}) = 50.78 \text{ m}^3$$

plus the 0.15 m of additional soil to bring the weekly cell cover to 0.30 m is

$$(3 \text{ cells/stack})(22.57 \text{ m}^2)(0.15 \text{ m}) = 10.16 \text{ m}^3$$

plus the additional 0.3 m to bring the final cover to 0.6 m,

$$(22.57 \text{ m}^2)(0.3 \text{ m}) = 6.77 \text{ m}^3$$

The total soil volume, including the 20.52 m³ for the sides of the stack, is

$$50.78 + 10.16 + 6.77 + 20.52 = 88.23 \text{ m}^3$$

The value for V_{sw} would then be

$$V_{sw} = (6.77 \text{ m}^3/\text{d})(15 \text{ d/stack}) = 101.55 \text{ m}^3/\text{stack}$$

The value for E would then be

$$E = \frac{101.55 + 88.23}{101.55} = 1.87$$

Thus, for this landfill, the separation wall will increase the volume by about 12 percent. This is not insignificant!

The estimated volume requirement for 20 years would be

$$V_{LF} = \frac{(780 \text{ Mg/y})(1.87)}{0.443 \text{ Mg/m}^3} \times 20 \text{ y} = 6.59 \times 10^4 \text{ m}^3$$

Since the average landfill depth will be three 2.4 m cells plus an additional 0.3 m final cover, the area will be

$$A_{LF} = \frac{6.59 \times 10^4}{(3)(2.4) + 0.3} = 8.78 \times 10^3 \text{ m}^2$$

An area approximately 100 m on a side would do very nicely.

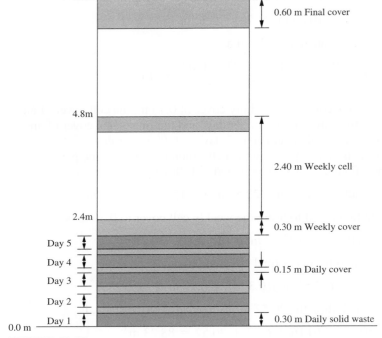

FIGURE 11-15

Schematic diagram of MSW landfill stack of three cells (Example 11-4).

Liner Selection.   In order to prevent groundwater contamination, strict leachate control measures are required. Under the 1991 Subtitle D rules promulgated by EPA, new landfills must be lined in a specific manner or meet maximum contaminant levels for the groundwater at the landfill boundary. The specified liner system includes a synthetic membrane (*geomembrane*) at least 30 mils (0.76 mm) thick supported by a compacted soil liner at least 0.6 m thick. The soil liner must have a hydraulic conductivity of no more than 1×10^{-7} cm/s. Flexible membrane liners consisting of high-density polyethylene (HDPE) must be at least 60 mils thick (40 CFR 257 and 258, and FR 9 OCT 1991). A schematic of the EPA specified liner system is shown in Figure 11-16.

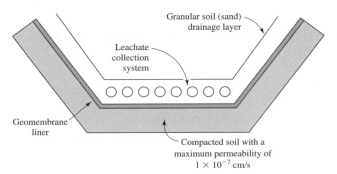

FIGURE 11-16

A composite liner and leachate collection system.

Several geomembrane materials are available. Some examples include polyvinyl chloride (PVC), high-density polyethylene (HDPE), chlorinated polyethylene (CPE), and ethylene propylene diene monomer (EPDM). Designers show a strong preference for PVC and especially for HDPE. Although the geomembranes are highly impermeable (hydraulic conductivities are often less than 1×10^{-12} cm/s), they can be easily damaged or improperly installed. Damage may occur during construction by construction equipment, by failure due to tensile stress generated by the overburden, tearing as a result of differential settling of the supporting soil, puncture from sharp objects in the overburden, puncture from coarse aggregate in the supporting soil, and tearing by landfill equipment during operation. Installation errors primarily occur during seaming when two pieces of geomembrane must be attached or when piping must pass through the liner. A liner placed with adequate quality control should have less than 3 to 5 defects per hectare.

The soil layer under the geomembrane acts as a foundation for the geomembrane and as a backup for control of leachate flow to the groundwater. Compacted clay generally meets the requirement for a hydraulic conductivity of less than 1×10^{-7} cm/s. In addition to having a low permeability, it should be: free of sharp objects greater than 1 cm in diameter, graded evenly without pockets or hillocks, compacted to prevent differential settlement, and free of cracks.

Leachate Breakthrough. Historically, landfill liners were constructed with only a single clay liner. Over time the leachate will pass through the liner. This is called *breakthrough*. The following equation may be used to estimate the time to breakthrough:

$$t = \frac{T^2 \eta}{K(H + T)} \qquad \text{🏳️} \ (11\text{-}12)$$

Where t = breakthrough time, y
T = thickness of the clay liner, m
η = clay liner porosity
K = hydraulic conductivity, m/y
H = depth of leachate above liner (also called "head"), m

Leachate Collection. Under the 1991 Subtitle D rules promulgated by EPA, the leachate collection system must be designed so that the depth of leachate above the liner does not exceed 0.3 m. The leachate collection system is designed by sloping the floor of the landfill to a grid of underdrain pipes* that are placed above the geomembrane. A 0.3-m-deep layer of granular material (for example, sand) with a high hydraulic conductivity (EPA recommends greater than 1×10^{-2} cm/s) is placed over the geomembrane to conduct the leachate to the underdrains. In addition to carrying the leachate, this layer also protects the geomembrane from mechanical damage from equipment and solid waste. In some instances a geonet (a synthetic matrix that resembles a miniature chain link fence), with a geofabric (an open-weave cloth) protective layer to keep out the sand, is placed under the sand and above the geomembrane to increase the flow of leachate to the pipe system.

---

*Underdrain pipes are perforated pipes designed to collect the leachate.

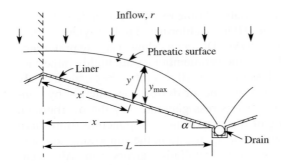

FIGURE 11-17
Geometry and symbols for calculating Y_{max}. (*Source:* McEnroe, 1993.)

Several different methods for estimating the steady-state maximum leachate depth have been proposed. EPA has proposed the following formula (refer to Figure 11-17 for an explanation of the notation) (U.S. EPA, 1989):

$$y_{max} = L\left(\frac{r}{2K}\right)^{0.5}\left[\frac{KS^2}{r} + 1 - \frac{KS}{r}\left(S^2 + \frac{r}{K}\right)^{0.5}\right] \qquad (11\text{-}13)$$

where y_{max} = maximum saturated depth, m
$\quad L$ = drainage distance, measured horizontal, m
$\quad r$ = vertical flow rate per unit horizontal area, $m^3/s \cdot m^2$
$\quad K$ = hydraulic conductivity of drainage layer, m/s
$\quad S$ = slope of liner ($= \tan \alpha$)

This formula may overestimate the value of y_{max} where the underdrain system has free drainage, that is, it is not undersized or clogged. Because this is commonly the case, McEnroe has proposed the following equations as a better approximation. (McEnroe, 1993):

$$Y_{max} = (R - RS + R^2S^2)^{0.5}\left[\frac{(1 - A - 2R)(1 + A - 2RS)}{(1 + A - 2R)(1 - A - 2RS)}\right]^{0.5A} \qquad (11\text{-}14)$$

for $R < 1/4$;

$$Y_{max} = \frac{R(1 - 2RS)}{1 - 2R}\exp\left[\frac{2R(S - 1)}{(1 - 2RS)(1 - 2R)}\right] \qquad (11\text{-}15)$$

for $R = 1/4$;
and

$$Y_{max} = (R - RS + R^2S^2)^{0.5}\exp\left[\frac{1}{B}\tan^{-1}\left(\frac{2RS - 1}{B}\right) - \frac{1}{B}\tan^{-1}\left(\frac{2R - 1}{B}\right)\right] \qquad (11\text{-}16)$$

for $R > 1/4$;

where $Y_{max} = y_{max}/(L \tan \alpha)$
$\quad R = r/(K \sin^2 \alpha)$
$\quad S$ = slope of liner ($= \tan \alpha$)
$\quad A = (1 - 4R)^{0.5}$
$\quad B = (4R - 1)^{0.5}$

The collected leachate must be treated because of the high concentration of pollutants it contains. In some instances on-site treatment is provided. This frequently is a biological treatment system. In other cases, the leachate may be pumped to a municipal treatment plant. In some recent designs, the leachate is recirculated through the landfilled waste. This provides moisture for the microbial population and accelerates the stabilization process. It also promotes the production of methane and provides some treatment for the biodegradable fraction of the constituents in the leachate.

Final Cover.   The major function of the final cover is to prevent moisture from entering the finished landfill. If no moisture enters, then at some point in time the leachate production will reach minimal proportions and the chance of groundwater contamination will be minimized.

Modern final cover design consists of a surface layer, biotic barrier, drainage layer, hydraulic barrier, foundation layer, and gas control. The surface layer is to provide suitable soil for plants to grow. This minimizes erosion. A soil depth of about 0.3 m is appropriate for grass. The biotic barrier is to prevent the roots of the plants from penetrating the hydraulic barrier. At this time, there does not seem to be a suitable material for this barrier. The drainage layer serves the same function here as in the leachate collection system—that is, it provides an easy flow path to a grid of perforated pipes. This collection piping system is subject to differential settling and may fail because of this settling. Some designers do not recommend installing it as they prefer to use the funds to develop a thicker hydraulic barrier. The hydraulic barrier serves the same function as the liner in that it prevents movement of water into the landfill. The EPA recommends a composite liner consisting of a geomembrane and a low hydraulic conductivity soil that also serves as the foundation for the geomembrane. This soil also protects the geomembrane from the rough aggregate in the gas control layer. The gas control layer is constructed of coarse gravel that acts as a vent to carry the gases to the surface. If the gas is to be collected for its energy value, a series of gas recovery wells is installed. A negative pressure is placed on these wells to draw the gas into the system.

Completed MSW Landfills

Completed landfills generally require maintenance because of uneven settling. Maintenance consists primarily of regrading the surface to maintain good drainage and filling in small depressions to prevent ponding and possible subsequent groundwater pollution. The final soil cover should be about 0.6 m deep.

Completed landfills have been used for recreational purposes such as parks, playgrounds, or golf courses. Parking and storage areas or botanical gardens are other final uses. Because of the characteristic uneven settling and gas evolution from landfills, construction of buildings on completed landfills should be avoided.

On occasion, one-story buildings and runways for light aircraft might be constructed. In such cases, it is important to avoid concentrated foundation loading, which can result in uneven settling and cracking of the structure. The designer must provide the means for the gas to dissipate into the atmosphere and not into the structure.

11-5 WASTE TO ENERGY

Utilization of the organic fraction of solid waste for fuel, while simultaneously reducing the volume, may be an important part of an integrated waste management plan. Specially designed power plants known as waste-to-energy facilities can produce energy through the combustion of municipal solid waste. In these facilities, trash volume is reduced by 90 percent and its weight by 75 percent. The remaining residue is disposed of in a MSW landfill. According to a 2004 Integrated Waste Services Association publication, 89 waste-to-energy facilities were in operation as of that time, disposing of 86 Gg of waste each day (IWSA, 2004). This waste was converted to approximately 2,500 megawatts of electric power.

Heating Value of Waste

The heating value of waste is measured in kilojoules per kilogram (kJ/kg), and is determined experimentally using a bomb calorimeter. A dry sample is placed in a chamber and burned. The heat released at a constant temperature of 25°C is calculated from a heat balance. Because the combustion chamber is maintained at 25°C, combustion water produced in the oxidation reaction remains in the liquid state. This condition produces the maximum heat release and is defined as the *higher heating value* (HHV).

In actual combustion processes, the temperature of the combustion gas remains above 100°C until the gas is discharged into the atmosphere. Consequently, the water from actual combustion processes is always in the vapor state. The heating value for actual combustion is termed the *lower heating value* (LHV). The following equation gives the relationship between HHV and LHV:

$$LHV = HHV - [(\Delta H_v)(9\,H)] \tag{11-17}$$

where ΔH_v = heat of vaporization of water
          = 2,420 kJ/kg
      H = hydrogen content of combusted material

The factor of 9 results because one gram mole of hydrogen will produce 9 gram moles of water (that is, 18/2). Note that this water is only that resulting from the combustion reaction. If the waste is wet, the free water must also be evaporated. The energy required to evaporate this water may be substantial. This results in a very inefficient combustion process from the point of view of energy recovery. The ash content also reduces the energy yield because it reduces the proportion of dry organic matter per kilogram of fuel and because it retains some heat when it is removed from the furnace.

Fundamentals of Combustion

Combustion is a chemical reaction where the elements in the fuel are oxidized. In waste-to-energy (WTE) plants, the fuel is, of course, the solid waste. The major oxidizable elements in the fuel are carbon and hydrogen. To a lesser extent sulfur and nitrogen are also present. With complete oxidation, carbon is oxidized to carbon dioxide, hydrogen to water, and sulfur to sulfur dioxide. Some fraction of the nitrogen may be oxidized to nitrogen oxides.

The combustion reactions are a function of oxygen, time, temperature, and turbulence (O, T, T, T). There must be a sufficient excess of oxygen to drive the reaction to completion in a short period of time. The oxygen is most frequently supplied by forcing air into the combustion chamber. Over 100 percent excess air may be provided to ensure a sufficient excess. Sufficient time must be provided for the combustion reactions to proceed. The amount of time is a function of the combustion temperature and the turbulence in the combustion chamber. Some minimum temperature must be exceeded to initiate the combustion reaction (that is, to ignite the waste). Higher temperatures also yield higher quantities of nitrogen oxide emissions, so there is a tradeoff in destroying the solid waste and forming air pollutants. Mixing of the combustion air and the combustion gases is essential for completion of the reaction.

As the solid waste enters the combustion chamber and its temperature increases, volatile materials are driven off as gases. Rising temperatures cause the organic components to thermally "crack" and form gases. When the volatile compounds are driven off, fixed carbon remains. When the temperature reaches the ignition temperature of carbon (700°C), it is ignited. To achieve destruction of all the combustible material (*burnout*), it is necessary to achieve 700°C throughout the bed of waste and ash (Pfeffer, 1992).

The flame zone is that area where the hot volatilized gases mix with oxygen. This reaction is very rapid. It goes to completion within 1 or 2 seconds if there is sufficient excess air and turbulence.

The evolution of solid waste combustion has led to higher temperatures both to destroy toxic compounds and to increase the opportunity to utilize the waste as an energy source by producing steam.

Conventional Incineration

The basic arrangement of the conventional incinerator is shown in Figure 11-18. Although the solid waste may have some heat value, it is normally quite wet and is not *autogenous* (self-sustaining in combustion) until it is dried. Conventionally, auxiliary fuel is provided for the initial drying stages. Because of the large amount of particulate matter generated in the combustion process, some form of air pollution control device is required. Normally, electrostatic precipitators or scrubbers are chosen. Bulk volume reduction in incinerators is about 90 percent. Thus, about 10 percent of the material still must be carried to a landfill.

Recovering Energy from Waste

In order to utilize the heat value of solid waste, most modern combustion devices are designed to recover the energy. The concept is more than 100 years old. The first refuse-to-electricity system was built in Hamburg, Germany, in 1896. In 1903, the first of several solid waste-fired electricity generating plants in the United States was installed in New York City.

There are now many WTE plants operating in the United States. They burn solid waste in a specially designed incinerator furnace jacketed with water-filled tubes to recover the heat as steam. The steam may be used directly for heating or to produce electricity.

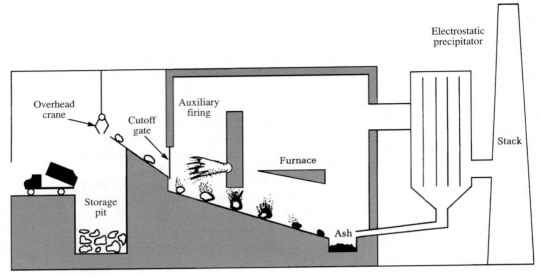

FIGURE 11-18
Schematic of a conventional traveling grate incinerator.

Many states require public utilities to buy the electricity produced at these plants. With efficient heat recovery and electric generators, WTE plants can produce about 600 kWh per mg of waste.

Refuse-Derived Fuel (RDF). Refuse-derived fuel is the combustible portion of solid waste that has been separated from the noncombustible portion through processes such as shredding, screening, and air classifying (Vence and Powers, 1980). By processing municipal solid waste (MSW), refuse-derived fuel containing 12 to 16 MJ/kg can be produced from between 55 and 85 percent of the refuse received. This system is also called a supplemental fuel system because the combustible fraction is typically marketed as a fuel to outside users (utilities or industries) as a supplement to coal or other solid fuels in their existing boilers.

In a typical system, MSW is fed into a trommel or rotating screen to remove glass and dirt, and the remaining fraction is conveyed to a shredder for size reduction. Shredded wastes may then pass through an air classifier to separate the "light fraction" (plastics, paper, wood, textiles, food wastes, and smaller amounts of light metals) from the "heavy fraction" (metals, aluminum, and small amounts of glass and ceramics).

The light fraction, after being routed through a magnetic system to remove ferrous metals, is ready for fuel use. The heavy fraction is conveyed to another magnetic removal system for recovery of ferrous metals. Aluminum may also be recovered. The remaining glass, ceramics, and other nonmagnetic materials from the heavy fraction are then sent to the landfill.

The first full-scale plant to prepare RDF has been in operation in Ames, Iowa, since 1975. Subsequently, other plants using similar technology have been designed and constructed. Figure 11-19 shows the process flow diagram for the Southeastern Virginia Public Service Authority's RDF plant.

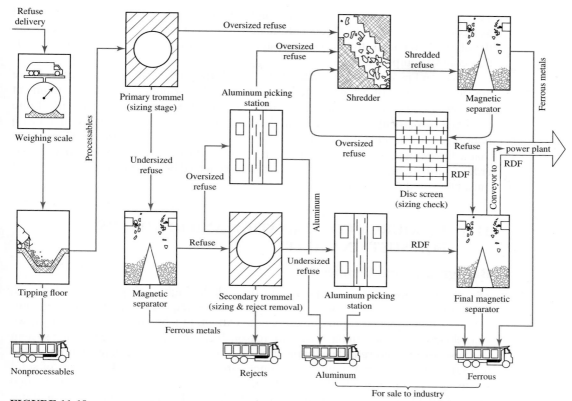

FIGURE 11-19
Southeastern Virginia Public Service Authority's refuse-derived fuel (RDF) plant.

Although there are a number of RDF production systems operating or starting up, they are still developmental in terms of process, equipment, and application. Data still are being gathered for prediction of performance and maintenance requirements.

Modular Incinerators. These units are available in various sizes. Their modularity enables them to be coupled with similar units to process available tonnage.

Most modular incinerators that produce energy incorporate a controlled air principle, use unprocessed MSW, and require a small amount of auxiliary fuel for startup. The waste is fed into a primary chamber where it is burned in the absence of sufficient oxygen for complete combustion. The resulting combustible gas passes through a second chamber, where excess air is injected, completing combustion. Auxiliary fuel may also be required in minimal quantities to maintain proper combustion temperatures.

After most of the particulate matter burns off, the hot effluent passes through a waste heat boiler to produce steam. The ash is water-quenched and disposed of at a landfill. The steam can be used directly or can be converted to electricity with the addition of a turbine generator.

The newer waste-to-energy plants are not without their problems. Serious concern has been raised about emissions of dioxins that result from the combustion process.

Two approaches are used to reduce the dioxin emission. Because the dioxin is formed as a combustion by-product from chlorinated plastics, it can be minimized by reducing the plastic in the feed stream. The second approach is to utilize sophisticated air pollution control equipment.

A second problem is associated with the ash from the combustion process. There are two categories of ash generated: fly ash from the air pollution control equipment and bottom ash from the furnace. Fly ash is of greater concern because the metals are adsorbed on particulates and are easily leached with water. When fly ash is mixed with bottom ash, the leachability of the metals is reduced. In 1994, the Supreme Court ruled that ash from municipal incinerators is not excluded from being considered as a hazardous waste (Chicago vs. EDF, 1994). It must be tested before it can be landfilled and must be treated if it fails the tests.

11-6 RESOURCE CONSERVATION AND RECOVERY FOR SUSTAINABILITY

Background and Perspective

The earth's prime mineral deposits are limited. As high-quality ores are depleted, lower-grade ores must be used. Lower-grade ores require proportionately greater amounts of energy and capital investment to extract. In a broad economic context, we should view with concern the long-term reasonableness of a market-accounting system that applies only current development costs to our use of depletable, nonrenewable natural resources such as aluminum, copper, iron, and petroleum. High rates of solid waste production imply high rates of virgin raw material extraction. In the United States, blatant mispricing—including the "depletion allowance" on minerals and unreasonably low rail rate fares on ores in contrast to scrap—is in no small way responsible for this state of affairs. Furthermore, our high-waste, low-recycle lifestyle is inherently wasteful of a bountiful endowment of natural resources.

Our renewable resources, primarily timber, are also under siege. Our prepackaged society, in combination with a wanton lack of care in our forests, has strained nature's capacity for growth and replenishment. Europe, India, and Japan have long been faced with a want of timber. We in the United States should learn from their predicaments.

The prevention of waste generation (resource conservation) and the productive use of waste material (resource recovery) represent means of alleviating some of the problems of solid waste management. At one time in our history, resource recovery played an important role in our industrial production. Until the mid-twentieth century, salvage (recovery and recycling) from household wastes was an important source of materials. In the five years preceding 1939, recycled copper, lead, aluminum, and paper supplied 44, 39, 28, and 30 percent, respectively, of the total raw materials shipments to fabricators in the United States (NCRR, 1974). Ultimately, it became more economical to process virgin materials than to use recovered materials.

In principle, processable municipal solid waste could provide 95 percent and 73 percent of our nation's needs in glass and paper, respectively. EPA estimates that overall, 30 percent of municipal solid waste was recovered in 2003. This represents an increasing trend. Table 11-12 shows the trend in recycling and reuse from 1960 to

TABLE 11-12

Generation, materials recovery, composting, and discards of municipal solid waste, 1960–2003[a, b]

	1960	1970	1980	1990	2000	2003
Generation	80.1	110.1	137.8	186.6	212.8	214.8
Recovery for recycling	5.1	7.3	13.2	26.4	47.6	50.4
Recovery for composting[c]				3.8	15.0	15.4
Total materials recovery	**5.1**	**7.3**	**13.2**	**30.2**	**62.6**	**65.7**
Discards after recovery	75.0	102.7	124.7	156.4	150.1	149.0

[a]*Source:* U.S. EPA, 2003.   [b]In teragrams (Tg).

[c]Composting of yard trimmings, food scraps, and other MSW organic material. Does not include backyard composting. Details may not add because of rounding.

2003 in millions of tons of waste. In 2003, 65.7 Tg million tons of waste were diverted from landfills by recycling and composting.

Table 11-13 shows a breakdown of recovered waste by 1=1 product in 2003. EPA estimates that during 2003, nearly 39 percent of containers and packaging were recycled. About 44 percent of aluminum beverage cans were recycled, as well as 48 percent of paper and paperboard, 22 percent of glass containers, and 8 percent of plastic packaging and containers. Newspapers, the most recycled product, were recycled at a rate of about 82 percent, while used telephone books were recycled at a rate of only 16 percent.

Recycling of municipal solid waste for profit or for energy recovery is rarely cost-effective. However, many communities have initiated recycling programs as a means of protecting the environment. Citizens have become increasingly aware of their role in protecting the natural environment, and so demand that communities offer recycling services. EPA has also set national goals to encourage active resource conservation and recovery programs.

Most states and the District of Columbia have enacted laws on recycling ranging from purchasing preferences to comprehensive recycling goals. Over 8,000 curbside recycling programs, 3,000 composting programs, and 200 municipal recycling facilities are in operation (Wolpin, 1994, and U.S. EPA, 2003). The recyclable market continues to fluctuate dramatically. For example, the price of old newsprint fell from $50/Mg in 1988 to less than $10/Mg in 1993 (Rogoff and Williams, 1995). It rose to over $100/Mg in 1995 (Paul, 1995).

The remainder of our discussion will be devoted to the technical details of several of the more promising resource conservation and recovery (RC & R) techniques. We have divided these into three broad categories entitled low technology, medium technology, and high technology. These categories refer to increasing degrees of sophistication in terms of implementation, equipment, and capital investment. No municipal government should be enticed into any one of these schemes with the hope of making money. The best that can be hoped for is defraying the additional costs over conventional landfilling and extending the life of the landfill by some modest amount. In some cases, even these modest goals may not be achieved.

TABLE 11-13
Generation and recovery of products in MSW by material 2003[a,b]

	Mass generated[c]	Mass recovered[c]	Recovery as a percent of generation
Durable goods			
Steel	10.16	3.06	30.2
Aluminum	0.96	Neg.[f]	Neg.
Other nonferrous metals[d]	1.44	0.96	66.7
Total metals	12.52	4.02	32.1
Glass	1.61	Neg.	Neg.
Plastics	7.61	0.30	3.9
Rubber and leather	5.36	1.00	18.6
Wood	4.78	Neg.	Neg.
Textiles	2.75	0.29	10.6
Other materials	1.18	0.89	75.4
Total durable goods	35.83	6.50	18.1
Nondurable goods			
Paper and paperboard	40.19	16.42	40.8
Plastics	5.76	Neg.	Neg.
Rubber and leather	0.80	Neg.	Neg.
Textiles	6.69	1.09	16.3
Other materials	2.96	Neg.	Neg.
Total nondurable goods	56.34	17.51	31.0
Containers and packaging			
Steel	2.58	1.56	60.6
Aluminum	1.76	0.63	35.6
Total metals	4.34	2.19	50.4
Glass	9.71	2.13	22.0
Paper and paperboard	35.20	19.87	56.4
Plastics	10.80	0.96	8.9
Wood	7.58	1.16	15.3
Other materials	0.20	Neg.	Neg.
Total containers and packaging	67.86	26.31	38.8
Other wastes			
Food, other[e]	25.04	0.68	2.7
Yard trimmings	25.95	14.61	56.3
Miscellaneous inorganic wastes	3.28	Neg.	Neg.
Total other wastes	54.25	15.33	28.2
Total MSW	214.28	65.59	30.6

[a]*Source:* U.S. EPA, 2003.
[b]Includes waste from residential, commercial, and institutional sources.
[c]In teragrams (Tg).
[d]Includes lead from lead-acid batteries.
[e]Includes recovery of other MSW organic material for composting.
[f]Neg. = negligible.

Low Technology RC & R

Returnable Beverage Containers.   The substitution of reusable products for single-use "disposable" products is a workable means of conserving natural resources. Legislation requiring mandatory refunds and/or deposits on both returnable and nonreturnable beverage containers has been and will continue to be hotly contested by the beverage and beverage container industries. States that have enacted mandatory refund and/or deposit legislation include California, Connecticut, Delaware, Hawaii, Maine, Massachusetts, Michigan, New York, Oregon, and Vermont as of 2002. The programs are successful in encouraging recycling of containers. Between 90 and 95 percent of the bottles are returned and between 80 and 85 percent of the cans are returned. In Oregon, a reduction in total roadside litter of 39 percent by item count and 47 percent by volume was reported after the second year of implementation of its law. Furthermore, for glass containers there is a significant energy savings in that a glass bottle reused 10 times consumes less than one-third of the energy of a single-use container. Average reuse cycles vary from 10 to 20 times per container.

Recycling.   The reprocessing of wastes to recover an original raw material was formerly called *salvage* and is now called *recycling*. At its lowest and most appropriate technological level, the materials are separated at the source by the consumer (*source-separation*). This is the most appropriate level because it requires the minimum expenditure of energy. With stringent goals for recycling, municipalities are looking at detailed recycling options.

Generally, the recycling options available to a municipality for residential use include:

Curbside collection

Drop-off centers

Material processing facility

Material transfer stations

Leaf/yard waste compost

Bulky waste collection and processing

Tire recovery

The primary method of recycling in the United States today is curbside collection. This has the advantage of being easier on the resident than having to drive to a recycling center. There are two basic types of curbside collection for recycling. In the first, the homeowner is given a number of bins or bags. The homeowner separates the refuse as it is used, placing it in the appropriate bin. On collection day the container is placed on the curb. The primary disadvantage of supplying home storage containers is the cost, which can represent a significant investment. A second method of curbside recycling is to provide the homeowner with only one bin, into which is placed all the recyclable materials. Curbside personnel then separate material as it is being picked up, placing each type of material into a separate compartment in the vehicle.

A second alternative is a drop-off center. Because recycling is a community-specific operation, a drop-off system must be designed around and in consideration of conditions particular to the area of involvement. To evaluate and select the most appropriate drop-off system, we must consider critical factors such as location, materials handled, population, number of centers, operation, and public information. When drop-offs are used to supplement curbside programs, fewer and smaller drop-off sites may be required. When drop-off sites are the only, or primary, recycling system in a community, the system must provide for increased capacity. Careful planning to accommodate traffic flow, as well as storage and collection of materials, must be part of the siting activity.

The convenience of a drop-off center will directly affect the amount of citizen participation. Strategically locating a drop-off center in an area of high traffic flow, where the center is highly visible, will encourage a greater level of participation. Even rural areas with widely scattered populations provide good locations for drop-offs. Rural homeowners have certain common travel patterns that bring them to a few locations at regular intervals—to a grocery store, church, or post office, to name a few. Figure 11-20 shows an example of a drive-through material recycling center.

A third major type of recycling is a materials recovery facility. In this case the recyclable material is taken by the municipality to a central facility where the material is separated via mechanical and labor-intensive means. Figure 11-21 shows an example layout of a separation facility and Figure 11-22 shows a mass balance of what can be expected at such a facility.

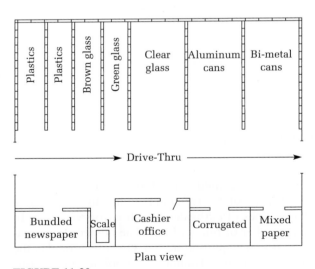

FIGURE 11-20
Enclosed drive-through drop-off center.

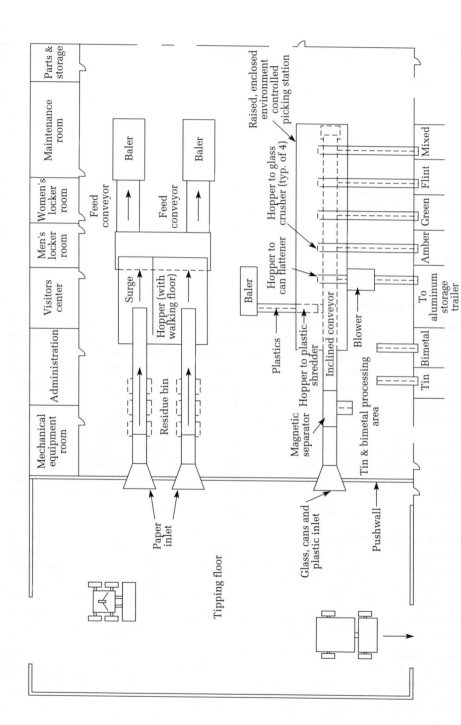

FIGURE 11-21
Material processing conceptual floor plan.

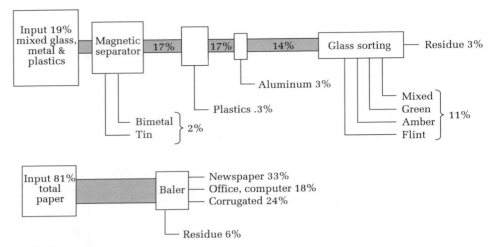

FIGURE 11-22
Material recovery facility process mass flow.

Medium Technology RC & R

Product Design. Simple changes in product configuration or packaging can result in conservation of resources. Three examples will suffice to illustrate the concept. In the mid-1970s several newspapers (for example, *Los Angeles Times, Washington Post,* and *New York Times*) switched from a traditional eight-column format to a new six-column format for news and nine-column format for advertising. This shift resulted in a 5 percent reduction in the amount of newsprint consumed. A large retail grocery store found that it could eliminate the custom of double bagging groceries by using a slightly heavier-weight bag with a reinforced bottom. This resulted in a 30 percent savings in the amount of fiber consumed. Many fast-food restaurants eliminated styrofoam containers for their sandwiches and now use paper wrapping, which is more readily biodegraded.

These kinds of changes are generally beyond the scope of the environmental engineer. However, their use can be encouraged, and purchases can be made that support those who use environmentally conservative packages and products.

Shredding and Separation. As a first step in a medium technology system or as an add-on to a landfill volume enhancement program, some materials may be reclaimed at a central processing point. The most likely candidates for recycling are paper, non-ferrous metals (for example, aluminum), and ferrous metals. Paper generally is removed by hand as the MSW passes along on a conveyor belt.* After passing through a shredder, ferrous metals can be removed using a magnetic separator. In large communities, where more than 1,000 Mg/wk of MSW is collected, some consideration may

---

*Depending upon the economy, hand sorting may be a losing proposition. An average worker can pick about 2.0 Mg of newspaper in an eight-hour day. At a wage of $5.50/h, a day's wages amount to $44.00, exclusive of overhead and fringe benefits. Using an overhead rate of 100 percent, the cost of sorting is $44.00/Mg. If the price for No. 6 newsprint (a grade of paper) is $22/Mg as it was in 1994, this is a loss of $22/Mg before transportation costs are deducted. Of course, in 1995, when the price was $116/Mg, it was a winning proposition.

be given to the separation and shredding of auto and truck tires. Asphaltic concrete plants may be able to use the shredded tires in their raw material feedstock. Because tires are troublesome at landfills (because no matter how deep they are buried, they often pop up to the surface), their recovery as a resource is doubly beneficial.

Composting. Compost is a humus-like material that results from the aerobic biological stabilization of the organic materials in solid waste. The most effective composting occurs when the waste stream is free of inorganic materials. Frequently, this makes source-separated yard waste ideal. For the biological process to be effective, the following conditions must be met (Tchobanoglous et al., 1993).

1. Particle size must be small (< 5 cm).

2. Aerobic conditions must be maintained by turning the compost pile or forcing air through it.

3. Adequate, but not excessive, moisture must be present (50 to 60 percent).

4. An adequate population of acclimated microorganisms must be present.

5. The carbon-to-nitrogen ratio must be in the range of 20–25 to 1.

The biodegradation process is exothermic and a well-operating compost will have a temperature between 55 and 60°C during the period of active degradation. These temperatures are effective in destroying pathogens. The processing cycle for composting is about 20 to 25 days with active degradation taking place over a 10- to 15-day period. One of the major drawbacks of composting is odors. Maintenance of aerobic conditions and a proper cure time minimize odor problems.

Compost is useful as a soil conditioner. In this role compost will: (1) improve soil structure, (2) increase moisture-holding capacity, (3) reduce leaching of soluble nitrogen, and (4) increase the buffer capacity of the soil. It should be emphasized that compost is not a valuable fertilizer. It contains only 1 percent or less of the major nutrients, such as nitrogen, phosphorus, and potash.

Composting is one of the fastest-growing aspects of ISWM. The driving force is legislation enacted to extend the life of landfills by removing yard waste from the waste stream. According to the EPA, recovery by composting was negligible in 1988. By 1990, EPA estimated that 2 percent of the nation's solid waste was being composted. The 2000 estimate was that 7 percent of the solid waste was being composted. In 1994, over 3,000 composting facilities were operating in the United States. Sludge composting facilities numbered over 180, and municipal solid waste composting was being practiced by 21 cities (Monk, 1994).

Methane Recovery. Methane is produced in sanitary landfills as a result of anaerobic decomposition of the organic fraction of the waste. In addition to gas extraction wells and a collection system, some gas processing equipment is employed. The minimum processing consists of dehydration, gas cooling, and, perhaps, removal of heavy hydrocarbons. The gas produced is a low-Joule gas having heating value of 18.6 MJ/m^3. In high-Joule processing systems, carbon dioxide and some hydrocarbons are removed to yield essentially pure methane. The resulting gas is of pipeline quality and has a heating value of approximately 37.3 MJ/m^3. The anticipated quantity of landfill gas (LFG) varies between 0.6 and 8.7 liters per kilogram of solid waste present per year (L/kg · y). The average production rate is 5 L/kg · y.

Although landfill sites as small as 11 ha have yielded substantial quantities of recoverable methane, the capital investment and complexity of the gas processing equipment will limit this technique to the larger sites (>65 ha). Otherwise, the technology is readily available and can make use of a resource that otherwise would dissipate into the atmosphere. According to EPA data, in 1999, 360 LFG-recovery projects nationwide produced the equivalent of 1,200 MW of power (Skinner, 1999).

High Technology RC & R

In the mid-1970s, under the auspices of the U.S. Environmental Protection Agency and with federal financing, several innovative high technologies for resource recovery were examined. At the end of the decade, a few workable systems and a large number of unworkable systems were identified.

Because the successful high technology systems depend, to a large measure, on the recovery of energy for their success, we will consider the worth of solid waste as a fuel. As illustrated in Table 11-14, MSW is not a very good fuel. On the other hand, its cost of $0.00/Mg may seem quite attractive. This is especially so when the price of anthracite coal may be $50/Mg and the price of No. 2 fuel oil is $250/Mg. Unfortunately, solid

TABLE 11-14
Net heating value of various materials

Material	Net heating value (MJ/kg)
Charcoal	26.3
Coal, anthracite	25.8
Coal, bituminous (hi volatile B)	28.5
Fuel oil, no. 2 (home heating)	45.5
Fuel oil, no. 6 (bunker C)	42.5
Garbage	4.2
Gasoline (regular, 84 octane)	48.1
Methane[a]	55.5
Municipal solid waste (MSW)	10.5
Natural gas[a]	53.0
Newsprint	18.6
Refuse derived fuel (RDF)	18.3
Rubber	25.6
Sewage gas[a]	21.3 to 26.6
Sewage sludge (dry solids)	23.3
Trash	19.8
Wood, oak	13.3 to 19.3
Wood, pine	14.9 to 22.3

[a]Densities taken as follows (all in kg/m^3): CH_4 = 0.680; natural gas = 0.756; sewage gas = 1.05.

waste, as a fuel, has a hidden cost. Unless the physical characteristics are upgraded by removing metals and glass and by reducing the particle size, MSW cannot be burned in conventional coal-fired power plants. The alternative is the construction of a special power plant that can handle the MSW as it is received. In either case, some cost is imposed.

It appears that if a high technology resource recovery facility is to be successful, it must meet the following criteria (Serper, 1980):

1. High technology resource recovery can only be economical in large metropolitan areas where landfill sites are unavailable or are very expensive, above $25/Mg, or in geographic locations where the water table makes safe landfilling impossible, as, for example, the city of New Orleans and its surrounding suburbs.

2. There must be an adequate refuse supply committed to the facility (a minimum of 1.8 Gg/d is needed). In general, this implies a population of 250,000 or more.

3. A customer must be obtained for the steam or the power generated by the plant and must be located close by. Firm contracts must be obtained for both the refuse supply and the sale of energy.

4. If the customer is totally dependent on the energy supplied by the facility, the combustion facility must be designed with the capacity to burn fossil fuel when refuse is unavailable or when the plant cannot process the raw refuse due to malfunctions of the processing equipment.

5. The logistics of delivering refuse to the resource recovery facility should be planned long in advance. It may be necessary to establish transfer stations and storage locations that will operate in conjunction with the resource recovery plant.

6. Systems that can dispose of both municipal refuse and sewage sludge will have economic advantages over systems that dispose of refuse only. With the ban of ocean dumping now in effect, local sewage districts are being forced to spend astronomical amounts of money to incinerate sludge. A co-disposal plant should reduce both the refuse and sludge disposal costs. In order to be economically competitive, sewage sludge must be dewatered to the maximum practical extent. A number of co-disposal plants are now in operation in Europe. Except for large installations, there will not be sufficient excess energy to warrant exporting it.

Many of the high technology systems have, as a common starting point, the medium technology materials recovery systems as their first process steps. These were discussed in a previous section.

11-7 CHAPTER REVIEW

When you have completed studying this chapter, you should be able to do the following without the aid of your textbook or notes:

1. State the average mass of solid waste produced per capita per day in the United States in 2003.

2. Differentiate between garbage, rubbish, refuse, and trash, based on their composition and source.

3. Compare the advantages and disadvantages of public and private solid waste collection systems.

4. List the three pickup methods (backyard, set-out/set-back, and curbside) and explain the advantages and disadvantages of each.

5. List the components of a time study for a waste collection system.

6. Compare the advantages and disadvantages of the four methods of collection truck routing.

7. Explain the four methods of integrating several crews.

8. Explain what a transfer station is and what purpose it serves.

9. List and discuss the factors pertinent to the selection of a landfill site.

10. Describe the two methods of constructing a MSW landfill.

11. Explain the purpose of daily cover in a MSW landfill and state the minimum desirable depth of daily cover.

12. Define leachate and explain why it occurs.

13. Sketch a MSW landfill that includes proper cover and a leachate collection system.

14. Define or explain the following terms: WTE, autogenous, HHV, LHV, RDF, source-separation.

15. Explain the relationship between oxygen, time, temperature, and turbulence in establishing efficient combustion reactions.

16. Explain the effect of source-separation on the heating value of solid waste and on the potential for hazardous air pollution emissions.

17. List two highly feasible methods of resource conservation and/or recovery in low technology and medium technology RC & R.

18. Describe and explain, in a basic manner, each of the two methods listed in number 17 above such that the average citizen could understand the method.

With the aid of this text, you should be able to do the following:

19. Determine the volume and mass of solid waste from various establishments.

20. Determine the required volume capacity of a solid waste collection truck, or conversely, determine the number of stops possible for a given truck volume, or the allowable mean time per collection.

21. Estimate the annual truck and labor cost for solid waste collection and the cost per service stop.

22. Lay out a truck route using the heuristic routing technique.

23. Determine the necessity and/or advisability of constructing a transfer station.

24. Estimate the volume and area requirements for a landfill.

25. Compute the LHV given the HHV and the chemical formula for a compound to be burned.

11-8 PROBLEMS

11-1. The student population of Metuchen High School is 881. The school has 30 standard classrooms. Assuming a 5-day school week with solid waste pickups on Wednesday and Friday before school starts in the morning, determine the size of storage container (dumpster) required. Assume waste is generated at a rate of 0.11 kg/cap · d plus 3.6 kg per room and that the density of uncompacted solid waste is 120.0 kg/m^3. Standard container sizes are as follows (all in m^3): 1.5, 2.3, 3.0, and 4.6.

    Answer: Select one 1.5-m^3 and one 4.6-m^3 container.

11-2. The Bailey Stone Works employs six people. Assuming that the density of uncompacted waste is 480 kg/m^3, determine the annual volume of solid waste produced by the stone works assuming a waste generation rate of 1 kg/cap · d.

11-3. As the supply of high-grade ores is used up, lower grade ores are used to produce minerals. Assuming that you are producing 100 kg of metal, use the mass balance method to calculate the kilograms of waste rock per kilogram of metal for ore containing 50, 25, 10, 5, and 2.5 percent metal.

11-4. Professor Green has made measurements of her household solid waste, shown in the table below. If the container volume is 0.0757 m^3, what is the average density of the solid waste produced in her household? Assume that the mass of each empty container is 3.63 kg.

Date	Can no.	Gross mass[a] (kg)
March 18	1	7.26
	2	7.72
March 25	1	10.89
	2	7.26
	3	8.17
April 8	1	6.35
	2	8.17
	3	8.62

[a]Container plus solid waste.

    Answer: Average density = 58.4 kg/m^3

11-5. The collection vehicle compacts the household solid waste in Problem 11-4 to 37 percent of its original volume. Estimate the density of the compacted waste in kg/m^3.

11-6. The typical composition of solid waste from Davis, California, is shown in Table 11-4. Calculate the density of this waste in kg/m^3 if the paper, cardboard, plastic, glass, and tin cans are removed.

11-7. Early Collection Systems is considering bidding on a solid waste management contract to collect all of the residential solid waste generated by Midden (population 44,000). The average solid waste generation rate is 1.17 kg/cap · d and the average uncompacted density is 144.7 kg/m^3. The request for bids specifies that each residence must have a minimum of two pickups per week (maximum of 4 days between pickups) and that there will be no rear-of-house pickups. Using the following assumptions, determine how many trucks of what size Early Collection Systems should plan on using. Assumptions for Midden:

> Average residential occupancy = 4/residence
>
> Average number of cans per stop = 3/wk at 0.0757 m^3/can
>
> Side loader compactor truck with a crew of one
>
> Truck compactor density rating = 475 kg/m^3
>
> Truck dump time = 7.50 min
>
> Delay time = 20.0 min
>
> Distance to disposal site = 24.0 km
>
> Number of trips to disposal site = 2/d
>
> Time between pickup stops = 18.00 s
>
> Dump time (regression coefficient a) = 12.60 s/can
>
> Standard side-loading compactor truck capacities (all in m^3): 9.0, 12.0, 15.0, 18.0, 19.0, 21, and 27
>
> *Answer:* Should have 12 trucks of 9.0 m^3 capacity.

11-8. The City of Forty Two (population 361,564) has requested your assistance in evaluating its solid waste collection system. Determine the mean time per collection stop plus the mean time to reach the next stop, the number of pickup locations per load, and the minimum number of trucks the city must own. Forty Two collection data:

> Average truck capacity = 18.0 m^3
>
> Average observed compaction ratio = 3.97
>
> Crew size = 2
>
> Number of pickups = 1 /wk (no rear-of-house service)
>
> Average number of cans per stop = 2.53/wk at 0.1136 m^3/can
>
> Average number of residents per stop = 4

Average uncompacted density = 100.76 kg/m³

Average transport time to disposal site including delays and dumping = 1.00 h/trip

Average number of trips to disposal site = 2/d

Rest breaks = 2 at 15.0 min

Average maintenance downtime = 24.0 min/d

Average workday = 8.00 h

Average percent of trucks out of service for major repairs = 15.0%

11-9. The City of Bon Chance (population 161,565) has requested your assistance in evaluating its solid waste collection system. Determine the mean time per collection stop plus the mean time to reach the next stop, the number of pickup locations per load, and the minimum number of trucks the city must own. Bon Chance collection data:

Average truck capacity = 18.0 m³

Average observed compaction ratio = 3.28

Crew size = 2

Number of pickups = 1 /wk (no rear-of-house service)

Average number of cans per stop = 2.95/wk at 0.0911 m³/can

Average number of residents per stop = 2.5

Average uncompacted density = 122.0 kg/m³

Average transport time to disposal site including delays and dumping = 1.50 h/trip

Average number of trips to disposal site = 2/d

Rest breaks = 2 at 15.0 min

Average maintenance downtime = 36.0 min/d

Average workday = 8.00 h

Average percent of trucks out of service for major repairs = 15.0%

11-10. Rework Example 11-3 assuming no rear-of-yard pickup and only one trip per day to the disposal site.

11-11. Rework Problem 11-8 using a time between pickup stops of 28.20 s and a dump time (regression coefficient a) of 12.80 s/can for a side-loading truck and a crew of one. Assume that the truck size remains the same but the number of trips to the disposal site is reduced to one per day.

11-12. Mr. Midas, owner and manager of Early Collection Systems, would like to make a 20 percent profit (before taxes) on the Midden collection system work (Problem 11-7). Using the data provided by Mr. Midas, shown in the table below, determine the annual cost per megagram (Mg) and the average weekly charge to each household in order for Mr. Midas to make a 20 percent profit before taxes.

Labor costs for Midden

Employee title	Number	Wage rate, $/h
Route supervisor[a]	1	29.60
Secretary/bookkeeper[a]	1	16.20
Mechanic[b]	1	20.61
Driver/collector	12	14.00
General laborer[a]	2	7.40

[a]paid by overhead.
[b]Mechanic is included in 0 & M cost.

Average workweek 40.0 h/wk, 5 d/wk
Overhead rate = 101.38% of total driver/collector wages

Truck data:
  Size = 9.0 m^3
  Capital cost = $117,000
  O&M cost = $6.46/km
  Anticipated life of truck = 5 y
  Interest rate = 8.75%
  Average annual distance for each truck = 16,412 km

11-13. Determine the annual cost per megagram and the average weekly charge per household for a system using a crew of two and for a system using a crew of one for the city of Nosleep (population 361, 564). Nosleep collection data:

Number of pickups = 1 /wk (no rear-of-house service)

Average number of cans per stop = 2.53/wk at 0.1136 m^3/can

Average number of residents per stop = 4

Mean time per collection stop plus mean time to reach next stop:
  For crew of one = 0.01180 h
  For crew of two = 0.00883 h

Average uncompacted density = 100.76 kg/m^3

Average observed compaction ratio = 3.97

Average transport time to disposal site including delays and dumping = 1.00 h/trip

Average number of trips to disposal site = 1/d

Rest breaks = 2 at 15.0 min

Average maintenance downtime = 24.0 min/d

Average workday = 8.00 h

Average percent of trucks out of service for major repairs = 15.0%

Labor costs for Nosleep

Employee title	Number	Wage rate, $/h
Director[a]	1	48.95
Secretary[a]	1	13.44
Bookkeeper[a]	1	23.02
Route supervisors	4	33.50
Senior mechanic[b]	1	37.68
Mechanic[b]	3	25.11
Crew of two:		
Driver (1/truck)	c	15.25
Collector (1/truck)	c	14.70
Crew of one:		
Driver/collector	c	16.00
General laborers[a]	4	7.40

[a]Paid by overhead.
[b]Included in O & M cost.
[c]Dependent on number and type of trucks required.

Average workweek = 40 h/wk, 5 d/wk

Overhead rate = 75.04% of total crew wages

Truck data:
Capital cost of 15.0-m^3 truck = $122,000
O & M for 15.0-m^3 truck = $5.75/km
Capital cost of 21.0-m^3 truck = $141,000
O & M cost for 21.0-m^3 truck = $6.55/km
Truck compactor density rating = 400 kg/m^3
Anticipated life = 5 y
Interest rate = 6.75%
Average annual distance for each truck = 11,797 km

11-14. The city manager of Bon Chance (population 161,565) has requested your services in analyzing three alternative city-managed schemes for collection of the city's solid waste. The three schemes are: (1) a system with crew-of-one trucks, (2) a system with crew-of-two trucks, and (3) a system with crew-of-three trucks. Using a spreadsheet program you have written, prepare an estimate of the annual cost per megagram (Mg) of each of these systems for the city manager. Bon Chance collection data:

Average observed compaction ratio = 3.28

Number of pickups = 1 /wk (no rear-of-house service)

Average number of cans per stop = 2.95/wk at 0.0911 m^3/can

Average number of people per stop = 2.5

Average uncompacted density = 122.0 kg/m^3

Mean time per collection stop plus mean time to reach next stop:

For crew of one = 0.88 min
For crew of two = 0.57 min
For crew of three = 0.37 min

Average transport time to disposal site including delays and dumping = 1.50 h/trip

Average number of trips to disposal site = 1/d

Rest breaks = 2 at 15.0 min

Average maintenance downtime = 36.0 min/d

Average workday = 8.00 h

Average percent of trucks out of service for major repairs = 15.0%

Bon Chance labor costs

Employee title	Number	Wage rate, $/h
Director[a]	1	48.95
Secretary[a]	1	13.44
Bookkeeper[a]	1	23.02
Route supervisors	4	33.50
Senior mechanic[b]	1	37.68
Mechanic[b]	3	25.11
Crew of one:		
Driver/collector	c	16.00
Crew of two:		
Driver (1/trusck)	c	15.25
Collector (1/truck)	c	14.70
Crew of three:		
Driver (1/truck)	c	15.25
Collector (2/truck)	c	14.70
General laborers[a]	4	7.40

[a]Paid by overhead.
[b]Included in O & M cost.
[c]Dependent on number and type of trucks required.

Average workweek = 40 h/wk, 5 d/wk

Overhead rate = 75.04% of total crew wages for crew of one.

Truck data:

Capital cost of 15.0 m^3 truck = $122,000
O & M for 15.0 m^3 truck = $5.75/km
Capital cost of 18.0 m^3 truck = $131,500
O & M cost for 18.0 m^3 truck = $6.55/km
Capital cost of 21.0 m^3 truck = $141,000
O & M cost for 21.0 m^3 truck = $7.60/km
Truck compactor density rating = 400 kg/m^3
Anticipated life = 5 y
Interest rate = 6.75%
Average annual distance for each truck = 15,260 km

11-15. Using the rules for heuristic routing, plan a collection route for the section of Redbud shown in Figure P-11-15. Assume that all streets are two-way and that the pattern is bounded by two-way streets on all four sides. Also assume that collection is on one side of the street at a time.

Answer: The solution has no dead distance and two left turns. Both left turns occur at the intersection of Simons and Garson.

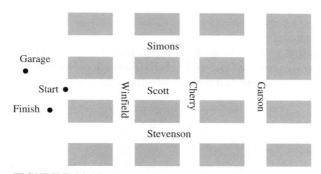

FIGURE P-11-15
Sketch map no. 1, Redbud.

11-16. Rework Problem 11-15 for the section of Mundy shown in Figure P-11-16. All of the streets are two-way and the pattern is bounded by two-way streets on all four sides. Collection is from one side of the street at a time.

11-17. Rework Problem 11-16 but assuming that West Zacks is one-way going north and that East Zacks is one-way going south.

Answer: The solution has three dead distances and 14 left turns. The dead distances are in the middle block of North and South Avenues. The left turns occur at the traffic signals.

North Avenue

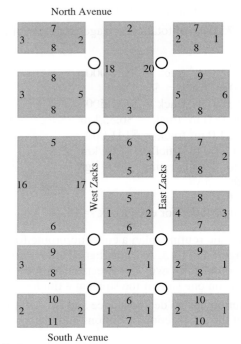

FIGURE P-11-16
Sketch map no. 2, Mundy. The numbers
refer to the number of stops in a block.
The circles denote traffic signals.

11-18. Using the rules for heuristic routing, plan a collection route for the section
of Travail shown in Figure P-11-18. Assume collection is on one side of
the street at a time and that all streets are two-way.

FIGURE P-11-18
Sketch map no. 3, Travail.

11-19. Divide the collection area shown in Figure P-11-19 into two approximately
equal collection routes with starting points at A(1) and A(2). The difference in the number of stops for each route should not exceed 25. Lay out

the collection route that begins at A(1). The collection route constraints are that there are to be no U-turns in streets and that collection is to be made from each side of the street with one driver/collector using a right-hand-drive collection vehicles. The preferred solution is one that minimizes overlaps.

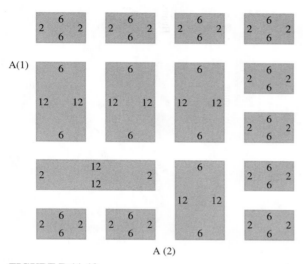

FIGURE P-11-19

Sketch map of Troublesome Creek collection area. 2, 6, 12 = number of residences along each block. (*Source:* Tchobanoglous, et al., 1993)

11-20. Divide the Olson collection area shown in Figure P-11-20 into two approximately equal collection routes with starting points at A(1) and A(2). The difference in the number of stops for each route should not exceed 25. Assuming that both sides of the street can be collected in one pass, lay out the collection route that begins at A(2). *HINTS: $N_p \approx 500$, Huntington Rd.* divides the two routes.

11-21. Repeat Problem 11-20 for the Masters collection area shown in Figure P-11-21 for the collection route that begins at A(1). *HINT: $N_p = 488$,* The route is roughly bounded by Highland Ave. and Concord Ave.

The following equations may be used to determine the speed as a function of the haul distance x for Problems 11-22, 11-23, and 11-24:

From 7.5 to 22 km: $s = -17.76 + \ln 2x$

From 22 to 40 km: $s = 10.36 + 0.86(2x)$

From 40 to 80 km: $s = 4.75 + 0.925(2x)$

Beyond 80 km: $s = 80$

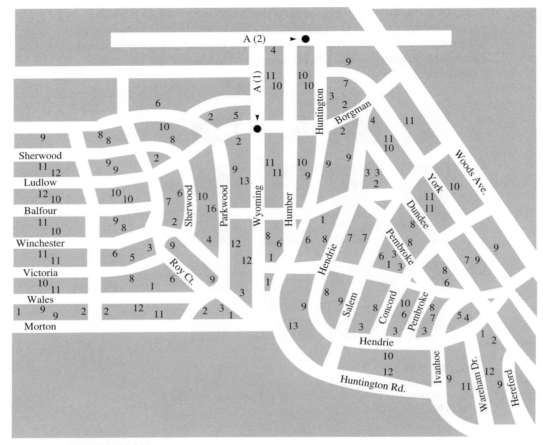

FIGURE P-11-20
Sketch map no. 4, Olson. The numbers refer to the number of stops in a block.

11-22. Write an equation for the relationship between the cost per megagram for hauling solid waste to a disposal site at a distance x, and the round trip time H_t it takes to travel to the disposal site for a crew of one in a 9.0-m^3 compactor truck (see Problems 11-7 and 11-12 for data). The haul speed is determined from the equations noted above.

Answer: TC = 13.29 + 1.51x + 4.18H_t

11-23. Repeat Problem 11-22 using a crew of one with 18-m^3 compactor trucks. (See Problem 11-14 data.)

11-24. Repeat Problem 11-22 using a crew of two with 18-m^3 compactor trucks. (See Problems 11-9 and 11-14 for data.)

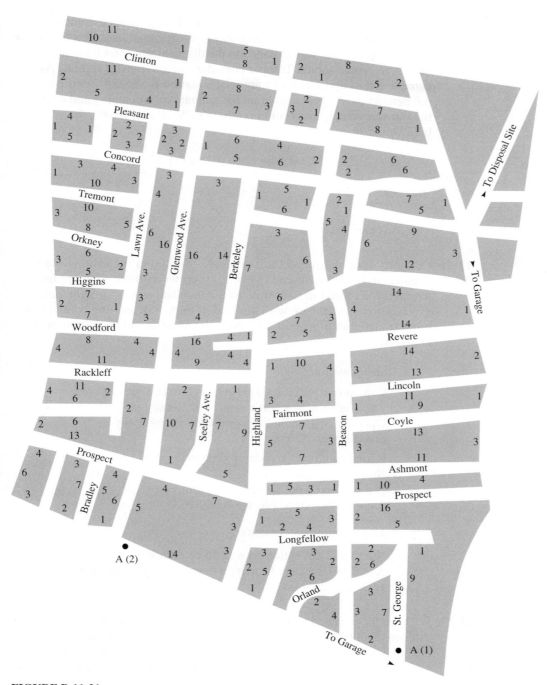

FIGURE P-11-21

Sketch map no. 5, Masters. The numbers refer to the number of stops in a block.

11-25. The town of Trooper (population 8,500) is closing its open dump and will transport its solid waste (9.53 Mg/wk) to the regional landfill at Tuppance Junction. If the one-way distance to the disposal site is 64.0 km and the crew size is one, should Trooper consider using a transfer station? Assume that the capital cost of the transfer station is $25,000 amortized at 6.0 percent over 5 years, and that the annual cost of owning and operating the transfer vehicle and station is $20,000. *Note:* A/P (6.0%, 5y) = 0.2374 and

$$V_T = 11.321 - 3.827\left(\frac{2x}{s}\right)$$

$$H_t = \frac{2x}{s}$$

Answer: No.

11-26. Calamity (population 35,000) generates 48,800 m³ of solid waste at a density of 425.0 kg/m³ each year. Four crews of three each are now averaging 1.08 h/d in haul time to the disposal site. Using the data given below, and the crew-of-three cost curves determine whether or not a transfer station should be considered based on an economical analysis.

Transfer station data:

Capital cost = $1,200,000 at 6.00% over 8 years
Transfer vehicle amortization = $55,000/y
Operator cost (including overhead) = $64,960/y
O & M = $6.55/km

Round trip travel and dump time to disposal site = 1.35 h
Number of trips = 5/d
Distance to disposal site = 46.7 km
$T_C = 13.22 + 0.6319(x) + 3.869(Hz)$

Savings from transfer station:

Number of collection crews will be reduced to two
Average daily round trip haul time to transfer station for each vehicle = 20 min

11-27. Estimate the area and volume of landfill to handle the solid waste from Midden (Problem 11-7) for 20 years. The Science Club at Midden High School has furnished the following data based on a 12-month survey. (One sample having a mass of 1.000 Mg was taken at the existing landfill during normal off-loading operations 1 day each month.) Assume a cell height of 2.40 m and that the recommended depths of cover will be used and that compaction will be normal.

Characterization of Midden solid waste

Component	Mass fraction
Food waste	0.0926
Paper	0.4954
Plastics, rubber, leather	0.0438
Textiles	0.0379
Metals	0.0741
Glass	0.1668
Miscellaneous	0.0894
Total	1.0000

11-28. Rework Problem 11-27 assuming that 50 percent of the paper is recycled.

11-29. A MSW landfill is being designed to handle solid waste generated by Binford at a rate of 50 Mg/d. It is expected that the waste will be delivered by compactor truck on a 5 d/week basis. The density as spread is 122 kg/m³. It will be spread in 0.50 m layers and compacted to 0.25 m. Assuming three such lifts per day and a daily cover of 0.15 m, determine the following: (a) annual volume of landfill consumed in m³, and (b) daily horizontal area covered by the solid waste. Ignore the soil volume between stacks.

11-30. Estimate the theoretical production of landfill gas (CH_4 only) from the degradation of 20.3 kg of rapidly decomposable MSW. Assume the density of methane is 0.7177 kg/m³.

11-31. Estimate the theoretical production of landfill gas (CH_4 plus CO_2) from the degradation of 3.3 kg of slowly decomposable MSW. Assume the density of methane is 0.7167 kg/m³ and that of carbon dioxide is 1.9768 kg/m³ at STP.

11-32. The city of Nosleep (Problem 11-13) is considering instituting a recycling program in which the residents presort the solid waste into four components: (1) mixed waste, (2) paper, (3) glass, and (4) metallics. From a research study report, we find that the mean time per collection stop plus the mean time to reach the next stop (t_p) can be estimated from the following equation (Tichenor, 1980):

$$t_p = 22.6 + 3.80R + 5.50S$$

where t = mean collection time, s
R = number of units of mixed waste per stop
S = sum of the number of units of separated paper, glass, and metallics per stop

Assuming that $S = 3.00$ and $R = 1.53$, rework Problem 11-13 to determine what savings in disposal cost is needed to offset the additional cost of collection for a crew of one.

11-33. Rework Example 11-4 assuming that 50 percent of the paper and 80 percent of the glass and metal are separated at the source and recycled.

11-34. Using the EPA method, estimate the maximum drainage distance L for a solid waste landfill with a rainfall of 4.0 cm/mo. Assume the hydraulic conductivity of the drainage layer is 2×10^{-2} cm/s and the slope of the liner is 1.0 percent.

11-35. Using a spreadsheet program you have written, rework Problem 11-34 with slopes of 0.5, 1.0, 2.0, and 3.0 percent. Estimate the maximum depth of leachate for wet season rainfall of 40.0 cm/mo for the case of the 1.0 percent slope.

11-36. The higher heating value for cellulose ($C_6H_{10}O_5$) is 32,600 kJ/kg. Compute the lower heating value.

11-37. The higher heating value for methane (CH_4) is 888,500 kJ/kg. Compute the lower heating value.

11-38. Typical residential food waste has a higher heating value of 4,500 kJ/kg on a dry mass basis. Compute the lower heating value if 6.0 percent of the waste by mass is hydrogen.

11-9   DISCUSSION QUESTIONS

11-1. What is the effect of crew size on the mean time per collection stop (t'_p)? How does the container location affect the mean time per collection stop?

11-2. Under what conditions would you recommend consideration of a transfer station?

11-3. Which of the following soil types would be suitable for (a) composite liner, (b) drainage layer, (c) gas venting:

   1. Gravel (>2.5 cm diameter)
   2. Glacial till
   3. Clay ($K = 1 \times 10^{-9}$ cm/s)
   4. Clay ($K = 1 \times 10^6$ cm/s)
   5. Sand ($K = 0.1$ cm/s)
   6. Sand ($K = 0.001$ cm/s)

11-4. A WTE plant is being proposed as part of an ISWM plan. The proponents of the WTE argue that recycling is not necessary and will have no effect on the performance of the plant. Do you agree or disagree? Explain.

11-5. Although the market value of compost is negligible, many communities have implemented yard waste composting systems. Explain why.

11-10   FE EXAM FORMATTED PROBLEMS

11-1. A field test of a clay liner resulted in the data shown below. Estimate the hydraulic conductivity of the liner.

Thickness of the clay liner = 0.60 m
Clay porosity = 0.55
Breakthrough time = 90.0 d
Depth of leachate above the liner = 3.0 m

a. 7.1×10^{-7} cm/s   b. 1.7×10^{-5} cm/s
c. 1.2×10^{-6} cm/s   d. 7.1×10^{-8} cm/s

11-2. Estimate the percent of landfill gas that is methane for a rapidly decomposing waste with the following composition: $C_{68}H_{111}O_{50}N$. assume the maximum theoretical landfill gas may be estimated with Equation 11-9.

a. 51%   b. 48%
c. 35%   d. 18%

11-3. Solid waste with a density of 110 kg/m^3 is spread in a 0.5 m layer at a landfill. If a steel-wheeled compactor can achieve a density of 400 kg/m^3, what thickness of waste will be achieved?

a. 1.8 m   b. 0.25 m
c. 0.14 m   d. 0.18 m

11-4. Estimate the weekly volume of solid waste being delivered to a municipal landfill if the average solid waste generation rate is 2.6 kg/person · d, the population is 17,200, and the density as delivered is 122 kg/m^3.

a. 1,800 m^3/wk   b. 2,600 m^3/wk
c. 1,900 m^3/wk   d. 370 m^3/wk

11-11   REFERENCES

Bader, C. (2001) "Where Are Collection Trucks Going?" *MSW Management: The Journal for Municipal Solid Waste Professionals,* vol. 12, no. 6, September/October.

CARB (1988) *The Landfill Gas Testing Program: A Report to the Legislature*, State of California Air Resources Board.

Chicago vs. EDF (1994) *City of Chicago, et al. vs. Environmental Defense Fund, et al.*, No. 92-1639, May.

Cossu, R., G. Andreottola, and A. Muntoni (1996) "Modeling Landfill Gas Production," in T. H. Christensen, R. Cossu, and R. Stegmann (eds.), E&FN Spon, Landfilling of Waste: Biogas London, pp. 237–250.

Heil (2006) at http://www.heil.com/products/starr.asp. and http://www.heil.com/products/pt1000.asp.

IIA (1968) *I.I.A. Standards*, Incinerator Institute of America, New York.

ISW (1970) *Municipal Refuse Disposal*, Institute for Solid Waste, American Public Works Association, Chicago.

IWSA (2004) *Waste-to-Energy, Clean, Reliable, Renewable Power*, Integrated Waste Services Association, Washington, DC.

Luken, K., and S. Bush (2002) "Automated Collection: Getting the Biggest Bang for Your Buck," *MSW Management: The Journal for Municipal Solid Waste Professionals*, vol. 12, no. 6, September/October.

McEnroe, B. M. (1993) "Maximum Saturated Depth Over Landfill Liner," *Journal of Environmental Engineering Division*, American Society of Civil Engineers, vol. 119, pp. 262–270.

Monk, R. B. (1994) "Digging in the Dirt, Unearthing Potential," *World Wastes*, vol. 37, no. 4 (April), cs1-cs-14.

NCRR (1974) *Resource Recovery from Municipal Solid Waste*, National Center for Resurce Recovery, Lexington Books, Lexington, MA.

O'Leary, P., and P. Walsh (2002) "Landfill Equipment and Operating Procedures," *Waste Age*, September, pp. 53–59.

Paul, S. (1995) "Reaching Equilibrium in Recycling Marketables," *World Wastes*, vol. 38, no. 8 (August), p. 52.

Pfeffer, J. T. (1992) *Solid Waste Management Engineering*, Prentice Hall, Upper Saddle River, NJ, p. 172.

Rogoff, M. J., and J. F. Williams (1995) "Marketing Efforts to Close Loop," *World Wastes*, vol. 38, no. 5 (May), p. 28.

Salvato, J. A. (1972) *Environmental Engineering and Sanitation*, Wiley-Interscience, New York, p. 427.

Schroeder, P. R., et al. (1984) *The Hydrologic Evaluation of Landfill Performance (HELP) Model Documentation, User's Guide*, U.S. Environmental Protection Agency Publication No. EPA 530 SW-84-009, Washington, DC.

Serper, A. (1980) "Resource Recovery Field Stands Poised Between Problems, Solutions," *Solid Waste Management/Resource Recovery Journal*, May, p. 86.

Shortsleeve, J., and R. Roche (1990) "Analyzing the Integrated Approach," *Waste Age*, March 1990, pp. 92–94.

Skinner, J. M. (1999) "Advancements in Reduction and Recovery," *MSW Management*.

Stone, R. (1969) *A Study of Solid Waste Collection Systems: Comparing One Man with Multi-Man Crews*, U.S. Department of Health, Education, and Welfare, Report No. SW-9C, Washington, DC, pp. 96–98.

Stone, R., and E. T. Conrad (1969) "Landfill Compaction Equipment Efficency," *Public Works*, May, pp. 111–113 and 160.

Tchobanoglous, G., H. Theisen, and R. Eliassen (1977) *Solid Wastes: Engineering Principles and Management Issues*, McGraw-Hill, New York, p. 95.

Tchobanoglous, G., H. Theisen, and S. Vigil (1993) *Integrated Solid Waste Management: Engineering Principles and Management Issues*, McGraw-Hill, New York, pp. 49, 214, 374, 388–391, 424, 686–695, 932–935.

Tichenor, Richard (1980) "Designing a Vehicle to Collect Source-Separated Recyclables," *Compost Science/Land Utilization*, vol. 21(l), January/February, pp. 36–4l.

U. Calif. (1952) *An Analysis of Refuse Collection and Sanitary Landfill Disposal,* University of California Technical Bulletin 8, Series 73, University of California Press, Berkeley, CA, p. 22.

U.S. EPA (1974) *Decision Makers Guide to Solid Waste Management,* U.S. Environmental Protection Agency, Washington, D.C.

U.S. EPA (1989) *Requirements for Hazardous Waste Landfill Design, Construction and Closure,* U.S. Environmental Protection Agency Publication No. EPA 625/4-89/022, Washington, DC, p. 89.

U.S. EPA (1995) *Decision Makers Guide to Solid Waste Management, Vol. II,* U.S. Environmental Protection Agency Publication No. EPA 530-R-95-023, Washington, DC.

U.S. EPA (1998) *Solid Waste Disposal Facility Criteria: Technical Manual,* U.S. Environmental Protection Agency Publication No. EPA 530-R-93-017, Washington, DC.

U.S. EPA (2003) *Municipal Solid Waste in the United States: Facts and Figures,* U.S. Environmental Protection Agency, Washington, DC.

U.S. EPA (2010) *Municipal Solid Waste Generation, Recycling, and Disposal in the United States: Facts and Figures* for 2009 U.S. Environmental Protection Agency, Washington, DC.

Vence, T. D., and D. L. Powers (1980) "Resource Recovery Systems, Part 1, Technological Comparison," *Solid Waste Management/Resource Recovery Journal,* May, pp. 26–28, 32, 34, 72, 92, 93.

Wolpin, B. (1994) "Go Figure," *World Wastes,* vol. 37, no. 10, October 1994, p. 4.

Zuena, A. J. (1987) "Snapshot of Small Transfer Station Costs," *Waste Age.*

CHAPTER
12

HAZARDOUS WASTE MANAGEMENT

12-1 INTRODUCTION

Applications

After a brief introduction to two hazardous wastes that have achieved national prominence, this chapter provides an introduction to the classification of hazardous waste, the major congressional acts and the EPA's implementation of the acts, hazardous waste management strategies, treatment technologies, disposal alternatives, and groundwater contamination and remediation. This chapter will provide you with tools to do the following:

- Determine whether or not a solid waste is hazardous

- Select a hazardous waste management strategy

- Select a treatment technology and assess its ability to comply with regulations

- Estimate the speed of movement of groundwater contaminants

- Assess the practicality of a pumping system to intercept a contaminated groundwater plume

Dioxins and Polychlorinated Biphenyls (PCBs)

Dioxins and PCBs have national prominence and notoriety. The following paragraphs present a summary of what these compounds are, where they come from, and their environmental impact.

Dioxins are found as over twenty different isomers of a basic chlorodioxin structure (Figure 12-1). The most common form, 2,3,7,8-tetrachlorodibenzo-p-dioxin (TCDD), has become recognized as probably the most poisonous of all synthetic chemicals. Dioxins are a contaminant by-product that may be thermally generated during the manufacture or burning of chlorophenols; pesticides such as 2,4,5-T; Agent Orange, a defoliant made of a 50/50 mix of 2,4-D and 2,4,5-T; algae-controlling herbicides; insecticides; and preservatives. Dioxins are not manufactured for any commercial purpose. They occur only as a contaminant by-product. To date no dioxin has been found to be formed naturally in the environment. Widespread TCDD contamination has been reported in particulate matter from commercial and domestic combustion processes. Additional background dioxin contamination (0.1 to 10 parts per million, ppm) may persist and bioaccumulate following the field application of herbicides.

TCDD is a crystalline solid at room temperature. It is only slightly soluble in water (0.2 to 0.6 parts per billion, ppb). TCDD is considered to be a highly stable compound. It is thermally degraded at temperatures over 700°C. It is photochemically degraded under ultraviolet light in the presence of a hydrogen-donating solvent such as a solution of olive oil in cyclohexanone.

TCDD contamination was found at ppm levels in 2,4,5-T and 2,4-D used for weed control in the United States and as a defoliant in Vietnam; in wastes at the Love Canal disposal sites; in orthochlorophenol crude spill residues in the Sturgeon, Missouri, train derailment; and in fallout from an explosion at a chlorophenol manufacturing plant spill in Seveso, Italy. It is at this last site that engineers and scientists were challenged to develop environmentally safe control strategies.

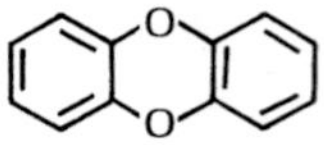

Unsubstituted dioxin

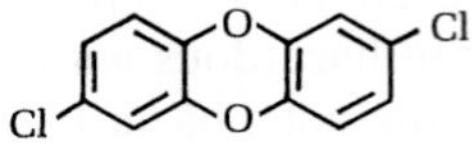

2, 7-DCDD

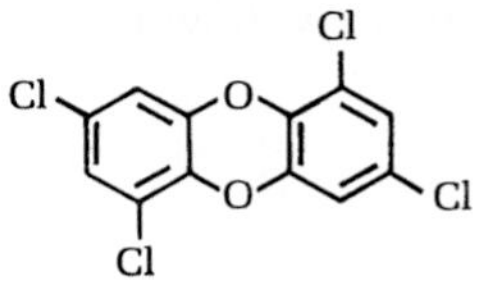

1, 3, 6, 8-TCDD

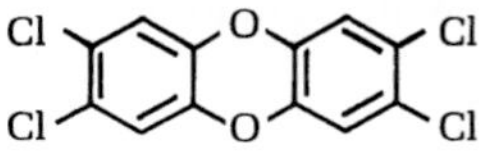

2, 3, 7, 8-TCDD

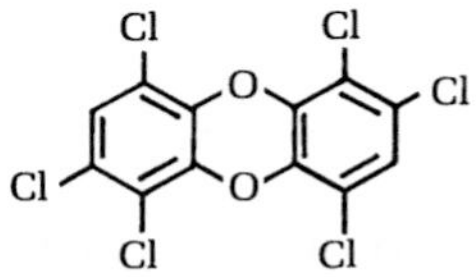

1, 2, 4, 6, 7, 9-HEXA-CDD

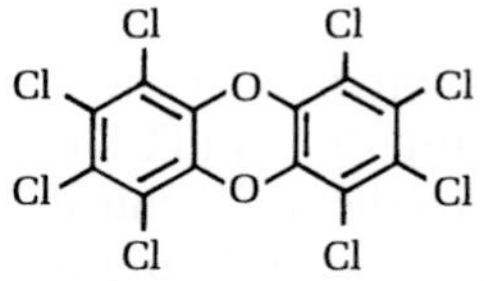

OCDD

FIGURE 12-1
Some examples of dioxins.

The environmental health effects of dioxin in people are not well documented. However, alleged birth defects in newborns in South Vietnam caused researchers to begin animal toxicological investigations. TCDD is known to cause severe skin disorders, such as chloracne. In test animals it is a carcinogen, teratogen, mutagen, and embryo-toxin, and is known to affect immune responses in mammals. It is considered persistent, and it bioaccumulates in aquatic organisms and people (U.S. EPA, 2005a). At this date (2011) no deaths have been directly correlated with low-level TCDD exposure. Nor have epidemiological findings shown any increased incidence of carcinogenesis, teratogenesis, mutagenesis or newborn defects, miscarriages, or similar adverse health effects in people. In 1994, the U.S. Environmental Protection Agency (EPA) released a report compiled by more than 100 scientists, including many not affiliated with EPA, that presents evidence that dioxin, even in trace amounts, may cause adverse human health effects (Hileman, 1994). EPA believes that dioxins are carcinogens and may cause a wide range of other effects including disruption of regulatory hormones, reproductive and immune system disorders, and abnormal fetal development (U.S. EPA, 2001 and 2005a). Levels of dioxins in the environment were negligible until about 1930, peaked about 1970, and have been declining since then. Concentrations of dioxins in human lipid tissue have declined since 1980.

The term PCB (polychlorinated biphenyls) refers to a class of organic chemicals produced by the chlorination of a biphenyl molecule. It is composed of ten possible forms and, theoretically, more than 200 isomers. These forms arise from a specified number of chlorine substitutions on the biphenyl molecule and correspond to the chemical nomenclatures monochlorobiphenyl, dichlorobiphenyl, trichlorobiphenyl, and so on. Several isomers for each PCB molecule are possible, the number depending on available substitution sites on each biphenyl portion (2–6, $2'-6'$) of the molecule. However, not all possible isomers are likely to be formed during the manufacturing processes. In general, the most common ones are those that have either an equal number of chlorine atoms on both rings or a difference of only one chlorine atom between rings. Some examples are shown in Figure 12-2.

Commercial PCB mixtures were manufactured under a variety of trade names. The chlorine content of any product varied from 18 to 79 percent, depending on the extent of chlorination during the manufacturing process or on the amount of isomeric mixing engaged in by individual producers. Each company had a specific system for identifying the chlorine content of its product. For example, Aroclor 1248, 1254, and 1260 indicate 48 percent, 54 percent, and 60 percent chlorine, respectively; Clophen A60, Phenochlor DP6, and Kaneclor 600 designate that these products contain mixtures of hexachlorobiphenyls.

The only important U.S. producer of PCBs was Monsanto Industrial Chemicals Co., which had plants at Anniston, Alabama, where production of PCBs ended in 1970; and Sauget, Illinois, where production ceased in 1977. Sold under Monsanto's registered trademark of Aroclors, mixtures of PCBs had been used originally as a coolant/dielectric for transformers and capacitors, as heat transfer fluids, and as protective coatings for woods

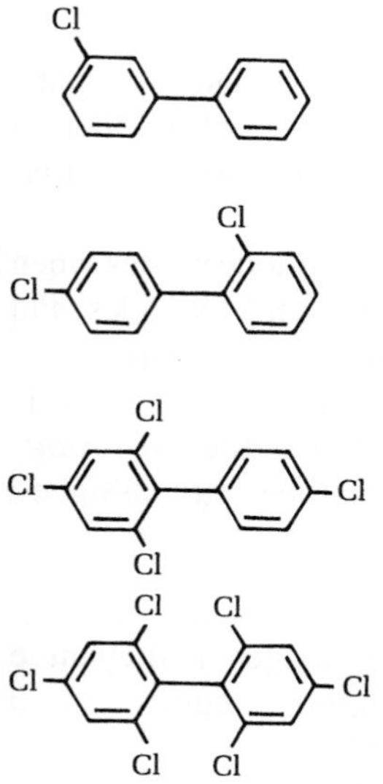

3-chlorobiphenyl

2,4′-dichlorobiphenyl

2,4,4′,6-tetrachlorobiphenyl

2,2′,4,4′,6,6′-hexachlorobiphenyl

FIGURE 12-2
Molecular structure and names of a few selected polychlorinated biphenyls.

when low flammability was essential or desirable. Producers and users alike, apparently unaware of any potential hazards from exposure to PCBs, initially operated in accordance with earlier results of toxicity tests that indicated no effects (Penning, 1930). The expansion of open-ended applications between 1930 and 1960, incorporating PCBs into such commodities as paints, inks, dedusting agents, and pesticides, led to the widespread dissemination of which we are now aware. By 1937, toxic effects were noted in occupationally exposed workers, and threshold limit values were imposed at manufacturing sites.

The general pattern of release of PCBs to the environment changed significantly during the early 1970s. Until then, essentially no restrictions were imposed either on the use or on the disposal of PCBs. After evidence became available in 1969 and 1970 that chronic exposure could result in hazards to human health and the environment, Monsanto voluntarily banned sales of PCBs, and the release rate from industrial use was reduced through stringent control measures. However, significant reservoirs of mobile PCBs (those available for transport among environmental media and biota) still exist along with even larger, currently immobile reservoirs. The latter include those materials containing PCBs that are still in service and those deposited in landfills and dumps. The major factor affecting future release of PCBs from these sources will be government regulations controlling storage and disposal of the chemical.

12-2 DEFINITION AND CLASSIFICATION OF HAZARDOUS WASTE

There are two ways a waste material is found to be hazardous (*Code of Federal Regulations,* 40 CFR 260): (1) by its presence on the EPA-developed lists, or (2) by evidence that the waste exhibits ignitable, corrosive, reactive, or toxic characteristics.

EPA's Hazardous Waste Designation System

The list of hazardous wastes includes spent halogenated and nonhalogenated solvents; electroplating baths; wastewater treatment sludges from many individual production processes; and heavy ends, light ends, bottom tars, and side-cuts from various distillation processes.

Some commercial chemical products are also listed as being hazardous wastes when discarded. These include "acutely hazardous" wastes such as arsenic acid, cyanides, and many pesticides, as well as "toxic" wastes such as benzene, toluene, and phenols.

EPA has designated five hazardous waste categories. Each hazardous waste is given an EPA Hazardous Waste Number. This is often referred to as the *Hazardous Waste Code.* Each of the five categories may be identified by the prefix letter assigned by EPA. The five categories may be described as follows:

1. Specific types of wastes from nonspecific sources; examples include halogenated solvents, nonhalogenated solvents, electroplating sludges, and cyanide solutions from plating batches. (There are 28 listings in this category. See 40 CFR 261.31.) These wastes have a waste code prefix letter F.

2. Specific types of wastes from specific sources; examples include oven residue from the production of chrome oxide green pigments and brine purification

muds from the mercury cell process in chlorine production where separated, prepurified brine is not used. (There are 111 listings in this category. See 40 CFR 261.32.) These wastes have a waste code prefix letter K.

3. Any commercial chemical product or intermediate, off-specification product, or residue that has been identified as an acute hazardous waste. Examples include potassium silver cyanide, toxaphene, and arsenic oxide. (There are approximately 203 listings in this category. See 40 CFR 261.33.) These wastes have a waste code prefix letter P.

4. Any commercial chemical product or intermediate, off-specification product, or residue that has been identified as hazardous waste. Examples include xylene, DDT, and carbon tetrachloride. (There are approximately 450 listings in this category. See 40 CFR 261.33.) These wastes have a waste code prefix letter U.

5. Characteristic wastes (40 CFR 261.21 through 40 CFR 261.27), which are wastes not specifically identified elsewhere, that exhibit properties of ignitability, corrosivity, reactivity, or toxicity. These wastes have a waste code prefix letter D.

The wastes that appear on one of the lists specified in items one through four are called *listed wastes*. The current list may be found at www.gpoaccess.gov/cfr.* Those wastes that are declared hazardous because of their general properties are called *characteristic wastes*. The characteristics of ignitability, corrosivity, and reactivity may be referred to as *ICR*. The toxicity characteristic may be referred to as *TC*.

Ignitability

A solid waste is said to exhibit the characteristic of ignitability if a representative sample of the waste has any of the following properties:

1. It is a liquid, other than an aqueous solution containing less than 24 percent alcohol by volume, and has a flash point less than 60°C†.

2. It is not a liquid and is capable, under standard temperature and pressure, of causing fire through friction, absorption of moisture, or spontaneous chemical changes; and, when ignited, burns so vigorously and persistently that it creates a hazard.

3. It is an ignitable, compressed gas.

4. It is an oxidizer.

A solid waste that exhibits the characteristic of ignitability is given an EPA Hazardous Waste Number of D001.

---

*In 2005, the **gpoaccess** search engine would locate only 40 CFR 261. To find a subparagraph such as 40 CFR 261.31, first search for 40 CFR 261, then scroll to the subparagraph of interest.

†Although it would seem to be a contradiction in terms, that is, calling a solid waste a liquid, Congress has done what was once only the province of the gods. In Section 1004 (27) of the Resource Conservation and Recovery Act of 1976, they saw fit to violate the laws of physics and make all of the physical states (liquids, gases, and solids) one and the same, that is, solid waste. By their definition, almost any discarded material is solid waste.

Corrosivity

A solid waste is said to exhibit the characteristic of corrosivity if a representative sample of the waste has either of the following properties:

1. It is aqueous and has a pH less than or equal to 2 or greater than or equal to 12.5.

2. It is a liquid that corrodes steel at a rate greater than 6.35 mm per year at a test temperature of 55°C.

A solid waste that exhibits the characteristic of corrosivity is given an EPA Hazardous Waste Number of D002.

Reactivity

A solid waste is said to exhibit the characteristic of reactivity if a representative sample of the waste has any of the following properties:

1. It is normally unstable and readily undergoes violent change without detonating.

2. It reacts violently with water.

3. It forms potentially explosive mixtures with water.

4. When mixed with water, it generates toxic gases, vapors, or fumes in a quantity sufficient to present a danger to human health or the environment.

5. It is a cyanide or sulfide-bearing waste that, when exposed to pH between 2 and 12.5, can generate toxic gases, vapors, or fumes in a quantity sufficient to present a danger to human health or the environment.

6. It is capable of detonation or explosive reaction if it is subjected to a strong initiating source or if heated under confinement.

7. It is readily capable of detonation or explosive decomposition or reaction at standard temperature and pressure.

8. It is a forbidden explosive, as defined in Department of Transportation regulations (49 CFR 173.51, 173.53, and 173.88).

A solid waste that exhibits the characteristic of reactivity is given an EPA Hazardous Waste Number of D003.

Toxicity

A solid waste is said to exhibit the characteristic of extraction procedure (EP) toxicity if, using the test methods described in Appendix II of the *Federal Register* (55 FR 11863 and 55 FR 26986), the extract from a representative sample of the waste contains any of the contaminants listed in Table 12-1 at a concentration equal to or greater than the respective value given in the table.

TABLE 12-1
Toxicity characteristic constituents and regulatory levels

EPA HW No.[a]	Constituent	Regulatory level (mg/L)
D004	Arsenic	5.0
D005	Barium	100.0
D018	Benzene	0.5
D006	Cadmium	1.0
D019	Carbon tetrachloride	0.5
D020	Chlordane	0.03
D021	Chlorobenzene	100.0
D022	Chloroform	6.0
D007	Chromium	5.0
D023	o-Cresol	200.0[b]
D024	m-Cresol	200.0[b]
D025	p-Cresol	200.0[b]
D026	Cresol	200.0[b]
D016	2,4-D	10.0
D027	1,4-Dichlorobenzene	7.5
D028	1,2-Dichloroethane	0.5
D029	1,1-Dichloroethylene	0.7
D030	2,4-Dinitrotoluene	0.13[c]
D012	Endrin	0.02
D031	Heptachlor (and its epoxide)	0.008
D032	Hexachlorobenzene	0.13[c]
D033	Hexachloro-1,3-butadiene	0.5
D034	Hexachloroethane	3.0
D008	Lead	5.0
D013	Lindane	0.4
D009	Mercury	0.2
D014	Methoxychlor	10.0
D035	Methyl ethyl ketone	200.0
D036	Nitrobenzene	2.0
D037	Pentachlorophenol	100.0
D038	Pyridine	5.0[c]
D010	Selenium	1.0
D011	Silver	5.0
D039	Tetrachloroethylene	0.7
D015	Toxaphene	0.5
D040	Trichloroethylene	0.5
D041	2,4,5-Trichlorophenol	400.0
D042	2,4,6-Trichlorophenol	2.0
D017	2,4,5-TP (Silvex)	1.0
D043	Vinyl chloride	0.2

[a]Hazardous waste number.
[b]If o-, m-, and p-cresol concentrations cannot be differentiated, the total cresol (D026) concentration is used. The regulatory level for total cresol is 200 mg/L.
[c]Quantitation limit is greater than the calculated regulatory level. The quantitation limit therefore becomes the regulatory level.

Figure 12-3 shows a generalized flow scheme for determining if a waste is hazardous according to EPA definitions. Of particular importance in the scheme are those things that are not included in the Resource Conservation and Recovery Act (RCRA) regulations. For example, domestic sewage, certain nuclear material, household wastes, including toxic and hazardous materials, and small quantities (less than 100 kg/mo) are excluded from the RCRA regulations. This does not mean that these wastes are not regulated at all. In fact they are regulated under other statutes and, thus, do not need to be regulated under RCRA.

Four important but controversial parts of the definition of a hazardous waste are the mixture rule, "contained in" policy, "derived from" rule, and waste-code carry through principle.

The *mixture rule* prevents dilution of waste for the purpose of escaping RCRA Subtitle C regulation. Under 40 CFR 261.3(a)(2), mixtures of a listed hazardous waste and other solid wastes become hazardous wastes. In certain instances for characteristic wastes and in those cases where the listed waste is not to be land disposed, dilution is permitted. The *dilution rules* are summarized in 56 FR 3875, 31 JAN 1991. When the dilution rules apply, the mixture of a hazardous waste with the diluent does not cause the diluent to become hazardous and may render the hazardous waste nonhazardous.

A corollary to the mixture rule is the *"contained in"* policy. Under this policy, media such as soil and water are treated as hazardous wastes if they "contain" listed hazardous wastes.

Any solid waste generated from the treatment, storage, or disposal of a hazardous waste, including any sludge, spill residue, ash, emission-control dust, or leachate (but not precipitation run-off) is a hazardous waste (40 CFR 261.3(c)). This is known as the " *derived from*" rule.

A corollary to the derived from and mixture rules is the *"waste-code carry through"* principle. The principle states that a solid waste derived from a hazardous waste or a mixture of hazardous and nonhazardous waste contains all of the same waste codes as the original waste (53 FR 31138, 31148).

Because of a legal suit and court order regarding the mixture rule, contained in policy, derived from rule, and waste-code carry through principle, EPA developed and proposed the *Hazardous Waste Identification Rule* (HWIR) (60 FR 66344, 12 DEC 1995). The proposed HWIR establishes exit levels for low-risk solid wastes that are designated as hazardous because they are listed, or have been mixed with, derived from, or contain a listed hazardous waste. If the constituents of the waste are below the exit levels, the waste may be disposed of as a nonhazardous (Subtitle D) waste.

The *Universal Waste Regulations* (40 CFR 273) were developed by EPA to streamline hazardous waste management standards for federal universal wastes (batteries, pesticides, thermostats, and lamps) that are widely generated. Batteries such as nickel-cadmium and small sealed lead-acid types, unused or banned pesticides, thermostats that contain mercury, and lamps such as fluorescent lights, neon, and mercury vapor that contain mercury or lead are considered *universal wastes*. The streamlining includes, for example, provisions to extend the amount of time that businesses can accumulate wastes, to increase the amount of waste that can be accumulated, and to eliminate the need for a manifest.

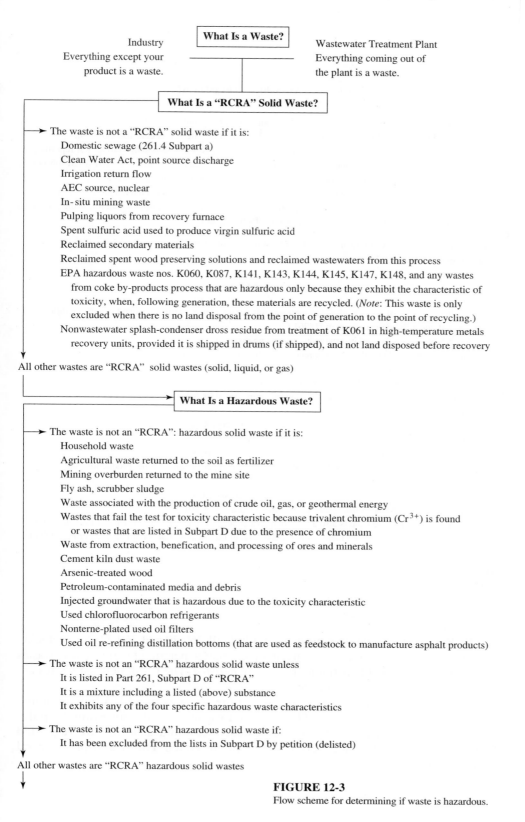

What Is a Waste?

Industry
Everything except your
product is a waste.

Wastewater Treatment Plant
Everything coming out of
the plant is a waste.

What Is a "RCRA" Solid Waste?

The waste is not a "RCRA" solid waste if it is:
  Domestic sewage (261.4 Subpart a)
  Clean Water Act, point source discharge
  Irrigation return flow
  AEC source, nuclear
  In-situ mining waste
  Pulping liquors from recovery furnace
  Spent sulfuric acid used to produce virgin sulfuric acid
  Reclaimed secondary materials
  Reclaimed spent wood preserving solutions and reclaimed wastewaters from this process
  EPA hazardous waste nos. K060, K087, K141, K143, K144, K145, K147, K148, and any wastes
    from coke by-products process that are hazardous only because they exhibit the characteristic of
    toxicity, when, following generation, these materials are recycled. (*Note*: This waste is only
    excluded when there is no land disposal from the point of generation to the point of recycling.)
  Nonwastewater splash-condenser dross residue from treatment of K061 in high-temperature metals
    recovery units, provided it is shipped in drums (if shipped), and not land disposed before recovery

All other wastes are "RCRA" solid wastes (solid, liquid, or gas)

What Is a Hazardous Waste?

The waste is not an "RCRA": hazardous solid waste if it is:
  Household waste
  Agricultural waste returned to the soil as fertilizer
  Mining overburden returned to the mine site
  Fly ash, scrubber sludge
  Waste associated with the production of crude oil, gas, or geothermal energy
  Wastes that fail the test for toxicity characteristic because trivalent chromium (Cr^{3+}) is found
    or wastes that are listed in Subpart D due to the presence of chromium
  Waste from extraction, benefication, and processing of ores and minerals
  Cement kiln dust waste
  Arsenic-treated wood
  Petroleum-contaminated media and debris
  Injected groundwater that is hazardous due to the toxicity characteristic
  Used chlorofluorocarbon refrigerants
  Nonterne-plated used oil filters
  Used oil re-refining distillation bottoms (that are used as feedstock to manufacture asphalt products)

The waste is not an "RCRA" hazardous solid waste unless
  It is listed in Part 261, Subpart D of "RCRA"
  It is a mixture including a listed (above) substance
  It exhibits any of the four specific hazardous waste characteristics

The waste is not an "RCRA" hazardous solid waste if:
  It has been excluded from the lists in Subpart D by petition (delisted)

All other wastes are "RCRA" hazardous solid wastes

FIGURE 12-3
Flow scheme for determining if waste is hazardous.

| **What Wastes Are Subject to Regulation?** |

→ The "RCRA" hazardous solid waste is currently not subject to Subtitle C regulations if:

    The total combined "RCRA" hazardous waste generated at the site is less than 100 kg/month

    It is intended to be legitimately reclaimed or reused, (261.6). However it is subject to RCRA reporting requirements regarding storage and transportion if it is a sludge or contains a Part 261 listed substance.

        The "RCRA" hazardous solid waste is temporarily exempt from certain regulations if:

    It is a hazardous waste that is generated in a product or raw material storage tank, a product or raw material transport vehicle or vessel, a product or raw material pipeline, or in a manufacturing process unit or an associated nonwaste-treatment-manufacturing unit.

All other "RCRA" hazardous solid wastes are subject to Subtitle C of RCRA regulation with respect to disposal, transport, and storage.

Requirements for Recyclable Materials

Hazardous wastes that are not subject to requirements for generator, transporters, and storage facilities:

Regulated under Subparts C through H (261.6):

    Recyclable materials used in a manner constituting disposal

    Hazardous wastes burned for energy recovery in boilers and industrial furnaces

    Recyclable materials from which precious metals are reclaimed

    Spent lead-acid batteries that are being reclaimed

Not subject to regulation or to the notification requirements of RCRA:

    Industrial ethyl alcohol that is reclaimed

    Used batteries returned to a battery manufacturer for regeneration

    Scrap metal

    Fuels produced from the refining of oil bearing hazardous wastes

    Oil reclaimed from hazardous waste resulting from normal petroleum refining, production, and transportation practices

    Hazardous waste fuel produced from oil-bearing hazardous wastes

    Petroleum coke produced from petroleum refinery hazardous wastes

    Used oil that is recycled and is also a hazardous waste solely because it exhibits a hazardous characteristic is not subject to the requirements of Parts 260–268 of this chapter, but is regulated under Part 279 of this chapter.

FIGURE 12-3
(*continued*)

RCRA provides a petition mechanism (40 CFR 260.20 and 260.22) for excluding a waste from nonspecific sources and at a particular generating facility. Those wastes that successfully pass the petition process are *delisted*. The list of delisted wastes appears in Appendix IX of 40 CFR 261.

Some waste streams do not come under the purview of RCRA but are, nonetheless, considered hazardous. These special wastes include, for example, polychlorinated biphenyls (PCBs) and asbestos. PCBs and asbestos are regulated under the Toxic Substances Control Act (abbreviated TSCA and pronounced "tas-kah").

12-3 RCRA AND HSWA

Congressional Actions on Hazardous Waste

In 1976 Congress passed the Resource Conservation and Recovery Act (abbreviated RCRA and pronounced "rick-rah") directing the U.S. Environmental Protection Agency to establish hazardous waste regulations. RCRA was amended in 1984 by the Hazardous and Solid Waste Amendments (abbreviated HSWA and pronounced "hiss-wah"). RCRA and HSWA were enacted to regulate the generation and disposal of hazardous wastes. These acts did not address abandoned or closed waste disposal sites or spills. The Comprehensive Environmental Response, Compensation, and Liability Act (abbreviated by CERCLA and pronounced "sir-klah"), commonly referred to as "Superfund," was enacted in 1980 to address these problems. SARA, the Superfund Amendments and Reauthorization Act of 1986, extended the provisions of CERCLA. In the following sections, we shall attempt to tell you about the who, what, where, and how of RCRA, HSWA, CERCLA, and SARA.

Cradle-to-Grave Concept

The EPA's cradle-to-grave hazardous waste management system is an attempt to track hazardous waste from its generation point (the "cradle") to its ultimate disposal point (the "grave"). The system requires generators to attach a manifest (itemized list describing the contents) form to their hazardous waste shipments. This procedure is designed to ensure that wastes are directed to, and actually reach, a permitted disposal site.

Generator Requirements

Generators of hazardous waste are the first link in the cradle-to-grave chain of hazardous waste management established under RCRA. Generators of more than 100 kilograms of hazardous waste or 1 kilogram of acutely hazardous waste per month must (with a few exceptions) comply with all of the generator regulations.

The regulatory requirements for hazardous waste generators include (U.S. EPA, 1986):

1. Obtaining an EPA ID number

2. Handling of hazardous waste before transport

3. Manifesting of hazardous waste

4. Recordkeeping and reporting

EPA assigns each generator a unique identification number. Without this number the generator is barred from treating, storing, disposing of, transporting, or offering for transportation any hazardous waste. Furthermore, the generator is forbidden from offering the hazardous waste to any transporter, or treatment, storage, or disposal (TSD) facility that does not also have an EPA ID number.

Pretransport regulations are designed to ensure safe transportation of a hazardous waste from origin to ultimate disposal. In developing these regulations, EPA adopted those used by the Department of Transportation (DOT) for transporting hazardous wastes (49 CFR Parts 172, 173, 178, and 179). These DOT regulations require:

1. Proper packaging to prevent leakage of hazardous waste during both normal transport conditions and in potentially dangerous situations, such as when a drum falls out of a truck.

2. Identification of the characteristics and dangers associated with the wastes being transported through labeling, marking, and placarding of the packaged waste.

These pretransport regulations only apply to generators shipping waste off-site.

In addition to adopting the DOT regulations outlined above, EPA also developed pretransport regulations that cover the accumulation of waste prior to transport. A generator may accumulate hazardous waste on site for 90 days or less as long as the following requirements are met:

1. *Proper storage:* The waste is properly stored in containers or tanks marked with the words "Hazardous Waste" and the date on which accumulation began.

2. *Emergency plan:* There is a contingency plan and emergency procedures for use in an emergency.

3. *Personnel training:* Facility personnel are trained in the proper handling of hazardous waste.

The 90-day period allows a generator to collect enough waste to make transportation more cost effective, that is, instead of paying to haul several small shipments of waste, the generator can accumulate waste until there is enough for one big shipment.

If the generator accumulates hazardous waste on-site for more than 90 days, it is considered an operator of a storage facility and must comply with requirements for such facilities. Under temporary, unforeseen, and uncontrollable circumstances, the 90-day period may be extended for up to 30 days by the EPA Regional Administrator on a case-by-case basis.

There is an exception to this 90-day accumulation period that applies to generators of between 100 and 1,000 kg/mo of hazardous waste who ship their waste off-site. People who fall in this category are called *small quantity generators* (SQG). HSWA required, and the EPA developed, regulations that allow such generators to accumulate waste for 180 days (or 270 days if the waste must be shipped over 320 km) before they are considered the operator of a storage facility.

The Uniform Hazardous Waste Manifest (the *manifest*) is the key to cradle-to-grave waste management (see Figure 12-4). Through the use of a manifest, generators can track the movement of hazardous waste from the point of generation to the point of ultimate treatment, storage, or disposal.

HSWA requires that each manifest certify that the generator has in place a program to reduce the volume and toxicity of the waste to the degree that is economically practicable, as determined by the generator, and that the treatment, storage, or disposal method chosen by the generator is the best practicable method currently available that minimizes the risk to human health and the environment.

Please print or type.   *(Form designed for use on elite (12-pitch) typewriter.)*

Form Approved. OMB No. 2000-0404. Expires 7-31-86

UNIFORM HAZARDOUS WASTE MANIFEST	1. Generator's US EPA ID No.	Manifest Document No.	2. Page 1 of	Information in the shaded areas is not required by Federal law.

3. Generator's Name and Mailing Address

A. State Manifest Document Number

B. State Generator's ID

4. Generator's Phone (   )

5. Transporter 1 Company Name	6.	US EPA ID Number	C. State Transporter's ID

D. Transporter's Phone

7. Transporter 2 Company Name	8.	US EPA ID Number	E. State Transporter's ID

F. Transporter's Phone

9. Designated Facility Name and Site Address	10.	US EPA ID Number	G. State Facility's ID

H. Facility's Phone

11. US DOT Description *(Including Proper Shipping Name, Hazard Class, and ID Number)*	12. Containers No.	Type	13. Total Quantity	14. Unit Wt/Vol	I. Waste No.
a.					
b.					
c.					
d.					

J. Additional Descriptions for Materials Listed Above

K. Handling Codes for Wastes Listed Above

15. Special Handling Instructions and Additional Information

16. GENERATOR'S CERTIFICATION: I hereby declare that the contents of this consignment are fully and accurately described above by proper shipping name and are classed, packed, marked, and labeled, and are in all respects in proper condition for transport by highway according to applicable international and national government regulations.

Unless I am a small quantity generator who has been exempted by statute or regulation from the duty to make a waste minimization certification under Section 3002(b) of RCRA, I also certify that I have a program in place to reduce the volume and toxicity of waste generated to the degree I have determined to be economically practicable and I have selected the method of treatment, storage, or disposal currently available to me which minimizes the present and future threat to human health and the environment.

Printed/Typed Name	Signature	Month	Day	Year

17. Transporter 1 Acknowledgement of Receipt of Materials

Printed/Typed Name	Signature	Month	Day	Year

18. Transporter 2 Acknowledgement of Receipt of Materials

Printed/Typed Name	Signature	Month	Day	Year

19. Discrepancy Indication Space

20. Facility Owner or Operator: Certification of receipt of hazardous materials covered by this manifest except as noted in Item 19.

Printed/Typed Name	Signature	Month	Day	Year

GENERATOR / *TRANSPORTER* / *FACILITY*

EPA Form 8700-22 (Rev. 4-85) Previous edition is obsolete.

FIGURE 12-4

Uniform hazardous waste manifest.

It is especially important for the generators to prepare the manifest properly because they are responsible for the hazardous waste they produce and its ultimate disposition.

The manifest is part of a controlled tracking system. Each time the waste is transferred, that is, from a transporter to the designated facility or from a transporter to another transporter, the manifest must be signed to acknowledge receipt of the waste. A copy of the manifest is retained by each link in the transportation chain. Once the waste is delivered to the designated facility, the owner or operator of the facility must send a copy of the manifest back to the generator. This system ensures that the generator has documentation that the hazardous waste has made it to its ultimate destination.

If 35 days pass from the date on which the waste was accepted by the initial transporter and the generator has not received a copy of the manifest from the designated facility, the generator must contact the transporter and/or the designated facility to determine the whereabouts of the waste. If 45 days pass and the manifest still has not been received, the generator must submit an exception report.

The recordkeeping and reporting requirements for generators provide EPA and states with a method for tracking the quantities of waste generated and the movement of hazardous wastes.

Transporter Regulations

Transporters of hazardous waste are the critical link between the generator and the ultimate off-site treatment, storage, or disposal of hazardous waste. The transporter regulations were developed jointly by EPA and the DOT to avoid contradictory requirements coming from the two agencies (U.S. EPA, 1986). Although the regulations are integrated, they are not contained under the same act. A transporter must comply with the regulations under 49 CFR 171-179 (The Hazardous Materials Transportation Act) as well as those under 40 CFR Part 263 (Subtitle C of RCRA).

Even if generators and transporters of hazardous waste comply with all appropriate regulations, transporting hazardous waste can still be dangerous. There is always the possibility that an accident will occur. To deal with this possibility, the regulations require transporters to take immediate action to protect health and the environment if a release occurs by notifying local authorities and/or diking off the discharge area.

The regulations also give officials special authority to deal with transportation accidents. Specifically, if a federal, state, or local official, with appropriate authority, determines that the immediate removal of the waste is necessary to protect human health or the environment, the official can authorize waste removal by a transporter who lacks an EPA ID and without the use of a manifest.

Treatment, Storage, and Disposal Requirements

Treatment, storage, and disposal facilities (TSDs) are the last link in the cradle-to-grave hazardous waste management system. All TSDs handling hazardous waste must obtain an operating permit and abide by the treatment, storage, and disposal regulations. The TSD regulations establish performance standards that owners and operators must apply to minimize the release of hazardous waste into the environment.

A TSD facility may perform one or more of the following functions (U.S. EPA, 1986):

1. *Treatment:* Any method, technique, or process, including neutralization, designed to change the physical, chemical, or biological character or composition of any hazardous waste so as to neutralize it or render it nonhazardous or less hazardous; to recover it; make it safer to transport, store, or dispose of; or make it amenable for recovery, storage, or volume reduction.

2. *Storage:* The holding of hazardous waste for a temporary period, at the end of which the hazardous waste is treated, disposed, or stored elsewhere.

3. *Disposal:* The discharge, deposit, injection, dumping, spilling, leaking, or placing of any solid waste or hazardous waste into or on any land or water so that any constituent thereof may enter the environment or be emitted into the air or discharged into any waters, including groundwaters.

The act establishes standards that consist of administrative-nontechnical requirements and technical requirements.

The purpose of the administrative/nontechnical requirements is to ensure that owners and operators of TSDs establish the necessary procedures and plans to operate a facility properly and to handle any emergencies or accidents. They cover the subject areas shown below:

Subpart	Subject
A	Who is subject to the regulations
B	General facility standards
	Waste analysis, security, inspections, training
	Ignitable, reactive, or incompatible wastes
	Location standards (permitted facilities)
C	Preparedness and prevention
D	Contingency plans and emergency procedures
E	Manifest system, recordkeeping, and reporting

The objective of the interim status technical requirements is to minimize the potential for threats resulting from hazardous waste treatment, storage, and disposal at existing facilities waiting to receive an operating permit. There are two groups of interim status requirements: general standards that apply to several types of facilities and specific standards that apply to a waste management method.

The general standards cover three areas:

1. Groundwater monitoring requirements

2. Closure, postclosure requirements

3. Financial requirements

Groundwater monitoring is only required of owners or operators of a surface impoundment, landfill, land treatment facility, or some waste piles used to manage hazardous waste. The purpose of these requirements is to assess the impact of a facility on the groundwater beneath it. Monitoring must be conducted for the life of the facility except at land disposal facilities, which must continue monitoring for up to 30 years after the facility has closed.

The groundwater monitoring program outlined in the regulations requires a monitoring system of four wells to be installed: one upgradient from the waste management unit and three downgradient. The downgradient wells must be placed so as to intercept any waste migrating from the unit, should such a release occur. The upgradient wells must provide data on groundwater that is not influenced by waste coming from the waste management unit (called background data). If the wells are properly located, comparison of data from upgradient and downgradient wells should indicate if contamination is occurring.

Once the wells have been installed, the owner or operator monitors them for one year to establish background concentrations for selected chemicals. These data form the basis for all future data comparisons. There are three sets of parameters for which background concentrations are established: drinking water parameters, groundwater quality parameters, and groundwater contamination parameters.

Closure is the period when wastes are no longer accepted, during which owners or operators of TSD facilities complete treatment, storage, and disposal operations, apply final covers to or cap landfills, and dispose of or decontaminate equipment, structures, and soil. Postclosure, which applies only to disposal facilities, is the 30-year period after closure during which owners or operators of disposal facilities conduct monitoring and maintenance activities to preserve and look after the integrity of the disposal system.

Financial requirements were established to ensure that funds are available to pay for closing a facility, for rendering postclosure care at disposal facilities, and to compensate third parties for bodily injury and property damage caused by sudden and nonsudden accidents related to the facility's operation (states and federal governments are exempted from abiding by these requirements). There are two kinds of financial requirements: financial assurance for closure/postclosure and liability coverage for injury and property damage.

Land Ban. The Hazardous and Solid Waste Amendments (HSWA) of 1984 significantly expanded the scope of the Resource Conservation and Recovery Act (RCRA). HSWA was created, in large part, in response to strongly voiced citizen concerns that existing methods of hazardous waste disposal, particularly land disposal, were not safe. Section 3004 of the act sets restrictions on land disposal of specific wastes. This is commonly called the "land ban," or *land disposal restrictions* (LDR). As specifically required by Section 3004(m), the agency established levels or methods of treatment, if any, which substantially reduce the likelihood of migration of hazardous constituents from waste so that short-term and long-term threats to human health and the environment are minimized. Congress established a stringent timetable for development of treatment standards. After the effective date of the promulgated standards, listed and characteristic wastes must be treated to meet the standards before the wastes can be placed in any form of land disposal facility. The only exception is where a special variance is approved based on a showing of no migration of hazardous constituents from the land disposal site for as long as the waste remains hazardous. The last set of congressionally mandated standards was promulgated in accordance with the timetable on May 8, 1990. EPA has subsequently published revisions to clarify and streamline the standards. The *Universal Treatment Standards* (UTS) are of particular note in this respect.

Prior to 1994, treatment facilities managing hazardous waste often had to meet LDR treatment standards established for many different listed and characteristic wastes. In some cases, a constituent regulated to a given concentration level for one waste was also regulated in another waste at a different concentration level. On September 18, 1994, EPA published the UTS to eliminate these differences (59 FR 47980, 18 SEP 1994, and 60 FR 242, 3 JAN 1995).

Underground Storage Tanks (UST)

A "UST system"* includes an underground storage tank, connected piping, underground ancillary equipment, and containment system, if any. On September 23, 1988, the EPA promulgated the final rules for underground storage tanks.

There are a number of exclusions to the new regulations, including:

Hazardous waste UST systems

Regulated wastewater treatment facilities

Any equipment or machinery that contains regulated substances for operational purposes such as hydraulic lift tanks and electrical equipment tanks

Any UST system of less than 415 liters

Any UST system containing a *de minimis* (negligible) concentration of regulated substances

Any emergency spill or overflow containment system that is expeditiously emptied after use

All UST systems must have corrosion protection. There are three ways to obtain corrosion protection for tanks: (1) construction of fiberglass-reinforced plastic, (2) steel- and fiberglass-reinforced plastic composite, or (3) a coated steel tank with cathodic protection. Cathodic protection systems must be regularly tested and inspected. All owners and operators must also provide spill and overfill prevention equipment and a certificate of installation to ensure that the methods of installation were in compliance with the regulations.

Release (leak) detection must be instituted for all UST systems. Several different methods are allowed for petroleum UST systems. However, some systems have specific requirements, for instance, a pressurized delivery system must be equipped with an automatic line leak detector and have an annual line tightness test. All new or upgraded UST systems storing hazardous substances must have secondary containment with interstitial monitoring.

---

*You can imagine the chagrin of regulators and others when the acronym for Leaking Underground Storage Tanks appears on meeting agendas and technical symposia!

When release is confirmed, owners and operators must begin corrective action. Immediate corrective action measures include mitigation of safety and fire hazards, removal of saturated soils and floating free product, and an assessment of further corrective action needed. As with any remediation situation, a corrective action plan may be required for long-term cleanups of contaminated soil and groundwater.

12-4 CERCLA AND SARA

The Superfund Law

The Comprehensive Environmental Response, Compensation, and Liability Act (CERCLA) of 1980, better known as "Superfund," became law "to provide for liability, compensation, cleanup and emergency response for hazardous substances released into the environment and the cleanup of inactive hazardous waste disposal sites." CERCLA was generally intended to give EPA authority and funds to clean up abandoned waste sites and to respond to emergencies related to hazardous waste. The law provides for both response and enforcement mechanisms. The four major provisions of the law establish:

1. A fund (the "superfund") to pay for investigations and remedies at sites where the responsible people cannot be found or will not voluntarily pay;

2. A priority list of abandoned or inactive hazardous waste sites for cleanup (the National Priority List);

3. The mechanism for action at abandoned or inactive sites (the National Contingency Plan);

4. Liability for those responsible for cleaning up.

Initially the trust fund was supported by taxes on producers and importers of petroleum and 42 basic chemicals. In its first five-year period, Superfund collected about $1.6 billion, with 86 percent of that money coming from industry and the remainder from federal government appropriations. In 1986 the Superfund Amendments and Reauthorization Act (SARA) greatly expanded the money available to remediate Superfund sites. The fund was raised to $8.6 billion for a five-year period by taxing petroleum products ($2.75 billion), business income ($2.5 billion), and chemical feedstocks ($1.4 billion). The remainder is from general revenues.

The National Priority List (NPL)

The NPL serves as a tool for the EPA to use in identifying sites that appear to present a significant risk to public health or the environment and that may merit use of Superfund money. First published in 1982, it is updated three times a year. In September 2005 the list contained 1,239 sites (U.S. EPA, 2005b). The first NPL was formulated from notification procedures and existing information sources. Subsequently, a numeric ranking system known as the *Hazard Ranking System* (HRS) was developed. Sites with high HRS scores may be added to the list. Sites on the NPL are eligible for Superfund money. Those with lower scores are not likely to be eligible.

The Hazard Ranking System (HRS)

The HRS is a procedure for ranking uncontrolled hazardous waste sites in terms of the potential threat based upon containment of the hazardous substances, route of release, characteristics and amount of the substances, and likely targets (40 CFR 300, Appendix A). The methodology of the HRS provides a quantitative estimate that represents the relative hazards posed by a site and takes into account the potential for human and environmental exposure to hazardous substances. The HRS score is based on the probability of contamination from four pathways—groundwater, surface water, soil, and air—on the site in question. The groundwater and air migration pathways are evaluated for ingestion and inhalation respectively. The surface water migration and soil exposure pathways are evaluated for multiple intake routes. Surface water is evaluated for (1) drinking water, (2) human food chain, and (3) environmental (contact) exposures. These exposures are evaluated for two separate migration components—overland/flood migration and groundwater to surface water migration. Soil is evaluated for potential exposure to the (1) resident population and (2) nearby population.

Use of the HRS requires considerable information about the site and its surroundings, the hazardous substances present, and the geology of the aquifers and the intervening strata. The factors that most affect an HRS site score are the proximity to a densely populated area or source of drinking water, the quantity of hazardous substances present, and the toxicity of those hazardous substances. The HRS methodology has been criticized for the following reasons:

1. There is a strong bias toward human health effects, with only a slight chance of a site in question receiving a high score if it represents only a threat or hazard to the environment.

2. Because of the human health bias, there is an even stronger bias in favor of highly populated affected areas.

3. The air emission migration route must be documented by an actual release, while groundwater and surface water routes have no such documentation requirement.

4. The scoring for toxicity and persistence of chemicals may be based on site containment, which is not necessarily related to a known or potential release of the toxic chemicals.

5. A high score for one migration route can be more than offset by low scores for the other migration routes.

6. Averaging of the route scores creates a bias against a site that has only one hazard, even though that one hazard may pose extreme threat to human health and the environment.

The HRS scores range from 0 to 100, with a score of 100 representing the most hazardous site. Occasional exceptions have been made in the HRS priority ranking to meet the CERCLA requirement that a site designated by a state as its top priority be included on the NPL.

The National Contingency Plan (NCP)

The *National Contingency Plan* (NCP) provides detailed direction on the action to be taken at a hazardous waste site, including initial assessment to determine if an emergency or imminent threat exists, emergency response actions, and a method to rank sites (the HRS) and establish priority for future action. When there is sufficient indication that a site poses a potential risk to the environment, a detailed study is required.

The NCP describes the steps to be taken for the detailed evaluation of the risks associated with a site. Such an evaluation is termed a *remedial investigation* (RI). The process of selecting an appropriate remedy is termed the *feasibility study* (FS). The remedial investigation and the feasibility study are often combined into a single measure, known popularly as a remedial investigation/feasibility study (RI/FS). The requirements of the RI/FS are usually outlined in a written work plan, which must be approved by the relevant federal and state agencies before it may be implemented.

A remedial investigation includes the development of detailed plans that address the following items (40 CFR 300.400):

1. *Site characterization:* A description of the hydrogeological and geophysical sampling and analytical procedures to be applied in order to discover the nature and extent of the waste materials, the physical characteristics of the site, and any receptors that could be affected by the wastes at the site.

2. *Quality control:* The guidelines to be enforced to ensure that all the data collected from the characterization program are valid and satisfactorily accurate.

3. *Health and safety:* The procedures to be employed to protect the safety of the individuals who will work at the site and perform the site characterization.

The RI activities and subsequent evaluation of the data gathered are termed a *risk assessment* or an endangerment assessment. The remedial investigation report documents the evaluation.

The remedial investigation report serves as a basis for the feasibility study, which evaluates various remedial alternatives. The review criteria include: overall protection of human health and the environment; compliance with applicable or relevant and appropriate regulations; long-term effectiveness; reduction in toxicity, mobility, or volume; short-term effectiveness; technical and administrative implementability; cost; state acceptance; and community acceptance. All the remedies selected must be capable of reducing the risk at the hazardous waste site to an acceptable level. And, in general, the lowest-cost alternative that achieves this objective is chosen as the course of action. The results of the feasibility study are presented in a written report, called the *record of decision* (ROD). This document serves as a preliminary basis for the design of the selected alternative.

One of the keys to the National Contingency Plan is that it specifies that the degree of cleanup be selected in accordance with several criteria, including the degree of hazard to the "public health, welfare and the environment." Therefore, there is no predetermined level of remediation that can be required or that must be achieved at any site. Rather, the degree of correction is established on a site-by-site basis. What is acceptable in one location may not necessarily be acceptable in another.

On completion and approval of the RI/FS, the next step is the preparation of plans and specifications for the selected remedy—the *remedial design* (RD). To complete the process, the actual construction and other activities are undertaken in accordance with the plans and specifications.

Liability

Perhaps the most far-reaching provision of CERCLA that has stood the test of the courts was the establishment of *strict, joint and several liability* for cleanup of an NPL site. Those identified by EPA as *potentially responsible parties* (PRPs) may include generators, present owners, or former owners of facilities or real property where hazardous wastes have been stored, treated, or disposed of, as well as those who accepted hazardous waste for transport and selected the facility. PRPs have strict liability; that is, liability without fault. Neither care nor negligence, neither good nor bad faith, neither knowledge nor ignorance, can be claimed as a defense. Congress correctly predicted that there would be instances where the PRPs would contest their contribution to the problem and would, then, be unwilling to share the costs or the responsibility. The strict liability provision orders that the PRP is liable even if the method of disposal was in accordance with prevailing standards, laws, and practice at the time of disposal. In other words, CERCLA is a "pay now, argue later" statute (O'Brien & Gere, 1988).

Although the language specific to "joint and several" liability was removed from CERCLA, the courts have interpreted the law as though the language were included. This means that if a PRP contributed any wastes to a site, that PRP can be held accountable for all costs associated with the cleanup. This concept was strongly reaffirmed in SARA. If the PRP refuses to pay, the federal government can sue to recover costs. These actions have been successful. In certain instances where those liable fail, without sufficient cause, to properly provide for cleanup, they may be liable for treble damages!

Superfund Amendments and Reauthorization Act (SARA)

SARA reaffirmed and strengthened many of the provisions and concepts of the CERCLA program. In SARA, Congress clearly expressed a preference, but not a requirement, for remedies such as incineration or chemical treatment that render a waste nonhazardous rather than transport to another disposal site or simple containment on site.

SARA directs that the level of cleanup should achieve compliance with *Applicable or Relevant and Appropriate Requirements* (ARARs). ARARs are environmental standards from programs other than CERCLA and SARA. For example, if a state has a regulation regarding atmospheric emissions from incinerators, then a Superfund cleanup using incineration must meet those *applicable* standards. Furthermore, if a similar standard appears to be *relevant* and *appropriate,* then EPA may elect to apply it. For example, if drums of waste found on an uncontrolled hazardous waste site have contents that appear to have the same constituents as F001-F005 spent solvent, then the UTS standards for RCRA waste may be considered relevant and appropriate even though there is no specific evidence to identify the origin of the waste.

SARA significantly strengthens the requirement to consider damages to natural resources, especially those off-site.

Title III.   SARA includes a major addition to the provisions of CERCLA, namely Title III—Emergency Planning and Community Right-to-Know. Under the Emergency Planning provisions, facilities must notify the State Emergency Response Commission if they have quantities of extremely hazardous substances that exceed EPA specified *Threshold Planning Quantities* (TPQ). In addition, communities must establish Local Emergency Planning Committees (LEPCs) to develop a chemical emergency response plan. This plan must include identification of regulated facilities, emergency response and notification procedures, training programs, and evacuation plans in case of a chemical release.

If a facility accidentally releases chemicals that are on one of two lists [that is, EPA's Extremely Hazardous Substance list or the CERCLA Section 103(a) list], in regulated quantities (RQ), and the release has the potential for exposure off-site, they must notify the LEPC immediately. The law also requires a report on response actions taken, known or anticipated health risks, and advice on medical attention for exposed individuals.

Perhaps the most revolutionary provision of Title III is the establishment of the *Community's Right-to-Know* amounts of chemicals and their location in facilities in their community. Thus, information about potential hazards from chemicals is available to the public. In addition, each year those facilities that release chemicals above specified threshold amounts must submit a *Toxic Release Inventory* (TRI) on an EPA-specified form ("Form R"). This inventory includes both accidental and routine releases, as well as off-site shipments of waste. The publication of these data has resulted in strenuous efforts by industry to control their previously unregulated and, hence, uncontrolled emissions because of the public outcry at the large quantities of materials being dumped into their environment.

12-5   HAZARDOUS WASTE MANAGEMENT

A logical priority in managing hazardous waste would be to:

1. Reduce the amount of hazardous wastes generated in the first place.

2. Stimulate "waste exchange." (One factory's hazardous wastes can become another's feedstock; for instance, acid and solvent wastes from some industries can be utilized by others without processing.)

3. Recycle metals, the energy content, and other useful resources contained in hazardous wastes.

4. Detoxify and neutralize liquid hazardous waste streams by chemical and biological treatment.

5. Reduce the volume of waste sludges generated in number four, above, by dewatering.

6. Destroy combustible hazardous wastes in special high-temperature incinerators equipped with proper pollution-control and monitoring systems.

7. Stabilize/solidify sludges and ash from numbers five and six to reduce leachability of metals.

8. Dispose of remaining treated residues in specially designed landfills.

Waste Minimization

The key elements necessary to the success of a waste-minimization program include (Fromm et al., 1986):

Top-level organizational commitment

Financial resources

Technical resources

Appropriate organization, goals, and strategy

The commitment of senior management is the first element that must be in place. Efforts to establish the other elements can follow. The organizational structure adopted should promote communication and feedback from participants. Often, the best ideas come from line operators who work with the processes day in and day out.

Some firms set quantitative waste-minimization goals. Other firms are more qualitative in their goal setting.

Waste Audit. An important first step in establishing a strategy for waste minimization is to conduct a waste audit. The audit should proceed stepwise:

1. Identify waste streams

2. Identify sources

3. Establish priority of waste streams for waste-minimization activity

4. Screen alternatives

5. Implement

6. Track

7. Evaluate progress

The key question that must be asked at the outset of a waste audit is "why is this waste being generated?" You must first establish the primary cause(s) of waste generation before attempting to find solutions. The audit should be waste stream–oriented in order to produce a list of specific waste-minimization options for additional evaluation or implementation. Once the causes are understood, solution options can be formulated. An efficient materials and waste tracking system that allows computation of mass balances is useful in establishing priorities. Knowing how much material is going in and how much of it is ending up as waste allows you to decide which process and which waste to address first.

Example 12-1.   A manufacturing company has, as part of their first audit, gathered the following data. Estimate the potential annual air emissions in kg of VOCs from the company.

Purchasing department records

Material	Purchase Quantity (barrels)
Methylene chloride (CH_2Cl_2)	228
Trichloroethylene (C_2HCl_3)	505

Wastewater treatment plant influent

Material	Average Concentration (mg/L)
CH_2Cl_2	4.04
C_2HCl_3	3.25

(Average flow into treatment plant is 0.076 m^3/s.)

Hazardous waste manifests

Material	Barrels	Concentration (%)
CH_2Cl_2	228	25
C_2HCl_3	505	80

Unused barrels at end of year

CH_2Cl_2	8
C_2HCl_3	13

Solution.   The materials balance diagram will be the same for each waste.

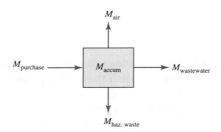

The mass balance equation would be

$$M_{\text{purchase}} = M_{\text{air}} + M_{\text{ww}} + M_{\text{hw}} + M_{\text{accum}}$$

Solving this equation for M_{air} gives us the estimated VOC emission.

First, we calculate the mass purchased. The density of each compound is found in Appendix A.

Mass Purchased

$$M(CH_2Cl_2) = (228 \text{ barrels/y})(0.12 \text{ m}^3/\text{barrel})(1,326 \text{ kg/m}^3)$$
$$= 36,279.36 \text{ kg/y}$$

$$M(C_2HCl_3) = (505 \text{ barrels/y})(0.12 \text{ m}^3/\text{barrel})(1,476 \text{ kg/m}^3)$$
$$= 89,445.60 \text{ kg/y}$$

Now we calculate the mass received at the wastewater treatment plant. (Note that $mg/L = g/m^3$.)

$$M(CH_2Cl_2) = (4.04 \text{ g/m}^3)(0.076 \text{ m}^3/\text{s})(86,400)(365)(10^{-3})$$
$$= 9,682.81 \text{ kg/y}$$

$$M(C_2HCl_3) = (3.25)(0.076)(86,400)(365)(10^{-3})$$
$$= 7,789.39 \text{ kg/y}$$

The mass shipped to the hazardous waste disposal facility is calculated next.

$$M(CH_2Cl_2) = (228)(0.12)(1,326)(0.25) = 9,069.84 \text{ kg/y}$$
$$M(C_2HCl_3) = (505)(0.12)(1,476)(0.80) = 71,556.48 \text{ kg/y}$$

Accumulated

$$M(CH_2Cl_2) = (8)(0.12)(1,326) = 1,272.96 \text{ kg/y}$$
$$M(C_2HCL_3) = (13)(0.12)(1,476) = 2,302.56 \text{ kg/y}$$

The estimated air emission for each compound is then

$$M(CH_2Cl_2) = 36,279.36 - 9,682.81 - 9,069.84 - 1,272.96$$
$$= 16,253.75, \text{ or } 16,000, \text{ kg/y}$$

$$M(C_2HCL_3) = 89,445.60 - 7,789.39 - 71,556.48 - 2,302.56$$
$$= 7,797.17, \text{ or } 7,800, \text{ kg/y}$$

Comments:

1. Note that we round to two significant figures because the volume of the barrels is known to only two significant figures.

2. From this analysis, to reduce the mass of air pollutants emitted, the company should attack the methylene chloride source first. We should also point out that simply counting "barrels in" from the purchasing record and "barrels out" from the hazardous waste manifest would give a highly erroneous picture of the environmental impact of this company's emissions.

3. From a waste minimization point of view, it is also apparent that C_2HCl_3, at 80 percent concentration in barrels going to hazardous waste disposal, is a candidate for recycling.

The first four steps of the waste audit allow you to generate a comprehensive set of waste management options following the hierarchy of source reduction first, waste exchange second, recycling third, and treatment last.

The screening of options begins with source control. The source control investigation should focus on (1) changes in input materials, (2) changes in process technology, and (3) changes in the human aspect of production. Input material changes can be classified into three separate elements: purification, substitution, and dilution.

Purification of input materials is performed in order to avoid the introduction of inerts or impurities into the production process. Such an introduction results in waste because the process inventory must be purged in order to prevent the undesirable accumulation of impurities. Examples of purification of feed materials to lower waste generation include the use of deionized rinse water in electroplating or the use of oxygen instead of air in oxychlorination reactors for production of ethylene dichloride.

Substitution is the replacement of a toxic material with one characterized by lower toxicity or higher environmental desirability. Examples include using phosphates in place of dichromates as cooling water corrosion inhibitors or the use of alkaline cleaners in place of chlorinated solvents for degreasing.

Dilution is a minor component of input material changes and is exemplified by use of more dilute plating solutions to minimize *dragout* (material carried out of one tank into another).

Technology changes are those made to the physical plant. Examples include process changes, equipment, piping or layout changes, changes to process operational settings, additional automation, energy conservation, and water conservation.

Procedural and/or institutional changes consist of improvements in the ways people affect the production process. Also referred to as "good operating practices" or "good housekeeping," these include operating procedures, loss prevention, waste segregation, and material handling improvements.

Waste Exchange

Waste minimization by consignment of excess unused materials to an independent party for resale to a third party, saves both in waste production and in the cost (environmental and financial) of production from new raw materials. In essence "one person's trash becomes another person's treasure." The difference between a manufacturing by-product, which is costly to treat or dispose, and a usable or salable by-product involves opportunity, knowledge of processes outside the generator's immediate production line, and comparative pricing of virgin material. Waste exchanges serve as information clearinghouses through which the availability and need for various types of materials can be established.

Recycling

Under RCRA and HSWA, EPA has carefully defined recycling to prohibit bogus recyclers that are really TSDs from taking advantage of more lenient rules for recycling. The definition says that a material is *recycled* if it is used, reused, or reclaimed (40 CFR 261.1 (c)(7)). A material is "used or reused" if it is either (1) employed as an ingredient (including its use as an intermediate) to make a product (however, a material will not satisfy this condition if distinct components of the material are recovered as separate end products, as when metals are recovered from metal-containing secondary materials); or (2) employed in a particular function as an effective substitute for a commercial product (40 CFR 261.1 (c)(5)). A material is *reclaimed* if it is processed to recover a useful product or if it is regenerated. Examples include the recovery of lead from spent batteries and the regeneration of spent solvents (40 CFR 261.1 (c)(4)) (U.S. EPA, 1988a).

Distillation processes can be utilized to recover spent solvent. The principal characteristics that determine the potential for recovery are the boiling points of the various useful constituents and the water content. The more dilute the waste solvent, the less economical it is to recover. Recovered solvents can be reused by the generator or sold for at least a substantial fraction of the cost of virgin material, and the credit for recovered solvent can more than offset the cost of recovery.

There are several technologies for recovery of metals from metal-plating rinse water. Most are applicable only to waste streams containing a single metal constituent. Examples include ion exchange, electrodialysis, evaporation, and reverse osmosis.

In October 1988 a federal appeals court struck down an EPA policy not to list used oil collected for recycling as a hazardous waste. Prior to that ruling, PCB-contaminated oils, petroleum industry sludges, and leaded tank bottoms were the only oils regulated. The majority of oil and oily wastes generated were not classified as hazardous under EPA regulations. These wastes are amenable either to recovery for use as fuel or to refinement for use as lubricants. Although all waste oil is now deemed hazardous, those oils that were formerly recovered may still be recovered, but the requirements for tracking them are more stringent.

12-6 TREATMENT TECHNOLOGIES

The wastes that remain after the implementation of waste minimization must be detoxified and neutralized. There are a large number of treatment technologies available to accomplish this. Many of these are applications of processes we have discussed in earlier chapters. Examples include: biological oxidation (Chapter 8), chemical precipitation, ion exchange, and oxidation-reduction (Chapter 6), and carbon adsorption (Chapter 9). Here we will discuss these as they apply to hazardous waste treatment, and we will introduce some new technologies.

Biological Treatment

In contrast to naturally occurring compounds, *anthropogenic compounds* (those created by human beings) are relatively resistant to biodegradation. One reason is that the organisms that are naturally present often cannot produce the enzymes necessary to bring about transformation of the original compound to a point at which the resultant intermediates can enter into common metabolic pathways and be completely mineralized.

Many environmentally important anthropogenic compounds are halogenated, and halogenation is often implicated as a reason for their persistence. The list of halogenated organic compounds includes pesticides, plasticizers, plastics, solvents, and trihalomethanes. Chlorinated compounds are the best known and most studied because of the highly publicized problems associated with DDT and other pesticides and numerous industrial solvents. Hence, chlorinated compounds serve as the basis for most of the information available on halogenated compounds.

Some of the characteristics that appear to confer persistence to halogenated compounds are the location of the halogen atom, the halide involved, and the extent of halogenation (Kobayashi and Rittman, 1982). The first step in biodegradation, then, is sometimes dehalogenation, for which there are several biological mechanisms.

Simple generalizations do not appear to be applicable. For example, until recently, oxidative pathways were mostly believed to be the typical means by which halogenated compounds were dehalogenated. Anaerobic, reductive dehalogenation, either biological or nonbiological, is now recognized as the critical factor in the transformation or biodegradation of certain classes of compounds. Compounds that require reductive dechlorination are common among the pesticides, as well as halogenated one- and two-carbon aliphatic compounds.

Reductive dehalogenation involves the removal of a halogen atom by oxidation-reduction. In essence, the mechanism involves the transfer of electrons from reduced organic substances via microorganisms or a nonliving (abiotic) mediator, such as inorganic ions (for example, Fe^{3+}) and biological products (for example, NAD(P), flavin, flavoproteins, hemoproteins, porphyrins, chlorophyll, cytochromes, and glutathione). The mediators are responsible for accepting electrons from reduced organic substances and transferring them to the halogenated compounds. The major requirements for the process are believed to be available free electrons and direct contact between the donor, mediator, and acceptor of electrons. Significant reductive dechlorination usually occurs only when the oxidation-reduction potential of the environment is 0.35 V and lower; the exact requirements appear to depend upon the compound involved (Kobayashi and Rittman, 1982).

Although simple studies using pure cultures of microorganisms and single substrates are valuable, if not essential, for determining biochemical pathways, they cannot always be used to predict biodegradability or transformation in more natural situations. The interactions among environmental factors, such as dissolved oxygen, oxidation-reduction potential, temperature, pH, availability of other compounds, salinity, particulate matter, competing organisms, and concentrations of compounds and organisms, often control the feasibility of biodegradation. The compound's physical or chemical characteristics, such as solubility, volatility, hydrophobicity, and octanol-water partition coefficient, contribute to the compound's availability in solution. Often compounds not soluble in the water are not readily available to organisms for biodegradation. There are some exceptions. For example, DDT, which is only slightly soluble in water, may be degraded by the white rot fungus found on decaying trees. This is because the enzymes involved in the white rot reaction are secreted from the cell (Kobayashi and Rittman, 1982).

Simple culture studies are similarly inadequate for predicting the fate of substances in the environment if there are many interactions between different organisms. First, substances that cannot be changed significantly in pure culture studies often will be degraded or transformed under mixed culture conditions. A good example of this type of interaction is *cometabolism,* in which a compound, the nongrowth substrate, is not metabolized as a source of carbon or energy, but is incidentally transformed by organisms using other compounds as growth substrates. The growth substrates provide the energy needed to cometabolize the nongrowth substrates. Second, products of the initial transformation by one organism may subsequently be broken down by a series of different organisms until compounds that can be metabolized by normal metabolic pathways are formed. An example is the degradation of DDT, which is reportedly mineralized directly by only one organism, a fungus; other organisms studied appear to degrade DDT only through cometabolism, resulting in numerous transformation products that subsequently can be used by other organisms. For example, *Hydrogenomonas* can metabolize DDT only as far as p-chlorophenylacetic acid (PCPA), while *Arthrobacter* species can then remove the PCPA (Kobayashi and Rittman, 1982).

Table 12-2 demonstrates that members of almost every class of anthropogenic compound can be degraded by some microorganism. The table also illustrates the wide variety of microorganisms that participate in environmentally significant biodegradation.

TABLE 12-2
Examples of anthropogenic compounds and microorganisms that can degrade them

Compound	Organism
Aliphatic (nonhalogenated)	
Acrylonitrile	Mixed culture of yeast mold, protozoan bacteria
Aliphatic (halogenated)	
Trichloroethane, trichloroethylene, methyl chloride, methylene chloride	Marine bacteria, soil bacteria, sewage sludge
Aromatic compounds (nonhalogenated)	
Benzene, 2,6-dinitrotoluene, creosol, phenol	*Pseudomonas* sp., sewage sludge
Aromatic compounds (halogenated)	Sewage sludge
1,2-; 2,3-; 1,4-dichlorobenzene, hexachlorobenzene, trichlorobenzene	
Pentachlorophenol	Soil microbes
Polycyclic aromatics (nonhalogenated)	
Benzo(a)pyrene, naphthalene	*Cunninghamella elegans*
Benzo(a)anthracene	*Pseudomonas*
Polycyclic aromatics (halogenated)	
PCBs	*Pseudomonas, Flavobacterium*
4-Chlorobiphenyl	Fungi
Pesticides	
Toxaphene	*Corynebacterium pyrogenes*
Dieldrin	Anacystic nidulans
DDT	Sewage sludge, soil bacteria
Kepone	Treatment lagoon sludge
Nitrosamines	
Dimethylnitrosamines	*Rhodopseudomonas*
Phthalate esters	Micrococcus 12B

(*Source:* Extracted from Table 1 of Kobayashi and Rittman, 1982.)

The metabolic capabilities of many microorganisms, in particular algae and oligo-trophic bacteria, are not well understood. Such knowledge is necessary if limiting reactions are to be determined and the proper types of organisms selected for specific applications. More information about appropriate types of microorganisms to be selected and maintained in "real-world" treatment systems is needed, especially for the more novel microbial cultures. In order to develop special-purpose organisms by genetic manipulation, major advances in the understanding of the genetic structure of the many different types of organisms in nature are needed.

Conventional biological treatment processes such as activated sludge and trickling filters have been used to treat hazardous wastes. The major modification to the activated sludge processes has been to extend the mean cell residence time from the conventional values of 4 to 15 days to much longer periods of 3 to 6 months. In a similar fashion, trickling filter loading rates are much lower than those employed in municipal treatment systems. One innovation that has been adopted by TSD facilities is the *sequencing batch reactor* (SBR). The SBR is a periodically operated, fill-and-draw reactor (Herzbron et al., 1985). Each reactor in an SBR system has five discrete periods in each cycle: fill, react, settle, draw, and idle. Biological reactions are initiated as the raw wastewater fills the tank. During the fill and react phase, the waste is aerated in the same fashion as an activated sludge unit. After the react phase, the mixed liquor suspended solids (MLSS) are allowed to settle. The treated supernatant is discharged during the draw phase. The idle stage, the time between the draw and fill, may be zero or may be a few days de-pending on wastewater flow demand. The SBR has a major advantage in that wastes may be tested for completeness of treatment before discharge.

Chemical Treatment

Chemical detoxification is a treatment technology, either employed as the sole treat-ment procedure or used to reduce the hazard of a particular waste prior to transport, incineration, and burial.

It is important to remember that a chemical procedure cannot magically make a toxic chemical disappear from the *matrix* (wastewater, sludge, etc.) in which it is found, but can only convert it to another form. Thus, it is vital to ensure that the products of a chemical detoxification step are less of a problem than the starting material. It is equally important to remember that the reagents for such a reaction can be hazardous.

The spectrum of chemical methods includes complexation, neutralization, oxida-tion, precipitation, and reduction. An optimum method would be fast, quantitative, in-expensive, and leave no residual reagent, which itself would be a pollution problem. The following paragraphs describe a few of these techniques.

Neutralization.   Solutions are neutralized by a simple application of the law of mass balance to bring about an acceptable pH. Sulfuric or hydrochloric acid is added to basic solutions, while caustic (NaOH) or slaked lime [$Ca(OH)_2$] is added to acidic solutions. Though a waste is hazardous at pH values less than 2 or greater than 12.5, and it would seem that simply bringing the pH into the range 2 to 12.5 would be adequate, good treat-ment practice requires that final pH values be in the range 6 to 8 to protect natural biota.

Oxidation. The cyanide molecule is destroyed by oxidation. Chlorine is the oxidizing agent most frequently used. Oxidation must be conducted under alkaline conditions to avoid the generation of hydrogen cyanide gas. This process is often referred to as alkaline chlorination. In chlorine oxidation, the reaction is carried out in two steps:

$$NaCN + 2NaOH + Cl_2 \rightleftharpoons NaCNO + 2NaCl + H_2O \qquad (12\text{-}1)$$

$$2NaCNO + 5NaOH + 3Cl_2 \rightleftharpoons 6NaCl + CO_2 + N_2 + NaHCO_3 + 2H_2O \quad (12\text{-}2)$$

In the first step, the pH is maintained above 10 and the reaction proceeds in a matter of minutes. In this step, great care must be taken to maintain relatively high pH values, because at lower pHs there is a potential for the evolution of highly toxic hydrogen cyanide gas. The second reaction step proceeds most rapidly around a pH of 8, but it is not as rapid as the first step. Higher pH values may be selected for the second step to reduce chemical consumption in the following precipitation steps. This increases the reaction time. Often the second reaction is not carried out because the CNO is considered nontoxic by current regulations.

Ozone also may be used as the oxidizing agent. Ozone has a higher redox potential than chlorine, thus there is a higher driving force toward the oxidized state. When ozone is used, the pH considerations are similar to those discussed for chlorine. Ozone cannot be purchased. It must be made on-site as part of the process.

This technology can be applied to a wide range of cyanide wastes: copper, zinc, and brass plating solutions; cyanide from cyanide salt heating baths; and passivating solutions. The process has been practiced on an industrial scale since the early 1940s. For extremely high cyanide concentrations (>1 percent), oxidation may not be desirable. Cyanide complexes of metals, particularly iron and to some extent nickel, cannot be decomposed easily by cyanide oxidation techniques.

Electrolytic oxidation of cyanide is carried out by anodic electrolysis at high temperatures. The theoretical basis of the process is that cyanide reacts with oxygen in solution in the presence of an electric potential to produce carbon dioxide and nitrogen gas. Normally, the destruction is carried out in a closed cell. Two electrodes are suspended in the solution and a DC current is applied to drive the reaction. The bath temperature must be maintained in the range of 50 to 95°C.

This technology is used for the destruction of cyanide in concentrated spent stripping solutions; in plating solutions for copper, zinc, and brass; in alkaline descalers; and in passivating solutions. It has been more successful for wastes containing high concentrations of cyanide (50,000 to 100,000 mg/L), but it has also been successfully used for concentrations as low as 500 mg/L.

Chemical oxidation methods for organic compounds in wastewater have received extensive study. In general they apply only to dilute solutions and often are considered expensive in comparison to the biological methods. Some examples include wet air oxidation, hydrogen peroxide, permanganate, chlorine dioxide, chlorine, and ozone oxidation. Of these, wet air oxidation and ozonation have shown promise as a pretreatment step for biological processes.

Wet air oxidation, also known as the Zimmerman process, operates on the principle that most organic compounds can be oxidized by oxygen given sufficient temperature and pressure. Wet air oxidation may be described as the aqueous phase oxidation of dissolved or suspended organic particles at temperatures of 175 to 325°C and sufficiently high pressure to prevent excessive evaporation. Air is bubbled through the liquid. The process is fuel efficient; once the oxidation reaction has started, it is usually self-sustaining. As this method is not limited by reagent cost, it is potentially the most widely applicable of all chemical oxidation methods. The method has been shown to be of use in destroying a wide range of organic compounds, including some pesticides. Although wet oxidation can provide acceptable levels of destruction for many hazardous compounds, it generally is not as complete as incineration. In many instances, the addition of metal salt catalysts can increase the destruction efficiency or allow the process to be run at lower temperature and/or pressure.

Precipitation. Metals are often removed from plating rinse waters by precipitation. This is a direct application of the solubility product principle (see Chapter 5). By raising the pH with lime or caustic, the solubility of the metal is reduced (Figure 12-5) and the metal hydroxide precipitates. Optimum removal is achieved by selecting the optimum pH as shown in Figure 12-5. Though there is an optimum for each metal, in many cases, the metals are mixed and the lowest value for an individual metal may not be achievable for the mixture.

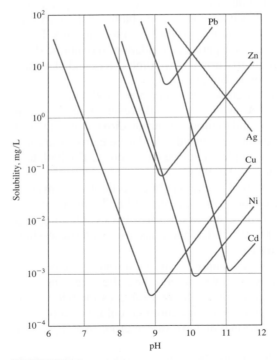

FIGURE 12-5
Solubilities of metal hydroxides as a function of pH.
(*Source:* U.S. EPA, 1981b.)

Example 12-2. A metal plating firm is installing a precipitation system to remove zinc. They plan to use a pH meter to control the feed of hydroxide solution to the mixing tank. What pH should the controller be set at to achieve a zinc effluent concentration of 0.80 mg/L? The K_{sp} of $Zn(OH)_2$ is 7.68×10^{-17}.

Solution. From Table A-9 in Appendix A we find that the zinc hydroxide reaction is

$$Zn^{2+} + 2OH^- \rightleftharpoons Zn(OH)_2$$

As shown in Chapter 5, we can write the solubility product equation as

$$K_{sp} = [Zn^{2+}][OH^-]^2$$

Because we want the zinc concentration to be no greater than 0.80 mg/L, we calculate the moles per liter of zinc.

$$[Zn^{2+}] = \frac{0.80 \text{ mg/L}}{(65.41 \text{ g/mol})(1,000 \text{ mg/g})} = 1.223 \times 10^{-5} \text{ moles/L}$$

Now we solve for the hydroxide concentration.

$$[OH^-]^2 = \frac{7.68 \times 10^{-17}}{1.223 \times 10^{-5}} = 6.28 \times 10^{-12}$$

$$= (6.28 \times 10^{-12})^{\frac{1}{2}} = 2.51 \times 10^{-6}$$

The pOH is

$$pOH = -\log(2.505 \times 10^{-6}) = 5.60$$

And the pH set point for the controller is

$$pH = 14 - pOH$$
$$= 14 - 5.60 = 8.4$$

Reduction. Although most heavy metals readily precipitate as hydroxides, hexavalent chromium used in plating solutions must be reduced to trivalent chromium before it will precipitate. Reduction is usually done with sulfur dioxide (SO_2) or sodium bisulfite ($NaHSO_3$). With SO_2 the reaction is

$$3SO_2 + 2H_2CrO_4 + 3H_2O \rightleftharpoons Cr_2(SO_4)_3 + 5H_2O \qquad (12\text{-}3)$$

Because the reaction proceeds rapidly at low pH, an acid is added to control the pH between 2 and 3.

Physical/Chemical Treatment

Several treatment processes are used to separate hazardous waste from aqueous solution. The waste is not detoxified but only concentrated for further treatment or recovery.

Carbon Adsorption. Adsorption is a mass transfer process in which gas vapors or chemicals in solution are held to a solid by intermolecular forces (for example, hydrogen bonding and van der Waals' interactions). It is a surface phenomenon. Pressure vessels having a fixed bed are used to hold the adsorbent. Activated carbon, molecular sieves, silica gel, and activated alumina are the most common adsorbents. The active sites become saturated at some point in time. When the organic material has commercial value, the bed is then regenerated by passing steam through it. The vapor-laden steam is condensed and the organic fraction is separated from the water. If the organic compounds have no commercial value, the carbon may be either incinerated or shipped to the manufacturer for regeneration. Carbon systems for recovery of vapor from degreasers and for polishing wastewater effluents have been in commercial application for over 20 years.

Distillation.   The separation of more volatile materials from less volatile materials by a process of vaporization and condensation is called *distillation*. When a liquid mixture of two or more components is brought to the boiling point of the mixture, a vapor phase is created above the liquid phase. If the vapor pressures of the pure components are different (which is usually the case), then the constituent(s) having the higher vapor pressure will be more concentrated in the vapor phase than the constituent(s) having the lower vapor pressure. If the vapor phase is cooled to yield a liquid, a partial separation of the constituents will result. The degree of separation depends on the relative differences in the vapor pressures. The larger the differences, the more efficient the separation. If the difference is large enough, a single separation cycle of vaporization and condensation is sufficient to separate the components. If the difference is not large enough, multiple cycles (stages) are required. Four types of distillation may be used: batch distillation, fractionation, steam stripping, and thin film evaporation.*

Both batch distillation and fractionation are well-proven technologies for recovery of solvents. Batch distillation is particularly applicable for wastes with high solids concentrations. Fractionation is applicable where multiple constituents must be separated and where the waste contains minimal suspended solids.

When the volatility of the organic compound is relatively high and the concentration relatively low, then some form of stripping may be appropriate. *Air stripping* has been used to purge large quantities of contaminated groundwater of small concentrations of volatile organic matter. The behavior of the process is the inverse of absorption discussed in Chapter 9. Air and contaminated liquid are passed countercurrently through a packed tower. The volatiles evaporate into the air, leaving a clean liquid stream. The contaminated air stream must then be treated to avoid an air pollution problem. Frequently this is accomplished by passing the air through an activated carbon column. The carbon is then incinerated. Air stripping has been used to remove tetrachloroethylene, trichloroethylene, and toluene from water (Gross and TerMaath, 1985; U.S. EPA, 1987).

The air stripper design equation may be developed in the same fashion as the absorber equation in Chapter 9. It is given here without that development:

$$Z_T = \frac{L}{A} \frac{\ln\left[\dfrac{C_1}{C_2} - \dfrac{LRT_g}{GH_c}\left(\dfrac{C_1}{C_2} - 1\right)\right]}{K_L a\left(1 - \dfrac{LRT_g}{GH_c}\right)} \qquad \text{(12-4)}$$

---

*Air stripping, though not strictly a distillation process because the condensation step is omitted, employs the same general principles of volatilization and, hence, is included in the discussion.

where
Z_T = depth of packing in tower, m
L = water flow, m^3/min
A = cross-sectional area of tower, m^2
G = air flow, m^3/min
H_c = Henry's constant, atm · m^3/mol
R = universal gas constant = 8.206×10^{-5} atm · m^3/mole · K
T_g = temperature of air, K
C_1, C_2 = influent and effluent organic concentration in the water, mol/m^3
K_L = overall mass transfer coefficient, mol/min · m^2 · mol/m^3
a = effective interfacial area of packing per unit volume for mass transfer, m^2/m^3

Another form of Equation 12-4 is shown in the end-of-chapter Problem 12-19.
Realistic values of the air-to-water ratio (G/L) range from 5 to several hundred. In an actual design, a safety factor of 20 percent would be added to Z_t. The column holding the packing would be somewhat larger to accommodate support structures and distribution piping (LaGrega et al., 2001).

Example 12-3. Well 12A at the City of Tacoma, WA, is contaminated with 350 μg/L of 1,1,2,2-tetrachloroethane. The water must be cleaned to the detection limit of 1.0 μg/L. Design a packed tower stripping column to meet this requirement using the following design parameters.

Henry's law constant = 5.0×10^{-4} atm · m^3/mol
$K_L a = 10 \times 10^{-3}$ s^{-1}
Air flow rate = 13.7 m^3/s
Liquid flow rate = 0.044 m^3/s
Temperature = 25°C
Column diameter may not exceed 4.0 m
Column height may not exceed 6.0 m

Solution. The Henry's law constants given in Appendix A are in kPa · m^3/moles. To convert these to atm · m^3/mole, divide by the atmospheric pressure at standard conditions, that is, 101.325 kPa/atm.

The stripper equation is then solved for $Z_T A$, the column volume.

$$Z_T A = (0.044) \frac{\ln\left[\dfrac{350}{1} - \dfrac{(0.044)(8.206 \times 10^{-5})(298)}{(13.7)(5.0 \times 10^{-4})} \left(\dfrac{350}{1} - 1 \right) \right]}{10 \times 10^{-3}\left[1 - \dfrac{(0.044)(8.206 \times 10^{-5})(298)}{(13.7)(5.0 \times 10^{-4})} \right]}$$

$$= (0.044)(6.75 \times 10^2)$$

$$= 29.7 \ m^3$$

Any number of solutions are now possible within the boundary conditions of 4 m diameter and 6 m height. For example, with the 20 percent safety factor and rounding,

Diameter (m)	Z_T (m)	Tower height (m)
4.00	2.36	3
3.34	3.39	5

For gases of lower volatility or higher concentration (>100 ppm) *steam stripping* may be employed. The physical arrangement of the process is much like that of an air stripper, except that steam is introduced instead of air. The addition of steam enhances the stripping process by decreasing the solubility of the organic in the aqueous phase and by increasing the vapor pressure. Steam stripping has been used to treat aqueous waste contaminated with chlorinated hydrocarbons, xylenes, acetone, methyl ethyl ketone, methanol, and pentachlorophenol. Concentrations treated range from 100 ppm to 10 percent organic compound (U.S. EPA, 1987).

Recovery of metals by evaporation is accomplished by boiling off sufficient water from the collected rinse stream to allow the concentrate to be returned to the plating bath. The condensed steam is recycled for use as rinse water. The boil-off rate, or evaporator duty, is set to maintain the water balance of the plating bath. Evaporation is usually performed under vacuum to prevent thermal degradation of additives and to reduce the amount of energy required for evaporation of the water.

There are four types of evaporators: rising film, flash evaporators using waste heat, submerged tube, and atmospheric pressure. Rising film evaporators are built so that the evaporative heating surface is covered by a wastewater film and does not lie in a pool of boiling wastewater. Flash evaporators are of similar configuration, but the plating solution is continuously recirculated through the evaporator along with the wastewater. This allows the use of waste heat in the plating bath to augment the evaporation process. In the submerged tube design, the heating coils are submerged in the wastewater. Atmospheric evaporators do not recover the distillate for reuse and they do not operate under vacuum.

Ion Exchange. Metals and ionized organic chemicals can be recovered by ion exchange. Ion exchange chemistry was discussed in Chapter 6. In ion exchange, the waste stream containing the ion to be removed is passed through a bed of resin. The resin is selected to remove either cations or anions. In the exchange process, ions of like charge are removed from the resin surface in exchange for ions in solution. Typically, either hydrogen or sodium is exchanged for cations (metal) in solution. When the bed becomes saturated with the exchanged ion, it is shut down and the resin is regenerated by passing a concentrated solution containing the original ion (hydrogen or sodium) back through the bed. The exchanged pollutant is forced off the bed in a concentrated form that may be recycled. A typical ion exchange column

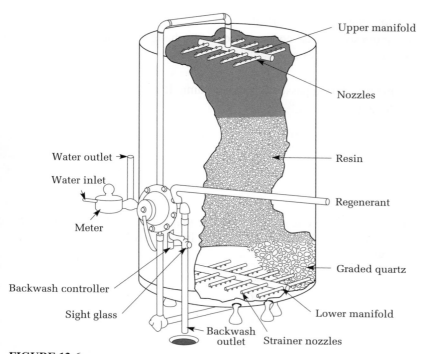

FIGURE 12-6
Typical ion exchange resin column.
(*Source:* U.S. EPA, 1981a)

is shown in Figure 12-6. A prefilter is required to remove suspended material that would hydraulically foul the column. It also removes organic contaminants and oils that would foul the resin.

As a rule, ion exchange systems are suitable for chemical recovery applications where the rinse water feed has a relatively dilute concentration ($< 1,000$ mg/L) and a relatively low concentration is required for recycle. Ion exchange has been demonstrated commercially for recovery of plating chemicals from acid-copper, acid-zinc, nickel, tin, cobalt, and chromium plating baths.

The breakthrough curves for an ion-exchange column and an adsorption column (Chapter 9) are similar. Thomas (1948) proposed a kinetic equation to describe the removal of the contaminant in the column:

$$\ln\left(\frac{C_o}{C} - 1\right) = \frac{(k)(q_o)(M)}{Q} - \frac{(k)(C_o)(\forall)}{Q} \tag{12-5}$$

where C_o = influent solute concentration, mg/L or milliequivalents/
L (meq/L)

C = effluent solute concentration, mg/L or milliequivalents/
L (meq/L)

k = rate constant, L/d · equivalent

q_o = maximum solid phase concentration of exchanged solute, equivalents/kg of resin

M = mass of resin, kg

$\mathcal{V}$ = volume of solution passed through column, L

Q = flow rate, L/d

This equation is of the form $y = mx + b$

where $y = \ln\left(\dfrac{C_o}{C} - 1\right)$

$x = \mathcal{V}$

This allows us to determine the rate constant and the maximum solid phase concentration from a plot of $\ln(C_o/C - 1)$ versus $\mathcal{V}$ as shown in Figure 12-7.

The slope of the line is equal to

$$\frac{kC_o}{Q}$$

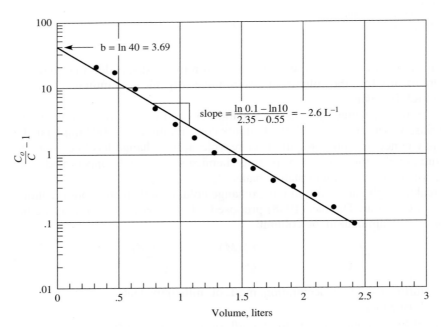

FIGURE 12-7
Plot of breakthrough data to estimate kinetic equation constants.
(Note: ordinate scale is logarithm to the base e.)

and the intercept is equal to

$$\frac{(k)(q_o)(M)}{Q}$$

Data from a laboratory or pilot scale breakthrough curve are required to obtain the plot. The same flow rate, in terms of bed volumes per unit time, should be used for both the pilot studies and the full-scale column.

Example 12-4. An electroplating rinse water containing 49 mg/L of zinc is to be treated by an ion exchange column to meet an allowable effluent concentration of 2.6 mg/L. A laboratory scale column has provided the breakthrough data shown in the first two columns of the table on page 906. The laboratory column data are as follows:

Inside diameter = 1.0 cm

Length = 10.0 cm

Mass of resin (moist basis) = 5.2 g

Water content = 17%

Density of dry resin = 0.65 g/cm^3

Liquid flow rate = 7.87 L/d

Initial concentration of zinc = 49 mg/L

The full-scale design must meet the following requirements:

Flow rate = 36,000 L/d

Hours of operation = 8 h/d

Regeneration is to be once every 5 days

Determine the mass of resin required.

Solution. The laboratory breakthrough data are converted to the form of Equation 12-5 in the following table. The initial concentration of zinc (C_o) is 49 mg/L. The meq/L is determined by first finding the equivalent weight (See Chapter 5) as

$$\frac{GMW}{n} = \frac{65.41 \text{ g/mole}}{2 \text{ eq/mole}} = 32.71 \text{ g/eq or mg/meq}$$

and dividing the concentration of zinc by its equivalent weight. The initial concentration (C_o) in meq/L is

$$\frac{49 \text{ mg/L}}{32.71 \text{ mg/meq}} = 1.50 \text{ meq/L}$$

Breakthrough Data

V, L	C, mg/L	C, meq/L	$\frac{C_0}{C} - 1$
0.32	2.25	0.06826	20.973
0.48	2.74	0.08313	17.044
0.64	4.56	0.13835	9.8421
0.80	8.32	0.25243	4.9423
0.96	12.74	0.38653	2.8807
1.12	17.70	0.53701	1.7932
1.28	23.54	0.71420	1.1003
1.44	27.48	0.83374	0.7991
1.60	30.58	0.92779	0.6167
1.76	35.34	1.07221	0.3990
1.92	37.02	1.12317	0.3355
2.08	39.38	1.19478	0.2555
2.24	42.50	1.28944	0.1632
2.40	45.10	1.36833	0.0962
2.56	44.10	1.33799	0.1211

The plot of these data is shown in Figure 12-7.
From the plot

$$k = (\text{slope})\left(\frac{Q}{C_0}\right) = (2.6 \text{ L}^{-1})\left(\frac{7.87 \text{ L/d}}{1.50 \text{ meq/L}}\right)$$

$$= 13.64 \text{ L/d} \cdot \text{meq}$$

and

$$q_0 = \frac{(b)(Q)}{(k)(M)} = \frac{(3.69)(7.87 \text{ L/d})}{(13.64 \text{ L/d} \cdot \text{meq})(4.316 \text{ g})}$$

$$= 0.4933 \text{ meq/g}$$

Note that the mass of the test column resin (M) is corrected for moisture, that is, the dry weight is

$$(5.2 \text{ g})(1 - 0.17) = 4.316 \text{ g}$$

Using these values of k and q_0 and reapplying Equation 12-5 we can determine the mass of resin for the full-scale column. Because the effluent concentration must not exceed 2.6 mg/L, we can solve the left-hand side of the equation as:

$$\ln\left(\frac{49}{2.6} - 1\right) = 2.882$$

The first term on the right-hand side contains the unknown (M). Using the constants determined above, and the daily flow rate, it may be simplified to

$$\frac{(13.64 \text{ L/d} \cdot \text{meq})(0.4933 \text{ meq/g})(M)}{36,000 \text{ L/d}} = 1.87 \times 10^{-4} (M)$$

Using a flow rate of 36,000 L/d and a 5 day operating cycle, the volume to treat (V) is

$$(36,000 \text{ L/d})(5 \text{ d}) = 180,000 \text{ L}$$

The second term on the right-hand side of the equation is then

$$\frac{(13.64 \text{ L/d} \cdot \text{meq})(1.50 \text{ meq/L})(180,000 \text{ L})}{36,000 \text{ L/d}} = 102.30$$

Setting the left-hand side of the equation equal to the right-hand side and solving for M yields

$$2.882 = 1.87 \times 10^{-4}(M) - 102.30$$
$$M = 5.6 \times 10^{5} \text{ g or } 560 \text{ kg}$$

In full-scale operation, the resin bed is not allowed to reach saturation because the concentration of the solute will exceed most discharge standards before this occurs. Normal operation then requires either an operating cycle that will allow regeneration of the spent resin during nonworking hours or, in the case of 24 hour, 7 day per week schedules, multiple beds so that one may be taken off-line.

The diameters of ion exchange columns may vary from centimeters to 6 m. Resin bed depths range from 1 to 3 m. Bed height-to-diameter ratios range from 1.5:1 to 1:3. The column shell is designed to allow for 100 percent expansion of the resin bed during backwashing (regeneration). Columns are normally prefabricated and shipped by truck. Column height generally does not exceed 4 m. Multiple columns in series are provided where the design height exceeds 4 m. The maximum column diameter is often controlled by the clearance under bridges passing over the highway.

During ion exchange, the normal flow pattern is downward through the bed. The hydraulic loading may range from 25 to 600 $\text{m}^3/\text{d} \cdot \text{m}^2$. Lower hydraulic loadings result in longer contact periods and better exchange efficiency. Because the surface of the bed acts like a filter, regeneration is often countercurrent, that is, the regenerating solution is pumped into the bottom of the column. This results in a cleansing of the column much like the backwashing of a rapid sand filter cleans it. Regeneration hydraulic loadings range from 60 to 120 $\text{m}^3/\text{d} \cdot \text{m}^2$.

Electrodialysis. The electrodialysis unit uses a membrane to selectively retain or transmit specific molecules. The membranes are thin sheets of ion exchange resin reinforced by a synthetic fiber backing. The construction of the unit is such that anion membranes are alternated with cation membranes in stacks of cells in series (Figure 12-8). An electric potential is applied across the membrane to provide the motive force for ion migration. Cation membranes permit passage of only positively charged ions, while anion membranes permit passage of only negatively charged ions. The flow is directed through the membrane in two hydraulic circuits (Figure 12-9). One circuit is ion-depleted and the other is ion-concentrated. The degree of purification achieved in the dilute circuit is set by the electric potential. The ability to pass the charge is proportional to the concentration of the ionic species in the dilute stream. Because ion migration is proportional to electric potential, the optimum system is a trade-off between energy requirements and degree of contaminant removal.

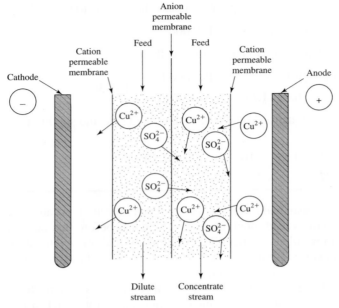

FIGURE 12-8

Electrodialysis. (Cations in the feed water show the same behavior as copper (Cu^{2+}) and anions show the same behavior as sulfate (SO_4^{2-}). Under the action of an electric field, cation-exchange membranes permit passage only of positive ions, while anion-exchange membranes permit passage only of negatively charged ions.) (*Source:* Davis and Masten, 2009.)

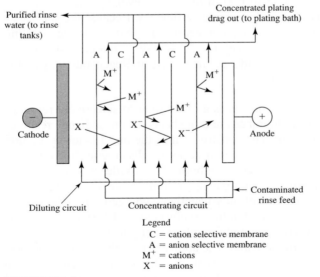

FIGURE 12-9

Electrodialysis unit flow schematic.

Electrodialysis has been in commercial operation for more than four decades in the production of potable water from brackish water. It has also been used in deashing of sugars, desalting of food products such as whey, and to recover waste developer in the photo processing industry and nickel from a metal-plating rinse water. Typically, electrodialysis can separate a waste stream containing 1,000 to 5,000 mg/L inorganic salts into a dilute stream that contains 100 to 500 mg/L salt and a concentrated stream that contains up to 10,000 mg/L salt.

Reverse Osmosis. Osmosis is defined as the spontaneous transport of a solvent from a dilute solution to a concentrated solution across an ideal semipermeable membrane that impedes passage of the solute but allows the solvent to flow. Solvent flow can be reduced by exerting pressure on the solution side of the membrane, as shown in Figure 12-10. If the pressure is increased above the osmotic pressure on the solution side, the flow reverses. Pure solvent will then pass from the solution into the solvent. As applied to metal finishing wastewater, the solute is the metal and the solvent is pure water.

Many configurations of the membrane are possible. The driving pressure is on the order of 1,000 to 5,500 kPa. No commercially available membrane polymer has demonstrated tolerance to all extreme chemical factors such as pH, strong oxidizing agents, and aromatic hydrocarbons. However, selected membranes have been demonstrated on nickel, copper, zinc, and chrome baths.

Solvent Extraction. Solvent extraction is also called *liquid extraction* and *liquid-liquid extraction*. Contaminants can be removed from a waste stream using liquid-liquid extraction if the wastewater is contacted with a solvent having a greater solubility for the target contaminants than the wastewater. The contaminants will tend to migrate from the wastewater into the solvent. Although predominately a method for separating organic materials, it may also be applied to remove metals if the solvent contains a material that will react with the metal. *Liquid ion exchange* is one kind of these reactions.

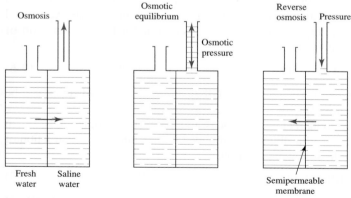

FIGURE 12-10
Direct and reverse osmosis.

In the solvent extraction process, the solvent and the waste stream are mixed to allow mass transfer of the constituent(s) from the waste to the solvent. The solvent, immiscible in water, is then allowed to separate from the water by gravity. The solvent solution containing the extracted contaminant is called the *extract*. The extracted waste stream with the contaminants removed is called the *raffinate*. As in distillation, the separation may need to be done in one or more stages. In general, more stages result in a cleaner raffinate. The degree of complexity of the apparatus varies from simple mixer/settlers to more exotic contacting devices. If the extract is sufficiently enriched, it may be possible to recover useful material. Distillation is often employed to recover the solvent and reusable organic chemicals. For metal recovery, the ion exchange material is regenerated by the addition of an acid or alkali. The process has found wide application in the ore processing industry, in food processing, in pharmaceuticals, and in the petroleum industry.

Incineration

In an incinerator, chemicals are decomposed by oxidation at high temperatures (800°C and greater). The waste, or at least its hazardous components, must be combustible in order to be destroyed. The primary products from combustion of organic wastes are carbon dioxide, water vapor, and inert ash. However, there are a multitude of other products that can be formed.

Products of Combustion.   The percentages of carbon, hydrogen, oxygen, nitrogen, sulfur, halogens, and phosphorus in the waste, as well as the moisture content, need to be known to determine stoichiometric combustion air requirements and to predict combustion gas flow and composition. Actual incineration conditions generally require excess oxygen to maximize the formation of *products of complete combustion* (POCs) and minimize the formation of *products of incomplete combustion* (PICs).

The incineration of halogenated organics results in the formation of halogenated acids, which require further treatment to ensure environmentally acceptable air emissions from the incineration process. Chlorinated organics are the most common halogenated hydrocarbons found in hazardous waste. The incineration of chlorinated hydrocarbons with excess air results in the formation of carbon dioxide, water, and hydrogen chloride. An example is the following reaction for the incineration of dichlorethane (Wentz, 1989):

$$2C_2H_4Cl_2 + 5O_2 \rightarrow 4CO_2 + 2H_2O + 4HCl \tag{12-6}$$

The hydrogen chloride must be removed before the carbon dioxide and steam can be safely exhausted into the atmosphere.

Hazardous waste may contain either organic or inorganic sulfur compounds. When these wastes are incinerated, sulfur dioxide is produced. For example, the destruction of ethyl mercaptan results in the following reaction:

$$2C_2H_5SH + 9O_2 \rightarrow 4CO_2 + 6H_2O + 2SO_2 \tag{12-7}$$

The sulfur dioxide produced by the incineration of sulfur-containing wastes must not exceed air quality standards.

Excess air must be provided to ensure complete combustion. However, the amount of the excess can only be determined empirically. For example, a highly volatile, clean, hydrocarbon waste would probably require much less excess air than would a heavy hydrocarbon sludge within a high solids content. Incineration of sludges and solids may require as much as two to three times excess air above stoichiometric equivalents. Too much excess air should be avoided because it increases the fuel required to heat the waste to destruction temperatures, reduces residence time for the hazardous wastes to be oxidized, and increases the volume of air emissions to be handled by the air pollution control equipment.

By-products from the incineration of hazardous wastes may also result from incomplete combustion as well as from the products of combustion. Products of incomplete combustion (PICs) include carbon monoxide, hydrocarbons, aldehydes, ketones, amines, organic acids, and polycyclic aromatic hydrocarbons (PAHs). In a well-designed incinerator, these products are insignificant in amount. However, in poorly designed or overloaded incinerators, PICs may pose environmental concerns. Polychlorinated biphenyls, for instance, decompose under such conditions into highly toxic chlorinated dibenzo furans (CDBF). The hazardous material, hexachlorocyclopentadiene (HCCPD), found in many hazardous wastes, is known to decompose into the even more hazardous compound hexachlorobenzene (HCB) (Oppelt, 1981).

Suspended particulate emissions are also produced during incineration. These include particles of mineral oxides and salts from the mineral constituents in the waste material, as well as fragments of incompletely burned combustibles.

Last, but not least, ash is a product of combustion. The ash is considered a hazardous waste. Metals not volatilized end up in the ash. Unburned organic compounds may also be found in the ash. When organic compounds remain, the ash may simply be incinerated. The metals must be treated prior to land disposal.

Design Considerations. The most important factors for proper incinerator design and operation are combustion temperature, combustion gas residence time, and the efficiency of mixing the waste with combustion air and auxiliary fuel.

Chemical and thermal dynamic properties of the waste that are important in determining its time/temperature requirements for destruction are its elemental composition, net heating value, and any special properties (for example, explosive properties) that may interfere with incineration or require special design considerations.

In general, higher heating values are required for solids versus liquids or gases, for higher operating temperatures, and for higher excess air rates if combustion is to be sustained without auxiliary fuel consumption. While sustained combustion (*autogenous combustion*) is possible with heating values as low as 9.3 MJ/kg, in the hazardous waste incineration industry it is common practice to blend wastes (and fuel oil, if necessary) to obtain an overall heating value of 18.6 MJ/kg or greater (Davis et al, 2000).

Blending is also used to limit the net chlorine content of chlorinated hazardous waste to a maximum of roughly 30 percent by weight to reduce chlorine concentrations in the combustion gas. The chlorine and, especially, hydrogen chloride that forms from the chlorine, are very corrosive. They oxidize the fire brick in the incinerator which causes it to fail.

Hazardous waste incinerators must be designed to achieve a 99.99 percent *destruction and removal efficiency* (DRE) of the *principal organic hazardous components* (POHCs) in the waste. This is commonly referred to as "four 9s DRE"; higher DREs may be referred to as five 9s, six 9s, that is, 99.999 and 99.9999 percent DRE, respectively. Because of the complexity of the wastes being burned, little success has been achieved in predicting the time and temperature requirements for achieving the 99.99 percent DRE. Empirical tests (*trial burns*) are required to demonstrate compliance. Experience has demonstrated that highly halogenated materials are more difficult to destroy than those with low halogen content.

Incinerator Types.   Two technologies dominate the incineration field: liquid injection and rotary kiln incinerators. Over 90 percent of all incineration facilities use one of these technologies. Of these, more than 90 percent are liquid injection units. Less commonly used incinerators include fluidized beds and starved air/pyrolysis systems.

Horizontal, vertical, and tangential liquid injection units are used. The majority of the incinerators for hazardous wastes inject liquid hazardous waste at 350 to 700 kPa through an atomizing nozzle into the combustion chamber. These liquid incinerators vary in size from 300,000 to 90 million Joules of heat released per second. An auxiliary fuel such as natural gas or fuel oil is often used when the waste is not autogenous. The liquid wastes are atomized into fine droplets as they are injected. A droplet size in the range 40 to 100 μm is obtained with atomizers or nozzles. The droplet volatilizes in the hot gas stream and the gas is oxidized. Efficient destruction of liquid hazardous wastes requires minimizing unevaporated droplets and unreacted vapors.

Residence time, temperature, and turbulence (often referred to as the "three Ts") are optimized to increase destruction efficiencies. Typical residence times are 0.5 to 2 seconds. Incinerator temperatures usually range between 800 and 1,600°C. A high degree of turbulence is desirable for achieving effective destruction of the organic chemicals in the waste. Depending on whether the liquid incinerator flow is axial, radial, or tangential, additional fuel burners and separate waste injection nozzles can be arranged to achieve the desired temperature, turbulence, and residence time. Vertical units are less likely to experience ash buildup. Tangential units have a much higher heat release and generally superior mixing.

The rotary kiln is often used in hazardous waste disposal systems because of its versatility in processing solid, liquid, and containerized wastes. Waste is incinerated in a refractory-lined rotary kiln, as shown in Figure 12-11. The shell is mounted at a slight incline from the horizontal plane to facilitate mixing the waste materials with circulating air. Solid wastes and drummed wastes are usually fed by a conveyor system or a ram. Liquids and pumpable sludges are injected through a nozzle. Noncombustible metal and other residues are discharged as ash at the end of the kiln.

Rotary kilns are typically 1.5 to 4 m in diameter and range in length from 3 to 10 m. Rotary kiln incinerators usually have a length-to-diameter ratio (*L/D*) of between two and eight. Rotational speeds range from 0.5 to 2.5 cm/s, depending on kiln periphery. High *L/D* ratios, along with slower rotational speeds, are used for wastes requiring longer residence times. The feed end of the kiln has airtight seals to adequately control the initial incineration reactions.

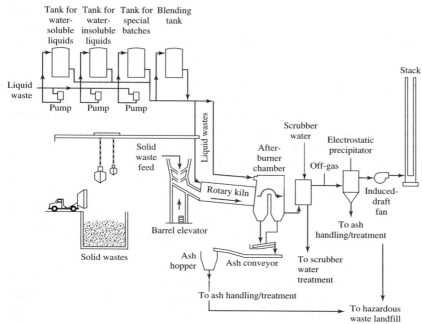

FIGURE 12-11
Rotary kiln incinerator.

Residence times for solid wastes are based on the rotational speed of the kiln and its angle. The residence time to volatilize waste is controlled by the gas velocity. The retention time of solids in the incinerator can be estimated from the following, where the coefficient 0.19 is based on limited experimental data:

$$\theta = \frac{0.19\,L}{NDS} \qquad (12\text{-}8)$$

where θ = retention time, min
L = kiln length, m
N = kiln rotational speed, rev/min
D = kiln diameter, m
S = kiln slope, m/m

Rotary kiln systems typically include secondary combustion chambers or afterburners to ensure complete destruction of the hazardous waste. Kiln operating temperatures range from 800 to 1,600°C. Afterburner temperatures range from 1,000 to 1,600°C. Liquid wastes are often injected into the secondary combustion chamber. The volatilized and combusted wastes leave the kiln and enter the secondary chamber, where additional oxygen is available and high heating value liquid wastes or fuel may be introduced. Both the secondary combustion chamber and the kiln are usually equipped with an auxiliary fuel firing system for startup.

Cement kilns are very efficient at destroying hazardous waste. Their long residence times and high operating temperatures exceed the requirements for destruction of most wastes. Hydrochloric acid generated from chlorinated hydrocarbon wastes is neutralized by the lime in the kiln while slightly lowering the alkalinity of the cement products. While cement plants can save energy by incinerating liquid wastes, the expense of obtaining permits and public resistance have inhibited use of this process.

Air Pollution Control (APC).   Typical APC equipment on an incinerator will include an afterburner, liquid scrubber, demister, and fine particulate control device. Afterburners are used to control emission of unburned organic by-products by providing additional combustion volume at an elevated temperature. Scrubbers are used to physically remove particulate matter, acid gases, and residual organics from the combustion gas stream. Metals, of course, are not destroyed in the incineration process. Some are volatilized and then collected in the air pollution control device. The large liquid droplets that escape from the scrubber are captured in a mist collector. The final stage in gas cleaning is to remove the fine particles that remain. Electrostatic precipitators have been used for this step. Scrubber water and residues from other APC devices are still considered hazardous and must be treated before ultimate land disposal.

Permitting of Hazardous Waste Incinerators.   The permitting of hazardous waste incinerators is a complex, multifaceted program conducted simultaneously on federal, state, and local levels. Because of the variety of state and local regulations for the handling, transportation, treatment, and disposal of hazardous wastes, as well as those concerning the operation of incinerators, each startup has a unique set of permit requirements.

Generally speaking, hazardous waste incinerators require at least the following permits: federal RCRA, state RCRA, for PCBs—the Toxic Substances and Control Act (TSCA), state and federal wastewater discharge, and state and federal air pollution control. A variety of local permits may also be necessary. Each of these require data substantiating an incinerator's operation at or above performance levels determined by environmental legislation. Each requires a public hearing and discussion of environmental impacts as well.

Hazardous waste incinerators must meet three performance standards (Theodore and Reynolds, 1987):

1. *Principal Organic Hazardous Constituents* (POHC). The DRE for a given POHC is defined as the mass percentage of the POHC removed from the waste. The POHC performance standard requires that the DRE for each POHC *designated* in the permit be 99.99 percent or higher. The DRE performance standard implicitly requires sampling and analysis to measure the amounts of the designated POHC(s) in both the waste stream and the stack effluent gas during a trial burn. (The term *designated POHC* is described in more detail later in this section.) The DRE is determined for each designated POHC from a mass balance of the waste introduced into the incinerator and in the stack gas*:

$$\text{DRE} = \frac{(W_{\text{in}} - W_{\text{out}})}{W_{\text{in}}} \times 100\% \qquad \qquad \text{PE} \ (12\text{-}9)$$

---

*Note that this is not a mass balance around the incinerator. Hazardous waste that ends up in the scrubber water, APC residue, and ash are not counted. Hence, the oxidation can be very poor and the incinerator can still meet the 99.99 percent rule if the scrubber is efficient and/or the waste ends up in the ash. This is one reason that residues are considered hazardous and must be treated before land disposal.

where W_{in} = mass feed rate of one POHC in the waste stream
W_{out} = mass emission rate of the same POHC present in exhaust emissions prior to release to the atmosphere

2. *Hydrochloric acid.* An incinerator burning hazardous waste and producing stack emissions of more than 1.8 kg/h of hydrogen chloride (HCl) must control HCl emissions such that the rate of emission is no greater than the larger of either 1.8 kg/h or 1 percent of the HCl in the stack gas prior to entering any pollution control equipment.

3. *Particulates.* Stack emissions of particulate matter are limited to 180 milligrams per dry standard cubic meter (mg/dscm) for the stack gas corrected to 7 percent oxygen. This adjustment is made by calculating a corrected concentration:

$$P_c = P_m \frac{14}{21 - Y} \qquad (12\text{-}10)$$

where P_c = corrected concentration of particulate, mg/dscm
P_m = measured concentration of particulate, mg/dscm
Y = percent oxygen in the dry flue gas

In this way, a decrease in the particulate concentration due solely to increasing air flow in the stack is not rewarded, and an increase in the particulate concentration due solely to reduction in the air flow in the stack is not penalized. Special rules for this calculation are being developed for oxygen-enriched combustion systems where the oxygen content is greater than the 21 percent found in the atmosphere.

Compliance with these performance standards is documented by a trial burn of the facility's waste streams. As part of the RCRA permit application, a trial burn plan detailing waste analysis, an engineering description of the incinerator, sampling and monitoring procedures, test schedule and protocol, as well as control information, must be developed. If EPA determines that the design is adequate, a temporary or draft permit is issued. This allows the owner or operator to build the incinerator and initiate the trial burn procedure.

The temporary permit covers four phases of operation. During the first phase, immediately following construction, the unit is operated for *shake-down* purposes to identify possible mechanical deficiencies and to ensure its readiness for the trial burn procedures. This phase of the permit is limited to 720 h of operation using hazardous waste feed. The trial burn is conducted during the second phase. This is the most critical component of the permitting process, because it demonstrates the incinerator's ability to meet the three performance standards. In addition, performance data collected during the trial burn phase are reviewed by the permitting official and become the basis for setting the conditions of the facility permit. These conditions are: (1) allowable waste analysis procedures, (2) allowable waste feed composition (including acceptable variations in the physical or chemical properties of the waste feed), (3) acceptable operating limits for carbon monoxide in the stack, (4) waste feed rate, (5) combustion temperature, (6) combustion gas flow rate, and (7) allowable variations in incinerator design and operating procedures (including a requirement for shutoff of waste feed during startup, shutdown, and at any time when conditions of the permit are violated).

To verify compliance with the POHC performance standard during the trial burn, it is not required that the incinerator DRE for every POHC identified in the waste be measured. The POHCs with the greatest potential for a low DRE, based on the expected difficulty of thermal degradation (incinerability) and the concentration of the POHC in the waste, become the *designated POHCs* for the trial burn. The EPA permit review personnel work with the owners/operators of the incinerator facility in determining which POHCs in a given waste should be designated for sampling and analysis during the trial burn.

If a wide variety of wastes are to be treated, a difficult-to-incinerate POHC at high concentration may be proposed for the trial burn. The substitute POHC is referred to as a *surrogate POHC*. The surrogate POHC does not have to be actually present in the normal waste. It does, however, have to be considered more difficult to incinerate than any POHC found in the waste.

The third phase consists of completing the trial burn and submitting the results. This phase can last several weeks to several months, during which the incinerator is allowed to operate under specified conditions. The data to be reported to regulatory agencies after the burn are: (1) a quantitative analysis of the POHCs in the waste feed, (2) a determination of the concentration of the particulates, POHCs, oxygen, and HCl in the exhaust gas, (3) a quantitative analysis of any scrubber water, ash residues, and other residues to determine the fate of the POHCs, (4) a computation of the DRE for the POHCs, (5) a computation of the HCl removal efficiency if the HCl emission rate exceeds 1.8 kg/h, (6) a computation of particulate emissions, (7) the identification of sources of fugitive emissions and their means of control, (8) a measurement of average, maximum, and minimum temperatures and combustion gas velocities (gas flows), (9) a continuous measurement of carbon monoxide (CO) in the exhaust gas, and (10) any other information EPA may require to determine compliance.

Provided that performance standards are met in the trial burn, the facility can begin its *fourth* and final phase, which continues through the duration of the permit. In the event that the trial burn results do not demonstrate compliance with standards, the temporary permit must be modified to allow for a second trial burn.

Example 12-5. A test burn waste mixture consisting of three designated POHCs (chlorobenzene, toluene, and xylene) is incinerated at 1,000°C. The waste feed rate and the stack discharge are shown in the following table. The stack gas flow rate is 375.24 dscm/min (dry standard cubic meters per minute). Is the unit in compliance?

Compound	Inlet (kg/h)	Outlet (kg/h)
Chlorobenzene (C_6H_5Cl)	153	0.010
Toluene (C_7H_8)	432	0.037
Xylene (C_8H_{10})	435	0.070
HCl	—	1.2
Particulates at 7% O_2	—	3.615

Outlet concentrations were measured in the stack after APC equipment.

Solution. We begin by calculating the DRE for each of the POHCs.

$$DRE = \frac{(W_{in}) - (W_{out})}{(W_{in})} \times 100$$

$$DRE_{chlorobenzene} = \frac{153 - 0.010}{153} \times 100 = 99.993\%$$

$$DRE_{toluene} = \frac{432 - 0.037}{432} \times 100 = 99.991\%$$

$$DRE_{xylene} = \frac{435 - 0.070}{435} \times 100 = 99.984\%$$

The DRE for each designated POHC must be at least 99.99 percent. In this case, the designated POHC xylene fails to meet the standard. The other POHCs exhibit a DRE of greater than 99.99 percent.

Now we check compliance for the HCl emission. The HCl emission may not exceed 1.8 kg/h or 1 percent of the HCl prior to the control equipment, whichever is greater. It is obvious that the 1.2 kg/h emission meets the 1.8 kg/h limit. This would be sufficient to demonstrate compliance, but we will calculate the mass emission rate prior to control for the purpose of comparison. To do this we assume all the chlorine in the feed is converted to HCl. The molar feed rate of chlorobenzene (M_{CB}) is

$$M_{CB} = \frac{W_{CB}}{(MW)_{CB}} = \frac{(153 \text{ kg/h})(1,000 \text{ g/kg})}{112.5 \text{ g/mole}}$$

$$= 1,360 \text{ mole/h}$$

where M_{CB} = molar flow rate of chlorobenzene
$(MW)_{CB}$ = molecular weight of chlorobenzene

Each molecule of chlorobenzene contains one atom of chlorine. Therefore,

$$M_{HCl} = M_{CB}$$
$$= 1,360 \text{ mole/h}$$
$$W_{HCl} = (GMW \text{ of } HCl)(\text{mole/h})$$
$$= (36.5 \text{ g/mole})(1,360 \text{ mole/h})$$
$$= 49,640 \text{ g/h or } 49.64 \text{ kg/h}$$

This is the HCl emission prior to control. The emission of 1.2 kg/h is greater than 1 percent of the uncontrolled emission, that is,

$$1\% \text{ of uncontrolled} = (0.01)(49.64)$$
$$= 0.4964 \text{ kg/h}.$$

However, the incinerator passes the HCl limits because the HCl emission is less than 1.8 kg/h.

The particulate concentration was measured at 7 percent O_2 and, therefore, does not need to be corrected. The outlet loading (W_{out}) of the particulates is

$$W_{out} = \frac{(3.615 \text{ kg/h})(10^6 \text{ mg/kg})}{(375.24 \text{ dscm/min})(60 \text{ min/h})}$$

$$= 160 \text{ mg/dscm}$$

Comment: This is less than the standard of 180 mg/dscm and is, therefore, in compliance with regard to particulates. However, because the incinerator fails the DRE for xylene, the unit is out of compliance.

Regulations for PCBs. Incineration of PCBs is regulated under the Toxic Substances Control Act (TSCA) rather than RCRA. Thus, some of the permit conditions for incineration of PCBs are different from other RCRA hazardous wastes.

The conditions for incineration of liquid PCBs may be summarized as follows (Wentz, 1989):

1. *Time and temperature.* Either of two conditions must be met. The residence time of the PCBs in the furnace must be 2 seconds at 1,200°C ± 100°C with 3 percent excess oxygen in the stack gas or, alternatively, the furnace residence time must be 1.5 seconds at 1,600°C ± 100°C with 2 percent excess oxygen in the stack gas.

   The EPA has interpreted these conditions to require a liquid PCB DRE ≥ 99.9999 percent.

2. *Combustion efficiency.* The combustion efficiency shall be at least 99.99 percent, computed as follows:

$$\text{Combustion efficiency} = \frac{C_{co_2}}{C_{co_2} + C_{co}} \times 100\% \qquad (12\text{-}11)$$

   where C_{co_2} = concentration of carbon dioxide in stack gas
   C_{co} = concentration of carbon monoxide in stack gas

3. *Monitoring and controls.* In addition to these permitted limits, owners or operators of incinerators are required to monitor and control the variables that affect performance. The rate and quantity of PCBs fed to the combustion system must be measured and recorded at regular intervals of no longer than 15 minutes. The temperatures of the incineration process must be continuously measured and recorded. The flow of PCBs to the incinerator must stop automatically whenever one of the following occurs: the combustion temperature drops below the temperatures specified, that is, 1,200 or 1,600°C; when there is a failure of monitoring operations; when the PCB rate and quantity measuring and recording equipment fails; or when excess oxygen falls below the percentage specified. Scrubbers must be used for HCl removal during PCB incineration.

In addition, a trial burn must be conducted and the following exhaust emissions must be monitored:

Oxygen (O_2)

Carbon monoxide (CO)

Oxides of nitrogen (NO_x)

Hydrogen chloride (HCl)

Total chlorinated organic content

PCBs

Total particulate matter

An incinerator used for incinerating nonliquid PCBs, PCB articles, PCB equipment, or PCB containers must comply with the same rules as those for liquid PCBs, and the mass air emissions from the incinerator must be no greater than 0.001 g PCB per kilogram of the PCB introduced into the incinerator, that is, a DRE of 99.9999 percent.

Stabilization/Solidification

Because of their elemental composition, some wastes, such as nickel, cannot be destroyed or detoxified by physical or chemical means. Thus, once they have been separated from aqueous solution and concentrated in ash or sludge, the hazardous constituents must be bound up in stable compounds that meet the LDR restrictions for leachability.

The terminology for this treatment technology has evolved in the last decade. In the early to mid 1980s "chemical fixation," "encapsulation," and "binding" were often used interchangeably with solidification and stabilization. With the promulgation of the LDR restrictions, the EPA established a more precise definition for solidification/ stabilization and discouraged the use of the other terms to describe the technology (U.S. EPA, 1988b). EPA linked solidification and stabilization because the resultant material from the treatment must be both stable and solid. "Stability" is determined by the degree of resistance of the mixture of the hazardous waste and additive chemical to leaching in the *Toxicity Characteristic Leaching Procedure* (TCLP) (55 FR 26986, JUN 29, 1990). In the EPA definition, then, solidification/stabilization refers to chemical treatment processes that chemically reduce the mobility of the hazardous constituent.

Reduced leachability is accomplished by the formation of a lattice structure and/or chemical bonds that bind the hazardous constituent and thereby limit the amount of constituent that can be leached when water or a mild acid solution comes into contact with the waste matrix. There are two principal solidification/stabilization processes: cement based and lime based. The cement or lime additive is mixed with the ash or sludge and water. It is then allowed to cure to form a solid. The correct mix proportions are determined by trial-and-error experiments on waste samples. In both techniques the stabilizing agent may be modified by other additives such as silicates. In general, this technology is applicable to wastes containing metals with little or no organic contamination, oil, or grease.

12-7 LAND DISPOSAL

Deep Well Injection

Deep well injection consists of pumping wastes into geologically secure formations. Pumping of wastes into these formations has been practiced primarily in Louisiana and Texas. In promulgating the final third of the LDR restrictions (55 FR 22530, 1 JUN 1990), the EPA allowed disposal of waste in Class I injection wells for wastes disposed under clean water act regulations.

Land Treatment

Land treatment is sometimes called "land farming" of the waste. In this practice, waste was incorporated with soil material in the manner that fertilizer or manure might be. Microorganisms in the soil degraded the organic fraction of the waste. Under the LDR restrictions, this practice is prohibited.

The Secure Landfill

Although far from ideal, the use of land for the disposal of hazardous wastes is a major option for the foreseeable future. Furthermore, we recognize that incinerator ash, scrubber bottoms, and the results of biological, chemical, and physical treatment leave residues of up to 20 percent of the original mass. These residues must be secured in an economical fashion. At this juncture, the secure landfill is the only option.

The basic physical problem with land disposal of hazardous waste stems from the movement of water. The dissolution of waste material results in contaminants being transported from the waste site to larger regions of the soil zone and, too often, to an underlying aquifer. Problems of groundwater pollution frequently lead to the condemnation of wells and to the contamination of surface water bodies fed by the associated aquifer. In many instances, well contamination is not detected until years after land disposal of waste has begun, because of the slow movement of the conveying ground water (Wood et al., 1984).

Water pollution, caused by a hazardous waste facility, may evolve in a variety of ways. Leachate from landfills may drain out of the side of the landfill and appear as surface runoff. It may seep down slowly through the unsaturated zone and enter an underlying aquifer. Fissures in liners lead to a downward migration of contaminants toward the water table.

Without the institution of remedial measures, buried waste usually acts as a continuing source of pollution. The waste constituents continue to be transported in the subsurface by infiltrating precipitation. Thus, sites that handle hazardous wastes are located above a natural barrier, as well as an applied liner. Moreover, the site is instrumented to continuously monitor the condition of any associated aquifers. In addition, a system for the collection and treatment of the leachate is required.

The technology of the secure landfill may be divided into two phases: siting and construction. The following discussion on siting is drawn primarily from E. F. Wood et al. (1984).

Landfill Siting. In siting a hazardous waste landfill, the four main considerations are air quality, groundwater quality, surface water quality, and subsurface migration of gases and leachates. Aside from the sociopolitical aspects, the last three components are the major factors to be considered in siting the landfill.

Air quality must be considered to prevent adverse effects to the air caused by volatilization, gas generation, gas migration, and wind dispersal of landfilled hazardous wastes. Generally, these can be controlled by proper construction techniques and do not inhibit the siting.

The hydrogeologic siting problem can be divided into four main areas: geology, soil, hydrology, and climate. Bedrock geology determines the structural framework that surfaces as landforms and the structural integrity of the landfill site.

Structural integrity of host rock is important in terms of seismic risk zones, dipping, and cleavage. Seismic risk zones indicate the presence of geologic faults and fractures. Faults and fractures provide a natural pathway for the flow of contaminants, even in low-permeability and low-porosity rock.

Transport capacity refers to a soil's ability to allow migration of contaminants. A soil with low permeability and porosity can lengthen the flow period and act as a natural defense by retarding the movement of contaminants. Glacial outwash plains and deltaic sands are both well-sorted sand and gravel beds with high permeability. Thus, they allow wastes to move faster and farther. Clays and silts have lower permeabilities and, thus, inhibit the movement of wastes.

Most contaminants will move either at the same rate or slower than the water. The relative speeds of the water and contaminant are a function of the contaminant and water characteristics. For example, organic contaminants that are relatively insoluble in water will be *retarded* more by the soil than organic contaminants that are relatively soluble in water. The pH of the water will also affect retardation. For example, at low pH (and in the absence of oxygen), iron will be present predominantly as ferrous iron (Fe^{2+}). This iron is quite soluble and will move with the water. If the pH is high (>6) and oxygen is present, the iron will be in the ferric (Fe^{3+}) form, which is much less soluble in water. The Fe^{3+} will precipitate and, therefore, will not move with the groundwater. The extent to which the chemicals are retarded is defined by the *retardation coefficient:*

$$R = \frac{v'_{water}}{v'_{contaminant}} \qquad \text{(12-12)}$$

where v'_{water} = linear speed of the water
     $v'_{contaminant}$ = linear speed of the contaminant

The retardation coefficient is a function of the hydrophobicity of the contaminant on a specified soil. For neutral organic chemicals, R is defined as:

$$R = 1 + \left(\frac{\rho_b}{\eta}\right) K_{oc} f_{oc} \qquad \text{(12-13)}$$

where ρ_b = bulk density of the soil
     η = porosity of soil as a fraction
     K_{oc} = partition coefficient into the organic carbon fraction of the soil
     f_{oc} = fraction of organic carbon in the soil

Some retardation coefficients for typical groundwater contaminants are given in Table 12–3.

TABLE 12-3
Retardation coefficients of typical groundwater contaminants[a]

Compound	Soil A[b]	Soil B[b]	Soil C[b]
Benzene	1.2	1.7	5.3
Toluene	1.7	3.6	17.0
Aniline	1.1	1.2	2.2
Di-n-propyl phthalate	5.6	19.0	110.0
Fluorene	23.0	86.0	500.0
n-Pentane	7.0	24.0	140.0

[a]Data and formulas used for computation of K_{oc} are from Schwarzenbach et al., 1993.
[b]For soil A: $\rho_b = 1.4$ g/cm^3; $\eta = 0.40$; $f_{oc} = 0.002$; for soil B: $\rho_b = 1.6$ g/cm^3;
$\eta = 0.30$; $f_{oc} = 0.005$; for soil C: $\rho_b = 1.75$ g/cm^3; $\eta = 0.55$; $f_{oc} = 0.05$.

An alternate expression for retardation is

$$R = 1 + \left(\frac{\rho_b}{\eta}\right)K_d \qquad \text{(12-14)}$$

Where the terms are the same as those given for Equation 12-13 with the exception of K_d, which is defined as the distribution coefficient.

Example 12-6.   An illegally buried drum of toluene has begun to leak into an unconfined drinking water aquifer. A homeowner's well is located 60 m down-gradient from the leaking drum. If this is a type C soil and the linear speed of the water in the aquifer is 4.7×10^{-6} m/s, how many years will it take for the toluene to reach the well?

Solution.   Using the value of R from Table 12–3, compute the linear speed of the contaminant as

$$v'_{contaminant} = \frac{v'_{water}}{R}$$

$$= \frac{4.7 \times 10^{-6} \text{ m/s}}{17.0} = 2.76 \times 10^{-7} \text{ m/s}$$

The travel time is then

$$\frac{\text{Distance}}{v'_{contaminant}} = \left(\frac{60 \text{ m}}{2.76 \times 10^{-7} \text{ m/s}}\right)\left(\frac{1}{86,400 \text{ s/d}}\right)\left(\frac{1}{365 \text{ d/y}}\right) = 6.88 \text{ or about 7 years}$$

Comment: We should point out two important notes: (1) this is an extreme simplification of a very complex problem that, in reality, may result in a very different answer from this computation, and (2) the concentration of the toluene is not addressed in this problem. The actual travel time may vary by an order of magnitude depending on the pumping rate of the well, precipitation patterns, and other undetermined hydrogeologic parameters. The concentration depends on the mass of toluene released, the quantity of water that dilutes it, the solubility of toluene in this water, and undetermined reactions in the soil.

Sorption capacity depends on the organic content, predominant minerals, pH, and soil. Sorption includes both absorption and adsorption of contaminants. Sorption is important in limiting the movement of metals, phosphorus, and organic chemicals. *Cation exchange capacity* (CEC) is a measure of the ability of the soil to trade cations in the soil for those in waste. The higher the CEC, the more metal will be retained. The capacity of soil to retard contaminant migration also depends on the presence of numerous hydrous oxides, particularly iron oxides, and other compounds such as phosphates and carbonates. These compounds precipitate heavy metals out of solution, making them unable to travel farther.

The hydrogen-ion concentration (pH) of soil influences the dominant removal mechanism for metal cations. The dominant removal mechanism for metal cations when pH < 5 is exchange or adsorption; when pH > 6, it is precipitation.

Hydrologic considerations in locating a hazardous waste landfill include distance to the groundwater table, the hydraulic gradient, the proximity of wells, and the proximity to surface waters.

When distance from the surface to the groundwater table is short, contaminant travel time is also short, allowing for little attenuation before pollutants disperse laterally in the saturated zone. It is desirable to have the average distance to the groundwater table large enough so that contaminants may be significantly attenuated. This also facilitates monitoring of the saturated zone. This will permit remedial action to be undertaken, if necessary.

A hydraulic gradient that slopes away from local groundwater supplies is desired. The steeper the hydraulic gradient, the lower the attenuation time and the faster the water movement. Therefore, a moderate hydraulic gradient may be most acceptable.

The distance from the disposal site to water-supply wells and surface waters must be as large as possible to protect them from potential contamination in case the landfill leaks. Furthermore, the proximity to surface waters must take into account the potential for flooding. Site flooding will weaken the structure of a land emplacement facility, causing it to fail and leak wastes. Therefore, it is essential that the facility not be built on a floodplain or area subject to local flooding. The facility should be designed so that it will not be flooded.

Climate is considered a driving force in contaminant migration, but we may exclude it when considering potential sites within the same region, where climate is unlikely to vary significantly.

Landfill Construction. A secure landfill means, in essence, that no leachate or other contaminant can escape from the fill and cause adverse impacts on the surface water or groundwater. Leakage from the site is not acceptable during or after operations. Neither is any external or internal displacement, which could be brought about by slumping, sliding, and flooding. Wastes must not be allowed to migrate from the site.

It is nearly impossible to create an impervious burial vault for hazardous wastes and guarantee its integrity forever. Landfill design and operation is regulated to minimize migration of wastes from the site. The current EPA rules (40 CFR 264.300) for hazardous waste landfills require a minimum of (1) two or

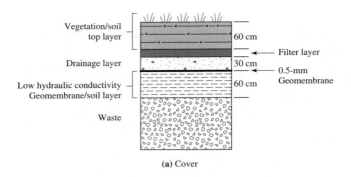

(a) Cover

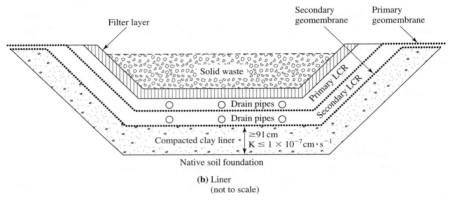

(b) Liner
(not to scale)

FIGURE 12-12
Minimum technology landfill liner, design and recommended final cover design. (*Source:* U.S. EPA, 1989b and 1991.)

more liners, (2) a leachate collection system above and between the liners, (3) surface run-on and run-off control to collect and control at least the water volume resulting from a 24-hour, 25-year storm, (4) monitoring wells, and (5) a "cap" (Figure 12-12).

The liner system must include (57 FR 3462, 29 JAN 1992):

1. A top liner designed and constructed of materials (for example, a geomembrane) to prevent migration of hazardous constituents into the liner during the active life and postclosure care period;

2. A composite bottom liner consisting of at least two components. The upper component must be designed and constructed of materials (for example, a geomembrane) to prevent migration of hazardous constituents into the liner during the active life and postclosure care period. The lower component must be designed and constructed of materials to minimize migration of hazardous constituents if a breach in the upper component were to occur. The lower component must be constructed of at least 91 cm of compacted soil material with a hydraulic conductivity of no more than 1×10^{-7} cm/s.

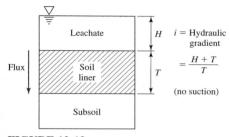

FIGURE 12-13
Definition of hydraulic gradient for landfill liner.

The leachate collection and removal system (LCR) immediately above the top liner must be designed, constructed, operated, and maintained to collect and remove leachate so that the leachate depth over the liner does not exceed 30 cm. The leachate collection and removal system between the liners and immediately above the bottom liner is also a leak detection system. The leachate collection system must, at a minimum, be:

1. Constructed with a bottom slope of 1 percent or more;

2. Constructed of a granular drainage material with a hydraulic conductivity of 1×10^{-2} cm/s or more and a thickness of 30 cm or more; or be constructed of synthetic or geonet drainage materials with a transmissibility of 3×10^{-5} m²/s or more;

3. Constructed of sufficient strength to prevent collapse and be designed to prevent clogging.

The design equations for the leachate collection system are the same as those used for a municipal landfill (Chapter 11). The leachate collection system must include pumps of sufficient size to remove the liquids to prevent leachate from backing up into the drainage layer. The leachate must be treated to meet discharge limits. The treated leachate may be discharged into the municipal wastewater treatment system or into a waterway.

The amount of leachate may be estimated using Darcy's law (Equation 4-27). The hydraulic gradient for a liner is defined as shown in Figure 12-13. The flow rate cannot exceed the amount of water available, that is, the product of the precipitation rate and the area of the landfill. The travel time of a contaminant through a soil layer may be estimated as the linear length of the flow path (T) divided by the seepage velocity (Equation 4-31).

Example 12-7. How long will it take for leachate to migrate through a 0.9 m clay liner with a hydraulic conductivity of 1×10^{-7} cm/s if the depth of leachate above the clay layer is 30 cm and the porosity of the clay is 55 percent?

Solution. The Darcy velocity is found using Equation 4-27.

$$v = K\left(\frac{dh}{dr}\right)$$

where the hydraulic gradient (dh/dr) is defined as in Figure 12-13:

$$\frac{dh}{dr} = \frac{0.30 \text{ m} + 0.9 \text{ m}}{0.9 \text{ m}} = 1.33$$

The Darcy velocity is then

$$v = (1 \times 10^{-7} \text{ cm/s})(1.33) = 1.33 \times 10^{-7} \text{ cm/s}$$

From Equation 4-31, the seepage velocity is

$$v' = \frac{K(dh/dr)}{\eta} = \frac{1.33 \times 10^{-7} \text{ cm/s}}{0.55} = 2.42 \times 10^{-7} \text{ cm/s}$$

The travel time is then

$$t = \frac{T}{v'} = \frac{(0.9 \text{ m})(100 \text{ cm/m})}{2.42 \times 10^{-7} \text{ cm/s}} = 3.71 \times 10^{8} \text{ s or about 12 years}$$

Comment: See Equation 11-12 for an alternative equation for estimating break-through time.

The site operator must keep careful records of the location and dimensions of each cell and must depict each cell on a map keyed to permanently surveyed vertical and horizontal markers. Records must show the contents of each cell and the approximate location of each hazardous waste type within the cell.

The purpose of groundwater monitoring is to ensure that programs for managing runon, runoff, and leachates are functioning properly so that groundwater remains un-contaminated. If contamination is occurring, early warning can be given and counter-measures taken. The site owner/operator has to place a sufficient number of monitoring wells around the limits of the facility to be able to describe the background (upgradient) and downgradient water quality. The regulations set forth, in detail, how the monitoring wells must be sunk, screened, sealed, sampled, and located, with special emphasis on location of the downgradient wells.

General groundwater quality, especially the suitability of the uppermost aquifer for use as a drinking water source, must meet EPA's primary drinking water standards. The flow rate for each sump must be calculated weekly during the active life and clo-sure period, and monthly during the postclosure care period. If the landfill is leaking to the groundwater, the site operator must file an assessment plan with the EPA that shows how the problem is to be remedied.

12-8 GROUNDWATER CONTAMINATION AND REMEDIATION

The Process of Contamination

Hazardous waste landfills are, of course, not the only source of groundwater contamination. Other sources include municipal landfills, septic tanks, mining and agricultural activities, "midnight dumping," and leaking underground storage tanks. It has been estimated that more than 35,000 underground storage tanks are leaking (U.S. EPA, 2004a).

The threat of contamination to groundwater depends on the specific geologic and hydrologic conditions of the site. The following paragraphs describe general considerations. All conditions may not exist at every site.

Leaking chemicals pass through several different hydrologic zones as they migrate through the soil to the groundwater system. The pore spaces in the unsaturated zone in the top soil layers are occupied by both air and water. Flow in this zone for liquid contaminants is downward by gravity. The upper region of the unsaturated zone is important for pollutant attenuation. Some chemicals are retained by adsorption onto organic material and chemically active soil particles. Some are trapped in the pore spaces and held by surface tension. These adsorbed and trapped chemicals may decompose through abiotic processes such as oxidation, reduction, and hydrolysis, as well as microbial activity, or they may simply remain sorbed onto the particles. Migration of precipitation may leach this sorbed and trapped material and carry it to the underlying aquifer for long periods of time after the source of contamination has been removed (Wentz, 1989).

In the capillary zone just above the saturated zone that marks the groundwater table, spaces between soil particles may be saturated by water rising from the water table by capillary action. Chemicals that are lighter than water will "float" on top of the water table in this zone and move in different directions and rates than dissolved contaminants.

The pore spaces between soil particles below the water table are saturated. Generally, the saturated zone is devoid of oxygen. The lack of dissolved oxygen limits the oxidation of chemicals.

Groundwater flow is laminar, with minimal mixing occurring as the groundwater moves. Dissolved chemicals will flow with groundwater and form distinct plumes. The shape and size of a contaminant plume depends upon the local hydrogeological setting, groundwater flow, the characteristics of the contaminants, and geochemistry. Solubility, adsorption characteristics, and degradation affect mobility. The density of the contaminant is important in determining the shape and movement of the plume. Lighter, less soluble chemicals, like gasoline, will tend to flow on top of the aquifer (see Figure 7-18). Water soluble contaminants tend to dissolve in and then flow with groundwater (see Figure 7-17). Dense, insoluble contaminants will sink to the bottom (see Figure 7-19) of the aquifer. *Volatile organic chemicals* (VOCs) in groundwater are extremely mobile. Polyvalent metal contaminants tend to adsorb onto clays and, hence, are not very mobile.

EPA's Groundwater Remediation Procedure

The federal program for cleanup of contaminated sites follows a procedural sequence as shown in Figure 12-14. Each of these steps is discussed in the following paragraphs.

Preliminary Assessment.   EPA involvement usually begins with the identification of a potential hazardous waste site. The initial information can come from a variety of sources, including local citizens and officials, state environmental agencies, the site owners themselves, or simply from awareness of potential problems associated with particular industries.

EPA has developed an inventory system called the Comprehensive Environmental Response, Compensation, and Liability Information System (CERCLIS) to document all

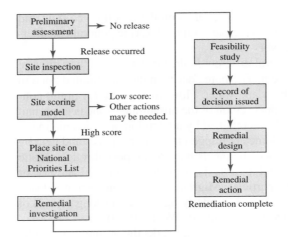

FIGURE 12-14
Steps involved in the Superfund cleanup process.

of the sites in the United States that may be candidates for remedial action. This is a continuing program that identifies sites as information about them becomes available. The growth in the number of CERCLIS sites has been dramatic and is expected to continue for the foreseeable future as additional abandoned and contaminated sites are discovered. As of September 2005, 12,031 sites were in the inventory. This list did not include an estimated 130,000 leaking underground storage tanks (U.S. EPA, 2004a and 2005b).

A *preliminary assessment* (PA) is the first step in identifying the potential for contamination from a particular site. The primary objectives of the PA are to determine if there has been a release of contaminant to the environment, if there is immediate danger to persons living or working near the site, and whether a site inspection is necessary. Samples for environmental analysis are generally not taken during the PA. Following the preliminary assessment, EPA or the designated state agency might determine that an immediate threat to residents or employees at the site requires an immediate removal action. Otherwise, on the basis of the preliminary assessment, the site is classified by EPA into one of the three following categories:

1. There is no further action needed, because there is no threat to human health or the environment.

2. Additional information is required to complete the preliminary assessment.

3. Inspection of the site is necessary.

Site Inspection.   Site inspection requires sampling to determine the types of hazardous substances that are present and to identify the extent of contamination and its migration. The actual site inspection includes preparation of a work plan and an on-site safety plan. The site assessment has three objectives:

1. To determine which releases pose no threat to public health and the environment;

2. To determine if there is any immediate threat to persons living or working near the release;

3. To collect data to determine whether or not a site should be included on the National Priorities List (NPL).

HRS, NPL, RI/FS, and ROD.   The next series of steps in the EPA's procedure include calculations to complete the HRS; inclusion on the NPL if the score is sufficiently high; conduct of a RI/FS; and issuance of an ROD. These steps were discussed in detail in Section 12-4.

Remedial Design and Remedial Action.   EPA-funded remedial actions may be taken only at those sites that are on the NPL. This ranking helps ensure that the Superfund dollars are used in the most cost-effective manner and where they will yield the greatest benefit.

Before a remedial action can be taken at a site, a number of questions must be answered. These can be classified as problem definition, design alternatives, and policy.

1. Problem definition questions: What are the contaminants and how much contamination is present? How large is the surface area of the contaminated site? What is the size of the contaminated groundwater plume? Where is the exact location of the plume and in what direction is it moving?

2. Design questions: Based on the alternatives available, what is the best way to clean up the site? How should these alternatives be implemented? What products will be produced during treatment? How long will it take to complete the remediation and what will it cost?

3. Policy questions: What level of protection is adequate? In other words, how clean is clean?

The answers to the first two sets of questions require scientific and engineering background that is supported by extensive sampling of the contaminated site area. The last question cannot be answered objectively; rather, it is a subjective and oftentimes political question.

The NCP defines three types of responses for incidents involving hazardous substances. In these responses *removal* is differentiated from *remediation*. Removal is, as its name suggests, the physical relocation of the waste—usually to a secure hazardous waste landfill. Remediation means that the waste is to be treated to make it less toxic and/or less mobile or the site is to be contained to minimize further release. Remediation can take place on-site or at a TSD facility. The three types of responses are:

1. *Immediate removal* is a prompt response to prevent immediate and significant harm to human health or the environment. Immediate removals must be completed within six months.

2. *Planned removal* is an expedited removal when some response, not necessarily an emergency response, is required. The same six-month limitation also applies to planned removal.

3. *Remedial response* is intended to achieve a site solution that is a permanent remedy for the particular problem involved.

Immediate removals are done to prevent an emergency involving hazardous substances. These emergencies might include fires; explosions; direct human contact with a hazardous substance; human, animal, or food-chain exposure; or contamination of drinking water sources. An immediate removal involves cleaning up the hazardous site to protect human health and life, containing the hazardous release, and minimizing the potential for damage to the environment. For example, a truck, train, or barge spill could involve an immediate removal determination by EPA to get the spill cleaned up.

Immediate removal responses may include activities such as sample collection and analysis, containment or control of the release, removal of the hazardous substances from the site, provision of alternate water supplies, installation of security fences, evacuation of threatened citizens, or general deterrent of the spread of the hazardous contaminants.

A planned removal involves a hazardous site that does not present an immediate emergency. Under Superfund, EPA may initiate a planned removal if the action will minimize the damage or risk and is consistent with a more effective long-term solution to the problem. Planned removals are carried out by EPA if the responsible party is either unknown or cannot or will not take timely and appropriate action. The state in which the cleanup is located must be willing to match at least 10 percent of the costs of the removal action, as well as agree to nominate the site in question for the National Priority List.

Mitigation and Treatment

Because the spread of contaminants is usually confined to a plume, only localized areas of an aquifer need to be reclaimed and restored. Cleanup of a contaminated aquifer, however, is often time-consuming and costly. The original source of contamination can be eliminated, but the complete restoration of the groundwater is fraught with additional problems, such as defining the site's subsurface soil and geologic composition, locating contamination sources, defining contaminant transport pathways, determining the extent and concentration of the contaminants, and choosing and implementing an effective remedial process (Griffin, 1988).

Cleanup methods for contaminated aquifers range from containment to destruction of the contaminants. Because, in the long run, containment does not really solve the problem, destruction of the contaminants is the preferred objective of a cleanup program. Examples of remedial methods include (LaGrega et al., 2001):

- Installing pumping wells to remove the contaminated water and then treating it with one of the technologies described in Section 12-6 (called *pump and treat*)

- Air sparging

- Soil vapor extraction

- Reactive treatment walls (also known as *permeable reactive barriers*) that allow the contaminated plume to pass through a reactive chemical or a an acclimated biomass.

In the next few paragraphs we will discuss the first of these systems. The other systems are discussed in detail in LaGrega et al. (2001).

Certainly, combinations of barriers and treatment methods should be considered. Source control (removal or remediation of the source), physical control, and treatment methods all will have their part in mitigating groundwater contamination problems. Legal implications may also dictate strategies that may be utilized (Griffin, 1988).

Pump and Treat.   The objectives of a pump-and-treat system include hydraulic containment of the contaminated plume and removal of the contaminant from the groundwater. The design of the well system for a pump-and-treat remediation is an application of well hydraulics described in Chapter 4.

The *capture zone* of an extraction well is that portion of the groundwater that will discharge into the well. The capture zone is not necessarily coincident with the cone of depression (Chapter 4) because the groundwater flow lines can be diverted by the influence of the pumping well without being captured by the well (Figure 12-15). Under steady-state conditions, the extent of the cone of depression

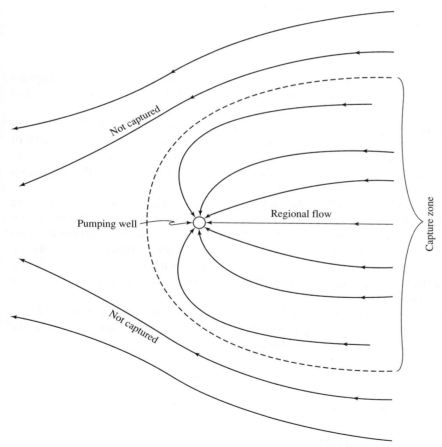

FIGURE 12-15
Groundwater flow lines influenced by pumping well.

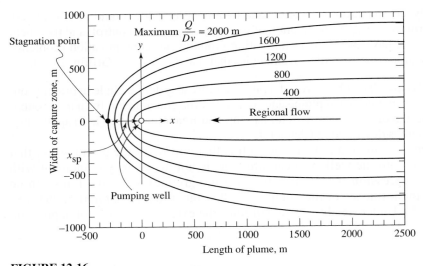

FIGURE 12-16
Type curve for analytical solution to capture zone analysis for a single extraction well.
(*Source:* Javandel and Tsang, 1986.)

largely depends on the transmissivity and pumping rate. The extent of the capture zone depends on the regional hydraulic gradient as well as the transmissivity and pumping rate.

Three parameters are used to delineate the capture zone: (1) the width of the capture zone at an infinite distance up-gradient from the pumping well, (2) the width of the capture zone at the location of the pumping well, and (3) the location of the down-gradient distance of the capture zone from the pumping well (called the *stagnation point*). These parameters are shown in Figure 12-16.

Javandel and Tsang (1986) developed a highly idealized model of the capture zone that can be used to examine the relationship between some of the important variables. The model assumes a homogeneous, isotropic aquifer uniform in cross section and infinite in width. The aquifer may be either confined or unconfined. However, in the case of the unconfined aquifer, drawdown must be insignificant with respect to the total thickness of the aquifer. The extraction wells are assumed to be fully penetrating.

With a single well located at the origin of the coordinate system shown in Figure 12-16, Javandel and Tsang (1986) developed the following equation to describe the y coordinate of the capture zone envelope:

$$y = \pm \frac{Q}{2Dv} - \frac{Q}{2\pi Dv}\tan^{-1}\frac{y}{x} \tag{12-15}$$

where x, y = distances from the origin m
  Q = well pumping rate, m^3/s
  D = aquifer thickness, m
  v = Darcy velocity, m/s

Note that the $\pm$ allows computation of the y coordinate above and below the x axis. Masters (1998) has shown that this equation may be rewritten in terms of the angle ϕ (in radians) drawn from the origin to the x, y coordinate of interest on the line describing the capture zone curve. That is,

$$\tan\phi = \frac{y}{x} \tag{12-16}$$

so that, for $0 \le \phi \ge 2\pi$, Equation 12-15 may be rewritten as

$$y = \pm \frac{Q}{2Dv} - \left(1 - \frac{\phi}{\pi}\right) \tag{12-17}$$

This equation allows us to examine some important fundamental relationships:

- The width of the capture zone is directly proportional to the pumping rate.

- The width of the capture zone is inversely proportional to the Darcy velocity.

- As x approaches infinity, $\phi = 0$ and $y = Q/(2Dv)$. This sets the maximum total width of the capture zone at $2\,[Q/(2Dv)] = Q/(Dv)$ as shown in Figure 12-16.

- For $\phi = \pi/2$, $x = 0$ and y is equal to $Q/(4Dv)$. Thus, the width of the capture zone at $x = 0$ is $2[Q/(4Dv)] = Q/(2Dv)$.

The distance to the stagnation point down-gradient of the extraction well (x_{sp}) may be estimated from the following equation (LeGrega et al., 2001):

$$x_{sp} = \frac{Q}{2\pi Dv} \tag{12-18}$$

Javandel and Tsang prepared a series of "type" curves for various well configurations (one to four wells) and several widths of the capture zone at $x = \infty$. The suggested approach to using the capture zone technique is summarized as follows:

1. Prepare a site map with the plume shape at the same scale as the type curves.

2. Superimpose the site map on the one-well type curve with the direction of regional flow parallel to the x axis. Place the leading edge of the plume just beyond the location of the extraction well. Select the capture zone curve that completely captures the plume. This defines the required value of Q/Dv at $x = \infty$.

3. Determine the required pumping rate by multiplying Q/Dv by Dv. If the required pumping rate can be achieved by the use of one well, then the problem is solved. If one well does not produce the required pumping rate, then go to step 4.

4. Repeat step 2 using the two, three, or four well–type curves as required to achieve an acceptable pumping rate. Each well in the multiple-well scenarios is assumed to pump at the same rate.

The capture zone of multiple extraction wells must overlap to prevent the ground-water flow from passing between them. If the distance between the extraction wells is less than or qual to $Q/\pi Dv$, the capture zones will overlap. Assuming that the wells are located symmetrically around the x axis, the optimum spacing may be calculated by using the following:

- For two wells space at $Q/\pi Dv$
- For three wells space at $1.26Q/\pi Dv$
- For four wells space at $1.2Q/\pi Dv$

The question of whether or not the required pumping rate can be achieved is deter-mined, in a confined aquifer, by the available drawdown that will not lower the piezometric surface into the aquifer. This can be calculated using the methods discussed in Chapter 4. For an unconfined aquifer, the restriction noted above, that the drawdown must be insig-nificant with respect to the total thickness of the aquifer, requires a judgment decision.

Example 12-8.   The drinking water well at village of Oh Six is threatened by a con-taminant plume in the aquifer. The confined aquifer is 28.7 m thick. It has a hydraulic conductivity of 1.5×10^{-4} m/s, a storage coefficient of 3.7×10^{-5}, and a regional hydraulic gradient of 0.003. The contaminant plume is 300 m wide at its widest point. The maximum allowable pumping rate based on the allowable drawdown is 0.006 m³/s. Locate a single extraction well so that the stagnation point is 100 m from the drinking water well and so that the capture zone encompasses the plume. A sketch map of the drinking water well and the plume relationship is shown below.

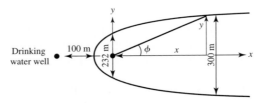

Solution.   Determine the Darcy velocity, using Equation 4-27:

$$v = K\frac{dh}{dr} = (1.5 \times 10^{-4}\text{ m/s})(0.003) = 4.5 \times 10^{-7}\text{ m/s}$$

The width of the capture zone at an infinite distance up-gradient is

$$\frac{Q}{Dv} = \frac{0.006\text{ m}^3/\text{s}}{(28.7\text{ m})(4.5 \times 10^{-7}\text{ m/s})} = 464.58\text{ or }465\text{ m}$$

The width of the capture zone at the extraction well will be

$$\frac{Q}{2Dv} = \frac{0.006\text{ m}^3/\text{s}}{2(28.7\text{ m})(4.5 \times 10^{-7}\text{ m/s})} = 232.29\text{ or }232\text{ m}$$

The stagnation point will be

$$x_{sp} = \frac{0.006 \text{ m}^3/\text{s}}{2\pi(28.7 \text{ m})(4.5 \times 10^{-7} \text{ m/s})} = 73.94 \text{ or } 74 \text{ m}$$

down-gradient from the extraction well.

The distance down-gradient from the lead edge of the plume that the extraction well must be placed is determined by using Equation 12-17. With $y = 150$ m,

$$150 \text{ m} = (232.29 \text{ m})\left(1 - \frac{\phi}{\pi}\right)$$

Solving for the angle (in radians) from the extraction well to the point where the plume just touches the capture zone is

$$\phi = 0.35 \; \pi \text{ rad}$$

Using the geometry shown in the sketch map, the distance is therefore

$$x = \frac{y}{\tan\phi} = \frac{150 \text{ m}}{\tan(0.35 \; \pi)} = \frac{150 \text{ m}}{1.96} = 76.4 \text{ or } 76 \text{ m}$$

Comment: This solution is, of course, highly idealized. The lead edge of the plume is conveniently of a geometry that allows us to locate it by using Equation 12-17. A more ellipsoidal plume geometry would project the lead edge in advance of the tangent point. This technique would then lead to a very erroneous positioning of the extraction well.

While the highly idealized situation used by Javandel and Tsang is useful in understanding the behavior of a well system to control the movement of a contaminant plume, in actual field sites the boundary conditions are rarely met. Computer models that have been calibrated to the local conditions will yield more reliable, although not perfect, understanding of the behavior of a proposed pump-and-treat system.

Because the rate of removal of contaminant decreases exponentially over time, and because the concentrations in the water rebound over time because of diffusion and desorption from the soil, pump-and-treat systems are limited in their value as a mass removal technology.

Non-Aqueous Phase Liquids (NAPLs) such as gasoline are referred to as "product" because their recovery may have some commercial value. When the NAPL floats on the groundwater table, special recovery techniques may be employed to recover it. Product recovery systems to recover NAPL use wells that terminate in the NAPL plume rather than in the aquifer. Because all hydrocarbons are slightly soluble in water, the product recovery system is usually accompanied by a groundwater pumping system to remove and treat the contaminated groundwater.

12-9   CHAPTER REVIEW

When you have completed studying this chapter, you should be able to do the following without the aid of your textbook or notes:

1. Sketch the chemical structure of 2,3,7,8-TCDD.

2. Explain how 2,3,7,8-TCDD occurs and/or when it is found in nature.

3. Sketch the chemical structure of the PCB 2,4′-dichlorobiphenyl.

4. Explain the origin of PCBs.

5. Define hazardous waste.

6. List the five ways a waste can be found to be hazardous and briefly explain each.

7. Explain why dioxin and PCB are hazardous wastes.

8. State how long generators may store their waste.

9. Explain what defines a small quantity generator and what "break" the rules give them.

10. Define the abbreviations CFR, FR, RCRA, HSWA, CERCLA, and SARA.

11. Explain the major difference (objective) between RCRA/HSWA and CERCLA/SARA.

12. Define/explain the terms "cradle-to-grave" and manifest system.

13. Explain what "land ban," or LDR, means.

14. Define the abbreviations TSD and UST.

15. Describe the three ways to meet corrosion protection standards for underground storage tanks.

16. List the four major provisions of CERCLA.

17. Define/explain the following abbreviations: NCP, NPL, HRS, RI, FS, ROD, and PRP.

18. Explain why it is important for a site to be placed on the NPL.

19. Explain the concept of "joint and several liability" and the implications to those with wastes found in an abandoned hazardous waste site.

20. List and explain four hazardous waste management techniques.

21. List the objectives of a waste audit.

22. Differentiate between waste minimization, waste exchange, and recycling.

23. List six disposal technologies for hazardous wastes.

24. Explain why seismic risk is important in landfill siting.

25. Explain how permeability, porosity, and sorption capacity of soil limit the migration of hazardous wastes.

26. Explain what hydrologic features are important in siting a landfill.

27. List the minimum EPA requirements for a hazardous waste landfill and sketch a landfill that meets these.

28. Explain the difference between deep well injection and land treatment.

29. Define the following acronyms: PIC, POC, POHC, and DRE, as they apply to incineration.

30. List the most important factors for proper incinerator design and operation.

31. List the two types of incinerators most commonly used for destroying hazardous waste.

32. Explain the terms "designated POHC" and "surrogate" as they apply to a trial burn.

33. Outline the steps in EPA's remediation procedures.

34. Differentiate between "remediation" and "removal" as they pertain to a CERCLA/SARA cleanup.

35. Explain why pump-and-treat remediation systems may take a very long time to clean up groundwater.

With the aid of this text, you should be able to do the following:

36. Determine whether or not a waste is an EPA hazardous waste based on its composition, source, or characteristics.

37. Perform a mass balance to identify waste sources or waste-minimization opportunities.

38. Write the reactions for oxidation or reduction of chemical contaminants to mineralized form.

39. Perform solubility product calculations to estimate treatment doses for precipitation or the concentration of contaminants that remain in solution.

40. Determine the dimensions of an air stripping column, air or liquid flow rate given the values for remaining variables.

41. Determine the mass of resin and column dimensions for an ion exchange column given laboratory or pilot breakthrough data.

42. Evaluate a chemical feed to an incinerator to determine whether or not the chlorine content is acceptable and design a mix of waste feeds to achieve a desired chlorine feed rate.

43. Evaluate the operating variables for an incinerator to determine regulatory compliance for DRE, HCl emissions, and particulate emissions.

44. Estimate the hydraulic conductivity of a liner material based on laboratory measurements.

45. Estimate the quantity of leachate given the precipitation rate, area, hydraulic gradient, and hydraulic conductivity.

46. Estimate the seepage velocity and travel time of a contaminant through a soil given the hydraulic gradient, hydraulic conductivity, porosity, and length of the flow path.

47. Locate one or more extraction wells in a contaminant plume for a specified pumping rate, aquifer thickness, hydraulic conductivity, and hydraulic gradient.

48. Estimate the required pumping rate for a single extraction well for a specified aquifer thickness, hydraulic conductivity, hydraulic gradient, and capture zone width.

12-10   PROBLEMS

12-1. Determine whether the following is a RCRA hazardous waste: Municipal wastewater containing 2.0 mg/L of selenium.

12-2. Determine whether the following is a RCRA hazardous waste: An empty pesticide container that a homeowner wishes to discard.

12-3. The town of What Cheer has set up a recycling center to collect old fluorescent light bulbs. They anticipate collecting about 250 kg/mo of fluorescent bulbs. What is the maximum time the fluorescent bulbs can be stored before they must be disposed? (*Hint:* use the internet to access the appropriate CFR.)

12-4. A vapor degreaser uses 590 kg/week of trichloroethylene (TCE). It is never dumped. The incoming parts have no TCE on them and the exiting parts drag out 3.8 L/h of TCE. The sludge removed from the bottom of the degreaser each week has 1.0 percent of the incoming TCE in it. The plant operates 8 h/d for 5 d/week. Draw the mass balance diagram for the degreaser and estimate the loss due to evaporation (in kg/week). The density of TCE is 1.460 kg/L.

Answer: $M_{evap} = 362.18$ or 360 kg/week

12-5. Using the following data and Figure P-12-5, use the mass balance technique to determine the mass flow rate (kg/d) of organic compounds to the condensate collection tank (sample location 4 in Figure P-12-5).

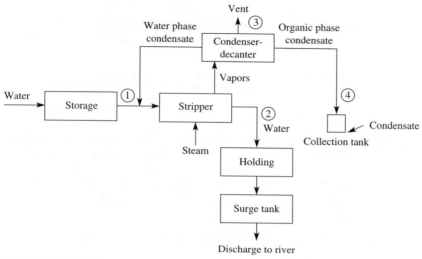

FIGURE P-12-5

Sample location	Flow rate, L/min	Total volatile organic	Temperature, °C
1	40.5	5,858. mg/L	25
2	44.8	0.037 mg/L	80
3	57.0 (vapor)	44.13%	20

Notes:

% is volume percent.

Vapor flow rate is corrected to 1 atm and 20°C.

Liquid organic density may be assumed to be 0.95 kg/L.

Assume the molecular weight of the organic vapor is equal to that of methylene chloride.

Steam mass flow rate is 252 kg/h at 106°C.

12-6. What is the efficiency of the condenser-decanter in Problem 12-5?

12-7. Given the waste constituent and concentration shown below, determine the quantity (in kg/d) of hydrated lime ($Ca(OH)_2$) required to neutralize the waste. Estimate the total dissolved solids (TDS) after neutralization. Report your answer in mg/L.

Constituent	Concentration, mg/L	Flow, L/min
HCl	100	5

Answers: Lime = 0.730 kg/d, TDS = 152 mg/L

12-8. Given the constituents and concentrations shown below, determine the quantities (in kg/d) of sulfuric acid required to neutralize the waste. Estimate the total dissolved solids (TDS) after neutralization. Report your answer in mg/L.

Constituent	Concentration, mg/L	Flow, L/min
NaOH	15	200

12-9. It has been proposed to mix a 1,500 L bath containing 5.00 percent by volume of H_2SO_4 with a 1,500 L bath containing 5.00 percent by weight of NaOH. The specific gravity of the acid added is 1.841 and its purity is 96 percent. The base added is 100 percent pure. Estimate the final pH (to two decimal places) and the final TDS (in mg/L) of the mixture of the two baths. (*Note:* the pH is very low.)

12-10. Write the reaction equation to oxidize sodium cyanide using sodium hypochlorite (NaOCl).

12-11. Write the reaction equation to oxidize sodium cyanide using ozone (O_3).

12-12. Write the reaction equation to reduce hexavalent chromium in chromic acid ($H_2Cr_2O_7$) to trivalent chromium using $NaHSO_3$.

12-13. A metal plating solution contains 50.00 mg/L of copper. Determine the concentration, in moles/L, to which the hydroxide concentration must be raised to precipitate all but 1.3 mg/L of the copper using lime. The K_{sp} of copper hydroxide is 2.00×10^{-19}. Estimate the final pH (report your answer to two decimal places).

12-14. A plating rinse water flowing at 100 L/min contains 50.0 mg/L of Zn. Calculate the theoretical pH required to achieve the EPA's pretreatment standard for existing dischargers of 2.6 mg/L and estimate the theoretical dose rate (g/min) of hydrated lime to remove only the required amount of Zn to achieve the standard (i.e., 50 mg/L minus the standard). Assume the lime is 100 percent pure.

12-15. A metal plating sludge as removed from a clarifier has a solids concentration of 4 percent. If the volume of sludge is 1.0 m^3/d, what volume will result if the sludge is processed in a filter press to a solids concentration of 30 percent? If the pressed sludge is dried to 80 percent solids, what volume will result?

> *Answer:* $V_1 = 0.133 \ m^3/d,$
> $V_2 = 0.05 \ m^3/d$

12-16. In Problem 12-15, ferrocyanide is found in the clarifier sludge at a concentration of 400 mg/kg (4 percent solids). Assuming that the ferrocyanide is part of the precipitate and that none escapes from the filter press, what concentration would be expected in filter cake? (*Hint: Set this up as a mass balance problem.*)

12-17. A drinking water supply at Oscoda, Michigan, has been contaminated by trichloroethylene. The average concentration in the water is estimated to be 6,000 μg/L. Using the following design parameters, design a packed-tower stripping column to reduce the water concentration to the state of Michigan discharge limit of 1.5 μg/L. Note that more than one column in series may be required for reasonable tower heights.

Henry's law constant = 6.74×10^{-3} m$^3 \cdot$ atm/mole

$K_L a = 0.720$ min^{-1}

Air flow rate = 60 m^3/min

G/L = 18

Temperature = 25°C

Column diameter may not exceed 4.0 m

Column height may not exceed 6.0 m

Answer: With an assumed height of 6 m, the diameter is 3.15 m

12-18. Well 13 at Watapitae is contaminated with 340 μg/L of tetrachloroethylene (perchloroethylene). The water must be remediated to achieve a concentration of 0.2 μg/L (the detection limit). Using a spreadsheet program you have written, design a packed-tower stripping column to meet this requirement using the following design parameters. (*Note:* more than one column in series may be required for reasonable tower heights.)

Henry's law constant = 100×10^{-4} m$^3 \cdot$ atm/mole

$K_L a = 14.5 \times 10^{-3}$ s^{-1}

Air flow rate = 15 m^3/s

Liquid flow rate = 0.22 m^3/s

Temperature = 20°C

Column diameter may not exceed 4.0 m

Column height may not exceed 6.0 m

12-19. An alternative form of Equation 12-4 uses the transfer unit concept discussed in Chapter 9. The relevant equations are (LaGrega, 2001):

1. Dimensionless Henry's law constant

$$H' = \frac{H_c}{RT_g}$$

2. Stripping factor

$$R_{sf} = \frac{(H')(G)}{L}$$

3. Height of transfer unit

$$HTU = \frac{L}{(A)(M_w)(K_L a)}$$

4. Number of transfer units

$$\text{NTU} = \left(\frac{R_{sf}}{R_{sf} - 1}\right) \ln\left[\frac{(C_1/C_2)(R_{sf} - 1) + 1}{R_{sf}}\right]$$

5. Height of packing in column

$$Z = (\text{NTU})(\text{HTU})$$

where
G = molar flow rate of air, moles/s
L = molar flow rate of water, moles/s
A = cross-sectional area of column, m^2
M_W = molar density of water
= 55,600 moles/m^3

Other terms are as defined for Equation 12-4.

Using the stripping factor equations, determine the height of packing for an air stripping column to reduce the concentration of ethylbenzene from 1.0 mg/L to 35 μg/L, using the following design parameters.

Henry's law constant = 6.44×10^{-3} m^3 · atm/mole
$K_La = 1.6 \times 10^{-2}$ s^{-1}
Liquid flow rate = 7.14 L/s
Temperature = 20°C
Column diameter may not exceed 4.0 m
Column height may not exceed 6.0 m

Because the air flow rate and diameter are not given, a trial-and-error solution is required. Use a spreadsheet program you have written to perform the trial-and-error solution. Use a 20 percent safety factor to estimate the final packing height.

12-20. An electroplating rinse water containing 55 mg/L of nickel is to be treated by an ion exchange column to meet an allowable effluent concentration of 2.6 mg/L. A laboratory-scale column has provided the breakthrough data shown in the table below. The laboratory column data are as follows:

Inside diameter = 1.0 cm
Length = 7.0 cm
Mass of resin (moist basis) = 5.2 g
Water content = 17%
Density of resin = 0.65 g/cm^3
Liquid flow rate = 7.68 L/d
Initial concentration = 55 mg/L

Breakthrough data

V, L	C, mg/L
0.160	4.23
0.320	5.14
0.480	10.03
0.640	16.65
0.800	23.62
0.960	29.54
1.120	35.46
1.280	39.04
1.440	44.04
1.600	49.54
1.760	53.32
1.920	54.14
2.080	53.22

The full-scale design must meet the following requirements:

Flow rate = 36,000 L/d

Hours of operation = 8 h/d

Regeneration is to be once every 5 days

Use a spreadsheet to plot the breakthrough data and determine the mass of resin required. (*Hint:* Some initial and final data points may be hard to plot and may need to be ignored to achieve a straight line on the semi-log plot.)

Answer: Mass of resin = 8.38×10^5 g or 840 kg

12-21. An electroplating rinse water containing 10 mg/L of silver is to be treated by an ion-exchange column to meet an allowable effluent concentration of 0.24 mg/L. A laboratory-scale column has provided the breakthrough data shown in the table below. The laboratory column data are as follows:

Inside diameter = 1.0 cm

Length = 14.85 cm

Mass of resin (moist basis) = 7.58 g

Water content = 34%

Density of resin = 0.65 g/cm^3

Liquid flow rate = 4.523 L/d

Initial concentration = 10 mg/L

Breakthrough data

V, L	C, mg/L	V, L	C, mg/L
0.1	0.00	1.1	2.00
0.2	0.00	1.2	3.33
0.3	0.01	1.3	5.00
0.4	0.02	1.4	6.67
0.5	0.04	1.5	8.00
0.6	0.08	1.6	8.89
0.7	0.16	1.7	9.41
0.8	0.31	1.8	9.69
0.9	0.61	1.9	9.84
1.0	1.15	2.0	9.92

The full-scale design must meet the following requirements:

Flow rate = 3,600 L/d

Hours of operation = 8 h/d

Regeneration is to be once every 5 days

Use a spreadsheet to plot the breakthrough data and determine the mass of resin required. (*Hint:* Some initial and final data points may be hard to plot and may need to be ignored to achieve a straight line on the semi-log plot.)

12-22. A very hard water is to be softened for use as an electroplating rinse water. The raw water analysis shows a calcium concentration of 107 mg/L and a magnesium concentration of 18 mg/L. The desired final hardness is 10 mg/L as $CaCO_3$. An ion-exchange column has been selected to achieve this hardness. A pilot-scale column has provided the breakthrough data shown in the table below. The laboratory column data are as follows (after Reynolds and Richards, 1996):

Inside diameter = 10.0 cm

Length = 91.5 cm

Mass of resin (moist basis) = 5.0 kg

Water content = 34%

Density of resin = 0.7 g/cm^3

Liquid flow rate = 2.25 L/h

Initial concentrations:

Ca = 107 mg/L as ion

Mg = 18 mg/L as ion

Breakthrough data

V, m³	C, meq/L
2.35	0.21
2.9	0.48
3.1	1.10
3.26	1.64
3.39	2.47
3.49	3.22
3.56	3.56
3.71	4.52
3.81	5.07
4.03	5.96
4.62	6.78

The full-scale design must meet the following requirements:

Flow rate = 570 m³/d

Regeneration is to be once every 60 days

Determine the mass of resin required. (*Hint:* see Chapter 6 for hardness equivalent weight and mg/L as $CaCO_3$ calculations and conversions.) Use a spreadsheet to plot the breakthrough data and determine the mass of resin required. (*Hint:* Some initial and final data points may be hard to plot and may need to be ignored to achieve a straight line on the semi-log plot.)

12-23. An incinerator operator receives the following shipments of waste for incineration. Can the operator mix these wastes to achieve 30 percent by mass of chlorine in the feed?

Trichloroethylene = 18.9 m³

1,1,1 Trichloroethane = 5.3 m³

Toluene = 213 m³

o-Xylene = 4.8 m³

12-24. An incinerator operator receives the following shipments of waste for incineration. What volume of methanol (CH_3OH) must the operator mix to achieve 30 percent by mass of chlorine in the feed? Assume the density of methanol is 0.7913 g/mL.

Carbon tetrachloride = 12.2 m³

Hexachlorobenzene = 153 m³

Pentachlorophenol = 2.5 m³

12-25. A hazardous waste incinerator is being fed methylene chloride at a concentration of 5,858 mg/L in an aqueous stream at a rate of 40.5 L/min. Calculate the mass flow rate of the feed in units of g/min.

12-26. Methylene chloride was measured in the flue gas of a hazardous waste incinerator at a concentration of 211.86 μg/m^3. If the flow rate of gas from the incinerator was 597.55 m^3/min, what was the mass flow rate of methylene chloride in g/min?

12-27. Assuming that the same incinerator is being evaluated in Problems 12-25 and 12-26, what is the DRE for the incinerator?

12-28. Xylene is fed into an incinerator at a rate of 481 kg/h. If the mass flow rate at the stack is 72.2 g/h, is the unit in compliance with the EPA rules?

12-29. 1,2-Dichlorobenzene is being burned in an incinerator under the following conditions:

> Operating temperature = 1,150°C
> Feed flow rate = 173.0 L/min
> Feed concentration = 13.0 g/L
> Residence time = 2.4 s
> Oxygen in stack gas = 7.0%
> Stack gas flow rate = 6.70 m^3/s at standard conditions
> Stack gas concentrations after APC equipment
> > Dichlorobenzene = 338.8 μg/dscm
> > HCl = 77.2 mg/dscm
> > Particulates = 181.6 mg/dscm

Assume all of the chlorine in the feed is converted to HCl. Does the incinerator comply with the EPA rules?

12-30. The POHCs from a trial burn are shown in the table below. The incinerator was operated at a temperature of 1,100°C. The stack gas flow rate was 5.90 dscm/s with 10.0 percent oxygen. Assuming that all the chlorine in the feed is converted to HCl, is the unit in compliance if the emissions are measured downstream of the APC equipment?

Compound	Inlet kg/h	Outlet kg/h
Benzene	913.98	0.2436
Chlorobenzene	521.63	0.0494
Xylenes	1,378.91	0.5670
HCl	n/a	4.85
Particulates	n/a	10.61

n/a = not applicable.

12-31. During a trial burn, an incinerator was fed a mixed feed containing trichloroethylene, 1,1,1-trichloroethane, and toluene in a aqueous solution. Each component accounted for 5.0 percent of the feed solution on a volume basis. The feed rate was 40 L/min. The incinerator was operated at a temperature of 1,200°C. The stack gas flow rate was 9.0 dscm/s with 7 percent oxygen. Assuming that all the chlorine in the feed is converted to HCl, is the unit in compliance with the following emissions measured after the APC equipment?

Trichloroethylene $= 170 \, \mu g/dscm$

1,1,1-Trichloroethane $= 353 \, \mu g/dscm$

Toluene $= 28 \, \mu g/dscm$

HCl $= 83.2$ mg/dscm

Particulates $= 123.4$ mg/dscm

12-32. During a trial burn, an incinerator was fed a mixed feed containing hexachlorobenzene (HCB), pentachlorophenol (PCP), and acetone (ACET) in an aqueous solution. Each component accounted for 9.3 percent of the feed solution on a volume basis, that is, HCB = 9.3 percent, PCP = 9.3 percent, and ACET = 9.3 percent. The feed rate was 140 L/min. The incinerator was operated at a temperature of 1,200°C. The stack gas flow rate was 28.32 dscm/s with 14 percent oxygen. Assuming that all the chlorine in the feed is converted to HCl, is the unit in compliance if the following emissions are measured downstream of the APC equipment?

Hexachlorobenzene $= 170 \, \mu g/dscm$

Pentachlorophenol $= 353 \, \mu g/dscm$

Acetone $= 28 \, \mu g/dscm$

HCl $= 83.2 \, \mu g/dscm$

Particulates $= 123.4$ mg/dscm

12-33. The permit for a rotary kiln hazardous waste incinerator specifies that the retention time for solids is 1 hour. The proposed dimensions and operating condition for the incinerator are:

Diameter $= 3.00$ m

Length $= 6.00$ m

Slope $= 2.00\%$

Peripheral speed $= 1.5$ m/min

Determine if the permit requirement will be met.

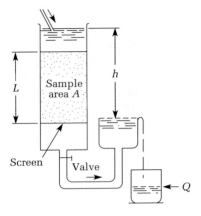

Standard constant head permeameter equation:

$$K = \frac{QL}{hAt}$$

where K = hydraulic conductivity
Q = quantity of discharge
L = length of sample
h = hydraulic head
A = cross-sectioned area of sample
t = time

FIGURE P-12-34

12-34. A standard permeameter is being considered for testing a clay for a hazardous waste landfill base. If the clay must have a hydraulic conductivity of 10^{-7} cm/s and the dimensions of the permeameter are as shown below, how long will the test take if a minimum of 100.0 milliliter of liquid must be collected for an accurate measurement? See Figure P-12-34 for notation and permeameter equation. Dimensions are:

$L = 10$ cm

$h = 1$ m

Diameter of sample $= 5.0$ cm

Answer: $t = 58.95$ or 60 d

12-35. A standard permeameter is being considered for testing a clay for a hazardous waste landfill base. If the clay must have a hydraulic conductivity of 10^{-7} cm/s and the dimensions of the permeameter are as shown in Figure P-12-34, the test will take 60 days if a minimum of 100.0 mL of liquid must be collected for an accurate measurement. You need the results in 30 days. What change in the design of the permeameter would you make to obtain results in 30 days? Show by calculation that your redesign would work.

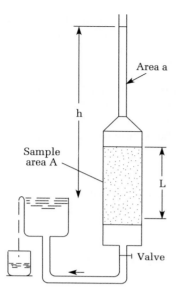

Area a

h

Sample
area A

L

Valve

Falling head permeameter equation:

$$K = 2.3\frac{a\,L}{A\,t}\;\log\left(\frac{h_0}{h_1}\right)$$

where K = hydraulic conductivity
a = cross-sectional area of stand pipe
A = cross-sectional area of sample
L = length of sample
t = time
h_0, h_1 = head at beginning of test
and at time t, respectively

FIGURE P-12-36

12-36. A soil sample has been tested to determine permeability using a falling-head permeameter. (See Figure P-12-36.) The following data were recorded:

Diameter of a = 1 mm

Diameter of A = 10 cm

Length, L = 25 cm

Initial head = 1.0 m

Final head = 25 cm

Duration of test = 14 days

From these data, calculate the hydraulic conductivity of the sample. Assuming the sample is representative of the landfill site, is this a good soil for a hazardous waste landfill base?

12-37. An old hazardous waste landfill was built on a 12-m-deep clay liner. An aquifer lies immediately below the clay layer. The clay layer through which the leachate must pass has a hydraulic conductivity of 1 3 1027 cm/s. If the liquid level (leachate) is 1.0 m deep above the clay layer, how much leachate (in m3/d) will reach the aquifer when the clay layer becomes saturated? Assume Darcy's law applies.

12-38. The three soil layers described below lie between the bottom of a hazardous waste landfill and the underlying aquifer. The depth of leachate above the top soil layer is 0.3 m. How long will it take (in years) for the leachate to migrate to an aquifer located at the bottom of soil C?

> Soil A
> > Depth = 3.0 m
> > Hydraulic conductivity = 1.8×10^{-7} cm/s
> > Porosity = 55%
>
> Soil B
> > Depth = 10 m
> > Hydraulic conductivity = 2.2×10^{-5} m/s
> > Porosity = 25%
>
> Soil C
> > Depth = 12.0 m
> > Hydraulic conductivity = 5.3×10^{-5} mm/s
> > Porosity = 35%

12-39. The practical quantitation limit (PQL) for the solvent trichloroethylene is 5 μg/L. If a barrel (approximately 0.12 m^3) of spent solvent leaked into an aquifer, approximately how many cubic meters of water would be contaminated at the PQL?

12-40. An aquifer has a hydraulic gradient of 8.6×10^{-4}, a hydraulic conductivity of 200 m/d, and a porosity of 0.23. A chemical with a retardation factor of 2.3 contaminates the aquifer. What is the linear velocity of the contaminant? How long will it take to travel 100 m in the aquifer?

12-41. An extraction well must be installed at the site of a leaking gasoline storage tank. The depth of the unconfined aquifer is 60.00 m and the hydraulic conductivity is 6.4×10^{-3} m/s. Measurements show that the plume does not extend more than 150 m from the center of the leak. At 130 m from the center of the leak, the plume is 0.1 m in depth. If the extraction well is 28 cm in diameter, what size pump (in m^3/s) is required so that the plume does not migrate any farther? (*Note:* this is an application of the well equations in Chapter 4.)

12-42. A drum (0.12 m^3) of carbon tetrachloride has leaked into a sandy soil. The soil has a hydraulic conductivity of 7×10^{-4} m/s and a porosity of 0.38. The groundwater table is 3 m below grade and has a hydraulic gradient of 0.002. The aquifer is 28 m thick. A single well intercept system is proposed, using a well pumping at 0.014 m^3/s. Estimate the width of the capture zone at the well.

12-43. If the leading edge of the plume in Problem 12-42 has spread to a width of 200 m, how far ahead of the plume must the well be located to intercept it?

12-11 DISCUSSION QUESTIONS

12-1. What was the outcome of the hazardous waste episode at Times Beach? (*Hint:* You will need to do an Internet search.)

12-2. It has been stated that, on the basis of LD_{50}, 2,3,7,8-TCDD is the most toxic chemical known. Why might this statement be misleading? How would you rephrase the statement to make it more scientifically correct?

12-3. A dry cleaner accumulates 10 kg per month of perchloroethylene (a hazardous waste solvent). To save shipping cost he would like to accumulate 6 months' worth before he ships it to a TSD facility. Can he do this? Explain. (*Hint*: search the applicable regulations in the CFR.)

12-4. Does the "land ban" actually ban the disposal of hazardous waste on the land? Explain.

12-5. A multimillion dollar company has just learned that one drum out of several hundred found at an abandoned waste disposal site has been identified as its property. Their attorney explains that the company may potentially be responsible for cleanup of all the drums at the site if no other former owners of the drums can be identified. Is this correct? Why or why not?

12-6. Your boss has proposed that your company institute a recycling program to minimize the generation of waste. Is recycling the best first step to investigate in a waste minimization program? If not, what others would you suggest and in what order?

12-7. A metal plater is proposing to treat waste sludge to recover the nickel from it. Would this be

a. recycling?

b. reusing?

c. reclaiming?

State the correct answer(s) and explain why you made your choice(s).

12-8. It is not necessary to measure every POHC in an incinerator trial burn. True or false? Explain your answer.

12-12   FE EXAM FORMATTED PROBLEMS

12-1. Calculate the incinerator destruction and removal efficiency for methylene chloride using the following measurements:

Aqueous influent	*Stack gas effluent*
Flow rate $= 53.0$ L/min	Flow rate $= 606.5$ m^3/min
Concentration $= 7,120$ mg/L	Concentration $= 188.3$ μg/m^3

a. 99.97%   b. 69.74%

c. 97.36%   d. 99.77%

12-2. What is the combustion efficiency of a PCB incinerator if the stack effluent concentration of $CO = 100$ ppm and the concentration of $CO_2 = 100,000$ ppm?

a. 0.999%   b. 99.0%

c. 10.0%   d. 99.9%

12-3. If the pH controller for feeding hydroxide to a zinc precipitation tank is set at pH $= 8.4$, what will the effluent concentration of zinc be after settling of the precipitate? The K_{sp} of $Zn(OH)_2$ is 7.68×10^{-17}.

a. 1.99 mg/L   b. 0.80 mg/L

c. 0.0013 mg/L   d. 0.082 mg/L

12-4. Estimate the soil partition coefficient of benzene for a soil having the following properties: bulk density $= 1.4$ g/cm^3, porosity $= 0.40$, organic carbon fraction $= 0.002$, and retardation coefficient $= 1.2$.

a. 29   b. 171

c. 100   d. 143

12-13   REFERENCES

Davis, M. L., C. R. Dempsey, and E. T. Oppelt (2000) "Waste Incineration Sources: Hazardous Waste," in W. T. Davis, *Air Pollution Engineering Manual*, Air Pollution Control Association, Pittsburgh, and John Wiley & Sons, New York, pp. 268–274.

Davis, M. L. and S. J. Masten (2009) *Principles of Environmental Engineering and Science,* McGraw-Hill, Boston, MA, p. 653.

Fromm, C. H., A. Bachrach, and M. S. Callahan (1986) "Overview of Waste Minimization Issues, Approaches and Techniques," in E. T. Oppelt, B. L. Blaney, and W. F. Kemner (eds.), *Transactions of an APCA International Specialty Conference on Peformance and Costs of Alternatives to Land Disposal of Hazardous Waste*, Air Pollution Control Association, Pittsburgh, pp. 6–20.

Griffin R. D. (1988) *Principles of Hazardous Materials Management,* Lewis Publishers, Ann Arbor, MI.

Gross, R. L., and S. G. TerMaath, (1985) "Packed Tower Aeration Strips Trichloroethylene from Groundwater," *Environmental Progress,* vol. 4, pp. 119–124.

Herzbron, P. A., R. L. Irvine, and K. C. Malinowski (1985) "Biological Treatment of Hazardous Waste in Sequencing Batch Reactors," *Journal of the Water Pollution Control Federation,* vol. 57, pp. 1163–1167.

Hileman, B. (1994) "EPA Reassesses Dioxins," *C&E News,* September 19, p. 6.

Javandel, I., and C. Tsang (1986) "Capture Zone Type Curves: A Tool for Cleanup," *Ground Water,* vol. 24, no. 5, pp. 616–625.

Kobayashi, H., and B. P. Rittman (1982) "Microbial Removal of Hazardous Organic Compounds," *Environmental Science and Technology,* vol. 16. pp. 170A–172A.

LaGrega, M. D., P. L. Buckingham, and J. C. Evans (2001) *Hazardous Waste Management,* McGraw–Hill, New York, pp. 471–473, 899–903, 1014–1016.

Masters, G. M. (1998) *Introduction to Environmental Engineering and Science,* Prentice Hall, Upper Saddle River, NJ, p. 240.

O'Brien & Gere Engineers Inc. (1988) *Hazardous Waste Site Remediation,* Van Nostrand Reinhold, New York, pp. 11–13.

Oppelt, E. T. (1981) "Thermal Destruction Options for Controlling Hazardous Wastes," *Civil Engineering ASCE,* pp. 72–75, September.

Penning, C. H. (1930) "Physical Characteristics and Commercial Possibility of Chlorinated Diphenyl," *Industrial & Engineering Chemistry,* vol. 22, pp. 1180–1183.

Reynolds, T. D., and P. A. Richards (1996) *Unit Operations and Processes in Environmental Engineering,* PWS Publishing, Boston, pp. 392–393.

Schwarzebach, R. P., P. M. Gschwend, and D. M. Imboden (1993) *Environmental Organic Chemistry,* John Wiley & Sons, New York, p. 274.

Theodore, L., and J. Reynolds (1987) *Introduction to Hazardous Waste Incineration,* John Wiley & Sons, New York, pp. 76–85.

Thomas, H. C. (1948) "Chromatography: A Problem of Kinetics," *Annals of the New York Academy of Science,* vol. 49, p. 161.

U.S. EPA (1981a) *Summary Report: Control and Treatment Technology for the Metal Finishing Industry—Ion Exchange,* U.S. Environmental Protection Agency Publication No. EPA 625/8-81-007.

U.S. EPA (1981b) *Development Document for Effluent Limitations: Guideline and Standards for the Metal Finishing Point Source Category,* U.S. Environmental Protection Agency Publication No. EPA/440/1-83-091, Washington, DC.

U.S. EPA (1986) *RCRA Orientation Manual,* U.S. Environmental Protection Agency, Publication No. EPA/530-SW-86-001, Washington, DC.

U.S. EPA (1987) *A Compendium of Technologies Used in the Treatment of Hazardous Waste,* U.S. Environmental Protection Agency Publication No. EPA/625/8-87/014), Cincinnati.

U.S. EPA (1988a) *Waste Minimization Opportunity Assessment Manual,* U.S. Environmental Protection Agency Publication No. EPA/625/7-88/003), Cincinnati, p. 2.

U.S. EPA (1988b) *Best Demonstrated Available Technology (BDAT) Background Document for FOO6,* U.S. Environmental Protection Agency Publication No. EPA/530-SW-88-009-I, Washington, DC.

U.S. EPA (1989b) *U.S. EPA Seminar Publication: Requirements for Hazardous Waste Landfill Design, Construction and Closure,* U.S. Environmental Protection Agency Publication No. EPA 625/4-89/022.

U.S. EPA (1991) *Design and Construction of RCRA/CERCLA Final Covers,* U.S. Environmental Protection Agency Publication No. EPA 625/4-89/022, Washington, DC.

U.S. EPA (2001) *Ninth Report on Carcinogens,* U.S. Environmental Protection Agency Report to Congress, Washington, DC.

U.S. EPA (2004a) *Underground Storage Tanks: Building on the Past to Protect the Future,* U.S. Environmental Protection Agency Publication No. EPA 512-R-04-001, Washington, DC.

U.S. EPA (2004b) *Risk Assessment Guidance Manual for Superfund, Volume I: Human Health Evaluation Manual,* U.S. Environmental Protection Agency Publication No. EPA/540/R/99/005, Washington, DC.

U.S. EPA (2005a) *Eleventh Report on Carcinogens,* U.S. Environmental Protection Agency Report to Congress, Washington, DC.

U.S EPA (2005b) CERCLIS Database, at http://cfpub.epa.gov/superrcpad/cursites.

Wentz, C. A. (1989) *Hazardous Waste Management,* McGraw-Hill, New York, pp. 206–207.

Wood, E. F., R. A. Ferrara, W. G. Gray, and G. F. Pinder (1984) *Groundwater Contamination from Hazardous Waste,* Prentice Hall, Englewood Cliffs, NJ, pp. 2–4, 145–158.

CHAPTER
13

SUSTAINABILITY AND GREEN ENGINEERING

13-1 INTRODUCTION

Sustainability

In the simplest dictionary style definition, sustainability is a method of harvesting or using a resource so that the resource is not depleted or permanently damaged. Beyond this simple definition, there are as many definitions as there are authors that write on the subject. This results from the many perspectives the authors bring to the subject. Here are several perspectives from which one might view sustainability: developing countries, developed countries, ecological, economic, social justice, worldwide, regional, national, local.

The conventional starting point for discussions on sustainability is that published by the World Commission on Environment and Development* (WCED, 1987): *Sustainable development is development that meets the needs of the present without compromising the ability of future generations to meet their own needs.* For our starting point, we prefer to define sustainability in terms of a sustainable economy. A *sustainable economy* is one that produces wealth and provides jobs for many human generations without degrading the environment. There are two fundamental principles of this definition of sustainability:

- Reduction in the use of both renewable and nonrenewable natural resources.

- Provision of solutions that are both long-term and market-based.

In the first instance, emphasis is placed on reduction of natural resources rather than end-of-pipe solutions. In a sustainable economy, development focuses on minimizing resource consumption through increased efficiency, reuse and recycling, and substitution of renewable resources for nonrenewable resources. *Renewable resources* are those that can be replaced within a few human generations. Some examples are timber, surface water, and alternative sources of power such as solar and wind. *Nonrenewable resources* are those that are replaceable only in geologic time scales. Groundwater, fossil fuels (coal, natural gas, and oil) and metal ores are examples of nonrenewable resources.

In the second instance, an effective and cost-efficient approach is for society to provide incentives to use alternatives or reduce the use of water, coal, gasoline, and other substances. Some of these incentives are in the form of technological advances that improve efficiency and some are in the form of sociopolitical changes.

The People Problem

An inherent problem with definitions of sustainability is "The People Problem," that is, the ability of future generations to meet their own needs. To understand the "The People Problem" we need to understand the characteristics of population growth.

---

*It is also known as the Brundtland Commission Report, after its chairman.

A simple way of viewing population growth is by assuming that growth is exponential:

$$P_t = P_o e^{kt}$$

FE (13-1)

where P_t = population at time t
 P_o = population at time $t = 0$
 k = rate of growth, individuals/individual · year
 t = time

The growth rate is a function of the crude birth rate (b), crude death rate (d), immigration rate (i), and emigration rate (e):

$$k = b - d + i - e \qquad (13\text{-}2)$$

Example 13-1 provides an illustration of a crude estimate of population growth.

Example 13-1. Estimate the percent growth of the global population from 2010 to 2050 using the following assumptions: crude birth rate ~20 per 1,000 people, crude death rate ~8 per 1,000 people, population ~6,892,000,000 or 6.892×10^9 (PRB, 2010).

Solution. Begin by estimating the rate of growth.

$$k = \frac{20}{1,000} - \frac{8}{1,000} = 0.020 - 0.008 = 0.012$$

The time interval is $2050 - 2010 = 40$ years and the population in 2050 is estimated to be

$$P_t = 6.892 \times 10^9 \{\exp(0.012 \times 40)\}$$
$$= 1.114 \times 10^{10} \text{ people}$$

This is an increase of

$$\left(\frac{1.114 \times 10^{10} - 6.892 \times 10^9}{6.892 \times 10^9} \right) (100\%) = 61.6, \text{ or about } 62\%$$

Comment: Historically, economic and social development has resulted in decreases in the rate of growth. For example, China's rate of growth in 1964 was about 31 per 1,000 people per year. In 1990, it was about 12 per 1,000 people per year. In 2010, it was about 5 per 1,000 people per year. As reference points, China's per capita income in 2000 was $950. Current (2010) growth projections estimate per capita income at $6,000 by 2015.

For comparison, the rate of growth in the United States was about 6 per 1,000 people in 2010; in the United Kingdom and France it was 4 per 1,000 people in 2010.

Projections of population using Equation 13-1 are quite crude. More sophisticated analyses take into account the *total fertility rate* (TFR). The TFR is the average number of children that would be born alive to a woman, assuming that current age-specific birth rates remain constant through the women's reproductive years. Four scenarios of

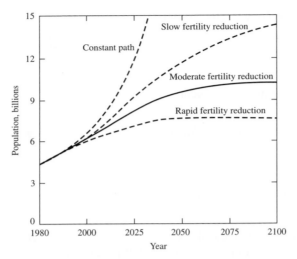

FIGURE 13-1
Four scenarios for world population growth. (*Source:* Haupt and Kane, 1985.)

global population growth for different TFRs are shown in Figure 13-1 (Haupt and Kane, 1985).

Recent studies (Bremner et al., 2010) reveal that the world's population has reached a transition point. The rapid growth of the second half of the 20th century has slowed. However, factors such as improving mortality and slower than expected declines in birth rate guarantee continued growth. Current projections of world population in 2050 range from 9.15 billion to 9.51 billion (Bremner, et al., 2010). This places the population growth on the "moderate fertility reduction" curve in Figure 13-1. This is a 38 percent global population increase rather than the 62 percent increase calculated in Example 13-1.

Given limited resources, both nonrenewable and renewable, there are long-term (> 100 years) implications for population growth that cannot be resolved by technological solutions (for example, food for starving people). We make this statement to place the remainder of the discussion in this chapter in context. We will focus on currently available technological improvements.

There Are No Living Dinosaurs

Over the millennia, the climate of the earth has changed. Natural processes such as variation in solar output, meteorite impacts, and volcanic eruptions cause climate change. These natural phenomena have resulted in ecosystem changes. Animal and plant species have evolved and adapted or have become extinct . . . there are no living dinosaurs. The rate of change has been a major factor in the ability of organisms to adapt successfully.

We inherently refer to the atmosphere when we speak of the climate. But there are also strong interactions with seas and oceans in terms of fluxes of energy, water, and carbon dioxide. While the atmosphere has only a small buffer capacity to resist change, the oceans provide a gigantic buffer capacity for changes in heat, water, and carbon dioxide. Thus, there is a considerable time lag between the causes of climate change and their effects. Because natural changes in climate occur relatively slowly, so also do their effects. Similarly, the effects of human impacts will take some time to be felt on a human time scale.

In this millennium, we are faced with the potential of significant global warming over a geologically short time period for the reasons discussed in Chapter 9. As noted in that chapter, the impacts of global warming on North America are forecast to be a mix of "good news" and "bad news." Exactly how these impacts will be addressed is a formidable challenge.

Vulnerability is a term used to assess the impacts of climate change in general and global warming in particular. To be vulnerable to the effects of global warming, human-environmental systems must not only be exposed to and sensitive to the changes but also unable to cope with the changes (Polsky and Cash, 2005). Conversely, human-environmental systems are **relatively sustainable** if they possess strong adaptive capacity and can employ it effectively. In the United States we have abundant cropland, sufficient natural resources, and a robust economy that gives us a strong adaptive capacity and, hence, a capacity for a relatively sustainable human-environmental system. Whether or not we can employ it effectively is another question. Many other countries do not have these key ingredients and, therefore, do not have a strong adaptive capacity . . . they are vulnerable.

We use the term *relatively sustainable* because there are limits to growth using 20th century economic models based on exploitation of nonrenewable resources. The time frame for adaptation on a global basis is on the order of a few generations or less for energy and minerals at current growth rates in consumption. Certain countries, the United States among them, are more richly blessed and have more time. But the time is not infinite.

Go Green

Green engineering is the design, commercialization, and use of processes and products that are feasible and economical while (U.S. EPA, 2010):

- reducing the generation of pollution at the source

- minimizing risk to human health and the environment

These are not new concepts. The code of ethics of the American Society of Civil Engineers and those of other professional engineering societies have incorporated these concepts for decades (see, for example, Figure 1-1 in this text and the model rules for professional conduct in the *Fundamentals of Engineering Supplied-Reference Handbook,* NCEES, 2011). In the decades of the 1980s and 1990s numerous publications were devoted to the concept of reducing pollution generation (A&WMA, 1988; Freeman, 1990; Freeman, 1995; Higgins, 1989; and Nemerow, 1995). What is "new" about green engineering is that it now has a catchy name *and* branches of civil engineering other than environmental engineering as well as other engineering disciplines have taken up the standard.

The concepts of green engineering are, in fact, a demonstration of adaptability to improve sustainability. Many of the changes that we have historically recognized as improvements in efficiency are steps in increasing sustainability. To be sure, there are new initiatives to take advantage of the catchword but they all contribute to sustainability . . . and they embody the two principles of sustainability:

- reducing society's use of natural resources and

- using market based solutions.

Of particular note for civil and environmental engineers is the emergence of whole building assessment systems like BRE Environmental Assessment Method (called BREEAM and used in the United Kingdom), Green Globes (used in Canada and the United States), and Leadership in Energy and Environmental Design (called LEED and used in the United States). These programs place considerable emphasis on the selection of green materials or products as an important aspect of sustainability.

In addition to these new assessment systems, an old system called life cycle assessment (LCA) has taken on a new perspective. Traditionally, LCA focused on the costs of building, operating, and closing a facility as a method of comparison of alternatives. The new LCA approach is a methodology for assessing the environmental performance of a product over its full life cycle (Trusty, 2009).

In the following sections of this chapter we will examine the parameters of sustainability for a renewable resource—water—and a nonrenewable resource—energy. In each of these cases we will give examples of green engineering to demonstrate the contribution of technology to sustainability. Obviously, these are only a sample of the current possibilities.

13-2   WATER RESOURCES

Water, Water, Everywhere

"Water: too much, too little, too dirty." (Loucks et al., 1981) This sentence succinctly summarizes the issues of sustainable water resource management. In Chapter 4, we discussed floods from the perspective of hydrologists and civil engineering applications. Here we will use these tools to discuss floods from the perspective of sustainability. While the discussion of droughts was not explicitly addressed in Chapter 4, the hydrologic equations for water balance and risk estimation discussed there also apply to droughts. We will use these in discussing droughts in the context of sustainability. A large portion of this text is devoted to environmental engineering measures to prevent water from becoming "too dirty" and cleaning water that is "too dirty."

Floods

There are two broad categories of floods: coastal and inland. Inland floods, also called *inundation floods*, as the name implies are the result of a combination of meteorological events that result in inundation of a floodplain. These may be seasonal, such as the historic flooding of the Nile and the rhythmic monsoons that flood great portions of India, or they may be highly irregular with long return periods. Coastal flooding most frequently results from exposure to cyclones or other intense storms. Prime examples are the Bay of Bengal and the Queensland coast of Australia exposed to cyclones, the Gulf and Atlantic coasts of the United States exposed to hurricanes, and the coasts of China and Japan affected by typhoons.

A special case of coastal flooding is that due to tsunamis. Earthquakes are a major cause of tsunamis. They most frequently occur along the Pacific rim because of crustal instability. Wave heights range from barely noticeable to over 10 m with run-ups from less than 5 m to over 500 m.

Floods by themselves are not disasters. Floods occurred long before civilizations arose and will continue long after they have disappeared. Inland floods bring nutrients and contribute to alterations in the river channel that provide nursery habitat for young fish, destroy existing riparian communities, and create new environments for new ecosystems. Both field data and model studies indicate that the most complex and diverse ecosystems are maintained only in riparian environments that fluctuate because of flooding (Power et al., 1995).

Estuaries and natural coastal areas are characterized by many different habitats, ranging from sandy beaches to salt marshes, mud flats, and tidal pools, that are inhabited by an extraordinary variety of animals and plants. In a similar fashion to inland floods, coastal floods alter the landscape to create new environments for new ecosystems.

Around 3000 B.C.E., the world's first agriculturally based urban civilizations appeared in Egypt and Mesopotamia and maize farming began in Central America. As populations grew, people moved to the floodplains to tap the natural resources of water and fish. Over the millennia these urban civilizations matured and grew such that today half of the world's 6.9 billion people live in an urban environment. There are between 18 and 25 major metropolitan cities with a population over 10 million (often cited as *megacities*). Most of these are located near water bodies that, at some point in time, will bring catastrophic floods.

Although only about 7 percent of the United States' total land area lies in floodplains, more than 20,800 communities are located in flood-prone areas (Hays, 1981). By the mid-1990s, 12 percent of the population occupied more than seven million structures in areas of periodic inundation (Gruntfest, 2000). About 50 percent of the population lives near the coast (Smith and Ward, 1998).

For every city like Chicago or Detroit that are relatively isolated from inland or costal flooding, there are cities like St. Louis, New Orleans, Los Angeles, Miami, and New York that will probably experience a major flood before the end of this century. As illustrated in Example 13-2, the probability is not small.

Example 13-2. Estimate the risk of a 100-year return period event occurring by the year 2100 if the current year is 2010.

Solution. Using Equation 4-14 with $T = 100$ years and $n = 2100 - 2010 = 90$

$$R = 1 - \left(1 - \frac{1}{T}\right)^n$$

$$= 1 - \left(1 - \frac{1}{100}\right)^{90} = 1 - 0.40 = 0.60$$

The estimate is that there is a 60 percent risk of a 100-year return period event being equaled or exceeded in the next 90 years.

In the United States inland floods tend to be repetitive. From 1972 to 1979, 1,900 communities were declared disaster areas by the federal government more than once, 351 were inundated at least three times, 46 at least four times, and 4 at least five times. Between 1900 and 1980 the coast of Florida experienced 50 major hurricanes, and even as far north as Maryland there is an average of one hurricane per year that has direct or fringe effects on the coast (Smith and Ward, 1998).

Floods and Climate Change.   The amount of precipitation falling in the heaviest 1 percent of the rain events in the United States increased 20 percent in the past 50 years with eastern events increasing by greater than 60 percent and western events increasing by 9 percent (Karl et al., 2009). During this time period, the greatest increases in heavy precipitation have occurred in the Northeast and Midwest. Using the middle 50 percent of the values from Global Circulation Models (GCMs), annual precipitation forecasts have been made for the time period 2080–2099. They are summarized as follows:

- western United States: 0 to 9 percent increase,

- central United States: precipitation projections cannot be distinguished from natural variability,

- eastern United States: increase by 5 to 10 percent.

These data (Figure 13-2) and projections imply that higher intensity rainfall is to be expected. With higher intensities comes high river flow rates and the potential for more floods.

As a result of melting of glaciers, global warming models predict an increase in sea level between 0.2 and 0.6 m by 2100 (IPCC, 2007). The sources of sea level rise include thermal expansion, melting of glaciers and ice caps, melting of the Antarctic ice sheet and the Greenland ice sheet. This will result in the loss of about 30 percent of global coast wetlands and an increase in coastal flooding.

Floods and Sustainability.   Vulnerability, in terms of flood planning, is a measure of the ability to anticipate, cope with, resist, and recover from harm caused by a flood hazard (Blaikie et al., 1994). As we noted in Section 13-1, human-environmental systems are relatively sustainable if they possess strong adaptive capacity and can employ it effectively. The Mississippi River floods regularly (notably so in 1927, 1937, 1947, 1965, 1993, 2008, and 2010). Florida has had hurricane flood surges in the range of 2 to 5 meters on at least five occasions (1935, 1960, 1992, and twice in 2004). The fact that the environs of the Mississippi River and Florida remain inhabited and prospering is evidence of sustainability and the adaptability of the people. As shown in Tables 13-1 and 13-2, this has not been without cost.

For a country like Bangladesh, the question of sustainability is moot. Because Bangladesh is a country of rivers, 66 percent of the country's 144,000 km^2 is either a floodplain or a delta. About 75 percent of the runoff from the Himalayan mountains drains through the country from July to October. The mean annual rainfall ranges from 1,500 mm in the west to over 3,000 mm in the southeast. Seventy-five to eighty

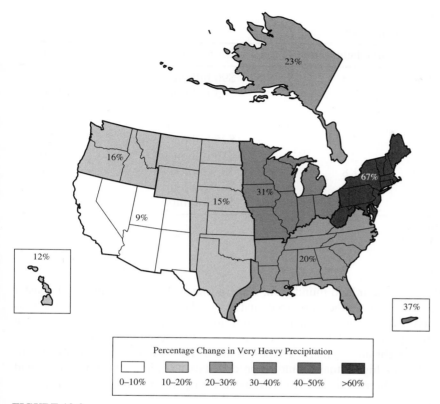

FIGURE 13-2
Increases in the amounts of very heavy precipitation from 1958–2007. (*Source:* Karl et al., 2009.)

percent of the rainfall comes from monsoons during the period from June to October. When the Brahmaputra, Ganges, and Meghna rivers peak in August or September approximately one-third of the country floods. Over one-third of the flooded area is under at least one meter of water (Wohl, 2000).

TABLE 13-1
Impact of 1927 and 1993 Mississippi floods

Parameter	1927 flood	1993 flood
Area flooded, millions of acres	12.8	20.1
Number of buildings damaged	137,000	70,000
Number of fatalities	246	52
Number of people evacuated	700,000	74,000
Property damage, billions of dollars[a]	21.6	22.3

[a]In 2012 dollars.
Data Source: Wright, 1996.

TABLE 13-2
Florida hurricane impacts

Year	Surge height, m	Economic loss	Fatalities
1926	4.6	$90 billion	373
1928	2.8	$25 million	1,836
1935	n/a	$6 million	408
1960	4	$387 million	50
1992	5	$25 billion	23
2004	2	$16 billion	10
2004	1.8	$8.9 billion	7
2004	1.8	$6.9 billion	3
2005	n/a	$16.8 billion	5

n/a = not available
Source: NOAA, 2010.

In addition to the inland inundation, Bangladesh suffers recurring lethal coastal storm surges. The surge effects are accentuated by a large astronomical tide, the seabed bathymetry, with shallow water extending to more than 300 km offshore in the northern part of the bay, and the coastal configuration that accentuates the funnel shape of the bay by a right-angle change in the coastline. This produces maximum storm-surge levels that are higher than would be produced on a straight coastline (Smith and Ward, 1998). The human devastation produced by the monsoons is summarized in Table 13-3.

The economic devastation is equally bad. In 1988–89, nearly half of the nation's development budget was spent to repair flood damage. In 1991, the storm surge caused damage estimated at one-tenth of the nation's gross domestic product.

TABLE 13-3
Bangladesh cyclone fatalities

Year	Fatalities
1822	40,000
1876	100,000
1897	175,000
1963	11,468
1965	19,279
1970	300,000
1985	11,000
1991	140,000

Data Source: Smith and Ward, 1998.

TABLE 13-4
Measures of adaptive capacity for Bangladesh and the United States[a]

Parameter	Bangladesh	United States
Population, million	164	310
Growth rate, %	1.5	0.6
Net migration, %	−1	+3
TFR	2.4	2.0
Infant mortality, deaths/1000	45	1.4
Population density, people/km^2	1,142	32
GNI[b], US$/capita	1,440	46,970

[a]PRB, 2010.
[b]Gross National Income, 2008.

The comparison of some fundamental parameters in Table 13-4 provides a basis for the very positive outlook for the ability of the United States to adapt in comparison to the daunting prospects for the Bengalis. Although Bangladesh's inundation brings desperately needed nutrients for the soil and replenishes ponds for fish and shrimp spawning, floods bring displacement. Historically, displaced households have moved to new lands. In a country with the world's highest population density, there is precious little "new" land. There is a collective fatalism to this cycle. The Bengalis cope astonishingly well with the floods they've always known. From a long-term perspective, it does not appear that Bangladesh has the resources to develop a *sustainable economy* that will produce wealth and provide jobs for many human generations without degrading the environment. They need outside help. Using funds from the National Science Foundation and the Georgia Tech Foundation to supplement sporadic funding from USAID (whose priorities changed from the Clinton to the Bush administration), Webster et al. (2010) were able to develop an exploratory project of 10-day flood forecasts to provide advance warning and allow implementation of an emergency response plan. The forecasts and response were successful for the 2009 flood season. The savings resulting from the forecast was estimated to be US$130–190 for fishery and agriculture income, US$500 per animal, and US$270 per household. Given that the average farmer's income is approximately US$470, the savings in the flooded regions were substantial. To continue this success, a better funding base and increased collaboration between federal and international agencies is required.

Floods and Green Engineering. In the last several decades it has become apparent that intensively engineered, densely populated, biologically impoverished river corridors and coastal areas are not a sustainable response to flood hazards. Unfortunately, most of the world's cultures have a tradition of aggressively interfering and altering natural systems and processes. We have selected two examples of alternatives that have been implemented in the United States. You will note that engineered structures play a minor role in sustainable flood protection. The "green engineering" component is recognizing that traditional structural solutions were bound to fail and that other solutions were required to minimize risk and to provide a *sustainable economy* for many human generations.

Flood-Warning Systems. Technological advances have made warning systems available to more than 1,000 communities in the United States. Coupled with advances in meteorological forecasting, these systems are lifesavers. The reduction in flood and hurricane fatalities can, in part, be directly related to implementation of these systems (see for example Tables 13-1 and 13-2). The lack of a warning system, as for example in Bangladesh, inevitably leads to great loss of life (Table 13-3). However, without a plan for disseminating the warning message and implementing an evacuation plan, the warning system is useless. The 1,835 fatalities in New Orleans that resulted from the Katrina hurricane in 2005 may be attributed to the lack of provision for the poor and handicapped that had no transportation before, during, and after the event as well as remote, safe facilities to accommodate them.

Given the difficulty in evacuating New Orleans, a metropolitan area with a population of 1,235,650 people, the prospects for successful evacuation of even a portion of one of the world's megacities in the event of a forecast 100- or 500-year storm or flood appear slim.

Of course, a warning system is only necessary after poor land use decisions have been made. The fact that the material resources and habitat are destroyed in a flood diminishes the potential for a sustainable economy that produces wealth and provides jobs.

Acquisition and Relocation. One of the most successful, sustainable, long-term solutions to repeated flooding is to move people and structures out of the floodplain. Almost 40 years ago the U.S. federal government authorized a cost-sharing program for relocation. One of the most dramatic uses of these funds was for the relocation of the village of Valmeyer, Illinois, from the Mississippi floodplain to a bluff 120 m above the river and two miles away. The village had been flooded in 1910, 1943, 1944, and 1947. Although the U.S. Army Corps of Engineers raised the levee to 14 m after the 1947 flood, it was topped by the 1993 flood, which destroyed 90 percent of the buildings. Relocation and rebuilding began almost immediately. The first business opened in May 1994 and the first home was occupied in April 1995. The town's population at the time of the 1993 flood was 900. In 2009, the population was 1,168.

Other examples of buyouts and relocation include Hopkinsville, Kentucky; Bismarck, North Dakota; Montgomery, Alabama; and Birmingham, Alabama. Since the 1993 Mississippi flood, nearly 20,000 properties in 36 states have been bought out (Gruntfest, 2000).

From a practical point of view, it is unlikely that more than a few of the 20,800 U.S communities that are flooded regularly are able or willing to move. However, it is frequently the case that only a portion of the town is flooded. This opens the opportunity for sections of communities to be relocated. The floodway can then become a resource for replenishment of the river ecosystem. In the last few decades, this type of activity has been a means of revitalization of downtown business districts. Predisaster mitigation by acquisition and relocation has resulted in benefit-cost ratios between 1.67

and 2.91 (Grimm, 1998). Given the immense cost of floods (Tables 13-1 and 13-2), this seems like an excellent market-based approach.

Droughts

Drought means something different for a climatologist, an agriculturalist, a hydrologist, a public water supply management official, and a wildlife biologist. There are more than 150 published definitions of drought. The literature of the 21st century appears to have adopted the classification by Wilhite and Glantz (1985). They address these different perspectives by classifying drought as a sequence of definitions related to the drought process and disciplinary perspectives: meteorological, agricultural, hydrological, and socioeconomical/political drought. Bruins (2000) adds another useful subdivision: pastoral drought.

In summary the definitions are as follows:

- *Meteorological (climatological) drought*: a deficiency of precipitation for an extended period of time that results from persistent large scale disruptions in the dynamic processes of the earth's atmosphere.

- *Agricultural drought*: a dry period where *arable* land that can be cultivated does not receive enough precipitation. This may occur even if the total annual precipitation reaches the average amount or above because the precipitation does not occur when the crop needs it. An example is in an early growth stage when seeds are germinating.

- *Pastoral drought*: land that is *nonarable* but is suitable for grazing may not receive enough moisture to keep natural, indigenous vegetation alive.

- *Hydrological drought*: this type is reflected in reduced stream flow, reservoir drawdown, dry lake beds, and lowering of the piezometric surface in wells.

- *Socioeconomical/political drought*: this type of drought occurs when the government or the private sector creates a demand for more water than is normally available. This may result from attempts to convert nonarable pasture to cultivation or population growth that outstrips the water supply.

The relationship between these various types of droughts, the duration of drought events, and the resultant socioeconomic impacts is shown schematically in Figure 13-3.

The temporal sequence in drought recovery begins when precipitation returns to normal and meteorological drought conditions have abated. Soil water reserves are replenished first, followed by increases in streamflow, reservoir and lake levels, followed by groundwater recovery. Drought impacts may diminish rapidly in the agricultural sector because of its reliance on soil water, but linger for months or years in other sectors dependent on stored surface or subsurface supplies. Groundwater users, often last to be affected by drought onset, may be last to return to normal conditions (Viau and Vogt, 2000).

Drought is a normal part of climate, rather than a departure from normal climate. Drought by itself is not a disaster. Whether it becomes a disaster depends on its impact on local people and the environment. The Nile River, which has been the source of life

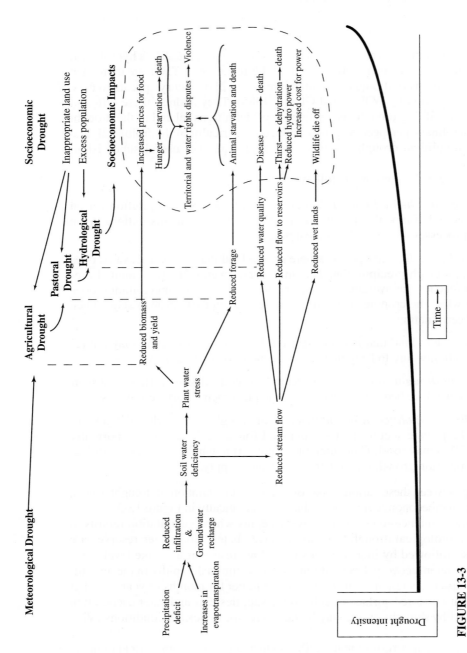

FIGURE 13-3
Relationship between various types of droughts, drought impacts, and drought duration.

for Egypt from its earliest inhabitation, is a source of long-term data that affirms this fact. The annual inundations brought moisture and nutrients to the soil. Low floods meant hardship because there was insufficient soil moisture and nutrients for crops. Hydrological drought in the Nile River has been documented for the period from 3000 B.C.E. to modern times. For the early Dynastic period and the Old Kingdom (circa 3000–2125 B.C.E.) there are 63 annual flood records from 11 different rulers. During this period there was a decrease of the mean discharge of about 30 percent (Bell, 1970; Butzer, 1976). It has been posited* that between 2250 and perhaps 1950 B.C.E. there was a series of catastrophic low floods (i.e., droughts) that lead to the demise of the Old Kingdom and the start of the First Intermediate period (Bell, 1971; Butzer, 1976). Again, in the late Ramessid period (circa 1170–1100 B.C.E.) there is economic evidence[†] of catastrophic failures of the annual flood (Butzer, 1976). Between 622 and 999 C.E. there were 102 years of poor floods (Bell, 1975).

Typically, we in the United States often associate drought with arid, semiarid, subhumid regions. In reality, drought occurs in most nations, in both dry and humid regions, and often on a yearly basis. As a basis for planning and management, the United Nations Educational, Scientific, and Cultural Organization (UNESCO) has developed a numerical representation of aridity based on the ratio of average precipitation (P) to average annual potential evapotranspiration (ET). This classification system is summarized in Table 13-5.

How Dry Is Dry?   Drought differs from other natural hazards. It is insidious. It takes months or, in some cases, years before it is recognized. This is because of natural aridity and the time it takes for the lack of precipitation to be manifest in agricultural drought, hydrological drought, or socioeconomic drought. Likewise, it is difficult to

TABLE 13-5
UNESCO definitions for bioclimatic aridity

Climatic zone	P/ET
Hyperarid	> 0.03
Arid	$0.03 - 0.20$
Semiarid	$0.20 - 0.50$
Subhumid	$0.50 - 0.75$

Source: UNESCO, 1979.

---

*Because there are no meteorological records of precipitation in ancient Egypt (and even if there were records they would invariably show little rain), the droughts were inferred using the definition of socioeconomic/political drought: lack of food implies a hydrological drought, which, in turn, implies an agricultural drought. Because the Nile floods are a result of monsoon rains in Ethiopia and Uganda, the meteorological drought had to occur there rather than in Egypt.

†The price of summer wheat with respect to metals rose to between eight and twenty-four times the standard price of earlier times. Prices recovered to predrought levels by about 1070 B.C.E. (Butzer, 1976). Again, this is a case of socioeconomic drought reflecting preceding hydrologic and agricultural droughts.

TABLE 13-6
Palmer drought severity index categories

Moisture category	PDSI
Extreme drought	≤ -4.00
Severe drought	-3.00 to -3.99
Moderate drought	-2.00 to -2.99
Mild drought	-1.00 to -1.99
Incipient drought	-0.50 to -0.99
Near normal	$+0.49$ to -0.49
Incipient wet	$+0.50$ to $+0.99$
Slightly wet	$+1.00$ to $+1.99$
Moderately wet	$+2.00$ to $+2.99$
Very wet	$+3.00$ to $+3.99$
Extremely wet	$\geq +4.00$

Source: Based on Palmer, 1965.

From this table we note that a PDSI less than -0.99 indicates drought.

determine when a drought has ended. Looking back, we have a better understanding of both the beginning and the end. One method of evaluation of a historic drought is the *Palmer Drought Severity Index* (PDSI). The PDSI is a single number representing precipitation, potential evapotranspiration, soil moisture, recharge, and runoff (Palmer, 1965). These measurements are discussed in Chapter 4. Implicitly, they are a measure of "agricultural drought."

Numerous authors have criticized the PDSI for a wide variety of reasons. Fundamentally, it is not an operational tool because the drought must end before it can be calculated. Moreover, it requires a substantial amount of sophisticated measurements and calculations to develop the index number. Nonetheless, it has has been used extensively to evaluate droughts in the United States. Even here it has limitations because Palmer used data from Iowa and Kansas that are not representative of the western states where snow accumulation and snow melt are an important component of the water balance. Because of its wide use in the United States and the extensive use of PDSI in comparative data with *dendrochronological* (tree-ring) reconstructions of historical records, we present it in summary form in Table 13-6.

Historical Reference Droughts.　For a simple reference frame, Fye et al. (2003) classify droughts in the United States into two broad categories: "Dust bowl-like droughts" and "1950's-like droughts." The "Dust bowl-like droughts" were put in this category because of the similarity of their spatial footprint with the 1930s drought. However, the 1930s drought was by far the worst drought over the last 500 years based on regional coverage, intensity, and duration. The 11-year drought from 1946 through 1956 was the second worst drought to impact the United States during the 20th century. Based on PDSI calculations for the 20th century and reconstructions of large-scale decadal tree ring reconstructions of summer PDSI, Fye et al. (2003) developed the maps reproduced

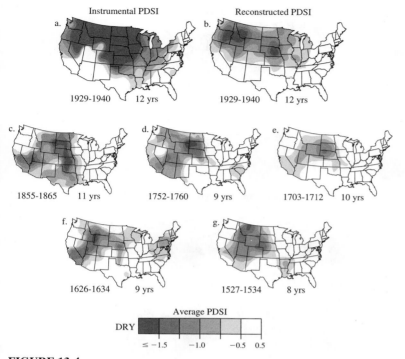

FIGURE 13-4

Dust bowl-like droughts. Instrumental PDSI (a) and PDSI analog constructed from tree-ring data (b) for the period 1929–1940. PDSI reconstructed, averaged, and mapped for the period 1527–1865 (c–g). (*Source:* Fye et al., 2003.)

in Figures13-4 and 13-5. Of special note is that these droughts lasted between 6 and 14 years and that shorter drought periods were not considered in the research. Earlier work using dendrochronology identified longer drought periods of severe drought (PDSI <-3.00) circa 900–1100 C.E. (200 years) and 1200–1350 C.E. (150 years) on the eastern slope of the Sierra Nevada (Stine, 1994). Fye et al. (2003) estimate that the 1950s-like decadal drought has a return period of about 45 years. Shorter drought periods, particularly in the southwest, have much shorter return periods.

Although the decadal droughts are predominantly located west of the Mississippi, there are instances of incipient and mild drought in areas that we consider to be humid in the dendrochronological analysis (i.e., the southeastern states and New England).

Regional Water Resource Limitations.   A comparison of the average regional consumptive use and renewable water supply in the United States is shown in Figure 13-6 on page 973. *Renewable supply* is the sum of precipitation and imports of water, minus the water that is not available for use through natural evaporation and exports. It is a simplified upper limit to the amount of water consumption that could occur in a region on a sustained basis. Because there are requirements to maintain minimum flows in streams

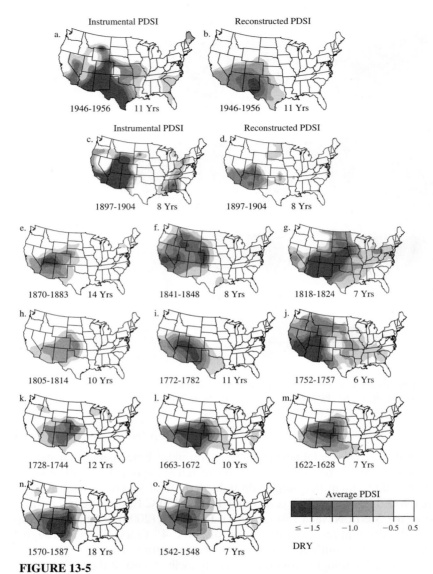

FIGURE 13-5

1950's-like droughts. Instrumental PDSI (a and c) and PDSI analogs constructed from tree-ring data (b and d). PDSI reconstructed, averaged, and mapped for the period 1542–1883 (e–o). (*Source:* Fye et al., 2003.)

and rivers for ecological, navigation, and hydropower, it is not possible to totally develop the renewable supply. However, the ratio of renewable supply to consumptive use is an index of the development of the resource (Metcalf & Eddy, Inc., 2007). Water resource issues for selected regions are summarized in the following paragraphs. These discussions are based on the *National Water Summary—1983* (USGS, 1984) and *Water 2025* (U.S. Department of Interior, 2003).

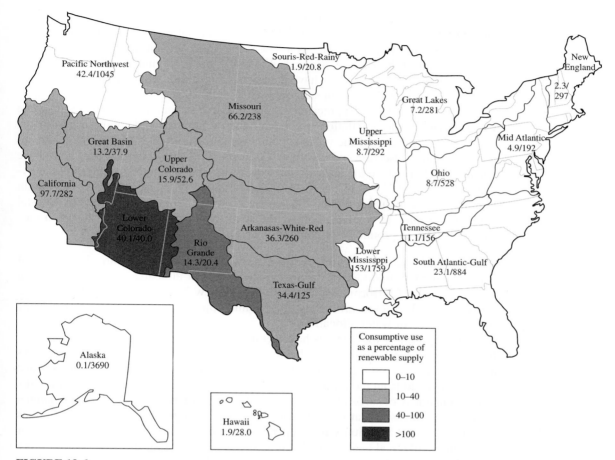

FIGURE 13-6

Comparison of average consumptive use and renewable water supply for the 20 water resources regions of United States (Adapted from USGS, 1984; updated using 1995 estimates of water use). The number in each water resource region is consumptive use/renewable water supply in 10^6 m^3/d, respectively, or consumptive use as a percentage of renewable supply as shown in the legend.

The West. Using the map in Figure 13-6, this region consists of the Pacific Northwest, California, and the Great Basin. While the index implies a consumptive use less than 40 percent of the renewable water supply, rather large local areas in these regions are already stressed to meet demand in regularly occurring meteorological droughts. The issues are (1) the explosive population growth in urban areas, (2) the emerging need for water for environmental and recreational uses, and (3) the national importance of the domestic production of food and fiber from western farms and ranches.

In the Pacific Northwest the majority of the precipitation falls on the western slopes of the Rocky Mountains. The Snake and Columbia are the major rivers. Fifteen large dams and more than 100 smaller dams have been built on these rivers and their tributaries. These dams harness 90 percent of the hydroelectric potential of the region and provide irrigation to over a million acres. Because of runoff, seepage, evaporation, and other losses, almost twice as much water must be delivered as is actually used. As much as 43 percent of the acreage is in low-value forage and pasture crops that account for 60 percent of the water consumed. While of great economic benefit to the region, these federally funded projects have decimated the salmon population with a loss of between 5 and 11 million adult salmonids (Feldman, 2007).

California has the most precarious water resource system in the United States. The vast majority of the population is concentrated in the semi-arid southern part of the state. Seventy percent of the water supplies are in the north and 80 percent of the demand is in the midsection and the south. California droughts of two or more years have occurred eight times in the last century. Record-breaking droughts occurred in the hydrologic years 1928–1934, 1976–1977, and 1987–1992. The most recent drought lasted from the hydrologic year 2007 to well into the 2010 hydrologic year. Coastal areas in the mid- to southern half of the state experience salt water intrusion (see Chapter 7) due to groundwater pumping.

The Colorado Basins.   The Upper and Lower Colorado basins are the source of water for the Colorado River Compact. The Compact allocates water among seven states: Arizona, California, Colorado, Nevada, New Mexico, Utah, and Wyoming. The Colorado Basin was arbitrarily divided into upper and lower basins at Lees Ferry for the purpose of water allocation.

Although all of these states have semi-arid to arid climates, Arizona is notable for the precariousness of its water balance.* As noted in Figure 13-6, the index shows a consumptive use greater than 100 percent of the renewable water supply. Surface water is supplied from the following rivers: Colorado, Verde, Gila, and Little Colorado. The vast majority of surface water comes from the Colorado River. The trend in annual flow volume measured at Lee's Ferry from 1895 through 2003 shows a continuous decline of about 0.5 million acre-feet per decade. This is due, in part, to upstream water use. Dendrocological analysis of Utah trees has revealed four droughts lasting more than 20 years, and nine lasting 15 to 20 years (USGS, 2004). Similar data from Mesa Verde, Arizona, reveal a drought lasting 23 years from circa 1276 to 1299 C.E. (Haury, 1935). Flow data from Lee's Ferry show three droughts lasting from four to eleven years and an ongoing drought from 2000 through 2003 (USGS, 2004).

Until the late 1970s, Arizona depended almost exclusively on groundwater to supply municipal water needs. The sparse precipitation means that the aquifer received virtually no recharge. As a consequence, the aquifer was literally being mined.

The Rio Grande.   Population density in the border lands of the Rio Grande River has quadrupled in Mexico and tripled in the United States since the 1950s (Mumme, 1995). Meanwhile, in-stream flows have fallen to 20 percent of historical levels.

---

*We have already addressed California's critical situation but would add the note that California has been using half a million acre-feet more water than was agreed to in the Compact and affirmed by the U.S. Supreme Court (i.e., 4.4 million acre-feet or 5.43×10^9 m^3). On October 10, 2003, a seven-state agreement called the California 4.4 Plan was signed that gave California until 2017 to reduce is draw on the river to the basic apportionment (Pulwarty et al., 2005).

The Central Great Plains. The Missouri, Arkansas-White-Red river basins and the Texas Gulf region are included in this region. A major trans-basin water diversion is from the Colorado River through tunnels drilled through the Rockies to supply Wyoming. As noted in "The West" discussion, the Upper Colorado basin is under severe stress to maintain flow rates to downstream compact members as agreed upon early in the last century.

Although there are major rivers that supply generous amounts of water to multiple communities, irrigated agriculture is a main end use in the region. Farmers in the "High Plains" states of Nebraska, Colorado, Kansas, Oklahoma, New Mexico, and Texas are drawing water from the Ogallala aquifer that underlies the region. In some instances the rate of withdrawal exceeds the average recharge rate (See Figures 13-7 and 13-8). At the current rate of withdrawal it is estimated that the aquifer will be unable to provide fresh water in another 50 to 100 years (KGS/KDA, 2010).

The Eastern Midwest. This region includes the Souris-Red-Rainy, Upper and Lower Mississippi, Ohio, and Tennessee river basins. This region can be said to have plenty of water. Seasonal meteorological droughts occur in some locations with resultant agricultural droughts.

Great Lakes. The Great Lakes hold 95 percent of the fresh surface water in the United States. This region includes the 121 watersheds in the states that border the Great Lakes. The surrounding states and Canadian provinces have ratified an agreement to prohibit export of water from this region. Local instances of groundwater depletion in urban areas are of concern.

New England. This region includes the traditional New England states of Maine, New Hampshire, Vermont, Massachusetts, Connecticut, and Rhode Island. Water resources in this region are generally abundant. Population pressure, especially in New Hampshire, Connecticut, and around the Boston metropolitan area, has resulted in dramatic stress on the drinking water supply and virtually irreparable damage to some surface water ecosystems. Even a short-term drought in this region would result in major socioeconomic impact.

Mid-Atlantic. This region includes the states of New York, New Jersey, Pennsylvania, Delaware, and Virginia. This region has experienced severe droughts in the past few decades. The major metropolitan areas rely on water systems that are highly sensitive to climate variation. Like the New England region, population pressure has resulted in dramatic stress on the drinking water supply and virtually irreparable damage to some surface water ecosystems. Even a short-term drought in this region would result in major negative socioeconomic impact.

South Atlantic-Gulf. This region includes North and South Carolina, Georgia, Alabama, and Florida. Development, population growth, demographic preference for the coastal areas, and large seasonal population swings present water management issues that are not easily resolved. A recent (2005–2008) drought in Alabama, Georgia, the Carolinas, and Tennessee has revealed major vulnerability. This was not an unusual event. For example, based on the climatological record, not including short-term agricultural droughts, Georgia can expect a meteorological drought of two or more years on the average about once in 25 years (Stooksbury, 2003).

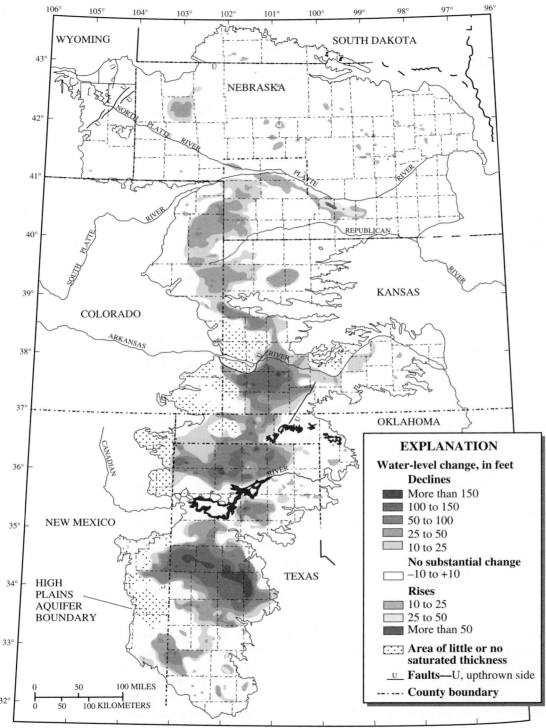

FIGURE 13-7
Ogallala aquifer. (*Source:* USGS 2007.)

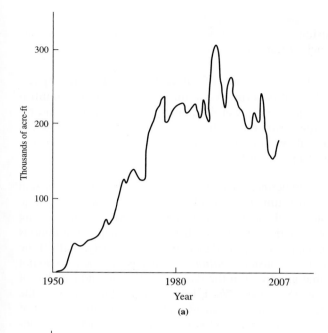

(a)

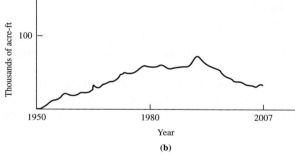

(b)

FIGURE 13-8
Water use for irrigation in Haskell County, Kansas. (a) Withdrawal, (b) Average recharge.

Florida has a large precipitation excess but it occurs over a three- to four-month period in the summer when water use is lower because the transient population leaves the state. During high demand in the winter, there is little precipitation.

All along the South Atlantic and Gulf coast, there is a burgeoning problem of salt water intrusion.

Droughts and Global Warming. As noted in the discussion of floods and climate change, GCM projections do not reveal an increase in meteorological drought. However, as noted in Chapter 9, forecast increases in temperature will result in

- drier crop conditions in the Midwest and Great Plains, requiring more irrigation

- warming in the western mountains that will decrease snowpack, cause more winter flooding, reduce summer flows, and increase agricultural, hydrological, and socioeconomic drought

- a rise in sea level between 0.18 and 0.57 m that would result in an increase in severity of saltwater intrusion into drinking water supplies in coastal areas particularly in Florida and much of the Atlantic coast

Drought and Sustainability. The determinants of vulnerability to drought impacts include: population growth, settlement patterns, economic development, health infrastructure, mitigation and preparedness, early warning, emergency assistance, and recovery assistance coupled with safe yield of available water resources. As we noted in Section 13-1, human-environmental systems are relatively sustainable if they possess strong adaptive capacity and can employ it effectively.

Safe Yield. Theoretically, safe yield defines the capability of a water source to sustain a stipulated level of water supply over time as river inflows and groundwater recharge varies both seasonally and annually (Dziegielewski and Crews, 1986). The engineering application of this definition is presented in Chapter 4. There are two major assumptions in the procedures described in Chapter 4: (1) the data exhibit stationarity, and (2) the required level of water supply remains constant. *Stationarity* means that the data do not exhibit trend or periodicity, or that any trend in the data is explicable and capable of reliable extrapolation into the future (Smith and Ward, 1998). The decline in flow of the Colorado River noted under "Regional Water Resource Limitations" and the forecast of reduction in snow melt runoff for California are just two examples of the lack of stationarity in the data. Population growth and increased saltwater intrusion because of rising sea levels are examples of the change in the required level of water supply that invalidates the second assumption. Nonetheless, the comparison of safe yield estimates and "current" water consumption shown in Table 13-7 provide an example of a convenient way to benchmark a municipal water supply.

From the perspective of drought mitigation, any water supply system with average demands below the safe yield should not experience water shortages during droughts that are less severe than a "design drought" used to derive the value of the safe yield.

TABLE 13-7
Water use and safe yield for selected cities

City	Current use, MGD[a]	Safe yield, MGD	Water use/ safe yield
Binghamton, NY	12.5	43	0.29
Denver, CO	198	266	0.74
Indianapolis, IN	102	112	0.91
Phoenix, AZ	272	257	1.06
Southern California	3,419	3,053	1.12
New York City, NY	1,533	1,290	1.19
Merrifield, VA	78.9	54	1.46

[a]MGD = million gallons per day.
Source: Dziegielewski et al., 1991.

Case Study—California.   Beginning at the end of the 19th century and continuing to this day, numerous audacious water projects have been developed to keep people coming to the state. Early in the 1900s, after shallow aquifers and seasonal rivers could no longer sustain Los Angeles, land was purchased in the Owens Valley, east of the Sierra Nevada. Completion of the Los Angeles Aqueduct sent the entire flow of the Owens River to Los Angeles. Within a decade the lake was a dust bowl and the San Fernado Valley was worth millions (Bourne and Burtynsky, 2010).

In the intervening years, California has built over 2,000 miles (3,218 km) of canals, pipelines, and aqueducts to transport water from the north to the south. Because of its dependence on snowmelt for much of its water, over 157 reservoirs have been built to hold the melt water for later distribution. Major pumping stations, located in the Sacramento–San Joaquin Delta, move the water to the south (Bourne and Burtynsky, 2010). The movement of water accounts for nearly 40 percent of the state's total energy supply (Feldman, 2007).

The Sacramento–San Joaquin Delta, a former 700,000 acre (283,300 ha) marsh, was drained and diked into islands that became prime farmland and exclusive residential property surrounded by waterways. Today the delta sits 20 feet (6 m) below sea level. Sea level rise combined with more severe storms threaten the dikes.

The delta sits just east of the Hayward Fault in one of the most dangerous earthquake zones in the United States. The forecast is that there is greater than a 60 percent chance of a major earthquake in the next 30 years. The average island in the delta now has a 90 percent chance of flooding in the next 50 years (Bourne and Burtynsky, 2010).

Los Angeles County has been pumping "fresh" water into the groundwater aquifer to prevent salt water intrusion into water supplies since 1947. About 10 billion gallons per year of recycled wastewater are used.

The population of southern California is increasing at a rate of more than 200,000 per year. In 2010, California had a budget deficit of $20 billion and a proposal for $11 billion in water projects.

Notwithstanding the implications of global warming, which exacerbate the existing precariousness of the water resource system, from a long-term perspective it does not appear that California has the water resources to develop a *sustainable economy* that will produce wealth and provide jobs for many human generations without degrading the environment. There are too many people for the available water resources and population growth is too high.

Droughts and Green Engineering.   As with floods, civil engineering structures play a minor role in sustainable drought protection. The "green engineering" component is recognizing that traditional structural solutions are bound to fail and that other solutions are required to minimize risk and to provide a *sustainable economy* for many human generations. These solutions fall into two broad categories: drought response planning and water conservation planning and implementation.

Drought Response Planning.   The basic goal of a response plan is to improve response efforts by enhancing monitoring and early warning, impact assessment, preparedness, and response. Activities in these areas are summarized in the following paragraphs.

- *Early warning:* Parameters that must be monitored to detect the early onset of drought include temperature, precipitation, stream flow, reservoir and groundwater levels, snowpack, and soil moisture.

- *Impact assessment:* Synthesis of the early warning data to describe the magnitude, duration, severity, and spatial extent of the drought requires some type of indicator. The PDSI may be appropriate to evaluate past droughts but it has been criticized for its inability to make "real time" assessments. Other tools such as the *Standardized Precipitation Index* (SPI), *Surface Water Supply Index* (SWSI), and the *U.S. Drought Monitor* (DM) maps have been proposed. Of these, the DM maps (http://www.drought.unl.edu/dm) are the most recent evolution. They are compiled with input from the National Drought Mitigation Center (NDMC), U.S. Department of Agriculture (USDA), National Oceanic and Atmospheric Administration's (NOAA) Climate Prediction Center (CPC), and the National Climatic Data Center (NCDC).

- *Preparedness:* Plans and facilities to implement them generally require legislative action. Examples include authorization to establish and activate water banks, mandatory conservation, drought surcharges, restrictions on outdoor residential use, and local emergency supply for municipalities and farm animals.

- *Response:* People and animals cannot drink plans. Fish cannot swim in plans. Agencies and individuals must be given **training and facilities** to implement the plans.

In the United States, drought response planning is primarily a state responsibility. At this date (2010) only five states (Alaska, Arkansas, Louisiana, Mississippi, and Vermont) are without a drought response plan.

California has delegated drought response planning to municipalities. San Diego has a vigorous, rigorous drought response plan that includes drought stages and required responses that are enforceable by code. Violators are subject to fines and potential water shutoff. Some examples are shown in Table 13-8. The demand reductions are based on "a reasonable probability that there will be a supply shortage" and that the specified reductions are required to "ensure that sufficient supplies will be available to meet anticipated demands."

The state of Georgia has a comprehensive drought response plan that is triggered by SPI readings for 3, 6, and 12 months. The plan includes several categories of users including municipal-industrial, agricultural, and water quality. Specific restrictions are applied for each of four drought levels. For example, in the municipal category, at the lowest level (Level 1), outdoor water use is limited to scheduled weekdays. At the highest level (Level 4), outdoor watering is banned (Georgia EPD, 2003). This is a proactive plan that should be able to prevent severe socioeconomic hardship and water rights lawsuits.

In contrast, Tennessee's drought response plan uses qualitative triggers: "Deteriorating water supplies or quality," "conflicts among users," and "very limited resource availability." Local actions are specified qualitatively rather than by prohibitions. *At no point are specific actions required.* For example, at the "drought alert level" public water suppliers are to "monitor water sources and water use." At the "mandatory restrictions level," restrictions *could* include "banning of some outdoor water uses, per capita quotas and

TABLE 13-8
San Diego's "Drought Watch Response Levels"

Level	Trigger, % demand reduction required	Typical Restrictions
1	<10	Voluntary restrictions become mandatory Prohibit excessive irrigation Washing of paved surfaces prohibited
2	<20	Landscape irrigation limited to 3 days/week Landscape irrigation is limited to 10 minutes
3	<40	Landscape irrigation limited to two days/week No new potable water services permitted
4	≥40	Landscape irrigation is stopped except for trees and shrubs

percent reductions of nonresidential users." At the "emergency management level" public water suppliers are to "provide bottled water" and "initiate hauling of water." This is a reactive plan that will not be able to prevent severe socioeconomic hardship and water rights lawsuits. The triggers will be too late to initiate preventive strategies. They are based on "hydrologic" drought. In the typical drought sequence, hydrologic drought is in the later stages of drought after meteorological drought and agricultural drought.

Water Conservation Planning and Implementation. The status of water conservation planning in the United States has been evaluated for each state. The status is reported in four categories (Rashid et al., 2010):

1. Mandate requiring comprehensive program
2. Recommend conservation planning as part of water supply
3. Provision of conservation tips
4. No provisions

All but 10 states fall into one of the first three categories. Those in the first three categories are shown by category number in Figure 13-9. Those in category 4 are Alabama, Alaska, Idaho, Kentucky, Mississippi, North Dakota, Ohio, South Dakota, West Virginia, and Wyoming.

Typical requirements include tiered rate structures, public conservation outreach, voluntary conservation initiatives, and mandatory conservation initiatives.

Case Study—Arizona. Arizona has, perhaps, the most ambitious and farsighted ongoing conservation plan. After the drought of 1976–1977, the legislature enacted the 1980 Groundwater Management Act (GMA). The act provided for *active management*

FIGURE 13-9
Status of state water conservation planning.

areas (AMAs) within the state. These are areas where the majority of the population and groundwater overdraft are located (Prescott, Phoenix, Pinal, Tucson, and Santa Cruz).The management goal for all of the AMAs is to develop a sustainable water supply. In the case of the major metropolitan areas, this goal is quantified as the safe yield. The AMAs include more than 80 percent of Arizona's population, 50 percent of the total water use, and 70 percent of the state's groundwater overdraft (Pulwarty et al., 2005).

Authorized by an act of Congress in 1969 and built by the U.S. Bureau of Reclamation, the Central Arizona Project (CAP) is the largest single source of renewable water supply in Arizona. The 540 km system of aqueducts, tunnels, pumping stations, and pipelines was designed to carry 1.4 million acre-feet per year of Arizona's Colorado River allocation from Lake Havasu to central and southern Arizona. The project cost $4 billion and took 22 years to complete. Under federal guidelines municipalities have a higher priority for CAP deliveries than agriculture. Deliveries to Phoenix began in 1985. Tucson began receiving water in 1992.

The GMA specifies mandatory reductions in demand for all sectors through conservation and transition to renewable supplies. A key component is the use of Colorado River water through the Assured Water Supply (AWS) program. The AWS requires that all new subdivisions in AMAs demonstrate a 100-year AWS based primarily on renewable water supplies before the subdivision is approved.

Recognizing the precariousness of Arizona's water supplies and the need to utilize its full allocation of the Colorado River flow, in 1996 the state created the Arizona Water Banking Authority (AWBA). The objectives of the AWBA are to:

1. Store water underground that can be recovered for municipal use.

2. Support the management goals of the AMAs.

3. Support Native American water rights.

4. Provide for interstate banking of Colorado River water to assist Nevada and California in meeting water supply while protecting Arizona's entitlement.

The AWS rules can be implemented because developers can access the AWBA through the Central Arizona Groundwater Replenishment District (CAGRD) by committing to replenish groundwater used by its members. With incentive pricing for agriculture and recharge, and the AWS rules in place, Arizona is now utilizing all of its Colorado River allocation. Arizona is banking on artificial recharge as a tool to offset future drought-related shortages.

Green Engineering in Metropolitan Areas. Surveys of water use in urban areas in the southwestern and western United States reveal that a majority of the per capita water use is out of doors (Figure 13-10). Conservation efforts in the metropolitan areas

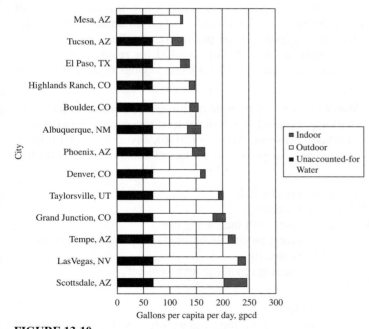

FIGURE 13-10
Per capita water use for single families.
(*Source:* Western Resource Advocates, 2003.)

focus first on outdoor use. Some example water conservation measures are summarized below.

- *Drip irrigation:* Using a small pipe with outlets for a plant, tree, or small area of plantings.

- *Smart irrigation:* Using meteorological data and/or soil moisture data to schedule timing and volume of water for irrigation. Real-time meteorological data are used to calculate evapotranspiration (ET is discussed in Chapter 4) and control irrigation. Soil moisture sensor (SMS) controllers measure the soil moisture at the root zone. While ET systems reduce overwatering, they may result in increases for those that have been underwatering (Green, 2010). SMS systems have an advantage over ET systems in that SMS systems can stop irrigation when enough has been applied by adjusting the run time to maintain the desired moisture (Philpott, 2008). Both systems are designed to avoid watering when it is raining.

- *Rainwater harvesting:* Methods for collecting, holding, and using rainwater (Waterfall, 2004).

- *Water irrigation auditor certification:* A program to train irrigation system installation and maintenance personnel on techniques to conserve water. Homes with in-ground sprinkler systems use 35 percent more water than those without in-ground systems (Feldman, 2007).

- *Water conservation incentives:* A program to credit water bills for a variety of conservation activities. Some examples include: irrigation conservation by replacing worn and leaking equipment ($25 per item), certified irrigation audit ($100), and turf grass removal ($0.25/ft^2 up to $400 for residential). For FY 2006–2010, Prescott, Arizona, estimates a total water savings of over 41 million gallons of water (Prescott, 2010).

- *Low flush toilets:* Water closets that operate on lower volumes of water. Typically a low system will use about 6 L/flush in contrast to a conventional system that uses about 15 L/flush. Prescott, Arizona (2010), saved about 10 million gallons of water over a four-year period by replacing standard units with low flush units.

- *Tiered rate structures:* Examples include
  —Uniform rate plus seasonal surcharge
  —Inclining block—rate increases with increasing use
  —Water budget rate—an inclining block rate tailored for individual customers
  —See Georgia EPD (2010) for details
  —A discussion of water pricing is given at the text website: www.mhhe.com/davis.

- *Water reuse:* Examples include
  —Using *gray water* that comes from bathing and washing facilities that does not contain concentrated human waste for irrigation
  —Using treated municipal wastewater for golf course irrigation

Green Engineering and Water Utilities.   Of the numerous suggestions for water conservation presented in *Water Conservation for Small- and Medium-Sized Utilities* (Green, 2010), two are highlighted here because of the magnitude of their impact on water conservation: leaks and water reuse. *Unaccounted for water* (UAW) is the term used to identify water produced but not otherwise sold. It includes leaks, unauthorized consumption, metering inaccuracies, and data handling errors. It is water that does not produce revenue for the utility.

In a 1997 audit, Lebanon, Tennessee, identified 45 percent of their water production was UAW (Leauber, 1997). Water audit and leak detection of 47 California water utilities found an average loss of 10 percent and a range from 5 to 45 percent. It is estimated that up to 700,000 acre-feet of leakage occurs in California each year (DWR, 2010). Community water systems in the United States lose about 6 billion gallons of water per day through leaks (Thornton et al., 2008). It is notable that 10 percent of all homes account for 58 percent of household leaks and that households with pools have 55 percent greater leakage than other households (Feldman, 2007).

Typically, UAW should not be more than 10 percent of production. Above this level, there is strong incentive to institute a leak detection and repair program. Advances in technology and expertise should make it possible to reduce losses in UAW to less than 10 percent (Georgia EPD, 2007).

Aside from the conservation aspect of leak detection and repair, there are other benefits:

- Repairing leaks with scheduled maintenance reduces overtime costs of unscheduled repairs.

- Leak detection and repair reduces energy consumption and power costs to deliver water.

- Leak detection and repair reduces chemical costs to treat water that will not be delivered to customers.

- Leak detection and repair reduces potential structural damage to roads.

- UAW audits, leak detection and repair have a very favorable return on investment.

- Little leaks get bigger with age . . . and ultimately result in pipe failure.

Water reuse, *water reclamation*, and *recycled water* are terms used for treated municipal wastewater that is given additional treatment and is distributed for specific, direct beneficial uses. The additional treatment is at a minimum tertiary treatment (Chapter 8) and often includes advanced wastewater treatment (for example, reverse osmosis discussed in Chapter 6). Water reuse has been practiced in California for over a century. In 1910 at least 35 communities were using wastewater for farm irrigation. By the end of 2001, reclaimed water use had reached over 648×10^6 m^3/y. Florida has been reclaiming wastewater for over 40 years. Approximately 834×10^6 m^3 of reclaimed wastewater was used in 2003 (Metcalf & Eddy, 2007). In 2008, Florida passed a law phasing out ocean outfall disposal of wastewater by 2025. A total of 416×10^6 m^3/y will be reclaimed to meet this goal (Greiner et al., 2009).

A list of typical applications is shown in Table 13-9.

TABLE 13-9
Example applications of reclaimed water

Category	Typical application
Agricultural irrigation	Crop irrigation
	Commercial nurseries
Groundwater recharge	Groundwater banking
	Groundwater replenishment
	Salt water intrusion control
	Subsidence control
Industrial recycling/reuse	Boiler feed water
	Cooling water
	Process water
Landscape irrigation	Cemeteries
	Golf courses
	Greenbelts
	Parks
Environmental enhancement	Fisheries
	Lake augmentation
	Marsh enhancement
	Stream flow augmentation
	Snowmaking
Nonpotable urban uses	Air conditioning where evaporative cooling is used
	Fire protection (sprinkler systems)
	Toilet flushing in recreational areas
Potable reuse	Blending in water supply reservoirs
	Blending in groundwater
	Direct pipe to water supply

Source: Metcalf & Eddy, 2007.

13-3 ENERGY RESOURCES

Until about 300 years ago, the majority of human energy needs were supplied by human and animal labor, water power, wind, wood and other burnable organic matter such as agricultural waste, dung and peat (now lumped together as *biomass*). With the advent of the Industrial Revolution in the late 18th and early 19th centuries, water power remained important but fossil fuels (primarily coal) began to be used on a large scale. Over the course of the 20th century petroleum products assumed a major role in supplying our energy needs.

The United States is both a major producer and major consumer of primary energy in the world. In 2006, the United States produced 21 percent of the world's primary energy and consumed 26 percent of the world's primary energy (EIA, 2010a and 2010b). Approximately 85 percent of our energy consumption is in the form of fossil fuels. Thirty-seven percent of our energy use is for transportation.

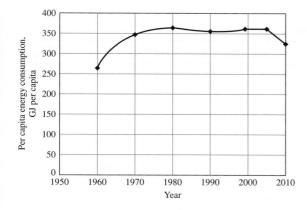

FIGURE 13-11
Per capita energy consumption in the United
States, GJ/capita-gigajoules per person.

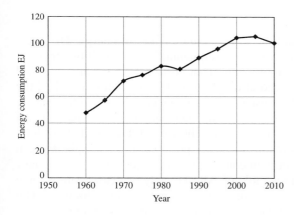

FIGURE 13-12
United States energy consumption,
EJ = exajoules.

Although our per capita energy consumption has stabilized in the last three decades, our total energy consumption has grown enormously in the last 40 years
(Figures 13-11 and 13-12).

Fossil Fuel Reserves

How do you compare a kg of coal with a liter of oil? Or, for that matter, a liter of oil
with a liter of natural gas? To make any useful comparison of fossil fuel reserves we
must use an energy basis. For fossil fuels, the net heating value (NHV) serves as a
common denominator for comparison. Typical values for NHV of some common fuels
are shown in Table 13-10.

Of course, comparison of energy value is not a comparison of the fuel's utility. For
example, you cannot replace gasoline with the equivalent NHV of coal or natural gas
and put it in the gas tank of your car. The fact that coal costs less than gasoline on a
dollar per joule ($/J) basis is irrelevant in comparing fuels for our current automobile
fleet.

TABLE 13-10
Typical values of net heating value

Material	Net heating value, MJ/kg
Charcoal	26.3
Coal, anthracite	25.8
Coal, bituminous	28.5
Fuel oil, no. 2 (home heating)	45.5
Fuel oil, no. 6 (bunker C)	42.5
Gasoline (regular, 84 octane)	48.1
Natural gas[a]	53.0
Peat	10.4
Wood, oak	13.3–19.3
Wood, pine	14.9–22.3

[a]Density take as 0.756 kg/m^3.

The question is not whether or not the world is running out of energy. The question is what are we willing to pay and what environmental impacts will be incurred. Let us first look at the question of "running out." That portion of the identified fossil fuels that can be economically recovered at the time of the calculation is called the *proven commercial energy reserve*. Because of changes in technology and the change in consumer willingness to pay, the reserve may change even though the mass of fuel does not. The estimated U.S. and world reserves are shown in Table 13-11. How long will the reserves last? Several types of estimates are possible. For example, we can estimate the length of time assuming current consumption rates (demand) with no new discoveries or changes in extraction technology. Alternatively, we can assume some growth in demand with no new discoveries or changes in extraction technology. The length of time that current reserves will last with a constant demand is expressed as

$$T_s = \frac{F}{A} \tag{13-3}$$

where T_s = time until exhaustion, (in years)
F = energy reserve, (in EJ*)
A = annual demand, (in EJ · year^{-1})

The length of time that current reserves will last with a growth in demand is expressed as

$$F = A\left[\frac{(1 + i)^n - 1}{i}\right] \tag{13-4}$$

where i = annual increase in demand as a fraction
n = the number of years to consume the reserve

*EJ = exajoule = 1×10^{18} J

TABLE 13-11
U.S. and world proven commercial energy reserves of fossil fuels[a]

Fuel	United States, EJ[b]	North America, EJ	World, EJ
Coal	5,600	5,800	19,800
Oil	130	1,300	8,000
Natural gas	230	300	6,700
Total Reserves	5,960	7,400	34,500

[a]Data for oil and gas are for January 2007; data for coal are for 2005.
[b]Exajoules = 10^{18} joules = 10^{12} megajoules.
Source: EIA, 2010c and 2010d.

In Example 13-3, we have shown how to make these two kinds of estimates.

Example 13-3. In 2004, the international consumption of coal for energy was 120.8 EJ (EIA, 2006). Assuming the 2004 demand remains constant, how long will world reserves last? The average world consumption of coal based energy increased 5.15% per year from 2000 to 2004. If that rate of increase remains constant, how long will world reserves last?

Solution. If the demand remains constant (Equation 13-3), the world reserve will last

$$\frac{19,800 \text{ EJ}}{120.8 \text{ EJ/year}} = 163.91 \text{ or } 160 \text{ years}$$

If the demand rises at a rate of 5.15 percent per year, then we can use the growth expression

$$19,800 \text{ EJ} = 120.8 \text{ EJ/y} \left[\frac{(1 + 0.0515)^n - 1}{0.0515} \right]$$

Solving for n

$$(163.91)(0.0515) = (1.0515)^n - 1$$
$$9.44 = (1.0515)^n$$

Taking the log of both sides

$$\log (9.44) = \log [(1.0515)^n] = n \log [1.0515]$$
$$0.975 = n (0.022)$$
$$n = 44.7 \text{ or } 45 \text{ years}$$

Long range forecasts of the supply of fossil fuels are illustrated in Figure 13-13. These estimates were based on data available in 2000. From this analysis it appears that a peak in fossil fuels will occur shortly. Recent advances in technology such as the improved recovery of oil from tar sands and natural gas from shale formations (LaCount, and Barcella, 2010) will push the peak further into the future. Thus, while

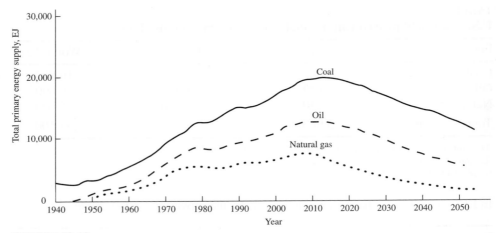

FIGURE 13-13
Peaking of world fossil fuel supply.

fossil fuels will still have a role to pay, by the end of this century it will be significantly smaller than it is today.

Nuclear Energy Resources

Fission and fusion are the two potential reactions that may be used to generate nuclear energy. During nuclear fission, a neutron penetrates the nucleus of a fissionable atom (a radioactive isotope of uranium or plutonium) and splits it into daughter products while simultaneously releasing energy. To generate energy by fusion, isotopes of a light element such as hydrogen are fused together to form a heavier element such as helium. In this process energy is released. All commercially operating nuclear reactors are based on fission reactions.

The two main isotopes of uranium are ^{235}U and ^{238}U. Of these, about 99.3 percent of the naturally occurring uranium is nonfissionable ^{238}U. To induce a sustained fission reaction, the probability that excess neutrons from one fission reaction will cause other fissionable atoms to be split is increased by increasing the concentration of fissionable material or by slowing down the neutrons so they are more likely to be captured by ^{235}U than ^{238}U.

In the middle of the 20th century, nuclear energy appeared to be a realistic energy alternative to fossil fuel. Many countries began building nuclear reactors. About 440 plants now generate 16 percent of the world's electric power. Some countries have made nuclear power their primary source of electricity. France, for example, gets 78 percent of its electricity from fission (Parfit and Leen, 2005).

There are, of course, pros and cons. On the plus side, fission provides abundant power, no global warming CO_2 emissions, and minimal impact on the landscape. On the minus side, in addition to the accidents at Three Mile Island in the United States and Chernobyl Nuclear Power Plant in Ukraine that have made the public skeptical at best, the capital investment in a nuclear power plant is huge compared to a fossil fuel fired power plant and radioactive waste has remained an unresolved problem for over

20 years in spite of the passage of the Nuclear Waste Policy Act of 1987. Although current technology exists to develop renewable nuclear energy, under current U.S. policies, the readily available uranium fuel will last only about 50 years (Parfit and Leen, 2005).

Although ^{238}U is not fissionable, a ^{238}U atom that captures a neutron will be transformed to fissionable plutonium (^{239}Pu). This is the basis of "breeder reactors." The National Academy of Science estimated that the transformation of abundant ^{238}U to ^{239}Pu could satisfy the U.S. electricity requirements for over a hundred thousand years (McKinney and Schoch, 1998). Yet, there are no breeder reactors operating in the United States Both environmental concerns and the potential for nuclear weapons proliferation have restricted the U.S. development of this energy source. Recently, well-known environmental advocates have come to regard the very real dangers of nuclear accidents and radioactive waste as less of a threat than the dangers of irreversible climate disruption. They now advocate the advancement of solutions to the economic, safety, waste storage, and proliferation problems because renewable nuclear energy is sustainable while firing fossil fuels is not (Hileman, 2006).

Sustainable Energy Sources

Sustainable energy sources, often called *renewables*, include hydropower, biomass or biofuels, geothermal, wave, and solar energy. Although photovoltaics, wind energy, and biofuels now provide only 6 percent of the world's energy, they are growing at annual rates of 17 to 29 percent (Hileman, 2006). Overall, renewable electricity supplied 9 percent of net generation in the United States in 2005 (computed from EIA, 2004, 2005b). Three of these alternative energy sources (hydropower, biomass or biofuels, geothermal, wave, and solar energy) along with renewable hydrogen are explored in the following paragraphs. It should be noted that consensus of virtually every energy expert is that there is no "silver bullet," that is, no single alternative is going to solve our long-term demand for energy. Society is going to need all of the energy it can get from all of the available alternatives.

Hydropower.   Hydropower exploits the most fundamental of physics principles by converting the potential energy of stored water to kinetic energy of falling water. The falling water passes through a turbine that drives an electrical generator. Unlike fossil fuels, hydropower is a renewable resource because the water is renewed through the hydrologic cycle. Although waterfalls have been used to generate hydropower, dams have been the major method of storing the water and rising the elevation to gain potential energy. From the definition of potential energy, the amount of energy available is a function of both the mass of water available and the elevation difference that can be created by damming up the water.

$$P.E. = mg(\Delta Z) \qquad (13\text{-}5)$$

where *P.E.* = potential energy, J

$\quad m$ = mass, kg

$\quad g$ = acceleration due to gravity

$\quad$ = 9.81 m/s^2

$\quad \Delta Z$ = head, the difference in elevation between the water surface at the top of the dam and the turbine, m

Ignoring the losses in energy from friction, the kinetic energy of the falling water is equal to the potential energy. This provides us with a method for estimating the velocity of the falling water. The definition of kinetic energy is

$$K.E. = \frac{1}{2}mv^2 \tag{13-6}$$

where v = velocity of water, m/s

If we set the two equations equal to one another we can solve for the velocity

$$v = [(2g)(\Delta Z)]^{0.5} \tag{13-7}$$

The power available may be estimated from the time rate at which work is done by the falling water. This is the mass flow rate of water traveling through a distance equal to the head.

$$Power = g(\Delta Z)\frac{dM}{dt} \tag{13-8}$$

where dM/dt = mass flow rate of water, kg/s
$\qquad$ = ρQ
$\qquad \rho$ = density of water, kg/m^3
$\qquad Q$ = flow rate of water, m^3/s

Example 13-4 illustrates how the potential energy and power may be estimated.

Example 13-4.   The Hoover Dam on the Colorado River at the Arizona/Nevada border is the highest dam in the United States It has a maximum height of 223 m and a storage capacity of about 3.7×10^{10} m^3. What is the potential energy of the Hoover Dam and Reservoir? If the maximum discharge is 950 m^3/s, what is the electrical capacity of the generating plant?

Solution.   Assuming the density of water is 1,000 kg/m^3 and applying the potential energy equation:

$$P.E. = (3.7 \times 10^{10} \text{ m}^3)(1,000 \text{ kg/m}^3)(9.81 \text{ m/s}^2)(111.5 \text{ m})$$
$$= 4.05 \times 10^{16} \text{ J or } 40.5 \text{ PJ}$$

Note that the final units are

$$\left(\frac{\text{kg} \cdot \text{m}}{\text{s}^2}\right)(\text{m}) = N \cdot \text{m} = \text{J}$$

and that the average head (1/2 of 223 m) is used for the calculation.

Of course, all of this energy cannot be recovered. The estimate assumes all of the water is at the average head, that the reservoir is filled to maximum capacity, and that the maximum height is equal to the maximum head when, in fact, the maximum head is a distance between the maximum elevation of the reservoir pool and the turbines. In addition, the efficiency of converting the flowing water to mechanical energy to turn the turbine and generator is less than 100 percent.

At a flow rate of 950 m^3/s, the electrical capacity is

Power $= (9.81 \text{ m/s}^2)(223 \text{ m})(1{,}000 \text{ kg/m}^3)(950 \text{ m}^3/\text{s}) = 2.08 \times 10^9 \text{ J/s}$
$= 2.08 \times 10^9 \text{ W} = 2{,}008 \text{ MW}$

Comment: The actual rating of the Hoover Dam is 2,000 MW. The average electrical power delivered may be considerably less than this because the average flow rate of water is less.

In 2005, the United States had a hydropower capacity of about 77 gigawatts (GW). This accounted for about 7 percent of the nation's electrical capacity (EIA, 2006). Most of this is developed at large dams such Hoover and Glen Canyon in Arizona. It is unlikely that any more large-scale plants will be built in the United States.

The combination of feasible topography for such dams and the negative environmental impacts prohibit further expansion of large-scale projects. Dams and the reservoirs they create, flood large tracts of land. For example, Lake Mead, the reservoir behind Hoover Dam, covers an area of 640 square kilometers. Natural habitats of a wide variety of biota are destroyed. In large projects, villages, homes, farms, and other natural resources are lost. Frequently, the reservoir traps sediments and nutrients normally supplied to downstream ecosystems while altering the dissolved oxygen, temperature, and mineral composition of the water.

An alternative source of hydropower is the 70,000 small dams already in existence in the United States. These provide an opportunity to develop so called "low-head" hydropower. Because the flow rates are low and the elevation differences are small, these projects are marginal economically.

Biofuels from Biomass. Biomass energy includes wastes, standing forests, and energy crops. In the United States, the "wastes" include wood scraps, pulp and paper scraps, and municipal solid wastes. In developing countries it may also include animal wastes. In developing countries, up to 90 percent of the energy may be supplied by biomass. Second only to hydropower in terms of renewable energy utilized in the United States, biomass burning provides about 3.8 percent of our nation's electrical capacity (computed from EIA, 2009). The environmental impacts are both positive and negative. Certainly, recovering the fuel value from wastes that would otherwise be buried is a positive aspect of burning biomass. Negative impacts result from the use of standing forests, especially those harvested in an unsustainable manner, that results in the creation of a wasteland. Soil erosion and deprivation of the replenishing nutrients have long-term consequences that are not repairable.

In the popular literature biofuel *is* ethanol. Other biofuels now in commercial use include alkyl esters and 1-butanol. In the United States, the primary feed stock for ethanol production is corn. Fermentation is the major method for production of ethanol for use as a fuel. The starch in the feed stock is hydrolyzed into glucose. The usual method of hydrolysis for fuels is by the use of dilute sulfuric acid and/or fungal amylase enzymes. Certain species of yeast (for example, *Saccharomyces cerevisiae*) anaerobically metabolize the glucose to form ethanol and carbon dioxide:

$$C_6H_{12}O_6 \rightleftharpoons 2\ CH_3CH_2OH + CO_2 \tag{13-9}$$

Ethanol is blended with gasoline. With the recent introduction of *flex-fuel* engine design the allowable ethanol content has been raised from about 15 percent to 85 percent—the so-called *E85*. In 2006 less than 2 percent of the total U.S. motor vehicle fleet was capable of running on E85 and there were only about 600 E85 service stations nationwide (Hess, 2006a).

Soybean oil serves as the primary feed stock for alkyl esters in the United States In global use, canola oil provides 84 percent of the feed stock. Alkyl esters are used as a diesel blend or substitute that is commonly referred to as *biodiesel*. In a modern diesel engine it can be blended in any percentage from B1 (1 percent biodiesel and 99 percent petrodiesel) to B99. Because of incentives from Congress, B20 is the popular blend (Pahl, 2005).

Biobutanol (1-butanol) is being brought to market as a competitor to ethanol. It has several advantages over ethanol. Ethanol attracts water and tends to corrode normal distribution pipelines. Thus, it must be transported by truck, rail, or barge to terminals where it is blended with gasoline. Butanol can be blended at higher concentrations than ethanol without having to retrofit automobile engines. It is also expected to have a better fuel economy than gasoline-ethanol blends (Hess, 2006b).

A lesser recognized but age-old source of biofuel is the methane generated from anaerobic decomposition of waste material. In Lansing, MI, for example the methane recovered from a municipal solid waste landfill is used to produce power for more than 4,500 homes each year.

There are three issues in the use of biofuels as a replacement for petrofuel: the energy balance, the environmental impact, and the availability of land. Critics have questioned the rationale behind the policies that promote ethanol for energy, stating that corn-ethanol has a negative energy value (Pimentel, 1991). That is, according to their estimates, the nonrenewable energy required to grow and convert corn into ethanol is greater than the energy value present in the ethanol fuel. Recent studies have concluded that changes in technology and increases in yields of corn and soybeans have resulted in net energy benefits (Farrell et al., 2006; Hammerschlag, 2006; Hill et al., 2006; and Shapouri et al., 2002). These studies revealed that the corn ethanol energy output:input ratios ranged from 1.25 to 1.34. The ratio for soybean biodiesel was even higher, ranging from 1.93 to 3.67.

The use of ethanol is estimated to reduce CO_2 emissions because, for example, of the uptake of CO_2 by corn. E16 could reduce CO_2 emissions from the light-duty fleet (all U.S. autos and small trucks) by 39 percent. Complete conversion to E85 would reduce the emissions on the order of 180 percent (Morrow et al., 2006). Conversely, the total life cycle emissions of five major air pollutants (carbon monoxide, fine particulate matter, volatile organic compounds, sulfur oxides, and nitrogen oxides) are higher with E85. Low-level biodiesel blends reduce these emissions while reducing greenhouse gas emissions that cause global warming (Hill et al., 2006).

The real issue for biofuels is the limit of agricultural production. "Even dedicating all U.S. corn and soybean production to biofuels would only meet 12 percent of gasoline demand and 6 percent of diesel demand." (Hill et al., 2006) The key to biofuels is to identify a nonfood crop that can be produced on marginal land. Switchgrass (*Panicum virgatum*) currently fulfills this requirement. It is a perennial warm-season grass native to the Midwest and Great Plains. It has been grown for decades as a pasture or hay crop on marginal land that is not well suited for conventional row crops. It is tolerant of both wet and dry conditions, requires less fertilizer and pesticides than corn or soybeans, and yields less agricultural waste. The major stumbling block is that the plant material is cellulosic . . . it cannot be economically converted to ethanol or butanol with our current (2010) technology.

Wind.   Even casual observation of hurricanes and tornadoes reveals that the wind has power. Sails and windmills have been used to harness the energy from lesser winds for centuries.

The force of the wind on a flat plate held normal to it is expressed as

$$F = \frac{1}{2}A\rho v^2 \qquad (13\text{-}10)$$

where F = force, N
A = area of plate, m^2
ρ = density of air, kg/m^3
v = wind speed, m/s

If the force moves the plate through a distance, then work is done. The product of (distance)$(A)(\rho)$ = mass, which is the definition of kinetic energy (Equation 13-6).

Two characteristics of Equation 13-10 are important in developing wind power. The first, well noted by sailors in "putting on more sail," is that the larger the area the greater the force captured. The second is less obvious. That is that the force is proportional to the square of the velocity, and power (kilowatts) is proportional to the cube of velocity (force × velocity). It is a characteristic of the wind that the velocity increases with distance above ground. Thus, a windmill will be more effective if it is placed at a higher elevation above the ground than a lower one. Modern wind turbines for generation of electrical power are placed on towers ranging from 30 to 200 m in height in "wind farms." Each wind turbine generates 750 kW to 5 MW.

While wind power is environmentally benign, it is not very dependable. Although the noise of 50 or more whirring propellers was initially a major concern in locating wind farms, current designs are quite benign in this respect. It has been forecast that up to 12 percent of the world demand for electricity could be provided by wind in the next two decades. Europe leads the world in wind power with almost 35 GW capacity. Denmark supplies about 20 percent of its electrical needs from wind power (Parfit and Leen, 2005). As of 2009, 34 GW have been installed in the United States. This represented about 3.3 percent of U.S. electric power (EIA, 2010e). In the last decade, the cost of wind power has dropped from 18–20 cents per kWh to 4–7 cents per kWh.

Solar.   In 20 days time, the earth receives energy from the sun equal to all of the energy stored as fossil fuels. This resource may be captured directly, as is done with greenhouses; passively, for example with thermal masses to absorb solar radiation; or actively, through water heaters, photovoltaic cells, or parabolic mirrors.

To be economically realistic, solar energy collection is limited to places with a lot of sunshine; Michigan is not an ideal setting. Furthermore, direct and passive systems are difficult to implement as a retrofit to existing structures. The system needs to be designed together. The active solar collection systems require large areas of land and are very expensive to install.

Photovoltaic (PV) systems for generating electricity have been on the market for about 40 years. Their high cost and the need for backup power or a large array of batteries when the sun doesn't shine have impeded their acceptance. The Japanese have been leaders in implementing improvements over the past 20 years. They have reduced the installed cost from $40–50 per W to $5–6 per W. This has resulted in a reduction in electricity cost to 11 to 12 cents per kWh. This is a very favorable rate in comparison to Japanese utility-generated electricity at 21 cents per kWh. In the United States, utility-generated electricity costs about 8.5 cents per kWh so PV systems are not competitive without subsidy. California and New Jersey are leaders in providing subsidies. These include not only cash rebates for installation but also regulatory programs that require utilities to purchase power from the PV users when they cannot use all the electricity they generate (Johnson, 2004). As of 2009, there was about 535 MW of solar capacity in the United States with about 450 MW installed in California (EIA, 2010e).

The worldwide growth of PV as an alternative has been over 29 percent per year for the last five years. In Japan, PV has grown at a rate of 43 percent annually since the early 1990s. They plan to have 4.8 GW of capacity installed by 2010.

Hydrogen. A cheap, robust fuel cell is the key to using hydrogen as a fuel. A membrane electrode assembly is the heart of the fuel cell (Figure 13-14). The proton-exchange membrane is a barrier to hydrogen but not protons. The membrane allows a catalyst to strip electrons, which power an electric motor, before the hydrogen reacts with oxygen to form water. The ultimate source of the hydrogen is water. Ideally, the

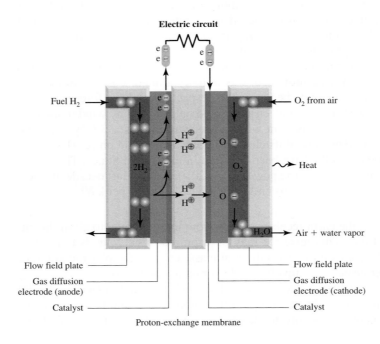

FIGURE 13-14

A typical membrane like this one shown between two flow field plates is the heart of the fuel cell. (*Source:* Davis & Masten, 2009.)

hydrogen is to be split from the water using natural systems such as photovoltaic cells to produce the power.

The automobile industry has focused a major research effort based on hydrogen as a fuel replacement for petroleum. Fuel cells and hydrogen fuel are a reality today but their application to automobiles has major hurdles to overcome:

- The cost of membranes must be reduced from the current price of $150 per m^2 to less than half that and preferably to about $35 per m^2.

- The life of the membranes must be doubled from 1,000 hours to 2,000 hours.

- The operating temperature range must be increased from 80°C to 100°C, and the relative humidity range must be increased from 25 percent to 80 percent.

- A suitable on-board hydrogen storage system must be devised.

- A hydrogen infrastructure comparable to the petroleum infrastructure must be developed.

These are not insurmountable obstacles, but it will be a while before hydrogen-powered vehicles replace the gasoline-fired engine.

Green Engineering and Energy Conservation

There are numerous examples that may be used to illustrate methods for energy conservation. Not all of these are practical and some have limited value because of the small segment of the energy demand that they impact. The several examples selected for discussion here were chosen because of their relevance to civil and environmental engineering practice.

Green Engineering and Building Construction. In the United States, buildings account for 42 percent of the energy consumption and 68 percent of the total electricity consumption (Janes, 2010). About 80 percent of this energy consumption is used in residential heating, cooling, and lighting systems. Examples of electricity consumption for common household appliances are shown in Table 13-12.

Building design teams have become increasingly aware of the need to incorporate green engineering in both new designs and retrofitting old ones. LEEDs and Green Globes provide a methodology for assessing the environmental performance of products selected for construction. Environmental performance is measured in terms of a wide range of potential impacts. Examples include human health respiratory effects, fossil fuel depletion, global warming potential, and ozone depletion. Both Green Globes and LEEDs have adopted a form of Life Cycle Assessment (LCA) that assigns credits for selecting prestudied building assemblies that are ranked in terms effects associated with making, transporting, using, and disposing of products. In the process of selecting material, the LCA of a product also is to include the use of other products required for cleaning or maintaining the product. As an interim aid to the design process, computer program (the Athena EcoCalculator—available at www.athenasmi.org) has been developed to assess alternatives. Ultimately, the EcoCalculator will serve as the basis for inputs to a separate computational system to evaluate whole building LCA (Trusty, 2009).

TABLE 13-12
Energy demand for common household appliances

Appliance	Average Demand, W	Comment
Air conditioning		
central	2,000–5,000	Function of size of house
room (window)	750–1,200	Function of size of room
Clothes dryer (electric)	4,400–5,000	
Clothes washer	500–1,150	
Computer	200–750	
Dishwasher	1,200–3,600	Function of water heater and dry cycle (1,200 W)
Freezer	335–500	
Furnace fan	350–875	Function of size of house
Lights		
incandesce equivalent	compact fluorescent	Note: CFL has has hazardous mercury component
40 W	11	
60 W	16	
75 W	20	
100 W	30	
Microwave oven	1,500	
Range		
oven	2,000–3,500	Temperature dependent
small burner element	1,200	
large burner element	2,300	
Refrigerator/freezer		
frost-free	400	
not frost-free	300	
side-by-side	780	
Television		
CRT 27-inch	170	
CRT 32-inch	200	
LCD 32-inch	125	
Plasma 42-inch	280	
Video game	100	
Water heater (electric)	4,000–4,500	

The first step in using the EcoCalculator is to select an assembly sheet from one of the following categories:

- Columns and beams
- Exterior walls
- Foundations and footings
- Interior walls

- Intermediate floors
- Roofs
- Windows

The number of assemblies in each category varies widely depending on the possible combinations of layers and materials. For example, the exterior wall category includes nine basic types, seven cladding types, three sheathing types, four insulation types, and two interior finish types (Athena Insititute, 2010).

Energy conservation is an important element in building design. For example, improved building insulation has a dramatic effect on reduction of energy consumption. From Equation 2-52, we may note that the effectiveness of insulation is a function of both its thermal conductivity and thickness. In the heating and air conditioning market it is more common to refer to the *resistance* (R) of the insulation than its thermal conductivity. The resistance is the reciprocal of the thermal conductivity:

$$R = \frac{1}{h_{tc}} \tag{13-11}$$

Larger values of R imply better insulating properties. When multiple layers of different materials are used, the combined resistance may be estimated as

$$R_T = R_1 + R_2 + \cdots R_i \tag{13-12}$$

The resistance form of Equation 2-52 is

$$\frac{dH}{dt} = \frac{1}{R_T}(A)(\Delta T) \tag{13-13}$$

where A = surface area, m^2
$\quad \Delta T$ = difference in temperature, K
$\quad R_T$ = resistance, m$^2 \cdot$ K/W

Example 13-5 illustrates the value of additional roofing insulation.

Example 13-5. A typical residential construction of the 1950s consisted of the layers shown below. Estimate the heat loss with the existing insulation scheme and with an additional 20 cm of organic bonded glass fiber insulation, if the indoor temperature is to be maintained at 20°C and the outdoor temperature is 0°C.

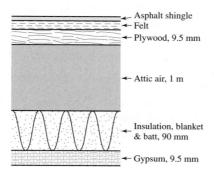

Asphalt shingle
Felt
Plywood, 9.5 mm

Attic air, 1 m

Insulation, blanket & batt, 90 mm

Gypsum, 9.5 mm

Solution. The resistance values are obtained from Table 13-13. Assuming a one square meter surface area, the total resistance for the original construction in units of $m^2 \cdot K/W$ is calculated as

$$R = \text{asphalt} + \text{felt} + \text{plywood} + \text{air} + \text{insulation} + \text{gypsum}$$

$$R = 0.077 + 0.21 + 0.10 + \frac{1{,}000 \text{ mm}}{90 \text{ mm}}(0.4) + 2.29 + 0.056 = 7.18 \text{ m}^2 \cdot K/W$$

where the ratio 1,000/90 is the number of 90 mm air spaces in 1 m of air in the attic. From Equation 13-13:

$$\frac{dH}{dt} = \frac{1}{7.18 \text{ m}^2 \cdot K/W}(1 \text{ m}^2)(20 - 0)$$

$$= 2.79 \text{ W}$$

The additional insulation will add resistance. To be in consistent units, we must multiply by the thickness of the insulation:

$$R = (27.7 \text{ m} \cdot K/W)(0.20 \text{ m}) = 5.54 \text{ m}^2 \cdot K/W$$

The new resistance is then

$$R_T = 7.18 + 5.54 = 12.72$$

and the heat loss is

$$\frac{dH}{dt} = \frac{1}{12.72 \text{ m}^2 \cdot K/W}(1 \text{ m}^2)(20 - 0)$$

$$= 1.57 \text{ W}$$

Comments:

1. This is a reduction of about 44 percent in the heat loss on a cold winter day.

2. The energy savings from additional insulation also applies to air conditioning.

3. To determine the annual savings, use the day-by-day temperature difference over a winter heating season and/or a summer cooling season. Heating/cooling degree days computed by the U.S. Weather Bureau may be used to facilitate the estimate.

Green Engineering and Building Operation. We tend to think of buildings as inert structures but they are in fact operating. For example, residential buildings have furnaces/air conditioners and refrigerators that operate whether or not anyone is home. Commercial and institutional buildings have heating and ventilating systems (HVAC), area lighting, and computers that operate 24 hours a day, seven days a week whether or not they are open for business.

Operations in residential buildings can be "engineered" to reduce energy consumption by the use of energy efficient appliances and regulating the heating/air conditioning systems with programmable thermostats that reduce operational times when the residence is not occupied. Turning off the lights, computers, and televisions when they are not in use also helps.

TABLE 13-13
Typical values of resistance for common building materials

Building material	R, m$^2 \cdot$ K/W	R for thickness shown, m$^2 \cdot$ K/W
Building board		
gypsum, 9.5 mm		0.056
particle board	7.35	
Building membrane		
2 layers, 0.73 kg/m^2 felt		0.21
Glass		
single glazing, 3 mm		0.16
double glazing, 6 mm air space		0.32
triple glazing, 6 mm air space		0.47
Insulating material		
Blanket and batt		
mineral fiber from glass,		
approx. 90 mm		2.29
approx. 150 mm		3.32
approx. 230 mm		5.34
approx. 275 mm		6.77
glass fiber, organic bonded	27.7	
loose fill milled paper	23	
spray applied polyurethane foam	40	
Roofing		
asphalt shingles		0.077
built-up, 10 mm		0.058
Masonry		
brick	1.15	
concrete	0.6	
Siding materials		
hardboard, 11 mm		0.12
plywood, 9.5 mm		0.10
Aluminum or steel, over sheathing,		
hollow-backed		0.11
insulating board backed, 9.5 mm		0.32
insulating board backed, 9.5 mm, foil		0.52
Soils	0.44	
Still air, 90 mm		0.4

Source: Data from ASHRAE (1993) *Handbook of Fundamentals,* American Society of Heating, Refrigerating and Air Conditioning Engineers, Atlanta.

Commercial/institutional "smart buildings" use computer control systems to regulate area lighting and HVAC systems based on occupancy. Regular preventive maintenance of the HVAC system is essential to conserve energy. Electric motors account for about one-half of the electricity used in the United States (Masters and Ela, 2008). Oversized motors and constant speed motors both contribute to inefficient use of electricity. Out-of-balance fans reduce the efficiency of the HVAC system. Personal computers should be turned off at the end of the workday. Computers draw virtually as much power when they are in the "sleep mode" as they do when they are active.

Installation of "smart" metering systems at residential and commercial buildings provides an opportunity for the owner to obtain real-time data on electricity and natural gas use (McNichol, 2011). These data can be used to develop and implement energy conservation plans on a building by building basis.

Green Engineering and Transportation. In the United States, transportation accounts for 28 percent of energy use (EIA, 2010f). The Corporate Average Fuel Economy (CAFE) standards were first enacted in 1975. The objective of the standards is to increase the motor vehicle fuel efficiency. The standards have been adjusted periodically to increase the miles per gallon (mpg). The standard for cars and light-duty trucks for model year 2011 is 24.1 mpg. The Energy Security Act (EISA) of 2007 requires that new passenger vehicles achieve 35 mpg in 2020. In addition to energy conservation, the achievement of the CAFE standards will result in the reduction of greenhouse gas carbon dioxide emissions.

Recycling of asphalt and concrete pavement materials not only saves raw material, it also saves energy. When recycled asphalt pavement is incorporated into new pavement, the asphalt cement in the old pavement is reactivated. Recycling in place (RIP) is another energy-saving step in asphaltic pavement rehabilitation. Other materials, including rubber from used tires, glass, and asphaltic roofing, can also be recycled into asphaltic pavement (NAPA, 2010).

Concrete pavement is also recyclable. In addition, supplementary cementitious material (SCM) may be used to replace Portland cement or as an additive. Common SCMs are fly ash (see Chapter 9), slag cement, ground blast furnace slag and, silica fume (ACPA, 2007).

Green Engineering and Water Supply and Wastewater Treatment. In some regions 30 to 50 percent of the total operating cost of the drinking water supply is for energy. At wastewater treatment plants, energy accounts for 25 to 40 percent of operating costs (Feldman, 2007). *Energy Conservation in Water and Wastewater Facilities* (Schroedel and Cavagnaro, 2010) provides numerous techniques to conserve energy. Several of these are highlighted in the following paragraphs.

Much of the cost for energy is for pumping. Pumps driven by electric motors are often regulated by adjusting valves instead of the speed of the motor itself. The excess electrical energy is wasted as heat. Variable frequency drives (vfd) adjust the speed and thus the pumping rate by adjusting the frequency of the electric current. This is a major means to reduce energy use and cost. Replacement of oversized motors and/or pump impellers is another method to reduce energy inefficiency. In wastewater plants, aerators driven by electric motors are another major energy consumer. Changing coarse bubble aerators to fine bubble aerators improves oxygen transfer efficiency and provides the opportunity to reduce motor sizes or implement vfd systems (Hebert, 2010). Using computer control systems to regulate the air flow in proportion to the wastewater flow and strength is another means to reduce energy consumption (Rogers, 2010).

Control of chemical dosing is another method to reduce energy consumption (Truax, 2010). For example, adjusting the dose of lime in softening plants (see Chapter 6) to achieve a final hardness of 130 mg/L as $CaCO_3$ instead of 80 mg/L as $CaCO_3$ will reduce mineral resource consumption, energy in production and transportation, and the energy and costs for sludge disposal.

13-4 CHAPTER REVIEW

When you have completed studying the chapter, you should be able to do the following without the aid of your textbook or notes:

1. Compare the WCED definition of sustainable development with the definition of sustainable economy.

2. Explain the difference between renewable and nonrenewable resources.

3. Explain the people problem in terms that a group of legislators can understand.

4. Define vulnerability in terms of sustainability under conditions of climate change.

5. List three key ingredients to a strong adaptive capacity for sustainability under conditions of climate change.

6. Define green engineering.

7. Explain why floods are a threat to sustainability.

8. Compare the ability of U.S. communities and those in Bangladesh to maintain a sustainable economy in the event of a flood.

9. List two green engineering programs to mitigate loss of life and flood damage.

10. Diagram and explain the relationship between the four definitions of drought.

11. Define PDSI and explain the significance of a PDSI of minus 4.

12. Explain why droughts are a threat to sustainability.

13. Explain the difference between drought response planning and water conservation planning and implementation.

14. Explain why leak detection and repair is an essential component of water conservation.

15. Describe four nonfossil fuel alternative energy sources.

With the aid of this text, you should be able to do the following:

16. Estimate a future population or growth rate given appropriate data.

17. Determine one of the following given the required data: the time until exhaustion of an energy reserve with a constant demand, the time until exhaustion of a mineral or energy reserve with a growth in demand, mass of mineral reserve, annual demand.

18. Perform an energy balance on a thermal power plant or other fossil fuel burning facility.

19. Perform an energy balance on a hydropower facility.

20. Perform an energy balance on a heated or cooled structure.

13-5   PROBLEMS

13-1. It has been estimated that at 2004 consumption rates, the world's petroleum reserve will last 37.5 years. Estimate the world consumption rate in 2004.

   Answer: 176 EJ/y

13-2. A house built in the 1950s has 14.86 m^2 of single-glazed windows. Estimate the heat loss with the existing single-glazed window and the loss if the windows are replaced with (a) double-glazed, (b) triple-glazed windows. Assume the indoor temperature is 20°C and the outdoor temperature is 0°C.

   Answer: Single-glaze $= 1.86 \times 10^3$ W; double-glaze $= 9.29 \times 10^2$ W

13-3. A 25 W compact fluorescent light bulb (CFL) produces the light equivalent of a 100 W incandescent bulb. The population of North America is estimated to be about 305 million. Estimate the amount of coal to light one 100 W incandescent light bulb for each person for one year and the amount of coal that would be saved if each incandescent bulb was replaced with a CFL. Assume the coal has a NHV of 28.5 MJ/kg and the power plants are 33 percent efficient.

13-4. A university computer lab has 32 machines. Each machine draws 400 W of power, most of which is given off as heat. Assuming that 100 percent of the electrical demand is given off as heat, estimate the amount of energy that is wasted each year if the computers are left running during the 8 hours at night that the lab is closed. (Note that the computers draw virtually as much power when they are in the "sleep mode" as they do when they are active.)

   Answer: 135 GJ

13-6   FE EXAM FORMATTED PROBLEMS

13-1. Estimate the growth rate of the population of China assuming a population of 1,311 million in 2005, a population of 1,338 million in 2010, and an exponential growth rate.

a. $0.00177 \ y^{-1}$ b. $5.4 \times 10^6 \ y^{-1}$

c. $0.20 \ y^{-1}$ d. $0.00408 \ y^{-1}$

13-2. Estimate the growth rate (individual/individual $\cdot$ y) if the crude birth rate is 14 per, 1000, the crude death rate is 8 per 1,000, and the net immigration is 3 per 1,000.

a. $0.0090 \ y^{-1}$ b. $0.0060 \ y^{-1}$

c. $0.0030 \ y^{-1}$ d. $9,000 \ y^{-1}$

13-7   REFERENCES

APCA (2007) *Green Highways*, American Concrete Pavement Association, Concrete Pavement Research and Technology Special Report, Skokie, IL.

ASHRAE (1993) *Handbook of Fundamentals*, American Society of Heating, Refrigerating and Air Conditioning Engineers, Atlanta, GA.

A&WMA (1988) *Waste Minimization*, Air &Waste Management Association, International Specialty Conference Proceedings, October, Baltimore, MD.

Athena Institute (2010) EcoCalculator, www.athenasmi.org/tools/ecoCalculator/index.html.

Bell, B. (1970) "The Oldest Records of the Nile Floods," *Geographical Journal*, vol. 136, pp. 569–573.

Bell, B. (1971) "The Dark Ages in Ancient Egypt: I The First Dark Age in Egypt," *American Journal of Archaeology*, vol. 75, pp. 1–26.

Bell, B. (1975) "Climate and History of Egypt," *American Journal of Archaeology*, vol. 79, pp. 223–269.

Blaikie, P., T. Cannon, I. Davis, and B. Wisner (1994) "Bangladesh—A 'Tech-fix' or People's Needs-Based Approach to Flooding?" in Blake, P.M. et al. (eds.) *At Risk: Natural Hazards, Peoples Vulnerability, and Disasters*, Routledge, Taylor & Francis Group, New York.

Bourne, J. K., and E. Burtynsky (2010) "California's Pipe Dream," *National Geographic*, vol. 217, no. 4, pp. 132–149.

Bremner, J., A. Frost, C. Haub, M. Mather, K. Ringheim, and E. Zuehlke (2010) "World Population Highlights: Key Findings from PRB's 2010 World Population Data Sheet," *Population Bulletin*, vol. 65, no. 2.

Bruins, H. J. (2000) "Drought Hazards in Israel and Jordan," in Wilhite, D. A. (ed.) *Drought: A Global Assessment, Volume II*, Routledge, Taylor & Francis Group, New York, pp.178–193.

Butzer, K. W. (1976) *Early Hydraulic Civilization in Egypt*, University of Chicago Press, Chicago, pp. 28–29, 55–56.

Davis, M. L. and S. J. Masten (2004) *Principles of Environmental Engineering and Science*, 2nd edition, McGraw-Hill Higher Education, Dubuque, Iowa, p. 292.

Downing, T. E. and K. Bakker (2000) "Drought Risk in a Changing Environment," in Vogt, J. V. and F. Somma (eds.) *Drought and Drought Mitigation in Europe*, Kluwer Academic Press, Boston, pp. 79–90.

DWR (2010) California Department of Water Resources, http://www.water.ca.gov/wateruseefficiency/leak.

Dziegielewski, B., G. D. Lynne, D. A. Wilhite, and D. P. Sheer (1991) *National Study of Water Management During Drought: A Research Assessment*, Institute of Water Resources Report, IWR Report 91-NDS-3, U.S. Army Corps of Engineers, Fort Belvoir, VA.

Dziegielewski, B. and J. E. Crews (1986) "Minimizing the Cost of Coping with Droughts: Springfield, IL," *Journal of Water Resource Management*, American Society of Civil Engineers, vol. 112, no. 4, pp. 419–438.

EIA (2004) *Historical Renewable Energy Consumption by Energy Use Sector and Source*, Energy Information Agency, U.S. Department of Energy, www.eia.doe.gov.

EIA (2005b) *Table 2.1a Energy Consumption by Sector, Selected Years, 1949–2005*, Energy Information Agency, U.S. Department of Energy, www.eia.doe.gov.

EIA (2005c) *Table 1. U.S. Energy Consumption by Energy Source, 2000–2004*, Energy Information Agency, U.S. Department of Energy, www.eia.doe.gov.

EIA (2005d) *Table 5b. Historical Renewable Energy Consumption by Energy Use Sector and Energy Source, 2000–2004*, Energy Information Agency, U.S. Department of Energy, www.eia.doe.gov.

EIA (2006) *Table 2.2 Existing Capacity by Energy Source*, Energy Information Agency, U.S. Department of Energy, www.eia.doe.gov.

EIA (2009) *U.S. Energy Consumption by Energy Source*, Energy Information Agency, U.S. Department of Energy, www.eia.doe.gov.

EIA (2010a) *Table E.1 World Primary Energy Consumption by Region*, Energy Information Agency, U.S. Department of Energy, www.eia.doe.gov.

EIA (2010b) *Table F.1 World Primary Energy Production by Region*, Energy Information Agency, U.S. Department of Energy, www.eia.doe.gov.

EIA (2010c) *Table 8.1. World Crude Oil and Natural Gas Reserves, January 1, 2007*, Energy Information Agency, U.S. Department of Energy, www.eia.doe.gov.

EIA (2010d) *Table 8.2. World Estimated Recoverable Coal, December 31, 2005*, Energy Information Agency, U.S. Department of Energy, www.eia.doe.gov.

EIA (2010e) *Electric Power Annual*, Energy Information Agency, U.S. Department of Energy, www.eia.doe.gov.

EIA (2010f) *Energy in Brief*, Energy Information Agency, U.S. Department of Energy, www.eia.doe.gov.

Farrell, A. E., R. J. Plevin, B. T. Turner, et al. (2006) "Ethanol Can Contribute to Energy and Environmental Goals," *Science*, vol. 311, pp. 506–508.

Feldman, D. L. (2007) *Water Policy for Sustainable Development*, The Johns Hopkins University Press, Baltimore, MD, pp. 145–147, 300–301, 304.

Finan, T., C. West, D. Austin, and T. McGuire (2002) "Processes of Adaptation to Climate Variability," *Climate Research*, vol. 21, no. 3, pp. 299–310.

Freeman, H. (1990) *Hazardous Waste Minimization*, McGraw-Hill, New York.

Freeman, H. (1995) *Industrial Pollution Prevention Handbook*, McGraw-Hill, New York.

Fye, F. K., D. W. Stahle, and E. R. Cook (2003) "Paleoclimatic Analogs to Twentieth-Century Moisture Regimes Across the United States," *Bulletin of the American Meteorological Society*, vol. 84, no. 7, pp. 901–909.

Georgia EPD (2003) *Georgia Drought Management Plan*, Department of Natural Resources, Environmental Protection Division, Atlanta, GA.

Georgia EPD (2007) *Water Leak Detection and Repair Program*, Georgia Department of Natural Resources, Environmental Protection Division, Atlanta, GA.

Georgia EPD (2010) *Georgia's Water Conservation Implementation Plan*, Georgia Department of Natural Resources, Environmental Protection Division, http://www.ConserveWaterGeorgia.net.

Glantz, M. (1994) *Drought Follows the Plow*, Cambridge University Press, Cambridge, U.K.

Glantz, M. (1994) "Drought Follows the Plow," in Wilhite, D. A. (ed.) *Drought: A Global Assessment, Volume II*, Routledge, Taylor & Francis Group, New York, pp. 285–291.

Green, D. (2010) *Water Conservation for Small- and Medium-Sized Utilities*, American Water works Association, Denver, CO.

Greiner, A. D, J. J Page, R. H. Cisterna, E. Vadiveloo, and P. A. Davis (2009) "In Search of a Sustainable Future," *Water Environment & Technology*, vol. 21, no. 11, pp. 35–39.

Grimm, M. (1998) "Floodplain Management," *Civil Engineering*, vol. 68, no. 3, pp. 62–64.

Gruntfest, E. (2000) "Nonstructural Mitigation of Flood Hazards," in Wohl, E. E. (ed.) *Inland Flood Hazards*, Cambridge University Press, Cambridge, U.K., pp. 394–395, 398.

Hammerschlag, R. (2006) "Ethanol's Energy Return on Investment: A Survey of the Literature 1990–Present, *Environmental Science & Technology*, vol. 40, no. 6, pp. 1744–1750.

Haupt, A. and T. T. Kane (1985) *Population Reference Handbook*, Population Reference Bureau, Washington, DC.

Haury, E. W. (1935) "Tree Rings—The Archeologist's Time Piece," *American Antiquity*, vol. 1, pp. 98–108.

Hays, W. W. (1981) *Gauging Geological and Hydrological Hazards: Earth-Science Considerations*, U.S. Geological Survey Professional Paper 1240-B, Washington, DC.

Hebert, J. (2010) "Process Aeration," Michigan Water Environment Association Process Seminar, East Lansing, MI, December 8.

Hess, G. (2006a) "Push for Biofuels Seen in Farm Bill," *Chemical & Engineering News*, 22 May, pp. 29–31.

Hess, G. (2006b) "BP and DuPont Plan 'Biobutanol," *Chemical & Engineering News*, June 26, p. 9.

Higgins, T. (1989) *Hazardous Waste Minimization Handbook*, Lewis Publishers, Chelsea, MI.

Hileman, B. (2006) "Heretical Position on Nuclear Powerr," *Chemical & Engineering News*, Aug 21, p. 43.

Hill, J., E. Nelson, D. Tilman, S. Polasku, and D. Tiffany (2006) "Environmental, Economic, and Energetic Costs and Benefits of Biodiesel and Ethanol Biofuels," *Proceedings National Academy of Science*, vol. 103, no. 30, pp. 11206–11210.

IPCC (2007) *Climate Change2007: Impacts, Adaptation annd Vulnerability—The Physical Science Basis, Summary for Policymakers*, Intergovernmental Panel on Climate Change, Cambridge University Press, Cambridge, U.K., pp. 1–18.

Janes, D. (2010) "Going for the Green," *EM*, January/February, pp. 24–25.

Johnson, J. (2004) "Power from the Sun," *Chemical & Engineering News*, June 21, pp. 25–28.

Karl, T. R., J. M. Melillo, and T. C. Peterson (2009) *Global Climate Change Impacts in the United States, U.S. Global Climate Change Research Program*, Cambridge University Press, Cambridge, U.K.

KGS/KDA (2010) "Water Use for Haskell County, Kansas," Kansas Geological Survey and Kansas Department of Agriculture,

LaCount, R. and M. L. Barcella (2010) "Shale Gas, A Game Changer for North American Energy and Environment," *EM*, Air & Waste Management Association, August, pp. 16–19.

Leauber, C. E. (1997) "Leak Detection Cost-Effective and Beneficial," *Journal of American Water Works Association*, vol. 89, no. 7, p. 10.

Loucks, D. P., J. R. Stedinger, and D. A. Haith (1981) *Water Resource Systems Planning and Analysis*, Prentice Hall, Englewood Cliffs, NJ, p. 3.

Masters, G. M., and W. P. Ela (2008) *Introduction to Environmental Engineering and Science*, 3rd.ed., Prentice Hall, Upper Saddle River, NJ, p. 612.

McKinney, M. L., and R. M. Schoch (1998) *Environmental Science Systems and Solutions*, Jones and Bartlett Publishers, Sudbury, MA, p. 10.

McNichol, T. (2011) "Rage Against the Machine," *Time*, January 10, p. 62.

Metcalf & Eddy, Inc. (2007) *Water Reuse*, McGraw-Hill, New York, pp. 20–24, 48, 54.

Morrow, W. R., W. M. Griffin, and H. S. Matthews (2006) "Modeling Switchgrass Derived Cellulosic Ethanol Distribution in the United States," *Environmental Science & Technology*, vol. 40, no. 9, pp. 2877–2886.

Mumme, S. P. (1995) " The New Regime for Managing U.S.–Mexican Water Resources," *Environmental Management*, vol. 19, no. 6, pp. 827–835.

NAPA (2010) *Benefits of Asphalt*, www.PaveGreen.com.

NCEES (2011) *Fundamentals of Engineering Supplied-Reference Handbook*, National Council of Examiners for Engineering and Surveying, Revised April 2011, p. 121.

Nemerow, N. L. (1995) *Zero Pollution for Industry*, John Wiley & Sons, Inc., New York.

NOAA (2010) http://www.nch.noaa.gov/HAW2/english/history.shtml.

Pahl, G. (2005) *Biodiesel, Growing a New Energy Economy*, Chelsea Green Publishing Company, White River Junction, VT, p. 47

Palmer, W. C. (1965) "Meteorological Drought," *Research Paper No. 45*, U.S. Weather Bureau, Washington, DC.

Parfit, M., and S. Leen (2005) "Powering the Future," *National Geographic*, vol. 208, no. 2, pp. 2–31.

Philpott, B. (2008) *Field Guide to Soil Moisture Sensor Use in Florida*, University of Florida, http://www.sjrwmd.com/floridawaterstar/pdfs/SMS_field_guide.pdf.

Pimentel, D. (1991) "Ethanol Fuels: Energy Security, Economics, and the Environment," *Journal of Agricultural and Environmental Ethics*, vol. 4, pp. 1–13.

Polsky, C., And D. W. Cash (2005) "Drought, Climate Change, and Vulnerability: The Role of Science and Technology in a Multi-Scale, Multi-Stressor World," in Wilhite, D.A. (ed.) *Drought and Water Crises: Science, Technology, and Management Issues*, Taylor & Francis, Boca Raton, FL, p. 218.

Power, M. E., G. Parker, W. E. Dietrich, and A. Sun (1995) "How Does a Floodplain Width Affect Floodplain River Ecology? A Preliminary Exploration Using Simulations," *Geomorphology*, vol. 13, pp. 301–307.

PRB (2010) *2010 World Population Data Sheet*, Population Reference Bureau Washington, DC.

Prescott (2010) *Water Smart*, at http://www.prescott-az.gov.

Pulwarty, R. S., K. L. Jacobs, and R. M. Dole (2005) "The Hardest Working River: Drought and Critical Water Problems in the Colorado River Basin," in Wilhite, D. A. (ed.) *Drought and Water Crises*, Tayor & Francis, New York, pp. 249–285.

Rashid, M. M., W. O. Maddaus, and M. L. Maddus (2010) "Progress in U.S. Water Conservation Planning and Implementation—1990–2009," *Journal of American Water Works Association*, vol. 102, no. 6, pp. 85–99.

Rogers, R. (2010) "Prime Mover Blower Operations," Michigan Water Environment Association Process Seminar, East Lansing, MI, December 8.

Schroedel, R. B., and P. V. Cavagnaro (2010) *Energy Conservation in Water and Wastewater Facilities*, Manual of Practice No. 32, Water Environment Federation, WEF Press, Alexandria, VA.

Shapouri, H., J. A. Duffield, and M. Wang (2002) *The Energy Balance of Corn Ethanol: An Update*, U.S. Department of Agriculture, Agricultural Economic Report No. 813, 15 pp.

Smith, K., and R. Ward (1998) *Floods: Physical Processes and Human Impacts*, John Wiley & Sons, New York, pp. 142, 153, 157–159, 191.

Stine, S. (1994) "Extreme and Persistent Drought in California and Patagonia During Medieval Time," *Nature*, vol. 369, pp. 546–549.

Stooksbury, D. E. (2003) "Historical Droughts in Georgia and Drought Assessment and Management," *Proceedings of the 2003 Georgia Water Resources Conference*, APR 23–24, Athens, GA.

Thornton, J., R. Sturm, and G. Kunkle (2008) *Water Loss Control*, 2nd edition, McGraw-Hill, New York.

Truax, T. (2010) "Chemical Metering Control Strategy," Michigan Water Environment Association Process Seminar, East Lansing, MI, December 8.

Trusty, W. (2009) "Incorporating LCA in Green Building Rating Systems," *EM*, Air & Waste Management Association, December, pp. 19–22.

UNESCO (1979) "Map of the World Distribution of Arid Regions, Explanatory Note," *Man and the Biosphere (MAB) Technical Notes 7*, Paris, UNESCO.

U.S. Department of Interior (2003) *Water 2025: Preventing Crises and Conflict in the West*, Washington, DC.

U.S. EPA (2010) http:www.epa.gov/oppt/greenengineering.

USGS (1984) *National Water Summary 1983*, U.S. Geological Survey Water Supply Paper 2250, Washington, DC.

USGS (2004) "Climatic Fluctuations, Drought, and Flow in the Colorado River Basin," fact sheet at http://pubs.usgs.gov/fs/2004/3062.

USGS (2007) *USGS Fact Sheet 2007–3029*, U. S. Geological Survey, Washington, DC.

Viau, A. A. and J. V. Vogt (2000) "Scale Issues in Drought Monitoring," in Vogt, J. V., and F. Somma (eds.), *Drought and Drought Mitigation in Europe*, Kulwer Academic Publishers, Boston, pp. 185–193.

Waterfall, P. (2004) *Harvesting Rainwater for Landscape Use*, University of Arizona Cooperative Extension, Pub. No. AZ1344.

WCED (1987) *Our Common Future*, World Commission on Environment and Development, Oxford University Press, Oxford, U.K.

Webster, P. J., J. Jian, T. M. Hopson, C. D. Hoyos, P. A. Agudelo, H. Chang, J. A. Curry, et al. (2010) "Extended-Range Probabilistic Forecasts of Ganges and Brahmaputra Floods in Bangladesh, *Bulletin of the American Meteorological Society*, vol. 91, no. 11, pp. 1493–1514.

Western Resources Advocates (2003) "Smart Water: A Comparative Study of Urban Water Use Efficiency Across the Southwest," Water Resource Advocates, Boulder, CO.

Wilhite, D. A., and W. E. Glantz (1985) "Understanding the Drought Phenomenon: The Role of Definitions," *Water International*, vol. 10, pp. 111–120.

Wilhite, D. A., and M. Buchanan-Smith (2005) "Drought as Hazard: Understanding the Natural and Social Context," in Wilhite, D.A. (ed.) *Drought and Water Crises*, Tayor & Francis, New York, p. 4.

Wohl, E. E. (ed.) (2000) *Inland Flood Hazards*, Cambridge University Press, Cambridge, U.K., pp. 25–26.

Wright, J. M. (1996) "Effects of the Flood on National Policy: Some Achievements, Major Challenges Remain," in Changnon, S. A. (ed.) *The Great Flood of 1993*, Westview Press, a Division of Harper Collins, Boulder, CO, p. 253.

Zonn, I., M. H. Glantz, and A. Rubenstein (2000) "The Virgin Lands Scheme in the Former Soviet Union," in Wilhite, D. A. (ed.) *Drought: A Global Assessment, Volume I*, Routledge, Taylor & Francis Group, New York, pp. 381–388.

APPENDIX

A

PROPERTIES OF AIR, WATER, AND SELECTED CHEMICALS

TABLE A-1
Physical properties of water at 1 atm

Temperature (°C)	Density, ρ (kg/m^3)	Specific weight, γ (kN/m^3)	Dynamic viscosity, μ (m(Pa $\cdot$ s))*	Kinematic viscosity, ν (μ(m^2/s))*
0	999.842	9.805	1.787	1.787
3.98	1,000.000	9.807	1.567	1.567
5	999.967	9.807	1.519	1.519
10	999.703	9.804	1.307	1.307
12	999.500	9.802	1.235	1.236
15	999.103	9.798	1.139	1.140
17	998.778	9.795	1.081	1.082
18	998.599	9.793	1.053	1.054
19	998.408	9.791	1.027	1.029
20	998.207	9.789	1.002	1.004
21	997.996	9.787	0.998	1.000
22	997.774	9.785	0.955	0.957
23	997.542	9.783	0.932	0.934
24	997.300	9.781	0.911	0.913
25	997.048	9.778	0.890	0.893
26	996.787	9.775	0.870	0.873
27	996.516	9.773	0.851	0.854
28	996.236	9.770	0.833	0.836
29	995.948	9.767	0.815	0.818
30	995.650	9.764	0.798	0.801
35	994.035	9.749	0.719	0.723
40	992.219	9.731	0.653	0.658
45	990.216	9.711	0.596	0.602
50	988.039	9.690	0.547	0.554
60	983.202	9.642	0.466	0.474
70	977.773	9.589	0.404	0.413
80	971.801	9.530	0.355	0.365
90	965.323	9.467	0.315	0.326
100	958.366	9.399	0.282	0.294

*Pa $\cdot$ s = (mPa $\cdot$ s) $\times 10^{-3}$
*m^2/s = (μm^2/s) $\times 10^{-6}$

TABLE A-2
Henry's law constants at 20°C

	H* (atm)	$H_u^\dagger$ (dimensionless)	$H_D^\dagger$ (atm · L/mg)	$H_m^\dagger$ (atm · m³/mol)
Oxygen	4.3×10^4	3.21×10	2.42×10^{-2}	7.73×10^{-1}
Methane	3.8×10^4	2.84×10	9.71×10^{-2}	6.38×10^{-1}
Carbon dioxide	1.51×10^2	1.13×10^{-1}	6.17×10^{-5}	2.72×10^{-3}
Hydrogen sulfide	5.15×10^2	3.84×10^{-1}	2.72×10^{-4}	9.26×10^{-3}
Vinyl chloride	3.55×10^5	2.65×10^2	1.02×10^{-1}	6.38
Carbon tetrachloride	1.29×10^3	9.63×10^{-1}	1.51×10^{-4}	2.32×10^{-2}
Trichloroethylene	5.5×10^2	4.1×10^{-1}	7.46×10^{-5}	9.89×10^{-3}
Benzene	2.4×10^2	1.8×10^{-1}	5.52×10^{-5}	4.31×10^{-3}
Chloroform	1.7×10^2	1.27×10^{-1}	2.55×10^{-5}	3.06×10^{-3}
Bromoform	3.5×10	2.61×10^{-2}	2.40×10^{-6}	6.29×10^{-4}
Ozone	5.0×10^3	3.71	1.87×10^{-3}	8.99×10^{-2}

*H values from Montgomery, 1985.
†H_u, H_D, and H_m calculated via Eqs. 5-49 to 5-51.

TABLE A-3

Saturation values of dissolved oxygen in freshwater exposed to a saturated atmosphere containing 20.9% oxygen under a pressure of 101.325 kPa[a]

Temperature (°C)	Dissolved oxygen (mg/L)	Saturated vapor pressure (kPa)
0	14.62	0.6108
1	14.23	0.6566
2	13.84	0.7055
3	13.48	0.7575
4	13.13	0.8129
5	12.80	0.8719
6	12.48	0.9347
7	12.17	1.0013
8	11.87	1.0722
9	11.59	1.1474
10	11.33	1.2272
11	11.08	1.3119
12	10.83	1.4017
13	10.60	1.4969
14	10.37	1.5977
15	10.15	1.7044
16	9.95	1.8173
17	9.74	1.9367
18	9.54	2.0630
19	9.35	2.1964
20	9.17	2.3373
21	8.99	2.4861
22	8.83	2.6430
23	8.68	2.8086
24	8.53	2.9831
25	8.38	3.1671
26	8.22	3.3608
27	8.07	3.5649
28	7.92	3.7796
29	7.77	4.0055
30	7.63	4.2430
31	7.51	4.4927
32	7.42	4.7551
33	7.28	5.0307
34	7.17	5.3200
35	7.07	5.6236
36	6.96	5.9422
37	6.86	6.2762
38	6.75	6.6264

[a]For other barometric pressures, the solubilities vary approximately in proportion to the ratios of these pressures to the standard pressures.

(*Source:* Calculated by G. C. Whipple and M. C. Whipple from measurements of C. J. J. Fox, *Journal of the American Chemical Society,* vol. 33, p. 362, 1911.)

TABLE A-4
Viscosity of dry air at approximately 100 kPa[a]

Temperature (°C)	Dynamic viscosity (μPa · s)
0	17.1
5	17.4
10	17.7
15	17.9
20	18.2
25	18.5
30	18.7
35	19.0
40	19.3
45	19.5
50	19.8
55	20.1
60	20.3
65	20.6
70	20.9
75	21.1
80	21.4
85	21.7
90	21.9
95	22.2
100	22.5
150	25.2

$\mu = 17.11 + 0.0536\,T + (P/8280)$ where T is in °C and P is in kPa.

TABLE A-5
Properties of air at standard conditions[a]

Molecular weight	M	28.97
Gas constant	R	287 J/kg · K
Specific heat at constant pressure	c_p	1,005 J/kg · K
Specific heat at constant volume	c_v	718 J/kg · K
Density	ρ	1.185 kg/m^3
Dynamic viscosity	μ	1.8515×10^{-5} Pa · s
Kinematic viscosity	ν	1.5624×10^{-5} m^2/s
Thermal conductivity	k	0.0257 W/m · K
Ratio of specific heats, c_p/c_v	k	1.3997
Prandtl number	Pr	0.720

[a] Measured at 101.325 kPa pressure and 298 K temperature.

TABLE A-6
Properties of saturated water at 298 K

Molecular weight	M	18.02
Gas constant	R	461.4 J/kg · K
Specific heat	c	4,181 J/kg · K
Prandtl number	Pr	6.395
Thermal conductivity	k	0.604 W/m · K

TABLE A-7
Frequently used constants

Standard atmospheric pressure	P_{atm}	101.325 kPa
Standard gravitational acceleration	g	9.8067 m/s^2
Universal gas constant	R_u	8,314.3 J/kg · mol · K
Electrical permittivity constant	ϵ_0	8.85×10^{-12} C/V · m
Electron charge	q_e	1.60×10^{-19} C
Boltzmann's constant	k	1.38×10^{-23} J/K

TABLE A-8
Properties of selected organic compounds

Name	Formula	M.W.	Density, g/mL	Vapor pressure, mm Hg	Henry's law constant kPa · m³/mol
Acetone	CH_3COCH_3	58.08	0.79	184	0.01
Benzene	C_6H_6	78.11	0.879	95	0.6
Bromodichloromethane	$CHBrCl_2$	163.8	1.971		0.2
Bromoform	$CHBr_3$	252.75	2.8899	5	0.06
Bromomethane	CH_3Br	94.94	1.6755	1,300	0.5
Carbon tetrachloride	CCl_4	153.82	1.594	90	3
Chlorobenzene	C_6H_5Cl	112.56	1.107	12	0.4
Chlorodibromomethane	$CHBr_2Cl$	208.29	2.451	50	0.09
Chloroethane	C_2H_5Cl	64.52	0.8978	700	0.2
Chloroethylene	C_2H_3Cl	62.5	0.912	2,550	4
Chloroform	$CHCl_3$	119.39	1.4892	190	0.4
Chloromethane	CH_3Cl	50.49	0.9159	3,750	1.0
1,2-Dibromoethane	$C_2H_2Br_2$	187.87	2.18	10	0.06
1,2-Dichlorobenzene	$1,2\text{-}Cl_2\text{-}C_6H_4$	147.01	1.3048	1.5	0.2
1,3-Dichlorobenzene	$1,3\text{-}Cl_2\text{-}C_6H_4$	147.01	1.2884	2	0.4
1,4-Dichlorobenzene	$1,4\text{-}Cl_2\text{-}C_6H_4$	147.01	1.2475	0.7	0.2
1,1-Dichloroethylene	$CH_2{=}CCl_2$	96.94	1.218	500	15
1,2-Dichloroethane	$ClCH_2CH_2Cl$	98.96	1.2351	60	0.1
1,1-Dichloroethane	CH_3CHCl_2	98.96	1.1757	180	0.6
Trans-1,2-Dichloroethylene	$CHCl{=}CHCl$	96.94	1.2565	300	0.6
Dichloromethane	CH_2Cl_2	84.93	1.327	350	0.3
1,2-Dichloropropane	$CH_3CHClCH_2Cl$	112.99	1.1560	50	0.4
Cis-1,3-Dichloropropylene	$ClCH_2CH{=}CHCl$	110.97	1.217	40	0.2
Ethyl benzene	$C_6H_5CH_2CH_3$	106.17	0.8670	9	0.8
Formaldehyde	HCHO	30.05	0.815		
Hexachlorobenzene	C_6Cl_6	284.79	1.5691		
Pentachlorophenol	Cl_5C_6OH	266.34	1.978		
Phenol	C_6H_5OH	94.11	1.0576		
1,1,2,2-Tetrachloroethane	$CHCl_2CHCl_2$	167.85	1.5953	5	0.05
Tetrachloroethylene	$Cl_2C{=}CCl_2$	165.83	1.6227	15	3
Toluene	$C_6H_5CH_3$	92.14	0.8669	28	0.7
1,1,1-Trichloroethane	CH_3CCl_3	133.41	1.3390	100	3.0
1,1,2-Trichloroethane	$CH_2ClCHCl_2$	133.41	1.4397	25	0.1
Trichloroethylene	$ClHC{=}CCl_2$	131.29	1.476	50	0.9
Vinyl chloride	$H_2C{=}CHCl$	62.50	0.9106	2,200	50
o-Xylene	$1,2\text{-}(CH_3)_2C_6H_4$	106.17	0.8802	6	0.5
m-Xylene	$1,3\text{-}(CH_3)_2C_6H_4$	106.17	0.8642	8	0.7
p-Xylene	$1,4\text{-}(CH_3)_2C_6H_4$	106.17	0.8611	8	0.7

Note: Ethene = ethylene; ethyl chloride = chloroethane; ethylene chloride = 1,2-dichloroethane; ethylidene chloride = 1,1-dichloroethane; methyl benzene = toluene; methyl chloride = chloromethane; methyl chloroform = 1,1,1-trichloroethane; methylene chloride = dichloromethane; tetrachloromethane = carbon tetrachloride; tribromomethane = bromoform.

TABLE A-9
Typical solubility product constants

Equilibrium equation	K_{sp} at 25°C
$AgCl \rightleftharpoons Ag^+ + Cl^-$	1.76×10^{-10}
$Al(OH)_3 \rightleftharpoons Al^{3+} + 3OH^-$	1.26×10^{-33}
$AlPO_4 \rightleftharpoons Al^{3+} + PO_4^{3-}$	9.84×10^{-21}
$BaSO_4 \rightleftharpoons Ba^{2+} + SO_4^{2-}$	1.05×10^{-10}
$Cd(OH)_2 \rightleftharpoons Cd^{2+} + 2OH^-$	5.33×10^{-15}
$CdS \rightleftharpoons Cd^{2+} + S^{2-}$	1.40×10^{-29}
$CdCO_3 \rightleftharpoons Ca^{2+} + CO_3^{2-}$	6.20×10^{-12}
$CaCO_3 \rightleftharpoons Ca^{2+} + CO_3^{2-}$	4.95×10^{-9}
$CaF_2 \rightleftharpoons Ca^{2+} + 2F^-$	3.45×10^{-11}
$Ca(OH)_2 \rightleftharpoons Ca^{2+} + 2OH^-$	7.88×10^{-6}
$Ca_3(PO_4)_2 \rightleftharpoons 3Ca^{2+} + 2PO_4^{3-}$	2.02×10^{-33}
$CaSO_4 \rightleftharpoons Ca^{2+} + SO_4^{2-}$	4.93×10^{-5}
$Cr(OH)_3 \rightleftharpoons Cr^{3+} + 3OH^-$	6.0×10^{-31}
$Cu(OH)_2 \rightleftharpoons Cu^{2+} + 2OH^-$	2.0×10^{-19}
$CuS \rightleftharpoons Cu^{2+} + S^{2-}$	1.0×10^{-36}
$Fe(OH)_3 \rightleftharpoons Fe^{3+} + 3OH^-$	2.67×10^{-39}
$FePO_4 \rightleftharpoons Fe^{3+} + PO_4^{3-}$	1.3×10^{-22}
$FeCO_3 \rightleftharpoons Fe^{2+} + CO_3^{2-}$	3.13×10^{-11}
$Fe(OH) \rightleftharpoons Fe^{2+} + 2OH^-$	4.79×10^{-17}
$FeS \rightleftharpoons Fe^{2+} + S^{2-}$	1.57×10^{-19}
$PbCO_3 \rightleftharpoons Pb^{2+} + CO_3^{2-}$	1.48×10^{-13}
$Pb(OH)_2 \rightleftharpoons Pb^{2+} + 2OH^-$	1.40×10^{-20}
$PbS \rightleftharpoons Pb^{2+} + S^{2-}$	8.81×10^{-29}
$Mg(OH)_2 \rightleftharpoons Mg^{2+} + 2OH^-$	5.66×10^{-12}
$MgCO_3 \rightleftharpoons Mg^{2+} + CO_3^{2-}$	1.15×10^{-5}
$MnCO_3 \rightleftharpoons Mn^{2+} + CO_3^{2-}$	2.23×10^{-11}
$Mn(OH)_2 \rightleftharpoons Mn^{2+} + 2OH^-$	2.04×10^{-13}
$NiCO_3 \rightleftharpoons Ni^{2+} + CO_3^{2-}$	1.45×10^{-7}
$Ni(OH)_2 \rightleftharpoons Ni^{2+} + 2OH^-$	5.54×10^{-16}
$NiS \rightleftharpoons Ni^{2+} + S^{2-}$	1.08×10^{-21}
$SrCO_3 \rightleftharpoons Sr^{2+} + CO_3^{2-}$	5.60×10^{-10}
$Zn(OH)_2 \rightleftharpoons Zn^{2+} + 2OH^-$	7.68×10^{-17}
$ZnS \rightleftharpoons Zn^{2+} + S^{2-}$	2.91×10^{-25}

(*Sources:* Linde, 2000; Sawyer, McCarty, and Parkin, 2003; Weast, 1983.)

TABLE A-10
Typical valences of elements and compounds in water

Element or compound	Valence
Aluminum	3^+
Ammonium (NH_4^+)	1^+
Barium	2^+
Boron	3^+
Cadmium	2^+
Calcium	2^+
Carbonate (CO_3^{2-})	2^-
Carbon dioxide (CO_2)	a
Chloride (*not* chlorine)	1^-
Chromium	$3^+, 6^+$
Copper	2^+
Fluoride (*not* fluorine)	1^-
Hydrogen	1^+
Hydroxide (OH^-)	1^-
Iron	$2^+, 3^+$
Lead	2^+
Magnesium	2^+
Manganese	2^+
Nickel	2^+
Oxygen	2^-
Nitrogen	$3^+, 5^+, 3^-$
Nitrate (NO_3^-)	1^-
Nitrite (NO_2^-)	1^-
Phosphorus	$5^+, 3^-$
Phosphate (PO_4^{3-})	3^-
Potassium	1^+
Silver	1^+
Silica	b
Silicate (SiO_4^{4-})	4^-
Sodium	1^+
Sulfate (SO_4^{2-})	2^-
Sulfide (S^{2-})	2^-
Zinc	2^+

[a]Carbon dioxide in water is essentially carbonic acid:

$$CO_2 + H_2O \rightleftharpoons H_2CO_3$$

As such, the equivalent weight = GMW/2.

[b]Silica in water is reported as SiO_2. The equivalent weight is equal to the gram molecular weight.

SOURCES

Linde, D. R. (2000) *CRC Handbook of Chemistry and Physics,* 81st ed., CRC Press, Boca Raton, FL, pp. 8-111–8-112.

Montgomery, J. M. (1985) *Water Treatment Principles and Design,* John Wiley & Sons, New York, p. 236.

Sawyer, C. N., P. L. McCarty, and G. F. Parkin (2003) *Chemistry for Environmental Engineering and Science,* 5th ed., McGraw-Hill, Boston, pp. 39–40

Weast, R. C. (1983) *CRC Handbook of Chemistry and Physics,* 64th ed., CRC Press, Boca Raton, FL, pp. B-219–B-220.

APPENDIX B

NOISE COMPUTATION TABLES AND NOMOGRAPHS

Project ____________________ Date ____________ Engineer ____________

Step			A	T_M	T_H	A	T_M	T_H	A	T_M	T_H	A	T_M	T_H	A	T_M	T_H	A	T_M	T_H
1	Traffic	Vehicle Volume, V(Vph)																		
2	Traffic	Vehicle Av. Speed, S(km/h)																		
3	Traffic	Combined Veh. Vol.*, V_C(Vph)																		
4	Prop.	Observer-Roadway Dist., D_C(m)																		
5	Shielding	Line-of-Sight Dist., L/S(m)																		
6	Shielding	Barrier Position Dist., P(m)																		
7	Shielding	Break in Barrier, B(m)																		
8	Shielding	Angle Subtended, θ (deg)																		
9	Prediction**	Unshield L_{10} Level (dBA)																		
10	Prediction**	Shielding Adjust. (dBA)																		
11	Prediction**	L_{10} at Observer (By Veh. Class)																		
12	Prediction**	L_{10} at Observer–Total																		

Code:

A = Automobiles, T_M = Medium Trucks, T_H = Heavy Trucks

* Applies only when automobile and medium truck average speeds are equal. $V_C = V_A + (10)V_{T_M}$.
Otherwise, multiply medium truck volume by 10 and use this volume to compute the Unshielded L_{10} level.

** If automobile-medium truck volume V_C is combined, use L_{10}. Nomograph prediction only once for these two vehicle classes.

FIGURE B-1

Blank noise prediction worksheet. (*Source*: National Cooperative Highway Research Program 174 (1976) *Highway Traffic Noise Prediction Model*, Federal Highway Administration, U.S. Department of Transportation, Washington, DC.)

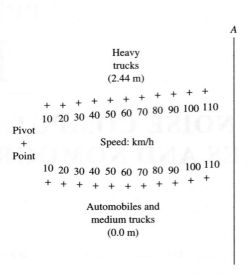

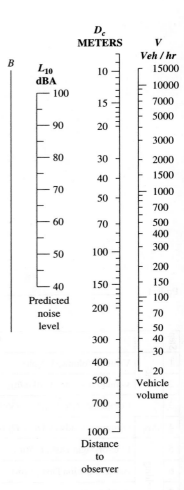

FIGURE B-2
Blank L_{10} nomograph. (*Source*: *NCHRP 174*,1976.)

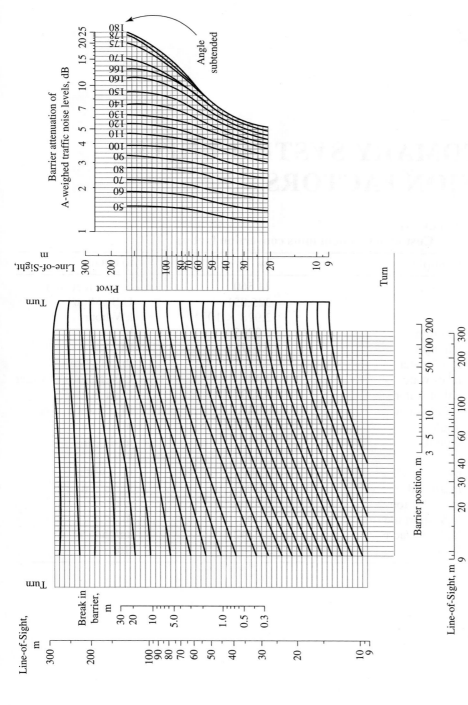

FIGURE B-3
Blank barrier nomograph. (*Source: NCHRP 174, 1976.*)

APPENDIX
C

U.S. CUSTOMARY SYSTEM UNITS CONVERSION FACTORS

TABLE C-1
U.S. Customary System units conversion factors

Multiply	by	to obtain
acre (ac)	43,560	square feet (ft^2)
acre-ft	325,851	U.S gallons
Btu	2.928×10^{-4}	kW-hour
Btu/min	0.02358	hp
Btu/min	0.01758	kW
ft^3 of water	62.4	lb_m of water
ft^3 of water	7.48	U.S. gallons of water
gal of water	0.1337	ft^3 of water
gal of water	8.34	lb_m of water
gpd/ft^2	0.04074	$m^3/d \cdot m^2$
gpm/ft^2	2.445	$m^3/h \cdot m^2$
hp	0.7457	kW
psi	2.307	ft of water
$lb_m/ft^2 \cdot d$	0.2048	$kg/m^2 \cdot d$
lb_m/U.S. ton	0.4999	g/kg
U.S. short tons	2,000	lb_m
U.S. tons/acre	0.2242	kg/ha
W-h	3.4144	Btu

INDEX

NOTE Bold, hyphenated numbers, for example 14–6, 14–25, 14–36, refer to Chapter 14 pages that may be found on the web at www.mhhe.com/davis

Useful conversion factors

Multiply	By	To Obtain
atmosphere (atm)	101.325	kilopascal (kPa)
Calorie (international)	4.1868	Joules (J)
centipoise	10^{-3}	Pa · s
centistoke	10^{-6}	m^2/s
cubic meter (m^3)	35.31	cubic feet (ft^3)
cubic meter	1.308	cubic yard (yd^3)
cubic meter	1,000.00	liter (L)
cubic meter/s	15,850.0	gallons/min (gpm)
cubic meter/s	22.8245	million gal/d (MGD)
cubic meter/m^2	24.545	gallons/sq ft (gal/ft^2)
cubic meter/d · m	80.52	gal/d · ft (gpd/ft)
cubic meter/d · m^2	24.545	gal/d · ft^2 (gpd/ft^2)
cubic meter/d · m^2	1.0	meters/d (m/d)
days (d)	24.00	hours (h)
days (d)	1,440.00	minutes (min)
days (d)	86,400.00	seconds (s)
dyne	10^{-5}	Newtons (N)
erg	10^{-7}	Joules (J)
grains (gr)	6.480×10^{-2}	grams (g)
grains/U.S. gallon	17.118	mg/L
grams (g)	2.205×10^{-3}	pounds mass (lb_m)
hectare (ha)	10^4	m^2
Hertz (Hz)	1	cycle/s
Joule (J)	1	N · m
J/m^3	2.684×10^{-5}	Btu/ft^3
kilogram/m^3 (kg/m^3)	8.346×10^{-3}	lb_m/gal
kilogram/m^3	1.6855	lb_m/yd^3
kilogram/ha (kg/ha)	8.922×10^{-1}	lb_m/acre
kilogram/m^2 (kg/m^2)	2.0482×10^{-1}	lb_m/ft^2
kilometers (km)	6.2150×10^{-1}	miles (mi)
kilowatt (kW)	1.3410	horsepower (hp)
kilowatt-hour	3.600	megajoules (MJ)
liters (L)	10^{-3}	cubic meters (m^3)
liters	1,000.00	milliliters (mL)
liters	2.642×10^{-1}	U.S. gallons
megagrams (Mg)	1.1023	U.S. short tons
meters (m)	3.281	feet (ft)
meters/d (m/d)	2.2785×10^{-3}	ft/min
meters/d	3.7975×10^{-5}	meters/s (m/s)
meters/s (m/s)	196.85	ft/min
meters/s	3.600	km/h
meters/s	2.237	miles/h (mph)
micron (μ)	10^{-6}	meters
milligrams (mg)	10^{-3}	grams (g)
milligrams/L	1	g/m^3
milligrams/L	10^{-3}	kg/m^3
Newton (N)	1	kg · m/s^2
Pascal (Pa)	1	N/m^2
Poise (P)	10^{-1}	Pa · s
square meter (m^2)	2.471×10^{-4}	acres
square meter (m^2)	10.7639	sq ft (ft^2)
square meter/s	6.9589×10^6	gpd/ft
Stoke (St)	10^{-4}	m^2/s
Watt (W)	1	J/s
Watt/cu meter (W/m^3)	3.7978×10^{-2}	hp/1,000 ft^3
Watt/sq meter · °C (W/m^2 · °C)	1.761×10^{-1}	Btu/h · ft^2 · °F